FUNCTIONS AND GRAPHS
A Precalculus Course

Raymond A. Barnett
Merritt College

McGRAW-HILL BOOK COMPANY
New York   St. Louis   San Francisco   Auckland   Bogotá   Hamburg
Johannesburg   London   Madrid   Mexico   Montreal   New Delhi   Panama
Paris   São Paulo   Singapore   Sydney   Tokyo   Toronto

FUNCTIONS AND GRAPHS: A PRECALCULUS COURSE

Copyright © 1985 by McGraw-Hill, Inc. All rights reserved. With the exception of Chapters 1 and 5, all material in this book is from *College Algebra, Trigonometry, and Analytic Geometry* by Raymond A. Barnett, copyright © 1984 by McGraw-Hill, Inc. All rights reserved. Printed in the United States of America. Except as permitted under the United States Copyright Act of 1976, no part of this publication may be reproduced or distributed in any form or by any means, or stored in a data base or retrieval system, without the prior written permission of the publisher.

2 3 4 5 6 7 8 9 0 D O C D O C 8 9 8 7 6 5

ISBN 0-07-003896-1

This book was set in Melior.
The editor was Peter R. Devine;
the production supervisor was Leroy A. Young.
New drawings were done by Carl Brown.
The cover was designed by Anne Canevari Green.
Project supervision was done by Phyllis Niklas.
R. R. Donnelley & Sons Company was printer and binder.

Cover Photo Credit

Ventral hind wing of banded purple admiral butterfly (Nymphalidae: *Limenitis arthemis*). Photo © by William C. Ferguson.

Chapter-Opening Photo Credits

Photos for Chapters 1 and 10 © by Peter Pearce. Photos for Chapters 2, 6, 7, 9, and 11 © by Anne Monk. Photo for Chapter 3 © by William C. Ferguson. Photo for Chapter 4 © by S. K. Webster, Monterey Bay Aquarium/BPS. Photo for Chapter 5 © by Glenn R. Steiner. Photo for Chapter 8 © by R. Humbert, Stanford University/BPS.

LIBRARY OF CONGRESS CATALOGING IN PUBLICATION DATA

Barnett, Raymond A.
    Functions and graphs.

    "With the exception of chapters 1 and 5, all material in this book is from College algebra, trigonometry, and analytic geometry"—T.p. verso.
    Includes index.
    1. Functions. 2. Algebra—Graphic methods.
I. Barnett, Raymond A. College algebra, trigonometry, and analytic geometry. II. Title.
QA331.3.B38   1985   512'.1   84-12519
ISBN 0-07-003896-1

Contents

Preface ix
To the Student xiii
Remarks on Calculator Use xv

*Chapter 1 Preliminaries — 1

    1-1 Sets   2
    1-2 Algebra and Real Numbers   9
    1-3 Linear Equations and Inequalities   19
    1-4 Absolute Value in Equations and Inequalities   35
    1-5 Nonlinear Inequalities   42
    1-6 Complex Numbers   50
    1-7 Quadratic Equations   57
    1-8 Equations Reducible to Quadratic Form   71
    1-9 Chapter Review   75

Chapter 2 Graphs and Functions — 81

    2-1 Rectangular Coordinate System; Graphing   82
    2-2 Relations and Functions   95
    2-3 Functions: Graphs and Properties   110
    2-4 Linear Relations and Functions   122
    2-5 Graphing Polynomial Functions   135
    2-6 Graphing Rational Functions   143
    2-7 Composite and Inverse Functions   154
   *2-8 Variation   164
    2-9 Chapter Review   175

* *Note to instructor:* May be omitted, depending on time, need, and interest. (Some topics are review topics and some are new topics.)

Chapter 3 Polynomial Functions and Theory of Equations 181

    3-1 Introduction 182
    3-2 Synthetic Division 184
    3-3 Remainder and Factor Theorems 188
    3-4 Fundamental Theorem of Algebra 194
    3-5 Isolating Real Zeros 200
    3-6 Finding Rational Zeros 206
*3-7 Approximating Irrational Zeros 213
*3-8 Partial Fraction Decomposition 218
    3-9 Chapter Review 226

Chapter 4 Exponential and Logarithmic Functions 229

4-1 Exponential Functions 230
4-2 Logarithmic Functions 238
4-3 Properties of Logarithmic Functions 244
4-4 Logarithms to Various Bases 248
4-5 Exponential and Logarithmic Equations 257
4-6 Chapter Review 265

Chapter 5 Trigonometric Functions 269

5-1 Introduction 270
5-2 The Wrapping Function 270
5-3 Circular Functions 279
5-4 Angles 290
5-5 Trigonometric Functions 299
5-6 Graphs of Trigonometric Functions 315
5-7 Graphing $y = A \sin(Bx + C)$ and $y = A \cos(Bx + C)$ 325
5-8 Inverse Trigonometric Functions 334
5-9 Chapter Review 343

Chapter 6 Trigonometric Identities and Conditional Equations 349

6-1 Introduction 350
6-2 Basic Identities and Their Use 350
6-3 Addition, Subtraction, and Cofunction Identities 358
6-4 Double-Angle and Half-Angle Identities 367
6-5 Product and Factor Identities 375
6-6 Trigonometric Equations 379
6-7 Chapter Review 389

Contents vii

*Chapter 7 Additional Topics in Trigonometry — 393

- 7-1 Solutions of Right Triangles — 394
- 7-2 Law of Sines — 399
- 7-3 Law of Cosines — 406
- 7-4 Vectors — 412
- 7-5 Polar and Rectangular Coordinates — 420
- 7-6 Sketching Polar Graphs — 426
- 7-7 Complex Numbers in Rectangular and Polar Forms — 432
- 7-8 De Moivre's Theorem — 439
- 7-9 Chapter Review — 443

Chapter 8 Additional Topics in Analytic Geometry — 447

- 8-1 Conic Sections — 448
- 8-2 Parabola — 449
- 8-3 Ellipse — 455
- 8-4 Hyperbola — 465
- 8-5 Translation of Axes — 477
- *8-6 Rotation of Axes — 484
- *8-7 Parametric Equations — 491
- 8-8 Chapter Review — 498

Chapter 9 Systems of Equations and Inequalities — 503

- 9-1 Systems of Linear Equations — A Review — 504
- 9-2 Systems and Augmented Matrices — An Introduction — 517
- 9-3 Gauss–Jordan Elimination — 525
- 9-4 Systems Involving Second-Degree Equations — 536
- *9-5 Systems of Linear Inequalities — 540
- 9-6 Chapter Review — 548

Chapter 10 Matrices and Determinants — 551

- 10-1 Matrix Addition; Multiplication of a Matrix by a Number — 552
- 10-2 Matrix Multiplication — 558
- *10-3 Inverse of a Square Matrix; Matrix Equations — 565
- 10-4 Determinant Functions — 578
- *10-5 Properties of Determinants — 585
- *10-6 Cramer's Rule — 592
- 10-7 Chapter Review — 597

Chapter 11 Sequences and Series — 603

    11-1 Sequences and Series — 604
  *11-2 Mathematical Induction — 610
    11-3 Arithmetic Sequences and Series — 619
    11-4 Geometric Sequences and Series — 625
  *11-5 Additional Applications — 631
    11-6 Binomial Formula — 634
    11-7 Chapter Review — 641

Appendixes — A1

  A-1 Significant Digits — A1
  A-2 Integer Exponents — A2
  A-3 Rational Exponents — A10
  A-4 Radicals — A16
  A-5 Algebraic Expressions, Basic Operations — A25
  A-6 Factoring — A32
  A-7 Fractions — A40
  A-8 Table Evaluation of Trigonometric Functions — A50

Tables — A55

    I Values of e^x and e^{-x} (0.00 to 3.00) — A56
   II Common Logarithms — A58
  III Natural Logarithms ($\ln x = \log_e x$) — A60
  IV Logarithms of Factorial n — A62
   V Trigonometric Functions — Radians or Real Numbers — A63
  VI Trigonometric Functions — Degrees and Minutes — A67

Answers — AA1

Index — I1

Preface

As indicated by the title, this is a precalculus text emphasizing functions and graphs. Prerequisite to its use is a course in intermediate algebra. Realizing that some prerequisite material may be forgotten, a concise review of some important topics from intermediate algebra can be found in Chapter 1 and in the appendixes. This review material may be treated systematically, reviewed as needed, or omitted altogether, depending on the background of a given class.

This text contains enough precalculus topics so that an instructor or department can design a course that will effectively meet the needs of their students and department. A diagram showing chapter dependencies and possible course outlines can be found on page xiv. In the table of contents we have also indicated those chapters and sections that may be omitted, depending on time, need, and interest.

- **Main Differences from the Other Books in the Author's Precalculus Series**

    1. In this text the review material is presented more concisely, and part has been moved to the appendixes.
    2. Trigonometric functions with real number domains are introduced through the use of the wrapping function, and trigonometric functions with angle domains follow. This order is reversed in the other books in the series containing trigonometry.
    3. The chapter titled "Additional Topics in Analytic Geometry" immediately follows the trigonometry chapters, whereas it was placed last in the author's *College Algebra, Trigonometry, and Analytic Geometry*.

- **Important Features**

    Readability. Every effort has been made to create a book students can read and understand. It is designed for students with varied backgrounds, interests, and abilities in mathematics. An average or below average student will experience success and a very capable student will be challenged. An informal style is used for exposition, statements of definitions, and proofs of theorems.

Examples. Each concept is illustrated with one or more examples, and following each worked out example is a parallel problem (with an answer given at the end of the section) so that students can immediately check their understanding of a concept. This arrangement results in active rather than passive student involvement in the learning process.

Exercises. Each exercise set contains a large number and variety of problems. In fact, over 4,000 carefully selected and graded problems are included in this text. Most exercise sets are divided into A, B, and C levels of difficulty.

Applications. The variety and number of significant applications distributed throughout the text should convince even the most skeptical student that mathematics is really useful. The applications, taken from many fields, are modern, topical, and interesting.

Topic Selection. The book contains a wide range of important precalculus topics which are carefully arranged for flexibility in coverage and continuity in transitions from one topic to another. The depth and completeness of coverage of most topics equals or exceeds that found in most precalculus texts. The function concept provides the unifying theme for the book.

Review. Each chapter ends with a complete review section that includes a summary of key terms and symbols, a comprehensive review exercise, and a practice test. (Also, as we pointed out earlier, Chapter 1 and the appendixes provide a review of intermediate algebra topics.)

 Calculator Use. Hand calculator use is encouragd throughout the text, including use in evaluating exponential, logarithmic, and trigonometric functions. (Optional table use is included for those still desiring that approach.) The author believes each student should have a scientific calculator with a user's manual for that calculator.

Student Aids (See Sections 1-2 and 1-3 for examples.)
1. **Think boxes** (dashed boxes) are used to enclose steps that are usually performed mentally.
2. **Common student errors** are identified where they naturally occur.
3. **Annotated examples** and developments are included.
4. **Functional use of a second color** clarifies the discussion.
5. **Comprehensive chapter reviews** with practice tests are provided.
6. **Answers** to odd-numbered problems and all chapter review problems are included in the back of the book.
7. **A solutions manual** is available at modest cost.

Instructor Aids

1. **Comprehensive test battery** with answer keys is provided in an instructor's manual. Two standard and two multiple-choice tests are given for most chapters.
2. **Answers** to even-numbered problems (not in the text) are in the instructor's manual.
3. **A solutions manual** is available without cost.

■ Coverage

Chapter 1. This chapter provides a review of important topics from intermediate algebra.

Chapter 2. The treatment of function, composite function, and inverse function is complete and well-motivated. Graphing techniques and the graphing of special functions are emphasized. A sound introduction to analytic geometry is contained in this chapter. Further topics on analytic geometry are found in other parts of the text and in Chapter 8.

Chapter 3. This chapter on polynomial functions and theory of equations provides a systematic approach to finding or approximating rational and irrational zeros of polynomial functions.

Chapter 4. In this chapter on exponential and logarithmic functions, properties and graphs are emphasized. The use of logarithmic functions as a computational tool has been deemphasized. On the other hand, calculator evaluation of exponential and logarithmic functions receives quite a bit of attention.

Chapters 5–7. These chapters cover trigonometric functions and their uses. Trigonometric functions with real number domains are introduced through the use of the wrapping function. A special section on trigonometric functions with angle domains follows and is presented in such a way that a student will be able to shift back and forth between the two approaches with relative ease, using the approach that is most enlightening and productive for a given purpose. Calculator evaluation of trigonometric functions is emphasized. Table evaluation, for those still desiring that approach, can be found in Appendix A-8. Exact values for multiples of special angles or multiples of special real numbers receive considerable attention, and these values are used extensively throughout all three chapters. Polar coordinates and polar graphing are included in Chapter 7.

Chapter 8. This chapter covers additional topics in analytic geometry. Discussions of conics, translation and rotation of axes, and parametric equations are included.

Chapter 9. Gauss–Jordan elimination is used to solve systems of linear equations. The presentation is gradual, well-motivated, and carefully done.

Chapter 10. In addition to the standard topics on matrices and determinants, inverses of square matrices and matrix equations are treated in some detail.

Chapter 11. This chapter on sequences and series includes a carefully motivated and executed presentation of mathematical induction.

- Error Check

Because of the careful checking and proofing by a number of very competent people (acting independently), the author and publisher believe this book to be substantially error-free. For any errors remaining, the author would be grateful if they were sent to: Mathematics Editor, College Division, 27th Floor, McGraw-Hill Book Company, 1221 Avenue of the Americas, New York, New York 10020.

- Acknowledgments

The preparation of a book requires the effort and skills of many people in addition to an author. I wish to thank the reviewers for their many helpful suggestions and comments. In particular, I wish to thank Thomas A. Atchison, Stephen F. Austin State University; Martin Broadwell, Jr., Florida College; Eddie J. Brown, St. Clair County Community College; Donald F. Devine, Western Illinois University; Leland Fry, Kirkwood Community College; Lotus Hershberger, Illinois State University; Sidney Katoni, New York City Technical College; Peter Lindstrom, North Lake College; Gary Ling, Golden Gate University; Stanley Lukawecki, Clemson University; Francis E. Masat, Glassboro State College; Ronald Prielipp, Bethany College; Bruce Reed, Virginia Polytechnic Institute; Donald G. Spencer, Northeast Louisiana University; and George Trytten, Luther College. A special thanks go to Margaret Barnett-Burnette, Fred Safier (City College of San Francisco), and Ward A. Soper (Walla Walla College) for their careful checking of all examples, matched problems, and exercise sets.

Raymond A. Barnett

To the Student

The following suggestions are made to help you get the most out of this book and your efforts.

As you study the text we suggest a five-step process. For each section:

> 1. Read a mathematical development.
> 2. Work through the illustrative example. } Repeat the 1-2-3 cycle until the section is finished.
> 3. Work the matched problem.
> 4. Review the main ideas in the section.
> 5. Work the assigned exercise at the end of the section.

All of this should be done with plenty of paper, pencils, and a wastebasket at hand. In fact, no mathematics text should be read without pencil and paper in hand; mathematics is not a spectator sport. Just as you cannot learn to swim by watching someone else swim, you cannot learn mathematics by simply reading worked examples — you must work problems, lots of them.

If you have difficulty with the course, then, in addition to doing the regular assignments, spend more time on the examples and matched problems and work more A exercises, even if they are not assigned. If the A exercises continue to be difficult for you, you probably should take an intermediate algebra course before attempting this one. If you find the course too easy, then work more C exercises, even if they are not assigned. If the C exercises are consistently easy for you, you are probably ready to start the calculus sequence.

Raymond A. Barnett

Suggested Course Outlines

Remarks on Calculator Use

Hand calculators are of two basic types relative to their internal logic (the way they compute): algebraic and reverse Polish notation (RPN). Throughout the book we will identify algebraic calculator steps with "A" and reverse Polish notation calculator steps with "P." Let us see how each type of calculator would compute

$$\frac{(5)(3)(2)-(7)(6)}{2(11)}$$

PRESS	DISPLAY
A: [5][×][3][×][2][−][7][×][6][=][÷][2][÷][11][=]	−0.54545455
P: [5][ENTER][3][×][2][×][7][ENTER][6][×][−][2][÷][11][÷]	−0.54545455

Some people prefer the algebraic logic and others prefer the Polish. Which is better is still being debated. The answer seems to rest with the type of problems one encounters and with individual preferences. The author owns both types and uses the one with Polish logic most frequently. However, he knows people who prefer the algebraic type, and they seem quite happy with their choice.

In any case, irrespective of the type of calculator that you own, it is essential that you read the user's manual for your own calculator. A large variety of calculators are on the market, and each is slightly different from the others. Therefore, it is important that you take the time to read the manual. Do not try to read and understand everything the calculator can do; this will only tend to confuse you. Read only those sections that pertain to the operations you are or will be using; then return to the manual as necessary when you encounter new operations.

In many places in the text calculator steps for new types of calculations will be shown (similar to those steps shown here). These are only aids. Try the calculation without the aid; then use the aid only if you get stuck.

It is important to remember that *a calculator is not a substitute for thinking*. It can save you a great deal of time in certain types of problems, but you still must know how and when to use it.

Raymond A. Barnett

Preliminaries

1-1 Sets
1-2 Algebra and Real Numbers
1-3 Linear Equations and Inequalities
1-4 Absolute Value in Equations and Inequalities
1-5 Nonlinear Inequalities
1-6 Complex Numbers
1-7 Quadratic Equations
1-8 Equations Reducible to Quadratic Form
1-9 Chapter Review

A natural design of mathematical interest. Can you guess the source? See the back of the book.

Chapter 1. Preliminaries

Section 1-1 Sets

- Set Notation
- Subsets and Equality
- Set Operations and Venn Diagrams

George Cantor (1845–1918), when about thirty, created a new mathematical concept, the set, and subsequently developed a theory of sets. This new theory, an outgrowth of his studies on infinity, has become a milestone in the development of mathematics. We will use only a few key ideas and symbols from this theory, ideas and symbols that will help us discuss certain mathematical developments with increased clarity and precision.

Set Notation

We can think of a **set** as any collection of objects that is **well defined**; that is, the collection is specified in such a way that we can tell whether any given object is or is not in the collection. In this course we will usually be interested in certain sets of numbers. Capital letters, such as A, B, and C, are often used to represent sets. For example,

$$A = \{1, 3, 5\} \qquad B = \{2, 5, 6\}$$

specify sets A and B. Each object in a set is called an **element** or **member** of the set. Symbolically:

$a \in A$ means "a is an element of set A."

$a \notin A$ means "a is not an element of set A."

Referring to sets A and B, we see that

$3 \in A$ 3 is an element of set A.

$4 \notin A$ 4 is not an element of set B.

1-1 Sets

A set without any elements is called the **empty** or **null set**. For example, the set of all solutions to the equation $x + 8 = x + 2$ is the empty set. Symbolically:

> $\emptyset$ represents the empty or null set

A set is usually described in one of two ways:

1. By **listing** the elements between braces $\{\ \}$: $\{1, 3, 5\}$.
2. By enclosing a **rule** within braces $\{\ \}$ that determines the elements of the set:

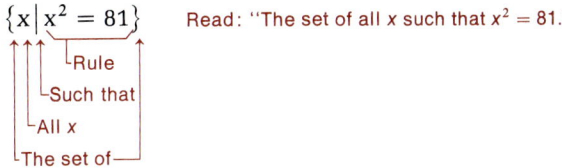

Read: "The set of all x such that $x^2 = 81$."

EXAMPLE 1   Let $D =$ The set of all numbers x such that $x^2 = 4$. Set D may be described by either the listing or the rule method:

Listing method: $D = \{-2, 2\}$
Rule method: $D = \{x \mid x^2 = 4\}$

PROBLEM 1*   Let $G =$ The set of all numbers x such that $x^2 = 64$.

(A) Write G using the listing method.
(B) Write G using the rule method.
(C) Indicate true (T) or false (F): $4 \in G$ $8 \in G$ $16 \notin G$

The letter x introduced above is a *variable*. In general, a **variable** is a symbol used as a placeholder for elements out of a set with two or more elements. A **constant**, on the other hand, is a symbol that names exactly one object.

■ Subsets and Equality

If each element of set A is also an element of set B, we say that A is a **subset** of set B. The set of all women in a class is a subset of the whole class. (Note that the definition of a subset allows a set to be a subset of itself.) If two sets have exactly the same elements (the order of listing does not matter), the sets are said to be **equal**. Set A is equal to set B if and only if A is a subset of B and B is a subset of A.

---

* Answers to matched problems in a given section are found near the end of the section, before the exercise set.

Symbolically,

Subsets

$A \subset B$ means "A is a subset of B." $\{3, 5\} \subset \{3, 5, 7\}$

$A = B$ means "A is equal to B." $\{4, 6\} = \{6, 4\}$

It is useful and interesting to note that

$\emptyset$ is a subset of every set.

It is certainly true that every element of $\emptyset$ is an element of any given set, since $\emptyset$ has no elements.

EXAMPLE 2 Let $A = \{-3, 0, 5\}$, $B = \{0, 5, -3\}$, and $C = \{0, 5\}$. Then each of the following statements is true.

$C \subset A$ $C \subset B$ $A = B$
$A \subset B$ $\emptyset \subset A$ $A \neq C$

PROBLEM 2 Let $M = \{-4, 6\}$, $N = \{6, -4\}$, and $P = \{-4\}$. Indicate true (T) or false (F).

(A) $M = N$ (B) $P \subset N$ (C) $N \neq P$
(D) $N \subset M$ (E) $\emptyset \subset P$ (F) $M \subset P$

■ **Set Operations and Venn Diagrams**

The **union** of sets A and B, denoted by $A \cup B$, is the set of all elements formed by placing all the elements of A and all the elements of B into one set (the same element is not repeated). Symbolically,

---
Union

$A \cup B = \{x \mid x \in A \;\; \textbf{or} \;\; x \in B\}$

Here we use the word *or* in the way it is most frequently used in mathematics; that is, x may be an element of set A or set B or both.

Venn diagrams are useful aids in visualizing set relationships. The union of two sets can be illustrated as shown in Figure 1.

The **intersection** of sets A and B, denoted by $A \cap B$, is the set of elements in set A that are also in set B. Symbolically,

---

Intersection

$$A \cap B = \{x \mid x \in A \text{ and } x \in B\}$$

This relationship is easily visualized in the Venn diagram shown in Figure 2.

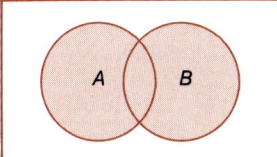

FIGURE 1 $A \cup B$ is the shaded region.

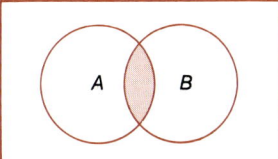

FIGURE 2 $A \cap B$ is the shaded region.

If $A \cap B = \emptyset$, then the sets A and B are said to be **disjoint**; this is illustrated in Figure 3.

The set of all elements under consideration is called the **universal set** U. Once the universal set is determined for a particular discussion, all other sets in that discussion must be subsets of U. We now define one more operation on sets called the *complement*. The **complement** of A (relative to U), denoted by A', is the set of elements in U that are not in A (see Fig. 4). Symbolically,

---

Complement

$$A' = \{x \in U \mid x \notin A\}$$

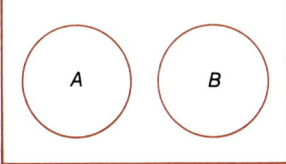

FIGURE 3 $A \cap B = \emptyset$

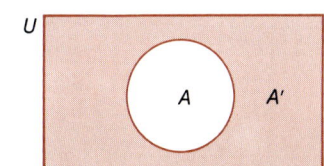

FIGURE 4 The complement of A is A'.

EXAMPLE 3   If $A = \{4, 5, 7\}$, $B = \{3, 6, 9\}$, and $C = \{3, 4, 5, 6, 7\}$, then

$B \cup C = \{3, 4, 5, 6, 7, 9\}$ Common elements are listed only once.

$B \cap C = \{3, 6\}$

$A \cap B = \emptyset$ A and B are disjoint.

A' relative to C is $\{3, 6\}$

PROBLEM 3   If $M = \{1, 2, 3, 4\}$, $N = \{1, 3, 5, 7\}$, and $Q = \{2, 4\}$, find:

(A) $M \cup N$ (B) $M \cap N$ (C) $N \cap Q$ (D) Q' relative to M

EXAMPLE 4   From a survey involving 100 college students, a marketing research company found that seventy-three students owned stereos, fifty-four owned bicycles, and forty-one owned bicycles and stereos.

(A) How many students owned either a stereo or a bicycle?
(B) How many students owned neither a bicycle nor a stereo?

Solution   Venn diagrams are very useful for this type of problem. If we let

U = Set of students in the sample (100)
S = Set of students who own stereos (73)
B = Set of students who own bicycles (54)
$B \cap S$ = Set of students who own bicycles and stereos (41)

then

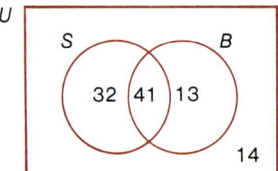

Place the number in the intersection first, then work outward.

(A) The number of students who own either a bicycle or a stereo is the number of students in the set $B \cup S$. You might be tempted to say that this is just the number of students in B plus the number of students in S, $54 + 73 = 127$, but this sum is larger than the sample we started with! What is wrong? We have actually counted the number in the intersection (41) twice. The correct answer, as seen in the Venn diagram, is

$13 + 41 + 32 = 86$ or $54 + 73 - 41 = 86$

(B) The number of students who own neither a bicycle nor a stereo is the number of students in the set $(B \cup S)'$—that is, 14.

PROBLEM 4 Refer to Example 4.

(A) How many students own a bicycle but not a stereo?
(B) How many students do not own both a bicycle and a stereo?

Answers to Matched Problems
1. (A) $\{-8, 8\}$ (B) $\{x | x^2 = 64\}$ (C) F, T, T
2. (A) T   (B) T   (C) T   (D) T   (E) T   (F) F
3. (A) $\{1, 2, 3, 4, 5, 7\}$ (B) $\{1, 3\}$ (C) $\emptyset$ (D) $\{1, 3\}$
4. (A) 13   (B) 59

Exercise 1-1

A *Indicate true (T) or false (F)*

1. $5 \in \{2, 3, 5\}$
2. $7 \notin \{2, 3, 5\}$
3. $\{3, 5\} \subset \{2, 3, 5\}$
4. $\{2, 3, 5\} = \{5, 2, 3\}$
5. $\{2, 3, 5\} \subset \{5, 2, 3\}$
6. $\{3, 5\} \in \{2, 3, 5\}$
7. $\emptyset \subset \{2, 3, 5\}$
8. $\emptyset = \{0\}$

In Problems 9–14 write the resulting sets using the listing method.

9. $\{3, 5, 7\} \cup \{4, 5, 6\}$
10. $\{1, 3, 5\} \cup \{2, 3, 4\}$
11. $\{3, 5, 7\} \cap \{4, 5, 6\}$
12. $\{1, 3, 5\} \cap \{2, 3, 4\}$
13. $\{3, 5, 7\} \cap \{4, 6, 8\}$
14. $\{3, 4, 7, 8\} \cap \{5, 6, 9\}$

B *In Problems 15–20 write the resulting sets using the listing method.*

15. $\{x | x - 8 = 0\}$
16. $\{x | x + 3 = 0\}$
17. $\{x | x^2 = 49\}$
18. $\{x | x^2 = 1\}$
19. $\{x | x \text{ is a prime number between 1 and 9, inclusive}\}$
20. $\{x | x \text{ is a composite number between 1 and 9, inclusive}\}$
21. For $U = \{-2, -1, 0, 1, 2\}$ and $M = \{-1, 0\}$, find M'.
22. For $U = \{1, 2, 3, 4, 5, 6, 7\}$ and $N = \{2, 4, 6\}$, find N'.

Problems 23–34 refer to the accompanying Venn diagram. How many elements are in the indicated sets?

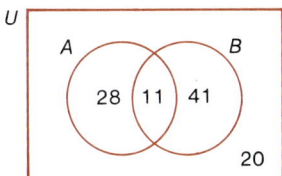

23. U
24. B
25. A'
26. B'
27. $A \cup B$
28. $A \cap B$
29. $A' \cap B$
30. $A \cap B'$

31. $(A \cap B)'$ **32.** $(A \cup B)'$ **33.** $A' \cap B'$ **34.** U'

35. If $M = \{2, 3, 4, 5\}$ and $N = \{4, 5, 6, 7\}$, find:
(A) $\{x \mid x \in M \text{ or } x \in N\}$ (B) $M \cup N$

36. For sets M and N in Problem 35, find:
(A) $\{x \mid x \in M \text{ and } x \in N\}$ (B) $M \cap N$

37. For $A = \{1, 2, 3, 4\}$, $B = \{2, 4, 6\}$, and $C = \{3, 4, 5, 6\}$, find $A \cup (B \cap C)$.

38. For sets A, B, and C in Problem 37, find $A \cap (B \cup C)$.

C Venn diagrams may be of help in Problems 39–44.

39. If $M \cap N = N$, can we always conclude that $N \subset M$?

40. If $M \cup N = N$, can we always conclude that $M \subset N$?

41. If $M \cap N = \emptyset$, can we always conclude that $M = \emptyset$ or $N = \emptyset$?

42. If M and N are arbitrary sets, can we always conclude that $M \cap N \subset N$?

43. If $M \subset N$ and $x \in N$, can we always conclude that $x \in M$?

44. If $M \subset N$ and $x \in M$, can we always conclude that $x \in N$?

45. How do the sets $\emptyset$, $\{0\}$, and $\{\emptyset\}$ differ from one another?

46. How many subsets does each of the following sets have?
(A) $\{a\}$ (B) $\{a, b\}$ (C) $\{a, b, c\}$
(D) A set with n elements

APPLICATIONS

Business and Economics

Problems 47–58 refer to the following marketing survey: From a random sample of 1,000 students on a college campus, it was found that 720 owned tape cassettes, 670 owned records, and 540 owned both. Let

C = Set of students in the sample who owned cassettes
R = Set of students in the sample who owned records

Following the procedures in Example 4, find the number of students in each set.

47. $C \cup R$ **48.** $C \cap R$ **49.** $(C \cup R)'$
50. $(C \cap R)'$ **51.** $C' \cap R$ **52.** $C \cap R'$

53. Set of students who owned either cassettes or records

54. Set of students who owned both cassettes and records

55. Set of students who owned neither cassettes nor records

56. Set of students who did not own both cassettes and records

57. Set of students who owned records but no cassettes

58. Set of students who owned cassettes but no records

Medicine—Blood Types Problems 59–66 refer to the following breakdown in blood types: When receiving a blood transfusion, a recipient must have all the antigens of the donor. A person may have one or more of the three antigens A, B, and Rh, or none at all. Eight blood types are possible, as indicated in the Venn diagram in the figure, where U is the set of all people under consideration. An A− person has A antigens but no B or Rh; an O+ person has Rh but neither A nor B; an AB− person has A and B antigens but no Rh; and so on. Using the figure, indicate which of the eight blood types are included in each set.

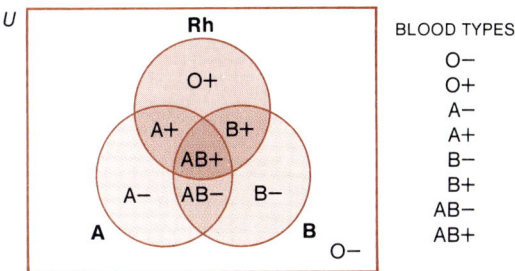

59. A ∩ B **60.** A ∩ Rh **61.** A ∪ B

62. A ∪ Rh **63.** (A ∪ B ∪ Rh)′ **64.** (A ∪ B)′

65. Rh′ ∩ A **66.** A′ ∩ B

Section 1-2 Algebra and Real Numbers

- The Real Number System
- The Real Number Line
- Basic Properties
- Further Properties
- Fraction Properties

In algebra we are interested in manipulating symbols in order to change or simplify algebraic expressions and to solve algebraic equations. Because many of these symbols represent real numbers, it is important to briefly review the real number system and some of its important properties. These properties provide the basic rules for much of the manipulation of symbols in algebra.

The Real Number System

The real number system is the number system in which you have worked most of your life. Table 1 describes the set of real numbers and some of the important types of numbers within the set of real numbers.

TABLE 1 The Set of Real Numbers

SYMBOL	NUMBER SYSTEM	DESCRIPTION	EXAMPLES
N	Natural numbers	Counting numbers (also called positive integers)	$1, 2, 3, \ldots$
Z	Integers	Set of natural numbers, their negatives, and 0	$\ldots, -2, -1, 0, 1, 2, \ldots$
Q	Rationals	Any number that can be represented as a/b, where a and b are integers and $b \neq 0$	$-4; \frac{-3}{5}; 0; 1; \frac{2}{3}; 3.67$
R	Reals	Set of all rational and irrational numbers (the irrational numbers are all the real numbers that are not rational)	$-4; \frac{-3}{5}; 0; 1; \frac{2}{3}; 3.67; \sqrt{2}; \pi; \sqrt[3]{5}$

Figure 5 illustrates how these sets of numbers are related to one another.

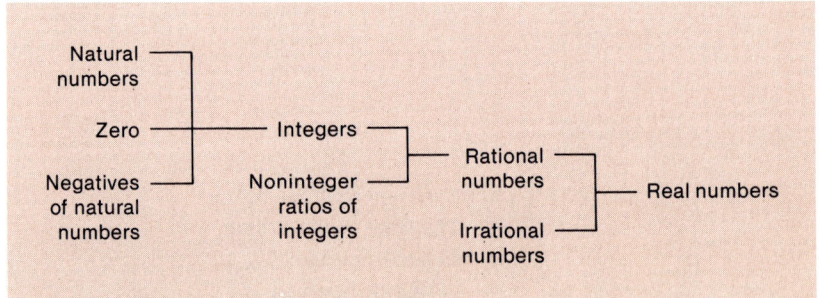

FIGURE 5 The real number system

The set of integers contains all the natural numbers and something else (their negatives and 0). The set of rational numbers contains all the integers and something else (noninteger ratios of integers). And the set of real numbers contains all the rational numbers and something else (irrational numbers).

Rational numbers have repeating decimal representations, whereas irrational numbers have infinite nonrepeating decimal representations.

1-2 Algebra and Real Numbers

For example, the decimal representations of the rational numbers 2, $\frac{4}{3}$, and $\frac{5}{11}$ are, respectively,

$$2 = 2.0000\ldots \qquad \frac{4}{3} = 1.333\ldots \qquad \frac{5}{11} = 0.454545\ldots$$

whereas those of the irrational numbers $\sqrt{2}$ and π are, respectively,

$$\sqrt{2} = 1.41421356\ldots \qquad \pi = 3.14159265\ldots$$

■ The Real Number Line

A one-to-one correspondence exists between the set of real numbers and the set of points on a line; that is, each real number corresponds to exactly one point, and each point to exactly one real number. A line with a real number associated with each point, and vice versa, as in Figure 6, is called a **real number line**, or simply a **real line**. Each number associated with a point is called the **coordinate** of the point. The point with coordinate 0 is called the **origin**. The arrow indicates a positive direction; the coordinates of all points to the right of the origin are called **positive real numbers** and those to the left of the origin are called **negative real numbers**.

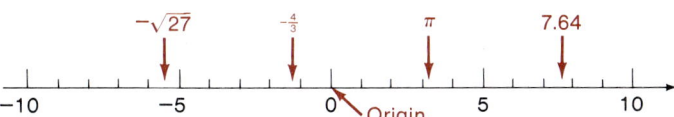

FIGURE 6   A real number line

■ Basic Properties

We now review (informally) a few basic real number properties. Other real number properties will be discussed as needed in other parts of the text. These properties of the real numbers become operational rules in the algebra of real numbers.

Real numbers can be added, subtracted, multiplied, and divided (except for division by 0). Does it matter in which order we perform addition, subtraction, multiplication, or division? In general, we may add in any order or multiply in any order (but this is not true for subtraction or division; for example, $4 - 2 \neq 2 - 4$ and $4 \div 2 \neq 2 \div 4$). This property is referred to as the **commutative property** (for addition and multiplication) for real numbers.

> **Commutative Property**
>
> For all real numbers a and b
>
> | Addition | Multiplication |
> |---|---|
> | $a + b = b + a$ | $ab = ba$ |
> | $7 + 2 = 2 + 7$ | $3 \cdot 5 = 5 \cdot 3$ |

When computing

$12 + 6 + 2$

$12 - 6 - 2$

$12 \cdot 6 \cdot 2$

$12 \div 6 \div 2$

does it matter how the numbers are grouped? That is, which of the following are true?

$(12 + 6) + 2 = 12 + (6 + 2)$

$(12 - 6) - 2 = 12 - (6 - 2)$

$(12 \cdot 6) \cdot 2 = 12 \cdot (6 \cdot 2)$

$(12 \div 6) \div 2 = 12 \div (6 \div 2)$

We see that the first and third are true, but the second and fourth are false. In general, we may group terms in addition and factors in multiplication in any way we please (but we are not free to group as we please in subtraction and division). This property is referred to as the **associative property** (for addition and multiplication) for real numbers.

> **Associative Property**
>
> For all real numbers a, b, and c,
>
> | Addition | Multiplication |
> |---|---|
> | $(a + b) + c = a + (b + c)$ | $(a \cdot b) \cdot c = a \cdot (b \cdot c)$ |
> | $(5 + 2) + 3 = 5 + (2 + 3)$ | $(4 \cdot 3) \cdot 2 = 4 \cdot (3 \cdot 2)$ |

> **Conclusion**
>
> In addition, commutativity and associativity permit us to change the order at will and insert or remove parentheses as we please. The same is true for multiplication, but not for subtraction and division.

1-2 Algebra and Real Numbers

What number added to a given number will give that number back again? What number times a given number will give that number back again? The answers are 0 and 1, respectively; thus, 0 and 1 are called the **identity elements** for the real numbers.

Identities

For each real number a,

$$a + 0 = a \qquad 1 \cdot a = a$$

$$(-3) + 0 = -3 \qquad 1 \cdot 7 = 7$$

0 is the additive identity.
1 is the multiplicative identity.

We now consider inverses. For each real number a, is there a real number that when added to a produces 0? For each real number $a \neq 0$, is there a real number that when multiplied times a produces 1? The answer in both cases is yes.

Inverses

For each real number a, there is a unique real number $-a$ such that

$$a + (-a) = 0$$

$$3 + (-3) = 0$$

For each real number $a \neq 0$, there is a unique real number $1/a$ such that

$$a\left(\frac{1}{a}\right) = 1$$

$$3\left(\frac{1}{3}\right) = 1$$

$-a$ is called the **additive inverse** of a or **negative** of a.*

$1/a$ is called the **multiplicative inverse** of a or **reciprocal** of a.

Zero has no multiplicative inverse.   What number times 0 is 1?

---

* $-a$ is not necessarily a negative number; it is positive if a is negative and negative if a is positive.

We now turn to an important real number property that involves both multiplication and addition. Consider the two computations:

$$3(4 + 2) = 3 \cdot 6 \qquad 3 \cdot 4 + 3 \cdot 2 = 12 + 6$$
$$= 18 \qquad\qquad\qquad\qquad = 18$$

Thus,

$$3(4 + 2) = 3 \cdot 4 + 3 \cdot 2$$

and we say that the factor 3 *distributes* over the sum $(4 + 2)$. In general, in the real number system multiplication always **distributes** over addition.

Distributive Property

For all real numbers a, b, and c,

$$a(b + c) = ab + ac \qquad\qquad (b + c)a = ba + ca$$
$$2(x + y) = 2x + 2y \qquad\qquad (3 + 5)x = 3x + 5x$$

All the properties listed in the preceding boxes are axioms (mathematical statements that are assumed true without proof). Many other real number properties (theorems) can be proved using these few axioms.

Throughout the rest of the book all variables represent real numbers unless stated to the contrary, and the properties given here will be assumed.

EXAMPLE 5 STATEMENT PROPERTY ILLUSTRATED
(A) $(7x)y = 7(xy)$ Associative ($\times$)
(B) $a(b + c) = (b + c)a$ Commutative ($\times$)
(C) $(2x + 3y) + 5y = 2x + (3y + 5y)$ Associative ($+$)
(D) $(x + y)(a + b) = (x + y)a + (x + y)b$ Distributive
(E) If $a + b = 0$, then $b = -a$. Inverse

PROBLEM 5 Which real number property justifies the indicated statement?

(A) $4 + (2 + x) = (4 + 2) + x$
(B) $(a + b) + c = c + (a + b)$
(C) $3x + 7x = (3 + 7)x$
(D) $(2x + 3y) + 0 = 2x + 3y$
(E) If $ab = 1$ and $a \neq 0$, then $b = 1/a$.

1-2 Algebra and Real Numbers

■ **Further Properties**

Subtraction and division can be defined in terms of addition and multiplication, respectively:

$$\text{SUBTRACTION:} \quad a - b = a + (-b) \qquad (-5) - (-3) = (-5) + (3) = -2$$

$$\text{DIVISION:} \quad b\overline{)a} = a \div b = \frac{a}{b} = a\left(\frac{1}{b}\right) \quad b \neq 0 \quad 3 \div 2 = 3\left(\frac{1}{2}\right)$$

Thus, to subtract b from a, add the negative (the additive inverse) of b to a. To divide a by b, multiply a by the reciprocal (the multiplicative inverse) of b. Note that division by 0 is not defined (one cannot divide by 0 ever!), since 0 does not have a reciprocal.

The following properties of negatives (called *theorems*) can be proved using the preceding axioms and definitions.

Properties of Negatives

For all real numbers a and b,

$$-(-a) = a$$

$$(-a)b = -(ab) = a(-b)$$

$$(-a)(-b) = ab$$

$$(-1)a = -a$$

$$\frac{-a}{b} = -\frac{a}{b} = \frac{a}{-b} \qquad b \neq 0$$

$$\frac{-a}{-b} = -\frac{-a}{b} = -\frac{a}{-b} = \frac{a}{b} \qquad b \neq 0$$

We now state two important properties (theorems) involving 0.

Zero Properties

For all real numbers a and b,

$$a \cdot 0 = 0$$

$$ab = 0 \quad \text{if and only if } a = 0 \text{ or } b = 0 \text{ or both.}$$

EXAMPLE 6

	STATEMENT	PROPERTY OR DEFINITION ILLUSTRATED
(A)	$3 - (-2) = 3 + [-(-2)]$	Subtraction
(B)	$-(-2) = 2$	Negatives
(C)	$-\dfrac{-3}{2} = \dfrac{3}{2}$	Negatives
(D)	$\dfrac{5}{-2} = -\dfrac{5}{2}$	Negatives
(E)	If $(x-3)(x+5) = 0$, then either $x - 3 = 0$ or $x + 5 = 0$.	Zero

PROBLEM 6 Which real number property or definition justifies each statement?

(A) $\dfrac{3}{5} = 3\left(\dfrac{1}{5}\right)$ (B) $(-5)(2) = -(5 \cdot 2)$ (C) $(-1)3 = -3$

(D) $\dfrac{-7}{9} = -\dfrac{7}{9}$ (E) If $(x+5) = 0$, then $(x-3)(x+5) = 0$.

■ **Fraction Properties**

Recall that the quotient $a \div b$, $b \neq 0$, written in the form a/b is called a **fraction**. The quantity a is called the **numerator** and the quantity b the **denominator**.

Fraction Properties

For all real numbers $a, b, c, d,$ and k (division by 0 excluded),

$\dfrac{a}{b} = \dfrac{c}{d}$ if and only if $ad = bc$

$\dfrac{4}{6} = \dfrac{6}{9}$ since $4 \cdot 9 = 6 \cdot 6$

$\dfrac{ka}{kb} = \dfrac{a}{b}$ $\dfrac{a}{b} \cdot \dfrac{c}{d} = \dfrac{ac}{bd}$ $\dfrac{a}{b} \div \dfrac{c}{d} = \dfrac{a}{b} \cdot \dfrac{d}{c}$

$\dfrac{7 \cdot 3}{7 \cdot 5} = \dfrac{3}{5}$ $\dfrac{3}{5} \cdot \dfrac{7}{8} = \dfrac{3 \cdot 7}{5 \cdot 8}$ $\dfrac{2}{3} \div \dfrac{5}{7} = \dfrac{2}{3} \cdot \dfrac{7}{5}$

$\dfrac{a}{b} + \dfrac{c}{b} = \dfrac{a+c}{b}$ $\dfrac{a}{b} - \dfrac{c}{b} = \dfrac{a-c}{b}$ $\dfrac{a}{b} + \dfrac{c}{d} = \dfrac{ad + bc}{bd}$

$\dfrac{3}{6} + \dfrac{5}{6} = \dfrac{3+5}{6}$ $\dfrac{7}{8} - \dfrac{3}{8} = \dfrac{7-3}{8}$ $\dfrac{2}{3} + \dfrac{3}{5} = \dfrac{2 \cdot 5 + 3 \cdot 3}{3 \cdot 5}$

1-2 Algebra and Real Numbers

Answers to Matched Problems

5. (A) Associative (B) Commutative (C) Distributive (D) Identity (E) Inverse
6. (A) Division (B) Negatives (C) Negatives (D) Negatives (E) Zero

Exercise 1-2

A In Problems 1–34 each statement illustrates the use of one of the following properties or definitions. Indicate which.

Commutative	Identity	Division
Associative	Inverse	Zero
Distributive	Subtraction	Negatives

1. $5 + 2x = 2x + 5$
2. $x + ym = x + my$
3. $7(3m) = (7 \cdot 3)m$
4. $(2w + 8) + 3 = 2w + (8 + 3)$
5. $x(y + z) = xy + xz$
6. $5(u + v) = 5u + 5v$
7. $-(-12) = 12$
8. $-\dfrac{3}{-5} = \dfrac{3}{5}$
9. $(-5) - (-2) = (-5) + [-(-2)]$
10. $8 - 12 = 8 + (-12)$
11. $\dfrac{-7}{9} = -\dfrac{7}{9}$
12. $\dfrac{-5}{-8} = \dfrac{5}{8}$
13. $3xyz + 0 = 3xyz$
14. $1 \cdot \left(-\dfrac{2}{3}\right) = -\dfrac{2}{3}$
15. $5 \div (-6) = 5\left(\dfrac{1}{-6}\right)$
16. $7 \div 9 = 7\left(\dfrac{1}{9}\right)$
17. $w + (-w) = 0$
18. $(-u) + [-(-u)] = 0$

B
19. $(7 + 12)x = 7x + 12x$
20. $8m + 5m = (8 + 5)m$
21. $4uv + 7uv = (4 + 7)uv$
22. $7x + 7y = 7(x + y)$
23. $(2x - 3)(x + 5) = 0$ if either $2x - 3 = 0$ or $x + 5 = 0$
24. $(u - 4)(3u - 7) = 0$ if either $u - 4 = 0$ or $3u - 7 = 0$
25. $(3x + 5) + 7 = 7 + (3x + 5)$
26. $(mn)p = p(mn)$
27. $(3x + 2) + (x + 5) = 3x + [2 + (x + 5)]$
28. $(5x)(7y) = 5[x(7y)]$
29. $(x + 3)(x + 5) = (x + 3)x + (x + 3)5$

30. $(m + n)(u + v) = m(u + v) + n(u + v)$
31. $x(x - y) + y(x - y) = (x + y)(x - y)$
32. $2x(x + 4) + 3(x + 4) = (2x + 3)(x + 4)$
33. $\dfrac{5}{-(x - 3)} = -\dfrac{5}{x - 3}$
34. $\dfrac{-7}{-(m + n)} = \dfrac{7}{m + n}$
35. If $ab = 0$, does either a or b have to be 0?
36. If $ab = 1$, does either a or b have to be 1?
37. Indicate which of the following are true.
    (A) All natural numbers are integers.
    (B) All real numbers are irrational.
    (C) All rational numbers are real numbers.
38. Indicate which of the following are true.
    (A) All integers are natural numbers.
    (B) All rational numbers are real numbers.
    (C) All natural numbers are rational numbers.
39. Give an example of a rational number that is not an integer.
40. Give an example of a real number that is not a rational number.
41. Given the sets of numbers: N (natural numbers), Z (integers), Q (rational numbers), and R (real numbers). Indicate to which set(s) each of the following numbers belong(s).
    (A) -3 (B) 3.14 (C) π (D) $\tfrac{2}{3}$
42. Given the sets of numbers N, Z, Q, and R (see Problem 41), indicate to which set(s) each of the following numbers belong(s).
    (A) 8 (B) $\sqrt{2}$ (C) -1.414 (D) $\dfrac{-5}{2}$
43. Indicate true (T) or false (F), and for each false statement find real number replacements for a and b that will illustrate its falseness. For all real numbers a and b,
    (A) $a + b = b + a$ (B) $a - b = b - a$
    (C) $ab = ba$ (D) $a \div b = b \div a$
44. Indicate true (T) or false (F), and for each false statement find real number replacements for a, b, and c that will illustrate its falseness. For all real numbers a, b, and c,
    (A) $(a + b) + c = a + (b + c)$ (B) $(a - b) - c = a - (b - c)$
    (C) $a(bc) = (ab)c$ (D) $(a \div b) \div c = a \div (b \div c)$

C 45. If $c = 0.151515\ldots$, then $100c = 15.1515\ldots$ and
$$100c - c = (15.1515\ldots) - (0.151515\ldots)$$
$$99c = 15$$
$$c = \tfrac{15}{99} = \tfrac{5}{33}$$

Proceeding similarly, convert the repeating decimal 0.090909...
into a fraction. (All repeating decimals are rational numbers, and
all rational numbers have repeating decimal representations.)

46. Repeat Problem 45 for 0.181818....

47. To see how the distributive property is behind the mechanics of long multiplication, compute each of the following and compare.

LONG MULTIPLICATION	USE OF THE DISTRIBUTIVE PROPERTY
23	$23 \cdot 12 = 23(2 + 10)$
$\times\ 12$	
───	$=$

48. For a and b real numbers, justify each step using a property in this section.

STATEMENT		REASON
1. $[a + b] + (-a) = (-a) + [a + b]$		**1.**
2.	$= [(-a) + a] + b$	**2.**
3.	$= 0 + b$	**3.**
4.	$= b$	**4.**

CALCULATOR PROBLEMS

Express each number as a decimal fraction to the capacity of your calculator. Observe the repeating decimal representation of the rational numbers and the apparent nonrepeating decimal representation of the irrational numbers.

49. (A) $\dfrac{8}{9}$ (B) $\dfrac{3}{11}$ (C) $\sqrt{5}$ (D) $\dfrac{11}{8}$

50. (A) $\dfrac{13}{6}$ (B) $\sqrt{21}$ (C) $\dfrac{7}{16}$ (D) $\dfrac{29}{111}$

Section 1-3   Linear Equations and Inequalities

- Equality
- Solving Linear Equations
- Inequality and Interval Notation
- Solving Linear Inequalities
- Applications

- **Equality**

The equal sign, =, used to join two expressions, asserts that the two expressions are names or descriptions of exactly the same object. Thus,

$a = b$

means a and b are names for the same object. Of course, $a \neq b$ means

a is not equal to b. It is interesting to note that the equality sign did not appear until rather late in history—the sixteenth century. It was introduced by the English mathematician Robert Recorde (1510–1558).

Several important properties of the equality symbol, =, follow directly from its logical meaning. These properties must hold any time the symbol is used.

Basic Properties of Equality

If a, b, and c are names of objects, then:

1. $a = a$ — REFLEXIVE PROPERTY
2. If $a = b$, then $b = a$. — SYMMETRIC PROPERTY
3. If $a = b$ and $b = c$, then $a = c$. — TRANSITIVE PROPERTY
4. If $a = b$, then either may replace the other in any statement without changing the truth or falsity of the statement. — SUBSTITUTION PRINCIPLE

The properties of equality are used extensively throughout mathematics. For example, using the symmetric property, we may reverse the left and right members of an equation any time we wish. That is,

if $\quad A = P + Prt, \quad$ then $\quad P + Prt = A$

Using the transitive property, we find that if

$$2x + 3x = (2 + 3)x \quad \text{and} \quad (2 + 3)x = 5x$$

then

$$2x + 3x = 5x$$

And, finally, if we know that

$$C = \pi D \quad \text{and} \quad D = 2R$$

then, using the substitution principle, D in the first formula may be replaced by $2R$ from the second formula to obtain

$$C = \pi(2R) = 2\pi R$$

■ Solving Linear Equations

We now turn our attention to methods of solving **first-degree** or **linear equations** in one variable—that is, to solving any equation that can be written in the following form:

LINEAR EQUATION $\quad ax + b = 0 \qquad a \neq 0$

---

where a and b are real constants and x is a variable.

The **replacement set** for a variable is defined to be the set of constants that are permitted to replace the variable, and the **solution set** for an equation is defined to be the set of elements from its replacement set that makes the equation a true statement. Any element of the solution set is called a **solution** or **root** of the equation. To **solve an equation** is to find the solution set for the equation.

Knowing what we mean by the solution set of an equation is one thing; finding it is another. To this end we introduce the idea of *equivalent equations*. After performing an operation on an equation, the resulting equation is said to be **equivalent** to the original if they both have the same solution set. A basic technique for solving equations is to perform operations on equations that produce simpler equivalent equations, and to continue the process until an equation is reached whose solution is obvious.

The properties of equality given in Theorem 1 produce equivalent equations when applied. These further properties follow directly from the basic properties of equality stated earlier.

THEOREM 1 | **Further Properties of Equality**

For a, b, and c any real numbers:

1. If $a = b$, then $a + c = b + c$.   ADDITION PROPERTY
2. If $a = b$, then $a - c = b - c$.   SUBTRACTION PROPERTY
3. If $a = b$, then $ca = cb$, $\quad c \neq 0$.   MULTIPLICATION PROPERTY
4. If $a = b$, then $\dfrac{a}{c} = \dfrac{b}{c}$, $\quad c \neq 0$.   DIVISION PROPERTY

EXAMPLE 7   Solve: $\dfrac{x+1}{3} - \dfrac{x}{4} = \dfrac{1}{2}$

Solution   If we could find a number that is exactly divisible by each denominator, then we could use the multiplication property in Theorem 1 to clear the equation of fractions. The smallest positive integer that is exactly divisible by each element in a set of positive integers is called the **least common multiple** (LCM) of the set. In this case, the LCM of 3, 4, and 2 is 12. Thus,

$$\frac{x+1}{3} - \frac{x}{4} = \frac{1}{2}$$ Multiply both sides by 12.

$$12\left(\frac{x+1}{3} - \frac{x}{4}\right) = 12 \cdot \frac{1}{2}$$ These steps can usually be done mentally.*

$$12 \cdot \frac{x+1}{3} - 12 \cdot \frac{x}{4} = 6$$

$$4(x+1) - 3x = 6$$ The equation is now free of fractions.

$$4x + 4 - 3x = 6$$

$$x = 2$$ Solution set = {2}

The check is left to the reader.

PROBLEM 7 Solve: $\dfrac{x}{5} - \dfrac{x-2}{2} = \dfrac{3}{4}$

Some equations involving variables in a denominator can be transformed into linear equations. We may proceed in essentially the same way as in the preceding example; however, the replacement set of the equation must exclude any value of the variable that will make a denominator 0. As long as we use the replacement set of the original equation, we may multiply through by the LCM of the denominators even though it contains a variable, and, according to Theorem 1, the new equation will be equivalent to the old.

EXAMPLE 8 Solve and check: $\dfrac{x}{2x-4} - \dfrac{2}{3} = \dfrac{7-2x}{3x-6}$

Solution

$$\frac{x}{2x-4} - \frac{2}{3} = \frac{7-2x}{3x-6}$$ Factor denominators.

$$\frac{x}{2(x-2)} - \frac{2}{3} = \frac{7-2x}{3(x-2)}$$ $x \neq 2$ (Why?)

*Dashed boxes indicate steps that are usually performed mentally.

1-3 Linear Equations and Inequalities

Multiply both sides by the LCM of the denominators, $6(x - 2)$, to obtain

$$6(x - 2)\frac{x}{2(x - 2)} - 6(x - 2)\frac{2}{3} = 6(x - 2)\frac{7 - 2x}{3(x - 2)}$$

$$3x - 4x + 8 = 14 - 4x$$

$$3x = 6$$

$$x = 2 \quad \text{No solution}$$

Since 2 is not in the replacement set of the original equation, the original equation has no solution. (Note that when $x = 2$ is replaced in the left and right members of the original equation, neither is defined.)

PROBLEM 8 Solve and check: $\dfrac{x - 3}{2x - 2} = \dfrac{1}{6} - \dfrac{1 - x}{3x - 3}$

EXAMPLE 9 Solve $A = P + Prt$ for P.

Solution

$A = P + Prt$ Think of P as a variable and A, r, and t as constants.

$P + Prt = A$ Symmetric property of equality

$P(1 + rt) = A$ Distributive property

$P = \dfrac{A}{1 + rt}$ Division property of equality ($rt \neq -1$)

PROBLEM 9 Solve $C = \tfrac{5}{9}(F - 32)$ for F.

It appears that any equation that can be written in the form

$$ax + b = 0 \qquad a \neq 0 \tag{1}$$

has exactly one solution. That this is true in general can be seen by solving equation (1) for x in terms of a and b.

$ax + b = 0$

$ax = -b$ Subtraction property of equality

$x = \dfrac{-b}{a}$ Division property of equality

Requiring $a \neq 0$ in equation (1) is an important restriction because without it we are able to write equations with first-degree members that have no solutions or have infinitely many solutions. For example,

$$2x - 3 = 2x + 5$$

has no solution. (Why?) And

$$3x - 4 = 5 + 3(x - 3)$$

has infinitely many solutions. (Why?) Try to solve each equation to see what happens.

■ Inequality and Interval Notation

We now define "less than" and "greater than" for the set of real numbers.

Definition of $a < b$ and $b > a$

For a and b real numbers, we say that **a is less than b** or **b is greater than a** and write

$$a < b \quad \text{or} \quad b > a$$

if there exists a positive real number p such that $a + p = b$ (or equivalently $b - a = p$).

We would certainly expect that if a positive number was added to any real number, the sum would be larger than the original. That is essentially what the definition states. When we write

a ≤ b

we mean **a is less than or equal to b**, and when we write

a ≥ b

we mean that **a is greater than or equal to b**.

The inequality symbols $<$ and $>$ have a very clear geometric interpretation on the real number line. If $a < b$, then a is to the left of b; if $c > d$, then c is to the right of d (Fig. 7).

FIGURE 7 $a < b, c > d$

It is an interesting and useful fact that for any two real numbers a and b, $a < b$, $a > b$, or $a = b$. This property (called the **trichotomy property**) is not shared with all number systems, as we will see later in this chapter when we extend the set of real numbers to the set of complex numbers.

1-3 Linear Equations and Inequalities

The double inequality $a < x \leq b$ means that $a < x$ and $x \leq b$; that is, x is between a and b, including b but not including a. Other variations on the theme, as well as a useful **interval notation**, are shown in Table 2.

TABLE 2

INTERVAL NOTATION	INEQUALITY NOTATION	LINE GRAPH
$[a, b]$	$a \leq x \leq b$	
$[a, b)$	$a \leq x < b$	
$(a, b]$	$a < x \leq b$	
(a, b)	$a < x < b$	
$[b, \infty)^*$	$x \geq b$	
(b, ∞)	$x > b$	
$(-\infty, a]$	$x \leq a$	
$(-\infty, a)$	$x < a$	

* The symbol ∞ (read "infinity") is not a number. When we write $[b, \infty)$, we are simply referring to the interval starting at b and continuing indefinitely to the right. We would never write $[b, \infty]$.

EXAMPLE 10 Write each of the following in inequality notation and graph on a real number line.

(A) $[-2, 3)$ (B) $(-4, 2)$ (C) $[-2, \infty)$ (D) $(-\infty, 3)$

Solution (A) $-2 \leq x < 3$

(B) $-4 < x < 2$

(C) $x \geq -2$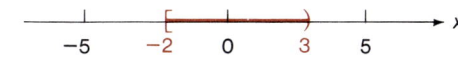

(D) $x < 3$

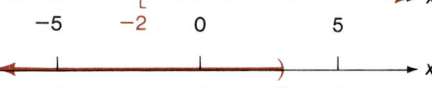

PROBLEM 10 Write each of the following in interval notation and graph on a real number line.

(A) $-3 < x \leq 3$ (B) $-1 \leq x \leq 2$
(C) $x > 1$ (D) $x \leq 2$

1 Preliminaries

■ Solving Linear Inequalities

We now turn to the problem of solving linear inequalities in one variable, such as

$$2(2x + 3) < 6(x - 2) + 10$$

and

$$-3 < 2x + 3 \leq 9$$

The **solution set** for an inequality is the set of elements from its replacement set that make the inequality a true statement. Any element of the solution set is called a **solution** of the inequality. To **solve an inequality** is to find its solution set. Two inequalities are **equivalent** if they have the same solution set. Just as with equations, we try to perform operations on inequalities that produce simpler equivalent inequalities, and to continue the process until an inequality is reached whose solution is obvious. The properties of inequalities given in Theorem 2 produce equivalent inequalities when applied.

THEOREM 2 — **Inequality Properties**

For a, b, and c any real numbers:

1. If $a < b$, then $a + c < b + c$.   ADDITION PROPERTY

     $-2 < 4$ $-2 + 3 < 4 + 3$

2. If $a < b$, then $a - c < b - c$.   SUBTRACTION PROPERTY

     $-2 < 4$ $-2 - 3 < 4 - 3$

3. If $a < b$ and c is positive, then $ca < ca$.

     $-2 < 4$ $3(-2) < 3(4)$

4. If $a < b$ and c is negative, then $ca > cb$.

     $-2 < 4$ $(-3)(-2) > (-3)(4)$

MULTIPLICATION PROPERTY (NOTE DIFFERENCE BETWEEN 3 AND 4)

5. If $a < b$ and c is positive, then $\dfrac{a}{c} < \dfrac{b}{c}$.

     $-2 < 4$ $\dfrac{-2}{2} < \dfrac{4}{2}$

6. If $a < b$ and c is negative, then $\dfrac{a}{c} > \dfrac{b}{c}$.

     $-2 < 4$ $\dfrac{-2}{-2} > \dfrac{4}{-2}$

DIVISION PROPERTY (NOTE DIFFERENCE BETWEEN 5 AND 6)

1-3 Linear Equations and Inequalities

Similar properties hold if each inequality sign is reversed, or if $<$ is replaced with $\leq$ and $>$ is replaced with $\geq$. Thus, we find that we can perform essentially the same operations on inequalities that we perform on equations. When working with inequalities, we have to be particularly careful of the use of the multiplication and division properties.

The sense of the inequality reverses if we multiply or divide both sides of an inequality statement by a negative number.

Let us sketch a proof of the multiplication property: If $a < b$, then by definition of $<$, there exists a positive number p such that $a + p = b$. If we multiply both sides of $a + p = b$ by a positive number c, we obtain $ca + cp = cb$, where cp is positive. (Why?) Thus, by definition of $<$, we see that $ca < cb$. Now if we multiply both sides of $a + p = b$ by a negative number c, we obtain $ca + cp = cb$ or $ca = cb - cp$ where cp is negative (Why?) and $-cp$ is positive. Hence, by definition of $<$, we see that $cb < ca$ or $ca > cb$.

Now let us see how the inequality properties are used to solve linear inequalities. Some examples will illustrate the process.

EXAMPLE 11 Solve and graph: $\dfrac{2x - 3}{4} + 6 \geq 2 + \dfrac{4x}{3}$

Solution

$\dfrac{2x - 3}{4} + 6 \geq 2 + \dfrac{4x}{3}$ Multiply both sides by 12, the LCM of 4 and 3.

$12 \cdot \dfrac{2x - 3}{4} + 12 \cdot 6 \geq 12 \cdot 2 + 12 \cdot \dfrac{4x}{3}$

$3(2x - 3) + 72 \geq 24 + 4 \cdot 4x$

$6x - 9 + 72 \geq 24 + 16x$

$6x + 63 \geq 24 + 16x$

$-10x \geq -39$

$x \leq 3.9$ or $(-\infty, 3.9]$ Sense reverses. (Why?)

<--------------]———→ x
             3.9

PROBLEM 11 Solve and graph: $\dfrac{4x - 3}{3} + 8 < 6 + \dfrac{3x}{2}$

EXAMPLE 12 Solve and graph: $-3 \leq 4 - 7x < 18$

Solution We proceed as in Example 11, but we try to isolate x in the middle with a coefficient of 1.

$$-3 \leq 4 - 7x < 18 \qquad \text{Subtract 4 from each member.}$$

$$-3 - 4 \leq 4 - 7x - 4 < 18 - 4$$

$$-7 \leq -7x < 14 \qquad \text{Divide each member by } -7.$$

$$\frac{-7}{-7} \geq \frac{-7x}{-7} > \frac{14}{-7} \qquad \text{Sense reverses. (Why?)}$$

$$1 \geq x > -2 \quad \text{or} \quad -2 < x \leq 1 \quad \text{or} \quad (-2, 1]$$

```
────(────────]────────► x
   -2        1
```

PROBLEM 12 Solve and graph: $-3 < 7 - 2x \leq 7$

■ **Applications**

A great many practical problems can be solved using algebraic techniques—so many, in fact, there is no one method of attack that will work for all. However, we can formulate a strategy that will help you organize your approach.

Strategy for Solving Word Problems

1. Read the problem carefully—several times if necessary; that is, until you understand the problem, know what is to be found, and know what is given.
2. Draw figures or diagrams and label known and unknown parts.
3. Look for formulas connecting the known quantities with the unknown quantities.
4. Let one of the unknown quantities be represented by a variable, say x, and try to represent all other unknown quantities in terms of x. This is an important step and must be done carefully.
5. Form an equation (or inequality) relating the unknown quantities with the known quantities.
6. Solve the equation (or inequality) and write answers to *all* parts of the problem requested.
7. Check and interpret all solutions in terms of the original problem and not just the equation (or inequality) found in step 5 (a mistake might have been made in setting up the equation or inequality in step 5).

1-3 Linear Equations and Inequalities

EXAMPLE 13 An excursion boat takes 1.5 times as long to go 360 miles up a river than to return. If the boat cruises at 15 miles/hour in still water, what is the rate of the current?

Solution Let

$$x = \text{Rate of current}$$
$$15 - x = \text{Rate of boat upstream}$$
$$15 + x = \text{Rate of boat downstream}$$

$$\text{Time upstream} = (1.5)(\text{Time downstream})$$

$$\frac{\text{Distance upstream}}{\text{Rate upstream}} = (1.5)\frac{\text{Distance downstream}}{\text{Rate downstream}}$$

$$\frac{360}{15 - x} = (1.5)\frac{360}{15 + x}$$

$$\frac{360}{15 - x} = \frac{540}{15 + x} \qquad \text{Multiply both sides by } (15 - x)(15 + x) \text{ to clear fractions.}$$

$$360(15 + x) = 540(15 - x)$$
$$5{,}400 + 360x = 8{,}100 - 540x$$
$$900x = 2{,}700$$
$$x = 3 \text{ miles/hour} \qquad \text{Rate of current}$$

Check

$$\text{Time upstream} = \frac{360}{15 - 3} = \frac{360}{12} = 30 \text{ hours}$$

$$\text{Time downstream} = \frac{360}{15 + 3} = \frac{360}{18} = 20 \text{ hours}$$

$$1.5(20) \stackrel{\checkmark}{=} 30 \text{ hours}$$

PROBLEM 13 A jet airliner takes 1.2 hours longer to fly from Paris to New York (3,600 miles) than to return. If the jet cruises at 550 miles/hour in still air, what is the average rate of the wind blowing in the direction of Paris from New York?

EXAMPLE 14 In a chemistry experiment a solution of hydrochloric acid is to be kept between 30° and 35° Celsius—that is, $30 \leq C \leq 35$. What is the range in temperature in degrees Fahrenheit? $[C = \frac{5}{9}(F - 32)]$

1 Preliminaries

Solution

$30 \leq C \leq 35$ Replace C with $\frac{5}{9}(F - 32)$.

$30 \leq \dfrac{5}{9}(F - 32) \leq 35$ Multiply each member by $\frac{9}{5}$. (Why?)

$\dfrac{9}{5} \cdot 30 \leq \dfrac{9}{5} \cdot \dfrac{5}{9}(F - 32) \leq \dfrac{9}{5} \cdot 35$

$54 \leq F - 32 \leq 63$ Add 32 to each member.

$54 + 32 \leq F - 32 + 32 \leq 63 + 32$

$86 \leq F \leq 95$ or $[86, 95]$

PROBLEM 14   A film developer is to be kept between 68° and 77° Fahrenheit—that is, $68 \leq F \leq 77$. What is the range in temperature in degrees Celsius? $(F = \frac{9}{5}C + 32)$

EXAMPLE 15   How many liters of a mixture containing 80% alcohol should be added to 5 liters of a 20% solution to yield a 30% solution?

Solution   Let x = Amount of 80% solution used.

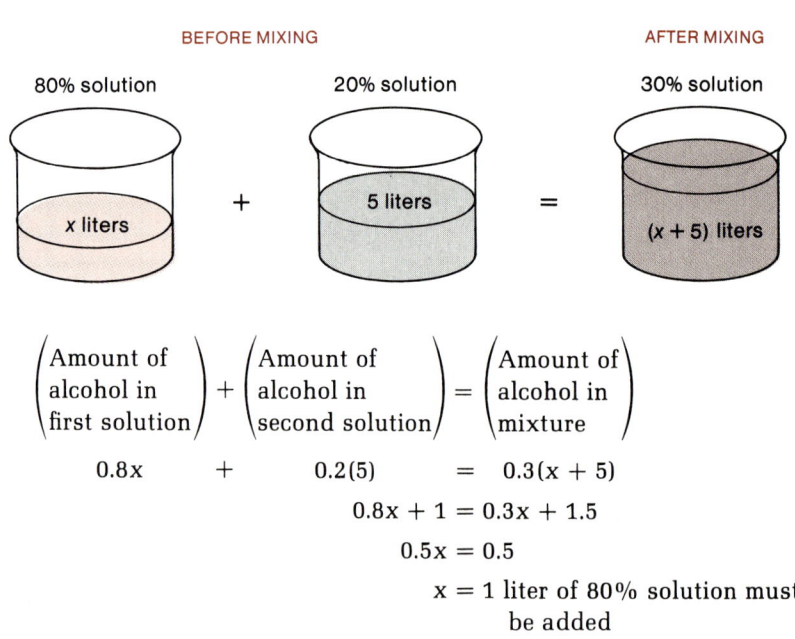

$$\begin{pmatrix} \text{Amount of} \\ \text{alcohol in} \\ \text{first solution} \end{pmatrix} + \begin{pmatrix} \text{Amount of} \\ \text{alcohol in} \\ \text{second solution} \end{pmatrix} = \begin{pmatrix} \text{Amount of} \\ \text{alcohol in} \\ \text{mixture} \end{pmatrix}$$

$0.8x \quad + \quad 0.2(5) \quad = \quad 0.3(x + 5)$

$0.8x + 1 = 0.3x + 1.5$

$0.5x = 0.5$

$x = 1$ liter of 80% solution must be added

Check $0.8(1) + 0.2(5) = 1.8$ Amount of alcohol before mixing

$0.3(1 + 5) = 1.8$ Amount of alcohol after mixing

PROBLEM 15   A chemical storeroom has a 90% acid solution and a 40% acid solution. How many centiliters must be taken from each to obtain 25 centiliters of a 50% acid solution?

Answers to Matched Problems

7. $x = \frac{5}{6}$ 8. No solution 9. $F = \frac{9}{5}C + 32$
10. (A) $(-3, 3]$
    (B) $[-1, 2]$
    (C) $(1, \infty)$
    (D) $(-\infty, 2]$
11. $x > 6$ or $(6, \infty)$
12. $5 > x \geq 0$ or $0 \leq x < 5$ or $[0, 5)$
13. 50 miles/hour
14. $20 \leq C \leq 25$ or $[20, 25]$
15. 5 centiliters of 90% solution, 20 centiliters of 40% solution

Exercise 1-3

Unless otherwise stated, the replacement set for all variables is the set of real numbers.

A Write in inequality notation and graph on a real number line.

1. $[-8, 7]$
2. $[-4, 8)$
3. $[-6, 6)$
4. $(-3, 3]$
5. $[-6, \infty)$
6. $(-\infty, 7)$

Write in interval notation and graph on a real number line.

7. $-2 < x \leq 6$
8. $-5 \leq x \leq 5$
9. $-7 < x < 8$
10. $-4 \leq x < 5$
11. $x \leq -2$
12. $x > 3$

Solve each equation and inequality, if possible. Graph the solution set for each inequality.

13. $5 + 4(t - 2) = 2(t + 7) + 1$
14. $5x - (7x - 4) - 2 = 5 - (3x + 2)$
15. $3 - m < 5(m - 3)$
16. $2(1 - u) \geq 5u$
17. $5 - \dfrac{2x - 1}{4} = \dfrac{x + 2}{3}$
18. $\dfrac{x + 3}{4} - \dfrac{x - 4}{2} = \dfrac{3}{8}$
19. $-2 - \dfrac{B}{4} \leq \dfrac{1 + B}{3}$
20. $\dfrac{y - 3}{4} - 1 > \dfrac{y}{2}$
21. $0.3x - 0.04(x + 1) = 2.04$
22. $0.02x - 0.5(x - 2) = 5.32$
23. $\dfrac{5x}{x + 5} = 2 - \dfrac{25}{x + 5}$
24. $\dfrac{3}{2x - 1} + 4 = \dfrac{6x}{2x - 1}$

25. $-4 < 5t + 6 \leq 21$ 26. $2 \leq 3m - 7 < 14$

B *Solve each equation and inequality, if possible. Graph the solution set for each inequality.*

27. $\dfrac{2x}{10} - \dfrac{3-x}{14} = \dfrac{2+x}{5} - \dfrac{1}{2}$

28. $\dfrac{2x-3}{9} - \dfrac{x+5}{6} = \dfrac{3-x}{2} + 1$

29. $\dfrac{q}{7} - 3 > \dfrac{q-4}{3} + 1$ 30. $\dfrac{p}{3} - \dfrac{p-2}{2} \leq \dfrac{p}{4} - 4$

31. $\dfrac{1}{3} - \dfrac{s-2}{2s+4} = \dfrac{s+2}{3s+6}$ 32. $\dfrac{n-5}{6n-6} = \dfrac{1}{9} - \dfrac{n-3}{4n-4}$

33. $\dfrac{5t-22}{t^2-6t+9} - \dfrac{11}{t^2-3t} - \dfrac{5}{t} = 0$

34. $\dfrac{5}{x-3} = \dfrac{33-x}{x^2-6x+9}$

35. $\dfrac{2x}{5} - \dfrac{1}{2}(x-3) \leq \dfrac{2x}{3} - \dfrac{3}{10}(x+2)$

36. $\dfrac{2}{3}(x+7) - \dfrac{x}{4} > \dfrac{1}{2}(3-x) + \dfrac{x}{6}$

37. $-12 < \tfrac{3}{4}(2-x) \leq 24$ 38. $-1 \leq \tfrac{2}{3}A + 5 \leq 11$

39. $16 < 7 - 3x \leq 31$ 40. $15 \leq 7 - \tfrac{2}{5}x \leq 21$

Solve for the indicated letter.

41. $a_n = a_1 + (n-1)d$ for d (arithmetic progressions)
42. $F = \tfrac{5}{9}C + 32$ for C (temperature scale)
43. $1/f = (1/d_1) + (1/d_2)$ for f (simple lens formula)
44. $1/R = (1/R_1) + (1/R_2)$ for R_1 (electric circuit)
45. $y = \dfrac{2x-3}{3x+5}$ for x

46. $x = \dfrac{3y+2}{y-3}$ for y

C 47. If both a and b are negative numbers and b/a is greater than 1, then is $a - b$ positive or negative?

48. If both a and b are positive numbers and b/a is greater than 1, then is $a - b$ positive or negative?

49. Prove the subtraction property of equality in Theorem 1.
50. Prove the multiplication property of equality in Theorem 1.

1-3 Linear Equations and Inequalities

51. Let m and n be real numbers with m larger than n. Then there exists a positive real number p such that $m = n + p$. Find the fallacy in the following argument:

$$m = n + p$$
$$(m - n)m = (m - n)(n + p)$$
$$m^2 - mn = mn + mp - n^2 - np$$
$$m^2 - mn - mp = mn - n^2 - np$$
$$m(m - n - p) = n(m - n - p)$$
$$m = n$$

52. Assume that $m > n > 0$; then

$$mn > n^2$$
$$mn - m^2 > n^2 - m^2$$
$$m(n - m) > (n + m)(n - m)$$
$$m > n + m$$
$$0 > n$$

But it was assumed $n > 0$. Can you find the error?

APPLICATIONS In Problems 53–56 set up appropriate inequalities and solve.

53. *Earth science.* As dry air moves upward it expands, and in so doing cools at a rate of about 5.5°F for each 1,000-foot rise up to about 40,000 feet. If the ground temperature is 70°F, then the temperature T at height h is given approximately by $T = 70 - 0.0055h$. For what range in altitude will the temperature be between 26° and −40°F?

54. *Energy.* If the power demands in a 110-volt electric circuit in a home vary between 220 and 2,750 watts, what is the range of current flowing through the circuit? ($W = EI$, where W = Power in watts, E = Pressure in volts, and I = Current in amperes.)

55. *Business and economics.* For a business to make a profit it is clear that revenue R must be greater than cost C; in short, a profit will result only if $R > C$. If a company manufactures records and its cost equation for a week is $C = 300 + 1.5x$ and its revenue equation is $R = 2x$, where x is the number of records sold in a week, how many records must be sold for the company to realize a profit?

56. *Psychology.* IQ is given by the formula

$$IQ = \frac{MA}{CA} 100$$

where MA is mental age and CA is chronological age. If

$$80 \leq IQ \leq 140$$

for a group of 12-year-old children, find the range of their mental ages.

In Problems 57–65 set up appropriate equations and solve.

57. *Business.* A stereo store marks up each item it sells 60% above wholesale price. What is the wholesale price on a record player that retails at $144?

58. *Business.* A person got a 10% raise one month and a 10% cut in salary the next month. What was the original monthly salary if the salary after the raise and cut was $1,980 per month?

59. *Chemistry.* A chemist has two solutions of sulfuric acid—a 20% solution and an 80% solution. How much of each should be used to obtain 100 liters of a 62% solution?

60. *Chemistry.* How many gallons of hydrochloric acid must be added to 12 gallons of a 30% solution to obtain a 40% solution?

61. *Rate-time.* You are at a river resort and rent a motor boat for 5 hours starting at 7 AM. You are told that the boat will travel at 8 miles/hour upstream and 12 miles/hour returning. You decide that you would like to go as far up the river as you can and still be back at noon. At what time should you turn back, and how far from the resort will you be at that time?

62. *Rate-time.* The cruising speed of an airplane is 150 miles/hour (relative to ground). You wish to hire the plane for a 3-hour sightseeing trip. You instruct the pilot to fly north as far as he can and still return to the airport at the end of the allotted time.
(A) How far north should the pilot fly if there is a 30 miles/hour wind blowing from the north?
(B) How far north should the pilot fly if there is no wind blowing?

63. *Earth science.* An earthquake emits a primary wave and a secondary wave. Near the surface of the earth the primary wave travels at about 5 miles/second, and the secondary wave at about 3 miles/second. From the time lag between the two waves arriving at a given seismic station, it is possible to estimate the distance to the quake. (The *epicenter* can be located by obtaining distance bearings at three or more stations.) Suppose a station measures a time difference of 12 seconds between the arrival of the two waves. How far is the earthquake from the station?

64. *Earth science.* A ship using sound-sensing devices above and below water recorded a surface explosion 39 seconds sooner on its underwater device than on its above-water device. If sound travels in air at about 1,100 feet/second, and in water at about 5,000 feet/second, how far away was the explosion?

65. *Puzzle.* After exactly 12 o'clock noon, what time will the hands of a clock be together again?

Section 1-4 Absolute Value in Equations and Inequalities

- Absolute Value and Distance
- Absolute Value in Equations and Inequalities

- Absolute Value and Distance

We start with a geometric definition of absolute value. If a is the coordinate of a point on a real number line, then the (nondirected) distance from the origin to a, a nonnegative quantity, is represented by $|a|$ and is referred to as the **absolute value** of a (Fig. 8). Thus, if $|x| = 5$, then x can be either -5 or 5.

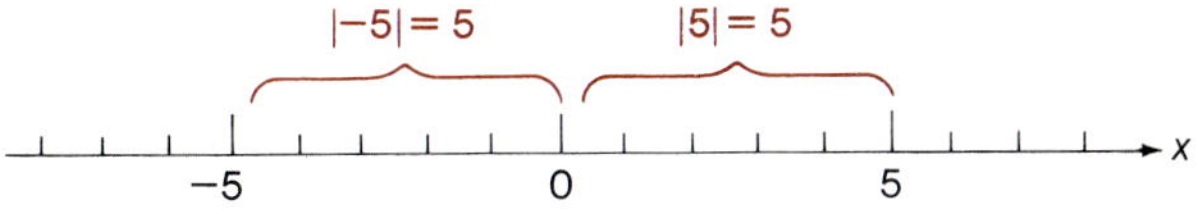

FIGURE 8 Absolute value

Symbolically, and more formally, we define absolute value as follows:

Absolute Value

$$|x| = \begin{cases} x & \text{if x is positive} \\ 0 & \text{if x is 0} \\ -x & \text{if x is negative} \end{cases}$$

[*Note:* $-x$ is positive if x is negative.]

Both the geometric and nongeometric definitions of absolute value are useful, as will be seen in the material that follows. Remember:

The absolute value of a number is never negative.

EXAMPLE 16 (A) $|7| = 7$
(B) $|\pi - 3| = \pi - 3$ Since $\pi - 3$ is nonnegative
(C) $|-7| = -(-7) = 7$
(D) $|3 - \pi| = -(3 - \pi) = \pi - 3$ Since $3 - \pi$ is negative

PROBLEM 16 Write without the absolute value sign.

(A) $|8|$ (B) $|\sqrt[3]{9} - 2|$ (C) $|-\sqrt{2}|$ (D) $|2 - \sqrt[3]{9}|$

Following the same reasoning used in Example 16B and D, it can be shown (see Problem 61 in Exercise 1-4) that:

For all real numbers a and b,

$$|b - a| = |a - b|$$

We use this result in defining the distance between two points on a real number line.

Distance between Points A and B

Let A and B be two points on a real number line with coordinates a and b, respectively. The **distance between A and B** (also called the **length of the line segment** joining A and B) is given by

$$d(A, B) = |b - a|$$

EXAMPLE 17 Find the distance between points A and B with coordinates a and b, respectively, as given.

(A) $a = 4$, $b = 9$ (B) $a = 9$, $b = 4$
(C) $a = 0$, $b = 6$ (D) $a = -3$, $b = 5$

Solution

(A) $d(A, B) = |9 - 4| = |5| = 5$

(B) $d(A, B) = |4 - 9| = |-5| = 5$

(C) $d(A, B) = |6 - 0| = |6| = 6$

(D) $d(A, B) = |5 - (-3)| = |8| = 8$

1-4 Absolute Value in Equations and Inequalities

It is clear, since $|b - a| = |a - b|$, that

d(A, B) = d(B, A)

Hence, in computing the distance between two points on a real number line, it does not matter how the two points are labeled—point A can be to the left or to the right of point B. Note also that if A is at the origin O, then

d(O, B) = |b − 0| = |b|

PROBLEM 17 Find the indicated distances given.

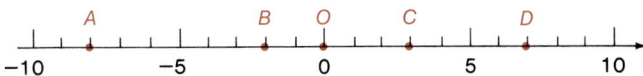

(A) $d(C, D)$ (B) $d(D, C)$ (C) $d(A, B)$
(D) $d(A, C)$ (E) $d(O, A)$ (F) $d(D, A)$

■ **Absolute Value in Equations and Inequalities**

Absolute value is frequently encountered in equations and inequalities. Some of these forms have immediate geometric interpretation.

EXAMPLE 18 Solve geometrically and graph.

(A) $|x - 3| = 5$ (B) $|x - 3| < 5$
(C) $0 < |x - 3| < 5$ (D) $|x - 3| > 5$

Solution   (A) Geometrically, $|x - 3|$ represents the distance between x and 3; thus, in $|x - 3| = 5$, x is a number whose distance from 3 is 5. That is,

$$x = 3 - 5 = -2 \quad \text{or} \quad x = 3 + 5 = 8$$

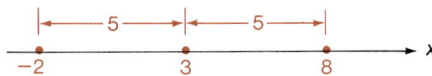

(B) Geometrically, in $|x - 3| < 5$, x is a number whose distance from 3 is less than 5; that is,

$$-2 < x < 8 \quad \text{or} \quad (-2, 8)$$

(C) The form $0 < |x - 3| < 5$ is encountered in calculus and more advanced mathematics. Geometrically, x is a number whose distance from 3 is less than 5, but x cannot equal 3. Thus,

$$-2 < x < 8 \qquad x \neq 3$$

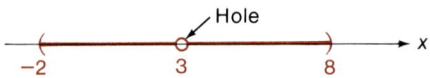

(D) Geometrically, in $|x - 3| > 5$, x is a number whose distance from 3 is greater than 5; that is,

$$x < -2 \quad \text{or} \quad x > 8$$

Note: This cannot be written as a double inequality.

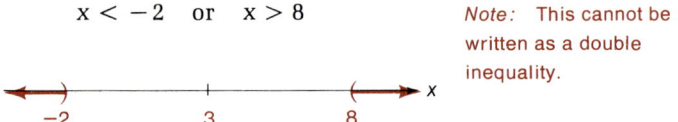

We summarize the preceding results in Table 3.

TABLE 3

FORM ($d > 0$)	GEOMETRIC INTERPRETATION	GRAPH
$\|x - c\| = d$	Distance between x and c is equal to d.	$\mid\leftarrow d \rightarrow\mid\leftarrow d \rightarrow\mid$ $c-d$ c $c+d$
$\|x - c\| < d$	Distance between x and c is less than d.	$c-d$ c $c+d$
$0 < \|x - c\| < d$	Distance between x and c is less than c, but $x \neq c$.	$c-d$ c $c+d$
$\|x - c\| > d$	Distance between x and c is greater than d.	$c-d$ c $c+d$

PROBLEM 18   Solve geometrically and graph.

(A) $|x + 2| = 6$ (B) $|x + 2| < 6$
(C) $0 < |x + 2| < 6$ (D) $|x + 2| > 6$

$\bigl[$*Hint:* $|x + 2| = |x - (-2)|.\bigr]$

Reasoning geometrically as before (noting that $|x| = |x - 0|$), we can establish Theorem 3.

1-4 Absolute Value in Equations and Inequalities

THEOREM 3

For $p > 0$,

1. $|x| = p$ is equivalent to $x = \pm p$.
2. $|x| < p$ is equivalent to $-p < x < p$.
3. $|x| > p$ is equivalent to $x < -p$ or $x > p$.

If we replace x in Theorem 3 with $ax + b$, we obtain the more general Theorem 4.

THEOREM 4

For $p > 0$,

1. $|ax + b| = p$ is equivalent to $ax + b = \pm p$.
2. $|ax + b| < p$ is equivalent to $-p < ax + b < p$.
3. $|ax + b| > p$ is equivalent to $ax + b < -p$ or $ax + b > p$.

EXAMPLE 19 Solve.

(A) $|3x + 5| = 4$ (B) $|x| < 5$
(C) $|2x - 1| < 3$ (D) $|7 - 3x| \leq 2$

Solution
(A) $|3x + 5| = 4$ (B) $|x| < 5$
$3x + 5 = \pm 4$ $\quad -5 < x < 5$
$3x = -5 \pm 4$
$x = \dfrac{-5 \pm 4}{3}$
$x = -3, -\tfrac{1}{3}$

(C) $|2x - 1| < 3$ (D) $|7 - 3x| \leq 2$
$-3 < 2x - 1 < 3$ $\quad -2 \leq 7 - 3x \leq 2$
$-2 < 2x < 4$ $\quad -9 \leq -3x \leq -5$
$-1 < x < 2$ $\quad 3 \geq x \geq \tfrac{5}{3}$
$\quad\quad\quad\quad\quad\quad\quad\quad \tfrac{5}{3} \leq x \leq 3$

PROBLEM 19 Solve.

(A) $|2x - 1| = 8$ (B) $|x| \leq 7$
(C) $|3x + 3| \leq 9$ (D) $|5 - 2x| < 9$

EXAMPLE 20 Solve.

(A) $|x| > 3$ (B) $|2x - 1| \geq 3$ (C) $|7 - 3x| > 2$

Solution (A) $|x| > 3$
$$x < -3 \quad \text{or} \quad x > 3$$

(B) $|2x - 1| \geq 3$
$$2x - 1 < -3 \quad \text{or} \quad 2x - 1 > 3$$
$$2x < -2 \quad \text{or} \quad 2x > 4$$
$$x < -1 \quad \text{or} \quad x > 2$$

(C) $|7 - 3x| > 2$
$$7 - 3x < -2 \quad \text{or} \quad 7 - 3x > 2$$
$$-3x < -9 \quad \text{or} \quad -3x > -5$$
$$x > 3 \quad \text{or} \quad x < \tfrac{5}{3}$$

PROBLEM 20 Solve.

(A) $|x| \geq 5$ (B) $|4x - 3| > 5$ (C) $|6 - 5x| > 16$

Answers to Matched Problems

16. (A) 8 (B) $\sqrt[3]{9} - 2$ (C) $\sqrt{2}$ (D) $\sqrt[3]{9} - 2$
17. (A) 4 (B) 4 (C) 6 (D) 11 (E) 8 (F) 15
18. (A) $x = -8, 4$

(B) $-8 < x < 4$ or $(-8, 4)$

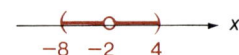

(C) $-8 < x < 4$, $x \neq -2$

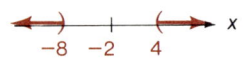

(D) $x < -8$ or $x > 4$

19. (A) $-\tfrac{7}{2}, \tfrac{9}{2}$ (B) $-7 \leq x \leq 7$ (C) $-4 \leq x \leq 2$
    (D) $-2 < x < 7$
20. (A) $x \leq -5$ or $x \geq 5$ (B) $x < -\tfrac{1}{2}$ or $x > 2$
    (C) $x < -2$ or $x > \tfrac{22}{5}$

Exercise 1-4 ■ **A** *Simplify, and write without absolute value signs. Leave radicals in simplest radical form.*

1. $|\sqrt{5}|$
2. $\left|-\tfrac{3}{4}\right|$
3. $|(-6) - (-2)|$
4. $|(-2) - (-6)|$
5. $|5 - \sqrt{5}|$
6. $|\sqrt{7} - 2|$
7. $|\sqrt{5} - 5|$
8. $|2 - \sqrt{7}|$

1-4 Absolute Value in Equations and Inequalities

Find the distance between points A and B with coordinates a and b, respectively, as given.

9. $a = -7$, $b = 5$
10. $a = 3$, $b = 12$
11. $a = 5$, $b = -7$
12. $a = 12$, $b = 3$
13. $a = -16$, $b = -25$
14. $a = -9$, $b = -17$

Find the indicated distances, given

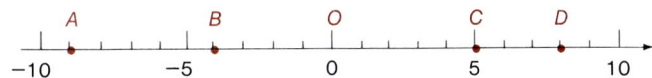

15. $d(B, O)$
16. $d(A, B)$
17. $d(O, B)$
18. $d(B, A)$
19. $d(B, C)$
20. $d(D, C)$

Solve and graph.

21. $|x| = 7$
22. $|x| = 5$
23. $|x| \leq 7$
24. $|t| \leq 5$
25. $|x| \geq 7$
26. $|x| \geq 5$
27. $|y - 5| = 3$
28. $|t - 3| = 4$
29. $|y - 5| < 3$
30. $|t - 3| < 4$
31. $|y - 5| > 3$
32. $|t - 3| > 4$
33. $|u + 8| = 3$
34. $|x + 1| = 5$
35. $|u + 8| \leq 3$
36. $|x + 1| \leq 5$
37. $|u + 8| \geq 3$
38. $|x + 1| \geq 5$

B *Solve.*

39. $|3x + 4| = 8$
40. $|2x - 3| = 5$
41. $|5x - 3| \leq 12$
42. $|2x - 3| \leq 5$
43. $|2y - 8| > 2$
44. $|3u + 4| > 3$
45. $|5t - 7| = 11$
46. $|6m + 9| = 13$
47. $|9 - 7u| < 14$
48. $|7 - 9M| < 15$
49. $|1 - \tfrac{2}{3}x| \geq 5$
50. $|\tfrac{3}{4}x + 3| \geq 9$
51. $|\tfrac{9}{5}C + 32| < 31$
52. $|\tfrac{5}{9}(F - 32)| < 40$

C *For what values of x does each of the following hold?*

53. $|x - 5| = x - 5$
54. $|x + 7| = x + 7$
55. $|x + 8| = -(x + 8)$
56. $|x - 11| = -(x - 11)$
57. $|4x + 3| = 4x + 3$
58. $|5x - 9| = (5x - 9)$
59. $|5x - 2| = -(5x - 2)$
60. $|3x + 7| = -(3x + 7)$

61. Show that $|b - a| = |a - b|$ for all real numbers a and b.
62. Prove that $|x|^2 = x^2$ for all real numbers x.

Section 1-5 Nonlinear Inequalities

- Algebraic Expressions and Polynomials
- Polynomial Inequalities
- Rational Inequalities

- Algebraic Expressions and Polynomials

Algebraic expressions are formed by using constants and variables and the algebraic operations of addition, subtraction, multiplication, division, and the taking of roots. For example,

$$\sqrt[3]{x^3 - 2x + 1} \qquad \frac{x - 5}{x^2 + 2x - 5} \qquad (3x^{-5} - 2x^{-3})^{2/3}$$

are all algebraic expressions. An algebraic expression involving only the operations of addition, subtraction, and multiplication on variables and constants, such as $x^3 - 2x^2 + 5x - 1$, is called a **polynomial**.

Polynomial in x

A **polynomial in x** is an algebraic expression of the form

$$a_n x^n + a_{n-1} x^{n-1} + \cdots + a_1 x + a_0$$

where the coefficients $a_0, a_1, \ldots, a_n$ are real numbers and n is a nonnegative integer.

Of course, we may consider polynomials in more than one variable. A polynomial in the two variables x and y is an algebraic expression formed by adding terms of the form $ax^m y^n$, where a is a real number and m and n are nonnegative integers. For example,

$$3x^3 - \sqrt{2}x^2 y + xy - \tfrac{1}{2}xy^2 + y^3 + 2x - 3$$

is a polynomial in two variables. Polynomials in three and more variables are defined in a similar way.

Polynomial forms are encountered frequently in mathematics, and for their more efficient study it is useful to classify them according to their degree. If a term in a polynomial has only one variable as a factor, then the **degree of that term** is the power of the variable. If two or more variables are present in a term as factors, then the **degree of the term** is the sum of the powers of the variables. The **degree of a polynomial** is the degree of the nonzero term with the highest degree in the polynomial. Any nonzero constant is defined to be a **polynomial of degree 0**. The number 0 is also a polynomial but is not assigned a degree.

1-5 Nonlinear Inequalities

EXAMPLE 21 (A) Polynomials in one variable:

$$x^2 - 3x + 2 \qquad 6x^3 - \sqrt{2}x - \tfrac{1}{3}$$

(B) Polynomials in several variables:

$$3x^2 - 2xy + y^2 \qquad 4x^3y^2 - \sqrt{3}xy^2z^5$$

(C) Nonpolynomials:

$$\sqrt{2x} - \frac{3}{x} + 5 \qquad \frac{x^2 - 3x + 2}{x - 3} \qquad \sqrt{x^2 - 3x + 1}$$

(D) The degree of the first term in $6x^3 - \sqrt{2}x - \tfrac{1}{3}$ is 3, the second term 1, the third term 0, and the whole polynomial 3.

(E) The degree of the first term in $4x^3y^2 - \sqrt{3}xy^2$ is 5, the second 3, and the whole polynomial 5.

PROBLEM 21 (A) Which of the following are polynomials?

$$3x^2 - 2x + 1 \qquad \sqrt{x - 3} \qquad x^2 - 2xy + y^2 \qquad \frac{x - 1}{x^2 + 2}$$

(B) Given the polynomial $3x^5 - 6x^3 + 5$, what is the degree of the first term? The second term? The whole polynomial?

(C) Given the polynomial $6x^4y^2 - 3xy^3$, what is the degree of the first term? The second term? The whole polynomial?

In addition to classifying polynomials by degree, we also call a single-term polynomial a **monomial**, a two-term polynomial a **binomial**, and a three-term polynomial a **trinomial**.

■ **Polynomial Inequalities**

You now know how to solve first-degree (linear) inequalities such as

$$3x - 7 \geq 5(x - 2) + 3$$

But how do we solve second-degree (quadratic) inequalities such as

$$x^2 - 12 < x$$

If, after collecting all nonzero terms on the left, we find that we are able to factor the left side in terms of first-degree factors, then we will be able to solve the inequality.

$$x^2 - 12 < x \quad \text{Move all nonzero terms to the left side.}$$
$$x^2 - x - 12 < 0 \quad \text{Factor left side.}$$
$$(x + 3)(x - 4) < 0$$

We are looking for values of x that will make the left side less than 0—that is, negative. What will the signs of (x + 3) and (x − 4) have to be so that their product is negative? They must have opposite signs.

Let us see whether we can determine where each of the factors is positive, negative, and 0. The point at which either factor is 0 is called a **critical point**. We will see why in a moment.

Sign analysis for (x + 3):

CRITICAL POINT	(x + 3) IS POSITIVE WHEN	(x + 3) IS NEGATIVE WHEN
x + 3 = 0	x + 3 > 0	x + 3 < 0
x = −3	x > −3	x < −3

It is useful to summarize these results on a real number line, as shown in Figure 9:

Sign of (x + 3) − − − − | + + + + + +
———————————•———————————→ x
−3
Critical point

FIGURE 9

Thus, (x + 3) is negative for values of x to the left of −3 and is positive for values of x to the right of −3.

Sign analysis for (x − 4):

CRITICAL POINT	(x − 4) IS POSITIVE WHEN	(x − 4) IS NEGATIVE WHEN
x − 4 = 0	x − 4 > 0	x − 4 < 0
x = 4	x > 4	x < 4

This is illustrated geometrically in Figure 10:

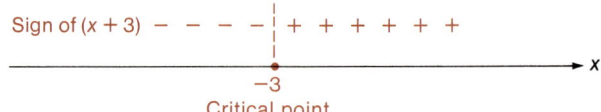

Sign of (x − 4) − − − − − − | + + +
———————————————•———————→ x
4
Critical point

FIGURE 10

Thus, (x − 4) is negative for values of x to the left of 4 and is positive for values of x to the right of 4.

1-5 Nonlinear Inequalities

Combining the results on a single real number line (Fig. 11) leads to a simple solution to the original problem.

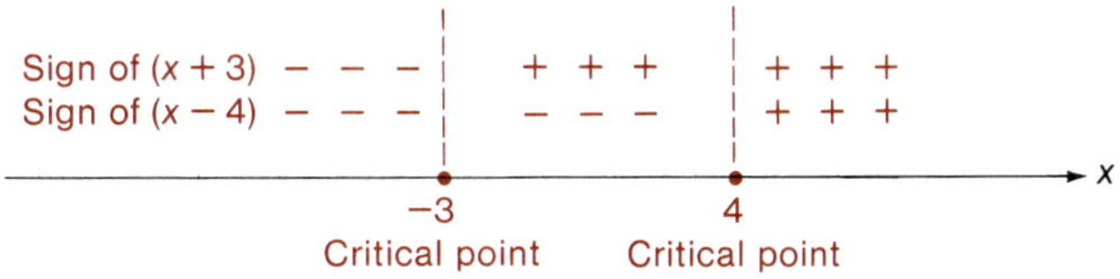

FIGURE 11

We see that the factors have opposite signs (thus their product is negative) for x between -3 and 4. We can now give the solution and graph for $x^2 - 12 \leq x$:

$-3 < x < 4$ Inequality notation

$(-3, 4)$ Interval notation

Proceeding as in the above example, we can easily prove Theorem 5, which is behind the sign-analysis method of solving second- and higher-degree inequalities as well as other types of inequalities.

THEOREM 5   The value of x at which $(ax + b)$ is 0 is called the **critical point** for $ax + b$. To the left of this critical point on a real number line $(ax + b)$ has one sign and to the right of this critical point $(ax + b)$ has the opposite sign $(a \neq 0)$.

EXAMPLE 22   Solve and graph: $3x^2 + 10x \geq 8$

Solution

$3x^2 + 10x \geq 8$ Move all nonzero terms to the left side.

$3x^2 + 10x - 8 \geq 0$ Factor the left side (if possible).

$(3x - 2)(x + 4) \geq 0$ Find critical points.

Critical points: $-4, \frac{2}{3}$

Locate the critical points on a real number line and determine the sign of each linear factor to the left and right of its critical point.

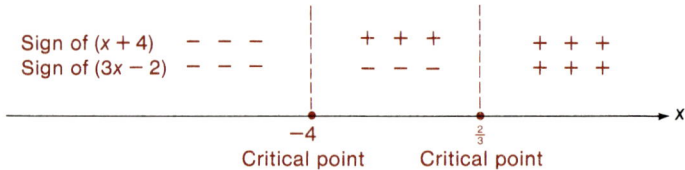

Note that the equality part of the inequality statement is satisfied at the critical points. The inequality part is satisfied when the product of the factors is positive—that is, when the factors have the same sign. From the figure we see that this happens to the left of -4 or to the right of $\frac{2}{3}$. We can now give the solution and graph.

$x \leq -4$ or $x \geq \frac{2}{3}$ Inequality notation

$(-\infty, -4] \cup [\frac{2}{3}, \infty)$ Interval notation

PROBLEM 22 Solve and graph: $2x^2 \geq 3x + 9$

EXAMPLE 23 Solve and graph: $x^3 - 4 \leq x - 4x^2$

Solution

$x^3 - 4 \leq x - 4x^2$ Move all nonzero terms to left side.
$x^3 + 4x^2 - x - 4 \leq 0$ Factor left side by grouping.
$(x^3 + 4x^2) - (x + 4) \leq 0$
$x^2(x + 4) - (x + 4) \leq 0$
$(x^2 - 1)(x + 4) \leq 0$
$(x - 1)(x + 1)(x + 4) \leq 0$ Find critical points.

Critical points: $-4, -1, 1$

Equality holds at all the critical points. The inequality holds when the left side is less than 0—that is, when the left side is negative. The left side is negative when $(x - 1)$, $(x + 1)$, and $(x + 4)$ are all negative or when one is negative and two are positive. We chart the sign of each factor on a real number line.

1-5 Nonlinear Inequalities

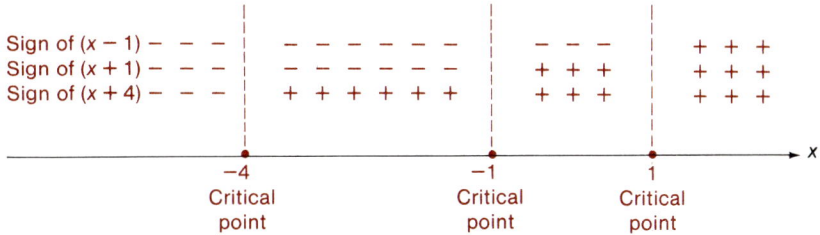

The solution is easily determined from the figure.

$x \leq -4$ or $-1 \leq x \leq 1$ Inequality notation

$(-\infty, -4] \cup [-1, 1]$ Interval notation

PROBLEM 23 Solve and graph: $x^3 + 12 > 3x^2 + 4x$

Remark: The key to solving polynomial inequalities is factoring. At this point we are able to factor only a few very special types of polynomials. In Chapter 3 procedures will be developed to help us factor polynomials of higher degree. We will then be able to apply the factoring technique of solving polynomial inequalities to a much wider class of problems.

■ Rational Inequalities

The sign-analysis technique described for solving polynomial inequalities can also be used to solve inequalities involving **rational forms** (quotients of polynomials) such as

$$\frac{x-3}{x+5} > 0 \qquad \frac{x^2+5x-6}{5-x} \leq 1$$

EXAMPLE 24 Solve and graph: $\dfrac{x^2 - x + 1}{2 - x} \geq 1$

Solution We might be tempted to start by multiplying both sides by $2 - x$ (as we would do if the inequality were an equation). However, since we do not know whether $2 - x$ is positive or negative, we do not know whether the sense of the inequality is to be changed.

We proceed instead as follows:

$$\frac{x^2 - x + 1}{2 - x} \geq 1 \quad \text{Move all nonzero terms to left side.}$$

$$\frac{x^2 - x + 1}{2 - x} - 1 \geq 0 \quad \text{Combine left side into a single fraction.}$$

$$\frac{x^2 - x + 1 - (2 - x)}{2 - x} \geq 0$$

$$\frac{x^2 - 1}{2 - x} \geq 0 \quad \text{Factor numerator.}$$

$$\frac{(x - 1)(x + 1)}{2 - x} \geq 0$$

Critical points: $-1, 1, 2$. Equality holds when $x = \pm 1$. The left side is not defined when $x = 2$.

The inequality holds when $(x - 1)$, $(x + 1)$, and $(2 - x)$ are all positive or two are negative and one is positive. We chart the sign of each on a real number line:

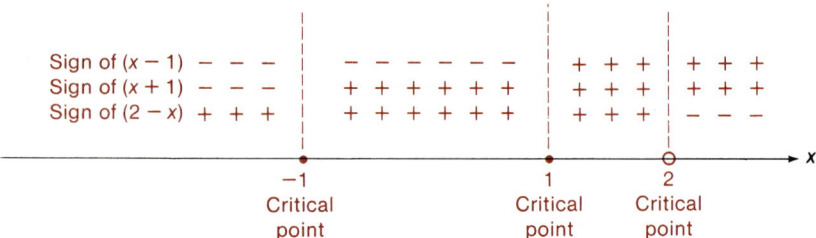

Note the sign pattern for $(2 - x)$. It is positive to the left of its critical point and negative to the right. From the figure it is easy to write the solution.

$x \leq -1 \quad \text{or} \quad 1 \leq x < 2 \quad \text{Inequality notation}$

$(-\infty, -1] \cup [1, 2) \quad \text{Interval notation}$

PROBLEM 24 Solve and graph: $\dfrac{3}{2 - x} \leq \dfrac{1}{x + 4}$

Answers to Matched Problems   **21.** (A) $3x^2 - 2x + 1, x^2 - 2xy + y^2$ (B) 5, 3, 5
(C) 6, 4, 6

22. $x \leq -\frac{3}{2}$ or $x \geq 3$
$(-\infty, -\frac{3}{2}] \cup [3, \infty)$

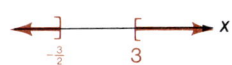

23. $-2 < x < 2$ or $x > 3$
$(-2, 2) \cup (3, \infty)$

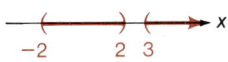

24. $-4 < x \leq -\frac{5}{2}$ or $x > 2$
$(-4, -\frac{5}{2}] \cup (2, \infty)$

Exercise 1-5 ■ Solve and graph. Express answers in both inequality and interval notation.

A
1. $x^2 - x - 12 < 0$
2. $x^2 - 2x - 8 < 0$
3. $x^2 - x - 12 \geq 0$
4. $x^2 - 2x - 8 \geq 0$
5. $x^2 < 10 - 3x$
6. $x^2 + x < 12$
7. $x^2 + 21 > 10x$
8. $x^2 + 7x + 10 > 0$
9. $x^2 \leq 8x$
10. $x^2 + 6x \geq 0$
11. $x^2 + 5x \leq 0$
12. $x^2 \leq 4x$
13. $x^2 > 4$
14. $x^2 \leq 9$

B
15. $x^2 + 9 \geq 6x$
16. $x^2 + 4 \geq 4x$
17. $x^3 + 5 \geq 5x^2 + x$
18. $x^3 + x^2 < 9x + 9$
19. $x^3 + 75 < 3x^2 + 25x$
20. $x^3 + 4x^2 \geq 4x + 16$
21. $\dfrac{x-2}{x+4} \leq 0$
22. $\dfrac{x+3}{x-1} \geq 0$
23. $\dfrac{x^2+5x}{x-3} \geq 0$
24. $\dfrac{x-4}{x^2+2x} \leq 0$
25. $\dfrac{x+4}{1-x} \leq 0$
26. $\dfrac{3-x}{x+5} \leq 0$
27. $\dfrac{1}{x} < 4$
28. $\dfrac{5}{x} > 3$
29. $\dfrac{2x}{x+3} \geq 1$
30. $\dfrac{2}{x-3} \leq -2$
31. $\dfrac{3x+1}{x+4} \leq 1$
32. $\dfrac{5x-8}{x-5} \geq 2$
33. $\dfrac{2}{x+1} \geq \dfrac{1}{x-2}$
34. $\dfrac{3}{x-3} \leq \dfrac{2}{x+2}$

C   35. $x^2 + 1 < 2x$ 36. $x^2 + 25 < 10x$
    37. $x^3 + 5x > 4x^2 + 20$ 38. $x^3 + 3x^2 + x + 3 < 0$
    39. $4x^4 + 4 \leq 17x^2$ 40. $x^4 + 36 \geq 13x^2$
    41. $|x^2 - 1| \leq 3$ 42. $\left|\dfrac{x+1}{x}\right| > 2$

Section 1-6 Complex Numbers

- Introductory Remarks
- The Complex Number System
- Complex Numbers and Radicals

- **Introductory Remarks**

The Pythagoreans (500–275 BC) found that the simple equation

$$x^2 = 2 \tag{1}$$

had no rational number solutions. If equation (1) were to have a solution, then a new kind of number had to be invented—the irrational numbers. The irrational numbers $\sqrt{2}$ and $-\sqrt{2}$ are both solutions to (1). Irrational numbers were not put on a firm mathematical foundation until the last century. The rational and irrational numbers together constitute the real number system.

Is there any need to extend the real number system further? Yes, since we find that another simple equation

$$x^2 = -1$$

has no real solutions (what real number squared is negative?). Once again, we are forced to invent a new kind of number, a number that has the possibility of being negative when it is squared. These new numbers are called **complex numbers**. The complex numbers evolved over a long period of time,* but, like the real numbers, it was not until the last century that they were placed on a firm mathematical foundation.

* BRIEF HISTORY OF COMPLEX NUMBERS

Approximate Date	Person	Event
50	Heron of Alexandria	First recorded encounter of a square root of a negative number
850	Mahavira of India	Said that a negative has no square root, since it is not a square
1545	Cardano of Italy	Solutions to cubic equations involved square roots of negative numbers.
1637	Descartes of France	Introduced the terms *real* and *imaginary*
1748	Euler of Switzerland	Used *i* for $\sqrt{-1}$
1832	Gauss of Germany	Introduced the term *complex number*

1-6 Complex Numbers

■ **The Complex Number System**

A **complex number** is a number of the form

$$a + bi$$

where a and b are real numbers and i is called the **imaginary unit**. Thus,

$3 - 2i \qquad \frac{1}{2} + 5i \qquad 2 - \frac{1}{3}i$

$0 + 3i \qquad 5 + 0i \qquad 0 + 0i$

are all complex numbers. Particular kinds of complex numbers are given special names as follows:

REAL NUMBERS: $\qquad a + 0i = a$

PURE IMAGINARY NUMBERS: $\quad 0 + bi = bi$

ZERO: $\qquad 0 + 0i = 0$

IMAGINARY UNIT: $\qquad 1i = i$

CONJUGATE OF $a + bi$: $\qquad a - bi$

Thus, we see that just as every integer is a rational number, every real number is a complex number; that is, the real numbers form a subset of the set of complex numbers. The complex number system is related to the other number systems that we have studied as shown in Figure 12.

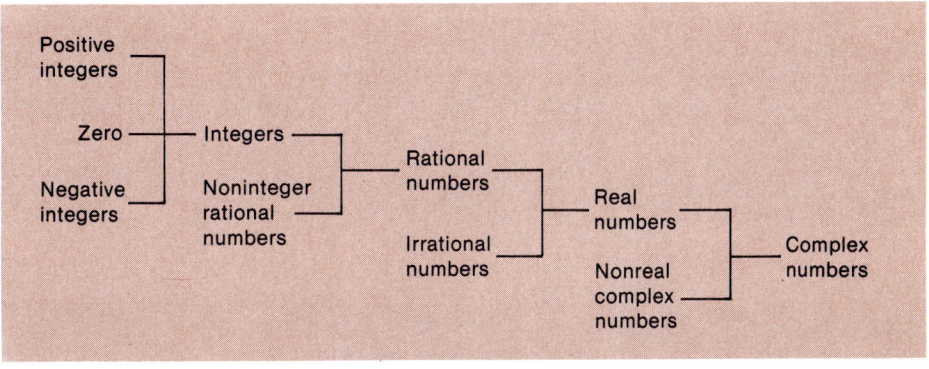

FIGURE 12

[*Note:* $a + bi$ is a real number if $b = 0$; $a + bi$ is a nonreal complex number if $b \neq 0$.]

To use complex numbers we must know how to add, subtract, multiply, and divide them. We start by defining equality, addition, and multiplication.

> EQUALITY: $a + bi = c + di$ if and only if $a = c$ and $b = d$
>
> ADDITION: $(a + bi) + (c + di) = (a + c) + (b + d)i$
>
> MULTIPLICATION: $(a + bi)(c + di) = (ac - bd) + (ad + bc)i$

The definitions, particularly the one for multiplication, may seem a little strange to you. But it turns out that if we want many of the properties for real numbers (commutative, associative, distributive, etc.) to continue to hold for complex numbers, and if we also want the possibility of having the square of a number negative, then we must define addition and multiplication in this way. Let us use the definition of multiplication to see what happens to i when it is squared:

$$i^2 = (\overset{a}{0} + \overset{b}{1}i)(\overset{c}{0} + \overset{d}{1}i)$$

$$= (\overset{a}{0} \cdot \overset{c}{0} - \overset{b}{1} \cdot \overset{d}{1}) + (\overset{a}{0} \cdot \overset{d}{1} + \overset{b}{1} \cdot \overset{c}{0})i$$

$$= -1 + 0i$$

$$= -1$$

Thus,

> $i^2 = -1$

and we have a number whose square is negative (and a solution to $x^2 = -1$). We choose to let

> $i = \sqrt{-1}$ and $-i = -\sqrt{-1}$

Fortunately, you do not have to memorize the definitions of addition and multiplication. We can show that the complex numbers under these definitions are associative and commutative and that multiplication

1-6 Complex Numbers

distributes over addition. As a consequence, we can manipulate complex numbers as if they were binomial forms in real number algebra, with the exception that i^2 is to be replaced with -1. The following example illustrates the mechanics of carrying out addition, subtraction, multiplication, and division.

EXAMPLE 25 Carry out the following operations and write each answer in the form $a + bi$.

(A) $(2 - 3i) + (6 + 2i)$ (B) $(7 - 3i) - (6 + 2i)$
(C) $(2 - 3i)(6 + 2i)$ (D) $(3 + 2i)/(5 + 3i)$

Solution

(A) $(2 - 3i) + (6 + 2i) = 2 - 3i + 6 + 2i$ Remove parentheses and combine like terms.
$= 8 - i$

(B) $(7 - 3i) - (6 + 2i) = 7 - 3i - 6 - 2i$ Remove parentheses and combine like terms.
$= 1 - 5i$

(C) $(2 - 3i)(6 + 2i) = 12 - 14i - 6i^2$ Multiply; then replace i^2 with -1.
$= 12 - 14i - 6(-1)$
$= 12 - 14i + 6$
$= 18 - 14i$

(D) To eliminate i from the denominator, we multiply the numerator and denominator by the conjugate of $5 + 3i$—that is, by $5 - 3i$:

$$\frac{3 + 2i}{5 + 3i} \cdot \frac{5 - 3i}{5 - 3i} = \frac{15 + i - 6i^2}{25 - 9i^2}$$

$$= \frac{15 + i + 6}{25 + 9}$$

$$= \frac{21 + i}{34} = \frac{21}{34} + \frac{1}{34}i \quad \text{Form } a + bi.$$

PROBLEM 25 Carry out the indicated operations and write each answer in the $a + bi$ form.

(A) $(3 + 2i) + (6 - 4i)$ (B) $(3 - 5i) - (1 - 3i)$

(C) $(2 - 4i)(3 + 2i)$ (D) $\dfrac{2 + 4i}{3 + 2i}$

EXAMPLE 26 Carry out the indicated operations and write each answer in the form $a + bi$.

(A) $(3 - 2i)^2 - 6(3 - 2i) + 13$ (B) $\dfrac{2 - 3i}{2i}$

1 Preliminaries

Solution (A) $(3 - 2i)^2 - 6(3 - 2i) + 13 = 9 - 12i + 4i^2 - 18 + 12i + 13$
$$= 9 - 12i - 4 - 18 + 12i + 13$$
$$= 0 + 0i \quad \text{or} \quad 0$$

(B) $\dfrac{2 - 3i}{2i} \cdot \dfrac{i}{i} = \dfrac{2i - 3i^2}{2i^2} = \dfrac{2i + 3}{-2} = -\dfrac{3}{2} - i$

PROBLEM 26 Carry out the indicated operations and write each answer in the form $a + bi$.

(A) $(3 + 2i)^2 - 6(3 + 2i) + 13$ (B) $\dfrac{4 - i}{3i}$

■ Complex Numbers and Radicals

Recall that we say that a is a square root of b if $a^2 = b$. It can be shown that if x is a positive or negative real number, then x has two square roots. We denote one by $\sqrt{x}$ and the other by $-\sqrt{x}$. If x is negative, the square roots of x are complex numbers. If we let $x = -a, a > 0$, then

$$\sqrt{-a} = i\sqrt{a} \qquad a > 0$$

This is readily verified by noting that $(i\sqrt{a})^2 = i^2 a = -a$. Hence, a square root of any negative real number can be written as the product of i and a square root of a positive real number. Thus,

$$\sqrt{-3} = i\sqrt{3} \qquad \sqrt{-4} = i\sqrt{4} = 2i$$

What are $\sqrt{-7}$ and $\sqrt{-9}$? [Answer: $i\sqrt{7}$ and $3i$]

EXAMPLE 27 Write in the form $a + bi$.

(A) $\sqrt{-4}$ (B) $4 + \sqrt{-5}$

(C) $\dfrac{-3 - \sqrt{-5}}{2}$ (D) $\dfrac{1}{1 - \sqrt{-9}}$

Solution (A) $\sqrt{-4} = i\sqrt{4} = 2i$

(B) $4 + \sqrt{-5} = 4 + i\sqrt{5}$

(C) $\dfrac{-3 - \sqrt{-5}}{2} = \dfrac{-3 - i\sqrt{5}}{2} = -\dfrac{3}{2} - \dfrac{\sqrt{5}}{2}i$

1-6 Complex Numbers

(D) $\dfrac{1}{1-\sqrt{-9}} = \dfrac{1}{1-3i} = \dfrac{1}{(1-3i)} \dfrac{(1+3i)}{(1+3i)}$

$= \dfrac{1+3i}{1-9i^2} = \dfrac{1+3i}{10} = \dfrac{1}{10} + \dfrac{3}{10}i$

PROBLEM 27 Write in the form $a + bi$.

(A) $\sqrt{-16}$ (B) $5 + \sqrt{-7}$ (C) $\dfrac{-5-\sqrt{-2}}{2}$ (D) $\dfrac{1}{3-\sqrt{-4}}$

Early resistance to these new numbers is suggested by the words used to name them: *complex* and *imaginary*. In spite of this early resistance, complex numbers have come into widespread use in both pure and applied mathematics. They are used extensively, for example, in electrical engineering, physics, chemistry, statistics, and aeronautical engineering. Our first use of them will be in connection with solutions of second-degree equations in the next section.

Answers to Matched Problems

25. (A) $9 - 2i$ (B) $2 - 2i$ (C) $14 - 8i$ (D) $\dfrac{14}{13} + \dfrac{8}{13}i$
26. (A) 0 (B) $-\dfrac{1}{3} - \dfrac{4}{3}i$
27. (A) $4i$ (B) $5 + i\sqrt{7}$ (C) $-\dfrac{5}{2} - (\sqrt{2}/2)i$ (D) $\dfrac{3}{13} + \dfrac{2}{13}i$

Exercise 1-6

A Perform the indicated operations and write each answer in the form $a + bi$.

1. $(2 + 4i) + (5 + i)$
2. $(3 + i) + (4 + 2i)$
3. $(-2 + 6i) + (7 - 3i)$
4. $(6 - 2i) + (8 - 3i)$
5. $(6 + 7i) - (4 + 3i)$
6. $(9 + 8i) - (5 + 6i)$
7. $(3 + 5i) - (-2 - 4i)$
8. $(8 - 4i) - (11 - 2i)$
9. $(4 - 5i) + 2i$
10. $6 + (3 - 4i)$
11. $(4i)(6i)$
12. $(3i)(8i)$
13. $-3i(2 - 4i)$
14. $-2i(5 - 3i)$
15. $(3 + 3i)(2 - 3i)$
16. $(-2 - 3i)(3 - 5i)$
17. $(2 - 3i)(7 - 6i)$
18. $(3 + 2i)(2 - i)$
19. $(7 + 4i)(7 - 4i)$
20. $(5 + 3i)(5 - 3i)$
21. $\dfrac{1}{2+i}$
22. $\dfrac{1}{3-i}$
23. $\dfrac{3+i}{2-3i}$
24. $\dfrac{2-i}{3+2i}$
25. $\dfrac{13+i}{2-i}$
26. $\dfrac{15-3i}{2-3i}$

B Convert square roots of negative numbers to complex form, perform the indicated operations, and express answers in the form $a + bi$.

27. $(2 - \sqrt{-4}) + (5 - \sqrt{-9})$
28. $(3 - \sqrt{-4}) + (-8 + \sqrt{-25})$
29. $(9 - \sqrt{-9}) - (12 - \sqrt{-25})$
30. $(-2 - \sqrt{-36}) - (4 + \sqrt{-49})$
31. $(3 - \sqrt{-4})(-2 + \sqrt{-49})$
32. $(2 - \sqrt{-1})(5 + \sqrt{-9})$
33. $\dfrac{5 - \sqrt{-4}}{7}$
34. $\dfrac{6 - \sqrt{-64}}{2}$
35. $\dfrac{1}{2 - \sqrt{-9}}$
36. $\dfrac{1}{3 - \sqrt{-16}}$

Write in the form $a + bi$.

37. $\dfrac{2}{5i}$
38. $\dfrac{1}{3i}$
39. $\dfrac{1 + 3i}{2i}$
40. $\dfrac{2 - i}{3i}$
41. $(2 - 3i)^2 - 2(2 - 3i) + 9$
42. $(2 - i)^2 + 3(2 - i) - 5$
43. Evaluate $x^2 - 2x + 2$ for $x = 1 - i$.
44. Evaluate $x^2 - 2x + 2$ for $x = 1 + i$.
45. Simplify: $i^2, i^3, i^4, i^5, i^6, i^7, i^8$
46. Simplify: $i^{12}, i^{13}, i^{14}, i^{15}, i^{16}$
47. For what real values of x and y will $(2x - 1) + (3y + 2)i = 5 - 4i$?
48. For what real values of x and y will $3x + (y - 2)i = (5 - 2x) + (3y - 8)i$?

C Perform the indicated operations and write each answer in the form $a + bi$.

49. $(a + bi) + (c + di)$
50. $(a + bi) - (c + di)$
51. $(a + bi)(a - bi)$
52. $(u - vi)(u + vi)$
53. $(a + bi)(c + di)$
54. $\dfrac{a + bi}{c + di}$
55. Show that $i^{4k} = 1, k \in N$.
56. Show that $i^{4k+1} = i, k \in N$.

Supply the reasons in the proofs for the following two theorems.

57. Theorem: The complex numbers are commutative under addition.
Proof: Let $a + bi$ and $c + di$ be two arbitrary complex numbers; then

STATEMENT	REASON
1. $(a + bi) + (c + di) = (a + c) + (b + d)i$	1.
2. $ = (c + a) + (d + b)i$	2.
3. $ = (c + di) + (a + bi)$	3.

58. *Theorem:* The complex numbers are commutative under multiplication.

Proof: Let $a + bi$ and $c + di$ be two arbitrary complex numbers; then

STATEMENT	REASON
1. $(a \cdot bi) \cdot (c + di) = (ac - bd) + (ad + bc)i$	**1.**
2. $ = (ca - db) + (da + cb)i$	**2.**
3. $ = (c + di)(a + bi)$	**3.**

Section 1-7 Quadratic Equations

- Solution by Factoring
- Solution by Square Root
- Solution by Completing the Square
- Solution by Quadratic Formula
- Applications

The next class of equations we will consider are the second-degree polynomial equations in one variable, called *quadratic equations*. A **quadratic equation** in one variable is any equation that can be written in the following form:

QUADRATIC EQUATION, STANDARD FORM: $\quad ax^2 + bx + c = 0 \qquad a \neq 0$

---

where x is a variable and a, b, and c are constants. We will refer to this form as the **standard form** for the quadratic equation.

- Solution by Factoring

If the coefficients a, b, and c are integers and are such that $ax^2 + bx + c$ can be written as the product of two first-degree factors with integral coefficients, then the quadratic equation can be quickly and easily solved. The method of solution by factoring rests on the zero property of real numbers.

Zero Property

If m and n are real numbers, then

$mn = 0$ if and only if $m = 0$ or $n = 0$ (or both).

EXAMPLE 28 Solve by factoring if possible.

(A) $6x^2 - 19x - 7 = 0$ (B) $2x^2 - 8x + 3 = 0$ (C) $2x^2 - 3x$

Solution (A) $6x^2 - 19x - 7 = 0$
$(2x - 7)(3x + 1) = 0$
$2x - 7 = 0$ or $3x + 1 = 0$
$x = \tfrac{7}{2}$ $x = -\tfrac{1}{3}$

(B) $2x^2 - 8x + 3$ cannot be factored using integers as coefficients. Another method must be used to solve this equation.

(C) $2x^2 = 3x$ Why shouldn't both members be divided by x?
$2x^2 - 3x = 0$
$x(2x - 3) = 0$
$x = 0$ or $2x - 3 = 0$
$2x = 3$
$x = \tfrac{3}{2}$

PROBLEM 28 Solve by factoring if possible.

(A) $3x^2 + 7x - 20 = 0$ (B) $2x^2 - 3x - 3 = 0$ (C) $4x^2 = 5x$

■ Solution by Square Root

We now turn our attention to quadratic equations of the form

$$ax^2 + c = 0 \qquad a \neq 0$$

that is, quadratic equations that have the first-degree term missing. The method of solution makes direct use of the definition of a square root of a number. The process is illustrated in the following example.

EXAMPLE 29 Solve by the square root method.

(A) $2x^2 - 3 = 0$ (B) $3x^2 + 27 = 0$ (C) $(x + \tfrac{1}{2})^2 = \tfrac{5}{4}$

Solution (A) $2x^2 - 3 = 0$
$x^2 = \tfrac{3}{2}$ What number squared is $\tfrac{3}{2}$?

$x = \pm\sqrt{\tfrac{3}{2}}$ or $\pm\dfrac{\sqrt{6}}{2}$

(B) $3x^2 + 27 = 0$

$\qquad x^2 = -9$ What number squared is -9?

$\qquad x = \pm\sqrt{-9}$ or $\pm 3i$

(C) $(x + \frac{1}{2})^2 = \frac{5}{4}$

$\qquad x + \frac{1}{2} = \pm\sqrt{\frac{5}{4}}$

$\qquad x = -\frac{1}{2} \pm \frac{\sqrt{5}}{2}$

$\qquad x = \dfrac{-1 \pm \sqrt{5}}{2}$

PROBLEM 29   Solve by the square root method.

(A) $3x^2 - 5 = 0$ (B) $2x^2 + 8 = 0$ (C) $(x + \frac{1}{3})^2 = \frac{2}{9}$

■ **Solution by Completing the Square**

The methods of square root and factoring are generally fast when they apply; however, there are equations, such as $2x^2 - 8x + 3 = 0$ (see Example 28B), that cannot be solved by these methods. A more general method must be developed to take care of this type of equation. The method of completing the square is such a method. This method is based on the process of transforming the standard quadratic equation

$$ax^2 + bx + c = 0$$

into the form

$$(x + A)^2 = B$$

where A and B are constants. The last equation can be easily solved by the square root method just discussed. But how do we transform the first equation into the second? The following brief discussion provides the key to the process.

What number must be added to $x^2 + bx$ so that the result is the square of a first-degree polynomial? There is an easy mechanical rule for finding this number, based on the square of the following binomials:

$$(x + m)^2 = x^2 + 2mx + m^2$$
$$(x - m)^2 = x^2 - 2mx + m^2$$

In either case, we see that the third term on the right is the square of one-half of the coefficient of x in the second term on the right. This observation leads directly to the rule for completing the square.

1 Preliminaries

> **Completing the Square**
>
> To complete the square of a quadratic of the form $x^2 + bx$, add the square of one-half the coefficient of x; that is, add $(b/2)^2$. Thus,
> $$x^2 + bx + \left(\frac{b}{2}\right)^2 = \left(x + \frac{b}{2}\right)^2$$

EXAMPLE 30 Complete the square for each of the following.

(A) $x^2 + 6x$ (B) $x^2 - 3x$ (C) $x^2 + bx$

Solution

(A) $x^2 + 6x$ Add $\left(\frac{6}{2}\right)^2$—that is, 9.

$x^2 + 6x + 9 = (x + 3)^2$

(B) $x^2 - 3x$ Add $\left(\frac{-3}{2}\right)^2$—that is, $\frac{9}{4}$.

$x^2 - 3x + \frac{9}{4} = (x - \frac{3}{2})^2$

(C) $x^2 + bx$ Add $\left(\frac{b}{2}\right)^2$—that is, $\frac{b^2}{4}$.

$x^2 + bx + \frac{b^2}{4} = \left(x + \frac{b}{2}\right)^2$

PROBLEM 30 Complete the square for each of the following.

(A) $x^2 + 10x$ (B) $x^2 - 5x$ (C) $x^2 + mx$

It is important to note that the rule for completing the square applies only to quadratic forms in which the coefficient of the second-degree term is 1. This causes little trouble, however, as you will see. We now solve two problems by the method of completing the square.

EXAMPLE 31 Solve by completing the square.

(A) $x^2 + 6x - 2 = 0$ (B) $2x^2 - 4x + 3 = 0$

Solution (A) $x^2 + 6x - 2 = 0$

$x^2 + 6x = 2$

$x^2 + 6x + 9 = 2 + 9$ Complete the square on the left by adding 9 to both members of the equation.

$(x + 3)^2 = 11$

$x + 3 = \pm\sqrt{11}$

$x = -3 \pm \sqrt{11}$

1-7 Quadratic Equations

(B) $2x^2 - 4x + 3 = 0$ Make the leading coefficient 1 by dividing by 2.

$x^2 - 2x + \tfrac{3}{2} = 0$

$x^2 - 2x = -\tfrac{3}{2}$ Complete the square.

$x^2 - 2x + 1 = -\tfrac{3}{2} + 1$

$(x - 1)^2 = -\tfrac{1}{2}$

$x - 1 = \pm\sqrt{-\tfrac{1}{2}}$

$x = 1 \pm \dfrac{\sqrt{2}}{2} i$

PROBLEM 31 Solve by completing the square.

(A) $x^2 + 8x - 3 = 0$ (B) $3x^2 - 12x + 13 = 0$

■ Solution by Quadratic Formula

Let us now consider the general quadratic equation, with unspecified coefficients,

$$ax^2 + bx + c = 0 \qquad a \neq 0$$

and solve it by completing the square exactly as we did in the preceding examples in which the coefficients were specified. To make the leading coefficient 1, multiply both members of the equation by $1/a$. Thus,

$$x^2 + \frac{b}{a}x + \frac{c}{a} = 0$$

Adding $-c/a$ to both members and then completing the square of the left member, we have

$$x^2 + \frac{b}{a}x + \frac{b^2}{4a^2} = \frac{b^2}{4a^2} - \frac{c}{a}$$

We now factor the left member and solve by the square root method.

$$\left(x + \frac{b}{2a}\right)^2 = \frac{b^2 - 4ac}{4a^2}$$

$$x + \frac{b}{2a} = \pm\sqrt{\frac{b^2 - 4ac}{4a^2}}$$

$$x = -\frac{b}{2a} \pm \frac{\sqrt{b^2 - 4ac}}{2a} \qquad \text{See Problem 69 in Exercise 1-7.}$$

1 Preliminaries

$$x = \frac{-b \pm \sqrt{b^2 - 4ac}}{2a} \qquad a \neq 0 \quad \text{QUADRATIC FORMULA}$$

The last equation is called the **quadratic formula**. It should be memorized and used to solve quadratic equations when all other methods fail.

EXAMPLE 32 Solve $2x + \frac{3}{2} = x^2$ by use of the quadratic formula. Leave the answer in simplest radical form.

Solution
$$2x + \frac{3}{2} = x^2 \qquad \text{Multiply both sides by 2.}$$

$$4x + 3 = 2x^2 \qquad \text{Write in standard form.}$$

$$2x^2 - 4x - 3 = 0 \qquad \text{Cannot be solved by factoring, so go directly to the quadratic formula.}$$

$$x = \frac{-b \pm \sqrt{b^2 - 4ac}}{2a} \qquad a = 2, \ b = -4, \ c = -3$$

$$x = \frac{-(-4) \pm \sqrt{(-4)^2 - 4(2)(-3)}}{2(2)}$$

$$x = \frac{4 \pm \sqrt{40}}{4} = \frac{4 \pm 2\sqrt{10}}{4} = \frac{2 \pm \sqrt{10}}{2}$$

$\left\{ \dfrac{2 + \sqrt{10}}{2}, \dfrac{2 - \sqrt{10}}{2} \right\}$ is the solution set for the equation.

$\left[\text{Note:} \quad \text{A common mistake is to cancel the 4's in } \dfrac{4 \pm 2\sqrt{10}}{4}. \text{ Also, some students carelessly write } 2 \pm \dfrac{\sqrt{10}}{2} \text{ for } \dfrac{2 \pm \sqrt{10}}{2}.\right]$

PROBLEM 32 Solve $x^2 - \frac{5}{2} = -3x$ by use of the quadratic formula. Leave the answer in simplest radical form.

EXAMPLE 33 Solve $5.37x^2 - 6.03x + 1.17 = 0$ to two decimal places using a hand calculator.

Solution
$$5.37x^2 - 6.03x + 1.17 = 0$$

$$x = \frac{6.03 \pm \sqrt{(-6.03)^2 - 4(5.37)(1.17)}}{2(5.37)}$$

$$x = 0.25, \ 0.87$$

CALCULATOR STEPS FOR THE SOLUTION INVOLVING THE NEGATIVE RADICAL

A: $\boxed{6.03}\ \boxed{x^2}\ \boxed{-}\ \boxed{(}\ \boxed{4}\ \boxed{\times}\ \boxed{5.37}\ \boxed{\times}\ \boxed{1.17}\ \boxed{)}\ \boxed{=}\ \boxed{\sqrt{x}}\ \boxed{+/-}\ \boxed{+}\ \boxed{6.03}\ \boxed{=}\ \boxed{\div}\ \boxed{2}\ \boxed{\div}\ \boxed{5.37}\ \boxed{=}$

P: $\boxed{6.13}\ \boxed{x^2}\ \boxed{\text{ENTER}}\ \boxed{4}\ \boxed{\text{ENTER}}\ \boxed{5.37}\ \boxed{\times}\ \boxed{1.17}\ \boxed{\times}\ \boxed{-}\ \boxed{\sqrt{x}}\ \boxed{+/-}\ \boxed{6.03}\ \boxed{+}\ \boxed{2}\ \boxed{\div}\ \boxed{5.37}\ \boxed{\div}$

1-7 Quadratic Equations

PROBLEM 33 Solve $2.79x^2 + 5.07x - 7.69 = 0$ to two decimal places using a hand calculator.

We conclude this part of the discussion by noting that $b^2 - 4ac$ in the quadratic formula is called the **discriminant**; this gives us useful information about the corresponding roots, as shown in Table 4.

TABLE 4   DISCRIMINANT AND ROOTS

Discriminant $b^2 - 4ac$	Roots of $ax^2 + bx + c = 0$ $a \neq 0$, a, b, and c real
Positive	Two distinct real roots
0	One real root
Negative	Two nonreal complex roots, one the conjugate of the other

■ Applications

We will now consider several applications that use quadratic equations in their solution. We restate the strategy for solving word problems presented in Section 1-3.

Strategy for Solving Word Problems

1. Read the problem carefully—several times if necessary; that is, until you understand the problem, know what is to be found, and know what is given.
2. Draw figures or diagrams and label known and unknown parts.
3. Look for formulas connecting the known quantities with the unknown quantities.
4. Let one of the unknown quantities be represented by a variable, say x, and try to represent all other unknown quantities in terms of x. This is an important step and must be done carefully.
5. Form an equation relating the unknown quantities with the known quantities.
6. Solve the equation and write answers to *all* parts of the problem requested.
7. Check and interpret all solutions in terms of the original problem and not just the equation found in step 5 (a mistake might have been made in setting up the equation in step 5).

EXAMPLE 34 The sum of a number and its reciprocal is $\frac{13}{6}$. Find all such numbers.

Solution Let x = The number, then

$$x + \frac{1}{x} = \frac{13}{6}$$ Multiply both sides by 6x. [*Note:* $x \neq 0$]

$$(6x)x + (6x)\frac{1}{x} = (6x)\frac{13}{6}$$

$$6x^2 + 6 = 13x$$
$$6x^2 - 13x + 6 = 0$$
$$(2x - 3)(3x - 2) = 0$$
$$2x - 3 = 0 \quad \text{or} \quad 3x - 2 = 0$$
$$x = \tfrac{3}{2} \qquad\qquad x = \tfrac{2}{3}$$

Check $\tfrac{3}{2} + \tfrac{2}{3} = \tfrac{13}{6} \qquad \tfrac{2}{3} + \tfrac{3}{2} = \tfrac{13}{6}$

PROBLEM 34 The sum of two numbers is 23 and their product is 132. Find the two numbers. [*Hint:* If one number is x, then the other number is 23 − x.]

EXAMPLE 35 An excursion boat takes 1.6 hours longer to go up a river than to return. If the rate of the current is 4 miles/hour, what is the rate of the boat in still water?

Solution Let x = Rate of boat in still water, then

x + 4 = Rate downstream

x − 4 = Rate upstream

$$\begin{pmatrix}\text{Time}\\ \text{upstream}\end{pmatrix} - \begin{pmatrix}\text{Time}\\ \text{downstream}\end{pmatrix} = 1.6$$

$$\frac{36}{x-4} - \frac{36}{x+4} = 1.6 \qquad\qquad T = \frac{D}{R}$$

$$36(x+4) - 36(x-4) = 1.6(x-4)(x+4)$$
$$36x + 144 - 36x + 144 = 1.6x^2 - 25.6$$
$$1.6x^2 = 313.6$$
$$x^2 = 196$$
$$x = \sqrt{196} = 14 \text{ miles/hour} \quad \text{Rate in still water}$$

[*Note:* $-\sqrt{196} = -14$ must be discarded, since it does not make sense in the problem.]

Check Time upstream $= \dfrac{D}{R} = \dfrac{36}{14-4} = 3.6$

Time downstream $= \dfrac{D}{R} = \dfrac{36}{14+4} = 2$

$$\dfrac{}{1.6} \quad \text{Difference of times}$$

1-7 Quadratic Equations

PROBLEM 35 Two boats travel at right angles to each other after leaving a dock at the same time. One hour later they are 25 miles apart. If one boat travels 5 miles/hour faster than the other, what is the rate of each? [*Hint:* Use the Pythagorean theorem,* remembering that distance equals rate times time.]

EXAMPLE 36 A tank can be filled in 4 hours by two pipes when both are used. How many hours are required for each pipe to fill the tank alone if the smaller pipe requires 3 hours more than the larger one? Compute answers to two decimal places.

Solution   Let

$$x = \text{Time for larger pipe to fill tank alone}$$
$$x + 3 = \text{Time for smaller pipe to fill tank alone}$$
$$4 = \text{Time for both pipes to fill tank together}$$

Then

$$\frac{1}{x} = \text{Rate for larger pipe} \left(\text{fills } \frac{1}{x} \text{ of the tank per hour}\right)$$

$$\frac{1}{x+3} = \text{Rate for smaller pipe} \left(\text{fills } \frac{1}{x+3} \text{ of the tank per hour}\right)$$

$$\frac{1}{4} = \text{Rate together} \left(\text{fills } \frac{1}{4} \text{ of the tank per hour}\right)$$

$$\begin{pmatrix}\text{Rate of}\\ \text{larger pipe}\end{pmatrix} + \begin{pmatrix}\text{Rate of}\\ \text{smaller pipe}\end{pmatrix} = \begin{pmatrix}\text{Rate}\\ \text{together}\end{pmatrix}$$

$$\frac{1}{x} + \frac{1}{x+3} = \frac{1}{4} \qquad \text{Multiply both sides by } 4x(x+3).$$

$$4(x+3) + 4x = x(x+3)$$
$$4x + 12 + 4x = x^2 + 3x$$
$$x^2 - 5x - 12 = 0$$

$$x = \frac{5 \pm \sqrt{73}}{2} \qquad \text{Why should we discard the negative answer?}$$

$$x = \frac{5 + \sqrt{73}}{2} \approx 6.77 \text{ hours} \qquad \text{Larger pipe}$$

$$x + 3 = 9.77 \text{ hours} \qquad \text{Smaller pipe}$$

* *Pythagorean theorem.* A triangle is a right triangle if and only if the square of the longest side is equal to the sum of the squares of the two shorter sides.

$$c^2 = a^2 + b^2$$

Check $\quad \dfrac{1}{6.77} + \dfrac{1}{9.77} \stackrel{?}{=} \dfrac{1}{4}$

$0.250\ 065 \stackrel{\checkmark}{\approx} 0.25$

[Note: We do not expect the check to be exact, since we rounded the answers to two decimal places. An exact check would be produced by using $x = (5 + \sqrt{73})/2$. The latter is left to the reader.]

PROBLEM 36 Two pipes can fill a tank in 3 hours when used together. Alone, one can fill the tank 2 hours faster than the other. How long will it take each pipe to fill the tank alone? Compute the answers to two decimal places.

Answers to Matched Problems

28. (A) $-4, \tfrac{5}{3}$ (B) Does not factor using integer coefficients (C) $0, \tfrac{5}{4}$
29. (A) $\pm\sqrt{\tfrac{5}{3}}$ or $\pm\sqrt{15}/3$ (B) $\pm 2i$ (C) $(-1 \pm \sqrt{2})/3$
30. (A) $x^2 + 10x + \mathbf{25} = (x + 5)^2$ (B) $x^2 - 5x + \tfrac{\mathbf{25}}{\mathbf{4}} = (x - \tfrac{5}{2})^2$
    (C) $x^2 + mx + \mathbf{(m^2/4)} = [x + (m/2)]^2$
31. (A) $-4 \pm \sqrt{19}$ (B) $(6 \pm i\sqrt{3})/3$ or $2 \pm (\sqrt{3}/3)i$
32. $(-3 \pm \sqrt{19})/2$
33. $-2.80, 0.98$
34. 11 and 12
35. 15 and 20 miles/hour
36. 5.16 and 7.16 hours

Exercise 1-7

■ *Leave all answers involving radicals in simplest radical form unless otherwise stated.*

A *Solve by factoring.*

1. $4u^2 = 8u$
2. $3A^2 = -12A$
3. $2d^2 + 15d = 8$
4. $3x^2 = 10x + 8$
5. $11x = 2x^2 + 12$
6. $8 - 10x = 3x^2$
7. $6x^2 + 5x = 4$
8. $6x^2 = 47x + 8$

Solve by the square root method.

9. $x^2 - 25 = 0$
10. $x^2 - 16 = 0$
11. $x^2 + 25 = 0$
12. $x^2 + 16 = 0$
13. $m^2 - 12 = 0$
14. $y^2 - 45 = 0$
15. $9y^2 - 16 = 0$
16. $4x^2 - 9 = 0$
17. $4x^2 + 25 = 0$
18. $16a^2 + 9 = 0$
19. $(n + 5)^2 = 9$
20. $(m - 3)^2 = 25$
21. $(d - 3)^2 = -4$
22. $(t + 1)^2 = -9$

1-7 Quadratic Equations

Solve using the quadratic formula.

23. $x^2 - 10x - 3 = 0$
24. $x^2 - 6x - 3 = 0$
25. $t^2 = 1 - t$
26. $u^2 = 1 - 3u$
27. $x^2 + 8 = 4x$
28. $y^2 + 3 = 2y$
29. $2x^2 + 1 = 4x$
30. $2m^2 + 3 = 6m$
31. $3q + 2q^2 = 1$
32. $p = 1 - 3p^2$
33. $5x^2 + 2 = 2x$
34. $7x^2 + 6x + 4 = 0$

B *Solve by completing the square.*

35. $x^2 - 6x - 3 = 0$
36. $y^2 - 10y - 3 = 0$
37. $2y^2 - 6y + 3 = 0$
38. $2d^2 - 4d + 1 = 0$
39. $3x^2 - 2x - 2 = 0$
40. $3x^2 + 5x - 4 = 0$
41. $x^2 + mx + n = 0$
42. $ax^2 + bx + c = 0, \quad a \neq 0$

Solve by any method.

43. $12x^2 + 7x = 10$
44. $9x^2 + 9x = 4$
45. $(2y - 3)^2 = 5$
46. $(3m + 2)^2 = -4$
47. $x^2 = 3x + 1$
48. $x^2 + 2x = 2$
49. $7n^2 = -4n$
50. $8u^2 + 3u = 0$
51. $2y = \dfrac{2}{y} + 3$
52. $L = \dfrac{15}{L - 2}$
53. $1 + \dfrac{8}{x^2} = \dfrac{4}{x}$
54. $\dfrac{2}{u} = \dfrac{3}{u^2} + 1$
55. $\dfrac{24}{10 + m} + 1 = \dfrac{24}{10 - m}$
56. $\dfrac{1.2}{y - 1} + \dfrac{1.2}{y} = 1$
57. $\dfrac{2}{x - 2} = \dfrac{4}{x - 3} - \dfrac{1}{x + 1}$
58. $\dfrac{3}{x - 1} - \dfrac{2}{x + 3} = \dfrac{4}{x - 2}$
59. $\dfrac{x + 2}{x + 3} - \dfrac{x^2}{x^2 - 9} = 1 - \dfrac{x - 1}{3 - x}$
60. $\dfrac{11}{x^2 - 4} + \dfrac{x + 3}{2 - x} = \dfrac{2x - 3}{x + 2}$

In Problems 61–64, solve for the indicated letters in terms of the other letters. Use positive square roots only.

61. $s = \tfrac{1}{2}gt^2$ for t
62. $a^2 + b^2 = c^2$ for a
63. $P = EI - RI^2$ for I
64. $A = P(1 + r)^2$ for r

C 65. Show that if r_1 and r_2 are the two roots of $ax^2 + bx + c = 0$, then $r_1 r_2 = c/a$.

66. For r_1 and r_2 in Problem 65, show that $r_1 + r_2 = -b/a$.

67. Show that if r_1 and r_2 are any nonzero numbers such that $r_1 r_2 = c/a$ and $r_1 + r_2 = -b/a$, then they are roots of $ax^2 + bx + c = 0$.

68. Use the results of Problems 65, 66, and 67 to check which of the following are roots of $2x^2 - 2x + 5 = 0$.
    (A) $-1, 2$ (B) $2 + \sqrt{3}, 2 - \sqrt{3}$ (C) $\frac{1}{2} - \frac{3}{2}i, \frac{1}{2} + \frac{3}{2}i$

69. In one stage of the derivation of the quadratic formula, we replaced $\pm\sqrt{(b^2 - 4ac)/4a^2}$ with $\pm\sqrt{b^2 - 4ac}/2a$. What justifies using $2a$ in place of $|2a|$?

70. Find the fallacy.

$$(n + 1)^2 = n^2 + 2n + 1$$

$$(n + 1)^2 - (2n + 1) = n^2$$

$$(n + 1)^2 - (2n + 1) - n(2n + 1) = n^2 - n(2n + 1)$$

$$(n + 1)^2 - (2n + 1) - n(2n + 1) + \frac{(2n + 1)^2}{4} = n^2 - n(2n + 1)$$

$$+ \frac{(2n + 1)^2}{4}$$

$$\left[(n + 1) - \left(\frac{2n + 1}{2}\right)\right]^2 = \left(n - \frac{2n + 1}{2}\right)^2$$

$$(n + 1) - \frac{2n + 1}{2} = n - \frac{2n + 1}{2}$$

$$n + 1 = n$$

CALCULATOR PROBLEMS Solve to two decimal places using a hand calculator.

71. $2.07x^2 - 3.79x + 1.34 = 0$ 72. $0.61x^2 - 4.28x + 2.93 = 0$
73. $4.83x^2 + 2.04x - 3.18 = 0$ 74. $5.13x^2 + 7.27x - 4.32 = 0$

Use the discriminant to determine which equations have real solutions.

75. $0.0134x^2 + 0.0414x + 0.0304 = 0$
76. $0.543x^2 - 0.182x + 0.00312 = 0$
77. $0.0134x^2 + 0.0214x + 0.0304 = 0$
78. $0.543x^2 - 0.182x + 0.0312 = 0$

APPLICATIONS *These problems are not grouped from easy (A) to difficult or theoretical (C). They are grouped according to type. However, the most difficult problems are double-starred (★★), moderately difficult problems are single-starred (★), and the easier problems are not marked.*

Numbers
79. Find two numbers such that their sum is 21 and their product is 104.

80. Find all numbers with the property that when the number is added to itself the sum is the same as when the number is multiplied by itself.

81. Find two consecutive positive even integers whose product is 168.

★82. The sum of a number and its reciprocal is $\frac{10}{3}$. Find the number.

Geometry
83. If the length and width of a 4 × 2-inch rectangle are each increased by the same amount, the area of the new rectangle will be twice the old. What are the dimensions to two decimal places of the new rectangle?

84. Find the base and height of a triangle with an area of 2 square feet if its base is 3 feet longer than its height. ($A = \frac{1}{2}bh$.)

★85. Approximately how far is the horizon of the earth from a balloon 4 miles high? Assume the radius of the earth is 4,000 miles. Estimate the answer to the nearest mile. [*Hint:* See the figure.]

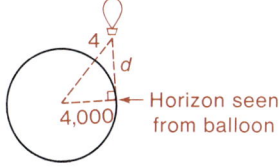

★★86. A flag has a white cross of uniform width on a red background. If the cross extends from edge to edge on a 4 × 3-feet flag, find its width if it takes up exactly half the total area of the flag.

Business and Economics
87. If P dollars are invested at r% compounded annually, at the end of 2 years the amount will be $A = P(1 + r)^2$. At what interest rate will $1,000 increase to $1,440 in 2 years? [*Note:* $A = \$1,440$ and $P = \$1,000$.]

★88. In a certain city the demand equation for stereo tapes is $q_d = 3,000/p$, where q_d is the quantity of tapes demanded on a given day if the selling price is $p per tape. (Notice that, as the price goes up, the number of tapes people are willing to buy goes down, and vice versa.) On the other hand, the supply equation is $q_s = 1,000p - 500$, where q_s is the quantity of tapes a supplier is willing to supply at $p per tape. (Notice that, as the price goes up, the number of tapes a supplier is willing to sell goes up, and vice versa.) At what price will supply equal demand; that is, at what price will $q_d = q_s$? In economic theory the price at which supply equals demand is called the *equilibrium point*, the point at which the price ceases to change.

Rate-Time

89. Two boats travel at right angles to each other after leaving the same dock at the same time. One hour later they are 13 miles apart. If one travels 7 miles/hour faster than the other, what is the rate of each?

90. A speedboat takes 1 hour longer to go 24 miles up a river than to return. If the boat cruises at 10 miles/hour in still water, what is the rate of the current?

★**91.** One pipe can fill a tank in 5 hours less than another; together they fill the tank in 5 hours. How long would it take each alone to fill the tank? Compute the answer to two decimal places.

★**92.** A new automatic machine can do a job in 1 hour less than an older machine. Together they can do the same job in 1.2 hours. How long would it take each to do the job alone?

Physics and Engineering

93. The pressure p in pounds per square foot of wind blowing at v miles/hour is $p = 0.003v^2$. If a pressure gauge on a bridge registers a wind pressure of 14.7 pounds per square foot, what is the velocity of the wind?

94. If a projectile is shot vertically into the air (from the ground) with an initial velocity of 176 feet/second, its distance y above the ground t seconds after it is shot (neglecting air resistance) is given by $y = 176t - 16t^2$.
  (A) Find the times when y is 0, and interpret the results physically.
  (B) Find the times when the projectile is 16 feet off the ground. Compute the answers to two decimal places.

★**95.** For a car traveling at a speed of v miles/hour, the least number of feet d under the best possible conditions necessary to stop it (including reaction time) is given by the empirical formula $d = 0.044v^2 + 1.1v$. Estimate the speed of a car that requires 165 feet to stop after the danger is realized.

★★**96.** A barrel 2 feet in diameter and 4 feet in height has a 1-inch diameter drainpipe in the bottom. It can be shown that the height h of the surface of the water above the bottom of the barrel at time t minutes after the drain has been opened is given by the formula

$h = [\sqrt{h_0} - (5t/12)]^2$, where h_0 is the water level above the drain at time $t = 0$. If the barrel is full and the drain opened, how long will it take to empty half the contents? [*Hint*: The problem is very easily solved if the right side of the equation is not squared.]

Section 1-8 Equations Reducible to Quadratic Form

- Equations Involving Radicals
- Equations Involving Rational Exponents

- **Equations Involving Radicals**

In solving an equation involving a radical such as

$$\sqrt{x - 1} = 2x - 3$$

it appears that we can remove the radical by squaring each side and then proceed to solve the resulting quadratic equation. Thus,

$$(\sqrt{x - 1})^2 = (2x - 3)^2$$
$$x - 1 = 4x^2 - 12x + 9$$
$$4x^2 - 13x + 10 = 0$$
$$(4x - 5)(x - 2) = 0$$
$$x = \tfrac{5}{4}, 2$$

Checking, we find that 2 is a solution, but $\tfrac{5}{4}$ is not. These results are a special case of Theorem 6.

THEOREM 6   If both sides of an equation are squared, then the solution set of the original equation is a subset of the solution set of the new equation.

Equation	Solution set
$x = 3$	$\{3\}$
$x^2 = 9$	$\{-3, 3\}$

This theorem provides us with a method of solving some equations involving radicals. It is important to remember that any new equation obtained by raising both members of an equation to the same power may have solutions (called **extraneous solutions**) that are not solutions of the original equation. On the other hand, any solution of the original equation must be among those of the new equation. Thus, every solution of the new equation must be checked in the original equation to eliminate so-called extraneous solutions.

EXAMPLE 37 Solve.

(A) $x + \sqrt{x - 4} = 4$ (B) $\sqrt{2x + 3} - \sqrt{x - 2} = 2$

Solution (A) $x + \sqrt{x - 4} = 4$

$\sqrt{x - 4} = 4 - x$ Isolate radical on one side.

$x - 4 = 16 - 8x + x^2$ Square both members.

$x^2 - 9x + 20 = 0$

$(x - 5)(x - 4) = 0$

$x = 5, 4$

Checking shows that 4 is good and 5 is extraneous; thus,

$$\{x \mid x + \sqrt{x - 4} = 4\} = \{4\}$$

(B) $\sqrt{2x + 3} - \sqrt{x - 2} = 2$

$\sqrt{2x + 3} = \sqrt{x - 2} + 2$

$2x + 3 = x - 2 + 4\sqrt{x - 2} + 4$

$x + 1 = 4\sqrt{x - 2}$

$x^2 + 2x + 1 = 16(x - 2)$

$x^2 - 14x + 33 = 0$

$(x - 11)(x - 3) = 0$

$x = 3, 11$

The work is made a little easier by having one radical on each side before squaring. After squaring, isolate the remaining radical on one side and square again.

Both solutions check; hence,

$$\{x \mid \sqrt{2x + 3} - \sqrt{x - 2} = 2\} = \{3, 11\}$$

PROBLEM 37 Solve.

(A) $x - 5 = \sqrt{x - 3}$ (B) $\sqrt{2x + 5} + \sqrt{x + 2} = 5$

■ **Equations Involving Rational Exponents**

If asked to solve the equation

$$x^{2/3} - x^{1/3} - 6 = 0$$

1-8 Equations Reducible to Quadratic Form

you might at first have trouble. But if you recognize that the equation is quadratic in $x^{1/3}$, you can solve for $x^{1/3}$ first and then solve for x. It may be convenient to make the substitution $u = x^{1/3}$, and then solve the equation

$$u^2 - u - 6 = 0$$
$$(u - 3)(u + 2) = 0$$
$$u = 3, -2$$

Replacing u with $x^{1/3}$, we obtain

$$x^{1/3} = 3 \qquad x^{1/3} = -2$$
$$x = 27 \qquad x = -8 \quad \text{Not } x = 3^{1/3} \text{ and } x = (-8)^{1/3}$$

In general, if an equation that is not quadratic can be transformed to the form

$$au^2 + bu + c = 0$$

where u is an expression in some other variable, then the equation is said to be in *quadratic form*. Once recognized as a quadratic form, an equation can often be solved using quadratic methods.

EXAMPLE 38 Solve as far as possible using techniques we have developed up to this point. Some equations may have additional complex solutions that you will not be able to find without further study in the theory of equations.

(A) $x^{10} + 6x^5 - 16 = 0$ (B) $4y^{-4} - 37y^{-2} + 9 = 0$

Solution (A) $x^{10} + 6x^5 - 16 = 0$

Let $u = x^5$ and solve:

$$u^2 + 6u - 16 = 0$$
$$(u + 8)(u - 2) = 0$$
$$u = -8, 2$$

Thus,

$$x^5 = -8 \qquad \text{or} \qquad x^5 = 2$$
$$x = \sqrt[5]{-8} = -\sqrt[5]{8} \qquad\qquad x = \sqrt[5]{2} \quad \text{Not } x = (-8)^5 \text{ and } x = 2^5$$

(B) $4y^{-4} - 37y^{-2} + 9 = 0$

Let $u = y^{-2}$, then

$$4u^2 - 37u + 9 = 0$$
$$(4u - 1)(u - 9) = 0$$
$$u = \tfrac{1}{4}, 9$$

$$y^{-2} = \tfrac{1}{4} \qquad y^{-2} = 9$$
$$\frac{1}{y^2} = \tfrac{1}{4} \qquad \frac{1}{y^2} = 9$$
$$y^2 = 4 \qquad y^2 = \tfrac{1}{9}$$
$$y = \pm 2 \qquad y = \pm \tfrac{1}{3}$$

PROBLEM 38 Solve as far as possible using techniques we have developed up to this point.

(A) $x^{2/3} - x^{1/3} - 12 = 0$ (B) $x^4 - 5x^2 + 4 = 0$
(C) $2x^{-2} - 5x^{-1} - 12 = 0$

Answers to Matched Problems

37. (A) 7   (B) 2
38. (A) 64, −27   (B) ±1, ±2   (C) $\tfrac{1}{4}, -\tfrac{2}{3}$

Exercise 1-8 ■

Find all solutions possible by the techniques that have been developed so far.

A
1. $\sqrt[3]{x+5} = 3$
2. $\sqrt[4]{x-3} = 2$
3. $\sqrt{5n+9} = n-1$
4. $m - 13 = \sqrt{m+7}$
5. $\sqrt{x+5} + 7 = 0$
6. $3 + \sqrt{2x-1} = 0$
7. $\sqrt{3x+4} = 2 + \sqrt{x}$
8. $\sqrt{3w-2} - \sqrt{w} = 2$
9. $y^4 - 2y^2 - 8 = 0$
10. $x^4 - 7x^2 - 18 = 0$
11. $x^{10} + 3x^5 - 10 = 0$
12. $x^{10} - 7x^5 - 8 = 0$
13. $2x^{2/3} + 3x^{1/3} - 2 = 0$
14. $x^{2/3} - 3x^{1/3} - 10 = 0$
15. $(m^2 - m)^2 - 4(m^2 - m) = 12$
16. $(x^2 + 2x)^2 - (x^2 + 2x) = 6$

B
17. $\sqrt{u-2} = 2 + \sqrt{2u+3}$
18. $\sqrt{3t+4} + \sqrt{t} = -3$
19. $\sqrt{3y-2} = 3 - \sqrt{3y+1}$
20. $\sqrt{2x-1} - \sqrt{x-4} = 2$
21. $\sqrt{7x-2} - \sqrt{x+1} = \sqrt{3}$
22. $\sqrt{3x+6} - \sqrt{x+4} = \sqrt{2}$
23. $3n^{-2} - 11n^{-1} - 20 = 0$
24. $6x^{-2} - 5x^{-1} - 6 = 0$
25. $9y^{-4} - 10y^{-2} + 1 = 0$
26. $4x^{-4} - 17x^{-2} + 4 = 0$
27. $y^{1/2} - 3y^{1/4} + 2 = 0$
28. $4x^{-1} - 9x^{-1/2} + 2 = 0$
29. $(m-5)^4 + 36 = 13(m-5)^2$
30. $(x-3)^4 + 3(x-3)^2 = 4$

C
31. $\sqrt{5-2x} - \sqrt{x+6} = \sqrt{x+3}$
32. $\sqrt{2x+3} - \sqrt{x-2} = \sqrt{x+1}$

Solve Problems 33–36 two ways: by squaring and by substitution.

33. $m - 7\sqrt{m} + 12 = 0$ **34.** $y - 6 + \sqrt{y} = 0$

35. $t - 11\sqrt{t} + 18 = 0$ **36.** $x = 15 - 2\sqrt{x}$

Section 1-9 Chapter Review

IMPORTANT TERMS AND SYMBOLS

1-1 Sets. Set; set notation: listing method, rule method; element; member; empty set; null set; variable; constant; subset; union; disjoint; intersection; universal set; complement; Venn diagram; $\in$; $\notin$; $\emptyset$; $A \subset B$; $A = B$; $A \cup B$; $A \cap B$; U; A'

1-2 Algebra and Real Numbers. The real number system, natural numbers, integers, rational numbers, irrational numbers, the real number line, coordinate, origin, positive real numbers, negative real numbers, commutative property, associative property, identities, inverses, distributive property, subtraction, division, properties of negatives, zero properties, fraction properties, numerator, denominator, N, Z, Q, R

1-3 Linear Equations and Inequalities. Equal to; not equal to; basic properties: reflexive property, symmetric property, transitive property, substitution principle; linear equation; first-degree equation; replacement set; solution; root; solution set; solving an equation; equivalent equations; further properties of equality: addition property, subtraction property, multiplication property, division property; least common multiple (LCM); less than; greater than; less than or equal to; greater than or equal to; trichotomy; interval notation; solving an inequality; solution; solution set; equivalent inequalities; inequality properties: addition property, subtraction property, multiplication property, division property; $=$; $\neq$; $<$; $>$; $\leq$; $\geq$; ∞

1-4 Absolute Value in Equations and Inequalities. Absolute value, distance between two points, absolute value in equations and inequalities, $|x|$, $|x - a|$, $d(A, B)$

1-5 Nonlinear Inequalities. Algebraic expression, polynomial, degree, monomial, binomial, trinomial, polynomial inequalities, rational inequalities, sign-analysis technique of solving

1-6 Complex Numbers. Complex number, pure imaginary number, imaginary unit, conjugate, equality, addition, subtraction, multiplication, division, complex numbers and radicals, $a + bi$, $i = \sqrt{-1}$, $i^2 = -1$, $a - bi$ is the conjugate of $a + bi$, $\sqrt{-a} = i\sqrt{a}$ $(a > 0)$

1-7 Quadratic Equations. Quadratic equation, solution by factoring, solution by square root, solution by completing the square, solution by quadratic formula, $ax^2 + bx + c = 0$ $(a \neq 0)$, $x = \left(-b \pm \sqrt{b^2 - 4ac}\right)/2a$

1-8 Equations Reducible to Quadratic Form. Equations involving radicals, equations involving rational exponents

Exercise 1-9 Chapter Review

Work through all the problems in this chapter review and check answers in the back of the book. (Answers to all problems are there, and following each answer is a number in italics indicating the section in which that type of problem is discussed.) Where weaknesses show up, review appropriate sections in the text. When you are satisfied that you know the material, take the practice test following this review.

A Problems 1 and 2 refer to the following sets:

$A = \{1, 2, 3, 4, 5, 6\}$ $\quad$ $B = \{1, 3, 5\}$
$C = \{3, 4, 5\}$ $\quad$ $D = \{2, 4, 6\}$ $\quad$ $E = \{5, 1, 3\}$

1. Find each of the following.
   - (A) $B \cup C$
   - (B) $B \cap C$
   - (C) $B \cap D$
   - (D) $A \cap C$
   - (E) B' relative to A

2. Indicate true (T) or false (F).
   - (A) $B \subset C$
   - (B) $D \subset A$
   - (C) $3 \in A$
   - (D) $3 \notin D$
   - (E) $B = E$
   - (F) $E \subset B$

3. $\{x \mid 6x^2 = 11x + 10\} = \{\text{List elements}\}$

4. Given the amount formula for simple interest $S = P + I$ and the simple interest formula $I = Prt$, what equality property permits us to write $S = P + Prt$?

Solve.

5. $0.05x + 0.25(30 - x) = 3.3$

6. $\dfrac{5x}{3} - \dfrac{4 + x}{2} = \dfrac{x - 2}{4} + 1$

Solve and graph Problems 7–11.

7. $3(2 - x) - 2 \leq 2x - 1$

8. $|y + 9| < 5$

9. $|3 - 2x| \leq 5$

10. $x^2 + x < 20$

11. $x^2 \geq 4x + 21$

12. Perform the indicated operations and write the answers in the form $a + bi$.
    - (A) $(-3 + 2i) + (6 - 8i)$
    - (B) $(3 - 3i)(2 + 3i)$
    - (C) $\dfrac{13 - i}{5 - 3i}$

Solve Problems 13–18.

13. $2x^2 - 7 = 0$

14. $2x^2 = 4x$

15. $2x^2 = 7x - 3$

16. $m^2 + m + 1 = 0$

17. $y^2 = \tfrac{3}{2}(y + 1)$

18. $\sqrt{5x - 6} - x = 0$

B 19. Let H be the set of all numbers x such that $10x + 11 = 6/x$.
  (A) Denote H by the rule method.
  (B) Denote H by the listing method.

20. Let $A = \{x \mid -1 \leq x < 2,\ x \text{ an integer}\}$
    $B = \{x \mid x < 4,\ x \text{ a natural number}\}$
  (A) Find $A \cup B$. (B) Find $A \cap B$. (C) Is $-3 \in B$?
  (D) Is $\emptyset \subset A$? (E) Is $B \subset A$? (F) Is $A \subset B$?

21. Which of the following sets is the empty set: $\emptyset$, $\{0\}$, $\{\emptyset\}$?

In Problems 22–27, each statement illustrates the use of one of the following real number properties or definitions. Indicate which.

  Commutative   Identity   Division
  Associative   Inverse   Zero
  Distributive   Subtraction   Negatives

22. $(-3) - (-2) = (-3) + [-(-2)]$
23. $3y + (2x + 5) = (2x + 5) + 3y$
24. $(2x + 3)(3x + 5) = (2x + 3)3x + (2x + 3)5$
25. $3 \cdot (5x) = (3 \cdot 5)x$
26. $\dfrac{a}{-(b-c)} = -\dfrac{a}{b-c}$
27. $3xy + 0 = 3xy$

28. Indicate true (T) or false (F):
  (A) An integer is a rational number and a real number.
  (B) An irrational number has a repeating decimal representation.

29. Give an example of an integer that is not a natural number.

Solve.

30. $\dfrac{7}{2-x} = \dfrac{10-4x}{x^2+3x-10}$
31. $\dfrac{u-3}{2u-2} = \dfrac{1}{6} - \dfrac{1-u}{3u-3}$

In Problems 32–35 solve and graph.

32. $\dfrac{x+3}{8} \leq 5 - \dfrac{2-x}{3}$
33. $|3x - 8| > 2$
34. $\dfrac{1}{x} < 2$
35. $\dfrac{3}{x-4} \leq \dfrac{2}{x-3}$

36. If the coordinates of A and B on a real number line are -8 and -2, respectively, find:
  (A) $d(A, B)$ (B) $d(B, A)$

37. Perform the indicated operations and write the final answers in the form $a + bi$.
  (A) $(3 + i)^2 - 2(3 + i) + 3$ (B) i^{27}

38. Convert to $a + bi$ forms, perform the indicated operations, and write the final answers in $a + bi$ form.

(A) $(2 - \sqrt{-4}) - (3 - \sqrt{-9})$

(B) $\dfrac{2 - \sqrt{-1}}{3 + \sqrt{-4}}$

(C) $\dfrac{4 + \sqrt{-25}}{\sqrt{-4}}$

Find all solutions possible using techniques we have developed so far.

39. $\left(u + \dfrac{5}{2}\right)^2 = \dfrac{5}{4}$

40. $1 + \dfrac{3}{u^2} = \dfrac{2}{u}$

41. $\dfrac{x}{x^2 - x - 6} - \dfrac{2}{x - 3} = 3$

42. $2x^{2/3} - 5x^{1/3} - 12 = 0$

43. $m^4 + 5m^2 - 36 = 0$

44. $\sqrt{y - 2} - \sqrt{5y + 1} = -3$

Solve Problems 45 and 46 for the indicated variable in terms of the other variables.

45. $P = M - Mdt$ for M (mathematics of finance)

46. $P = EI - RI^2$ for I (electrical engineering)

C 47. Is $A \cup B$ or $A \cap B$ defined by $\{x \mid x \in A$ and $x \in B\}$?

48. Indicate true (T) or false (F).
  (A) If $A \cap B = A$, then $A \subset B$.
  (B) If $A \cup B = A$, then $A \subset B$.
  (C) If $A \subset B$, then $A \cup B = B$.

49. Write $0.545\ 454\ 54\ldots$ in the form a/b, reduced to lowest terms, where a and b are positive integers.

50. Evaluate: $(a + bi)\left(\dfrac{a}{a^2 + b^2} - \dfrac{b}{a^2 + b^2}i\right)$, $a, b \neq 0$

Solve Problems 51–53.

51. $2x > \dfrac{x^2}{5} + 5$

52. $\dfrac{x^2}{4} + 4 \geq 2x$

53. $\left|x - \dfrac{8}{x}\right| \geq 2$

54. Solve by substitution and also by squaring: $x - 8\sqrt{x} + 15 = 0$

APPLICATIONS

55. *Chemistry.* A chemist has 1,200 milliliters of a 60% acid solution. How much should be drained off and replaced with pure acid to obtain the same amount of a 75% acid solution?

56. *Business.* From a survey of 1,000 residences, it was found that 850 had dead-bolt locks, 350 had alarm systems, and 275 had both.
  (A) How many residences had either dead bolts or an alarm system?
  (B) How many residences had neither dead bolts nor an alarm system?

57. *Numbers.* Find a number such that when its reciprocal is subtracted from the number the difference is $\frac{16}{15}$.

58. *Cost analysis.* Cost equations for manufacturing companies are often quadratic in nature. (At very high or very low outputs, the costs are more per unit because of inefficiency of plant operation at these extremes.) If the cost equation for manufacturing transistor radios is $C = x^2 - 10x + 31$, where C is the cost of manufacturing x units per week (both in thousands), find (A) the output for a $15,000 weekly cost and (B) the output for a $6,000 weekly cost.

59. *Break-even analysis.* The manufacturing company in Problem 58 sells its transistor radios for $3 each. Thus, its revenue equation is $R = 3x$, where R is revenue and x is the number of units sold per week (both in thousands). Find the break-even points for the company—that is, the output at which revenue equals cost.

Practice Test Chapter 1

Take this practice test as if it were a graded test. Allow yourself up to 50 minutes. Work the problems without looking back in the chapter. Correct your work using the answers (keyed to appropriate sections) in the back of the book.

Problems 1 and 2 refer to the following sets:

$A = \{-3, -2, -1, 0, 1, 2, 3\}$ $B = \{-2, 0, 2\}$
$C = \{1, 2, 3\}$ $D = \{-3, -1, 1, 3\}$ $E = \{-1, 1, -3, 3\}$

1. One of the following is false. Indicate which one.
 (A) $B \cup C = \{-2, 0, 1, 2, 3\}$ (B) $B \cap C = \{0, 2\}$
 (C) $B \cap D = \emptyset$ (D) D' relative to $A = B$

2. One of the following is false. Indicate which one.
 (A) $B \subset A$ (B) $\emptyset \subset D$
 (C) $\{1, 2, 3\} \in A$ (D) $E \subset D$

3. Each statement illustrates the use of one of the following real number properties or definitions. Indicate which.

Commutative	Identity	Division
Associative	Inverse	Zero
Distributive	Subtraction	Negatives

 (A) $(xy)z = z(xy)$
 (B) $(a + b)(x + y) = (a + b)x + (a + b)y$
 (C) $(-3) - (-7) = (-3) + [-(-7)]$
 (D) $1(7x + 3y) = 7x + 3y$
 (E) $(2x + 3) + (3x + 2) = 2x + [3 + (3x + 2)]$

4. Indicate true (T) or false (F).
   (A) A natural number is an integer and a real number.
   (B) A rational number has a nonrepeating decimal representation.

5. Evaluate $\dfrac{x^2 + 3}{x}$ for $x = 1 - 2i$. Write the final answer in the form $a + bi$.

Solve and graph. Write each solution using inequality notation and interval notation.

6. $0.05x + 0.25(20 - x) \geq 1.4$

7. $|3 - 2x| < 5$

8. $|x - 3| \geq 4$

9. $\dfrac{(x - 2)^2}{4 - x} \leq 1$

Solve.

10. $\dfrac{3}{x^2 + x} + \dfrac{1}{x} + \dfrac{3}{x + 1} = 0$

11. $1 + \dfrac{7}{x^2} = \dfrac{4}{x}$

12. $\dfrac{1}{x^2 - 4} + \dfrac{3x}{2 - x} = -2$

13. $x^{2/5} - x^{1/5} - 2 = 0$

14. $4x^{-4} - 7x^{-2} - 2 = 0$

15. $\sqrt{3x + 1} - \sqrt{x + 4} = 1$

16. An excursion boat takes 1 hour longer to go 24 miles up a river than to return. If the boat's speed in still water is 10 miles/hour, what is the rate of the current?

17. A chemical storeroom has an 80% alcohol solution and a 30% alcohol solution. How many milliliters of each should be used to obtain 50 milliliters of a 60% alcohol solution?

Graphs and Functions ■2

2-1 Rectangular Coordinate System; Graphing
2-2 Relations and Functions
2-3 Functions: Graphs and Properties
2-4 Linear Relations and Functions
2-5 Graphing Polynomial Functions
2-6 Graphing Rational Functions
2-7 Composite and Inverse Functions
2-8 Variation
2-9 Chapter Review

A natural design of mathematical interest. Can you guess the source? See the back of the book.

Chapter 2 ■ Graphs and Functions

Section 2-1 Rectangular Coordinate System; Graphing

- Cartesian Coordinate System
- Graphing: Point by Point
- Symmetry
- Distance between Two Points
- Circles

■ Cartesian Coordinate System

Just as we formed a **real number line** by establishing a one-to-one correspondence between the points on a line and the elements in the set of real numbers, we can form a **real plane** by establishing a one-to-one correspondence between the points in a plane and elements in the set of all ordered pairs of real numbers. This can be done by means of a Cartesian coordinate system.*

Recall that to form a **Cartesian (rectangular) coordinate system**, we select two real number lines, one vertical and one horizontal, and let them cross through their origins (0's) as indicated in Figure 1. Up and to

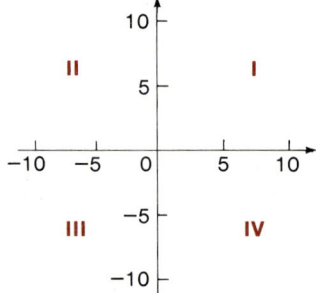

FIGURE 1 Cartesian coordinate system

---

* Named after René Descartes (1596–1650), the French philosopher-mathematician who is generally recognized as the founder of analytic geometry—the wedding of algebra and plane geometry.

2-1 Rectangular Coordinate System; Graphing

the right are the usual choices for the positive directions. These two number lines are called the **vertical axis** and the **horizontal axis**, or (together) the **coordinate axes**. The coordinate axes divide the plane into four parts called **quadrants**. The quadrants are numbered counterclockwise from I to IV.

Pick a point P in the plane at random (Fig. 2). Pass horizontal and vertical lines through the point. The vertical line will intersect the horizontal axis at a point with coordinate a, and the horizontal line will intersect the vertical axis at a point with coordinate b. These two numbers written as the ordered pair*

(a, b)

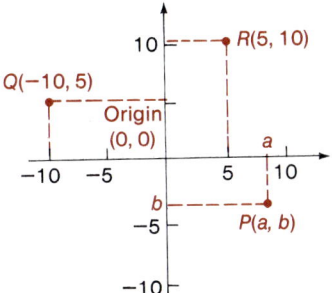

FIGURE 2

form the **coordinates** of the point P. In Figure 2, the coordinates of point Q are $(-10, 5)$ and of point R are $(5, 10)$. The first coordinate a of the coordinates of point P is also called the **abscissa** of P; the second coordinate b of the coordinates of point P is called the **ordinate** of P. The abscissa of Q in Figure 2 is -10, and the ordinate of Q is 5. The point with coordinates $(0,0)$ is called the **origin**.

We know that the coordinates (a, b) exist for each point in the plane, since every point on each axis has a real number associated with it. Hence, by the procedure just described, each point located in the plane can be labeled with a unique pair of real numbers. Conversely, by reversing the process, each pair of real numbers can be associated with a unique point in the plane. Thus, we have established *a one-to-one correspondence between the points in a plane and the elements in the*

---

* An **ordered pair** of real numbers is a pair of numbers in which the order is specified. We have now used (a, b) as the coordinates of a point and as an interval on a real number line. The context in which (a, b) is used will determine its meaning.

set of all ordered pairs of real numbers. This result is often referred to as the **fundamental theorem of analytic geometry**.

■ Graphing: Point by Point

Because of the fundamental theorem of analytic geometry, we are now in a position to look at algebraic forms geometrically and to look at geometric forms algebraically. We start by considering an equation in two variables, say

$$y = x^2 - 4 \tag{1}$$

A **solution** to equation (1) is an ordered pair of numbers (a, b) such that

$$b = a^2 - 4$$

The **solution set** of equation (1) is the set of all of its solutions. More formally,

Solution set of equation (1) = $\{(x, y) | y = x^2 - 4\}$

that is, the set of all ordered pairs (x, y) such that $y = x^2 - 4$.

To find the solutions of equation (1), we simply replace one of the variables with a number and solve for the other variable. For example, if $x = 2$, then $y = 2^2 - 4 = 0$, and the ordered pair $(2, 0)$ is a solution. Continuing in the same way, assigning different values for x and solving for y (or vice versa), we can obtain as many solutions in the solution set as we please. Now, each solution in the solution set (since it is an ordered pair of real numbers) is the coordinates of a point in a Cartesian coordinate system. The set of all points in a Cartesian coordinate system that have coordinates from the solution set form the **graph** of the given equation. Thus, **to graph an equation in two variables** is to graph its solution set.

Returning to equation (1), we find that its solution set has infinitely many elements and its graph will extend off any paper we might choose, no matter how large. Thus, **to sketch a graph of an equation**, we include enough points from its solution set so that what remains is apparent.

EXAMPLE 1   Sketch a graph of $y = x^2 - 4$.

Solution   We make up a table of ordered pairs (solutions) of numbers that satisfy the given equation:

x	−4	−3	−2	−1	0	1	2	3	4
y	12	5	0	−3	−4	−3	0	5	12

If, after plotting these, there remain certain regions of ambiguity in the graph, we plot enough additional solutions to resolve this ambiguity. These solutions are plotted and joined with a smooth curve.

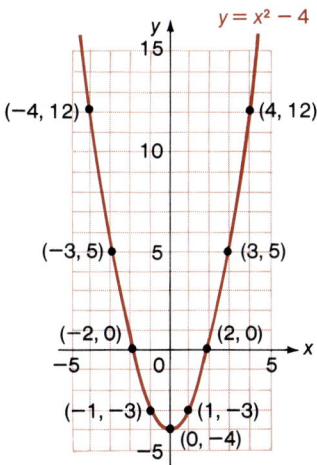

The resulting figure is called a **parabola**. Notice that if we fold the paper along the y axis, the right side will match the left side. We say that the graph is **symmetric with respect to the y axis** and call the y axis the **axis of the parabola**. (More will be said about parabolas later in the text.)

PROBLEM 1 Sketch a graph of $y^2 = x$.

The procedure used to sketch the graph of $y = x^2 - 4$ in Example 1 is called **point-by-point plotting**. As equations get more involved, this basic approach to graphing is substantially aided by the use of hand calculators. In addition to calculators, there are mathematical aids to graphing that can speed up the process significantly. A number of mathematical aids will be discussed in this book, and additional powerful aids to graphing are developed in courses on calculus.

■ Symmetry

We noticed that the graph of $y = x^2 - 4$ in Example 1 is *symmetric with respect to the y axis*; that is, the two parts of the graph coincide if the paper is folded along the y axis. Similarly, we say that a graph is *symmetric with respect to the x axis* if the parts above and below the x axis coincide when the paper is folded along the x axis. In general, we define symmetry with respect to the y axis, x axis, and origin as follows:

Symmetry

A graph is **symmetric with respect to**:

1. **The y axis** if $(-a, b)$ is on the graph whenever (a, b) is on the graph.
2. **The x axis** if $(a, -b)$ is on the graph whenever (a, b) is on the graph.
3. **The origin** if $(-a, -b)$ is on the graph whenever (a, b) is on the graph.

Figure 3 illustrates these three types of symmetry.

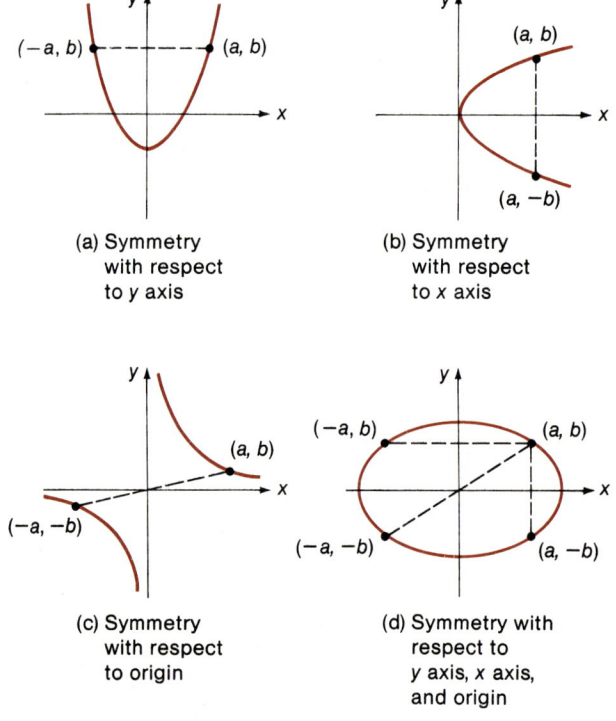

(a) Symmetry with respect to y axis

(b) Symmetry with respect to x axis

(c) Symmetry with respect to origin

(d) Symmetry with respect to y axis, x axis, and origin

FIGURE 3 Symmetry

2-1 Rectangular Coordinate System; Graphing

Given an equation, if we could determine the symmetry properties of its graph ahead of time, we could save a lot of time and energy in sketching the graph. For example, if we knew that the graph of $y = x^2 - 4$ in Example 1 were symmetric with respect to the y axis, we would have to carefully sketch only the right side of the graph; then reflect the result across the y axis to obtain the whole sketch—the point-by-point plotting would be cut in half!

The tests for symmetry (based on the preceding discussion) are given in Table 1. These tests are easily applied and are very helpful aids to graphing.

TABLE 1 TESTS FOR SYMMETRY

Symmetry with Respect to the	Equation Remains Unchanged if
y axis	x is replaced with $-x$
x axis	y is replaced with $-y$
Origin	x and y are replaced with $-x$ and $-y$

Using the tests on $y = x^2 - 4$ from Example 1, we replace x with $-x$ and observe that the equation does not change:

$$y = (-x)^2 - 4$$
$$= x^2 - 4$$

Thus, according to the test, the graph is symmetric with respect to the y axis. Note that if we replace y with $-y$, the equation will change:

$$-y = x^2 - 4$$

or

$$y = 4 - x^2$$

EXAMPLE 2 Test for symmetry and graph.

(A) $y = x^3$ (B) $x^2 + 4y^2 = 36$

Solution (A) Substitute $-x$ for x and $-y$ for y in $y = x^3$, and observe that the equation remains unchanged:

$$-y = (-x)^3$$
$$-y = -x^3$$
$$y = x^3$$

Thus, the graph is symmetric with respect to the origin. Note that positive values of x produce positive values for y, and negative values of x produce negative values for y; hence, the graph occurs in the first and third quadrants. We make a careful sketch in the first quadrant; then reflect these points through the origin to obtain the complete sketch.

x	0	1	2
y	0	1	8

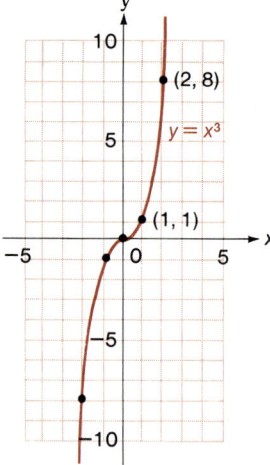

At a glance, the graph shows us how y varies as x varies. A graph is a visual aid and should be constructed to impart the maximum amount of information using the least amount of effort on the part of the observer. Label coordinate axes and indicate scales on both axes.

(B) Since both x and y occur to only even powers in $x^2 + 4y^2 = 36$, the equation will remain unchanged if x is replaced with $-x$ or if y is replaced with $-y$. Consequently, the graph is symmetric with respect to the y axis, x axis, and origin. We need to make a careful sketch in only the first quadrant, reflect this graph across the y axis, and then reflect everything across the x axis. To find first-quadrant solutions, we solve the equation for either y in terms of x or x in terms of y. We choose the latter because the result is simpler to work with.

$$x^2 + 4y^2 = 36$$
$$x^2 = 36 - 4y^2$$
$$x = \pm\sqrt{36 - 4y^2}$$

To obtain the first-quadrant portion of the graph, we sketch $x = \sqrt{36 - 4y^2}$ for $0 \leq y \leq 3$. Note that y cannot be larger than 3. (Why?)

x	6	$\sqrt{32} \approx 5.7$	$\sqrt{20} \approx 4.5$	0	Choose values for
y	0	1	2	3	y and solve for x.

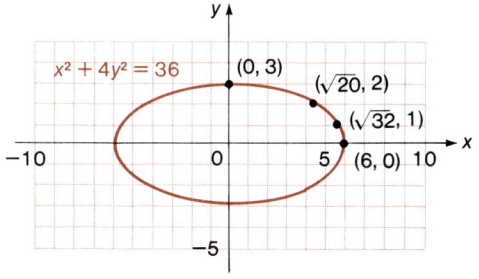

This figure is called an *ellipse*.

PROBLEM 2 Test for symmetry and graph.

(A) $y = x$ (B) $9x^2 + y^2 = 36$

■ **Distance between Two Points**

Analytic geometry is concerned with two basic problems:

1. Given an equation, find its graph.
2. Given a figure (line, circle, parabola, ellipse, etc.) in a coordinate system, find its equation.

So far we have concentrated on the first problem. We now introduce a basic tool that is used extensively in solving the second problem. This basic tool is the *distance-between-two-points formula*, which is easily derived using the Pythagorean theorem. Let $P_1(x_1, y_1)$ and $P_2(x_2, y_2)$ be two points in a rectangular coordinate system (the scale on each axis is assumed to be the same). Then referring to Figure 4, we see that

$$[d(P_1, P_2)]^2 = |x_2 - x_1|^2 + |y_2 - y_1|^2$$
$$= (x_2 - x_1)^2 + (y_2 - y_1)^2 \quad \text{Since } |N|^2 = N.$$

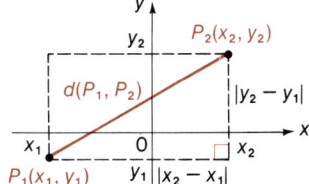

FIGURE 4

Thus:

Distance between $P_1(x_1, y_1)$ and $P_2(x_2, y_2)$

$$d(P_1, P_2) = \sqrt{(x_2 - x_1)^2 + (y_2 - y_1)^2}$$

EXAMPLE 3 Find the distance between $(-3, 5)$ and $(-2, -8)$.*

Solution It does not matter which point we designate P_1 and P_2 because of the squaring in the formula. Let $(x_1, y_1) = (-3, 5)$ and $(x_2, y_2) = (-2, -8)$. Then

$$d = \sqrt{[(-2) - (-3)]^2 + [(-8) - (5)]^2} = \sqrt{170}$$

PROBLEM 3 Find the distance between $(6, -3)$ and $(-7, -5)$.

■ Circles

The distance-between-two-points formula would still be helpful if its only use were to find actual distances between points, such as in Example 3. However, its more important use is in finding equations of figures in a rectangular coordinate system. We will use it to derive the general equation of a circle. We start with a coordinate-free definition of a circle.

Definition of a Circle

A **circle** is the set of all points in a plane equidistant from a fixed point. The fixed distance is called the **radius**, and the fixed point is called the **center**.

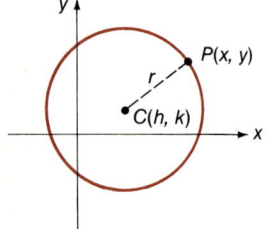

FIGURE 5

Let us find the equation of a circle with radius r ($r > 0$) and center at (h, k) in a rectangular coordinate system (Fig. 5). The point $P(x, y)$ is on the circle if and only if $d(P, C) = r$; that is, if and only if

$$\sqrt{(x - h)^2 + (y - k)^2} = r \qquad r > 0$$

---

* We often speak of the point (a, b) when we are referring to the point with coordinates (a, b). This shorthand, though not accurate, causes little trouble, and we will continue the practice.

or, equivalently,

$$(x - h)^2 + (y - k)^2 = r^2$$

Standard Equations of a Circle

1. Circle with radius r $(r > 0)$ and center at (h, k):

$$(x - h)^2 + (y - k)^2 = r^2$$

2. Circle with radius r $(r > 0)$ and center at $(0, 0)$:

$$x^2 + y^2 = r^2$$

EXAMPLE 4 Find the equation of a circle with radius 4 and center at (A) $(-3, 6)$ and (B) $(0, 0)$. Graph each equation.

Solution (A) $(h, k) = (-3, 6)$ and $r = 4$

$$(x - h)^2 + (y - k)^2 = r^2$$
$$[x - (-3)]^2 + (y - 6)^2 = 4^2$$
$$(x + 3)^2 + (y - 6)^2 = 16$$

To graph the equation, locate the center $C(-3, 6)$ and draw a circle of radius 4.

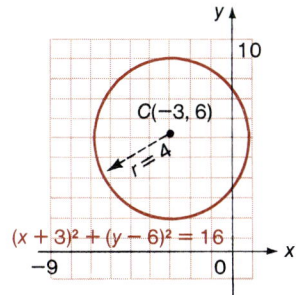

(B) $(h, k) = (0, 0)$ and $r = 4$

$$x^2 + y^2 = r^2$$
$$x^2 + y^2 = 4^2$$
$$x^2 + y^2 = 16$$

To graph the equation, locate the center at the origin and draw a circle of radius 4.

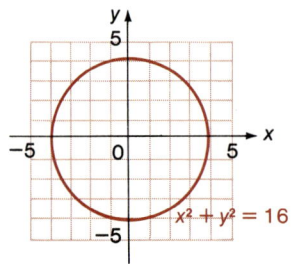

PROBLEM 4 Find the equation of a circle with radius 3 and center at (A) (3, −2) and (B) (0, 0). Graph each equation.

EXAMPLE 5 Find the center and radius of the circle with equation $x^2 + y^2 + 6x - 4y = 23$.

Solution We try to transform the equation into the form $(x - h)^2 + (y - k)^2 = r^2$ by completing the square relative to x and relative to y. From this standard form we can determine the center and radius.

$$x^2 + y^2 + 6x - 4y = 23$$
$$(x^2 + 6x \quad) + (y^2 - 4y \quad) = 23 \qquad \text{Complete the squares}$$
$$(x^2 + 6x + 9) + (y^2 - 4y + 4) = 23 + 9 + 4$$
$$(x + 3)^2 + (y - 2)^2 = 36$$
$$[x - (-3)]^2 + (y - 2)^2 = 6^2$$

Center: $C(h, k) = C(-3, 2)$
Radius: $r = \sqrt{36} = 6$

PROBLEM 5 Find the center and radius of the circle with equation $x^2 + y^2 - 8x + 10y = -25$.

Answers to Matched Problems 1.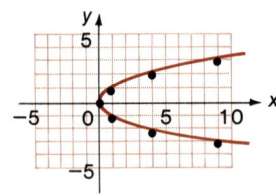

2. (A) Symmetric with respect to the origin

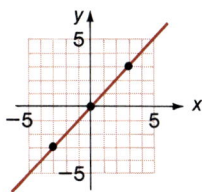

(B) Symmetric with respect to the x axis, y axis, and origin

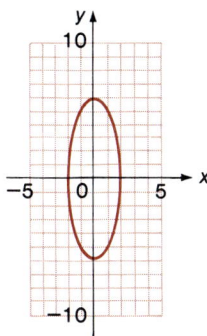

3. $d = \sqrt{173}$

4. (A) $(x - 3)^2 + (y + 2)^2 = 9$ (B) $x^2 + y^2 = 9$

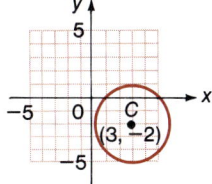 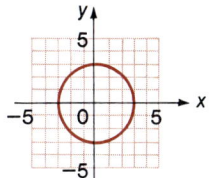

5. $(x - 4)^2 + (y + 5)^2 = 16$, radius: 4, center: $(4, -5)$

Exercise 2-1 ■ **A** *Determine symmetry with respect to the x axis, y axis, or origin, if any exists, and graph.*

1. $y = 2x - 4$
2. $y = \frac{1}{2}x + 1$
3. $y = |x|$
4. $y = -|x|$
5. $|y| = x$
6. $|y| = -x$
7. $|x| = |y|$
8. $y = -x$

2 Graphs and Functions

Find the distance between the indicated points. Leave the answer in radical form.

9. $(-6, -4)$, $(3, 4)$
10. $(-5, 4)$, $(6, -1)$
11. $(6, 6)$, $(4, -2)$
12. $(5, -3)$, $(-1, 4)$

Write the equation of a circle with the indicated center and radius.

13. $C(0, 0)$, $r = 7$
14. $C(0, 0)$, $r = 5$
15. $C(2, 3)$, $r = 6$
16. $C(5, 6)$, $r = 2$
17. $C(-4, 1)$, $r = \sqrt{7}$
18. $C(-5, 6)$, $r = \sqrt{11}$
19. $C(-3, -4)$, $r = \sqrt{2}$
20. $C(4, -1)$, $r = \sqrt{5}$

B *Determine symmetry with respect to the x axis, y axis, or origin, if any exists, and graph.*

21. $y^2 = x + 2$
22. $y^2 = x - 2$
23. $y = x^2 + 1$
24. $y + 2 = x^2$
25. $9x^2 + y^2 = 9$
26. $4x^2 + y^2 = 4$
27. $x^2 + 4y^2 = 4$
28. $x^2 + 9y^2 = 9$
29. $y^3 = x$
30. $y = x^4$

In Problems 31 and 32 determine whether the given points are vertices of a right triangle. (Recall, a triangle is a right triangle if and only if the square of the longest side is equal to the sum of the squares of the shorter sides.)

31. $(-3, 2)$, $(1, -2)$, $(8, 5)$
32. $(-4, -1)$, $(0, 7)$, $(6, -6)$

33. Find x such that $(x, 8)$ is thirteen units from $(2, -4)$.
34. Find y such that $(-2, y)$ is five units from $(-6, 6)$.

Find the center and radius of the circle with the given equation:

35. $(x - 3)^2 + (y - 5)^2 = 49$
36. $(x - 6)^2 + (y - 1)^2 = 64$
37. $(x + 4)^2 + (y - 2)^2 = 7$
38. $(x - 5)^2 + (y + 7)^2 = 15$
39. $x^2 + y^2 - 6x - 4y = 36$
40. $x^2 + y^2 - 2x - 10y = 55$
41. $x^2 + y^2 + 8x - 6y + 8 = 0$
42. $x^2 + y^2 + 4x + 10y + 15 = 0$

C *In Problems 43–48 determine symmetry with respect to the x axis, y axis, or origin, if any exists, and graph.*

43. $y^3 = |x|$
44. $|y| = x^3$
45. $xy = 1$
46. $xy = -1$
47. $y = 6x - x^2$
48. $y = x^2 - 6x$

49. Find the equation of the perpendicular bisector of the line segment joining $(-6, -2)$ and $(4, 4)$.

50. Show that

$$\left(\frac{x_1 + x_2}{2}, \frac{y_1 + y_2}{2}\right)$$

is the **midpoint formula** for the line segment joining (x_1, y_1) and (x_2, y_2) by using the distance-between-two-points formula.

Find the equation of a circle that has a diameter with the indicated end points. [*Hint:* See Problem 50.]

51. $(7, -3)$, $(1, 7)$ **52.** $(-3, 2)$, $(7, -4)$

CALCULATOR PROBLEMS

Find the perimeter to two decimal places of the triangle with the indicated vertices.

53. $(-3, 1)$, $(1, -2)$, $(4, 3)$ **54.** $(-2, 4)$, $(3, 1)$, $(-3, -2)$

Graph, using symmetry properties where appropriate and a hand calculator.

55. $y = 0.6x^2 - 4.5$ **56.** $x = 0.8y^2 - 3.5$
57. $y = \sqrt{17 - x^2}$ **58.** $y = \sqrt{100 - 4x^2}$
59. $y = x^{2/3}$ **60.** $y^{2/3} = x$

Section 2-2 Relations and Functions

- Introduction
- Relations and Functions
- Relations Specified by Equations
- Function Notation
- A Brief History of Function

- Introduction

Relations among various sets of objects abound in one's daily activities. For example,

To each person there corresponds an age.

To each item in a drugstore there corresponds a price.

To each automobile there corresponds a license number.

To each circle there corresponds an area.

To each number there corresponds its cube.

To each nonzero real number there corresponds two square roots.

One of the most important aspects of science is establishing relations among various types of phenomena. Once a relation is known, predictions can be made. A chemist can use a gas law to predict the pressure of an enclosed gas given its temperature; an engineer can use a formula to predict the deflections of a beam subject to different loads; an economist would like to be able to predict interest rates given the rate of change of the money supply; and so on.

Establishing and working with relations are so fundamental to both pure and applied science that people have found it necessary to describe them in the precise language of mathematics. Special relations called *functions* represent one of the most important concepts in all of mathematics. Effort made to understand and use this concept correctly right from the beginning will be rewarded many times.

■ Relations and Functions

What do all the examples of relations given here have in common? Each deals with the matching of elements from a first set, called the **domain** of the relation, with elements in a second set, called the **range** of the relation.

Consider the accompanying table showing three relations involving the cube, square, and square root. (The choice of small domains enables us to introduce two important concepts in a relatively simple setting. Shortly, we will consider relations with infinite domains.) The first two relations are examples of functions. The third is not a function. These two very important terms, *relation and function*, are defined in the box at the top of the next page.

RELATION 1		RELATION 2		RELATION 3	
Domain (Number)	Range (Cube)	Domain (Number)	Range (Square)	Domain (Number)	Range (Square Root)
0 → 0		−2 ↘	4	0 → 0	
1 → 1		−1 ↘↗	1	1 ⟨	1, −1
2 → 8		0 ↗↘	0	4 ⟨	2, −2
		1 ↗		9 ⟨	3, −3
		2			

2-2 Relations and Functions

Definition of a Relation and of a Function: Rule Form

A **relation** is a rule (process or method) that produces a correspondence between a first set of elements called the **domain** and a second set of elements called the **range** such that to each element in the domain there corresponds *one or more* elements in the range.

A **function** is a relation with the added restriction that to each domain element there corresponds *one and only one* range element.

(All functions are relations, but some relations are not functions.)

In the cube, square, and square root examples, we see that all three are relations according to the definition.* Relations 1 and 2 are also functions, since to each domain value there corresponds exactly one range value (for example, the square of -2 is 4 and no other number). On the other hand, relation 3 is not a function, since to at least one domain value there corresponds more than one range value (for example, to the domain value 9 there corresponds -3 and 3, both square roots of 9).

Since in a relation (or function) elements in the range are paired with elements in the domain by some rule or process, this correspondence (pairing) can be illustrated using ordered pairs of elements where the first component represents a domain element and the second component a corresponding range element. Thus, we can write relations 1–3 as

Relation 1 = $\{(0, 0), (1, 1), (2, 8)\}$

Relation 2 = $\{(-2, 4), (-1, 1), (0, 0), (1, 1), (2, 4)\}$

Relation 3 = $\{(0, 0), (1, 1), (1, -1), (4, 2), (4, -2), (9, 3), (9, -3)\}$

This suggests an alternative but equivalent way of defining relations and functions that provides additional insight into these concepts.

Definition of a Relation and of a Function: Set Form

A **relation** is *any* set of ordered pairs of elements.

A **function** is a relation with the added restriction that no two distinct ordered pairs can have the same first component.

The set of first components in a relation (or function) is called the **domain** of the relation, and the set of second components is called the **range**.

---

*We have used the word *relation* earlier as a word from our ordinary language. After the formal definition, the word *relation* becomes part of our technical mathematical vocabulary. From now on when we use the word *relation* in a mathematical context, it will have the meaning as specified.

According to this definition, we see (as before) that relation 3 is not a function, since there are two distinct ordered pairs [(1, 1) and (1, −1), for example] that have the same first component (more than one range element is associated with a given domain element).

The rule form of the definition of a relation and a function suggests a formula or a "machine" operating on domain values to produce range values—a dynamic process. On the other hand, the set definition of these concepts is closely related to graphs in a Cartesian coordinate system—a static form. Each approach has its advantages in certain situations.

One of the main objectives of this section is to expose you to the more common ways relations and functions are specified (including special notation) and to provide you with experience in determining whether a given relation is or is not a function.

As a consequence of the definitions, we find that a relation (or function) can be specified in many different ways: by an equation, by a table, by a set of ordered pairs of elements, and by a graph, to name a few of the more common ways (Table 2). All that matters is that we are given a set of elements called the domain and a rule (method or process) for obtaining corresponding range values for each domain value.

TABLE 2   COMMON WAYS OF SPECIFYING RELATIONS AND FUNCTIONS

Method	Illustration	Example
Equation	$y = x^2 - x, \quad x \in R^*$	$x = -1$ corresponds to $y = 2$
Table	$\begin{array}{c\|ccc} m & 1 & 2 & 3 \\ \hline n & 1 & 8 & 27 \end{array}$	$m = 2$ corresponds to $n = 8$
Sets of ordered pairs of elements	(a) $\{(1, 1), (2, 8), (3, 27)\}$ (b) $\{(x, y) \mid y = x^3, x \in R\}$	3 corresponds to 27 $x = -2$ corresponds to $y = -8$
Graph	(graph of sideways parabola on u-v axes, vertex near origin opening right, passing through approximately $(0, \pm 2)$)	$u = 0$ corresponds to $v = \pm 2$

* Recall that R is the set of real numbers.

Which relation in Table 2 is not a function? The relation specified by the graph is not a function, since it is possible for a domain value to correspond to more than one range value. (What does $u = -5$ correspond to?)

It is very easy to determine whether a relation is a function if one has its graph.

Vertical Line Test for a Function

A relation is a function if each vertical line in the coordinate system passes through *at most* one point on the graph of the relation. (If a vertical line passes through two or more points on the graph of a relation, then the relation is not a function.)

■ **Relations Specified by Equations**

Most of the domains and ranges included in this text will be sets of numbers, and the rules associating range values with domain values will be equations in two variables.

Consider the equation

$$y = x^2 - x \qquad x \in R$$

For each **input** x we obtain one **output** y. For example,

If $x = 3$, then $y = 3^2 - 3 = 6$.

If $x = -\frac{1}{2}$, then $y = (-\frac{1}{2})^2 - (-\frac{1}{2}) = \frac{1}{4} + \frac{1}{2} = \frac{3}{4}$.

The input values are domain values and the output values are range values. The equation (a rule) assigns each domain value x a range value y. The variable x is called an *independent variable* (since values are "independently" assigned to x from the domain), and y is called a *dependent variable* (since y's value "depends" on the value assigned to x). In general, any variable used as a placeholder for domain values is called an **independent variable**; any variable that is used as a placeholder for range values is called a **dependent variable**.

Unless stated to the contrary, we shall adhere to the following convention regarding domains and ranges for relations and functions specified by equations.

> **Agreement on Domains and Ranges**
>
> If a relation or function is specified by an equation and the domain is not indicated, then we shall assume that the domain is the set of all real number replacements of the independent variable (inputs) that produce real values for the dependent variable (outputs). The range is the set of all outputs corresponding to input values.

Most equations in two variables specify relations, but when does an equation specify a function?

> **Equations and Functions**
>
> If, in an equation in two variables, there corresponds exactly one value of the dependent variable (output) for each value of the independent variable (input), then the equation specifies a function. If there is more than one output for at least one input, then the equation does not specify a function.

EXAMPLE 6 (A) Is the relation specified by the equation $y^2 = x + 1$ a function, assuming x is the independent variable?
(B) What is the domain of the relation?

Solution (A) The relation is not a function, since, for example, if $x = 3$, then $y = \pm 2$.
(B) The domain of the relation (since it is not explicitly given) is the set of all real x that produce real y. Solving for y in terms of x, we obtain

$$y = \pm\sqrt{x + 1}$$

For y to be real, $x + 1$ must be greater than or equal to 0; that is,

$$x + 1 \geq 0$$
$$x \geq -1$$

Thus,

Domain: $x \geq -1$ or $[-1, \infty)$

PROBLEM 6 (A) Is the relation specified by the equation $x^2 + y^2 = 25$ a function, assuming x is the independent variable?
(B) What is the domain of the relation?

Function Notation

We have just seen that a function involves two sets of elements, a domain and a range, and a rule of correspondence that enables one to assign each element in the domain to exactly one element in the range. We use different letters to denote names for numbers; in essentially the same way, we will now use different letters to denote names for functions. For example, f and g may be used to name the two functions

$f: \quad y = 2x + 1$

$g: \quad y = x^2 + 2x - 3$

If x represents an element in the domain of a function f, then we will often use the symbol

$f(x)$

in place of y to designate the number in the range of the function f to which x is paired (Fig. 6).

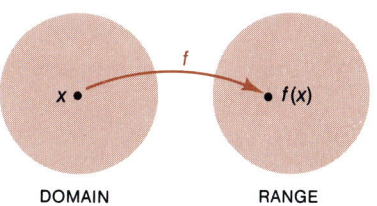

FIGURE 6   DOMAIN   RANGE

The function f "maps" the domain value x into the range value $f(x)$.

It is important not to think of $f(x)$ as the product of f and x. The symbol $f(x)$ is read "f of x," or "the value of f at x." The variable x is an independent variable; both y and $f(x)$ are dependent variables.

This function notation is extremely important, and its correct use should be mastered as early as possible. For example, in place of the more formal representation of the functions f and g, we can now write

$f(x) = 2x + 1 \quad \text{and} \quad g(x) = x^2 + 2x - 3$

The function symbols $f(x)$ and $g(x)$ have certain advantages over the variable y in certain situations. For example, if we write $f(3)$ and $g(5)$, then each symbol indicates in a concise way that these are range values of particular functions associated with particular domain values. Let us find $f(3)$ and $g(5)$.

To find $f(3)$, we replace x by 3 wherever x occurs in

$f(x) = 2x + 1$

and evaluate the right side:

$$f(3) = 2 \cdot 3 + 1$$
$$= 6 + 1$$
$$= 7$$

Thus,

$f(3) = 7$ The function f assigns the range value 7 to the domain value 3; the ordered pair (3, 7) belongs to f.

To find $g(5)$, we replace x by 5 wherever x occurs in

$$g(x) = x^2 + 2x - 3$$

and evaluate the right side:

$$g(5) = 5^2 + 2 \cdot 5 - 3$$
$$= 25 + 10 - 3$$
$$= 32$$

Thus,

$g(5) = 32$ The function g assigns the range value 32 to the domain value 5; the ordered pair (5, 32) belongs to g.

It is very important to understand and remember the definition of $f(x)$:

The Function Symbol $f(x)$

For any element x in the domain of the function f, the function symbol

$$f(x)$$

represents the element in the range of f corresponding to x in the domain of f. [If x is an input value, then $f(x)$ is an output value, or symbolically, $f: \quad x \to f(x)$.] The ordered pair $(x, f(x))$ belongs to the function f.

Figure 7, illustrating a "function machine," may give you additional insight into the nature of functions and the function symbol $f(x)$. We can think of a function machine as a device that produces exactly one output (range) value for each input (domain) value based on a set of instructions such as those found in an equation, graph, or table. (If more than one output value is produced for an input value, then the machine would be a "relation machine" and not a "function machine.")

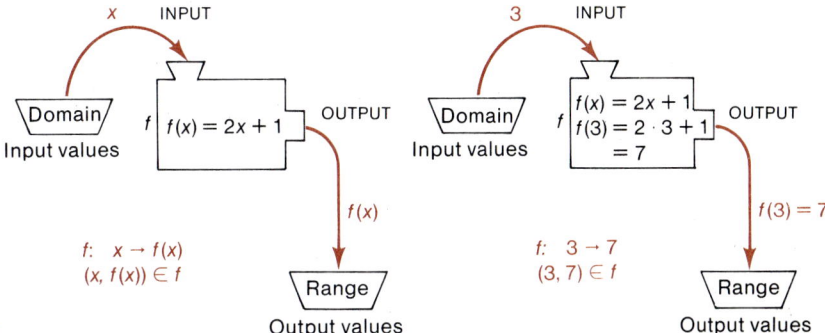

FIGURE 7 "Function Machine"— exactly one output for each input

For the function $f(x) = 2x + 1$, the machine takes each domain value (input), multiplies it by 2, then adds 1 to the result to produce the range value (output). Different rules inside the machine result in different functions.

EXAMPLE 7 Let $f(x) = |x| - 1$, $g(x) = 1 - x^2$, $I(x) = x$, and $F(x) = 5$. Then

(A) $f(-2) = |-2| - 1 = 2 - 1 = 1$

(B) $g(-3) = 1 - (-3)^2 = 1 - 9 = -8$

(C) $F(7) = 5$ *F is a constant function with domain the set of real numbers R and range {5}.*

(D) $\dfrac{2f(-1) - 3I(3)}{g(-1)} = \dfrac{2[|-1| - 1] - 3[3]}{1 - (-1)^2} = \dfrac{2(0) - 9}{0} = \dfrac{-9}{0}$; not defined

(E) $f(a) + g(a) = [|a| - 1] + [1 - a^2] = |a| - a^2$

(F) $\dfrac{f(a+h) - f(a)}{h} = \dfrac{[|a+h| - 1] - [|a| - 1]}{h} = \dfrac{|a+h| - |a|}{h}$

PROBLEM 7 For the functions f, g, I, and F in Example 7, find:

(A) $f(-6)$ (B) $g(-4)$ (C) $\dfrac{f(0) - g(0)}{2g(0)}$

(D) $4F(-2) + 2I(0)$ (E) $g(m) + F(m)$ (F) $\dfrac{g(a+h) - g(a)}{h}$

EXAMPLE 8 (A) Find the domain and range for the function

$$f = \{(-2, 3), (-1, 3), (0, 2), (1, 2)\}$$

Solution   Domain $= \{-2, -1, 0, 1\}$ *Set of first components*

Range $= \{2, 3\}$ *Set of second components*

(B) Find the domain for the function g where

$$g(x) = \frac{2x}{x^2 - 4}$$

Solution   Since the domain is not explicitly stated, it is understood to be the set of all real x that produce real g(x). The expression $2x/(x^2 - 4)$ represents a real number for all x except for

$$x^2 - 4 = 0$$
$$x^2 = 4$$
$$x = \pm 2$$

Domain: All real numbers except ± 2.

(C) Find the domain for the function H where

$$H(u) = \sqrt{25 - u^2}$$

Solution   The domain is the set of all real u such that $\sqrt{25 - u^2}$ is a real number—that is, such that $25 - u^2 \geq 0$. Solving the inequality $25 - u^2 = (5 - u)(5 + u) \geq 0$ by methods discussed in Section 1-5, we find

Domain: $-5 \leq u \leq 5$ or $[-5, 5]$

PROBLEM 8   (A) Find the domain and range for the function

$$h = \{(x, y) \mid y = x^2, \quad x \in \{-2, 0, 2\}\}$$

(B) Find the domain of the function G where

$$G(t) = \frac{2t + 1}{t^2 - t - 6}$$

(C) Find the domain of the function f where

$$f(x) = \sqrt{\frac{x - 2}{x + 3}}$$

Remark: It is useful to summarize the different ways a function can be specified. All the following specify the same function f:

$$f(x) = \sqrt{25 - x^2}$$
$$f: \quad y = \sqrt{25 - x^2}$$
$$f: \quad x^2 + y^2 = 25 \qquad y \geq 0$$
$$f: \quad x \to \sqrt{25 - x^2} \qquad \text{Read ``x is mapped into } \sqrt{25 - x^2}\text{.''}$$
$$f = \{(x, y) \mid y = \sqrt{25 - x^2}\}$$

$$f = \{(x, y) \mid x^2 + y^2 = 25, \ y \geq 0\}$$

f:

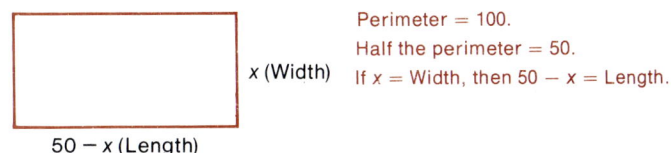

EXAMPLE 9 A rectangular feeding pen for cattle is to be made with 100 meters of fencing.

(A) If x represents the width of the pen, express its area $A(x)$ in terms of x.

(B) What is the domain of the function A (determined by the physical restrictions)?

Solution (A) Draw a figure and label the sides.

Perimeter = 100.
Half the perimeter = 50.
If x = Width, then $50 - x$ = Length.

x (Width)

$50 - x$ (Length)

$$A(x) = (\text{Width})(\text{Length}) = x(50 - x) \quad \text{Area depends on width } x.$$

(B) To have a pen, x must be positive, but x must also be less than 50 (or the length will not exist). Thus,

Domain: $0 < x < 50$ Inequality notation

$(0, 50)$ Interval notation

PROBLEM 9 Work Example 9 with the added assumption that a large barn is to be used as one side of the pen.

■ A Brief History of Function

In reviewing the history of function, we are made aware of the tendency of mathematicians to extend and generalize a concept. The word *function* appears to have been first used by Leibniz in 1694 to stand for any quantity associated with a curve. By 1718, Johann Bernoulli considered a function any expression made up of constants and a variable. Later in the same century, Euler came to regard a function as any equation made up of constants and variables. Euler made extensive use of the extremely important notation $f(x)$, although its origin is generally attributed to Clairaut (1734).

The form of the definition of function that has been used until well into this century (many texts still contain this definition) was formulated by Dirichlet (1805–1859). He stated that, if two variables x and y are so related that for each value of x there corresponds exactly one value of y, then y is said to be a (single-valued) function of x. He called x, the variable to which values are assigned at will, the independent variable, and y, the variable whose values depend on the values assigned to x, the dependent variable. He called the values assumed by x the domain of the function, and the corresponding values assumed by y the range of the function.

Now, since set concepts permeate almost all mathematics, we have the more general definitions of function presented in this section in terms of sets of ordered pairs of elements. The function concept is one of the most important concepts in mathematics, and as such it plays a central and natural role as a guide for the selection and development of material in many mathematics courses (look at the title of this book and chapter titles following this one; also look at the section titles following this one).

Answers to Matched Problems

6. (A) No (B) Domain: $-5 \leq x \leq 5$ or $[-5, 5]$

7. (A) 5 (B) -15 (C) -1 (D) 20 (E) $6 - m^2$
(F) $-2a - h$

8. (A) Domain $= \{-2, 0, 2\}$; range $= \{0, 4\}$
(B) Domain: All real numbers except -2 and 3
(C) Domain $x < -3$ or $x \geq 2$ Inequality notation
$(-\infty, -3) \cup [2, \infty)$ Interval notation

9. (A) $A(x) = x(100 - 2x)$
(B) Domain: $0 < x < 50$ Inequality notation
$(0, 50)$ Interval notation

Exercise 2-2 ■ A

Indicate whether each relation in Problems 1–12 is or is not a function.

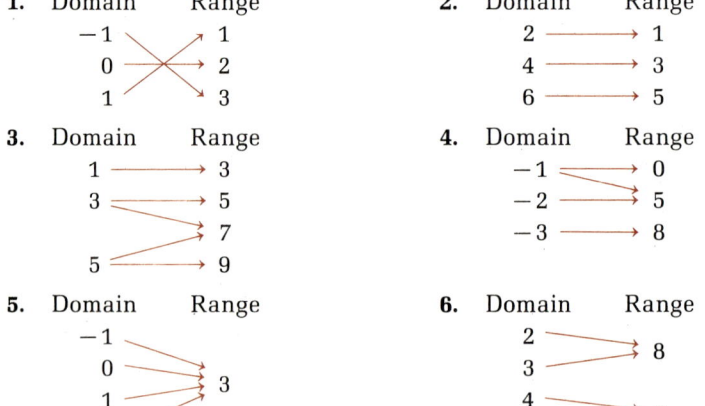

2-2 Relations and Functions

7.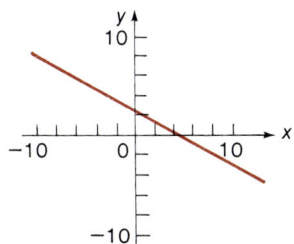

8.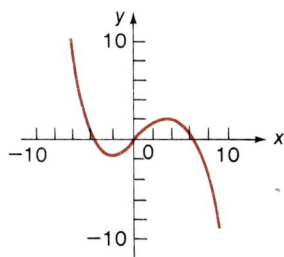

9.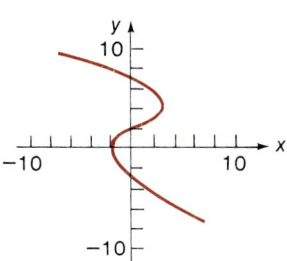

10.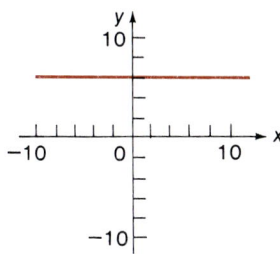

11.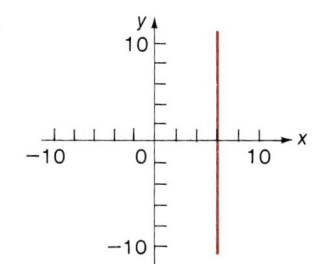

12.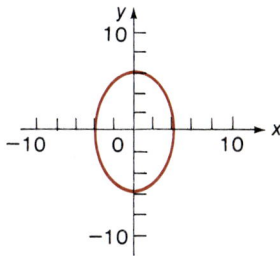

Problems 13–24 refer to

$$f(x) = 3x - 5 \qquad g(t) = 4 - t \qquad F(m) = 3m^2 \qquad G(u) = u - u^2$$

Evaluate as indicated.

13. $f(-1)$
14. $g(6)$
15. $G(-2)$
16. $F(-3)$
17. $F(-1) + f(3)$
18. $G(2) - g(-3)$
19. $2F(-2) - G(-1)$
20. $3G(-2) + 2F(-1)$
21. $\dfrac{f(0) \cdot g(-2)}{F(-3)}$
22. $\dfrac{g(4) \cdot f(2)}{G(1)}$
23. $\dfrac{f(2 + h) - f(2)}{h}$
24. $\dfrac{g(3 + h) - g(3)}{h}$

B *Each equation specifies a relation. Which specify a function given that x is an independent variable?*

25. $y = 3 - x$
26. $y = 2x + 3$
27. $y = 2x^2 - 3x + 5$
28. $y = (2 - x)^3$

29. $y^2 - x = 2$
30. $y - x^2 = 2$
31. $y = |x - 2|$
32. $|y| = x + 2$
33. $x^2 + y^2 = 81$
34. $16x^2 + y^2 = 16$
35. $y^3 = x$
36. $y^5 = x$
37. $y = \dfrac{x+3}{x-2}$
38. $y = \dfrac{x-4}{2x^2 + 7x - 4}$
39. $y = \sqrt{x+1}$
40. $y^2 = x + 1$
41. $y = \sqrt{\dfrac{x+3}{x-2}}$
42. $y = \sqrt{x^2 + x - 12}$

Given each relation in Problems 43–50, state its domain and range and indicate which are functions. The variable x is independent.

43. $f = \{(2, 4), (4, 2), (2, 0), (4, -2)\}$
44. $g = \{(-1, 3), (0, 1), (1, 3)\}$
45. $G = \{(-4, 1), (0, 1), (4, 1)\}$
46. $H = \{(-6, 3), (-4, 5), (-6, 0)\}$
47. $P = \{(x, y) \mid y^2 = x, \; x \in \{0, 1, 4\}\}$
48. $Q = \{(x, y) \mid |y| = x, \; x \in \{0, 1, 2\}\}$
49. $R = \{(x, y) \mid y = x^2 - x, \; x \in \{-2, 2\}\}$
50. $h = \{(x, y) \mid x^2 + y^2 = 25, \; x \in \{-5, 5\}\}$

For Problems 51–62, determine the domain of the relation in the indicated problem.

51. Problem 25
52. Problem 26
53. Problem 27
54. Problem 28
55. Problem 32
56. Problem 31
57. Problem 33
58. Problem 34
59. Problem 37
60. Problem 38
61. Problem 41
62. Problem 42

63. If $g(x) = 2 - x^2$, find: $\dfrac{g(3+h) - g(3)}{h}$

64. If $f(x) = x^2$, find: $\dfrac{f(2+h) - f(2)}{h}$

65. If $Q(t) = t^2 - 2t + 1$, find: $\dfrac{Q(1+h) - Q(1)}{h}$

66. If $P(m) = 2m^2 + 3$, find: $\dfrac{P(2+h) - P(2)}{h}$

C 67. If $f(x) = x^2 - 1$, find: $\dfrac{f(a+h) - f(a)}{h}$

68. If $g(x) = x^2 + x - 1$, find: $\dfrac{g(a+h) - g(a)}{h}$

69. If $f(x) = x^2 - 1$, find: $\dfrac{f(x+h) - f(x)}{h}$

70. If $g(x) = x^2 + x - 1$, find: $\dfrac{g(x+h) - g(x)}{h}$

71. If $h(x) = x^3$, find: $\dfrac{h(x+h) - h(x)}{h}$

72. If $g(x) = x^3 + x$, find: $\dfrac{g(x+h) - g(x)}{h}$

APPLICATIONS Each relationship in Problems 73–78 can be described by a function. Write an equation that specifies the function.

73. *Cost function.* The cost per day $C(x)$ for renting a car at $10 per day plus 12¢ per mile for x miles. (The cost depends on the number of miles driven.)

74. *Cost function.* The cost per day $C(x)$ of manufacturing x pairs of skis if fixed costs are $375 per day and variable costs are $68 per pair of skis manufactured. (The cost per day depends on the number of skis manufactured per day.)

75. *Area function.* The area $A(r)$ of a circle is π times the square of the radius r. (The area depends on the radius.)

76. *Earth science.* The pressure $P(d)$ in the ocean in pounds per square inch is found by dividing the depth d by 33, adding 1 to the quotient, and multiplying the final result by 15. (The pressure below sea level depends on the depth.)

77. *Manufacturing.* A candy box is to be made out of a piece of cardboard 8 by 12 inches. Equal-sized squares, x inches on a side, will be cut from each corner, and then the ends and sides will be folded up. Find a formula for the volume of the box $V(x)$ in terms of x. From practical considerations, what is the domain of the function V?

78. *Construction.* A rancher has 20 miles of fencing to fence a rectangular piece of grazing land along a straight river. If no fence is required along the river and the sides perpendicular to the river are x miles long, find a formula for the area $A(x)$ of the rectangle in terms of x. From practical considerations, what is the domain of the function A?

79. *Physics—rate.* The distance in feet that an object falls in a vacuum is given by $s(t) = 16t^2$, where t is time in seconds. Find:

(A) $s(0), s(1), s(2), s(3)$

(B) $\dfrac{s(2+h) - s(2)}{h}$

(C) What happens in (B) when h tends to 0? Interpret physically.

Section 2-3   Functions: Graphs and Properties

- Graphs of Functions
- Function Properties
- More Aids to Graphing

Each function that has a real number domain and range has a graph—the graph of the ordered pairs of real numbers that constitute the function. In this section we will identify several basic functions by name and sketch their graphs. Special function properties will be introduced through graphs and definitions, and additional aids to graphing will be presented.

- **Graphs of Functions**

When functions are graphed, domain values are usually associated with the horizontal axis and range values with the vertical axis. Thus, if we graph

$$y = f(x)$$

then x would be the independent variable and the abscissa of a point on the graph of the function f; y and $f(x)$ would be dependent variables and either the ordinate of a point on the graph of f (Fig. 8).

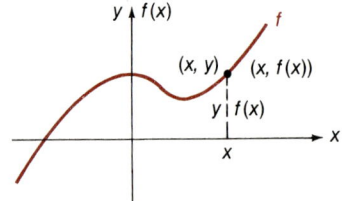

FIGURE 8

The **graph of a function f** is the same as the graph of the equation $y = f(x)$. Figure 9 illustrates the graphs of several basic functions with which you have had some experience in equation form.

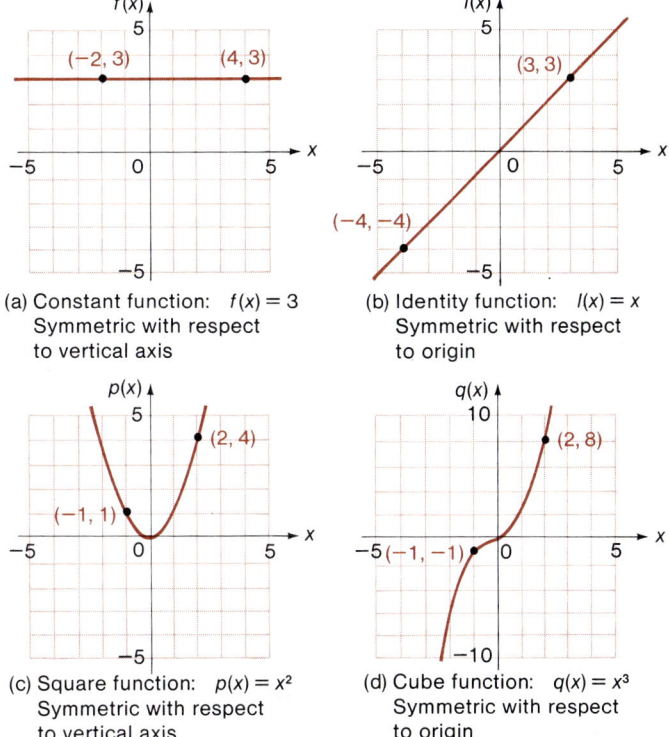

(a) Constant function: $f(x) = 3$
Symmetric with respect to vertical axis

(b) Identity function: $l(x) = x$
Symmetric with respect to origin

(c) Square function: $p(x) = x^2$
Symmetric with respect to vertical axis

(d) Cube function: $q(x) = x^3$
Symmetric with respect to origin

FIGURE 9

■ Function Properties

A function whose graph is symmetric with respect to the vertical axis is called an **even function**. A function whose graph is symmetric with respect to the origin is called an **odd function**. (Is it possible for a function with independent variable x to be symmetric with respect to the x axis?) The constant and square functions in Figure 9 are even functions, and the identity and cube functions are odd functions. Of course, if the graph of a function is not symmetric with respect to either the vertical axis or the origin, then it is neither even nor odd. As a consequence of the tests for symmetry discussed in Section 2-1, we have the following tests for even and odd functions:

Even and Odd Functions

If $f(-x) = f(x)$, then f is an **even function**.
If $f(-x) = -f(x)$, then f is an **odd function**.

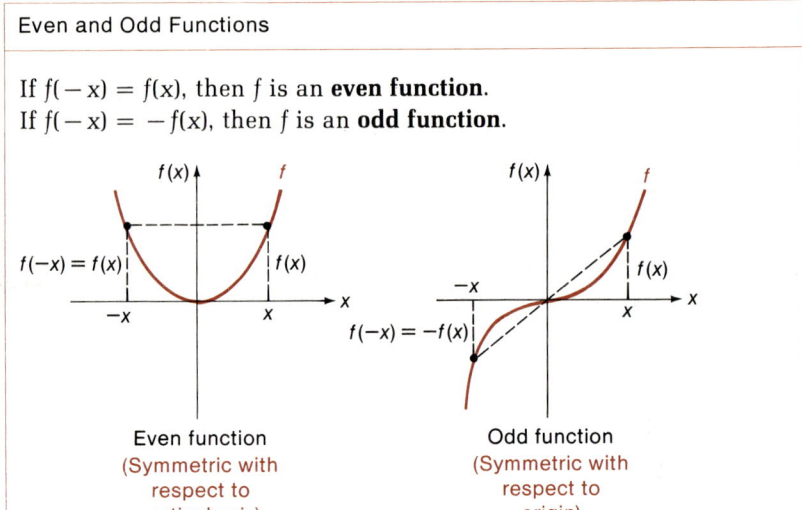

Even function
(Symmetric with respect to vertical axis)

Odd function
(Symmetric with respect to origin)

EXAMPLE 10 Without graphing, determine whether the functions f, g, and h are even, odd, or neither.

(A) $f(x) = |x|$ (B) $g(x) = x^3 + 1$ (C) $h(x) = \sqrt[3]{x}$

Solution (A) $f(-x) = |-x| = |x| = f(x)$; therefore, f is even.

(B) $g(x) = x^3 + 1$
$g(-x) = (-x)^3 + 1 = -x^3 + 1$ $g(-x) \neq g(x)$
$-g(x) = -(x^3 + 1) = -x^3 - 1$ $g(-x) \neq -g(x)$

Therefore, g is neither even nor odd.

(C) $h(-x) = \sqrt[3]{-x} = -\sqrt[3]{x} = -h(x)$; therefore, h is odd.

PROBLEM 10 Without graphing, determine whether the functions F, G, and H are even, odd, or neither.

(A) $F(x) = x^3 + x$ (B) $G(x) = x^2 + 1$ (C) $H(x) = 2x + 4$

Why are we interested in knowing whether a function is even or odd? If we want to graph a function specified by an equation, then the even–odd test in the box provides a useful aid for graphing: If the function is even, then its graph is symmetric with respect to the vertical axis; if it is odd, then its graph is symmetric with respect to the origin.

In addition, certain problems and developments in calculus and more advanced mathematics are simplified if one recognizes the presence of either an even or odd function.

We now take a look at increasing–decreasing properties of functions. Intuitively, a function is increasing over an interval I in its domain if its graph rises as the independent variable increases over I; a function is decreasing over I if its graph falls as the independent variable increases over I (Fig. 10).

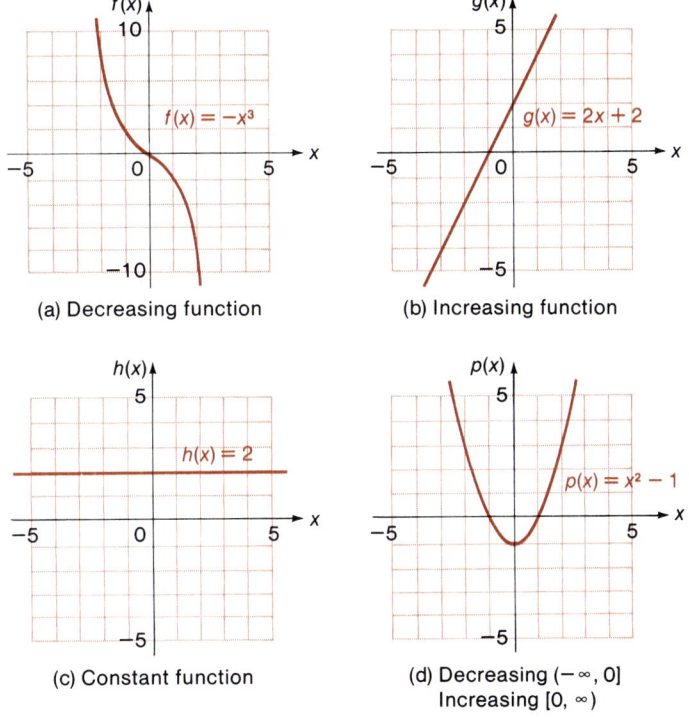

FIGURE 10

(a) Decreasing function
(b) Increasing function
(c) Constant function
(d) Decreasing $(-\infty, 0]$ Increasing $[0, \infty)$

More formally, we define increasing, decreasing, and constant functions as follows:

Increasing, Decreasing, and Constant Functions

Let I be an interval in the domain of a function f, then

1. **f is increasing** on I if $f(b) > f(a)$ whenever $b > a$ in I.
2. **f is decreasing** on I if $f(b) < f(a)$ whenever $b > a$ in I.
3. **f is constant** on I if $f(a) = f(b)$ for all a and b in I.

Another important property of functions is the **continuity property**. Our discussion of continuity must remain informal at this point and will rely heavily on geometric considerations. (A precise presentation of continuity can be found in calculus texts.) Nevertheless, it is useful to have an intuitive idea of the concept of continuity before a formal presentation is made. Let us introduce the idea through an interesting function called the *greater integer function*.

The **greatest integer** of a real number x, denoted by $[\![x]\!]$, is the integer n such that $n \leq x < n + 1$ (that is, $[\![x]\!]$ is the largest integer less than or equal to x). For example,

$$[\![3.45]\!] = 3$$
$$[\![7]\!] = 7$$
$$[\![0]\!] = 0$$
$$[\![-2.13]\!] = -3 \quad \text{Not } -2$$
$$[\![-8]\!] = -8$$

The **greatest integer function** f (also called a **step function**) is determined by the equation $f(x) = [\![x]\!]$. The domain of f is the set of all real numbers and the range of f is the set of integers. A sketch of the graph of f is shown in Figure 11.

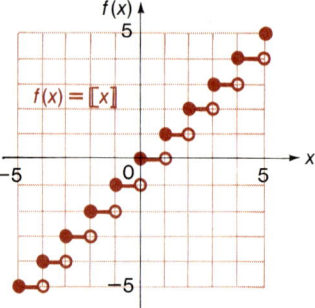

FIGURE 11 Greatest integer function

For x in the interval	$[\![x]\!]$ has the value
⋮	⋮
$[-2, -1)$	-2
$[-1, 0)$	-1
$[0, 1)$	0
$[1, 2)$	1
$[2, 3)$	2
⋮	⋮

[*Note*: A solid dot on the graph indicates the point is part of the graph; a hollow dot indicates the point is not part of the graph.]

We notice in Figure 11 that at each integer value for x there is a break in the graph, and between integer values for x there is no break. If the graph of a function is not broken (disconnected) at a point, then the function is said to be **continuous** at that point. A function whose graph is broken (disconnected) at a certain point is said to be **discontinuous** at that point. A function is **continuous over an interval** if its graph is continuous (not broken) at each value on the interval. Thus, we see that the greatest integer function is discontinuous at each integer but is continuous in each interval that does not contain an integer.

EXAMPLE 11 Let a function f be defined as follows:

$$f(x) = \begin{cases} 0 & \text{for } x < 0 \\ -x + 2 & \text{for } 0 \leq x < 2 \\ 2 & \text{for } x \geq 2 \end{cases}$$

Graph f and indicate points of discontinuity.

Solution Note that f is defined by different formulas for different parts of its domain. This is a perfectly acceptable way to define a function and is used in many applications. (All that matters is that we have a way of determining a range value for each domain value.) The graph of f is given here. We see from the graph that the function f is discontinuous at $x = 0$ and $x = 2$.

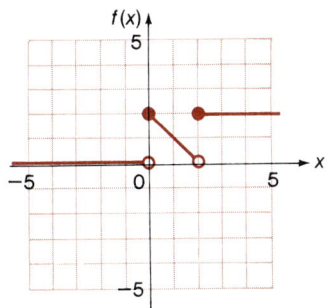

PROBLEM 11 Let a function g be defined as follows:

$$g(x) = \begin{cases} 1 & \text{for } x < -1 \\ x^2 & \text{for } -1 \leq x < 2 \\ 0 & \text{for } x \geq 2 \end{cases}$$

Graph g and indicate points of discontinuity.

Remarks: In calculus it can be shown that:

1. Polynomial functions [functions defined by equations of the form $f(x) = $ (Polynomial in x)] are continuous for all real numbers; that is, polynomial functions have no breaks in their graphs. Thus, f where $f(x) = 2x^3 - x^2 + 3x - 1$ is continuous for all real numbers. [*Note:* Domain is the set of all real numbers.]

2. Rational functions [functions defined by equations of the form $f(x) =$ (Polynomial in x)/(Polynomial in x)] are continuous for all real numbers except values of x that make a denominator 0. Thus, f where

$$f(x) = \frac{x-1}{x^2 - x - 6} = \frac{x-1}{(x-3)(x+2)}$$

is continuous for all real numbers, except $x = -2$ and $x = 3$; that is, f is discontinuous (there is a break in the graph) at $x = -2$ and at $x = 3$. [Note: Domain is the set of all real numbers except -2 and 3.]

We will have more to say about these remarks and their consequences in later sections.

■ **More Aids to Graphing**

There are situations in which it is possible to use known graphs of basic functions to graph related functions. Involved here are vertical and horizontal shifting, reflecting with respect to an axis, and expanding and contracting. These graphing aids will be illustrated through examples.

EXAMPLE 12   Graph $f(x) = x^2$, $F(x) = x^2 + 3$, and $G(x) = x^2 - 2$.

Solution   We already know that the graph of $y = x^2$ is a parabola opening upward; it passes through the origin and has as its axis the y axis [see figure (a)]. To graph $y = x^2 + 3$, we add 3 to each ordinate value for the graph of $y = x^2$ [see figure (b)]. To graph $y = x^2 - 2$, we subtract 2 from each ordinate value [see figure (c)]. The net result is that the graph of $y = x^2 + 3$ is just the graph of $y = x^2$ shifted upward three units, and the graph of $y = x^2 - 2$ is just the graph of $y = x^2$ shifted downward two units.

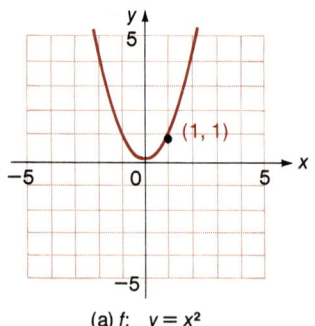

(a) f: $y = x^2$

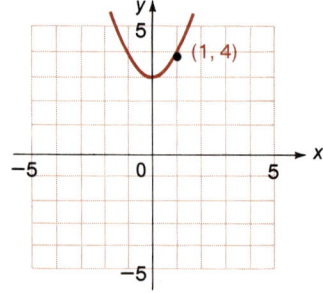

(b) F: $y = x^2 + 3$
The graph of $y = x^2 + 3$ is the same as the graph of $y = x^2$ shifted up three units.

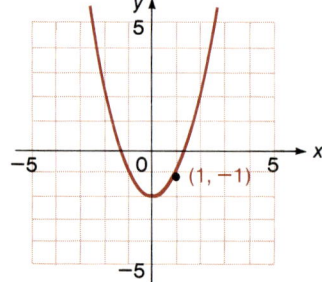

(c) G: $y = x^2 - 2$
The graph of $y = x^2 - 2$ is the same as the graph of $y = x^2$ shifted down two units.

Following the same line of reasoning as in Example 12, the results can be generalized as given in Table 3.

TABLE 3 VERTICAL SHIFTING (TRANSLATION)

To Graph	Shift the Graph of $y = f(x)$
$y = f(x) + k$, $k > 0$	Up k units
$y = f(x) + k$, $k < 0$	Down $|k|$ units

PROBLEM 12 Graph $f(x) = |x| + k$ for $k = 0$, $k = 2$, and $k = -3$.

We now turn to horizontal shifting.

EXAMPLE 13 Graph $f(x) = x^2$, $P(x) = (x + 2)^2$, and $Q(x) = (x - 3)^2$.

Solution Observe the following:

$$f(x) = x^2 \qquad P(x) = (x + 2)^2 \qquad Q(x) = (x - 3)^2$$
$$f(a) = a^2 \qquad P(a - 2) = (a - 2 + 2)^2 \qquad Q(a + 3) = (a + 3 - 3)^2$$
$$(a, a^2) \in f \qquad \qquad = a^2 \qquad \qquad = a^2$$
$$\qquad \qquad (a - 2, a^2) \in P \qquad (a + 3, a^2) \in Q$$

Thus, the point with abscissa a on the graph of $y = f(x) = x^2$ has the same ordinate value, a^2, as the point with abscissa $a - 2$ on the graph of $y = P(x) = (x + 2)^2$ and the point with abscissa $a + 3$ on the graph of $y = Q(x) = (x - 3)^2$. We conclude that the graph of $y = (x + 2)^2$ is the same as the graph of $y = x^2$ shifted to the left two units [see figure (b)]. And the graph of $y = (x - 3)^2$ is the same as the graph of $y = x^2$ shifted to the right three units [see figure (c)].

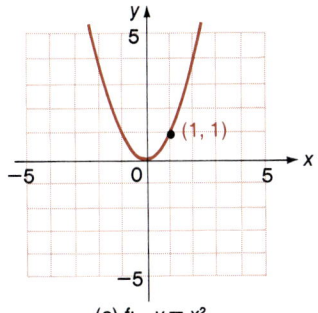

(a) f: $y = x^2$

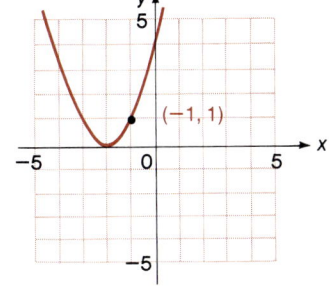

(b) P: $y = (x + 2)^2$

The graph of $y = (x + 2)^2$ is the same as the graph of $y = x^2$ shifted to the left two units.

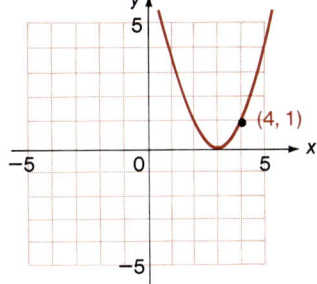

(c) Q: $y = (x - 3)^2$

The graph of $y = (x - 3)^2$ is the same as the graph of $y = x^2$ shifted to the right three units.

Following the same line of reasoning as in Example 13, the results can be generalized as given in Table 4.

TABLE 4   HORIZONTAL SHIFTING (TRANSLATION)

To Graph	Shift the Graph of $y = f(x)$
$y = f(x + h)$, $h > 0$	To the left h units
$y = f(x + h)$, $h < 0$	To the right $|h|$ units

Remark: The horizontal shift is just the opposite of what many people expect. A positive h is associated with a shift to the left; a negative h is associated with a shift to the right. On the other hand, relative to vertical shifts, a positive k [in $y = f(x) + k$] is associated with a shift upward, and a negative k is associated with a shift downward.

PROBLEM 13   Graph $f(x) = |x + h|$ for $h = 0$, $h = 2$, and $h = -3$.

We conclude this section by considering reflections, expansions, and contractions of graphs.

EXAMPLE 14   Graph $f(x) = x^2$, $R(x) = -x^2$, $S(x) = 2x^2$, and $T(x) = \frac{1}{2}x^2$.

Solution   If we take the negative of each ordinate value on the graph of $y = x^2$

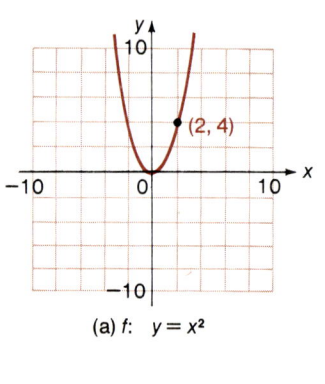

(a) f: $y = x^2$

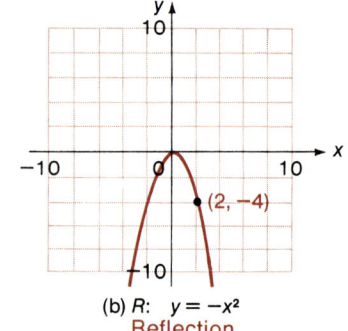

(b) R: $y = -x^2$
Reflection

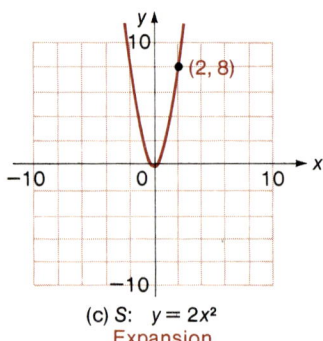

(c) S: $y = 2x^2$
Expansion

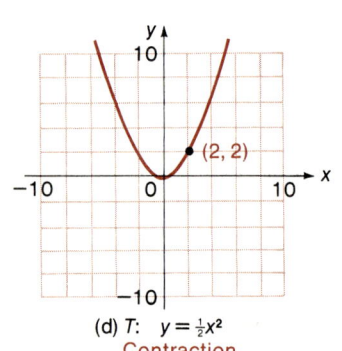

(d) T: $y = \frac{1}{2}x^2$
Contraction

2-3 Functions: Graphs and Properties   119

[figure (a)], we will obtain the graph of $y = -x^2$ [figure (b)]. If we double each ordinate value on the graph of $y = x^2$, we will obtain the graph of $y = 2x^2$ [figure (c)]. And if we cut each ordinate value in half on the graph of $y = x^2$, we will obtain the graph of $y = \frac{1}{2}x^2$ [figure (d)].

Following the same line of reasoning as in Example 14, the results can be generalized as given in Table 5.

TABLE 5   REFLECTION, EXPANSION, CONTRACTION

Graph of	Relationship to the Graph of $y = f(x)$
$y = -f(x)$	Reflection of the graph of $y = f(x)$ across the x axis
$y = Cf(x)$, $C > 1$	All ordinate values are expanded by a factor of C
$y = Cf(x)$, $0 < C < 1$	All ordinate values are contracted by a factor of C

PROBLEM 14   Graph $f(x) = C|x|$ for $C = 1$, $C = -1$, $C = 2$, and $C = \frac{1}{2}$.

Answers to Matched Problems
10. (A) Odd   (B) Even   (C) Neither
11. Discontinuous at $x = 2$

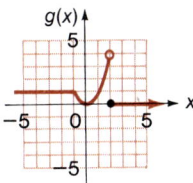

12.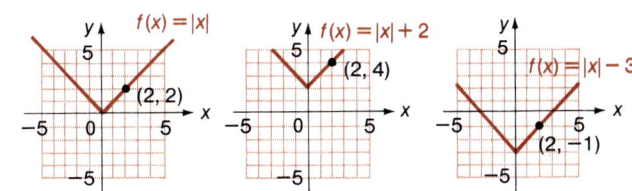

13.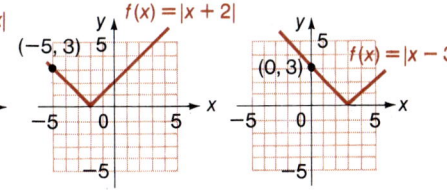

14.

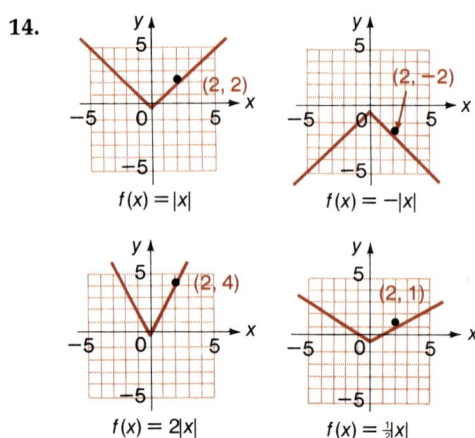

Exercise 2-3 ■ A

Problems 1–12 refer to functions f, g, p, and q given by the following graphs. (Assume the graphs continue as indicated beyond the parts shown.)

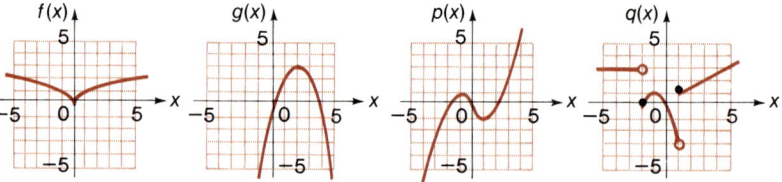

1. Which functions are symmetric with respect to the vertical axis?
2. Which functions are symmetric with respect to the origin?
3. Which functions are even?
4. Which functions are odd?
5. Indicate intervals over which f is:
   (A) Constant   (B) Increasing   (C) Decreasing
6. Indicate intervals over which g is:
   (A) Constant   (B) Increasing   (C) Decreasing
7. Indicate intervals over which q is:
   (A) Constant   (B) Increasing   (C) Decreasing
8. Indicate intervals over which p is:
   (A) Constant   (B) Increasing   (C) Decreasing
9. Which functions are continuous on the interval $(-3, 3)$?
10. Which functions are continuous on the interval $(2, 4)$?
11. Which functions have points of discontinuity? Name them.
12. Is function f discontinuous at $x = 0$?

B *Without graphing, indicate whether each function is even, odd, or neither.*

13. $g(x) = x^3 + x$
14. $f(x) = x^5 - x$
15. $m(x) = x^4 + 3x^2$
16. $h(x) = x^4 - x^2$
17. $F(x) = x^5 + 1$
18. $f(x) = x^5 - 3$
19. $G(x) = x^4 + 2$
20. $P(x) = x^4 - 4$
21. $q(x) = x^2 + x - 3$
22. $n(x) = 2x - 3$

Graph as indicated.

23. $f(x) = \sqrt{x} + k$ for $k = 0$, $k = 2$, $k = -3$
24. $g(x) = -x^2 + k$ for $k = 0$, $k = 4$, $k = -1$
25. $F(x) = \sqrt{x + h}$ for $h = 0$, $h = 4$, $h = -1$
26. $G(x) = -(x + h)^2$ for $h = 0$, $h = 1$, $h = -2$
27. $p(x) = C\sqrt{x}$ for $C = 1$, $C = -1$, $C = 2$, $C = \frac{1}{2}$
28. $Q(x) = Cx$ for $C = 1$, $C = -1$, $C = 2$, $C = \frac{1}{2}$

Indicate how the graph of each function is related to the graph of $y = x^2$, $y = |x|$, or $y = \sqrt{x}$. Graph each function.

29. $g(x) = -(x + 2)^2$
30. $h(x) = -(x - 3)^2$
31. $G(x) = -|x + 2|$
32. $H(x) = -|x - 3|$
33. $P(x) = -\sqrt{x - 1}$
34. $Q(x) = -\sqrt{x + 2}$

Graph each function using the aids to graphing discussed in this section. Indicate if a function is either even or odd. Indicate any points of discontinuity.

35. $g(x) = -4$
36. $f(x) = -2$
37. $S(x) = -2|x + 2|$
38. $f(x) = -\frac{1}{2}(x - 2)^2$
39. $g(x) = \frac{1}{2}(x + 1)^2 - 2$
40. $h(x) = -2|x - 2| + 2$
41. $y = \dfrac{1}{|x|}$
42. $y = \dfrac{1}{x}$
43. $f(x) = \dfrac{|x|}{x}$
44. $g(x) = x + \dfrac{|x|}{x}$

45. $f(x) = \begin{cases} 1 & \text{if } x < 0 \\ x + 1 & \text{if } 0 \leq x < 2 \\ 2 & \text{if } x \geq 2 \end{cases}$

46. $g(x) = \begin{cases} -x & \text{if } x < 0 \\ 2 & \text{if } 0 \leq x < 2 \\ x - 2 & \text{if } x \geq 2 \end{cases}$

47. $p(x) = \begin{cases} 0 & \text{if } x < 0 \\ 4 - x^2 & \text{if } 0 \leq x < 2 \\ 0 & \text{if } x \geq 2 \end{cases}$

48. $T(x) = \begin{cases} 0 & \text{if } x < 0 \\ x^2 & \text{if } 0 \leq x < 2 \\ 0 & \text{if } x \geq 2 \end{cases}$

49. $p(x) = -[\![x]\!]$
50. $q(x) = [\![x]\!] + 2$
51. $m(x) = -[\![x - 1]\!]$
52. $g(x) = [\![x + 3]\!]$

C 53. $m(x) = x^2 - 2|x|$
54. $n(x) = 2|x| - x^2$
55. $f(x) = x^3 - 3x$
56. $T(x) = 3x - x^3$
57. $r(x) = x - [\![x]\!]$
58. $S(x) = [\![x]\!] - x$

Section 2-4 Linear Relations and Functions

- Linear Functions
- Slope of a Line
- Equations of Lines—Standard Forms
- Parallel and Perpendicular Lines
- Concluding Remarks

We now turn to special relations and functions called *linear relations and functions*. Both are used extensively in mathematical developments and applications.

■ Linear Functions

We start the discussion by defining a linear function:

Linear Function

A function f is a **linear function** if

$f(x) = ax + b \qquad a \neq 0$

where a and b are real numbers.

The algebraic expression $ax + b$, $a \neq 0$, is a *first-degree polynomial*; hence, a linear function is called a **first-degree polynomial function**. (In subsequent sections and chapters we will discuss second- and higher-degree polynomial functions.)

The word *linear* is used in naming the function, since its graph is a straight line. To see this, we will show that any three points having coordinates that satisfy

$$f(x) = ax + b, \qquad a \neq 0 \tag{1}$$

are **collinear** (that is, they lie on the same line). Pick three arbitrary but distinct values for x, and label them so that $x_1 < x_2 < x_3$. Then, using equation (1), we obtain

$$P_1(x_1, f(x_1)) = P_1(x_1, ax_1 + b)$$
$$P_2(x_2, f(x_2)) = P_2(x_2, ax_2 + b)$$
$$P_3(x_3, f(x_3)) = P_3(x_3, ax_3 + b)$$

These three points have coordinates that satisfy equation (1). We now show that P_1, P_2, and P_3 are collinear by showing that

$$d(P_1, P_2) + d(P_2, P_3) = d(P_1, P_3)$$

To accomplish this, we compute each of the distances and add the first two to obtain the third.*

$$\begin{aligned} d(P_1, P_2) &= \sqrt{(x_2 - x_1)^2 + [(ax_2 + b) - (ax_1 + b)]^2} \\ &= \sqrt{(x_2 - x_1)^2 + a^2(x_2 - x_1)^2} \\ &= \sqrt{(x_2 - x_1)^2(1 + a^2)} = |x_2 - x_1|\sqrt{1 + a^2} \\ &= (x_2 - x_1)\sqrt{1 + a^2} \quad \text{Since } x_2 > x_1 \end{aligned}$$

Similarly,

$$d(P_2, P_3) = (x_3 - x_2)\sqrt{1 + a^2}$$
$$d(P_1, P_3) = (x_3 - x_1)\sqrt{1 + a^2}$$

Thus,

$$\begin{aligned} d(P_1, P_2) + d(P_2, P_3) &= (x_2 - x_1)\sqrt{1 + a^2} + (x_3 - x_2)\sqrt{1 + a^2} \\ &= (x_2 - x_1 + x_3 - x_2)\sqrt{1 + a^2} \\ &= (x_3 - x_1)\sqrt{1 + a^2} = d(P_1, P_3) \end{aligned}$$

We conclude that P_1, P_2, and P_3 are collinear. We have just proved Theorem 1 (a result you no doubt guessed to be true from the graphing that was done earlier).

THEOREM 1 The graph of a linear function f,

$$f(x) = ax + b \qquad a \neq 0$$

is a straight line.

---

* Recall from plane geometry that the shortest distance between two points is a straight line joining the two points.

Now that we know this theorem as a fact, graphing linear functions is very easy. Since two points determine a line, we have only to find two solutions to $y = ax + b$, plot them, and then draw a line through the two points using a straightedge. (Sometimes it is useful to find a third solution to $y = ax + b$ as a check point.)

EXAMPLE 15 Graph the linear function f given by

$$f(x) = -\tfrac{2}{3}x + 4$$

Solution

x	f(x)
0	4
3	2
6	0

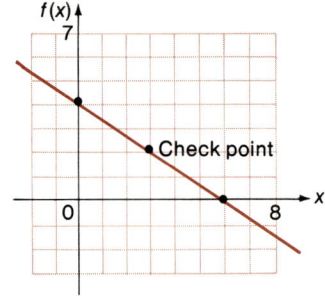

PROBLEM 15 Graph $g(x) = \tfrac{3}{2}x - 6$.

■ Slope of a Line

We next turn to a measure of the "steepness" of a line called *slope*. (We first note that a **vertical line** is a line parallel to the y axis, and a **horizontal line** is a line parallel to the x axis.)

Slope of a Line

If $P_1(x_1, y_1)$ and $P_2(x_2, y_2)$ are two distinct points on a nonvertical line ($x_1 \neq x_2$), then the **slope** of the line is given by

$$m = \frac{y_2 - y_1}{x_2 - x_1} \qquad x_1 \neq x_2$$

Interpreted geometrically, the slope is the ratio of the change in y (**rise**) to the change in x (**run**) as we move from P_1 to P_2. If the points are labeled so that $x_2 > x_1$, then m is positive if $y_2 > y_1$, 0 if $y_2 = y_1$, and negative if $y_2 < y_1$. On a vertical line $x_1 = x_2$; thus, its slope is not defined. All four cases are illustrated in Figure 12.

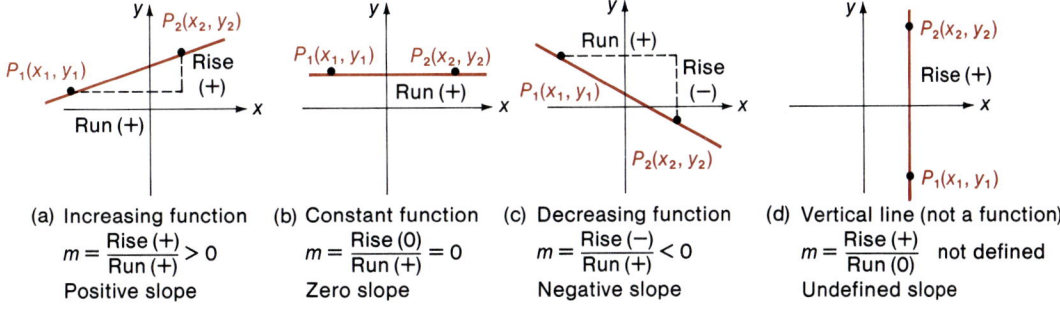

FIGURE 12 Slope of a line

In using the formula to find the slope of a line through two points, it does not matter which point is labeled P_1 or P_2, since

$$\frac{y_2 - y_1}{x_2 - x_1} = \frac{y_1 - y_2}{x_1 - x_2}$$

In addition, it is important to note that the definition of slope does not depend on the two points chosen on the line as long as they are distinct. This follows from the fact that the ratios of corresponding sides of similar triangles are equal. (The details of a proof are left to the reader.)

EXAMPLE 16 Sketch a line through each pair of points and find the slope of each line.

(A) $(-3, -4), (3, 2)$ (B) $(-2, 3), (1, -3)$
(C) $(-4, 2), (3, 2)$ (D) $(2, 4), (2, -3)$

Solution (A)

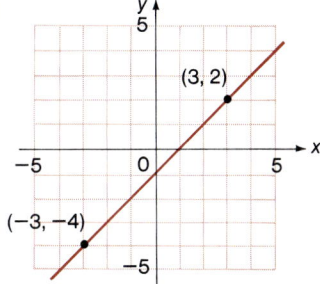

(B)

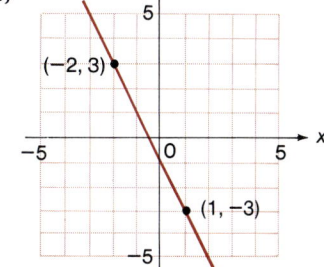

$$m = \frac{2 - (-4)}{3 - (-3)} = \frac{6}{6} = 1 \qquad m = \frac{-3 - 3}{1 - (-2)} = \frac{-6}{3} = -2$$

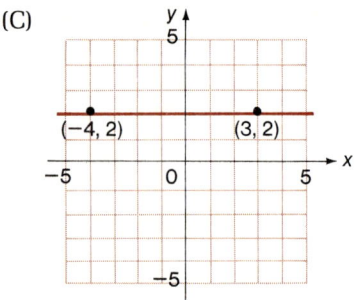

$$m = \frac{2-2}{3-(-4)} = \frac{0}{7} = 0 \qquad m = \frac{-3-4}{2-2} = \frac{-7}{0};$$

slope not defined

PROBLEM 16 Find the slope of the line through each pair of points. Do not graph.
(A) $(-3, -3), (2, -3)$ (B) $(-2, -1), (1, 2)$
(C) $(0, 4), (2, -4)$ (D) $(-3, 2), (-3, -1)$

■ **Equations of Lines—Standard Forms**

There are several standard equations of lines that we will now identify. (Each has a particular advantage for certain situations.) We will start with the simplest equations, those for horizontal and vertical lines. Consider the two equations

$x + 0y = a$ or $x = a$ (2)

$0x + y = b$ or $y = b$ (3)

In equation (2), y can be any number as long as $x = a$. Thus, the graph of $x = a$ is a vertical line crossing the x axis at $(a, 0)$. In equation (3), x can be any number as long as $y = b$. Thus, the graph of $y = b$ is a horizontal line crossing the y axis at $(0, b)$. We summarize these results as follows:

Vertical and Horizontal Lines	
EQUATION	GRAPH
$x = a$ (short for $x + 0y = a$)	Vertical line through $(a, 0)$
$y = b$ (short for $0x + y = b$)	Horizontal line through $(b, 0)$

EXAMPLE 17 Find the equations of the horizontal and vertical lines that pass through $(-7, 12)$.

Solution Horizontal line: $y = 12$

Vertical line: $x = -7$

PROBLEM 17 Find the equations of the horizontal and vertical lines that pass through $(5, -2)$.

Now suppose a nonvertical line passes through $P_1(x_1, y_1)$ and has slope m. What is its equation? Choose $P(x, y)$, $x \neq x_1$, as an arbitrary point in the coordinate system. The point P will be on the line if and only if the slope of the line through P_1 and P is m—that is, if and only if

$$\frac{y - y_1}{x - x_1} = m \tag{4}$$

or

$$y - y_1 = m(x - x_1) \tag{5}$$

Note that (x_1, y_1) also satisfies equation (5); hence, *any* point having coordinates that satisfy equation (5) is on the line that passes through (x_1, y_1) with slope m.

We have just obtained the **point–slope form** of the equation of a line.

Point–Slope Form

An equation of a line through $P_1(x_1, y_1)$ with slope m is

$$y - y_1 = m(x - x_1)$$

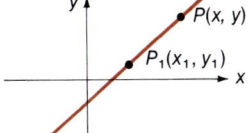

Remember that $P(x, y)$ is a variable point and $P_1(x_1, y_1)$ is fixed.

EXAMPLE 18 Find the equation of the line that passes through $(4, -1)$ and $(-8, 5)$. Write the resulting equation in the form $y = ax + b$.

Solution First find the slope:

$$m = \frac{y_2 - y_1}{x_2 - x_1} = \frac{5 - (-1)}{-8 - 4} = \frac{6}{-12} = -\frac{1}{2}$$

Then use either point for (x_1, y_1).

$(x_1, y_1) = (4, -1)$ or $(x_1, y_1) = (-8, 5)$

$y - y_1 = m(x - x_1)$ $\qquad$ $y - y_1 = m(x - x_1)$

$y - (-1) = -\frac{1}{2}(x - 4)$ $\qquad$ $y - 5 = -\frac{1}{2}[x - (-8)]$

$y + 1 = -\frac{1}{2}x + 2$ $\qquad$ $y - 5 = -\frac{1}{2}x - 4$

$y = -\frac{1}{2}x + 1$ $\qquad$ $y = -\frac{1}{2}x + 1$

Note that we obtained the same equation for both choices of (x_1, y_1).

PROBLEM 18 Find the equation of the line that passes through $(-2, 4)$ and $(1, -2)$. Write the answer in the form $y = mx + b$.

EXAMPLE 19 A sporting goods store sells a tennis racket that cost $60 for $82 and a pair of ski boots that cost $80 for $106.

(A) If the markup policy of the store for items that cost more than $30 is assumed to be linear and is reflected in the pricing of these two items, write an equation that relates retail price R to cost C.

(B) Use the equation to find the retail price for a pair of running shoes that cost $40.

Solution (A) If the retail price R is assumed linearly related to cost C, then we are looking for an equation whose graph passes through $(60, 82)$ and $(80, 106)$. We find the slope, and then use the point–slope form to find the equation.

$$m = \frac{106 - 82}{80 - 60} = \frac{24}{20} = 1.2$$

$R - R_0 = m(C - C_0)$

$R - 82 = 1.2(C - 60)$

$R - 82 = 1.2C - 72$

$R = 1.2C + 10$

(B) $R = 1.2(40) + 10 = \$58$

PROBLEM 19 The management of a company that manufactures ball-point pens estimates costs for running the company to be $200 per day at zero output, and $700 per day at an output of 1,000 pens.

(A) Assuming total cost per day C is linearly related to total output per day x, write an equation relating these two quantities.

(B) What is the total cost per day for an output of 5,000 pens?

We now observe that if the point–slope form is solved for y in terms of x, then we obtain another useful form:

$y - y_1 = m(x - x_1)$

$y - y_1 = mx - mx_1$

$y = mx + (y_1 - mx_1)$

2-4 Linear Relations and Functions

which is of the form

$$y = mx + b$$

where $b = y_1 - mx_1$ is a constant. Geometrically, the coefficient of x represents the slope and b is the y coordinate of the point of crossing of the line with the y axis (generally called the **y intercept**). Thus, we have the slope–intercept form.

Slope–Intercept Form

$y = mx + b$

$m = \dfrac{\text{Rise}}{\text{Run}} = \text{Slope}$

$b = y$ intercept

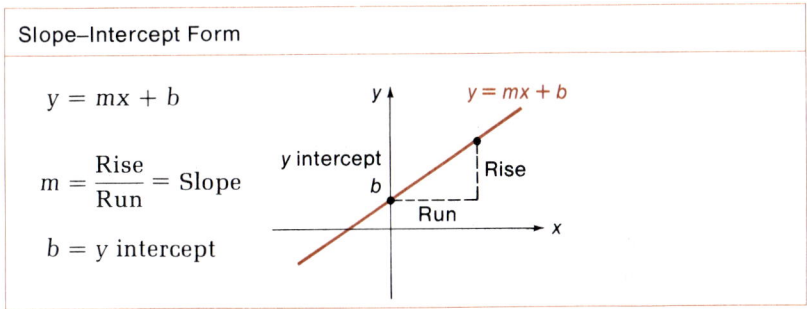

Let us look at the equation

$$Ax + By = C \qquad A \text{ and } B \text{ not both } 0 \tag{6}$$

We now have enough information to conclude that the graph of this equation is always a line. If $B \neq 0$, then solving for y in terms of x, we obtain

$$y = -\frac{A}{B}x + \frac{C}{B}$$

which is the equation of a line with slope $-A/B$ and y intercept C/B. If $B = 0$ and $A \neq 0$, then equation (6) can be written in the form

$$x = \frac{C}{A}$$

which is the equation of a vertical line crossing the x axis at $(C/A, 0)$. Thus, we have sketched a partial proof of Theorem 2.

THEOREM 2   The graph of the equation

$$Ax + By = C \quad \text{Standard form}$$

where A, B, and C are constants (A and B not both 0) and x and y are variables, is a straight line. Any line in a rectangular coordinate system has an equation of this form.

130 2 Graphs and Functions

A graph of an equation of the form $Ax + By = C$, where $A, B, C \neq 0$, can be quickly made by plotting the x and y intercepts of the graph. (Recall, the **x intercept** is the x coordinate of the point of intersection of the graph and the x axis, and the **y intercept** is the y coordinate of the point of intersection of the graph and the y axis.) To find the x intercept, set $y = 0$ and solve for x; to find the y intercept, set $x = 0$ and solve for y.

EXAMPLE 20   Graph $2x - 3y = 9$.

Solution   Find the x and y intercepts and draw a line through these two points with a straightedge.

x	y
0	−3
4.5	0

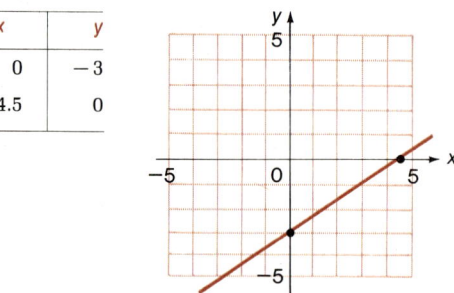

PROBLEM 20   Graph $4x + 5y = 12$.

We conclude our identification of special equations of lines with an **intercept form**.

Intercept Form

The equation

$$\frac{x}{a} + \frac{y}{b} = 1$$

is the equation of a line with x intercept a and y intercept b.

To see this is the case, let $x = 0$ and solve for y; then let $y = 0$ and solve for x. The intercept form is particularly useful if it is desired to find the equation of a line with given intercepts.

EXAMPLE 21   Find the equation of a line with x intercept 4 and y intercept -2. Write the final answer in the form $Ax + By = C$, where $A, B,$ and C are integers.

2-4 Linear Relations and Functions

Solution

$$\frac{x}{4} + \frac{y}{-2} = 1 \quad \text{Multiply both sides by 4.}$$

$$x - 2y = 4$$

PROBLEM 21   Find the equation of a line with x intercept 3.5 and y intercept 7. Write the final answer in the form $Ax + By = C$, where A, B, and C are integers.

■ Parallel and Perpendicular Lines

From geometric considerations, we know that two vertical lines are parallel to each other and that a horizontal line and a vertical line are perpendicular to each other. How can we tell when two nonvertical lines are parallel or perpendicular to each other? Theorem 3 (which we state without proof) provides a convenient test.

THEOREM 3

> Given two (nonvertical) lines L_1 and L_2 with slopes m_1 and m_2, respectively, then
>
> $L_1 \parallel L_2$ if and only if $\quad m_1 = m_2.$
> $L_1 \perp L_2$ if and only if $\quad m_1 m_2 = -1.$

The symbols $\parallel$ and $\perp$ mean, respectively, "is parallel to" and "is perpendicular to." In the case of perpendicularity, the condition $m_1 m_2 = -1$ can also be written as

$$m_2 = -\frac{1}{m_1} \quad \text{or} \quad m_1 = -\frac{1}{m_2}$$

Thus, **two (nonvertical) lines are perpendicular if and only if their slopes are the negative reciprocals of each other.**

EXAMPLE 22   Given the line L: $3x - 2y = 5$ and the point $P(-3, 5)$, find an equation of a line through P that is (A) parallel to L, (B) perpendicular to L. Write the final answers in the form $y = mx + b$.

Solution   First find the slope of L by writing $3x - 2y = 5$ in the equivalent slope–intercept form $y = mx + b$:

$$3x - 2y = 5$$
$$-2y = -3x + 5$$
$$y = \tfrac{3}{2}x - \tfrac{5}{2}$$

Thus, the slope of L is $\frac{3}{2}$. The slope of a line parallel to L is the same, $\frac{3}{2}$, and that of a line perpendicular to L, $-\frac{2}{3}$. We can now find the equations of the two lines in parts (A) and (B) using the point–slope formula (which needs to be memorized).

(A) $y - y_1 = m(x - x_1)$
$\quad y - 5 = \frac{3}{2}(x + 3)$
$\quad y - 5 = \frac{3}{2}x + \frac{9}{2}$
$\quad y = \frac{3}{2}x + \frac{19}{2}$

(B) $y - y_1 = m(x - x_1)$
$\quad y - 5 = -\frac{2}{3}(x + 3)$
$\quad y - 5 = -\frac{2}{3}x - 2$
$\quad y = -\frac{2}{3}x + 3$

PROBLEM 22 Given the line L: $4x + 2y = 3$ and the point $P(2, -3)$, find an equation of a line through P that is (A) parallel to L, (B) perpendicular to L. Write the final answers in the form $y = mx + b$.

■ **Concluding Remarks**

Relative to lines in a rectangular coordinate system, linear equations in two variables, and linear functions, we have answered two of the fundamental questions in analytic geometry:

1. Given an equation, what is its graph?
2. Given a graph, what is its equation?

We now know how to graph linear equations on sight and how to write equations of lines given certain information about the lines.

Answers to Matched Problems

15.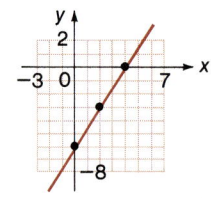

16. (A) $m = 0$ (B) $m = 1$ (C) $m = -4$
    (D) m not defined

17. Horizontal line: $y = -2$;
    vertical line: $x = 5$

18. $y = -2x$

19. (A) $C = 0.5x + 200$ (B) $\$2{,}700$

20.

x	0	3
y	$\frac{12}{5}$	0

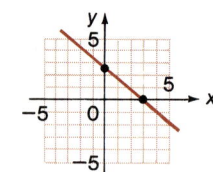

2-4 Linear Relations and Functions

21. $2x + y = 7$
22. (A) $y = -2x + 1$ (B) $y = \frac{1}{2}x - 4$

Exercise 2-4

A *Graph each equation and indicate the slope of the graph if it exists.*

1. $f(x) = 2x - 4$
2. $g(x) = -\frac{1}{2}x + 6$
3. $h(x) = -\frac{3}{5}x + 4$
4. $p(x) = -\frac{3}{2}x + 6$
5. $y = -\frac{3}{4}x$
6. $y = \frac{2}{3}x - 3$
7. $2x - 3y = 15$
8. $4x + 3y = 24$
9. $4x - 5y = -24$
10. $6x - 7y = -49$
11. $\frac{x}{2} - \frac{y}{3} = 1$
12. $\frac{x}{4} + \frac{y}{5} = 1$
13. $\frac{y}{8} - \frac{x}{4} = 1$
14. $\frac{y}{6} - \frac{x}{5} = 1$
15. $x = -3$
16. $y = -2$
17. $y = 3.5$
18. $x = 2.5$

B *Write an equation of the line that contains the indicated point(s) and/or has the indicated slope and/or has the indicated intercepts. Write the final equation in the form $y = mx + b$ or $x = c$.*

19. $(0, 4)$; $m = -3$
20. $(2, 0)$; $m = 2$
21. $(-5, 4)$; $m = -\frac{2}{5}$
22. $(3, -3)$; $m = -\frac{1}{3}$
23. $(5, 5)$; $m = 0$
24. $(-4, -2)$; $m = \frac{1}{2}$
25. $(1, 6)$, $(5, -2)$
26. $(-3, 4)$, $(6, 1)$
27. $(-4, 8)$, $(2, 0)$
28. $(2, -1)$, $(10, 5)$
29. $(-3, 4)$, $(5, 4)$
30. $(0, -2)$, $(4, -2)$
31. $(4, 6)$, $(4, -3)$
32. $(-3, 1)$, $(-3, -4)$
33. x intercept 6
    y intercept 2
34. x intercept 3
    y intercept 4
35. x intercept -4
    y intercept 3
36. x intercept -4
    y intercept -5

In Problems 37–48 write an equation of the line that contains the indicated point and meets the indicated condition(s). Write the final answer in the form $Ax + By = C$, $A > 0$.

37. $(-3, 4)$; parallel to $y = 3x - 5$
38. $(-4, 0)$; parallel to $y = -2x + 1$
39. $(2, -3)$; perpendicular to $y = -\frac{1}{3}x$

40. $(-2, -4)$; perpendicular to $y = \frac{2}{3}x - 5$
41. $(2, 5)$; parallel to y axis
42. $(7, 3)$; parallel to x axis
43. $(3, -2)$; vertical
44. $(-2, -3)$; horizontal
45. $(5, 0)$; parallel to $3x - 2y = 4$
46. $(3, 5)$; parallel to $3x + 4y = 8$
47. $(0, -4)$; perpendicular to $x + 3y = 9$
48. $(-2, 4)$; perpendicular to $4x + 5y = 0$
49. Graph $f(x) = mx + 2$ for $m = 2$, $m = \frac{1}{2}$, $m = 0$, $m = -\frac{1}{2}$, and $m = -2$, all on the same coordinate system.
50. Graph $g(x) = -\frac{1}{2}x + b$ for $b = -3$, $b = 0$, and $b = 3$, all on the same coordinate system.

Problems 51–56 refer to the quadrilateral with vertices $A(0, 2)$, $B(4, -1)$, $C(1, -5)$, and $D(-3, 2)$.

51. Show that $AB \parallel DC$.
52. Show that $DA \parallel CB$.
53. Show that $AB \perp BC$.
54. Show that $AD \perp DC$.

C 55. Find an equation of the perpendicular bisector of AD. [Hint: First find the midpoint of AD.]

56. Find an equation of the perpendicular bisector of AB.
57. If the graph of a linear function g has slope $-\frac{2}{3}$ and passes through $(-6, 4)$, find $g(x)$.
58. If the graph of a linear function f passes through $(-6, 2)$ and $(-4, -4)$, find $f(x)$.
59. If $f(-1) = 3$ and $f(3) = 5$ for a linear function f, find $f(x)$.
60. If $h(0) = 4$ and $h(-3) = -2$ for a linear function h, find $h(x)$.

APPLICATIONS 61. *Physics—spring stretch.* It is known from physics (Hooke's law) that the relationship between the stretch s of a spring and the weight w causing the stretch is linear (a principle upon which all spring scales are constructed). A 10-pound weight stretches a spring 1 inch and with no weight the stretch of the spring is zero.
   (A) Find a linear function f: $s = f(w) = mw + b$, that represents this relationship. [Hint: Both points $(10, 1)$ and $(0, 0)$ are on the graph of f.]
   (B) Find $f(15)$ and $f(30)$—that is, the stretch of the spring for 15-pound and 30-pound weights.

(C) What is the slope of the graph of f? (The slope indicates the increase in stretch for each pound increase in weight.)

(D) Graph f for $0 \leq w \leq 40$.

62. *Business—depreciation.* An electronic computer was purchased by a company for $20,000 and is assumed to have a salvage value of $2,000 after 10 years (for tax purposes). Its value is depreciated linearly from $20,000 to $2,000.

(A) Find the linear function f: $V = f(t)$, that relates value V in dollars to time t in years.

(B) Find $f(4)$ and $f(8)$, the values of the computer after 4 and 8 years, respectively.

(C) Find the slope of the graph of f. (The slope indicates the decrease in value per year.)

(D) Graph f for $0 \leq t \leq 10$.

63. *Biology—nutrition.* A biologist needs to prepare a special diet for a group of experimental animals. Two food mixes, M and N, are available. If mix M contains 20% protein and mix N contains 10% protein, what combinations of each mix will provide exactly 20 grams of protein? Let x be the amount of M used and y the amount of N used. Then write a linear equation relating x, y, and 20. Graph this equation for $x \geq 0$ and $y \geq 0$.

64. *Psychology—motivation.* In an experiment on motivation (see J. S. Brown, *Journal of Comparative Physiology and Psychology*, 1948, **41**:450–465), J. S. Brown trained a group of rats to run down a narrow passage to receive food in a goal box. He then connected the rats, using a harness, to an overhead wire that was attached to a spring scale. A rat was placed at different distances d (in centimeters) from the goal box, and the pull p (in grams) of the rat toward the food was measured. Brown found that the relationship between the two variables was close to being linear and could be approximated by the linear function f: $p = f(d) = -\frac{1}{5}d + 70$, $30 \leq d \leq 175$.

(A) Find $f(30)$ and $f(175)$, the pull toward the goal box at 30 centimeters and 175 centimeters, respectively.

(B) What is the slope of the graph of f? (The slope indicates the change in pull per unit change in position away from the goal box.)

(C) Graph f.

Section 2-5 Graphing Polynomial Functions

- Graphing by "Nested Factoring"
- Quadratic Functions—Special Properties

2 Graphs and Functions

A very important class of functions is polynomial functions. A function f defined by an equation of the form

$$f(x) = a_n x^n + a_{n-1} x^{n-1} + \cdots + a_1 x + a_0 \qquad a_n \neq 0$$

where the coefficients, a_i, are constants and n is a nonnegative integer, is called an **nth-degree polynomial function**. In the last section we discussed first-degree polynomial functions. The following equations define polynomial functions of various degrees:

$$f(x) = 2x - 3 \qquad\qquad g(x) = 2x^2 - 3x + 2$$
$$P(x) = x^3 - 2x^2 + x - 1 \qquad Q(x) = x^4 - 5$$

In Section 2-3 we discussed continuity properties of functions and noted that **polynomial functions are continuous everywhere**. Unless otherwise restricted, the domain of a polynomial function is the set of real numbers. (In the next chapter we will extend the domain to the set of complex numbers.)

In this section we will consider a particularly efficient way of graphing many polynomial functions; we will also consider some special properties of quadratic functions.

■ Graphing by "Nested Factoring"

Graphing polynomial functions by "nested factoring" is a device used to speed up the process of point-by-point plotting. It is particularly well suited for hand calculator use but is also effective for hand or mental calculations. An example will illustrate the process.

EXAMPLE 23   Graph: $P(x) = x^3 + 3x^2 - x - 3, \quad -4 \leq x \leq 2$

Solution   We first write $P(x)$ in a "nested factored" form as follows:

$P(x) = x^3 + 3x^2 - x - 3$ Factor the first two terms, and repeat until you
$\quad = (x + 3)x^2 - x - 3$ cannot go any further.
$\quad = [(x + 3)x - 1]x - 3$

This "nested factored" form is particularly convenient for evaluating $P(x)$ for various values of x by hand, and even more convenient for use with a hand calculator when x is not a small whole number. When using a hand calculator, store the chosen value of x and recall it as necessary as you proceed from left to right. In this case, all the calculations involving integers from -4 to 2 can be done mentally. If decimal values between the integers are desired (for increased graph clarity), then a hand calculator will be a considerable help.

2-5 Graphing Polynomial Functions

Proceed (mentally from the "inside out") using

$$P(x) = [(x + 3)x - 1]x - 3$$

$P(-4) = [((-4) + 3)(-4) - 1](-4) - 3 = -15$
$P(-3) = [((-3) + 3)(-3) - 1](-3) - 3 = 0$
and so on

CALCULATOR COMPUTATION

A: [4] [+/−] [STO] [1] [CLR] [(] [(] [RCL] [1] [+] [3] [)] [×] [RCL] [1] [−] [1] [)] [×] [RCL] [1] [−] [3] [=]

P: [4] [+/−] [STO] [1] [ENTER] [3] [+] [RCL] [1] [×] [1] [−] [RCL] [1] [×] [3] [−]

[*Note:* Here we can see the power of the RPN logic.]

We construct a table of ordered pairs of numbers belonging to the function P. We then plot these points and join them with a smooth curve. (It is important to plot enough points so that it is clear what happens between the points when the points are joined by a smooth curve.)

x	P(x)
−4	−15
−3	0
−2	3
−1	0
0	−3
1	0
2	15

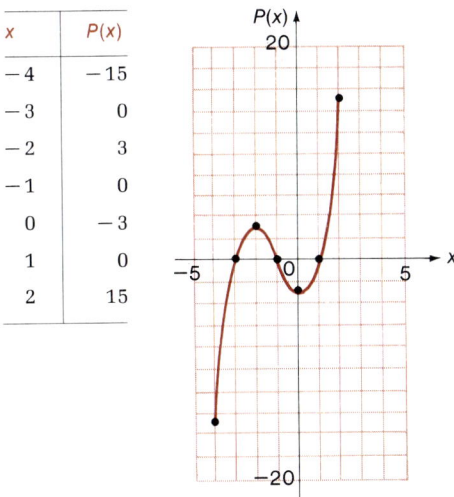

PROBLEM 23 Graph $P(x) = x^3 - 4x^2 - 4x + 16$, $-3 \leq x \leq 5$, using the nested factoring method.

Two nested factorings are shown here for polynomials with terms missing.

$P(x) = x^3 - 2x^2 - 5$
$ = (x - 2)x^2 - 5$
$ = [(x - 2)x]x - 5$

$Q(x) = 2x^3 - 4x + 3$
$ = (2x^2 - 4)x + 3$

Quadratic Functions—Special Properties

Let us now consider a special second-degree polynomial function, called a **quadratic function**, defined by an equation of the form

$$Q(x) = ax^2 + bx + c \qquad a \neq 0 \tag{1}$$

In earlier sections we graphed special cases of equation (1) and found, in particular, that the graph of $f(x) = x^2$ is as indicated in Figure 13. The graph is a parabola.

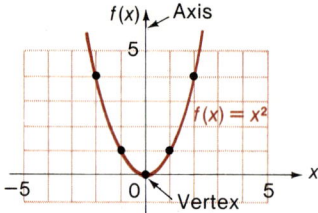

FIGURE 13 A parabola

It can be shown that the graph of any quadratic function is a parabola. In fact, the graph of $Q(x) = ax^2 + bx + c$, where $a \neq 0$, is the same as the graph of $f(x) = x^2$ modified by shifting it left, right, up, or down, and/or reflecting, and/or by expansion or contraction, depending on the values of the constants a, b, and c. The line of symmetry is called the **axis** of the parabola, and the point of intersection of the parabola and its axis is called the **vertex**.

To get a quick sketch of equation (1) we could proceed by the nested factoring method just discussed, and this is exactly what we would do in many cases. However, we can uncover some interesting and useful properties of quadratic functions by transforming equation (1) into the form

$$Q(x) = a(x + h)^2 + k \qquad a, h, k \text{ constants} \tag{2}$$

by completing a square. We look at the process through an example. Starting with

$$f(x) = 2x^2 - 8x + 5$$

we transform it into form (2) by completing a square as follows:

$f(x) = 2x^2 - 8x + 5$ Factor the coefficient of *x* out of the first two terms.
$ = 2(x^2 - 4x) + 5$
$ = 2(x^2 - 4x + ?) + 5$ Complete the square within the parentheses.

2-5 Graphing Polynomial Functions

$$= 2(x^2 - 4x + 4) + 5 - 8$$ We added 4 to complete the square inside the parentheses. But because of the 2 outside of the parentheses, we have actually added 8, so we must subtract 8.

$$= 2(x - 2)^2 - 3$$ The transformation is complete. The graph is the same as the graph of $y = 2x^2$ shifted to the right two units and down three units.

Thus

$$f(x) = \underbrace{2(x - 2)^2}_{\text{Never negative (Why?)}} - 3$$

When $x = 2$, the first term on the right is 0, and we add 0 to -3 to obtain $f(2) = -3$. For *any* other value of x, we will add a positive number to -3, thus making $f(x)$ larger. Therefore,

$$f(2) = -3$$

is the minimum value of $f(x)$ for *all* x—a very important result! The vertical line $x = 2$ is the axis of the parabola and $(2, -3)$ are the coordinates of its vertex.

We plot the vertex and the axis and a couple of points on either side of the axis to complete the graph (Fig. 14).

x	f(x)
2	-3
1	-1
3	-1
0	5
4	5

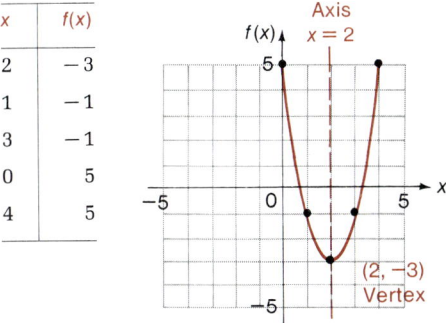

FIGURE 14

Note the important results we have obtained with this approach. We have found:

The axis of the parabola

Its vertex

The minimum value of $f(x)$

The graph of $y = f(x)$

If we had started with the general quadratic function defined by $f(x) = ax^2 + bx + c$, $a \neq 0$, following the same process as in the preceding example, we would have obtained the following general result.

2 Graphs and Functions

Properties of $f(x) = ax^2 + bx + c$, $a \neq 0$

1. **Axis (of symmetry):** $x = -\dfrac{b}{2a}$

2. **Vertex:** $\left(-\dfrac{b}{2a},\ f\left(-\dfrac{b}{2a}\right)\right)$

3. **Maximum or minimum value of $f(x)$:**

$$f\left(-\dfrac{b}{2a}\right) \begin{cases} \text{Minimum} & \text{if } a > 0 \\ \text{Maximum} & \text{if } a < 0 \end{cases}$$

4. **Graph:**

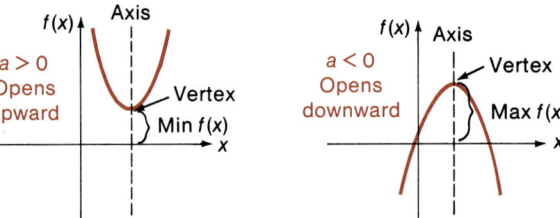

To graph a quadratic function using the method of completing the square, we can either actually complete the square as in the preceding example or use the properties in the box; some people can more readily remember a formula, others a process. We will use the properties in the box in the next example.

EXAMPLE 24 Graph, finding the axis, vertex, and maximum or minimum of $f(x)$:

Solution $f(x) = 12x - 2x^2$

$f(x) = -2x^2 + 12x$ Write in standard form, $f(x) = ax^2 + bx + c$, and note that $a = -2$, $b = 12$, and $c = 0$.

Axis of symmetry:

$$x = -\dfrac{b}{2a} = -\dfrac{12}{2(-2)} = 3$$

$x = 3$ Axis of symmetry

Vertex:

$$\left(-\dfrac{b}{2a},\ f\left(-\dfrac{b}{2a}\right)\right) = (3, f(3)) = (3, 18)$$

2-5 Graphing Polynomial Functions

Maximum value of $f(x)$ (since $a = -2 < 0$):

Max $f(x) = f(3) = 18$

Graph of $y = f(x)$: To graph f, locate the axis and vertex, then locate a couple of points on either side of the axis.

x	f(x)
3	18
2	16
4	16
1	10
5	10
0	0
6	0

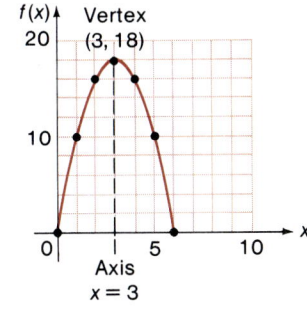

PROBLEM 24 Graph, finding the axis, vertex, and maximum or minimum of $f(x)$.

$f(x) = x^2 - 2x - 3$

Answers to Matched Problems **23.** 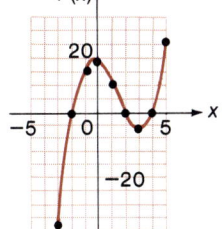 **24.** Min $f(x) = f(1) = -4$

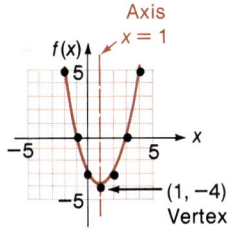

Exercise 2-5

Even though most of the problems in this exercise can be done by hand, a hand calculator will relieve much of the drudgery. The graphing problems involving nested factoring are particularly suited for calculator use.

A Graph each quadratic function using the nested factoring method.

1. $P(x) = x^2 - 4x - 5$
2. $P(x) = x^2 - 6x + 5$
3. $P(x) = -x^2 + 6x$
4. $P(x) = -x^2 + 2x + 8$

Graph, finding the axis, vertex, and maximum or minimum of $f(x)$. [Hint: Look at pages 138 and 139 carefully.]

5. $f(x) = (x - 3)^2 + 2$
6. $f(x) = \frac{1}{2}(x + 2)^2 - 4$
7. $f(x) = -(x + 3)^2 - 2$
8. $f(x) = -(x - 2)^2 + 4$

B   Graph by completing the square. Indicate the axis, vertex, and maximum or minimum value of $f(x)$.

9. $f(x) = x^2 + 6x + 11$
10. $f(x) = x^2 - 8x + 14$
11. $f(x) = -x^2 + 6x - 6$
12. $f(x) = -x^2 - 10x - 24$

Graph each polynomial function using the nested factoring method.

13. $P(x) = x^3 - 5x^2 + 2x + 8$, $-2 \leq x \leq 5$
14. $P(x) = x^3 + 2x^2 - 5x - 6$, $-4 \leq x \leq 3$
15. $P(x) = x^3 + 4x^2 - x - 4$, $-5 \leq x \leq 2$
16. $P(x) = x^3 - 2x^2 - 5x + 6$, $-3 \leq x \leq 4$

C   Graph by completing the square. Indicate the axis, vertex, and maximum or minimum value of $f(x)$.

17. $f(x) = \frac{1}{2}x^2 + 2x$
18. $f(x) = 2x^2 - 12x + 14$
19. $f(x) = -2x^2 - 8x - 2$
20. $f(x) = -\frac{1}{2}x^2 + 4x - 4$

Graph each polynomial function using the nested factoring method.

21. $P(x) = x^4 - 2x^3 - 2x^2 + 8x - 8$
22. $P(x) = x^4 - 2x^2 + 16x - 15$
23. $P(x) = x^4 + 4x^3 - x^2 - 20x - 20$
24. $P(x) = x^4 - 4x^2 - 4x - 1$

APPLICATIONS

25. *Construction.* A rectangular dog pen is to be made with 100 feet of fence wire.
    (A) If x represents the width of the pen, express its area $A(x)$ in terms of x.
    (B) Considering the physical limitations, what is the domain of the function A?
    (C) Graph the function for this domain.
    (D) Determine the dimensions of the rectangle that will make the area maximum.

26. *Construction.* Work Problem 25 with the added assumption that an existing property fence will be used for one side of the pen. (Let x equal the width—see the figure.)

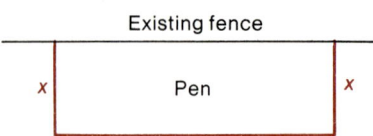

2-6 Graphing Rational Functions   143

27. *Packaging.* A candy box is to be made out of a rectangular piece of cardboard that measures 8 by 12 inches. Equal-sized squares (x by x inches) will be cut out of each corner, and then the ends and sides will be folded up to form a rectangular box.
   (A) Write the volume of the box V(x) in terms of x.
   (B) Considering the physical limitations, what is the domain of the function V?
   (C) Graph the function for this domain.
   (D) From the graph, estimate to the nearest half-inch the size square that must be cut from each corner to yield a box with the maximum volume. What is the maximum volume?

28. *Packaging.* A parcel delivery service will deliver only packages with length plus girth (distance around) not exceeding 108 inches. A packaging company wishes to design a box with a square base (x by x inches) that will have a maximum volume and will meet the delivery service's restrictions.
   (A) Write the volume of the box V(x) in terms of x.
   (B) Considering the physical limitations imposed by the delivery service, what is the domain of the function V?
   (C) Graph the function for this domain.
   (D) From the graph estimate to the nearest inch the dimensions of the box with the maximum volume. What is the maximum volume?

Section 2-6 Graphing Rational Functions

- Rational Functions
- Vertical and Horizontal Asymptotes
- Graphing Rational Functions

Rational Functions

Just as rational numbers are defined in terms of quotients of integers, rational functions are defined in terms of quotients of polynomials. The following equations define rational functions:

$$f(x) = \frac{x-1}{x^2 - x - 6} \qquad g(x) = \frac{1}{x} \qquad h(x) = \frac{x^3 - 1}{x}$$

$$p(x) = 2x^2 - 3 \qquad q(x) = 3 \qquad r(x) = 0$$

In general, a function R is a **rational function** if

$$R(x) = \frac{P(x)}{Q(x)} \qquad Q(x) \neq 0$$

and P(x) and Q(x) are polynomials. The **domain of R** is the set of all real numbers x such that $Q(x) \neq 0$. In Section 2-3 we noted that rational functions are continuous (no breaks or holes in the graph) for all values of x except those for which the denominator Q(x) is 0. If $x = a$ and $Q(a) = 0$, then R is said to be **discontinuous at x = a**.

■ Vertical and Horizontal Asymptotes

Even though a rational function R may be discontinuous at $x = a$ (no graph for $x = a$), it is still useful to know what happens to the graph of R when x is close to a. Let us make this discussion concrete through a very simple rational function f defined by

$$f(x) = \frac{1}{x}$$

It is clear that the function f is discontinuous at $x = 0$. But what happens to f(x) when x approaches 0 from either side of 0? A few table values will give us an idea of what happens to f(x) when x gets close to 0 (Tables 6 and 7). As x approaches 0 from the right (denoted by $x \to 0^+$), x stays positive and 1/x gets larger and larger; that is, 1/x increases without bound (denoted by $1/x \to \infty$).* As x approaches 0 from the left (denoted by $x \to 0^-$), x stays negative and 1/x decreases without bound (denoted by $1/x \to -\infty$). Thus, the graph of $f(x) = 1/x$ approaches the y axis (but never touches it) as x gets closer to 0. The graph of $f(x) = 1/x$ for

TABLE 6
x APPROACHES 0
FROM THE RIGHT ($x \to 0^+$)

x	1	0.1	0.01	0.001	0.0001	0.000 01	0.000 001	...
1/x	1	10	100	1,000	10,000	100,000	1,000,000	...

TABLE 7
x APPROACHES 0
FROM THE LEFT ($x \to 0^-$)

x	−1	−0.0	−0.01	−0.001	−0.0001	−0.000 01	−0.000 001	...
1/x	−1	−10	−100	−1,000	−10,000	−100,000	−1,000,000	...

* The symbol ∞, called **infinity**, does not represent a real number. When we write $1/x \to \infty$, we mean that 1/x exceeds any given number N no matter how large N is chosen.

$-4 \leq x \leq 4$ is shown in Figure 15. Note that f is an odd function and the graph is symmetric with respect to the origin.

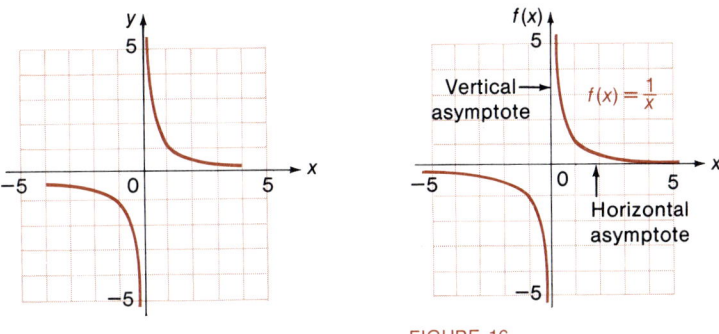

FIGURE 15

FIGURE 16

Now let us look at the behavior of $f(x) = 1/x$ as $|x|$ gets very large—that is, as $x \to \infty$ and as $x \to -\infty$. Consider Tables 8 and 9. As x increases without bound ($x \to \infty$), $1/x$ stays positive and approaches 0 from above ($1/x \to 0^+$). As x decreases without bound ($x \to -\infty$), $1/x$ stays negative and approaches 0 from below ($1/x \to 0^-$). The graph never touches the x or y axes; that is, there are no x or y intercepts. A sketch of the graph of $f(x) = 1/x$ is completed in Figure 16. The curve is an example of a plane curve called a *hyperbola*.

TABLE 8

$x \to \infty$

x	1	10	100	1,000	10,000	100,000	1,000,000	...
$1/x$	1	0.1	0.01	0.001	0.0001	0.000 01	0.000 001	...

TABLE 9

$x \to -\infty$

x	-1	-10	-100	$-1,000$	$-10,000$	$-100,000$	$-1,000,000$	...
$1/x$	-1	-0.1	-0.01	-0.001	-0.0001	$-0.000\ 01$	$-0.000\ 001$	...

The fixed lines that the graph approaches are called *asymptotes*. In this case the y axis is a *vertical asymptote* and the x axis is a *horizontal asymptote*. In general, a line $x = a$ is a **vertical asymptote** for the graph of $y = f(x)$ if $f(x)$ either increases or decreases without bound as $x \to a^+$ or $x \to a^-$. A line $y = b$ is a **horizontal asymptote** if $f(x)$ approaches b as $x \to -\infty$ or $x \to \infty$. Graphing certain kinds of functions is considerably aided by locating vertical and horizontal asymptotes first, if they exist. Using the same kind of reasoning as in the preceding example, we have the following general method of locating vertical asymptotes for rational functions.

Vertical Asymptotes

Let R be a rational function defined by

$$R(x) = \frac{P(x)}{Q(x)}$$

where $P(x)$ and $Q(x)$ are polynomials. If a is a real number such that $Q(a) = 0$ and $P(a) \neq 0$, then the line $x = a$ is a vertical asymptote of the graph $y = R(x)$.

To gain added insight in locating horizontal asymptotes for the graph of $R(x) = P(x)/Q(x)$, consider the following three examples.

1. Degree of numerator less than degree of denominator:

$$f(x) = \frac{2x - 1}{3x^2 - 2x}$$

Divide each term in numerator and denominator by x^2, the highest power of x that appears in the numerator and denominator.

$$= \frac{\dfrac{2x}{x^2} - \dfrac{1}{x^2}}{\dfrac{3x^2}{x^2} - \dfrac{2x}{x^2}}$$

Reduce internal fractions.

$$= \frac{\dfrac{2}{x} - \dfrac{1}{x^2}}{3 - \dfrac{2}{x}}$$

As $x \to \infty$ or $-\infty$, $2/x \to 0$ and $1/x^2 \to 0$; hence, $f(x) \to 0$. Thus, the line $y = 0$ is a horizontal asymptote.

2. Degree of numerator equal to degree of denominator:

$$g(x) = \frac{2x^2 - 1}{3x^2 - 2x}$$

Divide each term in numerator and denominator by x^2, the highest power of x that appears in the numerator and denominator.

$$= \frac{\dfrac{2x^2}{x^2} - \dfrac{1}{x^2}}{\dfrac{3x^2}{x^2} - \dfrac{2x}{x^2}}$$

Reduce internal fractions.

2-6 Graphing Rational Functions

$$= \frac{2 - \dfrac{1}{x^2}}{3 - \dfrac{2}{x}}$$

As $x \to \infty$ or $-\infty$, $1/x^2 \to 0$ and $2/x \to 0$; hence, $g(x) \to \frac{2}{3}$. Thus, $y = \frac{2}{3}$ is a horizontal asymptote.

3. Degree of numerator greater than degree of denominator:

$$h(x) = \frac{2x^3 - 1}{3x^2 - 2x}$$ Divide each term in numerator and denominator by x^3, the highest power of x that appears in the numerator and denominator.

$$= \frac{\dfrac{2x^3}{x^3} - \dfrac{1}{x^3}}{\dfrac{3x^2}{x^3} - \dfrac{2x}{x^3}}$$ Reduce internal fractions.

$$= \frac{2 - \dfrac{1}{x^3}}{\dfrac{3}{x} - \dfrac{2}{x^2}}$$

As $x \to \infty$ or $-\infty$, the numerator approaches 2 and the denominator approaches 0. Thus, $|h(x)|$ increases without bound and there is no horizontal asymptote.

Reasoning in the same way as in these examples, one can establish the general method of locating horizontal asymptotes shown in the box.

Horizontal Asymptotes

Let R be a rational function defined by the quotient of two polynomials as follows:

$$R(x) = \frac{a_m x^m + \cdots + a_1 x + a_0}{b_n x^n + \cdots + b_1 x + b_0} \qquad a_m, b_n \neq 0$$

1. For $m < n$, the x axis ($y = 0$) is a horizontal asymptote.
2. For $m = n$, the line $y = a_m/b_n$ is a horizontal asymptote.
3. For $m > n$, there are no horizontal asymptotes.

2 Graphs and Functions

■ **Graphing Rational Functions**

We now use these new aids to graphing (along with other aids discussed earlier) to graph several rational functions. First, we outline a systematic approach to the problem of graphing rational functions:

To Graph a Rational Function

$$y = R(x) = \frac{P(x)}{Q(x)}$$

1. Determine symmetry with respect to the vertical axis and origin. (Is R even or odd?)
2. Find and plot x and y intercepts.
3. Find points of discontinuity.
4. Determine vertical and horizontal asymptotes and sketch them using broken lines.
5. Find where the graph of $y = R(x)$ is above and below the x axis. [Solve $R(x) > 0$ and $R(x) < 0$.]
6. Determine the behavior of the graph when it is close to its asymptotes. [What happens to $f(x)$ when $x \to \infty$ or $-\infty$? If the line $x = a$ is a vertical asymptote, what happens to $f(x)$ when $x \to a^+$ or a^-?]
7. Complete the sketch of the graph by plotting additional points as necessary and joining these points with a smooth curve. (Do not cross points of discontinuity.)

EXAMPLE 25 Graph: $y = f(x) = \dfrac{2x}{x - 3}$

Solution 1. Symmetry: Compute

$$f(-x) = \frac{-2x}{-x - 3}$$

f is neither even nor odd, since $f(-x) \neq f(x)$ and $f(-x) \neq -f(x)$.

2. Intercepts:

When $x = 0$, $y = \dfrac{2 \cdot 0}{0 - 3} = 0$ y intercept

When $y = 0$, $0 = \dfrac{2x}{x - 3}$ x intercept

$x = 0$

The graph crosses the coordinate axes only at the origin. Sketch the intercepts [figure (a)].

3. Points of discontinuity: f is discontinuous where $Q(x) = x - 3 = 0$—that is, at $x = 3$.

4. Asymptotes:

   Vertical: $x = 3$

   Horizontal: $y = \dfrac{2}{1} = 2$ Degrees of numerator and denominator are the same.

Sketch the asymptotes [figure (a)].

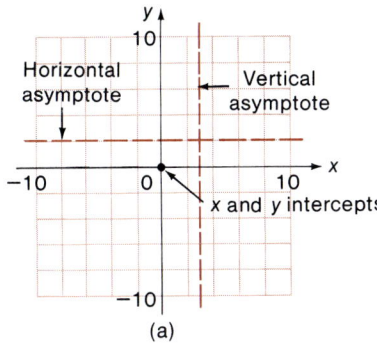

(a)

5. Where is the graph above and below the x axis? (Review Section 1-5.)

   Critical points: $x = 0, 3$

   Sign of $2x$ $- -$ | $+ + + +$ | $+ +$
   Sign of $(x - 3)$ $- -$ | $- - - -$ | $+ +$

   $\qquad\qquad\qquad\qquad 0 \qquad\qquad 3 \qquad\qquad\qquad x$

   $f(x) > 0$ for $x < 0$ or $x > 3$ Graph above x axis
   $f(x) < 0$ for $0 < x < 3$ Graph below x axis

6. Asymptotic behavior: Vertical asymptote ($x = 3$): With the help of steps 4 and 5, we see that

   $f(x) \to -\infty$ as $x \to 3^-$
   $f(x) \to \infty$ as $x \to 3^+$

   Horizontal asymptote ($y = 2$): With the help of steps 4 and 5, we see that

   $f(x) \to 2$ as $x \to -\infty$ and as $x \to \infty$

7. **Complete the sketch:** By plotting a few additional points, we obtain the graph as in figure (b).

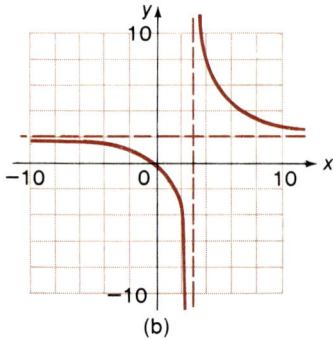

(b)

As one gains experience in graphing, many of the steps in Example 25 can be done mentally (or on scratch paper) and the process can be speeded up considerably.

PROBLEM 25 Proceed as in Example 25 and graph: $y = f(x) = \dfrac{3x}{x+2}$

EXAMPLE 26 Graph: $y = f(x) = \dfrac{x-1}{x^2-1}$

Solution Factor the denominator:

$$y = f(x) = \frac{x-1}{(x-1)(x+1)} = \frac{1}{x+1} \qquad x \neq 1$$

1. **Symmetry:** f is neither even nor odd.
2. **Intercepts:**

$$\text{When } x = 0, \quad y = \frac{1}{0+1} = 1 \quad \text{y intercept}$$

No x intercept

Graph the intercept [figure (a)].

3. **Points of discontinuity:** $x = -1$ and $x = 1$.
4. **Asymptotes:** Vertical: Find values of x such that the denominator $x^2 - 1 = 0$, but the numerator $x - 1 \neq 0$. This is true for $x = -1$. Thus, the line $x = -1$ is a vertical asymptote.
Horizontal: x axis (since the degree of the numerator is less than the degree of the denominator).

2-6 Graphing Rational Functions

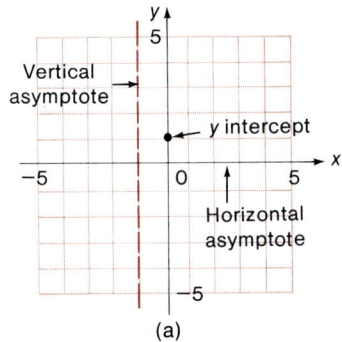

(a)

Sketch in asymptotes [figure (a)].

5. Where is graph above and below the x axis?   By inspection,

$$f(x) = \frac{1}{x+1} > 0 \quad \text{when } x > -1 \quad \text{Graph above } x \text{ axis}$$

$$f(x) = \frac{1}{x+1} < 0 \quad \text{when } x < -1 \quad \text{Graph below } x \text{ axis}$$

6. Asymptotic behavior: Vertical asymptote ($x = 1$): With the help of steps 4 and 5, we see that

$$f(x) \to -\infty \quad \text{as } x \to 1^-$$
$$f(x) \to \infty \quad \text{as } x \to 1^+$$

Horizontal asymptote ($y = 0$): With the help of steps 4 and 5, we see that

$$f(x) \to 0 \quad \text{as } x \to -\infty \quad \text{and} \quad \text{as } x \to \infty$$

7. Complete the sketch: By plotting a few additional points we obtain the graph in figure (b). Note what happens at the two points of discontinuity $x = -1$ and $x = 1$. At $x = -1$ we have a huge break in the graph; at $x = 1$ there is just a point missing, which, if replaced, would make the function continuous there.

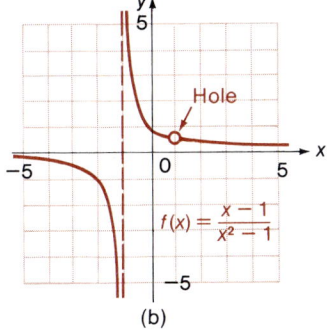

(b)

PROBLEM 26   Graph: $y = f(x) = \dfrac{x+1}{x^2-1}$

EXAMPLE 27   Graph: $y = g(x) = \dfrac{x^2+4}{x^2-4}$

Solution   Factor the denominator:

$$y = f(x) = \dfrac{x^2+4}{(x-2)(x+2)}$$

1. Symmetry: $f(-x) = f(x)$; hence, f is even and its graph is symmetric with respect to the y axis.
2. Intercepts:

   When $x = 0$, $y = \dfrac{0+4}{0-4} = -1$ y intercept

   No x intercept $x^2 + 4 \neq 0$ for all x

   Graph the intercepts [figure (a)].
3. Points of discontinuity: $x = -2$ and $x = 2$.
4. Asymptotes:

   Vertical: $x = -2$ and $x = 2$
   Horizontal: $y = 1$ Degrees of numerator and denominator are the same.

   Sketch the asymptotes on graph [figure (a)].

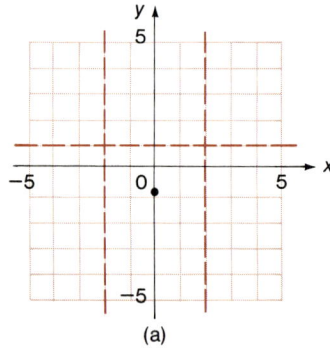

(a)

5. Where is the graph above and below the x axis?

$$f(x) = \dfrac{x^2+4}{(x-2)(x+2)}$$

Critical points: $x = -2$ and $x = 2$

2-6 Graphing Rational Functions   153

Sign of $(x^2 + 4)$	+ +	+ + + +	+ +	Positive for all x
Sign of $(x - 2)$	− −	− − − −	+ +	
Sign of $(x + 2)$	− −	+ + + +	+ +	

                  −2   2

$f(x) > 0$ for $x < -2$ and $x > 2$ Graph above x axis

$f(x) < 0$ for $-2 < x < 2$ Graph below x axis

6. Asymptotic behavior: Vertical asymptotes ($x = -2$ and $x = 2$): With the help of steps 4 and 5, we see that

$f(x) \to \infty$ as $x \to -2^-$

$f(x) \to -\infty$ as $x \to -2^+$

$f(x) \to -\infty$ as $x \to 2^-$

$f(x) \to \infty$ as $x \to 2^+$

Horizontal asymptotes: With the help of steps 4 and 5, we see that

$f(x) \to 1$ as $x \to -\infty$ and as $x \to \infty$

7. Complete the sketch: By plotting a few additional points we obtain the graph in figure (b).

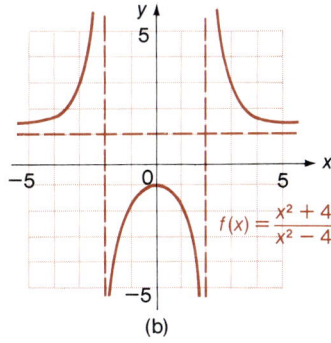

(b)

PROBLEM 27   Graph: $f(x) = \dfrac{x^2}{x^2 - 1}$

Answers to Matched Problems

25.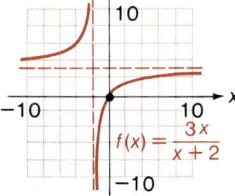

26.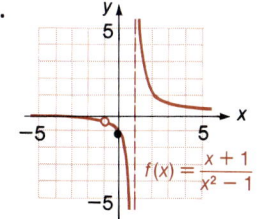

27.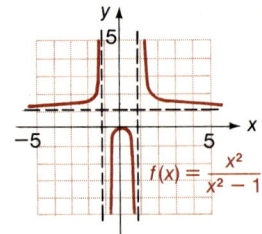

$f(x) = \dfrac{x^2}{x^2 - 1}$

Exercise 2-6 ■ Graph each function. Sketch in any horizontal or vertical asymptotes.

A 1. $f(x) = \dfrac{1}{x - 4}$ 2. $g(x) = \dfrac{1}{x + 3}$

3. $p(x) = \dfrac{-1}{x - 4}$ 4. $m(x) = \dfrac{-1}{x + 3}$

5. $f(x) = \dfrac{x}{x + 1}$ 6. $f(x) = \dfrac{3x}{x - 3}$

B 7. $q(x) = \dfrac{2x - 1}{x}$ 8. $r(x) = \dfrac{1 - x}{x}$

9. $h(x) = \dfrac{x}{2x - 2}$ 10. $p(x) = \dfrac{3x}{4x + 4}$

11. $g(x) = \dfrac{1 - x^2}{x^2}$ 12. $f(x) = \dfrac{x^2 + 1}{x^2}$

13. $f(x) = \dfrac{9}{x^2 - 9}$ 14. $g(x) = \dfrac{6}{x^2 - x - 6}$

15. $f(x) = \dfrac{x}{x^2 - 1}$ 16. $p(x) = \dfrac{x}{1 - x^2}$

17. $g(x) = \dfrac{2}{x^2 + 1}$ 18. $f(x) = \dfrac{x}{x^2 + 1}$

19. $h(x) = \dfrac{2x^2}{x^2 + 1}$ 20. $f(x) = \dfrac{-2x^4}{x^4 + 1}$

C 21. $m(x) = \dfrac{x^2 - 1}{x^2 + x - 2}$ 22. $g(x) = \dfrac{x^2 - x - 6}{x^2 - 4}$

Section 2-7 Composite and Inverse Functions

- Composite Functions
- Inverse Relations and Functions
- One-to-One Correspondence and Inverses

2-7 Composite and Inverse Functions

In this section we will discuss two important methods of obtaining new relations and functions from known functions.

■ Composite Functions

Consider the function h given by the equation

$$h(x) = \sqrt{2x + 1}$$

Inside the radical is a first-degree polynomial that defines a linear function. So the function h is really a combination of a square root function and a linear function. We see this clearly as follows:

Let $\quad u = 2x + 1 = g(x)$
$\qquad y = \sqrt{u} = f(u)$

Then $\quad h(x) = f[g(x)]$

The function h is said to be the **composite** of the two simpler functions f and g. (Loosely speaking, we can think of h as a function of a function.) What can we say about the domain of h given the domains of f and g? In forming the composite

$$h(x) = f[g(x)]$$

x must be restricted so that x is in the domain of g and g(x) is in the domain of f. Since the domain of f, where $f(u) = \sqrt{u}$, is the set of nonnegative real numbers,* we see that $g(x)$ must be nonnegative; that is,

$\qquad g(x) \geq 0$
$\qquad 2x + 1 \geq 0$
$\qquad\quad x \geq -\frac{1}{2}$

Thus, the domain of h is this restricted domain of g.

A special function symbol is often used to represent the **composite of two functions**, which we define in general terms in the box on page 156.

---

* Recall from Section 2-2 we said that if a function is specified by an equation and the domain is not indicated, then we shall assume that the domain is the set of all real replacements of the independent variable that produce real values for the dependent variable. The range is the set of all dependent variable values corresponding to independent variable values.

Composite Functions

Given functions f and g, then f ∘ g is called their **composite** and is defined by the equation

$(f \circ g)(x) = f[g(x)]$

The domain of f ∘ g is the set of all numbers x such that x is in the domain of g and g(x) is in the domain of f.

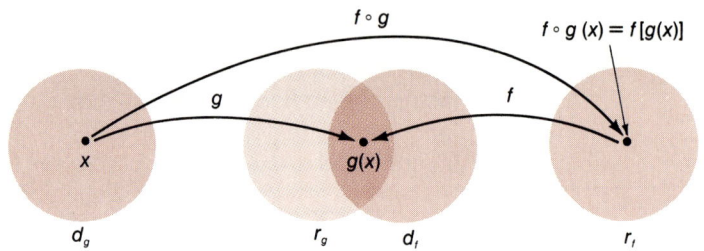

EXAMPLE 28 Find (f ∘ g)(x) and (g ∘ f)(x) and their domains for $f(x) = x^{10}$ and $g(x) = 3x^4 - 1$.

Solution $(f \circ g)(x) = f[g(x)] = [g(x)]^{10} = (3x^4 - 1)^{10}$
$(g \circ f)(x) = g[f(x)] = 3[f(x)]^4 - 1 = 3(x^{10})^4 - 1 = 3x^{40} - 1$

The domain of f and the domain of g are the set of all real numbers R. Thus, for all real x, x is in the domain of g and g(x) is in the domain of f, and x is in the domain of f and f(x) is in the domain of g. Hence, the domain of f ∘ g and the domain of g ∘ f are R.

PROBLEM 28 Find (f ∘ g)(x) and (g ∘ f)(x) and their domains for $f(x) = 2x + 1$ and $g(x) = (x - 1)/2$.

■ Inverse Relations and Functions

We now turn to a second way of obtaining new functions from given functions. The process is used to obtain logarithmic functions from exponential functions and inverse trigonometric functions from trigonometric functions.

Given a relation G, if we interchange the order of the components in each ordered pair belonging to G, we obtain a new relation G^{-1} called the **inverse of G**. [Note: G^{-1} is a relation–function symbol; it does not mean 1/G.] For example, if

$G = \{(2, 4), (-1, 3), (0, 4)\}$

then by reversing the components in each ordered pair in G, we obtain

$$G^{-1} = \{(4, 2), (3, -1), (4, 0)\}$$

It follows from the definition (and is evident from the example) that the domains and ranges of G and G^{-1} are interchanged.

Inverse of G

If G is a relation, the inverse of G, denoted by G^{-1}, is given by

$$G^{-1} = \{(b, a) | (a, b) \in G\}$$
Domain of G^{-1} = Range of G
Range of G^{-1} = Domain of G

If a relation G is specified by an equation, say

$$G: \quad y = 2x - 1 \tag{1}$$

then how do we find G^{-1}? The answer is easy: We interchange the variables in equation (1). Thus,

$$G^{-1}: \quad x = 2y - 1 \tag{2}$$

or, solving for y,

$$G^{-1}: \quad y = \frac{x + 1}{2} \tag{3}$$

Any ordered pair of numbers that satisfies equation (1), when reversed in order, will satisfy equations (2) and (3). For example, (3, 5) satisfies equation (1) and (5, 3) satisfies equations (2) and (3), as can easily be checked.

If we sketch a graph of G, G^{-1}, and y = x on the same coordinate system (Fig. 17), we will observe something interesting. If we fold the

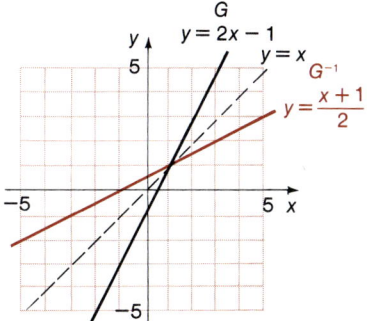

FIGURE 17

paper along the line $y = x$, then the graphs of G and G^{-1} match. Actually, we can graph G^{-1} by drawing G with wet ink and folding the paper along $y = x$ before the ink dries; G will then print G^{-1}. [To prove this in general, one has to show that the line $y = x$ is the perpendicular bisector of the line joining (a, b) and (b, a).] Knowing that the graphs of G and G^{-1} are symmetric relative to the line $y = x$ makes it easy to graph G^{-1} if G is known, and vice versa.

In Figure 17 observe that G and G^{-1} are both functions. This is not always the case, however. Inverses of some functions may not be functions. Consider the following example.

EXAMPLE 29 The relation f is given by $y = x^2$.

(A) Find f^{-1}.
(B) Graph f, f^{-1}, and $y = x$. Is either f or f^{-1} a function?
(C) Indicate the domain and range of f and f^{-1}.

Solution (A) f^{-1}: $x = y^2$ or $y = \pm\sqrt{x}$

(B)
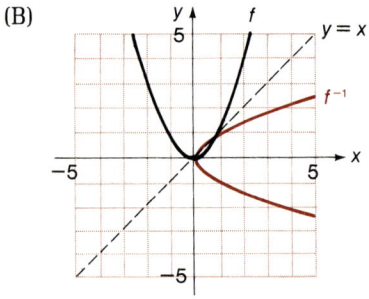

f is a function

f^{-1} is not a function

(C) Domain of $f = R =$ Range of f^{-1}

Range of $f = [0, \infty) =$ Domain of f^{-1}

PROBLEM 29 The relation f is given by $y = |x| + 2$.

(A) Find f^{-1}.
(B) Graph f, f^{-1}, and $y = x$. Is either f or f^{-1} a function?
(C) Indicate the domain and range of f and f^{-1}.

■ One-to-One Correspondence and Inverses

If we are given a function f, how can we tell in advance whether its inverse f^{-1} will be a function? The answer is contained in the concept of one-to-one correspondence. A **one-to-one correspondence** exists

2-7 Composite and Inverse Functions

between two sets if each element in the first set corresponds to exactly one element in the second set, and each element in the second set corresponds to exactly one element in the first set. Consider the two functions f and g and their inverses:

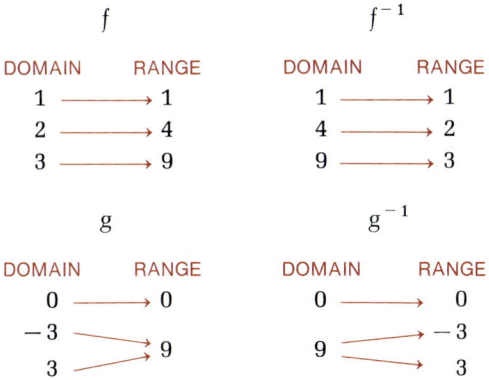

Function f has a one-to-one correspondence between domain and range values (notice that f^{-1} is also a function). Function g does not have a one-to-one correspondence between domain and range values (notice that g^{-1} is not a function).

THEOREM 4

Inverses

A function f has an inverse that is a function if and only if there exists a one-to-one correspondence between domain and range values of f. In this case,

$$(f \circ f^{-1})(y) = y \quad \text{and} \quad (f^{-1} \circ f)(x) = x$$

Theorem 4 is interpreted schematically in Figure 18. Figure 19 illustrates some functions that are one to one,* and Figure 20 illustrates

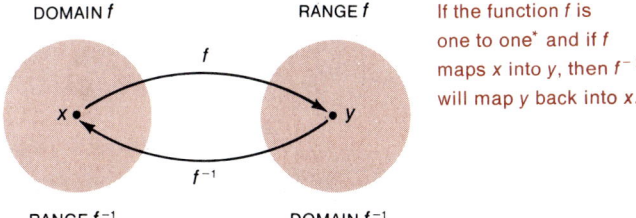

If the function f is one to one* and if f maps x into y, then f^{-1} will map y back into x.

FIGURE 18

---

* When we refer to a function as being *one to one*, we mean there exists a one-to-one correspondence between its domain and range values.

some that are not. The following observation can be made (and proved in general): **All increasing functions and all decreasing functions are one to one and hence have inverses that are functions.**

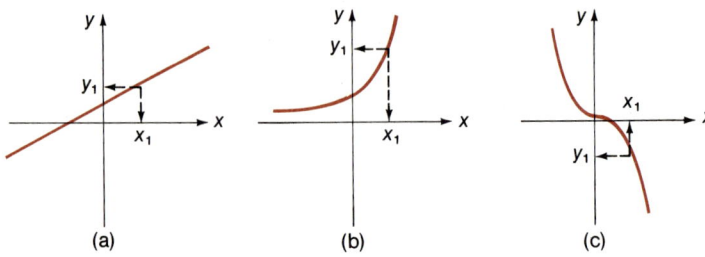

FIGURE 19 Functions that are one to one (each has an inverse that is a function).

(a) (b) (c)

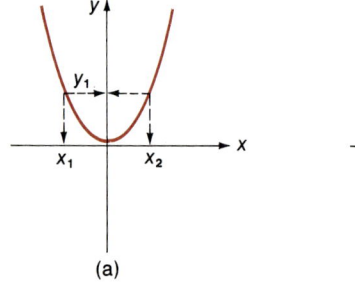 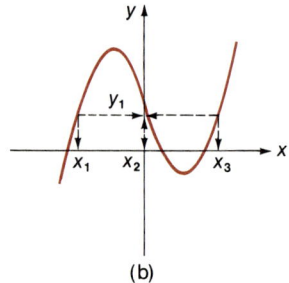

FIGURE 20 Functions that are not one to one (neither has an inverse that is a function).

(a) (b)

When a function is defined by an equation and the function is either increasing or decreasing (thus, it is one to one), we can often find its inverse in terms of an equation.

EXAMPLE 30 Given $f(x) = 3x + 2$, find:

(A) $f^{-1}(x)$ (B) $f^{-1}(5)$
(C) $(f^{-1} \circ f)(5)$ (D) $(f^{-1} \circ f)(x)$

Solution Function f is linear with slope 3 and hence is increasing and has an inverse that is a function.

(A) $f: \quad y = 3x + 2$ Replace $f(x)$ with y in $f(x) = 3x + 2$.

$f^{-1}: \quad x = 3y + 2$ Interchange variables x and y to obtain f^{-1}.

$$y = \frac{x-2}{3}$$ Solve for y in terms of x.

Thus,

$$f^{-1}(x) = \frac{x-2}{3}$$ Replace y with $f^{-1}(x)$.

(B) $f^{-1}(5) = \dfrac{5-2}{3} = \dfrac{3}{3} = 1$

(C) $(f^{-1} \circ f)(5) = f^{-1}[f(5)]$ We are just verifying the results of Theorem 4; that is, if f and f^{-1} are both functions and 5 is in the domain of $f^{-1} \circ f$, then $(f^{-1} \circ f)(5) = 5$, or, in general, $(f^{-1} \circ f)(x) = x$ for all x in the domain of $f^{-1} \circ f$.

$= \dfrac{f(5) - 2}{3}$

$= \dfrac{17 - 2}{3}$

$= \dfrac{15}{3} = 5$

(D) $(f^{-1} \circ f)(x) = f^{-1}[f(x)]$ See comment in part C.

$= \dfrac{f(x) - 2}{3}$

$= \dfrac{(3x + 2) - 2}{3} = x$

PROBLEM 30   Given $g(x) = \dfrac{x}{3} - 2$, find:

(A) $g^{-1}(x)$ (B) $g^{-1}(-2)$
(C) $(g^{-1} \circ g)(3)$ (D) $(g^{-1} \circ g)(x)$

There are many cases where we start with a function that is not one to one and restrict its domain so that the function is either increasing or decreasing and hence becomes one to one. We do this so that we can obtain an inverse that is a function. Suppose we start with $f(x) = x^2$. Because f is not one to one, its inverse will not be a function [Fig. 21(a)]. There are many ways in which the domain of f can be restricted to obtain either an increasing or a decreasing function. Figures 21(b) and 21(c) illustrate two such restrictions.

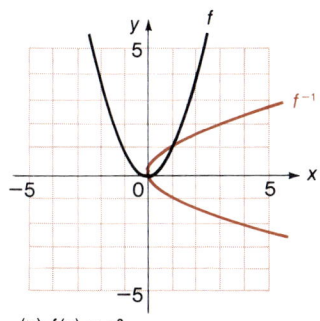
(a) $f(x) = x^2$
f^{-1} is not a function

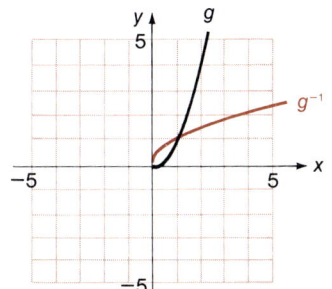
(b) $g(x) = x^2$, $x \geq 0$
g^{-1} is a function

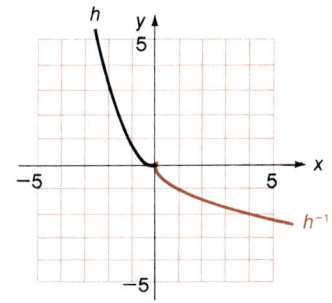
(c) $h(x) = x^2$, $x \leq 0$
h^{-1} is a function

FIGURE 21

EXAMPLE 31 Given $f(x) = x^2$, $x \geq 0$, find:

(A) The domain and range of f and f^{-1}
(B) $f^{-1}(x)$ (C) $f^{-1}(4)$ (D) $(f \circ f^{-1})(3)$ (E) $(f \circ f^{-1})(x)$

Solution (A) Domain of $f = [0, \infty) =$ Range of f^{-1}
Range of $f = [0, \infty) =$ Domain of f^{-1}

(B) $f:$ $y = x^2$ $x \geq 0$ Replace $f(x)$ with y in $f(x) = x^2$.
     $f^{-1}:$ $x = y^2$ $y \geq 0$ Interchange variables x and y to obtain f^{-1}.
          $y = \sqrt{x}$ Solve for y in terms of x. [Note: $y \neq -\sqrt{x}$. Why?]
          $f^{-1}(x) = \sqrt{x}$ Replace y with $f^{-1}(x)$.

(C) $f^{-1}(4) = \sqrt{4} = 2$

(D) $(f \circ f^{-1})(3) = 3$ Direct use of Theorem 4.

(E) $(f \circ f^{-1})(x) = x$ Direct use of Theorem 4.

PROBLEM 31 Given $g(x) = x^2 - 1$, $x \geq 0$, find:

(A) The domain and range of g and g^{-1}
(B) $g^{-1}(x)$ (C) $g^{-1}(8)$ (D) $(g^{-1} \circ g)(5)$ (E) $(g^{-1} \circ g)(x)$

Answers to Matched Problems
28. $(f \circ g)(x) = x$, Domain $= R$; $(g \circ f)(x) = x$, Domain $= R$
29. (A) $f^{-1}:$ $x = |y| + 2$ or $|y| = x - 2$ or $y = \pm(x - 2)$

(B)

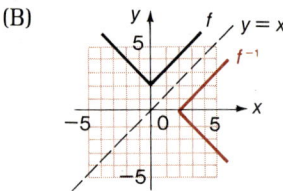

f is a function
f^{-1} is not a function

(C) Domain of $f = R =$ Range of f^{-1}
Range of $f = [2, \infty) =$ Domain of f^{-1}

30. (A) $3x + 6$ (B) 0   (C) 3   (D) x
31. (A) Domain of $g = [0, \infty) =$ Range of g^{-1}
        Range of $g = [-1, \infty) =$ Domain of g^{-1}
(B) $\sqrt{x + 1}$ (C) 3   (D) 5   (E) x

Exercise 2-7 ■ **A** Find $(f \circ g)(x)$ and $(g \circ f)(x)$ for functions f and g as indicated.

1. $f(x) = x^7$, $g(x) = x^2 - x + 1$
2. $f(x) = x^{12}$, $g(x) = 2x^3 - 5$

3. $f(x) = \sqrt{x}$, $g(x) = 2x + 5$
4. $f(x) = \sqrt{x}$, $g(x) = x - 4$
5. $f(x) = x^{1/2}$, $g(x) = 1 - x^2$
6. $f(x) = x^{1/4}$, $g(x) = 4 - x^2$
7. $f(x) = |x|$, $g(x) = 3x - 2$
8. $f(x) = |x - 1|$, $g(x) = \sqrt{x}$
9. $f(x) = x^{2/3}$, $g(x) = x^3 - 4$
10. $f(x) = x^{3/5}$, $g(x) = 3x^{10} + x^5$
11. $f(x) = |x|$, $g(x) = -7$
12. $f(x) = \sqrt{x}$, $g(x) = 4$

Graph each relation, its inverse, and $y = x$ on the same coordinate system. Identify all functions in addition to $y = x$. [Use the graph of the original relation to find the graph of the inverse. Remember, if $(a, b) \in f$, then $(b, a) \in f^{-1}$.]

13. $H = \{(-2, \frac{1}{4}), (0, 1), (1, 2), (2, 4)\}$
14. $G = \{(-4, 2), (-1, 2), (2, 2), (4, 2)\}$
15. g: $y = 2x - 2$ 16. f: $y = -\frac{1}{2}x + 2$
17. p: $y = x^2 + 1$ 18. q: $y = (x + 2)^2$

B For the following functions f that are one to one, find:
(A) The domain and range of f and f^{-1}
(B) $f^{-1}(x)$ (C) $f^{-1}(4)$ (D) $(f^{-1} \circ f)(2)$ (E) $(f^{-1} \circ f)(x)$

19. $f(x) = 2x - 2$ 20. $f(x) = -\frac{1}{2}x + 2$
21. $f(x) = 3x - 5$ 22. $f(x) = 2x - 4$
23. $f(x) = x^2 + 1$ 24. $f(x) = (x + 2)^2$
25. $f(x) = x^2 + 1$, $x \geq 0$ 26. $f(x) = (x + 2)^2$, $x \geq -2$
27. $f(x) = (x - 1)^2$ 28. $f(x) = x^2 - 2$
29. $f(x) = (x - 1)^2$, $x \geq 1$ 30. $f(x) = x^2 - 2$, $x \geq 0$

Find the domain of $f \circ g$ and of $g \circ f$, given functions f and g in:

31. Problem 1   32. Problem 2   33. Problem 3
34. Problem 4   35. Problem 5   36. Problem 6
37. Problem 7   38. Problem 8   39. Problem 9
40. Problem 10 41. Problem 11 42. Problem 12

C For f as indicated in Problems 43–46, find:
(A) $(f \circ f)(x)$ (B) $f^{-1}(x)$ (C) $(f \circ f^{-1})(x)$
(D) Any conclusions?

43. $f(x) = 4 - x$ 44. $f(x) = \sqrt{2 - x}$

45. $f(x) = 1/x$ **46.** $f(x) = \sqrt{9 - x^2}, \quad x \geq 0$

47. For $f(x) = (x + 1)/(x - 2)$, find a function g so that $(g \circ f)(x) = x$.

48. For $f(x) = (3x + 1)/(x - 2)$, find a function g so that $(g \circ f)(x) = x$.

49. Given two functions F and G with common domain X, then $F = G$ if and only if $F(x) = G(x)$ for all x in X. Using this definition, show that $f^{-1} \circ f = I$, where I is the identity function that has the same domain as $f^{-1} \circ f$. This suggests why the notation f^{-1} is used for the inverse.

50. Does a constant function have an inverse that is a function? Explain.

APPLICATIONS **51.** *Geometry.* For a circle with circumference C, diameter D, and radius R, we have

$$C = f(D) = \pi D \quad \text{and} \quad D = g(R) = 2R$$

Find $(f \circ g)(R)$ and interpret.

52. *Cost analysis.* The cost C to produce x units of a given product per month is given by

$$C = f(x) = 19{,}200 + 160x$$

If the demand x each month at a selling price of $p per unit is given by

$$x = g(p) = 200 - \frac{p}{4}$$

find $(f \circ g)(p)$ and interpret.

Section 2-8 Variation

- Direct Variation
- Inverse Variation
- Joint Variation
- Combined Variation

In reading scientific material, one is likely to come across statements such as "The pressure of an enclosed gas varies directly as the absolute temperature," or "The frequency of vibration of air in an organ pipe varies inversely as the length of the pipe," or even more complicated statements such as "The force of attraction between two bodies varies jointly as their masses and inversely as the square of the distance between the two bodies." These statements have precise mathematical meaning in that they represent particular types of functions. The purpose of this section is to investigate these special functions.

2-8 Variation

■ Direct Variation

The statement **y varies directly as x** means

$$y = kx \qquad k \neq 0$$

where k is a constant called the **constant of variation**. Similarly, the statement "y varies directly as the square of x" means

$$y = kx^2 \qquad k \neq 0$$

and so on. The first equation defines a linear function, and the second a quadratic function.

Direct variation is illustrated by the familiar formulas

$$C = \pi D \quad \text{and} \quad A = \pi r^2$$

where the first formula asserts that the circumference of a circle varies directly as the diameter, and the second that the area of a circle varies directly as the square of the radius. In both cases, π is the constant of variation.

EXAMPLE 32 Translate each statement into an appropriate equation, and find the constant of variation if $y = 16$ when $x = 4$.

(A) y varies directly as x.

Solution $y = kx$ Do not forget *k*.

To find the constant of variation k, substitute $x = 4$ and $y = 16$ and solve for k.

$16 = k \cdot 4$

$k = \frac{16}{4} = 4$

Thus, $k = 4$ and the equation of variation is

$y = 4x$

(B) y varies directly as the cube of x.

Solution $y = kx^3$ Do not forget *k*.

To find k, substitute $x = 4$ and $y = 16$.

$16 = k \cdot 4^3$

$k = \frac{16}{64} = \frac{1}{4}$

Thus, the equation of variation is

$y = \frac{1}{4}x^3$

PROBLEM 32   If y = 4 when x = 8, find the equation of variation for each statement.

(A) y varies directly as x.
(B) y varies directly as the cube root of x.

■ Inverse Variation

The statement **y varies inversely as x** means

$$y = \frac{k}{x} \qquad k \neq 0$$

where k is a constant (the constant of variation). As in the case of direct variation, we also discuss y varying inversely as the square of x, and so on.

An illustration of inverse variation is given in the distance-rate-time formula $d = rt$ in the form $t = d/r$ for a fixed distance d. In driving a fixed distance, say $d = 400$ miles, time varies inversely as the rate; that is,

$$t = \frac{400}{r}$$

where 400 is the constant of variation—as the rate increases, the time decreases, and vice versa.

EXAMPLE 33   Translate each statement into an appropriate equation, and find the constant of variation if y = 16 when x = 4.

(A) y varies inversely as x.

Solution

$$y = \frac{k}{x} \qquad \text{Do not forget } k.$$

To find k, substitute x = 4 and y = 16.

$$16 = \frac{k}{4}$$

$$k = 64$$

2-8 Variation

Thus, the equation of variation is

$$y = \frac{64}{x}$$

(B) y varies inversely as the square root of x.

Solution

$$y = \frac{k}{\sqrt{x}}$$

To find k, substitute $x = 4$ and $y = 16$.

$$16 = \frac{k}{\sqrt{4}}$$

$$k = 32$$

Thus, the equation of variation is

$$y = \frac{32}{\sqrt{x}}$$

PROBLEM 33 If $y = 4$ when $x = 8$, find the equation of variation for each statement.

(A) y varies inversely as x.
(B) y varies inversely as the square of x.

■ Joint Variation

The statement **w varies jointly as x and y** means

$$w = kxy \qquad k \neq 0$$

where k is a constant (the constant of variation). Similarly, if

$$w = kxyz^2 \qquad k \neq 0$$

we would say that "w varies jointly as x, y, and the square of z," and so on. For example, the area of a rectangle varies jointly as its length and width (recall $A = lw$), and the volume of a right circular cylinder varies jointly as the square of its radius and its height (recall $V = \pi r^2 h$). What is the constant of variation in each case?

Combined Variation

The basic types of variation introduced above are often combined. For example, the statement "w varies jointly as x and y and inversely as the square of z" means

$$w = k\frac{xy}{z^2} \qquad k \neq 0 \qquad \text{We do not write:} \quad w = \frac{kxy}{kz^2}$$

Thus the statement, "The force of attraction F between two bodies varies jointly as their masses m_1 and m_2 and inversely as the square of the distance d between the two bodies," means

$$F = k\frac{m_1 m_2}{d^2} \qquad k \neq 0$$

If (assuming k is positive) either of the two masses is increased, the force of attraction increases; on the other hand, if the distance is increased, the force of attraction decreases.

EXAMPLE 34 The pressure P of an enclosed gas varies directly as the absolute temperature T and inversely as the volume V. If 500 cubic feet of gas yields a pressure of 10 pounds per square foot at a temperature of 300 K (absolute temperature*), what will be the pressure of the same gas if the volume is decreased to 300 cubic feet and the temperature increased to 360 K?

Solution *Method 1.* Write the equation of variation $P = k(T/V)$, and find k using the first set of values:

$$10 = k(\tfrac{300}{500})$$
$$k = \tfrac{50}{3}$$

Hence, the equation of variation for this particular gas is $P = \tfrac{50}{3}(T/V)$. Now find the new pressure P using the second set of values:

$$P = \tfrac{50}{3}(\tfrac{360}{300}) = 20 \text{ pounds per square foot}$$

Method 2 (generally faster than Method 1). Write the equation of variation $P = k(T/V)$; then convert to the equivalent form:

$$\frac{PV}{T} = k$$

---

* A Kelvin (absolute) and a Celsius degree are the same size, but 0 on the Kelvin scale is $-273°$ on the Celsius scale. This is the point at which molecular action is supposed to stop and is called *absolute zero*.

2-8 Variation

If P_1, V_1, and T_1 are the first set of values for the gas, and P_2, V_2, and T_2 are the second set, then

$$\frac{P_1 V_1}{T_1} = k \quad \text{and} \quad \frac{P_2 V_2}{T_2} = k$$

Hence

$$\frac{P_1 V_1}{T_1} = \frac{P_2 V_2}{T_2}$$

Since all values are known except P_2, substitute and solve. Thus,

$$\frac{(10)(500)}{300} = \frac{P_2(300)}{360}$$

$$P_2 = 20 \text{ pounds per square foot}$$

PROBLEM 34 The length L of skid marks of a car's tires (when brakes are applied) varies directly as the square of the speed v of the car. If skid marks of 20 feet are produced at 30 miles/hour, how fast would the same car be going if it produced skid marks of 80 feet? Solve in two ways (see Example 34).

EXAMPLE 35 The frequency of pitch f of a given musical string varies directly as the square root of the tension T and inversely as the length L. What is the effect on the frequency if the tension is increased by a factor of 4 and the length is cut in half?

Solution   Write the equation of variation:

$$f = \frac{k\sqrt{T}}{L} \quad \text{or equivalently} \quad \frac{f_2 L_2}{\sqrt{T_2}} = \frac{f_1 L_1}{\sqrt{T_1}}$$

We are given that $T_2 = 4T_1$ and $L_2 = 0.5L_1$. Substituting in the second equation, we have

$$\frac{f_2 0.5 L_1}{\sqrt{4T_1}} = \frac{f_1 L_1}{\sqrt{T_1}} \quad \text{Solve for } f_2.$$

$$\frac{f_2 0.5 L_1}{2\sqrt{T_1}} = \frac{f_1 L_1}{\sqrt{T_1}}$$

$$f_2 = \frac{2\sqrt{T_1} f_1 L_1}{0.5 L_1 \sqrt{T_1}} = 4f_1$$

Thus, the frequency of pitch is increased by a factor of 4.

PROBLEM 35 The weight w of an object on or above the surface of the earth varies inversely as the square of the distance d between the object and the center of the earth. If an object on the surface of the earth is moved into space so as to double its distance from the earth's center, what effect will this move have on its weight?

Answers to Matched Problems
32. (A) $y = \tfrac{1}{2}x$ (B) $y = 2\sqrt[3]{x}$
33. (A) $y = 32/x$ (B) $y = 256/x^2$
34. $v = 60$ miles/hour
35. It will be one-fourth as heavy.

Exercise 2-8 ■ A

Translate each problem into an equation using k as the constant of variation.

1. F varies directly as the square of v.
2. u varies directly as v.
3. The pitch or frequency f of a guitar string of a given length varies directly as the square root of the tension T of the string.
4. Geologists have found in studies of earth erosion that the erosive force (sediment-carrying power) P of a swiftly flowing stream varies directly as the sixth power of the velocity v of the water.
5. y varies inversely as the square root of x.
6. I varies inversely as t.
7. The biologist Reaumur suggested in 1735 that the length of time t that it takes fruit to ripen during the growing season varies inversely as the sum T of the average daily temperatures during the growing season.
8. In a study on urban concentration, F. Auerbach discovered an interesting law. After arranging all the cities of a given country according to their population size, starting with the largest, he found that the population P of a city varied inversely as the number n indicating its position in the ordering.
9. R varies jointly as S, T, and V.
10. g varies jointly as x and the square of y.
11. The volume of a cone V varies jointly as its height h and the square of the radius r of its base.
12. The amount of heat put out by an electrical appliance (in calories) varies jointly as time t, resistance R in the circuit, and the square of the current I.

2-8 Variation

Solve using either of the two methods illustrated in Example 34.

13. u varies directly as the square root of v. If $u = 2$ when $v = 2$, find u when $v = 8$.
14. y varies directly as the square of x. If $y = 20$ when $x = 2$, find y when $x = 5$.
15. L varies inversely as the square root of M. If $L = 9$ when $M = 9$, find L when $M = 3$.
16. I varies inversely as the cube of t. If $I = 4$ when $t = 2$, find I when $t = 4$.

B *Translate each problem into an equation using k as the constant of variation.*

17. U varies jointly as a and b and inversely as the cube of c.
18. w varies directly as the square of x and inversely as the square root of y.
19. The maximum safe load L for a horizontal beam varies jointly as its width w and the square of its height h, and inversely as its length l.
20. Joseph Cavanaugh, a sociologist, found that the number of long-distance phone calls n between two cities in a given time period varied (approximately) jointly as the populations P_1 and P_2 of the two cities, and inversely as the distance d between the two cities.

Solve using either of the two methods illustrated in Example 34.

21. Q varies jointly as m and the square of n, and inversely as P. If $Q = -4$ when $m = 6$, $n = 2$, and $P = 12$, find Q when $m = 4$, $n = 3$, and $P = 6$.
22. w varies jointly as x, y, and z and inversely as the square of t. If $w = 2$ when $x = 2$, $y = 3$, $z = 6$, and $t = 3$, find w when $x = 3$, $y = 4$, $z = 2$, and $t = 2$.
23. The weight w of an object on or above the surface of the earth varies inversely as the square of the distance d between the object and the center of the earth. If a girl weighs 100 pounds on the surface of the earth, how much would she weigh (to the nearest pound) 400 miles above the earth's surface? (Assume the radius of the earth is 4,000 miles.)
24. A child was struck by a car in a crosswalk. The driver of the car had slammed on his brakes and left skid marks 160 feet long. He told the police he had been driving at 30 miles/hour. The police know that the length of skid marks L (when brakes are applied) varies directly as the square of the speed of the car v, and that at 30 miles/hour (under ideal conditions) skid marks would be 40 feet long. How fast was the driver actually going before he applied his brakes?

25. Ohm's law states that the current I in a wire varies directly as the electromotive force E and inversely as the resistance R. If $I = 22$ amperes when $E = 110$ volts and $R = 5$ ohms, find I if $E = 220$ volts and $R = 11$ ohms.

26. Anthropologists, in their study of race and human genetic groupings, often use an index called the *cephalic index*. The cephalic index C varies directly as the width w of the head and inversely as the length l of the head (both when viewed from the top). If an Indian in Baja California (Mexico) has measurements of $C = 75$, $w = 6$ inches, and $l = 8$ inches, what is C for an Indian in northern California with $w = 8.1$ inches and $l = 9$ inches?

C 27. If the horsepower P required to drive a speedboat through water varies directly as the cube of the speed v of the boat, what change in horsepower is required to double the speed of the boat?

28. The intensity of illumination E on a surface varies inversely as the square of its distance d from a light source. What is the effect on the total illumination on a book if the distance between the light source and the book is doubled?

29. The frequency of vibration f of a musical string varies directly as the square root of the tension T and inversely as the length L of the string. If the tension of the string is increased by a factor of 4 and the length of the string is doubled, what is the effect on the frequency?

30. In an automobile accident the destructive force F of a car varies (approximately) jointly as the weight w of the car and the square of the speed v of the car. (This is why accidents at high speed are generally so serious.) What would be the effect on the destructive force of a car if its weight were doubled and its speed were doubled?

ADDITIONAL APPLICATIONS *The following problems include significant applications from many different areas and are arranged according to subject area. The more difficult problems are marked with two stars (**), the moderately difficult problems are marked with one star (*), and the easier problems are not marked.*

Astronomy 31. The square of the time t required for a planet to make one orbit around the sun varies directly as the cube of its mean (average) distance d from the sun. Write the equation of variation, using k as the constant of variation.

★32. The centripetal force F of a body moving in a circular path at constant speed varies inversely as the radius r of the path. What happens to F if r is doubled?

2-8 Variation

33. The length of time t a satellite takes to complete a circular orbit of the earth varies directly as the radius r of the orbit and inversely as the orbital velocity v of the satellite. If $t = 1.42$ hours when $r = 4{,}050$ miles and $v = 18{,}000$ miles/hour (Sputnik I), find t for $r = 4{,}300$ miles and $v = 18{,}500$ miles/hour.

Life Science

34. The number N of gene mutations resulting from x-ray exposure varies directly as the size of the x-ray dose r. What is the effect on N if r is quadrupled?

35. In biology there is an approximate rule, called the *bioclimatic rule* for temperate climates, which states that the difference d in time for fruit to ripen (or insects to appear) varies directly as the change in altitude h. If $d = 4$ days when $h = 500$ feet, find d when $h = 2{,}500$ feet.

Physics—Engineering

36. Over a fixed distance d, speed r varies inversely as time t. Police use this relationship to set up speed traps. (The graph of the resulting function is a hyperbola.) If in a given speed trap $r = 30$ miles/hour when $t = 6$ seconds, what would be the speed of a car if $t = 4$ seconds?

★37. The length L of skid marks of a car's tires (when the brakes are applied) varies directly as the square of the speed v of the car. How is the length of skid marks affected by doubling the speed?

38. The time t required for an elevator to lift a weight varies jointly as the weight w and the distance d through which it is lifted, and inversely as the power P of the motor. Write the equation of variation, using k as the constant of variation.

39. The total pressure P of the wind on a wall varies jointly as the area of the wall A and the square of the velocity of the wind v. If $P = 120$ pounds when $A = 100$ square feet and $v = 20$ miles/hour, find P if $A = 200$ square feet and $v = 30$ miles/hour.

★★40. The thrust T of a given type of propeller varies jointly as the fourth power of its diameter d and the square of the number of revolutions per minute n it is turning. What happens to the thrust if the diameter is doubled and the number of revolutions per minute is cut in half?

Psychology

41. In early psychological studies on sensory perception (hearing, seeing, feeling, and so on), the question was asked: "Given a certain level of stimulation S, what is the minimum amount of added stimulation ΔS that can be detected?" A German physiologist, E. H. Weber (1795–1878) formulated, after many experiments, the famous law that now bears his name: "The amount of change ΔS

that will be just noticed varies directly as the magnitude S of the stimulus."

(A) Write the law as an equation of variation.

(B) If a person lifting weights can just notice a difference of 1 ounce at the 50-ounce level, what will be the least difference she will be able to notice at the 500-ounce level?

(C) Determine the just noticeable difference in illumination a person is able to perceive at 480 candlepower if he is just able to perceive a difference of 1 candlepower at the 60-candlepower level.

42. Psychologists in their study of intelligence often use an index called IQ. IQ varies directly as mental age MA and inversely as chronological age CA (up to the age of 15). If a 12-year-old boy with a mental age of 14.4 has an IQ of 120, what will be the IQ of an 11-year-old girl with a mental age of 15.4?

Music 43. The frequency of vibration of air in an open organ pipe varies inversely as the length of the pipe. If the air column in an open 32-foot pipe vibrates 16 times per second (low C), then how fast would the air vibrate in a 16-foot pipe?

44. The frequency of pitch f of a musical string varies directly as the square root of the tension T and inversely as the length l and the diameter d. Write the equation of variation using k as the constant of variation. (It is interesting to note that if pitch depended on only length, then pianos would have to have strings varying from 3 inches to 38 feet.)

Photography 45. The f-stop numbers N on a camera, known as focal ratios, vary directly as the focal length F of the lens and inversely as the diameter d of the diaphragm opening (effective lens opening). Write the equation of variation using k as the constant of variation.

★46. In taking pictures using flashbulbs, the lens opening (f-stop number) N varies inversely as the distance d from the object being photographed. What adjustment should you make on the f-stop number if the distance between the camera and the object is doubled?

Chemistry ★47. Atoms and molecules that make up the air constantly fly about like microscopic missiles. The velocity v of a particular particle at a fixed temperature varies inversely as the square root of its molecular weight w. If an oxygen molecule in air at room temperature has an average velocity of 0.3 mile/second, what will be the average velocity of a hydrogen molecule, given that the hydrogen molecule is one-sixteenth as heavy as the oxygen molecule?

48. The Maxwell–Boltzmann equation says that the average velocity v of a molecule varies directly as the square root of the absolute temperature T and inversely as the square root of its molecular weight w. Write the equation of variation using k as the constant of variation.

Business **49.** The amount of work A completed varies jointly as the number of workers W used and the time t they spend. If ten workers can finish a job in eight days, how long will it take four workers to do the same job?

50. The simple interest I earned in a given time varies jointly as the principal p and the interest rate r. If \$100 at 4% interest earns \$8, how much will \$150 at 3% interest earn in the same period?

Geometry ★**51.** The volume of a sphere varies directly as the cube of its radius r. What happens to the volume if the radius is doubled?

★**52.** The surface area S of a sphere varies directly as the square of its radius r. What happens to the area if the radius is cut in half?

Section 2-9 Chapter Review

IMPORTANT TERMS AND SYMBOLS

2-1 Rectangular Coordinate System; Graphing. Cartesian coordinate system; coordinate axes; quadrants; coordinates of a point; abscissa; ordinate; point-by-point graphing; graph of an equation in two variables; symmetry with respect to the x axis, the y axis, and the origin; distance-between-two-points formula; circle; equation of a circle

2-2 Relations and Functions. Relation; function; domain; range; relations specified by equations, tables, ordered pairs, and graphs; independent variable; dependent variable; function notation; $f(x)$; f: $y = f(x)$

2-3 Functions: Graphs and Properties. Graph of a function f, constant function, identity function, square function, cube function, even function, odd function, increasing function, decreasing function, greatest integer function $[\![x]\!]$, discontinuity at a point, continuity at a point, continuity over an interval, vertical and horizontal shifting, reflecting with respect to an axis, expanding and contracting

2-4 Linear Relations and Functions. Linear function, first-degree polynomial function, graph of a linear function, slope, vertical line, horizontal line, point–slope form, slope–intercept form, intercept form, parallel lines, perpendicular lines

2-5 Graphing Polynomial Functions. Polynomial function, continuity, graphing by nested factoring, quadratic functions, graphs of quadratic functions, axis, vertex, maximum or minimum

2-6 Graphing Rational Functions. Rational function, points of discontinuity, infinity, horizontal asymptote, vertical asymptote, graphing rational functions, ∞

2-7 Composite and Inverse Functions. Composite function, inverse relation, one-to-one correspondence, f^{-1}

2-8 Variation. Direct variation, inverse variation, joint variation, combined variation, constant of variation

Exercise 2-9 Chapter Review

Work through all the problems in this chapter review and check answers in the back of the book. (Answers to all problems are there, and following each answer is a number in italics indicating the section in which that type of problem is discussed.) Where weaknesses show up, review appropriate sections in the text. When you are satisfied that you know the material, take the practice test following this review.

A
1. How are the graphs of the following related to the graph of $y = x^2$?
   (A) $y = -x^2$ (B) $y = x^2 - 3$ (C) $y = (x + 3)^2$
2. Given the points $A(-2, 3)$ and $B(4, 0)$, find:
   (A) The distance between A and B
   (B) The slope of AB
   (C) The slope of a line perpendicular to AB
3. Write the equations of the vertical and horizontal lines passing through $(-3, 4)$. What is the slope of each?

Problems 4–8 refer to the following functions:

$$f(x) = 3x + 5 \qquad g(x) = 4 - x^2 \qquad h(x) = 5 \qquad m(x) = 2|x| - 1$$

Find the indicated quantities or expressions.

4. $f(2) + g(-2) + h(0)$

5. $\dfrac{m(-2) + 1}{g(2) + 4}$

6. $\dfrac{f(2 + h) - f(2)}{h}$

7. $\dfrac{g(a + h) - g(a)}{h}$

8. $(f \circ g)(x)$ and $(g \circ f)(x)$

9. Given $f(x) = 4x - 1$, find:
   (A) $f^{-1}(x)$ (B) $f^{-1}(7)$ (C) $(f^{-1} \circ f)(x)$

10. Graph $3x + 2y = 9$ and indicate its slope.

11. Write an equation of a line with x intercept 6 and y intercept 4. Write the final answer in the form $Ax + By = C$, where A, B, and C are integers.

12. Graph $f(x) = 1/(x + 2)$. Indicate any vertical or horizontal asymptotes with broken lines.

13. y varies directly as x, and inversely as z.
    (A) Write the equation of variation.
    (B) If $y = 4$ when $x = 6$ and $z = 2$, find y when $x = 4$ and $z = 4$.

B   **14.** Write an equation of a line through $A(-6, -4)$ and $B(4, 1)$. Write the final answer in the form $Ax + By = C$, where A, B, and C are integers.

15. Find the equation of a circle with radius 2 and center at $(3, -2)$.

16. If the slope of a line is negative, is the function represented by the graph increasing, decreasing, or constant?

17. Discuss the graph of $4x^2 + 9y^2 = 36$ relative to symmetry with respect to the x axis, y axis, and origin.

18. Given:

$$f(x) = \sqrt[3]{x} \qquad g(x) = \frac{x^2}{x^2 - 1} \qquad h(x) = x + 1$$

    (A) Which are odd?
    (B) Which are even?
    (C) Which are neither even nor odd?

19. Graph: $f(x) = -|x + 1| - 1$

20. Write an equation of the line (A) parallel to, or (B) perpendicular to the line $6x + 3y = 5$ and passing through the point $(-2, 1)$. Write final answers in the form $y = mx + b$.

21. Graph $f(x) = x^2 - 6x + 5$. Show the axis and vertex, and find the maximum or minimum value of $f(x)$.

22. Given: $f(x) = \dfrac{x - 1}{x + 2}$

    (A) Is f even or odd?
    (B) Find the x and y intercepts.
    (C) Find the points of discontinuity.
    (D) Find the equations of horizontal and vertical asymptotes.
    (E) Where is the graph above and below the x axis?
    (F) Sketch a graph of f (include asymptotes).

23. What are the coordinates of the center and the radius of the circle given by $x^2 + y^2 - 6x + 8y = 0$?

24. Write $P(x) = x^3 - 2x^2 - 5x + 6$ in a nested factored form and graph for $-3 \leq x \leq 4$.

25. Each of the following equations defines a function. Which have inverses that are functions?

(A) $f(x) = x^3$ (B) $g(x) = (x - 2)^2$
(C) $h(x) = 2x - 3$ (D) $F(x) = (x + 3)^2$, $x \geq -3$

26. Given: f: $y = 2^x$, $x \in \{-2, -1, 0, 1, 2\}$
    (A) Find the domain and range of f and f^{-1}.
    (B) Graph f, f^{-1}, and $y = x$ on the same coordinate system.
    (C) Which of f and f^{-1} is a function?

27. Given $f(x) = x^2 - 1$, $x \geq 0$, find:
    (A) The domain and range of f and f^{-1}
    (B) $f^{-1}(x)$ (C) $f^{-1}(3)$ (D) $(f^{-1} \circ f)(4)$
    (E) $(f^{-1} \circ f)(x)$

28. The time t required for an elevator to lift a weight varies jointly as the weight w and the distance d through which it is lifted, and inversely as the power P of the motor. Write the equation of variation using k as the constant of variation.

C 29. Find the equations for the horizontal and vertical asymptotes for the graphs of:

(A) $f(x) = \dfrac{3x}{2x + 3}$ (B) $g(x) = \dfrac{5x}{x^2 - x - 6}$

30. For what values of x are f and g discontinuous in Problem 29?

31. For $f(x) = (x + 2)/(x - 3)$, find:
    (A) $f^{-1}(x)$ (B) $f^{-1}(3)$ (C) $(f^{-1} \circ f)(x)$

32. Find the equation of the set of points equidistant from $(3, 3)$ and $(6, 0)$. What is the name of the geometric figure formed by this set?

33. For what values of x is the graph of $f(x) = (x - 1)/(x + 2)$ below its horizontal asymptote? Above its horizontal asymptote?

34. The total force F of a wind on a wall varies jointly as the area of the wall A and the square of the velocity of the wind. How is the total force on the wall affected if the area is cut in half and the velocity is doubled?

Practice Test   Chapter 2

Take this practice test as if it were a graded test. Allow yourself up to 50 minutes. Work the problems without looking back in the chapter. Correct your work using the answers (keyed to appropriate sections) in the back of the book.

1. (A) Find an equation of the line through $P(-4, 3)$ and $Q(0, -3)$. Write the final answer in the form $Ax + By = C$, where A, B, and C are integers with $A > 0$.
   (B) Find $d(P, Q)$.

Practice Test Chapter 2

2. Given the line $L: 3x + 4y = 10$, find an equation of the line through $(4, -2)$ that is (A) parallel to L, or (B) perpendicular to L. Write the final answers in the form $y = mx + b$.

3. Find the center and radius of the circle given by $x^2 + y^2 + 4x - 6y = 3$.

Problems 4 and 5 refer to the following functions:

$$f(x) = 3x \qquad g(x) = 2|x| - 1 \qquad F(x) = 3 \quad 2x^2$$

4. (A) $\dfrac{f(2) + g(0)}{h(2)} = ?$ (B) $\dfrac{F(a+h) - F(a)}{h} = ?$

5. (A) List the even functions.
   (B) Which have graphs that are symmetric with respect to the origin?
   (C) Which are one to one and thus have inverses that are functions?

6. Find the maximum or minimum value of $f(x) = x^2 - 6x + 11$ without graphing. What are the coordinates of the vertex of the graph?

Problems 7–9 refer to the function f defined by: $f(x) = \dfrac{x-1}{2x+2}$

7. (A) Is f even or odd?
   (B) Where is f discontinuous?
   (C) Find the x and y intercepts for the graph of f.
   (D) Find the equations of the horizontal and vertical asymptotes.

8. For what values of x is the graph of f above the x axis? Below the x axis?

9. Sketch a graph of f. Draw vertical and horizontal asymptotes with broken lines.

10. Given $f(x) = \sqrt{x-8}$ and $g(x) = |x|$.
    (A) Find $(f \circ g)(x)$ and $(g \circ f)(x)$.
    (B) Find the domains for $f \circ g$ and $g \circ f$.

11. Given $f(x) = 3x - 7$, find:
    (A) $f^{-1}(x)$ (B) $f^{-1}(5)$ (C) $(f^{-1} \circ f)(x)$
    (D) Is f a decreasing or increasing function?

12. Given $f(x) = \sqrt{x-1}$.
    (A) Find the domain and range for f and f^{-1}.
    (B) Graph f, f^{-1}, and $y = x$ on the same coordinate system and indicate which of f and f^{-1} is a function.
    (C) Find $f^{-1}(x)$, if it exists.

13. Suppose H varies directly as n and inversely as the square of m.
    (A) Write the equation of variation.
    (B) If $H = 4$ when $n = 2$ and $m = 3$, find H when $n = 4$ and $m = 2$.
    (C) What happens to H if both n and m are doubled?

14. How is the graph of $f(x) = -(x - 2)^2 - 1$ related to the graph of $g(x) = x^2$?

15. (A) Write $f(x) = x^3 - 3x^2 - x + 3$ in a nested factored form.
    (B) Using (A), graph f for $-2 \leq x \leq 4$.

Polynomial Functions and Theory of Equations ▪3

3-1 Introduction
3-2 Synthetic Division
3-3 Remainder and Factor Theorems
3-4 Fundamental Theorem of Algebra
3-5 Isolating Real Zeros
3-6 Finding Rational Zeros
3-7 Approximating Irrational Zeros
3-8 Partial Fraction Decomposition
3-9 Chapter Review

A natural design of mathematical interest. Can you guess the source? See the back of the book.

Chapter 3 ▪ Polynomial Functions and Theory of Equations

Section 3-1 Introduction

We know how to solve first- and second-degree polynomial equations (linear and quadratic equations). For example, for the linear equations

$$ax + b = 0 \qquad a \neq 0$$

$$x = -\frac{b}{a}$$

and for the quadratic equation

$$ax^2 + bx + c = 0 \qquad a \neq 0$$

$$x = \frac{-b \pm \sqrt{b^2 - 4ac}}{2a}$$

What about third- and higher-degree polynomial equations? For example, how do we solve equations such as

$$2x^3 - 3x^2 + x - 5 = 0$$

and

$$x^7 - 6x^4 + 3x - 1 = 0$$

It turns out that there are direct methods (though complicated) for finding all solutions for any third- or fourth-degree polynomial equation. However, Evariste Galois (1811–1832), a Frenchman, proved at the age of 20 that for polynomial equations of degree greater than four there was no finite step-by-step process that would always yield all solutions.* This does not mean that we give up looking for solutions to higher-degree polynomials. In this chapter you will find that solutions always exist for all polynomial equations of degree greater than or equal to one.

---

* Galois' contribution, using the new concept of "group," was of the highest mathematical significance and originality. However, his contemporaries hardly read his papers, dismissing them as "almost unintelligible." At the age of 21, involved in political agitation, Galois met an untimely death in a duel. A short but fascinating account of Galois' tragic life can be found in E. T. Bell's *Men of Mathematics* (New York: Simon & Schuster, 1937), pp. 362–377.

3-1 Introduction

We will develop methods for finding or approximating all real solutions of polynomials with real coefficients.

To aid this endeavor, we will find it helpful to switch our emphasis from polynomial equations to polynomial functions. We will then uncover some important properties of polynomial functions that will lead directly to solutions of certain polynomial equations. The following definitions make a useful connecting link between polynomial functions and polynomial equations.

For the **nth-degree polynomial function** P given by

$$P(x) = a_n x^n + a_{n-1} x^{n-1} + \cdots + a_1 x + a_0 \qquad a_n \neq 0$$

where the coefficients are real or complex, r is said to be a **zero of the function P**, or a **zero of the polynomial P(x)**, or a **solution or root of the equation P(x) = 0**, if

$$P(r) = 0$$

A zero of a polynomial may or may not be 0; a zero of a polynomial is *any* number (real or complex) that makes a polynomial 0. If we consider the graph of $y = P(x)$, then a real zero of $P(x)$ is simply an x intercept. Consider the polynomial

$$P(x) = x^2 - 4x + 3$$

The graph of P is shown in Figure 1.

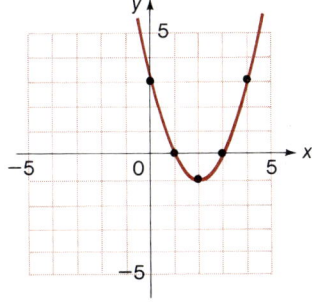

FIGURE 1

The x intercepts 1 and 3 are zeros of $P(x) = x^2 - 4x + 3$, since $P(1) = 0$ and $P(3) = 0$. The x intercepts 1 and 3 are also solutions or roots for the equation $x^2 - 4x + 3 = 0$.

In general:

Zeros and Roots

The x intercepts of the graph of $y = P(x)$ are real **zeros** of P and $P(x)$ and real solutions or **roots** for the equation $P(x) = 0$.

Section 3-2 Synthetic Division

- Algebraic Long Division
- Synthetic Division

We now digress for a moment to discuss algebraic long division. This apparent digression actually provides us with a very useful tool that we will use throughout most of this chapter.

- **Algebraic Long Division**

We can find quotients of polynomials by a long-division process similar to that used in arithmetic. An example will illustrate the process.

EXAMPLE 1 Divide $5 + 4x^3 - 3x$ by $2x - 3$.

Solution

$$
\begin{array}{r}
2x^2 + 3x + 3 \\
2x - 3 \overline{\smash{\big)}\,4x^3 + 0x^2 - 3x + 5} \\
\underline{4x^3 - 6x^2} \\
6x^2 - 3x \\
\underline{6x^2 - 9x} \\
6x + 5 \\
\underline{6x - 9} \\
14 = R
\end{array}
$$

Remainder

Arrange the dividend and the divisor in descending powers of the variable. Insert, with 0 coefficients, any missing terms of degree less than three. Divide the first term of the divisor into the first term of the dividend. Multiply the divisor by $2x^2$, line up like terms, subtract as in arithmetic, and bring down $-3x$. Repeat the process until the degree of the remainder is less than that of the divisor.

Thus,

$$\frac{4x^3 - 3x + 5}{2x - 3} = 2x^2 + 3x + 3 + \frac{14}{2x - 3}$$

Check

$$(2x - 3)\left[(2x^2 + 3x + 3) + \frac{14}{2x - 3}\right] = (2x - 3)(2x^2 + 3x + 3) + 14$$

$$= 4x^3 - 3x + 5$$

PROBLEM 1 Divide $6x^2 - 30 + 9x^3$ by $3x - 4$.

- **Synthetic Division**

Being able to divide a polynomial $P(x)$ by a linear polynomial of the form $x - r$ quickly and accurately will be of great aid to us (as strange as it may seem now) in the search for zeros of higher-degree polynomial

3-2 Synthetic Division

functions. This kind of division can be carried out efficiently by a method called **synthetic division**. The method is most easily understood through an example. Let us start by dividing $P(x) = 2x^4 + 3x^3 - x - 5$ by $x + 2$, using ordinary long division. The critical parts of the process are indicated in color.

$$
\begin{array}{r}
2x^3 - 1x^2 + 2x - 5 \quad \text{Quotient} \\
x + 2 \overline{\smash{\big)}\, 2x^4 + 3x^3 + 0x^2 - 1x - 5} \quad \text{Dividend} \\
\underline{2x^4 + 4x^3} \\
-1x^3 + 0x^2 \\
\underline{-1x^3 - 2x^2} \\
2x^2 - 1x \\
\underline{2x^2 + 4x} \\
-5x - 5 \\
\underline{-5x - 10} \\
5 \quad \text{Remainder}
\end{array}
$$

Divisor is $x + 2$.

The numerals printed in color, which represent the essential part of the division process, are arranged more conveniently as:

$$
\begin{array}{r|rrrrr}
 & \multicolumn{5}{c}{\text{Dividend coefficients}} \\
 & 2 & 3 & 0 & -1 & -5 \\
 &   & 4 & -2 & 4 & -10 \\
\hline
2 & 2 & -1 & 2 & -5 & 5 \\
\end{array}
$$

Quotient coefficients · Remainder

Mechanically, we see that the second and third rows of numerals are generated as follows. The first coefficient 2 of the dividend is brought down and multiplied by 2 from the divisor, and the product 4 is placed under the second dividend coefficient 3 and subtracted. The difference -1 is again multiplied by the 2 from the divisor, and the product is placed under the third coefficient from the dividend and subtracted. This process is repeated until the remainder is reached. The process can be made a little faster, and less prone to sign errors, by changing $+2$ from the divisor to -2 and adding instead of subtracting. Thus,

$$
\begin{array}{r|rrrrr}
 & \multicolumn{5}{c}{\text{Dividend coefficients}} \\
 & 2 & 3 & 0 & -1 & -5 \\
 &   & -4 & 2 & -4 & 10 \\
\hline
-2 & 2 & -1 & 2 & -5 & 5 \\
\end{array}
$$

Quotient coefficients · Remainder

Key Steps in the Synthetic Division Process

1. Arrange the coefficients of P(x) in order of descending powers of x (write 0 as the coefficient for each missing power).
2. After writing the divisor in the form x − r, use r to generate the second and third rows of numbers as follows. Bring down the first coefficient of the dividend and multiply it by r; then add the product to the second coefficient of the dividend. Multiply this sum by r, and add the product to the third coefficient of the dividend. Repeat the process until a product is added to the constant term of P(x). [Note: This process is well suited to hand calculator use. Store r; then proceed from left to right recalling r and using it as indicated.]
3. The last number in the third row of numbers is the remainder; the other numbers in the third row are the coefficients of the quotient, which is of degree 1 less than P(x).

EXAMPLE 2 Use synthetic division to find the quotient and remainder resulting from dividing $P(x) = 4x^5 - 30x^3 - 50x - 2$ by $x + 3$. Write the answer in the form $Q(x) + R/(x - r)$, where R is a constant.

Solution $x + 3 = x - (-3)$; therefore, $r = -3$.

$$\begin{array}{c|cccccc} & 4 & 0 & -30 & 0 & -50 & -2 \\ & & -12 & 36 & -18 & 54 & -12 \\ \hline -3 & 4 & -12 & 6 & -18 & 4 & -14 \end{array}$$

The quotient is $4x^4 - 12x^3 + 6x^2 - 18x + 4$ with a remainder of -14. Thus,

$$\frac{P(x)}{x + 3} = 4x^4 - 12x^3 + 6x^2 - 18x + 4 + \frac{-14}{x + 3}$$

PROBLEM 2 Repeat Example 2 with $P(x) = 3x^4 - 11x^3 - 18x + 8$ and divisor $x - 4$.

Answers to Matched Problems

1. $3x^2 + 6x + 8 + \dfrac{2}{3x - 4}$

2. $\dfrac{P(x)}{x - 4} = 3x^3 + x^2 + 4x - 2 + \dfrac{0}{x - 4}$

   $= 3x^3 + x^2 + 4x - 2$

3-2 Synthetic Division

Exercise 3-2 ■ **A** *Divide, using algebraic long division. Write the quotient and indicate the remainder.*

1. $(4m^2 - 1) \div (2m - 1)$
2. $(y^2 - 9) \div (y + 3)$
3. $(6 - 6x + 8x^2) \div (2x + 1)$
4. $(11x - 2 + 12x^2) \div (3x + 2)$
5. $(x^3 - 1) \div (x - 1)$
6. $(a^3 + 27) \div (a + 3)$
7. $(3y - y^2 + 2y^3 - 1) \div (y + 2)$
8. $(3 + x^3 - x) \div (x - 3)$

Use algebraic long division and synthetic division to write the quotient $P(x) \div (x - r)$ in the form $P(x)/(x - r) = Q(x) + R/(x - r)$ where R is a constant.

9. $(x^2 + 3x - 7) \div (x - 2)$
10. $(x^2 + 3x - 3) \div (x - 3)$
11. $(4x^2 + 10x - 9) \div (x + 3)$
12. $(2x^2 + 7x - 5) \div (x + 4)$
13. $(2x^3 - 3x + 1) \div (x - 2)$
14. $(x^3 + 2x^2 - 3x - 4) \div (x + 2)$

B *Divide, using synthetic division. Write the quotient and indicate the remainder. As coefficients get more involved, a hand calculator will be very helpful.*

15. $(3x^4 - x - 4) \div (x + 1)$
16. $(5x^4 - 2x^2 - 3) \div (x - 1)$
17. $(x^5 + 1) \div (x + 1)$
18. $(x^4 - 16) \div (x - 2)$
19. $(2x^3 + 4x^2 - 9x - 11) \div (x + 3)$
20. $(x^4 - 3x^3 - 5x^2 + 6x - 3) \div (x - 4)$
21. $(2x^4 - 13x^3 + 14x^2 + 15) \div (x - 5)$
22. $(x^5 + 10x^2 + 5x + 2) \div (x + 2)$
23. $(4x^4 + 2x^3 - 6x^2 - 5x + 1) \div (x + \frac{1}{2})$
24. $(2x^3 - 5x^2 + 6x + 3) \div (x - \frac{1}{2})$
25. $(4x^3 + 4x^2 - 7x - 6) \div (x + \frac{3}{2})$
26. $(3x^3 - x^2 + x + 2) \div (x + \frac{2}{3})$
27. $(x^3 - 2x^2 + 3x - 1) \div (x - 0.3)$
28. $(2x^3 + 3x^2 - 2x + 1) \div (x - 0.2)$

29. $(3x^3 + 2x - 4) \div (x - 0.2)$
30. $(4x^3 - 2x^2 - 1) \div (x - 0.3)$

C *Divide, using algebraic long division. Write the quotient and indicate the remainder.*

31. $(16x - 5x^3 - 8 + 6x^4 - 8x^2) \div (2x - 4 + 3x^2)$
32. $(8x^2 - 7 - 13x + 24x^4) \div (3x + 5 + 6x^2)$

Divide, using synthetic division. Write the quotient and indicate the remainder. A hand calculator will be very helpful in Problems 33–36.

33. $(2x^3 - 5x^2 - 8x + 6) \div (x - 3.3)$
34. $(2x^4 - x^2 + 3x - 2) \div (x - 0.6)$
35. $(2x^3 - 5x^2 - 8x + 6) \div (x + 1.4)$
36. $(x^3 + 2x - 4) \div (x - 1.2)$
37. $(x^3 - 3x^2 + x - 3) \div (x - i)$
38. $(x^3 - 2x^2 + x - 2) \div (x + i)$
39. (A) Divide $P(x) = a_2 x^2 + a_1 x + a_0$ by $x - r$, using synthetic division and the long-division process, and compare the coefficients of the quotient and the remainder produced by each method.
    (B) Expand the expression representing the remainder. What do you observe?
40. Repeat Problem 39 for $P(x) = a_3 x^3 + a_2 x^2 + a_1 x + a_0$.

Section 3-3 Remainder and Factor Theorems

- Division Algorithm
- Remainder Theorem
- Graphing Polynomials
- Factor Theorem

- **Division Algorithm**

If we divide $P(x) = 2x^4 - 5x^3 - 4x^2 + 13$ by $x - 3$, we obtain

$$\frac{2x^4 - 5x^3 - 4x^2 + 13}{x - 3} = 2x^3 + x^2 - x - 3 + \frac{4}{x - 3} \qquad x \neq 3$$

If we multiply both members by $x - 3$, then

$$2x^4 - 5x^3 - 4x^2 + 13 = (x - 3)(2x^3 + x^2 - x - 3) + 4$$

3-3 Remainder and Factor Theorems

This last equation is an identity in that the left side is equal to the right side for *all* replacements of x by real or complex numbers, including $x = 3$. This example suggests the important **division algorithm**, which we state as Theorem 1 without proof.

THEOREM 1 | Division Algorithm

For each polynomial $P(x)$ of degree one or greater and each number r, there exists a unique polynomial $Q(x)$ of degree 1 less than $P(x)$ and a unique number R (which may be 0) such that

$$P(x) = (x - r)Q(x) + R$$

The polynomial $Q(x)$ is called the **quotient**, $x - r$ the **divisor**, and R the **remainder**.

■ Remainder Theorem

We now use the division algorithm in Theorem 1 to prove the important and useful remainder theorem.

The equation in Theorem 1

$$P(x) = (x - r)Q(x) + R$$

is an identity; that is, it is true for all real or complex replacements for x. In particular, if we let $x = r$, then we observe a very interesting and extremely useful relationship:

$$P(r) = (r - r)Q(r) + R$$
$$= 0 \cdot Q(r) + R$$
$$= 0 + R$$
$$= R$$

In words, the value of a polynomial $P(x)$ at $x = r$ is the same as the remainder R one obtains by dividing $P(x)$ by $x - r$. We have proved the well-known remainder theorem (Theorem 2).

THEOREM 2 | Remainder Theorem

If R is the remainder after dividing the polynomial $P(x)$ by $x - r$, then

$$P(r) = R$$

EXAMPLE 3 If $P(x) = 4x^4 + 10x^3 + 19x + 5$, find $P(-3)$ by (A) using the remainder theorem and synthetic division, and (B) evaluating $P(-3)$ directly.

Solution (A)

$$\begin{array}{r|rrrrr}
 & 4 & 10 & 0 & 19 & 5 \\
 &   & -12 & 6 & -18 & -3 \\
\hline
-3 & 4 & -2 & 6 & 1 & 2 = R = P(-3)
\end{array}$$

(B) $P(-3) = 4(-3)^4 + 10(-3)^3 + 19(-3) + 5$
$ = 2$

PROBLEM 3 Repeat Example 3 for $P(x) = 3x^4 - 16x^2 - 3x + 7$ and $x = -2$.

■ **Graphing Polynomials**

The remainder theorem and synthetic division provide us with an efficient way of graphing polynomials. In terms of the mechanics, the process is equivalent to the "nested factoring" method of graphing discussed in Section 2-5 (see Problems 39–42 in Exercise 3-3). The following example illustrates the process.

EXAMPLE 4 Graph: $P(x) = x^3 + 3x^2 - x - 3$, $-4 \le x \le 2$

Solution We evaluate $P(x)$ from $x = -4$ to $x = 2$, for selected values of x, using synthetic division and the remainder theorem. The process is speeded by forming a synthetic division table. The second row is left blank and the computation for succeeding rows is done either mentally or on a handheld calculator—the hand calculator becomes increasingly useful as the coefficients become more numerous or complicated. The table also provides other important information, as will be seen in subsequent sections.

$$\begin{array}{r|rrrrl}
 & 1 & 3 & -1 & -3 & \\
\hline
-4 & 1 & -1 & 3 & -15 & = P(-4) \\
-3 & 1 & 0 & -1 & 0 & = P(-3) \\
-2 & 1 & 1 & -3 & 3 & = P(-2) \\
-1 & 1 & 2 & -3 & 0 & = P(-1) \\
0 & 1 & 3 & -1 & -3 & = P(0) \\
1 & 1 & 4 & 3 & 0 & = P(1) \\
2 & 1 & 5 & 9 & 15 & = P(2)
\end{array}$$

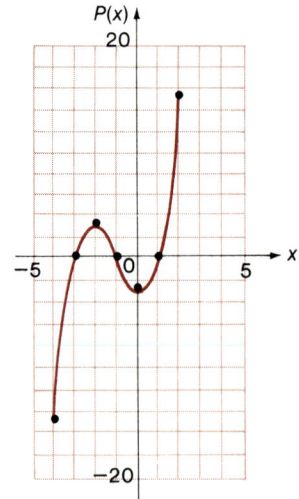

PROBLEM 4 Graph $P(x) = x^3 - 4x^2 - 4x + 16$, $-3 \leq x \leq 5$. Find points using synthetic division and the remainder theorem.

■ **Factor Theorem**

The equation $P(x) = (x - r)Q(x) + R$ in Theorem 1 may, because of the remainder theorem, be written in a form where R is replaced by $P(r)$. Thus,

$$P(x) = (x - r)Q(x) + P(r)$$

It is easy to see that $x - r$ is a factor of $P(x)$ if and only if $P(r) = 0$—that is, if and only if r is a zero of the polynomial $P(x)$ [or a root or solution of the polynomial equation $P(x) = 0$]. This result is known as the **factor theorem** (Theorem 3).

THEOREM 3

Factor Theorem

If r is a zero of the polynomial $P(x)$, then $x - r$ is a factor of $P(x)$; conversely, if $x - r$ is a factor of $P(x)$, then r is a zero of $P(x)$.

If we can find a zero of a polynomial, then we can find one of its factors. On the other hand, if we can find a linear factor of a polynomial, we can find a zero of the polynomial.

EXAMPLE 5 (A) Use the factor theorem to show that $x + 1$ is a factor of $P(x) = x^{25} + 1$.
(B) What are the zeros of $P(x) = 3(x - 5)(x + 2)(x - 3)$?

Solution (A) $x + 1 = x - (-1)$; thus, $r = -1$.

$$P(r) = P(-1) = (-1)^{25} + 1 = -1 + 1 = 0$$

Hence, -1 is a zero of $P(x) = x^{25} + 1$. Thus, $x - (-1) = x + 1$ is a factor of $x^{25} + 1$.

(B) 5, -2, and 3 are zeros of $P(x)$, since $(x - 5)$, $(x + 2)$, and $(x - 3)$ are all factors of $P(x)$.

PROBLEM 5 (A) Use the factor theorem to show that $x - 1$ is a factor of $P(x) = x^{54} - 1$.
(B) What are the zeros of the polynomial

$$P(x) = 2(x + 3)(x + 7)(x - 8)(x + 1)?$$

Answers to Matched Problems   **3.** $P(-2) = -3$ for both parts, as it should.

4.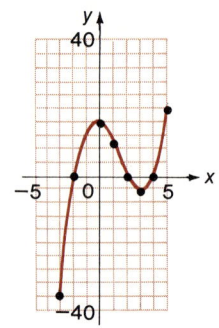

5. (A) $r = 1$ and $P(1) = 1^{54} - 1 = 1 - 1 = 0$; therefore, $x - r = x - 1$ is a factor of $P(x) = x^{54} - 1$.
   (B) $-3, -7, 8, -1$

Exercise 3-3 ■

A   *A hand calculator will prove helpful in some of the following problems.*

Use synthetic division and the remainder theorem in each of the following problems.

1. Find $P(-2)$, given $P(x) = 3x^2 - x - 10$.
2. Find $P(-3)$, given $P(x) = 4x^2 + 10x - 8$.
3. Find $P(2)$, given $P(x) = 2x^3 - 5x^2 + 7x - 7$.
4. Find $P(5)$, given $P(x) = 2x^3 - 12x^2 - x + 30$.
5. Find $P(-4)$, given $P(x) = x^4 - 10x^2 + 25x - 2$.
6. Find $P(-7)$, given $P(x) = x^4 + 5x^3 - 13x^2 - 30$.

Find the zeros for the following polynomials using the factor theorem.

7. $P(x) = (x - 3)(x + 5)$
8. $P(x) = (x + 2)(x - 7)$
9. $P(x) = 2(x + \frac{1}{2})(x - 8)(x + 2)$
10. $P(x) = 3(x - \frac{2}{3})(x - 5)(x + 7)$

Determine whether the second polynomial is a factor of the first polynomial without dividing or using synthetic division. [Hint: Evaluate directly and use the factor theorem.]

11. $x^{18} - 1$; $x - 1$
12. $x^{18} - 1$; $x + 1$
13. $3x^3 - 7x^2 - 8x + 2$; $x + 1$
14. $3x^4 - 2x^3 + 5x - 6$; $x - 1$

3-3 Remainder and Factor Theorems

B *Use synthetic division and the remainder theorem in each of the following problems.*

15. Find $P(\frac{1}{2})$, given $P(x) = 4x^3 - 8x^2 + 5x - 4$.
16. Find $P(\frac{1}{3})$, given $P(x) = 6x^3 + 4x^2 - 5x - 4$.
17. Find $P(0.3)$ for $P(x) = x^3 - 2x + 1$.
18. Find $P(0.7)$ for $P(x) = 2x^3 + 3x^2 - 5x + 2$.

Graph each polynomial function using synthetic division and the remainder theorem.

19. $P(x) = x^3 - 5x^2 + 2x + 8$, $-2 \leq x \leq 5$
20. $P(x) = x^3 + 2x^2 - 5x - 6$, $-4 \leq x \leq 3$
21. $P(x) = x^3 + 4x^2 - x - 4$, $-5 \leq x \leq 2$
22. $P(x) = x^3 - 2x^2 - 5x + 6$, $-3 \leq x \leq 4$

Find three solutions for each equation.

23. $(x + 4)(x + 8)(x - 1) = 0$
24. $(x - 2)(x + 5)(x - 3) = 0$
25. $7(x - \frac{1}{8})(x + \frac{3}{5})(x + 4) = 0$
26. $4(x + \frac{3}{4})(x - 5)(x - \frac{2}{3}) = 0$

Use the quadratic formula and the factor theorem to factor each polynomial.

27. $P(x) = x^2 - 3x + 1$
28. $P(x) = x^2 - 4x - 2$
29. $P(x) = x^2 - 6x + 10$
30. $P(x) = x^2 - 4x + 5$

Determine whether the second polynomial is a factor of the first polynomial without dividing or using synthetic division.

31. $x^n - a^n$; $x - a$
32. $x^n - a^n$, n even; $x + a$
33. $4x^7 - 2x^6 + x^2 + 2x + 5$; $x - 1$
34. $2x^5 - 5x^2 - x + 4$; $x + 1$

C *Graph each polynomial function in Problems 35–38 using synthetic division and the remainder theorem. A hand calculator may prove useful.*

35. $P(x) = x^4 - 2x^3 - 2x^2 + 8x - 8$
36. $P(x) = x^4 - 2x^2 + 16x - 15$
37. $P(x) = x^4 + 4x^3 - x^2 - 20x - 20$
38. $P(x) = x^4 - 4x^2 - 4x - 1$

39. Recall from Section 2-5 that polynomials can also be conveniently evaluated using a nested factoring scheme. For example, to evaluate $P(x) = x^4 - 6x^3 + 19x^2 - 26x + 18$ for $x = 2$, we would first

write $P(x) = \{[(x - 6)x + 19]x - 26\}x + 18$. Find $P(2)$ using this "factored" form, and by synthetic division using the remainder theorem. How do the two methods compare step for step?

40. Repeat Problem 39 for the equation $P(x) = 3x^4 - 10x^2 + 5x - 2$ and $x = -2$.

41. (A) Write $P(x) = a_2x^2 + a_1x + a_0$ in the form $P(x) = (a_2x + a_1)x + a_0$ and find $P(r)$ using the latter.
(B) Find $P(r)$ using synthetic division and the remainder theorem, and compare with part (A).

42. Repeat Problem 41 for $P(x) = a_3x^3 + a_2x^2 + a_1x + a_0$.

Section 3-4 Fundamental Theorem of Algebra

- Fundamental Theorem of Algebra
- n Zeros Theorem
- Complex Zeros
- Remarks

In our search for zeros of polynomial functions it would be useful to know at the outset how many zeros to expect for a given function. The following two theorems tell us exactly how many zeros exist for a polynomial function of a given degree. Even though the theorems do not tell us how to find the zeros, it is still very helpful to know that what we are looking for exists. These theorems were first proved in 1797 by Carl Friedrich Gauss, one of the greatest mathematicians of all time, at the age of 20.

- Fundamental Theorem of Algebra

Theorem 4, often referred to as the **fundamental theorem of algebra**, requires a proof that is beyond the scope of this book, so we state it without proof.

THEOREM 4   Fundamental Theorem of Algebra

Every polynomial $P(x)$ of degree $n \geq 1$, with real or complex coefficients, has at least one real or complex zero.

3-4 Fundamental Theorem of Algebra

■ *n Zeros Theorem*

If $P(x) = a_n x^n + a_{n-1} x^{n-1} + \cdots + a_1 x + a_0$ is a polynomial of degree $n \geq 1$, then, according to this theorem, it has at least one zero, say r_1. According to the factor theorem, $x - r_1$ is a factor of $P(x)$. Thus,

$$P(x) = (x - r_1) Q_1(x)$$

where $Q_1(x)$ is a polynomial of degree $n - 1$. If $n - 1 = 0$, then $Q_1(x) = a_n$. If $n - 1 \geq 1$, then, by the fundamental theorem of algebra, $Q_1(x)$ has at least one zero, say r_2. And

$$Q_1(x) = (x - r_2) Q_2(x)$$

where $Q_2(x)$ is a polynomial of degree $n - 2$. Thus,

$$P(x) = (x - r_1)(x - r_2) Q_2(x)$$

If $n - 2 = 0$, then $Q_2(x) = a_n$. If $n - 2 \geq 1$, then $Q_2(x)$ has at least one zero, say r_3. And

$$Q_2(x) = (x - r_3) Q_3(x)$$

where $Q_3(x)$ is a polynomial of degree $n - 3$.

We continue in this way until $Q_k(x)$ is of degree zero—that is, until $k = n$. At this point, $Q_n(x) = a_n$ and we have

$$P(x) = (x - r_1)(x - r_2) \cdots (x - r_n) a_n$$

Thus, $r_1, r_2, \ldots, r_n$ are n zeros (not necessarily distinct) of $P(x)$. Is it possible for $P(x)$ to have more than these n zeros? Let us assume that r is a number different from the zeros above; then

$$P(r) = a_n (r - r_1)(r - r_2) \cdots (r - r_n) \neq 0$$

since r is not equal to any of the zeros. Hence, r is not a zero and we conclude that $r_1, r_2, \ldots, r_n$ are the only zeros of $P(x)$. We have just sketched a proof of Theorem 5.

THEOREM 5 | *n Zeros Theorem*

Every polynomial $P(x)$ of degree $n \geq 1$, with real or complex coefficients, can be expressed as the product of n linear factors (hence, has exactly n zeros—not necessarily distinct).

If $P(x)$ is represented as the product of linear factors and $x - r$ occurs m times, then r is called a **zero of multiplicity m**. For example, if

$$P(x) = 4(x - 5)^3 (x + 1)^2 (x - i)(x + i)$$

then this seventh-degree polynomial has seven zeros, not all distinct. Five is a zero of multiplicity 3 (or a triple zero); -1 is a zero of multiplicity 2 (or a double zero). Thus, this seventh-degree polynomial has exactly seven zeros if we count 5 and -1 with their respective multiplicities.

EXAMPLE 6 If -2 is a double zero of $P(x) = x^4 - 7x^2 + 4x + 20$, write $P(x)$ as a product of first-degree factors.

Solution Since -2 is a double zero of $P(x)$, we can write

$$P(x) = (x + 2)^2 Q(x)$$
$$= (x^2 + 4x + 4) Q(x)$$

and find $Q(x)$ by dividing $P(x)$ by $x^2 + 4x + 4$. Carrying out the division, we obtain

$$Q(x) = x^2 - 4x + 5$$

The zeros of $Q(x)$ are found, using the quadratic formula, to be $2 - i$ and $2 + i$. Thus, $P(x)$ written as a product of linear factors is

$$P(x) = (x + 2)^2 [x - (2 - i)][x - (2 + i)]$$

PROBLEM 6 If 3 is a double zero of $P(x) = x^4 - 12x^3 + 55x^2 - 114x + 90$, write $P(x)$ as a product of first-degree factors.

■ Complex Zeros

Something interesting happens if we restrict the coefficients of a polynomial to real numbers. Let us use the quadratic formula to find the zeros of the polynomial

$$P(x) = x^2 - 6x + 10$$

To find the zeros of $P(x)$, we solve $P(x) = 0$:

$$x^2 - 6x + 10 = 0$$

$$x = \frac{6 \pm \sqrt{36 - 40}}{2}$$

$$= \frac{6 \pm \sqrt{-4}}{2} = \frac{6 \pm 2i}{2} = 3 \pm 2i$$

The zeros of $P(x)$ are $3 - 2i$ and $3 + 2i$, conjugate complex numbers (see Section 1-6). Also observe that the complex zeros in Example 6 are the conjugate complex numbers $2 - i$ and $2 + i$.

3-4 Fundamental Theorem of Algebra

In general, one can prove the following theorem:

THEOREM 6

Complex Zeros Theorem

Nonreal complex zeros of polynomials with real coefficients, if they exist, occur in conjugate pairs.

As a consequence of Theorems 5 and 6, we immediately know (think this through) that:

Real Zeros and Odd-Degree Polynomials

A polynomial of odd degree with real coefficients always has at least one real zero.

EXAMPLE 7 Let $P(x)$ be a third-degree polynomial with real coefficients. One of the following statements is false; indicate which one.

(A) $P(x)$ has at least one real zero.
(B) $P(x)$ has three zeros.
(C) $P(x)$ can have two real zeros and one complex zero.

Solution Statement (C) is false, since complex zeros of polynomials with real coefficients *must* occur in conjugate pairs. If $P(x)$ has two real zeros, then we know that the third zero must also be real.

PROBLEM 7 Let $P(x)$ be a polynomial of fourth degree with real coefficients. One of the following statements is false; indicate which one.

(A) $P(x)$ has four zeros.
(B) $P(x)$ has at least two real zeros.
(C) If we know $P(x)$ has three real zeros, then the fourth zero must be real.

■ Remarks

The fundamental theorem of algebra tells us that in the set of complex numbers not only $x^2 + 1 = 0$ has a solution, but every polynomial equation with real or complex coefficients has a solution.

3 Polynomial Functions and Theory of Equations

This important and useful result does not come free. In extending the real numbers to a number system that provides solutions for all polynomial equations we have to give up something—namely, an ordering of the number system. The complex numbers cannot be ordered; that is, in general, we cannot say that one complex number is less than or greater than another.

Answers to Matched Problems

6. $P(x) = (x - 3)^2[x - (3 - i)][x - (3 + i)]$

7. (B) is false. According to the three theorems in this section, the possible combinations of real and complex zeros for $P(x)$ are as follows: (1) four complex, (2) two real and two complex, (3) four real. So $P(x)$ may not have any real zeros.

Exercise 3-4

A *Write the zeros of each polynomial, and indicate the multiplicity of each if over one. What is the degree of the polynomial?*

1. $P(x) = (x + 8)^3(x - 6)^2$
2. $P(x) = (x - 5)(x + 7)^2$
3. $P(x) = 3(x + 4)^3(x - 3)^2(x + 1)$
4. $P(x) = 5(x - 2)^3(x + 3)^2(x - 1)$

Find a polynomial $P(x)$ of lowest degree, with leading coefficient 1, that has the indicated set of zeros. (Leave the answer in a factored form.) Indicate the degree of the polynomial.

5. 3 (multiplicity 2) and -4
6. -2 (multiplicity 3) and 1 (multiplicity 2)
7. -7 (multiplicity 3), $\frac{2}{3}$, and -5
8. $\frac{1}{3}$ (multiplicity 2), 5, and -1
9. $(2 - 3i)$, $(2 + 3i)$, -4 (multiplicity 2)
10. $i\sqrt{3}$ (multiplicity 2), $-i\sqrt{3}$ (multiplicity 2), and 4 (multiplicity 3)

B *Given the indicated polynomials, what are the possible combinations of real and complex zeros?*

11. $P(x) = 2x^3 - 3x^2 + x - 5$
12. $P(x) = 2x^4 - 2x^3 + x - 8$
13. $P(x) = 3x^6 - 5x^5 + 3x^2 - 4$
14. $P(x) = x^5 - 2x^3 + 5x^2 - 6$

Write $P(x)$ as a product of first-degree factors.

15. $P(x) = x^3 + 9x^2 + 24x + 16$; -1 is a zero
16. $P(x) = x^3 - 4x^2 - 3x + 18$; 3 is a double zero
17. $P(x) = x^4 - 1$; 1 and -1 are zeros

3-4 Fundamental Theorem of Algebra

18. $P(x) = x^4 + 2x^2 + 1$; i is a double zero
19. $P(x) = 2x^3 - 17x^2 + 90x - 41$; $\frac{1}{2}$ is a zero
20. $P(x) = 3x^3 - 10x^2 + 31x + 26$; $-\frac{2}{3}$ is a zero

Given the indicated equations, what are the possible combinations of real and complex solutions?

21. $x^4 - 3x^3 + 5x - 6 = 0$
22. $2x^3 - 4x^2 - x + 3 = 0$
23. $x^6 - 3x^4 + x^3 - x - 7 = 0$
24. $4x^5 + x^4 - 5x^2 - x + 3 = 0$

Multiply.

25. $[x - (4 - 5i)][x - (4 + 5i)]$
26. $[x - (2 - 3i)][x - (2 + 3i)]$
27. $[x - (a + bi)][x - (a - bi)]$
28. $(x - bi)(x + bi)$

C *In Problems 29–32 find two other zeros of P(x), given the indicated zeros.*

29. $P(x) = x^3 - 5x^2 + 4x + 10$; $3 - i$ is one zero
30. $P(x) = x^3 + x^2 - 4x + 6$; $1 + i$ is one zero
31. $P(x) = x^3 - 3x^2 + 25x - 75$; $-5i$ is one zero
32. $P(x) = x^3 + 2x^2 + 16x + 32$; $4i$ is one zero
33. The solutions to the equation $x^3 - 1 = 0$ are all the cube roots of 1.
    (A) How many cube roots of 1 are there?
    (B) 1 is obviously a cube root of 1; find all others.
34. The solutions to the equation $x^3 - 8 = 0$ are all the cube roots of 8.
    (A) How many cube roots of 8 are there?
    (B) 2 is obviously a cube root of 8; find all others.
35. If P is a polynomial function of degree n, with n odd, then what is the maximum number of times the graph of $y = P(x)$ can cross the x axis? What is the minimum number of times?
36. Answer the questions in Problem 35 for n even.
37. Given $P(x) = x^2 + 2ix - 5$ with $2 - i$ a zero, show that $2 + i$ is not a zero of P(x). Does this contradict Theorem 6? Explain.
38. If P(x) and G(x) are two polynomials of degree n, and if $P(x) = G(x)$ for more than n values of x, then how are P(x) and Q(x) related?

Section 3-5 Isolating Real Zeros

- Descartes' Rule of Signs
- Bounding Real Zeros
- Sign Changes in $P(x)$

For the rest of this chapter we will focus on the problem of finding real zeros of polynomials with real coefficients. Three theorems will help us greatly in this regard. The first theorem gives us useful information about the possible number of real zeros of a given polynomial; the second theorem tells us how to determine a finite interval that contains all the real zeros, if they exist; and the third theorem will help us isolate particular zeros further within this interval.

Descartes' Rule of Signs

When the terms of a polynomial with real coefficients are arranged in order of descending powers, we say that a **variation in sign** occurs if two successive terms have opposite signs. Missing terms (terms with 0 coefficients) are ignored. For a given polynomial $P(x)$ we are going to be interested in the total number of variations in sign in both $P(x)$ and $P(-x)$.

EXAMPLE 8 If $P(x) = 3x^4 - 2x^3 + 3x - 5$, how many variations in sign are in $P(x)$ and in $P(-x)$?

Solution

$P(x) = 3x^4 - 2x^3 + 3x - 5$ Three variations in sign

$P(-x) = 3x^4 + 2x^3 - 3x - 5$ One variation in sign

PROBLEM 8 If $P(x) = 2x^5 - x^4 - x^3 + x + 5$, how many variations in sign are in $P(x)$ and in $P(-x)$?

The number of variations in sign for $P(x)$ and for $P(-x)$ gives us useful information about the number of real zeros of a polynomial with real coefficients. In 1636, René Descartes, a French philosopher-mathematician, gave the first proof of a simplified version of a theorem that now bears his name. We state Theorem 7 without proof, since a proof is beyond the scope of this book.

3-5 Isolating Real Zeros

THEOREM 7	Descartes' Rule of Signs
	Given a polynomial $P(x)$ with real coefficients:
	1. *Positive real zeros.* The number of positive real zeros of $P(x)$ is never greater than the number of variations in sign in $P(x)$ and, if less, then always by an even number.
	2. *Negative real zeros.* The number of negative real zeros of $P(x)$ is never greater than the number of variations in sign in $P(-x)$ and, if less, then always by an even number.

EXAMPLE 9 What can you say about the number of positive and negative real zeros of:

(A) $P(x) = 3x^4 - 2x^3 + 3x - 5$
(B) $Q(x) = 2x^6 + x^4 - x + 3$

Solution (A) $P(x) = 3x^4 - 2x^3 + 3x - 5$ Three variations in sign
$P(x) = 3x^4 + 2x^3 - 3x - 5$ One variation in sign

Positive real zeros: three or one

Negative real zeros: one

(B) $Q(x) = 2x^6 + x^4 - x + 3$ Two variations in sign
$Q(-x) = 2x^6 + x^4 + x + 3$ No variations in sign

Positive real zeros: two or zero

Negative real zeros: none

PROBLEM 9 What can you say about the number of positive and negative real zeros of:

(A) $P(x) = 4x^5 + 2x^4 - x^3 + x - 5$ (B) $Q(x) = x^3 + 3x^2 + 5$

■ Bounding Real Zeros

Any number that is greater than or equal to the largest zero of a polynomial is called an **upper bound of the zeros** of the polynomial; any number that is less than or equal to the smallest zero of a polynomial is called a **lower bound of the zeros** of the polynomial. Theorem 8 enables us to determine upper and lower bounds of all real zeros of a polynomial with real coefficients.

THEOREM 8	Upper and Lower Bounds of Real Zeros
	Given a polynomial $P(x)$ with real coefficients, degree $n \geq 1$, and the coefficient of the nth-degree term positive. Let $P(x)$ be divided by $x - r$ using synthetic division. 1. *Upper bound.* If $r > 0$ and all numbers in the quotient row of the synthetic division are nonnegative, then r is an upper bound of the zeros of $P(x)$. 2. *Lower bound.* If $r < 0$ and all numbers in the quotient row of the synthetic division alternate in sign, then r is a lower bound of the zeros of $P(x)$. [*Note:* In this lower-bound test, if 0 appears in one or more places in the quotient row, the sign in front of it can be considered either positive or negative.]

We sketch a proof of part 1 of Theorem 8. The proof of part 2 is similar, only a little more difficult.

If all the numbers in the quotient row of the synthetic division are nonnegative after dividing $P(x)$ by $x - r$, then

$$P(x) = (x - r)Q(x) + R$$

where the coefficients of $Q(x)$ are nonnegative and R is nonnegative. If $x > r > 0$, then $x - r > 0$ and $Q(x) > 0$; hence,

$$P(x) = (x - r)Q(x) + R > 0$$

Thus, $P(x)$ cannot be 0 for any x greater than r, and r is an upper bound for the real zeros of $P(x)$.

EXAMPLE 10   Find the smallest positive integer and the largest negative integer that, by Theorem 8, are upper and lower bounds, respectively, for the real zeros of

$$P(x) = x^3 - 3x^2 - 18x + 4$$

Solution   An easy way to locate these upper and lower bounds, particularly if the coefficients of $P(x)$ are not too large, is to test $r = 1, 2, 3, \ldots$ until the quotient row turns nonnegative; then test $r = -1, -2, -3, \ldots$ until the quotient row alternates in sign. The resulting table will provide side benefits as we will see later.

3-5 Isolating Real Zeros

		1	−3	−18	4
	1	1	−2	−20	−16
	2	1	−1	−20	−36
	3	1	0	−18	−50
	4	1	1	−14	−52
	5	1	2	−8	−36
UB	6	1	3	0	4
	−1	1	−4	−14	18
	−2	1	−5	−8	20
	−3	1	−6	0	4
LB	−4	1	−7	10	−36

← {This quotient row is nonnegative; hence, 6 is an upper bound (UB).

← {This quotient row alternates in sign; hence, −4 is a lower bound (LB).

Because of Theorem 8, we now know that all real zeros of $P(x) = x^3 - 3x^2 - 18x + 4$ (or all real solutions of $x^3 - 3x^2 - 18x + 4 = 0$) lie between −4 and 6.

PROBLEM 10 Repeat Example 10 for $P(x) = x^3 - 4x^2 - 5x + 8$.

■ Sign Changes in $P(x)$

Observing sign changes in a polynomial $P(x)$ with real coefficients as x is replaced with different real numbers leads to the further isolation of real zeros of $P(x)$. Recall that a polynomial function P (with real coefficients) is continuous everywhere; that is, the graph of $y = P(x)$ has no holes or breaks (Fig. 2). This property of polynomial functions is the basis of Theorem 9.

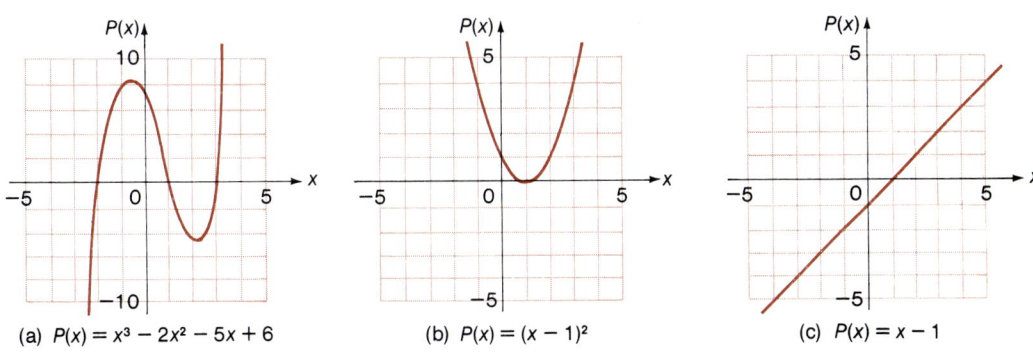

(a) $P(x) = x^3 - 2x^2 - 5x + 6$ (b) $P(x) = (x - 1)^2$ (c) $P(x) = x - 1$

FIGURE 2 The x intercepts of the graph of $y = P(x)$ are the real zeros of $P(x)$

THEOREM 9	Location Theorem
	If $P(x)$ is a polynomial with real coefficients, and if $P(a)$ and $P(b)$ are of opposite sign, then there is at least one real zero between a and b.

Notice in Figure 2(a) that $P(-3) < 0$ and $P(4) > 0$ and there are three real zeros between -3 and 4. Also, in Figure 2(c) $P(-3) < 0$ and $P(4) > 0$ and there is one zero between -3 and 4. Since the graph of a polynomial function P with real coefficients is continuous, and if $P(a)$ and $P(b)$ are of opposite sign, the graph of $y = P(x)$ must cross the x axis at least once for x between a and b.

The converse of Theorem 9 is false; that is, if $P(x)$ has at least one real zero, then $P(x)$ may or may not change sign as x crosses a zero. Compare Figure 2(b) and (c). Both functions have zeros at $x = 1$, but the first is never negative, whereas the second is negative for $x < 1$ and positive for $x > 1$.

EXAMPLE 11 Show that there is at least one real zero of

$$P(x) = x^4 - 2x^3 - 6x^2 + 6x + 9$$

between 1 and 2.

Solution Show that $P(1)$ and $P(2)$ have opposite signs.

```
        1    -2    -6     6     9
    1 | 1   -1    -7    -1     8 = P(1)
    2 | 1    0    -6    -6    -3 = P(2)
```

Since $P(1)$ and $P(2)$ have opposite signs, there is at least one real zero between 1 and 2 (Theorem 9).

PROBLEM 11 Show that there is at least one real zero of

$$P(x) = 2x^4 - 3x^3 - 3x - 4$$

between 2 and 3.

Answers to Matched Problems

8. Two in $P(x)$ and three in $P(-x)$
9. (A) Positive real zeros: three or one
       Negative real zeros: two or none
   (B) Positive real zeros: none
       Negative real zeros: one
10. Lower bound: -2; upper bound: 5
11. $P(2) = -2$ and $P(3) = 68$, and the conclusion follows from Theorem 9.

3-5 Isolating Real Zeros

Exercise 3-5

A Using Descartes' rule of signs, what can you say about the number of positive and negative zeros of each of the following polynomials?

1. $P(x) = 2x^2 + x - 4$
2. $Q(x) = 3x^2 - x - 5$
3. $M(x) = 7x^2 + 2x + 4$
4. $N(x) = -3x^2 - 2x - 1$
5. $Q(x) = 2x^3 - 4x^2 + x - 3$
6. $P(x) = x^3 + 7x^2 - x + 2$

Find the smallest positive integer and the largest negative integer that, by Theorem 8, are upper and lower bounds, respectively, for the real zeros of each of the following polynomials.

7. $P(x) = x^2 - 2x + 3$
8. $Q(x) = x^2 - 3x - 2$
9. $M(x) = x^3 - 3x + 5$
10. $R(x) = x^3 - 2x^2 + 3$
11. $M(x) = x^4 - x^2 + 3x + 2$
12. $N(x) = x^4 - 2x^3 + 4x - 3$

Show, using Theorem 9, that for each polynomial there is at least one real zero between the given values of a and b.

13. $P(x) = x^2 - 3x - 2;\quad a = 3,\quad b = 4$
14. $Q(x) = x^2 - 3x - 2;\quad a = -1,\quad b = 0$
15. $P(x) = x^3 - 3x + 5;\quad a = -3,\quad b = -2$
16. $P(x) = x^3 - 2x^2 - 4;\quad a = 2,\quad b = 3$
17. $Q(x) = x^3 - 3x^2 - 3x + 9;\quad a = 1,\quad b = 2$
18. $G(x) = x^3 - 3x^2 - 3x + 9;\quad a = -2,\quad b = -1$

B For each polynomial P(x),

(A) Discuss the possible number of real zeros using Descartes' rule of signs.

(B) Find the smallest and largest integers that are, respectively, upper and lower bounds of the zeros of P(x) according to Theorem 8.

(C) Discuss the location of real zeros within the lower and upper bound interval by applying Theorem 9 to integer values of x within this interval.

19. $P(x) = x^3 - x^2 - 6x + 6$
20. $P(x) = x^3 - 3x^2 - 2x + 6$
21. $P(x) = x^3 - 2x - 6$
22. $P(x) = x^3 - 3x^2 - 5$
23. $P(x) = x^4 + 4x^3 - 2x^2 - 12x - 3$
24. $P(x) = x^4 - 4x^3 + 8x - 4$
25. $P(x) = x^5 - 3x^3 + 2x - 5$
26. $P(x) = 2x^5 - 5x^4 - 2x + 5$

27. Prove that $P(x) = x^4 + 3x^2 - x - 5$ has two complex and two real zeros, without finding the zeros.

28. Prove that $P(x) = x^3 + 3x^2 + 5$ has one negative real zero and two complex zeros, without finding the zeros.

29. Prove that the graph of $P(x) = x^5 + 3x^3 + x$ crosses the x axis only once without graphing $y = P(x)$.

30. Prove that the graph of $P(x) = x^4 + 3x^2 + 7$ does not cross the x axis at all. Do not graph $y = P(x)$.

Section 3-6 Finding Rational Zeros

- Rational Zero Theorem
- Strategy for Finding Rational Zeros

Rational Zero Theorem

We start our investigation with a quadratic function whose zeros can be found easily by factoring. From this example, we will point out a relationship that generally holds for all polynomials with integer coefficients.

$$P(x) = 6x^2 - 13x - 5 = (2x - 5)(3x + 1)$$

Zeros of $P(x)$: $\quad \dfrac{5}{2}, \; -\dfrac{1}{3} \; \text{or} \; \dfrac{-1}{3}$

Notice that the numerator of each zero (5 and -1) is a factor of -5, the constant term in $P(x)$. The denominator of each zero (2 and 3) is a factor of 6, the coefficient of the highest-degree term in $P(x)$. These observations are generalized in Theorem 10.

THEOREM 10 | **Rational Zero Theorem**

If the rational number b/c, in lowest terms, is a zero of the polynomial

$$P(x) = a_n x^n + a_{n-1} x^{n-1} + \cdots + a_1 x + a_0$$

with integer coefficients, then b must be a factor of a_0 [the constant term in $P(x)$] and c must be a factor of a_n [the coefficient of the highest-degree term in $P(x)$].

The proof of Theorem 10 is not difficult, and is instructive, so we sketch it here. Since b/c is a zero of $P(x)$,

$$a_n \left(\dfrac{b}{c}\right)^n + a_{n-1} \left(\dfrac{b}{c}\right)^{n-1} + \cdots + a_1 \left(\dfrac{b}{c}\right) + a_0 = 0 \tag{1}$$

3-6 Finding Rational Zeros

If we multiply both members of equation (1) by c^n, we obtain

$$a_n b^n + a_{n-1} b^{n-1} c + \cdots + a_1 b c^{n-1} + a_0 c^n = 0 \tag{2}$$

which can be written in the form

$$a_n b^n = c(-a_{n-1} b^{n-1} - \cdots - a_0 c^{n-1}) \tag{3}$$

Thus c is a factor of $a_n b^n$, since the expression in parentheses is an integer. (Why?) And since b and c are **relatively prime** (that is, have no common factors other than ± 1), b^n and c must be relatively prime; hence c must be a factor of a_n. [Divide both sides of equation (3) by c to see why.]

Now, if we solve equation (2) for $a_0 c^n$ and factor b out of the right side, we have

$$a_0 c^n = b(-a_n b^{n-1} - \cdots - a_1 c^{n-1})$$

we see that b is a factor of $a_0 c^n$ and hence a factor of a_0, since b and c are relatively prime.

We emphasize that **Theorem 10 does not say that a polynomial with integers as coefficients has rational zeros; it simply states that if it does, then they must meet the conditions stated in the theorem. In short, it enables us to list a set of rational numbers that must include all rational zeros if they exist.**

EXAMPLE 12 List all possible rational zeros for

$$P(x) = 2x^4 - 3x^3 + x - 9$$

Solution If b/c (in lowest terms) is a rational zero of $P(x)$, then b must be a factor of -9 and c must be a factor of 2.

Possible values of b (factors of -9): $\pm 1, \pm 3, \pm 9$ (4)

Possible values of c (factors of 2): $\pm 1, \pm 2$ (5)

Thus, writing all possible fractions b/c where b is from (4) and c is from (5), we have

Possible rational zeros for $P(x)$: $\pm 1, \pm 3, \pm 9, \pm \frac{1}{2}, \pm \frac{3}{2}, \pm \frac{9}{2}$ (6)

$$\left[\text{Note:} \quad \frac{\pm 1}{\pm 1} = \pm 1, \quad \frac{\pm 3}{\pm 1} = \pm 3, \quad \text{etc.} \right]$$

Thus, if $P(x)$ has rational zeros, they must be in list (6).

PROBLEM 12 List all possible rational zeros for

$$P(x) = x^3 + 2x^2 - 5x - 6$$

■ **Strategy for Finding Rational Zeros**

If a polynomial is of first or second degree, then we can always find *all* of its zeros using methods discussed in Chapter 1. With all the tools we have developed in this chapter, we are now ready to state a strategy that will efficiently lead to all rational zeros of polynomials of degree $n \geq 3$ with integer coefficients, if they exist. Of course, we could just test each of the possible rational zeros that result from the rational zero theorem. However, we can make the process more efficient by using many of the other properties of polynomials discussed earlier in this chapter. In addition, some of these same properties and procedures will help us locate irrational zeros in the next section.

Strategy for Finding Rational Zeros

Assume $P(x)$ is a polynomial of degree $n \geq 3$ with integer coefficients.

Step 1. List the possible rational zeros of $P(x)$ using Theorem 10.

Step 2. List the possible number of positive and negative real zeros using Descartes' rule of signs.

Step 3. Test the possible rational zeros from the list in step 1 being guided by the results of step 2 and the following steps a–e:

[*Note:* If a rational zero r is found in any of the steps a through e below, write

$P(x) = (x - r)Q(x)$

and immediately proceed to find the rational zeros for $Q(x)$, the reduced polynomial relative to $P(x)$. If $Q(x)$ is of degree $n \geq 3$, return to step 1 using $Q(x)$ in place of $P(x)$. If $Q(x)$ is quadratic, find *all* of its zeros using standard methods for solving quadratic equations.]

a. Use the results of Descartes' rule of signs (step 2) in conjunction with steps b through e to further eliminate possible rational zeros from the list in step 1.
b. Form a synthetic division table by testing the possible integer zeros from the list in step 1.
c. Watch for sign changes in $P(x)$ and test any fractions from the list in step 1 that are between integers that produce sign changes in $P(x)$.
d. Watch for test values that are lower or upper bounds for the real zeros of $P(x)$ — add more integer values if necessary to locate lower and upper bounds. Eliminate any rational numbers from the list in step 1 that are below the lower bound or above the upper bound.
e. Test the remaining possible rational zeros from step 1. If none of the possible rational zeros from step 1 are zeros, we conclude that $P(x)$ has no rational zeros. The zeros must then be irrational or nonreal complex.

3-6 Finding Rational Zeros

Let us see how the strategy works in several concrete examples.

EXAMPLE 13 Find all rational zeros of $P(x) = 2x^3 - x^2 - 8x + 4$.

Solution Step 1. List the possible rational zeros.

$\pm 1, \quad \pm 2, \quad \pm 4, \quad \pm \frac{1}{2}$

Step 2. List the possible number of positive and negative real zeros.

$P(x) = 2x^3 - x^2 - 8x + 4$ Two variations in sign
$P(-x) = -2x^3 - x^2 + 8x + 4$ One variation in sign

Positive real zeros: two or none

Negative real zeros: one

Step 3. Test the possible rational zeros listed in step 1.

$$\begin{array}{c|cccc} & 2 & -1 & -8 & 4 \\ \hline 1 & 2 & 1 & -7 & -3 \\ 2 & 2 & 3 & -2 & 0 \end{array}$$ 2 is a real zero

We have found a real zero, so we write:

$P(x) = (x - r)Q(x)$
$\quad\quad = (x - 2)(2x^2 + 3x - 2)$

The zeros of $Q(x)$, since it is quadratic, can be found by solving $Q(x) = 0$:

$2x^2 + 3x - 2 = 0$
$(2x - 1)(x + 2) = 0$
$2x - 1 = 0 \quad \text{or} \quad x + 2 = 0$
$\quad\quad x = \frac{1}{2} \quad\quad\quad\quad x = -2$

Thus, the rational zeros of $P(x)$ are ± 2 and $\frac{1}{2}$.

PROBLEM 13 Find all rational zeros for $P(x) = 3x^3 + 10x^2 + x - 6$.

EXAMPLE 14 Find *all* zeros for $P(x) = 2x^3 - 5x^2 - 8x + 6$.

Solution Step 1. List the possible rational zeros.

$\pm 1, \quad \pm 2, \quad \pm 3, \quad \pm 6, \quad \pm \frac{1}{2}, \quad \pm \frac{3}{2}$

Step 2. List the possible number of positive and negative real zeros.

$P(x) = 2x^3 - 5x^2 - 8x + 6$ Two variations in sign
$P(-x) = -2x^3 - 5x^2 + 8x + 6$ One variation in sign

Positive real zeros: two or none

Negative zeros: one

Step 3. Test the possible rational zeros listed in step 1.

[*Note:* We may include other numbers as needed to isolate zeros further.]

		2	−5	−8	6	
	1	2	−3	−11	−5	
	2	2	−1	−10	−14	
	3	2	1	−5	−9	There is an irrational zero between 3 and 4
UB	4	2	3	4	22	(Why?); also 4 is an upper bound.
	−1	2	−7	−1	7	Real zero between −1 and 1
	0	2	−5	−8	6	Real zero between 0 and 1
	$\frac{1}{2}$	2	−4	−10	1	Irrational zero between $\frac{1}{2}$ and 1 (Why?)
LB	−2	2	−9	10	−14	Real zero between −1 and −2; also, −2 is a lower bound
	$-\frac{3}{2}$	2	−8	4	0	$-\frac{3}{2}$ is a rational zero

We now know that $P(x)$ has three real zeros: $-\frac{3}{2}$ is the negative zero, and the table tells us that there are two positive irrational zeros, one between $\frac{1}{2}$ and 1 and the other between 3 and 4. We write

$$P(x) = (x - r)Q(x) = (x + \tfrac{3}{2})(2x^2 - 8x + 4)$$

and find the zeros of the reduced polynomial $Q(x)$ by solving $Q(x) = 0$:

$$2x^2 - 8x + 4 = 0$$
$$x^2 - 4x + 2 = 0$$

$$x = \frac{4 \pm \sqrt{16 - 4(1)(2)}}{2}$$

$$= \frac{4 \pm 2\sqrt{2}}{2} = 2 \pm \sqrt{2}$$

The zeros of $P(x)$ are $-\frac{3}{2}$ and $2 \pm \sqrt{2}$ or, to two decimal places:

−1.5, 0.59, 3.41   Notice that 0.59 is between $\frac{1}{2}$ and 1, and 3.41 is between 3 and 4, as was predicted from the synthetic division table.

PROBLEM 14 Find *all* zeros for $P(x) = 2x^3 - 7x^2 + 6x + 5$.

EXAMPLE 15 Find all rational zeros for $P(x) = x^4 - 7x^3 + 17x^2 - 17x + 6$.

Solution *Step 1.* List the possible rational zeros.

±1, ±2, ±3, ±6

3-6 Finding Rational Zeros

Step 2. List the possible number of positive and negative real zeros.

$P(x) = x^4 - 7x^3 + 17x^2 - 17x + 6$ Four variations in sign

$P(-x) = x^4 + 7x^3 + 17x^2 + 17x + 6$ No variation in sign

Positive real zeros: four, two, or none

Negative real zeros: none

We can eliminate all the negative numbers from the list in step 1, since there are no negative real zeros.

Step 3. Test the possible rational zeros listed in step 1.

$$\begin{array}{r|rrrrr} & 1 & -7 & 17 & -17 & 6 \\ \hline 1 & 1 & -6 & 11 & -6 & 0 \end{array}$$ 1 is a zero

Write

$$P(x) = (x - r)Q(x)$$
$$= (x - 1)(x^3 - 6x^2 + 11x - 6)$$

and return to step 1 for the reduced polynomial

$$Q(x) = x^3 - 6x^2 + 11x - 6$$

Step 1. List the possible rational zeros.

1, 2, 3, 6 The negatives of these were eliminated in step 2 above.

We now go directly to step 3, since step 2 does not add a lot of additional information.

Step 3. Test the possible rational zeros listed above.

$$\begin{array}{r|rrrr} & 1 & -6 & 11 & -6 \\ \hline 1 & 1 & -5 & 6 & 0 \end{array}$$ 1 is a zero

Write

$$Q(x) = (x - r)Q_1(x)$$
$$= (x - 1)(x^2 - 5x + 6)$$

and find the zeros of the reduced polynomial $Q_1(x)$ by solving $Q_1(x) = 0$:

$$x^2 - 5x + 6 = 0$$
$$(x - 2)(x - 3) = 0$$
$$x = 2 \quad \text{or} \quad 3$$

The zeros of $P(x)$ are 1 (multiplicity 2), 2, and 3.

PROBLEM 15 Find all rational zeros for $P(x) = x^4 + 8x^3 + 23x^2 + 28x + 12$.

3 Polynomial Functions and Theory of Equations

With a little practice and ingenuity (educated guessing) you will be able to reduce the number of steps and effort required to find rational zeros, if they exist. To develop this efficiency, you must work problems yourself. The more you work, the easier and faster the process will become.

Answers to Matched Problems

12. ± 1, ± 2, ± 3, ± 6 **13.** $-3, -1, \frac{2}{3}$
14. $-\frac{1}{2}, 2 \pm i$ **15.** $-3, -2$ (multiplicity 2), -1

Exercise 3-6

A   For each polynomial:
(A) List all possible rational zeros (Theorem 10).
(B) Find all rational zeros. If there are no rational zeros, say so.

1. $P(x) = x^3 - 2x^2 - 5x + 6$
2. $P(x) = x^3 + 3x^2 - 6x - 8$
3. $P(x) = 3x^3 - 11x^2 + 8x + 4$
4. $P(x) = 2x^3 + x^2 - 4x - 3$
5. $P(x) = 12x^3 - 16x^2 - 5x + 3$
6. $P(x) = 2x^3 - 9x^2 + 14x - 5$
7. $P(x) = 3x^3 + 7x^2 - 10x - 4$
8. $P(x) = 2x^3 - 5x^2 - 2x + 15$

B
9. $P(x) = x^3 - 3x^2 + 6$
10. $P(x) = x^3 - 3x + 1$
11. $P(x) = x^4 - 2x^3 - 2x^2 + 8x - 8$
12. $P(x) = 2x^4 + 5x^3 - 7x^2 - 6x + 4$
13. $P(x) = 3x^4 - 8x^3 - 6x^2 + 17x + 6$
14. $P(x) = 12x^4 - 8x^3 - 37x^2 + 7x + 6$

Find all roots (rational, irrational, and complex) for each polynomial equation.

15. $2x^3 - 5x^2 + 1 = 0$
16. $2x^3 - 10x^2 + 12x - 4 = 0$
17. $x^4 + 4x^3 - x^2 - 20x - 20 = 0$
18. $x^4 - 4x^2 - 4x - 1 = 0$
19. $2x^5 - 3x^4 - 2x + 3 = 0$
20. $x^4 - 2x^2 - 16x - 15 = 0$

C *Write each polynomial as a product of linear factors.*

21. $P(x) = 6x^3 + 13x^2 - 4$
22. $P(x) = 6x^3 - 17x^2 - 4x + 3$
23. $P(x) = x^3 + 2x^2 - 9x - 4$
24. $P(x) = x^3 - 8x^2 + 17x - 4$

Show that each of the following real numbers is not rational by writing an appropriate polynomial and making use of Theorem 10.

25. $\sqrt{6}$ 26. $\sqrt{12}$ 27. $\sqrt[3]{5}$ 28. $\sqrt[5]{8}$

Solve each inequality using the factoring method discussed in Section 1-5. [Hint: Find the zeros of the corresponding polynomial function; then use the factor theorem.]

29. $x^2 \leq 4x - 1$
30. $x^2 > 2x + 1$
31. $2x^3 + 6 \geq 13x - x^2$
32. $5x^3 - 3x^2 < 10x - 6$

Section 3-7   Approximating Irrational Zeros

How do we find irrational zeros of polynomials of degree greater than two? If the polynomial has no rational zeros but has irrational zeros, then the rational zero theorem discussed in the last section will be of little help. Since general methods for finding zeros of third- and fourth-degree polynomials are long and involved, and no general method exists for finding zeros of polynomials of degree higher than four, we introduce the *method of successive approximation*, which will enable us to approximate irrational zeros to any decimal accuracy desired.

    Practically speaking, this method is used only to approximate irrational zeros to a couple of decimal places, since more efficient methods are available, particularly after you have had some calculus. The method of successive approximation does have the advantage, however, of being easily understood and easily remembered, and of providing a foundation for understanding the more refined methods that you are likely to encounter later. The tedious aspect of its application is reduced sharply by use of a hand calculator, even an inexpensive one.

    We outline a general strategy for finding all real zeros (rational zeros exactly and irrational zeros approximately).

> **Strategy for Finding All Real Zeros of a Polynomial P(x)**
>
> Step 1. Find all rational zeros by the methods of Section 3-6 and set them aside.
>
> Step 2. Write the final reduced polynomial $Q(x)$ resulting from step 1.
>
> Step 3. Use Descartes' rule of signs to determine the possible number of positive and negative real zeros left after step 1.
>
> Step 4. Isolate real zeros further. Form a synthetic division table using $Q(x)$ to:
>   a. Locate lower and upper bounds for irrational zeros.
>   b. Isolate irrational zeros, if possible, between successive integers by observing sign changes in $Q(x)$.
>   c. A graph of $y = Q(x)$ between lower and upper bounds may be useful.
>
> Step 5. Approximate irrational zeros (located approximately in step 4b) to desired accuracy using the *method of successive approximation* (described in Example 16).

EXAMPLE 16 Find all real zeros for $P(x) = 2x^4 + x^3 + 4x^2 - 6x - 4$. (Approximate irrational zeros to two decimal places.)

Solution *Step 1.* Find rational zeros, if any.

Using methods of the preceding section, we find $-\frac{1}{2}$ to be the only rational zero:

$$\begin{array}{r|rrrrr} & 2 & 1 & 4 & -6 & -4 \\ -\frac{1}{2} & 2 & 0 & 4 & -8 & 0 \end{array}$$

We write

$P(x) = (x + \frac{1}{2})(2x^3 + 4x - 8)$ Factor 2 out of the second factor.
$\quad\quad = 2(x + \frac{1}{2})(x^3 + 2x - 4)$

Step 2. Write the final reduced polynomial from step 1.

$Q(x) = x^3 + 2x - 4$

Step 3. Determine the possible positive and negative real zeros for $Q(x)$.

$Q(x) = x^3 + 2x - 4$ One variation in sign
$Q(-x) = -x^3 - 2x - 4$ No variation in sign

Positive real zeros: one
Negative real zeros: none

Now we know for certain that there is one positive irrational zero.

Step 4. Isolate real zeros further. We form a synthetic division table

3-7 Approximating Irrational Zeros

to try to isolate the irrational zero between two integers by observing sign changes in P(x). We start with x = 0, since the irrational zero is positive (thus, 0 is a lower bound).

$$
\begin{array}{r|rrrr}
 & 1 & 0 & 2 & -4 \\
\hline
0 & 1 & 0 & 2 & -4 \\
1 & 1 & 1 & 3 & -1 \\
2 & 1 & 2 & 6 & 8 \quad \text{2 is an upper bound}
\end{array}
$$

Real zero $\begin{cases} 1 \\ 2 \end{cases}$

From the table, we see that the irrational zero must be between 1 and 2. To obtain a clearer picture of where the zero lies in this interval, we sketch a graph of y = Q(x) for the interval [0, 2]—see figure (a). (This graphing step is optional in practice.)

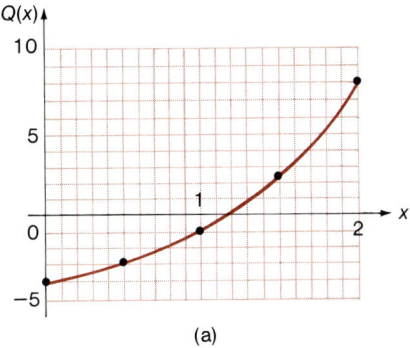

(a)

Step 5. Approximate the irrational zero.

We approximate this irrational zero to two decimal places using the *method of successive approximation*.

The Method of Successive Approximation

The first step is to divide the unit interval containing the zero into tenths. To determine the first decimal of the zero, we locate the interval in this subdivision within which P(x) changes sign. We repeat this process, dividing this subinterval into ten parts, to locate the second decimal place, and so on. The process can be repeated as long as desired (barring fatigue) to produce a decimal approximation of an irrational zero to any accuracy desired. To obtain an accuracy of two decimal places, we go to the third and round back to the second. In general, we go to one more place than the accuracy desired, and then round back one place.

The irrational zero of $Q(x)$ is between 1 and 2. We divide the interval [1, 2] into tenths, locate $(1, P(1))$ and $(2, P(2))$, and join these two points with a straight line to determine approximately where the zero of $Q(x)$ lies [figure (b)]. [Generally we draw only the straight line; figure (b) contains the approximating straight line and the actual graph of $y = Q(x)$ for comparison.]

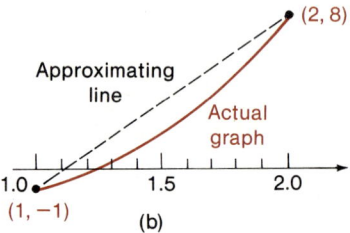

(b)

From figure (b), it appears that the zero is closer to 1 than to 2, so we start from that end. Using synthetic division and a hand calculator, we find $P(1.0)$, $P(1.1)$, $P(1.2)$, and so on, until a sign change occurs:

		1	0	2	−4	
	1.0	1	1	3	−1	
Zero {	1.1	1	1.1	3.21	−0.47	} Sign change
	1.2	1	1.2	3.44	0.13	

The zero is between 1.1 and 1.2. We now divide the interval from 1.1 to 1.2 into tenths, locate $(1.1, P(1.1))$ and $(1.2, P(1.2))$ approximately, and join these two points with a straight line to determine approximately where the graph of $y = Q(x)$ crosses the x axis [figure (c)].

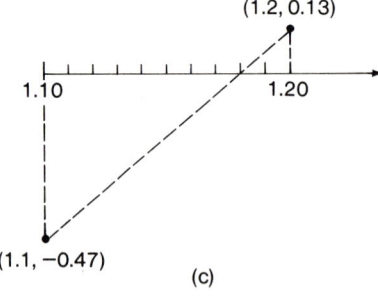

(c)

Test values close to the point of intersection to determine the sign change in $Q(x)$.

3-7 Approximating Irrational Zeros

	1	0	2	-4
1.16	1	1.16	3.346	-0.119
Zero { 1.17	1	1.17	3.369	-0.058 } Sign change
1.18	1	1.18	3.392	0.003

We now have the zero between 1.17 and 1.18. Continuing, we divide this interval into tenths and proceed as above [figure (d)].

	1	0	2	-4
1.178	1	1.178	3.3877	-0.0093
Zero { 1.179	1	1.179	3.3900	-0.0031 } Sign change
1.180	1	1.180	3.3924	0.0030

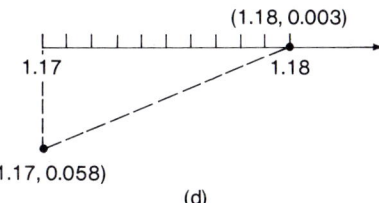

(d)

It is now clear that the irrational zero to two decimal places is 1.18, so the real zeros for the original polynomial

$$P(x) = 2x^4 + x^3 + 4x^2 - 6x - 4$$

are -0.5 and 1.18.

[Note: It is useful to observe that the method of successive approximations can be used to find real zeros for functions other than polynomial functions.]

PROBLEM 16 Find all real zeros for $P(x) = x^3 + 2x - 7$. Approximate irrational zeros to one decimal place.

Answers to Matched Problems   **16.**   1.6

Exercise 3-7

A hand calculator will prove useful in most of the problems in this exercise.

A   *Find the irrational zero to one decimal place in the indicated interval.*

1. $P(x) = x^3 - 5x^2 + 3;$ $[4, 5]$
2. $P(x) = x^3 - 5x + 3;$ $[0, 1]$
3. $P(x) = x^3 + x - 1;$ $[0, 1]$
4. $P(x) = x^3 - x^2 - x - 1;$ $[1, 2]$

B Find all real zeros of each polynomial. Approximate irrational zeros to one-decimal-place accuracy.

5. $P(x) = x^4 - 6x - 7$
6. $P(x) = x^4 - x^3 + 10x^2 - 28x + 18$
7. $P(x) = 2x^5 - 5x^4 - 7x^3 + 4x^2 + 21x + 9$
8. $P(x) = 2x^4 + 3x^3 + 6x^2 - x - 15$

C In Problems 9 and 10, isolate each irrational root between successive integers; then approximate the largest root to two decimal places.

9. $x^3 - 5x^2 + 3 = 0$ 10. $x^3 + x^2 - 6x - 2 = 0$

11. Show that even though P(2) and P(3) have the same sign, $P(x) = x^4 - 4x^3 + x^2 + 6x + 2$ has at least two real zeros between 2 and 3.

12. Approximate the largest zero of P(x) in Problem 11 in the interval [2, 3] to one decimal place.

APPLICATIONS

13. *Construction.* An open metal container is to be made from a rectangular piece of sheet metal, 11 × 9 inches, by cutting out squares of the same size from each corner and bending up the sides. If the volume of the container is to be 72 cubic inches, how large a square should be cut from each corner?

14. *Construction.* A rectangular box has dimensions of 1 × 1 × 2 feet. If each dimension is increased by the same amount, how much should this amount be to triple the volume of the box? Approximate the answer to one decimal place.

15. *Physics—engineering.* In physics it can be shown that a solid buoy in the form of a sphere, with radius r and specific gravity s, $0 < s < 1$, will sink in water to a depth of x as given by the equation

$$x^3 - 3rx^2 + 4r^3s = 0$$

How far will a plastic buoy of radius 1 foot and specific gravity $s = 0.1$ sink? Give the answer accurate to one decimal place.

Section 3-8 Partial Fraction Decomposition

- Preliminaries
- Partial Fraction Decomposition

- Preliminaries

You have now had some experience in combining two or more algebraic fractions into single fractions by addition or subtraction. For example,

3-8 Partial Fraction Decomposition

problems such as

$$\frac{2}{x+5} + \frac{3}{x-4} = \frac{5x+7}{x^2+x-20}$$

should be routine. There are several places in more advanced courses, particularly in calculus and differential equations, where it is a great advantage to be able to reverse the process—that is, to be able to express the quotient of two polynomials as the sum of two or more simpler quotients called **partial fractions**. This process of decomposing a quotient into partial fractions, like many reverse processes, is more difficult than the original.

We confine our attention to quotients of the form $P(x)/D(x)$, where $P(x)$ and $D(x)$ are real polynomials. In addition, we will assume that the degree of $P(x)$ is less than the degree of $D(x)$. If the degree of $P(x)$ is greater than or equal to that of $D(x)$, we have only to divide $P(x)$ by $D(x)$ to obtain

$$\frac{P(x)}{D(x)} = Q(x) + \frac{R(x)}{D(x)}$$

where the degree of $R(x)$ is less than that of $D(x)$. For example,

$$\frac{x^4 - 3x^3 + 2x^2 - 5x + 1}{x^2 - 2x + 1} = x^2 - x - 1 + \frac{-6x + 2}{x^2 - 2x + 1}$$

If the degree of $P(x)$ is less than that of $D(x)$, then $P(x)/D(x)$ is called a **proper fraction**. Our task now is to figure out a systematic way to decompose proper fractions into the sum of two or more partial fractions. The following three theorems take care of the problem completely. The first and third theorems are stated without proof.

THEOREM 11 | Two polynomials are equal to each other if and only if the coefficients of like-degree terms are equal.

For example, if

$$5x - 3 = (A + 2B)x + B$$

then

$$B = -3$$
$$A + 2B = 5$$
$$A + 2(-3) = 5$$
$$A = 11$$

THEOREM 12 For a polynomial with real coefficients, there always exists a complete factoring involving only prime linear and/or quadratic factors relative to the set of real numbers.

That Theorem 12 is true can be seen as follows: From earlier theorems in this chapter, we know that an nth-degree polynomial $P(x)$ has n zeros and n linear factors. If the coefficients of $P(x)$ are real, complex zeros occur in conjugate pairs. Thus, if we multiply the factors corresponding to each pair of conjugate complex zeros (when they exist), we will obtain quadratic factors with real coefficients, as can be readily seen as follows (where a and b are real numbers):

Let $a \pm bi$ be two conjugate complex zeros of $P(x)$, then $[x - (a + bi)]$ and $[x - (a - bi)]$ are two linear factors of $P(x)$. Multiplying these two factors, we have

$$[x - (a + bi)][x - (a - bi)] = x^2 - 2ax + (a^2 + b^2)$$

The quadratic is a polynomial with real coefficients.

■ Partial Fraction Decomposition

We are now ready to state Theorem 13, which forms the basis for partial fraction decompositions.

THEOREM 13 Partial Fraction Decomposition

Any reduced proper fraction $P(x)/D(x)$ can be decomposed into the sum of partial fractions as follows:

1. If $D(x)$ has a nonrepeating linear factor of the form $ax + b$, then the partial fraction decomposition of $P(x)/D(x)$ contains a term of the form

$$\frac{A}{ax + b} \qquad A \text{ a constant}$$

2. If $D(x)$ has a k-repeating linear factor of the form $(ax + b)^k$, then the partial fraction decomposition of $P(x)/D(x)$ contains terms of the form

$$\frac{A_1}{ax + b} + \frac{A_2}{(ax + b)^2} + \cdots + \frac{A_k}{(ax + b)^k} \qquad A_1, A_2, \ldots, A_k \text{ constants}$$

3-8 Partial Fraction Decomposition

---

3. If $D(x)$ has a nonrepeating quadratic factor of the form $ax^2 + bx + c$, the partial fraction decomposition of $P(x)/D(x)$ contains a term of the form

$$\frac{Ax + B}{ax^2 + bx + c} \qquad A \text{ and } B \text{ constants}$$

4. If $D(x)$ has a k-repeating quadratic factor of the form $(ax^2 + bx + c)^k$, then the partial fraction decomposition of $P(x)/D(x)$ contains terms of the form

$$\frac{A_1 x + B_1}{ax^2 + bx + c} + \frac{A_2 x + B_2}{(ax^2 + bx + c)^2} + \cdots + \frac{A_k x + B_k}{(ax^2 + bx + c)^k}$$

$$A_1, \ldots, A_k, \; B_1, \ldots, B_k \text{ constants}$$

Let us see how the theorem is used to obtain partial fraction decompositions in several examples.

EXAMPLE 17 Decompose $\dfrac{5x + 7}{x^2 + 2x - 3}$ into partial fractions.

Solution We first try to factor the denominator. If it is irreducible in the real numbers, then we will not be able to go further. In this example the denominator factors, so we apply step 1 from Theorem 13:

$$\frac{5x + 7}{(x - 1)(x + 3)} = \frac{A}{x - 1} + \frac{B}{x + 3} \tag{1}$$

To find the constants A and B we combine the right member of equation (1) to form a single fraction:

$$\frac{A(x + 3) + B(x - 1)}{(x - 1)(x + 3)}$$

and equate the numerator to $5x + 7$. Thus,

$$5x + 7 = A(x + 3) + B(x - 1) \tag{2}$$

We could multiply the right member and find A and B by using Theorem 11, but in this case it is easier to take advantage of the fact that equation (2) is an identity; that is, it must hold for all values of x. In particular, we note that if we let $x = 1$, then the second term of the right member drops out and we can solve for A.

$$5 \cdot 1 + 7 = A(1 + 3) + B(1 - 1)$$
$$12 = 4A$$
$$A = 3$$

Similarly, if we let $x = -3$, the first term will drop out and we find

$$-8 = -4B$$
$$B = 2$$

Hence,

$$\frac{5x + 7}{x^2 + 2x - 3} = \frac{3}{x - 1} + \frac{2}{x + 3}$$

as can easily be checked.

PROBLEM 17 Decompose $\dfrac{7x + 6}{x^2 + x - 6}$ into partial fractions.

EXAMPLE 18 Decompose $\dfrac{6x^2 - 14x - 27}{(x + 2)(x - 3)^2}$ into partial fractions.

Solution Using steps 1 and 2 from Theorem 13, we write

$$\frac{6x^2 - 14x - 27}{(x + 2)(x - 3)^2} = \frac{A}{x + 2} + \frac{B}{x - 3} + \frac{C}{(x - 3)^2}$$

$$= \frac{A(x - 3)^2 + B(x + 2)(x - 3) + C(x + 2)}{(x + 2)(x - 3)^2}$$

Thus for all x,

$$6x^2 - 14x - 27 = A(x - 3)^2 + B(x + 2)(x - 3) + C(x + 2)$$

If $x = 3$, then

$$-15 = 5C$$
$$C = -3$$

If $x = -2$, then

$$25 = 25A$$
$$A = 1$$

If $x = 0$, then

$$-27 = 9 - 6B - 6$$
$$B = 5$$

Thus,

$$\frac{6x^2 - 14x - 27}{(x + 2)(x - 3)^2} = \frac{1}{x + 2} + \frac{5}{x - 3} - \frac{3}{(x - 3)^2}$$

PROBLEM 18 Decompose $\dfrac{x^2 + 11x + 15}{(x - 1)(x + 2)^2}$ into partial fractions.

3-8 Partial Fraction Decomposition

EXAMPLE 19 Decompose $\dfrac{5x^2 - 8x + 5}{(x-2)(x^2 - x + 1)}$ into partial fractions.

Solution First we see that the quadratic in the denominator is irreducible in the real numbers and then use steps 1 and 3 from Theorem 13 to write

$$\frac{5x^2 - 8x + 5}{(x-2)(x^2 - x + 1)} = \frac{A}{x - 2} + \frac{Bx + C}{x^2 - x + 1}$$

$$= \frac{A(x^2 - x + 1) + (Bx + C)(x - 2)}{(x-2)(x^2 - x + 1)}$$

Thus, for all x,

$$5x^2 - 8x + 5 = A(x^2 - x + 1) + (Bx + C)(x - 2)$$

If $x = 2$, then

$$9 = 3A$$
$$A = 3$$

If $x = 0$, then

$$5 = 3 - 2C$$
$$C = -1$$

If $x = 1$, then

$$2 = 3 + (B - 1)(-1)$$
$$B = 2$$

Hence,

$$\frac{5x^2 - 8x + 5}{(x-2)(x^2 - x + 1)} = \frac{3}{x - 2} + \frac{2x - 1}{x^2 - x + 1}$$

PROBLEM 19 Decompose $\dfrac{7x^2 - 11x + 6}{(x-1)(2x^2 - 3x + 2)}$ into partial fractions.

EXAMPLE 20 Decompose $\dfrac{x^3 - 4x^2 + 9x - 5}{(x^2 - 2x + 3)^2}$ into partial fractions.

Solution Since $x^2 - 2x + 3$ is irreducible in the real numbers, we proceed to use step 4 from Theorem 13 to write

$$\frac{x^3 - 4x^2 + 9x - 5}{(x^2 - 2x + 3)^2} = \frac{Ax + B}{x^2 - 2x + 3} + \frac{Cx + D}{(x^2 - 2x + 3)^2}$$

$$= \frac{(Ax + B)(x^2 - 2x + 3) + Cx + D}{(x^2 - 2x + 3)^2}$$

Thus, for all x,

$$x^3 - 4x^2 + 9x - 5 = (Ax + B)(x^2 - 2x + 3) + Cx + D$$

Multiplying out and rearranging the right member, we obtain

$$x^3 - 4x^2 + 9x - 5 = Ax^3 + (B - 2A)x^2 + (3A - 2B + C)x + (3B + D)$$

Now we use Theorem 11 to equate coefficients of like-powered terms.

$$A = 1$$
$$B - 2A = -4$$
$$3A - 2B + C = 9$$
$$3B + D = -5$$

From these equations we easily find that $A = 1$, $B = -2$, $C = 2$, and $D = 1$. And now we can write

$$\frac{x^3 - 4x^2 + 9x - 5}{(x^2 - 2x + 3)^2} = \frac{x - 2}{x^2 - 2x + 3} + \frac{2x + 1}{(x^2 - 2x + 3)^2}$$

PROBLEM 20 Decompose $\dfrac{3x^3 - 6x^2 + 7x - 2}{(x^2 - 2x + 2)^2}$ into partial fractions.

It should be clear that one of the key problems in decomposing quotients of polynomials into partial fractions is factoring the denominator into linear and quadratic factors with real coefficients. The material in the earlier parts of this chapter can be put to effective use in this regard.

Answers to Matched Problems

17. $\dfrac{4}{x - 2} + \dfrac{3}{x + 3}$

18. $\dfrac{3}{x - 1} - \dfrac{2}{x + 2} + \dfrac{1}{(x + 2)^2}$

19. $\dfrac{2}{x - 1} + \dfrac{3x - 2}{2x^2 - 3x + 2}$

20. $\dfrac{3x}{x^2 - 2x + 2} + \dfrac{x - 2}{(x^2 - 2x + 2)^2}$

Exercise 3-8 ■ **A** Find constants A, B, C, and D so that the right member is equal to the left.

1. $\dfrac{7x - 14}{(x - 4)(x + 3)} = \dfrac{A}{x - 4} + \dfrac{B}{x + 3}$

2. $\dfrac{9x + 21}{(x + 5)(x - 3)} = \dfrac{A}{x + 5} + \dfrac{B}{x - 3}$

3. $\dfrac{17x - 1}{(2x - 3)(3x - 1)} = \dfrac{A}{2x - 3} + \dfrac{B}{3x - 1}$

4. $\dfrac{x - 11}{(3x + 2)(2x - 1)} = \dfrac{A}{3x + 2} + \dfrac{B}{2x - 1}$

3-8 Partial Fraction Decomposition

5. $\dfrac{3x^2 + 7x + 1}{x(x+1)^2} = \dfrac{A}{x} + \dfrac{B}{x+1} + \dfrac{C}{(x+1)^2}$

6. $\dfrac{x^2 - 6x + 11}{(x+1)(x-2)^2} = \dfrac{A}{x+1} + \dfrac{B}{x-2} + \dfrac{C}{(x-2)^2}$

7. $\dfrac{3x^2 + x}{(x-2)(x^2+3)} = \dfrac{A}{x-2} + \dfrac{Bx+C}{x^2+3}$

8. $\dfrac{5x^2 - 9x + 19}{(x-4)(x^2+5)} = \dfrac{A}{x-4} + \dfrac{Bx+C}{x^2+5}$

9. $\dfrac{2x^2 + 4x - 1}{(x^2+x+1)^2} = \dfrac{Ax+B}{x^2+x+1} + \dfrac{Cx+D}{(x^2+x+1)^2}$

10. $\dfrac{3x^3 - 3x^2 + 10x - 4}{(x^2-x+3)^2} = \dfrac{Ax+B}{x^2-x+3} + \dfrac{Cx+D}{(x^2-x+3)^2}$

B *Decompose into partial fractions.*

11. $\dfrac{-x+22}{x^2-2x-8}$

12. $\dfrac{-x-21}{x^2+2x-15}$

13. $\dfrac{3x-13}{6x^2-x-12}$

14. $\dfrac{11x-11}{6x^2+7x-3}$

15. $\dfrac{x^2-12x+18}{x^3-6x^2+9x}$

16. $\dfrac{5x^2-36x+48}{x(x-4)^2}$

17. $\dfrac{5x^2+3x+6}{x^3+2x^2+3x}$

18. $\dfrac{6x^2-15x+16}{x^3-3x^2+4x}$

19. $\dfrac{2x^3+7x+5}{x^4+4x^2+4}$

20. $\dfrac{-5x^2+7x-18}{x^4+6x^2+9}$

21. $\dfrac{x^3-7x^2+17x-17}{x^2-5x+6}$

22. $\dfrac{x^3+x^2-13x+11}{x^2+2x-15}$

C 23. $\dfrac{4x^2+5x-9}{x^3-6x-9}$

24. $\dfrac{4x^2-8x+1}{x^3-x+6}$

25. $\dfrac{x^2+16x+18}{x^3+2x^2-15x-36}$

26. $\dfrac{5x^2-18x+1}{x^3-x^2-8x+12}$

27. $\dfrac{-x^2+x-7}{x^4-5x^3+9x^2-8x+4}$

28. $\dfrac{-2x^3+12x^2-20x-10}{x^4-7x^3+17x^2-21x+18}$

29. $\dfrac{4x^5+12x^4-x^3+7x^2-4x+2}{4x^4+4x^3-5x^2+5x-2}$

30. $\dfrac{6x^5-11x^4+x^3-10x^2-2x-2}{6x^4-7x^3+x^2+x-1}$

Section 3-9 Chapter Review

IMPORTANT TERMS AND SYMBOLS

3-1 Introduction. Zero of a function P, zero of a polynomial P(x), solution or root of the equation $P(x) = 0$

3-2 Synthetic Division. Algebraic long division, synthetic division.

3-3 Remainder and Factor Theorems. Division algorithm, remainder theorem, graphing polynomials, factor theorem

3-4 Fundamental Theorem of Algebra. Fundamental theorem of algebra, n zeros theorem, complex zeros

3-5 Isolating Real Zeros. Variation in sign, Descartes' rule of signs, upper and lower bounds of real zeros, sign changes in $P(x)$

3-6 Finding Rational Zeros. Rational zero theorem, strategy for finding rational zeros

3-7 Approximating Irrational Zeros. Strategy for finding all real zeros of a polynomial $P(x)$ with real coefficients, method of successive approximation

3-8 Partial Fraction Decomposition. Partial fractions, partial fraction decomposition

Exercise 3-9 Chapter Review

Work through all the problems in this chapter review and check answers in the back of the book. (Answers to all problems are there, and following each answer is a number in italics indicating the section in which that type of problem is discussed.) Where weaknesses show up, review appropriate sections in the text. When you are satisfied that you know the material, take the practice test following this review.

A
1. Use synthetic division to divide $P(x) = 2x^3 + 3x^2 - 1$ by $D(x) = x + 2$, and write the answer in the form $P(x) = D(x)Q(x) + R$.
2. If $P(x) = x^5 - 4x^4 + 9x^2 - 8$, find $P(3)$ using the remainder theorem and synthetic division.
3. What are the zeros of $P(x) = 3(x - 2)(x + 4)(x + 1)$?
4. If $P(x) = x^2 - 2x + 2$ and $P(1 + i) = 0$, find another zero of $P(x)$.
5. Using Descartes' rule of signs, what can you say about the number of positive and negative zeros of
    (A) $P(x) = x^3 - x^2 - x + 3$ (B) $P(x) = x^5 + x^3 + 4$
6. According to the upper and lower bound theorem in this chapter, which of the following are upper or lower bounds of zeros of $P(x) = x^3 - 4x^2 + 2$: $-2, -1, 3, 4$?

7. How do you know that $P(x) = 2x^3 - 3x^2 + x - 5$ has at least one real zero between 1 and 2?
8. Write the possible rational zeros for $P(x) = x^3 - 4x^2 + x + 6$.
9. Find all rational zeros for $P(x) = x^3 - 4x^2 + x + 6$.
10. Decompose $(7x - 11)/(x - 3)(x + 2)$ into partial fractions.

B 11. Use synthetic division to divide $P(x) = 3x^3 + 4x^2 - 7x - 3$ by $x - \frac{2}{3}$, and write the answer in the form $P(x) = D(x)Q(x) + R$.
12. If $P(x) = 4x^3 - 8x^2 - 3x - 3$, find $P(-\frac{1}{2})$ using the remainder theorem and synthetic division.
13. Use the quadratic formula and the factor theorem to factor $P(x) = x^2 - 2x - 1$.
14. Is $x + 1$ a factor of $P(x) = x^{25} + 1$? Explain without dividing or using synthetic division.
15. For $P(x) = 2x^4 - 3x^3 - 14x^2 + 2x + 4$,
    (A) Using Descartes' rule of signs, discuss the possible number of real zeros.
    (B) Find the smallest and largest integers that are, respectively, upper and lower bounds of zeros of $P(x)$ according to Theorem 8.
    (C) Discuss the location of real zeros within the lower and upper bound interval.
16. Determine all rational zeros of $P(x) = 2x^3 - 3x^2 - 18x - 8$.
17. Factor the polynomial in Problem 16 into linear factors.
18. Find all rational zeros of $P(x) = x^3 - 3x^2 + 5$.
19. Find all zeros (rational, irrational, and complex) for $P(x) = 2x^3 - 3x^2 + 3x - 1$.
20. Factor the polynomial in Problem 19 into linear factors.
21. Find the real zero of $P(x) = x^4 - x^2 - 2$ between 1 and 2 to one-decimal-place accuracy.
22. Decompose $\dfrac{-x^2 + 3x + 4}{x(x - 2)^2}$ into partial fractions.
23. Decompose $\dfrac{8x^2 - 10x + 9}{2x^3 - 3x^2 + 3x}$ into partial fractions.

C 24. Use synthetic division to divide $P(x) = x^3 + 3x + 2$ by $[x - (1 + i)]$, and write the answer in the form $P(x) = D(x)Q(x) + R$.
25. Find a polynomial of lowest degree with leading coefficient 1 that has zeros $-\frac{1}{2}$ (multiplicity 2), -3, and 1 (multiplicity 3). (Leave the answer in factored form.) What is the degree of the polynomial?
26. Repeat Problem 25 for a polynomial $P(x)$ with zeros -5, $2 - 3i$, and $2 + 3i$.

27. Find all real roots of $2x^4 - x^3 - 12x^2 - 14x + 10 = 0$ (irrational roots to one-decimal-place accuracy).

28. Decompose $\dfrac{5x^2 + 2x + 9}{x^4 - 3x^3 + x^2 - 3x}$ into partial fractions.

Practice Test Chapter 3

Take this practice test as if it were a graded test. Allow yourself up to 50 minutes. Work the problems without looking back in the chapter. Correct your work using the answers (keyed to appropriate sections) in the back of the book.

1. If $P(x) = 8x^4 - 14x^3 - 13x^2 - 4x + 7$, find $Q(x)$ and R such that $P(x) = (x - \frac{1}{4})Q(x) + R$. What is $P(\frac{1}{4})$?

2. Is $x + 1$ a factor of $P(x) = 9x^{26} - 11x^{17} + 8x^{11} - 5x^4 - 7$? Explain.

Problems 3–6 refer to $P(x) = 2x^3 - 7x^2 + 2x + 6$.

3. (A) List all possible rational zeros of $P(x)$.
   (B) Discuss the possible number of positive and negative real zeros of $P(x)$ using Descartes' rule of signs.

4. (A) Find all intervals of the form $[a, b]$, where a and b are successive integers, that contain at least one real zero of $P(x)$.
   (B) Find the smallest and largest integers that are, respectively, lower and upper bounds of zeros for $P(x)$.

5. Find all zeros of $P(x)$.

6. Factor $P(x)$ as a product of first-degree factors.

7. How do we know that $x^3 + 3x - 5 = 0$ has exactly one real root and that root is positive?

8. Find the real root of the equation in Problem 7 to one decimal place.

9. Decompose $\dfrac{x^2 - 2x + 10}{(x + 2)(x - 1)^2}$ into partial fractions.

10. Decompose $\dfrac{3x^2 + 2x + 4}{(x + 1)(x^2 + 4)}$ into partial fractions.

Exponential and Logarithmic Functions   4

4-1 Exponential Functions
4-2 Logarithmic Functions
4-3 Properties of Logarithmic Functions
4-4 Logarithms to Various Bases
4-5 Exponential and Logarithmic Equations
4-6 Chapter Review

A natural design of mathematical interest. Can you guess the source? See the back of the book.

Chapter 4 ▪ Exponential and Logarithmic Functions

Most of the functions we have considered have been **algebraic functions**—that is, functions defined by means of the basic algebraic operations on variables and constants. In this chapter we will define and investigate the properties of two new and important classes of functions: exponential and logarithmic functions.

Section 4-1 Exponential Functions

- Exponential Functions
- Graphing an Exponential Function
- Typical Exponential Graphs
- Base e
- Basic Exponential Properties

■ **Exponential Functions**

In this and the next section we will consider two new kinds of functions that use variable exponents in their definitions. To start, note that

$$f(x) = 2^x \quad \text{and} \quad g(x) = x^2$$

are not the same function. The function g is a quadratic function, which we have already discussed; the function f is a new function called an *exponential function*. An **exponential function** is a function defined by an equation of the form:

Exponential Function

$$f(x) = b^x \qquad b > 0, \quad b \neq 1$$
---

where b is a constant, called the **base**, and the exponent x is a variable. The replacement set for the exponent, the **domain of f**, is the set of real numbers R. The **range of f** is the set of positive real numbers. We require b to be positive to avoid complex numbers such as $(-2)^{1/2}$.

- Graphing an Exponential Function

Many students, if asked to graph an exponential function such as $f(x) = 2^x$, would not hesitate at all. They would likely make up a table by assigning integers to x, plot the resulting points, and then join these points with a smooth curve (Fig. 1). The only catch is that 2^x has not

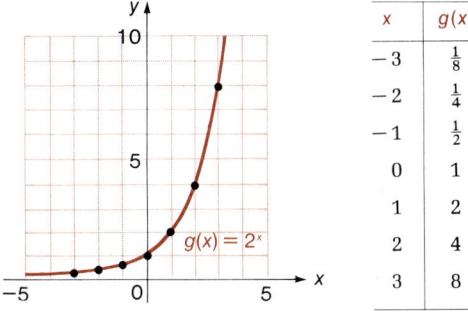

FIGURE 1

been defined at this point for all real numbers. We know what 2^5, 2^{-3}, $2^{2/3}$, $2^{-3/5}$, $2^{1.4}$, and $2^{-3.15}$ all mean (that is, 2^p, where p is a rational number), but what does

$$2^{\sqrt{2}}$$

mean? The question is not easy to answer at this time. In fact, a precise definition of $2^{\sqrt{2}}$ must wait for more advanced courses, where we can show that

$$b^x$$

names a real number for b a positive real number and x any real number, and that the graph of $g(x) = 2^x$ is as indicated in Figure 1. We can also show that for x irrational, b^x can be approximated as closely as we like by using rational number approximations for x. Since $\sqrt{2} = 1.414213\ldots$, for example, the sequence

$$2^{1.4}, 2^{1.41}, 2^{1.414}, \ldots$$

approximates $2^{\sqrt{2}}$, and as we move to the right the approximation improves.

■ **Typical Exponential Graphs**

It is useful to compare the graphs of $y = 2^x$ and $y = (\frac{1}{2})^x = 2^{-x}$ by plotting both on the same coordinate system [Fig. 2(a)]. The graph of

$$f(x) = b^x \qquad b > 1 \text{ [Fig. 2(b)]}$$

will look very much like the graph of $y = 2^x$, and the graph of

$$f(x) = b^x \qquad 0 < b < 1 \text{ [Fig. 2(b)]}$$

will look very much like the graph of $y = (\frac{1}{2})^x$. Note in both cases that the x axis is a horizontal asymptote and the graphs will never touch it.

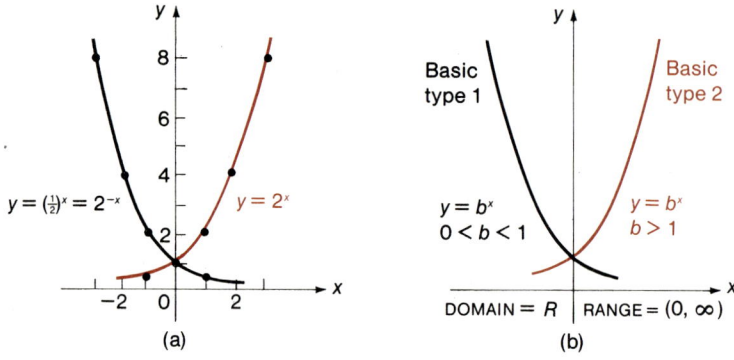

FIGURE 2

[*Note:* An exponential function is either increasing or decreasing, and hence is one to one and has an inverse that is a function. This fact will be important to us in the next section when we define a logarithmic function as an inverse of an exponential function.]

EXAMPLE 1 Graph $y = \frac{1}{2} 4^x$ for $-3 \leq x \leq 3$.

Solution

x	y
−3	0.01
−2	0.03
−1	0.13
0	0.50
1	2.00
2	8.00
3	32.00

A: $\boxed{4}\ \boxed{y^x}\ \boxed{3}\ \boxed{+/-}\ \boxed{=}\ \boxed{\div}\ \boxed{2}\ \boxed{=}$

P: $\boxed{4}\ \boxed{\text{ENTER}}\ \boxed{3}\ \boxed{+/-}\ \boxed{y^x}\ \boxed{2}\ \boxed{\div}$

4-1 Exponential Functions

PROBLEM 1   Graph $y = \frac{1}{2}4^{-x}$ for $-3 \leq x \leq 3$.

Exponential functions are often referred to as *growth functions* because of their widespread use in describing different kinds of growth phenomena. These functions are used to describe population growth of people, animals, and bacteria; radioactive decay (negative growth); growth of a new chemical substance in a chemical reaction; increase or decline in the temperature of a substance being heated or cooled; growth of money at compound interest; light absorption (negative growth) as it passes through air, water, or glass; decline of atmospheric pressure as altitude is increased; and growth of learning a skill such as swimming or typing relative to practice.

■ **Base e**

For introductory purposes, the bases 2 and $\frac{1}{2}$ were convenient choices; however, a certain irrational number, denoted by e, is by far the most frequently used exponential base for both theoretical and practical purposes. In fact,

$$f(x) = e^x$$

is often referred to as *the* exponential function because of its widespread use. The reasons for the preference for e as a base are made clear in more advanced courses. And at that time, it is shown that e is approximated by $(1 + 1/n)^n$ to any decimal accuracy desired by making n (an integer) sufficiently large. The irrational number e to eight decimal places is

$$e \approx 2.718\ 281\ 83$$

Similarly, e^x can be approximated by using $(1 + 1/n)^{nx}$ for sufficiently large n. Because of the importance of e^x and e^{-x} tables for their evaluation are readily available. In fact, many hand calculators can evaluate these functions directly. A short table (Table I) for e^x and e^{-x} can be found in the back of this book for those not using a calculator.

The important constant e along with two other important constants $\sqrt{2}$ and π are shown on the number line in Figure 3.

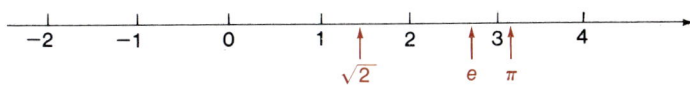

FIGURE 3

EXAMPLE 2   Graph $y = 10e^{-0.5x}$, $-3 \leq x \leq 3$, using a hand calculator or Table I.

4 Exponential and Logarithmic Functions

Solution

x	y
−3	44.82
−2	27.18
−1	16.49
0	10.00
1	6.07
2	3.68
3	2.23

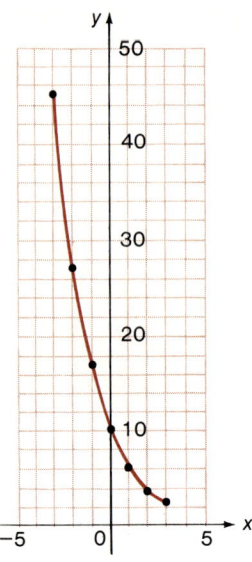

A: $\boxed{3}\,\boxed{+/-}\,\boxed{\times}\,\boxed{0.5}\,\boxed{+/-}\,\boxed{=}\,\boxed{e^x}\,\boxed{\times}\,\boxed{10}\,\boxed{=}$

P: $\boxed{3}\,\boxed{+/-}\,\boxed{\text{ENTER}}\,\boxed{0.5}\,\boxed{+/-}\,\boxed{\times}\,\boxed{e^x}\,\boxed{10}\,\boxed{\times}$

PROBLEM 2 Graph $y = 10e^{0.5x}$, $-3 \leq x \leq 3$, using a hand calculator or Table I.

EXAMPLE 3 If $P are invested at 100r% compounded continuously, then the amount A in the account at the end of t years is given by (from mathematics of finance):

$$A = Pe^{rt}$$

If $100 is invested at 12% compounded continuously, graph the amount in the account relative to time for a period of 10 years.

Solution We wish to graph

$$A = 100e^{0.12t} \qquad 0 \leq t \leq 10$$

We make up a table of values using a calculator or Table I, graph the points from the table, and join the points with a smooth curve. The table and graph are shown at the top of the next page.

PROBLEM 3 Repeat Example 3 with $5,000 being invested at 20% compounded continuously.

4-1 Exponential Functions

t	A
0	100
1	113
2	127
3	143
4	162
5	182
6	205
7	232
8	261
9	294
10	332

■ **Basic Exponential Properties**

It can be shown that the five laws for rational exponents (Appendix A-3) hold for irrational exponents. Thus, we now assume that all five laws of exponents hold for any real exponents as long as the bases involved are positive.

As a consequence of exponential functions being either increasing or decreasing and thus one to one, we have:

$$b^m = b^n \quad \text{if and only if} \quad m = n, \, b > 0, \, b \neq 1$$

Thus, if $2^{15} = 2^{3x}$, then $3x = 15$ and $x = 5$.

Answers to Matched Problems

1. $y = \frac{1}{2} 4^{-x}$

x	y
-3	32.00
-2	8.00
-1	2.00
0	0.50
1	0.13
2	0.03
3	0.01

2. $y = 10e^{0.5x}$

x	y
−3	2.23
−2	3.68
−1	6.07
0	10.00
1	16.49
2	27.18
3	44.82

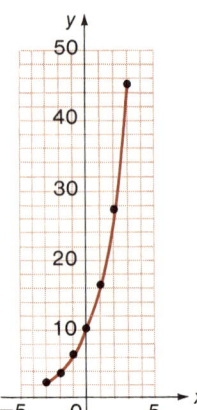

3. $A = 5{,}000e^{0.2t}$

t	A
0	5,000
1	6,107
2	7,459
3	9,111
4	11,128
5	13,591
6	16,601
7	20,276
8	24,765
9	30,248
10	36,945

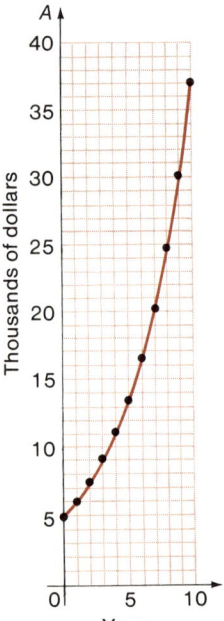

Exercise 4-1 ■ A

Graph each exponential function for $-3 \leq x \leq 3$. Plot points using integers for x, and join the points with a smooth curve.

1. $y = 3^x$
2. $y = 2^x$
3. $y = (\frac{1}{3})^x = 3^{-x}$
4. $y = (\frac{1}{2})^x = 2^{-x}$
5. $y = 5 \cdot 3^x$ [Note: $4 \cdot 3^x \neq 12^x$]

4-1 Exponential Functions

6. $y = 5 \cdot 2^x$

B **7.** $y = 2^{x+3}$ **8.** $y = 3^{x+1}$

9. $y = 7(\frac{1}{2})^{2x} = 7 \cdot 2^{-2x}$ **10.** $y = 11 \cdot 2^{-2x}$

Graph Problems 11–14 for $-3 \leq x \leq 3$. Use a calculator or Table I.

11. $y = e^x$ **12.** $y = e^{-x}$

13. $y = 10e^{-0.12x}$ **14.** $y = 100e^{0.25x}$

C **15.** Graph $y = 10 \cdot 2^{-x^2}$ for $-2 \leq x \leq 2$.

16. Graph $y = e^{-x^2}$ for $x = -1.5, -1.0, -0.5, 0, 0.5, 1.0, 1.5$, and join these points with a smooth curve.

17. Graph $y = y_0 2^x$, where y_0 is the value of y when $x = 0$. (Express the vertical scale in terms of y_0.)

18. Graph $y = y_0 e^{-0.22x}$, where y_0 is the value of y when $x = 0$. (Express the vertical scale in terms of y_0.)

19. Graph $y = 2^x$ and $x = 2^y$ on the same coordinate system.

20. Graph $f(x) = 10^x$ and $y = f^{-1}(x)$ on the same coordinate system.

APPLICATIONS **21.** If we start with 2¢ and double the amount each day, at the end of n days we will have 2^n¢. Graph $f(n) = 2^n$ for $1 \leq n \leq 10$. (Pick the scale on the vertical axis so that the graph will not go off the paper.)

22. *Compound interest.* If a certain amount of money P, called the principal, is invested at $100r\%$ interest compounded annually, the amount of money A after t years is given by

$$A = P(1 + r)^t$$

Graph this equation for $P = \$10$, $r = 0.10$, and $0 \leq t \leq 10$.

23. *Earth science.* The atmospheric pressure P, in pounds per square inch, can be calculated approximately using the formula

$$P = 14.7e^{-0.21x}$$

where x is altitude relative to sea level in miles. Graph the equation for $-1 \leq x \leq 5$.

24. *Bacterial growth.* If bacteria in a certain culture double every hour, write a formula that gives the number of bacteria N in the culture after n hours, assuming the culture has N_0 bacteria to start with.

25. *Radioactive decay.* Radioactive strontium-90 has a half-life of 28 years; that is, in 28 years one-half of any amount of strontium-90 will change to another substance because of radioactive decay. If we place a bar containing 100 milligrams of strontium-90 in a nuclear reactor, the amount of strontium-90 that will be left after

t years is given by $A = 100(\frac{1}{2})^{t/28}$. Graph this exponential function for $t = 0$, 28, 2(28), 3(28), 4(28), 5(28), and 6(28), and join these points with a smooth curve.

26. *Radioactive decay.* Radioactive argon-39 has a half-life of 4 minutes; that is, in 4 minutes one-half of any amount of argon-39 will change to another substance because of radioactive decay. If we start with A_0 milligrams of argon-39, the amount left after t minutes is given by $A = A_0(\frac{1}{2})^{t/4}$. Graph this exponential function for $A_0 = 100$ and $t = 0$, 4, 8, 12, 16, and 20, and join these points with a smooth curve.

27. *Sociology—small-group analysis.* Sociologists Stephan and Mischler found that, when the members of a discussion group of ten were ranked according to the number of times each participated, the number of times $N(i)$ the ith-ranked person participated was given approximately by the exponential function

$$N(i) = N_1 e^{-0.11(i-1)} \qquad 1 \leq i \leq 10$$

where N_1 is the number of times the top-ranked person participated in the discussion. Graph the exponential function using $N_1 = 100$.

Section 4-2 Logarithmic Functions

- Logarithmic Functions
- From Logarithmic to Exponential and Vice Versa
- Finding x, b, or y in $y = \log_b x$
- Logarithmic–Exponential Identities

- **Logarithmic Functions**

We now define a new class of functions, called **logarithmic functions**, as inverses of exponential functions. (Since exponential functions are one to one, their inverses are functions.) Here you will see why we placed special emphasis on the general concept of inverse functions in Section 2-7. If you know quite a bit about a function, then (knowing about inverses in general) you will automatically know quite a bit about its inverse. For example, the graph of f^{-1} is the graph of f reflected across the line $y = x$, and the domain and range of f^{-1} are, respectively, the range and domain of f.

If we start with the exponential function

$f: \quad y = 2^x$

4-2 Logarithmic Functions

and interchange the variables x and y, we obtain the inverse of f:

f^{-1}: $x = 2^y$

The graphs of f and f^{-1} (along with $y = x$) are shown in Figure 4. This new function is given the name **logarithmic function with base 2**, and is symbolized as follows (since we cannot "algebraically" solve $x = 2^y$ for y):

$y = \log_2 x$

Thus,

$y = \log_2 x$ is equivalent to $x = 2^y$

that is, $\log_2 x$ is the power to which 2 must be raised to obtain x. (Symbolically, $x = 2^y = 2^{\log_2 x}$.)

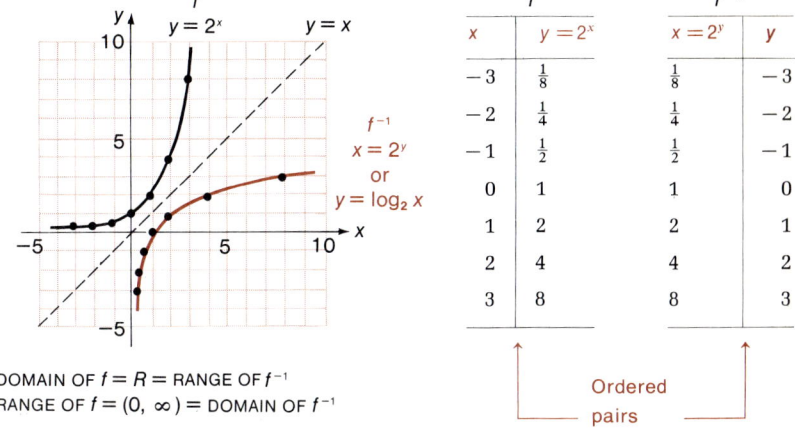

FIGURE 4
DOMAIN OF $f = R$ = RANGE OF f^{-1}
RANGE OF $f = (0, \infty)$ = DOMAIN OF f^{-1}

In general, we define the **logarithmic function with base b** to be the inverse of the exponential function with base b ($b > 0$, $b \neq 1$).

Definition of Logarithmic Function

For $b > 0$ and $b \neq 1$,

$y = \log_b x$ is equivalent to $x = b^y$

(The log to the base b of x is the power to which b must be raised to obtain x.)

$y = \log_{10} x$ is equivalent to $x = 10^y$

$y = \log_e x$ is equivalent to $x = e^y$

It is very important to remember that $y = \log_b x$ and $x = b^y$ define the same function, and as such can be used interchangeably.

Since the domain of an exponential function includes all real numbers and its range is the set of positive real numbers, the **domain** of a logarithmic function is the set of all positive real numbers and its **range** is the set of all real numbers. Thus, $\log_{10} 3$ is defined, but $\log_{10} 0$ and $\log_{10}(-5)$ are not defined (3 is a logarithmic domain value, but 0 and -5 are not). Typical logarithmic curves are shown in Figure 5.

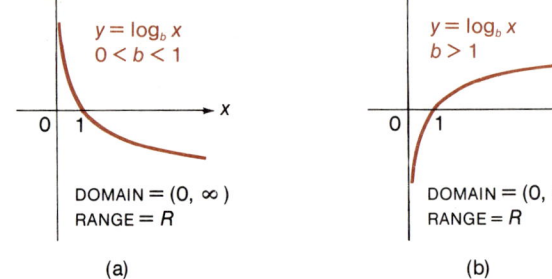

FIGURE 5 Typical logarithmic graphs

■ **From Logarithmic to Exponential and Vice Versa**

We now look into the matter of converting logarithmic forms to equivalent exponential forms, and vice versa.

EXAMPLE 4 From logarithmic form to exponential form:

(A) $\log_2 8 = 3$ is equivalent to $8 = 2^3$
(B) $\log_{25} 5 = \frac{1}{2}$ is equivalent to $5 = 25^{1/2}$
(C) $\log_2 \frac{1}{4} = -2$ is equivalent to $\frac{1}{4} = 2^{-2}$

PROBLEM 4 Change to equivalent exponential form.

(A) $\log_3 27 = 3$ (B) $\log_{36} 6 = \frac{1}{2}$ (C) $\log_3 (\frac{1}{9}) = -2$

EXAMPLE 5 From exponential form to logarithmic form:

(A) $49 = 7^2$ is equivalent to $\log_7 49 = 2$
(B) $3 = \sqrt{9}$ is equivalent to $\log_9 3 = \frac{1}{2}$
(C) $\frac{1}{5} = 5^{-1}$ is equivalent to $\log_5 (\frac{1}{5}) = -1$

PROBLEM 5 Change to equivalent logarithmic form.

(A) $64 = 4^3$ (B) $2 = \sqrt[3]{8}$ (C) $\frac{1}{16} = 4^{-2}$

4-2 Logarithmic Functions

■ **Finding x, b, or y in $y = \log_b x$**

To gain a little deeper understanding of logarithmic functions and their relationship to the exponential functions, we will look at a few problems where one is to find x, b, or y in $y = \log_b x$, given the other two values. All values were chosen so that the problems can be solved without tables or a calculator.

EXAMPLE 6 Find x, b, or y as indicated.

(A) Find y: $y = \log_4 8$.

Solution   Write $y = \log_4 8$ in equivalent exponential form:

$8 = 4^y$ *Write each number to the same base 2.*
$2^3 = 2^{2y}$ *Recall $b^m = b^n$ if and only if $m = n$.*
$2y = 3$
$y = \frac{3}{2}$

Thus, $\frac{3}{2} = \log_4 8$.

(B) Find x: $\log_3 x = -2$.

Solution   Write $\log_3 x = -2$ in equivalent exponential form:

$x = 3^{-2}$

$x = \frac{1}{3^2} = \frac{1}{9}$

Thus, $\log_3 (\frac{1}{9}) = -2$.

(C) Find b: $\log_b 1{,}000 = 3$.

Solution   Write $\log_b 1{,}000 = 3$ in equivalent exponential form:

$1{,}000 = b^3$ *Write 1,000 as a third power.*
$10^3 = b^3$
$b = 10$

Thus, $\log_{10} 1{,}000 = 3$.

PROBLEM 6   Find x, b, or y as indicated.

(A) $y = \log_9 27$ (B) $\log_2 x = -3$ (C) $\log_b 100 = 2$

■ **Logarithmic–Exponential Identities**

Recall from Section 2-7 that if f and f^{-1} are both functions (that is, if f is one to one), then

$f^{-1}[f(x)] = x$ and $f[f^{-1}(x)] = x$

Applying these general properties to $f(x) = b^x$ and $f^{-1}(x) = \log_b x$, we see that

$$f^{-1}[f(x)] = x \qquad f[f^{-1}(x)] = x$$
$$\log_b [f(x)] = x \qquad b^{f^{-1}(x)} = x$$
$$\log_b b^x = x \qquad b^{\log_b x} = x$$

Thus, we have the useful logarithmic–exponential identities:

Logarithmic–Exponential Identities

For $b > 0$, $b \neq 1$,

1. $\log_b b^x = x$
2. $b^{\log_b x} = x \qquad x > 0$

EXAMPLE 7
(A) $\log_{10} 10^5 = 5$
(B) $\log_{10} 0.01 = \log_{10} 10^{-2} = -2$
(C) $\log_e e^{2x+1} = 2x + 1$
(D) $\log_4 1 = \log_4 4^0 = 0$
(E) $10^{\log_{10} 7} = 7$
(F) $e^{\log_e x^2} = x^2$

PROBLEM 7 Find each of the following:

(A) $\log_{10} 10^{-5}$
(B) $\log_5 25$
(C) $\log_{10} 1$
(D) $\log_e e^{m+n}$
(E) $10^{\log_{10} 4}$
(F) $e^{\log_e (x^4 + 1)}$

Answers to Matched Problems

4. (A) $27 = 3^3$ (B) $6 = 36^{1/2}$ (C) $\frac{1}{9} = 3^{-2}$
5. (A) $\log_4 64 = 3$ (B) $\log_8 2 = \frac{1}{3}$ (C) $\log_4 (\frac{1}{16}) = -2$
6. (A) $y = \frac{3}{2}$ (B) $x = \frac{1}{8}$ (C) $b = 10$
7. (A) -5 (B) 2 (C) 0 (D) $m + n$ (E) 4
   (F) $x^4 + 1$

Exercise 4-2 ■ A

Rewrite in equivalent exponential form.

1. $\log_3 9 = 2$
2. $\log_2 4 = 2$
3. $\log_3 81 = 4$
4. $\log_5 125 = 3$
5. $\log_{10} 1{,}000 = 3$
6. $\log_{10} 100 = 2$
7. $\log_e 1 = 0$
8. $\log_8 1 = 0$

Rewrite in equivalent logarithmic form.

9. $64 = 8^2$
10. $25 = 5^2$
11. $10{,}000 = 10^4$
12. $1{,}000 = 10^3$
13. $u = v^x$
14. $a = b^c$
15. $9 = 27^{2/3}$
16. $8 = 4^{3/2}$

4-2 Logarithmic Functions

Find each of the following.

17. $\log_{10} 10^5$
18. $\log_5 5^3$
19. $\log_2 2^{-4}$
20. $\log_{10} 10^{-7}$
21. $\log_6 36$
22. $\log_3 9$
23. $\log_{10} 1{,}000$
24. $\log_{10} 0.001$

Find x, y, or b as indicated.

25. $\log_2 x = 2$
26. $\log_3 x = 2$
27. $\log_4 16 = y$
28. $\log_8 64 = y$
29. $\log_b 16 = 2$
30. $\log_b 10^{-3} = -3$

B *Rewrite in equivalent exponential form.*

31. $\log_{10} 0.001 = -3$
32. $\log_{10} 0.01 = -2$
33. $\log_{81} 3 = \frac{1}{4}$
34. $\log_4 2 = \frac{1}{2}$
35. $\log_{1/2} 16 = -4$
36. $\log_{1/3} 27 = -3$
37. $\log_a N = e$
38. $\log_k u = v$

Rewrite in equivalent logarithmic form.

39. $0.01 = 10^{-2}$
40. $0.001 = 10^{-3}$
41. $1 = e^0$
42. $1 = (\frac{1}{2})^0$
43. $\frac{1}{8} = 2^{-3}$
44. $\frac{1}{8} = (\frac{1}{2})^3$
45. $\frac{1}{3} = 81^{-1/4}$
46. $\frac{1}{2} = 32^{-1/5}$
47. $7 = \sqrt{49}$
48. $11 = \sqrt{121}$

Find each of the following.

49. $\log_b b^u$
50. $\log_b b^{uv}$
51. $\log_e e^{1/2}$
52. $\log_e e^{-3}$
53. $\log_2 \sqrt{8}$
54. $\log_5 \sqrt[3]{5}$
55. $\log_{23} 1$
56. $\log_{17} 1$
57. $\log_4 8$
58. $\log_4 (\frac{1}{4})$

Find x, y, or b as indicated.

59. $\log_4 x = \frac{1}{2}$
60. $\log_{25} x = \frac{1}{2}$
61. $\log_{1/3} 9 = y$
62. $\log_{49} (\frac{1}{7}) = y$
63. $\log_b 1{,}000 = \frac{3}{2}$
64. $\log_b 4 = \frac{2}{3}$
65. $\log_b 1 = 0$
66. $\log_b b = 1$

C 67. For $f = \{(x, y) | y = 1^x\}$, discuss the domain and range for f and f^{-1}. Are both relations functions?

68. Why is 1 not a suitable logarithmic base? [*Hint:* Try to find $\log_1 5$.]

69. (A) For $f = \{(x, y) | y = 10^x\}$, graph f and f^{-1} using the same coordinate axes.

(B) Discuss the domain and range of f and f^{-1}.
(C) What other name could you use for the inverse of f?

70. Prove that $\log_b (1/x) = -\log_b x$.

Find the inverse of:

71. $f(x) = 5^{3x-1} + 4$ **72.** $g(x) = 3^{2x-3} - 2$

73. $g(x) = 3 \log_b (5x - 2)$ **74.** $f(x) = 2 + \log_b (2x - 3)$

Section 4-3 Properties of Logarithmic Functions

- Basic Logarithmic Properties
- Use of the Logarithmic Properties

Basic Logarithmic Properties

Logarithmic functions have several very useful properties that follow directly from the fact that they are inverses of exponential functions. These properties will enable us to convert multiplication problems into addition problems, division problems into subtraction problems, and power and root problems into multiplication problems. In addition, we will be able to solve exponential equations such as $2 = 10^x$.

THEOREM 1 | **Properties of Logarithmic Functions**

If b, M, and N are positive real numbers, $b \neq 1$, and p is a real number, then

1. $\log_b b^u = u$
2. $\log_b MN = \log_b M + \log_b N$
3. $\log_b \dfrac{M}{N} = \log_b M - \log_b N$
4. $\log_b M^p = p \log_b M$
5. $\log_b 1 = 0$

The first property in Theorem 1 follows directly from the definition of a logarithmic function. The proof of the second property is based on the laws of exponents. To bring exponents into the proof, we let

$$u = \log_b M \quad \text{and} \quad v = \log_b N$$

4-3 Properties of Logarithmic Functions

and convert these to the equivalent exponential forms

$$M = b^u \quad \text{and} \quad N = b^v$$

Now, see if you can provide the reasons for each of the following steps:

$$\log_b MN = \log_b b^u b^v = \log_b b^{u+v} = u + v = \log_b M + \log_b N$$

The other properties are established in a similar manner.

■ Use of the Logarithmic Properties

We now see how logarithmic properties can be used to convert multiplication problems into addition problems, division problems into subtraction problems, and power and root problems into multiplication problems.

EXAMPLE 8

(A) $\log_{10} 10^5 = 5$ $\log_b b^u = u$

(B) $\log_b 3x = \log_b 3 + \log_b x$ $\log_b MN = \log_b M + \log_b N$

(C) $\log_b \dfrac{x}{5} = \log_b x - \log_b 5$ $\log_b \dfrac{M}{N} = \log_b M - \log_b N$

(D) $\log_b x^7 = 7 \log_b x$ $\log_b M^p = p \log_b M$

(E) $\log_b \dfrac{mn}{pq} = \log_b mn - \log_b pq$ $\log_b \dfrac{M}{N} = \log_b M - \log_b N$

$\quad = \log_b m + \log_b n$ $\log_b MN = \log_b M + \log_b N$
$\quad\quad - (\log_b p + \log_b q)$
$\quad = \log_b m + \log_b n$
$\quad\quad - \log_b p - \log_b q$

(F) $\log_b (mn)^{2/3} = \tfrac{2}{3} \log_b mn$ $\log_b M^p = p \log_b M$
$\quad = \tfrac{2}{3}(\log_b m + \log_b n)$ $\log_b MN = \log_b M + \log_b N$

(G) $\log_b \dfrac{x^8}{y^{1/5}} = \log_b x^8 - \log_b y^{1/5}$ $\log_b \dfrac{M}{N} = \log_b M - \log_b N$

$\quad = 8 \log_b x - \tfrac{1}{5} \log_b y$ $\log_b M^p = p \log_b M$

PROBLEM 8 Write in terms of simpler logarithmic forms, as in Example 8.

(A) $\log_b \left(\dfrac{r}{uv}\right)$ (B) $\log_b \left(\dfrac{m}{n}\right)^{3/5}$ (C) $\log_b \left(\dfrac{u^{1/3}}{v^5}\right)$

EXAMPLE 9 If $\log_e 3 = 1.10$ and $\log_e 7 = 1.95$, find

(A) $\log_e \left(\tfrac{7}{3}\right)$

Solution $\log_e (\frac{7}{3}) = \log_e 7 - \log_e 3 = 1.95 - 1.10 = 0.85$

(B) $\log_e \sqrt[3]{21}$

Solution $\log_e \sqrt[3]{21} = \log_e (21)^{1/3} = \frac{1}{3} \log_e (3 \cdot 7) = \frac{1}{3}(\log_e 3 + \log_e 7)$
$= \frac{1}{3}(1.10 + 1.95) = 1.02$

PROBLEM 9   If $\log_e 5 = 1.609$ and $\log_e 8 = 2.079$, find:

(A) $\log_e \left(\dfrac{5^{10}}{8}\right)$ (B) $\log_e \sqrt[4]{\dfrac{8}{5}}$

Finally, we note that since logarithmic functions are one to one,

$$\log_b m = \log_b n \quad \text{if and only if} \quad m = n$$

Thus, if $\log_{10} x = \log_{10} 32.15$, then $x = 32.15$.

The following example and problem, though somewhat artificial, will give you additional practice in using the properties in Theorem 1.

EXAMPLE 10   Find x so that $\log_b x = \frac{2}{3} \log_b 27 + 2 \log_b 2 - \log_b 3$ without using a calculator or table.

Solution $\log_b x = \frac{2}{3} \log_b 27 + 2 \log_b 2 - \log_b 3$ Express right side in terms of a single log.

$= \log_b 27^{2/3} + \log_b 2^2 - \log_b 3$ Property 4

$= \log_b 9 + \log_b 4 - \log_b 3$ $27^{2/3} = 9, 2^2 = 4$

$= \log_b \dfrac{9 \cdot 4}{3} = \log_b 12$ Properties 2 and 3

Thus,

$\log_b x = \log_b 12$

Hence,

$x = 12$

PROBLEM 10   Find x so that $\log_b x = \frac{2}{3} \log_b 8 + \frac{1}{2} \log_b 9 - \log_b 6$ without using a calculator or table.

Answers to Matched Problems

8. (A) $\log_b r - \log_b u - \log_b v$ (B) $\frac{3}{5}(\log_b m - \log_b n)$
   (C) $\frac{1}{3} \log_b u - 5 \log_b v$

9. (A) 14.01 (to four significant digits)   10. $x = 2$
   (B) 0.1175 (to four significant digits)

4-3 Properties of Logarithmic Functions

Exercise 4-3 ■ **A** *Write in terms of simpler logarithmic forms (going as far as you can with logarithmic properties—see Example 8).*

1. $\log_b uv$
2. $\log_b rt$
3. $\log_b (A/B)$
4. $\log_b (p/q)$
5. $\log_b u^5$
6. $\log_b w^{25}$
7. $\log_b N^{3/5}$
8. $\log_b u^{-2/3}$
9. $\log_b \sqrt{Q}$
10. $\log_b \sqrt[5]{M}$
11. $\log_b uvw$
12. $\log_b (u/vw)$

Write each expression in terms of a single logarithm with a coefficient of 1.

Example: $\log_b u^2 - \log_b v = \log_b (u^2/v)$.

13. $\log_b A + \log_b B$
14. $\log_b P + \log_b Q + \log_b R$
15. $\log_b X - \log_b Y$
16. $\log_b x^2 - \log_b y^3$
17. $\log_b w + \log_b x - \log_b y$
18. $\log_b w - \log_b x - \log_b y$

If $\log_b 2 = 0.69$, $\log_b 3 = 1.10$, and $\log_b 5 = 1.61$, find the logarithm to the base b of each of the following numbers.

19. $\log_b 30$
20. $\log_b 6$
21. $\log_b (\tfrac{2}{5})$
22. $\log_b (\tfrac{5}{3})$
23. $\log_b 27$
24. $\log_b 16$

B *Write in terms of simpler logarithmic forms (going as far as you can with logarithmic properties—see Example 8).*

25. $\log_b u^2 v^7$
26. $\log_b u^{1/2} v^{1/3}$
27. $\log_b \left(\dfrac{1}{a}\right)$
28. $\log_b \left(\dfrac{1}{M^3}\right)$
29. $\log_b \left(\dfrac{\sqrt[3]{N}}{p^2 q^3}\right)$
30. $\log_b \left(\dfrac{m^5 n^3}{\sqrt{p}}\right)$
31. $\log_b \sqrt[4]{\dfrac{x^2 y^3}{\sqrt{z}}}$
32. $\log_b \sqrt[5]{\left(\dfrac{x}{y^4 z^9}\right)^3}$

Write each expression in terms of a single logarithm with a coefficient of 1.

33. $2 \log_b x - \log_b y$
34. $\log_b m - \tfrac{1}{2} \log_b n$
35. $3 \log_b x + 2 \log_b y - 4 \log_b z$
36. $\tfrac{1}{3} \log_b w - 3 \log_b x - 5 \log_b y$
37. $\tfrac{1}{5}(2 \log_b x + 3 \log_b y)$
38. $\tfrac{1}{3}(\log_b x - \log_b y)$

If $\log_b 2 = 0.69$, $\log_b 3 = 1.10$, and $\log_b 5 = 1.61$, find the logarithm to the base b of the following numbers.

39. $\log_b 7.5$
40. $\log_b 1.5$
41. $\log_b \sqrt[3]{2}$
42. $\log_b \sqrt{3}$
43. $\log_b \sqrt{0.9}$
44. $\log_b \sqrt[3]{\tfrac{3}{2}}$

C **45.** Find x so that $\frac{3}{2} \log_b 4 - \frac{2}{3} \log_b 8 + 2 \log_b 2 = \log_b x$.
46. Find x so that $3 \log_b 2 + \frac{1}{2} \log_b 25 - \log_b 20 = \log_b x$.
47. Write $\log_b y - \log_b c + kt = 0$ in exponential form free of logarithms.
48. Write $\log_e x - \log_e 100 = -0.08t$ in exponential form free of logarithms.
49. Prove that $\log_b (M/N) = \log_b M - \log_b N$ under the hypotheses of Theorem 1.
50. Prove that $\log_b M^p = p \log_b M$ under the hypotheses of Theorem 1.
51. Prove that $\log_b MN = \log_b M + \log_b N$ by starting with $M = b^{\log_b M}$ and $N = b^{\log_b N}$.
52. Prove that $\log_b (M/N) = \log_b M - \log_b N$ by starting with $M = b^{\log_b M}$ and $N = b^{\log_b N}$.

Section 4-4 Logarithms to Various Bases

- Common and Natural Logarithms—Calculator Evaluation
- Common Logarithms—Table Evaluation (Optional)
- Natural Logarithms—Table Evaluation (Optional)
- Change-of-Base Formula

John Napier (1550–1617) is credited with the invention of logarithms. They evolved out of an interest in reducing the computational strain in astronomy research. This new computational tool was immediately accepted by the scientific world. Now, with the availability of inexpensive hand calculators, logarithms have lost most of their importance as a computational device. However, the logarithmic concept has been greatly generalized since its conception, and logarithmic functions are used widely in both theoretical and applied sciences. For example, even with a very good scientific hand calculator, we still need logarithmic functions to solve the simple-looking exponential equation from population growth studies and the mathematics of finance:

$2 = 1.08^x$

Of all possible logarithmic bases, the base e and the base 10 are used almost exclusively. Before we can use logarithms in certain practical problems, we need to be able to approximate the logarithm of any number to either base 10 or base e. And conversely, if we are given the logarithm of a number to base 10 or base e, we need to be able to approximate the number. Historically, tables such as Table II and Table III at the back of the book were used for this purpose, but now with inexpensive scientific hand calculators readily available, most people will choose a calculator, since it is faster and far more accurate than any table you might use.

■ Common and Natural Logarithms—Calculator Evaluation

Common logarithms (also called **Briggsian logarithms**) are logarithms with base 10. **Natural logarithms** (also called **Napierian logarithms**) are logarithms with base e. Most scientific calculators have a button labeled "log" (or "LOG") and a button labeled "ln" (or "LN"). The former represents a common (base 10) logarithm and the latter a natural (base e) logarithm. In fact, "log" and "ln" are both used extensively in mathematical literature, and whenever you see either used in this book without a base indicated, they will be interpreted as follows:

Logarithmic Notation

$$\log x = \log_{10} x$$
$$\ln x = \log_e x$$

To find the common or natural logarithm using a scientific calculator is very easy: You simply enter a number from the domain of the function and push the log or ln button.

EXAMPLE 11 Use a scientific calculator to find each to six decimal places.

(A) log 3,184   (B) ln 0.000 349   (C) log (−3.24)

Solution

	ENTER	PRESS	DISPLAY
(A)	3,184	$\boxed{\log}$	3.502973
(B)	0.000 349	$\boxed{\ln}$	−7.960439
(C)	−3.24	$\boxed{\log}$	Error

Why is an error indicated in (C)? Because −3.24 is not in the domain of the log function.

PROBLEM 11 Use a scientific calculator to find each to six decimal places.

(A) log 0.013 529   (B) ln 28.693 28   (C) ln (−0.438)

EXAMPLE 12 Use a scientific calculator to evaluate each to three decimal places.

(A) $n = \dfrac{\log 2}{\log 1.1}$ (B) $n = \dfrac{\ln 3}{\ln 1.08}$

Solution (A) First note that $(\log 2)/(\log 1.1) \neq \log 2 - \log 1.1$. Recall (see Section 4-3) that $\log_b (M/N) = \log_b M - \log_b N$, which is, of course, not the same as $(\log_b M)/(\log_b N)$.

$n = \dfrac{\log 2}{\log 1.1} = 7.273$

A: $\boxed{2}\,\boxed{\log}\,\boxed{\div}\,\boxed{1.1}\,\boxed{\log}\,\boxed{=}$
P: $\boxed{2}\,\boxed{\log}\,\boxed{\text{ENTER}}\,\boxed{1.1}\,\boxed{\log}\,\boxed{\div}$

(B) $n = \dfrac{\ln 3}{\ln 1.08} = 14.275$

A: $\boxed{3}\,\boxed{\ln}\,\boxed{\div}\,\boxed{1.08}\,\boxed{\ln}\,\boxed{=}$
P: $\boxed{3}\,\boxed{\ln}\,\boxed{\text{ENTER}}\,\boxed{1.08}\,\boxed{\ln}\,\boxed{\div}$

PROBLEM 12 Use a scientific calculator to evaluate each to two decimal places.

(A) $n = \dfrac{\ln 2}{\ln 1.1}$ (B) $n = \dfrac{\log 3}{\log 1.08}$

We now turn to the second problem: Given the logarithm of a number, find the number. We make direct use of the logarithmic–exponential relationships that were discussed in Section 4-2.

Logarithmic–Exponential Relationships

$\log x = y$ is equivalent to $x = 10^y$
$\ln x = y$ is equivalent to $x = e^y$

EXAMPLE 13 Find x to three significant digits, given the indicated logarithms.

(A) $\log x = -9.315$ (B) $\ln x = 2.386$

Solution (A) $\log x = -9.315$ Change to equivalent exponential form.
$x = 10^{-9.315}$ $\boxed{9.315}\,\boxed{+/-}\,\boxed{10^x}$
$x = 4.84 \times 10^{-10}$ Notice the answer is displayed in scientific notation in the calculator.

(B) $\ln x = 2.386$ Change to equivalent exponential form.
$x = e^{2.386}$ $\boxed{2.386}\,\boxed{e^x}$
$x = 10.9$

4-4 Logarithms to Various Bases

PROBLEM 13 Find x to four significant figures, given the indicated logarithms.

(A) $\ln x = -5.062$ (B) $\log x = 12.0821$

■ **Common Logarithms—Table Evaluation (Optional)**

We now show how Table II in the back of this book can be used to approximate common logarithms. Recalling that any decimal fraction can be written in scientific notation (see Appendix A-2), we see that

$$\log_{10} 33{,}800 = \log_{10} (3.38 \times 10^4)$$
$$= \log_{10} 3.38 + \log_{10} 10^4$$
$$= \log_{10} 3.38 + 4$$

and that

$$\log_{10} 0.003\ 51 = \log_{10} (3.51 \times 10^{-3})$$
$$= \log_{10} 3.51 + \log_{10} 10^{-3}$$
$$= \log_{10} 3.51 - 3$$

In general:

If a number N is written in scientific notation

$$N = r \times 10^k \quad 1 \leq r < 10, \quad k \text{ an integer}$$

then

$$\log N = \log (r \times 10^k)$$
$$= \log r + \log 10^k$$
$$= \underbrace{\log r}_{\text{Mantissa}} + \underbrace{k}_{\text{Characteristic}}$$

Thus, if common logarithms of r, $1 \leq r < 10$, are given in a table, we will be able to approximate the common logarithm of any positive decimal fraction to the accuracy of the table.

Using methods of advanced mathematics, a table of common logarithms of numbers from 1 to 10 can be computed to any decimal accuracy desired. Table II in the back of this book is such a table to four-decimal-place accuracy. It is useful to remember that if x is between 1 and 10, then log x is between 0 and 1 (Fig. 6).

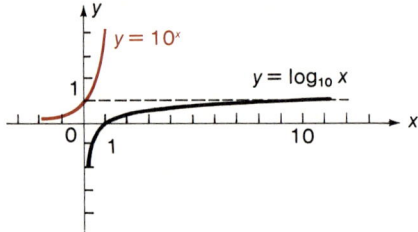

FIGURE 6

To illustrate the use of Table II, a small portion of it is reproduced in Table 1. To find log 3.47, for example, we first locate 3.4 under the x heading; then we move across to the column headed 7, where we find .5403. Thus, log 3.47 = 0.5403.*

TABLE 1

x	0	1	2	3	4	5	6	7	8	9
3.2	.5051	.5065	.5079	.5092	.5105	.5119	.5132	.5145	.5159	.5172
3.3	.5185	.5198	.5211	.5224	.5237	.5250	.5263	.5276	.5289	.5302
3.4	.5315	.5328	.5340	.5353	.5366	.5378	.5391	**.5403**	.5416	.5428
3.5	.5441	.5453	.5465	.5478	.5490	.5502	.5514	.5527	.5539	.5551

Now let us finish finding the common logarithms of 33,800 and 0.003 51.

EXAMPLE 14 (A) $\log 33{,}800 = \log (3.38 \times 10^4)$
$= \log 3.38 + \log 10^4$ Use Table 1.
$= 0.5289 + 4$
$= 4.5289$

(B) $\log 0.003\ 51 = \log (3.51 \times 10^{-3})$
$= \log 3.51 + \log 10^{-3}$ Use Table 1.
$= 0.5453 - 3$
$= -2.4547$

PROBLEM 14 Use Table 1 to find:

(A) log 328,000   (B) log 0.000 342

---

* Throughout the rest of this chapter we will use = in place of ≈ in many places, realizing that values are only approximately equal. Occasionally, we will use ≈ when a special emphasis is desired.

4-4 Logarithms to Various Bases

Now let us reverse the problem; that is, given the log of a number, find the number. To find the number, we first write the log of the number in the form

$$m + c$$

where m (the mantissa) is a nonnegative number between 0 and 1, and c (the characteristic) is an integer; then reverse the process illustrated in Example 14.

EXAMPLE 15 (A) If $\log x = 2.5224$, find x.

Solution
$$\log x = 2.5224$$
$$= 0.5224 + 2$$
$$= \log 3.33 + \log 10^2$$
$$= \log (3.33 \times 10^2)$$

Write 2.5224 in the form $m + c$, $0 \leq m < 1$ and c an integer. Look for 0.5224 in the body of Table 1. Thus, we see that $0.5224 = \log 3.33$.

Thus,
$$x = 3.33 \times 10^2 \quad \text{or} \quad 333$$

(B) If $\log x = 0.5172 - 4$, find x.

Solution
$$\log x = 0.5172 - 4$$
$$= \log 3.29 + 10^{-4}$$
$$= \log (3.29 \times 10^{-4})$$

Use Table 1.

Thus,
$$x = 3.29 \times 10^{-4} \quad \text{or} \quad 0.000\ 329$$

(C) If $\log x = -4.4685$, find x.

Solution
$$\log x = -4.4685$$
$$= 0.5315 - 5$$
$$= \log 3.40 + \log 10^{-5}$$
$$= \log (3.40 \times 10^{-5})$$

Convert -4.4685 to $m + c$ form with $0 \leq m < 1$ by adding and subtracting 5:

$$5.0000 - 5$$
$$-4.4685$$
$$\overline{0.5315 - 5}$$

Thus,
$$x = 3.40 \times 10^{-5} \quad \text{or} \quad 0.000\ 0340$$

PROBLEM 15 Use Table 1 to find x if

(A) $\log x = 5.5378$ (B) $\log x = 0.5289 - 3$ (C) $\log x = -2.4921$

What if a number has more significant digits than are included in a table? We then use the nearest table value, or, if more accuracy is desired, we use a calculator.

EXAMPLE 16 (A) $\log 32{,}683 \approx \log (3.27 \times 10^4)$ Round 3.2683 to nearest table value—that is, to 3.26.

$$\boxed{= \log 3.27 + \log 10^4}$$
$$= 0.5145 + 4$$
$$= 4.5145$$

(B) To find x if $\log x = 0.5241 - 3$, we observe (in Table 1) that 0.5241 is between 0.5237 and 0.5250, but is closer to 0.5237. Thus, we write

$$\log x = 0.5241 - 3$$
$$\approx 0.5237 - 3$$

Since 0.5241 is not in the body of the table, select the value in the table that is closest; that is, select 0.5237.

$$\boxed{\begin{aligned}&= \log 3.34 + \log 10^{-3}\\ &= \log (3.34 \times 10^{-3})\end{aligned}}$$

Thus,

$$x = 3.34 \times 10^{-3} \quad \text{or} \quad 0.003\,34$$

PROBLEM 16 Find:
(A) $\log 0.034\,319$ (B) x if $\log x = 6.5473$
(C) x if $\log x = -4.4942$

■ **Natural Logarithms—Table Evaluation (Optional)**

Approximating natural logarithms using Table III in the back of this book proceeds in much the same way as finding common logarithms using Table II, except the arithmetic is a little more complicated. A couple of examples will illustrate the process.

EXAMPLE 17 (A) $\ln 52{,}400 = \ln (5.24 \times 10^4)$
$$= \ln 5.24 + 4 \ln 10$$

[ln 5.24 is read out of the main table (Table III), and 4 ln 10 is obtained from the list at the top of the table.]

$$= 1.6563 + 9.2103$$
$$= 10.8666$$

(B) $\ln 0.002\,78 = \ln (2.78 \times 10^{-3})$
$$= \ln 2.78 - 3 \ln 10$$
$$= 1.0225 - 6.9078$$
$$= -5.8853$$

PROBLEM 17 Use Table III to find:
(A) $\ln 0.000\,683$ (B) $\ln 328{,}000$ (C) $\ln 23{,}582$

4-4 Logarithms to Various Bases

■ Change-of-Base Formula

If we have a means (through either a calculator or table) of finding logarithms of numbers to one base, then by means of the change-of-base formula we can find the logarithm of a number to any other base.

Change-of-Base Formula

$$\log_b N = \frac{\log_a N}{\log_a b}$$

Can you supply the reasons for each step in the following derivation of this formula?

$$y = \log_b N$$
$$N = b^y$$
$$\log_a N = \log_a b^y$$
$$\log_a N = y \log_a b$$
$$y = \frac{\log_a N}{\log_a b}$$
$$\log_b N = \frac{\log_a N}{\log_a b} \quad \text{Since } y = \log_b N$$

EXAMPLE 18 Find $\log_5 14$ using common logarithms.

Solution $\log_5 14 = \dfrac{\log_{10} 14}{\log_{10} 5} = 1.640$ A: $\boxed{14}\ \boxed{\log}\ \boxed{\div}\ \boxed{5}\ \boxed{\log}\ \boxed{=}$

P: $\boxed{14}\ \boxed{\log}\ \boxed{\text{ENTER}}\ \boxed{5}\ \boxed{\log}\ \boxed{\div}$

PROBLEM 18 Find $\log_7 729$.

Answers to Matched Problems
11. (A) $-1.868\ 734$ (B) $3.356\ 663$ (C) Not possible
12. (A) 7.27 (B) 14.27
13. (A) $0.006\ 333$ (B) 1.21×10^{12}
14. (A) 5.5159 (B) $0.5340 - 4 = -3.4660$
15. (A) 3.45×10^5 (B) 3.38×10^{-3} (C) 3.22×10^{-3}
16. (A) -1.4647 (B) 3.53×10^6 (C) 3.20×10^{-5}
17. (A) -7.2890 (B) 12.7008
    (C) Approx. $\ln 23{,}600 = 10.0690$
18. 3.3874

Exercise 4-4

Hand calculators are used extensively in this exercise set.

A   Use a calculator to find each to four decimal places.

1. log 82,734
2. log 843,250
3. log 0.001 439
4. log 0.035 604
5. ln 43.046
6. ln 2,843,100
7. ln 0.081 043
8. ln 0.000 0324

Use Table II to find each. (Optional)

9. log 7.29
10. log 6.37
11. log 2,040
12. log 327
13. log 0.0413
14. log 0.000 927

B   Use a calculator to find x to four significant digits, given:

15. $\log x = 5.3027$
16. $\log x = 1.9168$
17. $\log x = -3.1773$
18. $\log x = -2.0411$
19. $\ln x = 3.8655$
20. $\ln x = 5.0884$
21. $\ln x = -0.3916$
22. $\ln x = -4.1083$

Use the nearest values in Table II to approximate each of the following. (Optional)

23. log 304,918
24. log 82,734
25. log 0.004 769
26. log 0.061 94

Find x using the nearest values in Table II. (Optional)

27. $\log x = 7.437\ 15$
28. $\log x = 9.113\ 64$
29. $\log x = -4.8013$
30. $\log x = -3.4128$

Evaluate each of the following to three decimal places using a calculator.

31. $n = \dfrac{\log 2}{\log 1.15}$
32. $n = \dfrac{\log 2}{\log 1.12}$
33. $n = \dfrac{\ln 3}{\ln 1.15}$
34. $n = \dfrac{\ln 4}{\ln 1.2}$
35. $x = \dfrac{\ln 0.5}{-0.21}$
36. $t = \dfrac{\log 200}{2 \log 2}$

Use the change-of-base formula and a calculator with either log or ln to find each to four decimal places.

37. $\log_5 372$
38. $\log_4 23$
39. $\log_8 0.0352$
40. $\log_2 0.005\ 439$
41. $\log_3 0.1483$
42. $\log_{12} 435.62$

4-5 Exponential and Logarithmic Equations 257

Find each of the following using the nearest values in Table III. (Optional)

43. ln 2.35 **44.** ln 7.02 **45.** ln 603,517
46. ln 5,233 **47.** ln 0.003 1687 **48.** ln 0.071 33

C *Find x to five significant digits using a calculator.*

49. $x = \log(5.3147 \times 10^{12})$ **50.** $x = \log(2.0991 \times 10^{17})$
51. $x = \ln(6.7917 \times 10^{-12})$ **52.** $x = \ln(4.0304 \times 10^{-8})$
53. $\log x = 32.068\ 523$ **54.** $\log x = -12.731\ 64$
55. $\ln x = -14.667\ 13$ **56.** $\ln x = 18.891\ 143$

Section 4-5 Exponential and Logarithmic Equations

- Exponential Equations
- Logarithmic Equations

Equations involving exponential and logarithmic functions, such as

$$2^{3x-2} = 5 \quad \text{and} \quad \log(x + 3) + \log x = 1$$

are called **exponential** and **logarithmic equations**, respectively. Logarithmic properties play a central role in their solution.

■ Exponential Equations

The following examples illustrate the use of logarithmic properties in solving exponential equations.

EXAMPLE 19 Solve $2^{3x-2} = 5$ for x to four decimal places.

Solution $2^{3x-2} = 5$ How can we get *x* out of the exponent? Use logs! If two positive quantities are equal, their logs are equal.

$\log 2^{3x-2} = \log 5$ Use $\log_b N^p = p \log_b N$ to get $3x - 2$ out of
$(3x - 2)\log 2 = \log 5$ the exponent position.

$3x - 2 = \dfrac{\log 5}{\log 2}$ Remember: $\log 5/\log 2 \neq \log 5 - \log 2$

$x = \dfrac{1}{3}\left(2 + \dfrac{\log 5}{\log 2}\right)$

A: ⬚5⬚ ⬚log⬚ ⬚÷⬚ ⬚2⬚ ⬚log⬚ ⬚=⬚ ⬚+⬚ ⬚2⬚ ⬚=⬚ ⬚÷⬚ ⬚3⬚ ⬚=⬚

or

A: [3] [1/x] [×] [(] [2] [+] [(] [5] [log] [÷] [2] [log] [)] [)] [=]

P: [5] [log] [ENTER] [2] [log] [÷] [2] [+] [3] [÷]

$$x = 1.4406 \qquad \text{To four decimal places}$$

PROBLEM 19 Solve $35^{1-2x} = 7$ for x to four decimal places.

EXAMPLE 20 If a certain amount of money P (principal) is invested at $100r\%$ interest compounded annually, then the amount of money A in the account after n years, assuming no withdrawals, is given by

$$A = P(1 + r)^n$$

How long will it take the money to double if it is invested at 6% compounded annually?

Solution To find the doubling time, we replace A in $A = P(1.06)^n$ with 2P and solve for n.

$2P = P(1.06)^n$ Divide both sides by P.

$2 = 1.06^n$ Take the common or natural log of both sides.

$\log 2 = \log 1.06^n$

$\log 2 = n \log 1.06$ Note how log properties are used to get n out of the exponent position.

$$n = \frac{\log 2}{\log 1.06}$$

$= 12$ years   To the nearest year

PROBLEM 20 Repeat Example 20 changing the interest rate from 6% compounded annually to 9% compounded annually.

EXAMPLE 21 The atmospheric pressure P (in pounds per square inch) at x miles above sea level is given approximately by

$$P = 14.7e^{-0.21x}$$

At what height will the atmospheric pressure be half of the sea-level pressure? Compute the answer to two significant digits.

Solution Sea-level pressure is the pressure at $x = 0$. Thus,

$$P = 14.7e^0 = 14.7$$

One-half of sea-level pressure is $14.7/2 = 7.35$. Now our problem is to find x so that $P = 7.35$; that is, we solve $7.35 = 14.7e^{-0.21x}$ for x.

4-5 Exponential and Logarithmic Equations

$$7.35 = 14.7e^{-0.21x}$$ Divide both sides by 14.7 to simplify.

$$0.5 = e^{-0.21x}$$ Take the natural log of both sides.

$$\ln 0.5 = \ln e^{-0.21x}$$ Why use natural logs? Compare with common log to see why.

$$\ln 0.5 = -0.21x$$

$$x = \frac{\ln 0.5}{-0.21}$$ Use a hand calculator (or Table III).

$$= 3.3 \text{ miles}$$ To two significant digits

PROBLEM 21 Using the formula in Example 21, find the altitude in miles to two significant digits so that the atmospheric pressure will be one-eighth that at sea level.

The graph of

$$y = \frac{e^x + e^{-x}}{2} \tag{1}$$

is a curve called a *catenary* (Fig. 7). A uniform cable suspended between two fixed points is a physical example of such a curve.

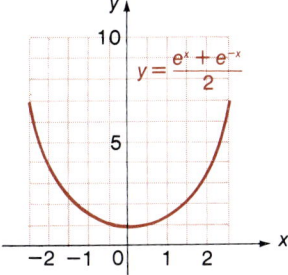

FIGURE 7 Catenary

EXAMPLE 22 Given equation (1), find x for y = 2.5. Compute the answer to four decimal places.

Solution
$$y = \frac{e^x + e^{-x}}{2}$$

$$2.5 = \frac{e^x + e^{-x}}{2}$$

$$5 = e^x + e^{-x}$$ Multiply both sides by e^x.

$$5e^x = e^{2x} + 1$$

$$e^{2x} - 5e^x + 1 = 0$$ This is a quadratic in e^x.

Let $u = e^x$, then

$$u^2 - 5u + 1 = 0$$

$$u = \frac{5 \pm \sqrt{25 - 4(1)(1)}}{2}$$

$$= \frac{5 \pm \sqrt{21}}{2}$$

Replace u with e^x and solve for x.

$$e^x = \frac{5 \pm \sqrt{21}}{2}$$

Take the natural log of both sides (both values on the right are positive).

$$\ln e^x = \ln \frac{5 \pm \sqrt{21}}{2}$$

$$x = \ln \frac{5 \pm \sqrt{21}}{2}$$

$$= -1.5668, \; 1.5668$$

PROBLEM 22 Given $y = (e^x - e^{-x})/2$, find x for $y = 1.5$. Compute the answer to three decimal places.

■ Logarithmic Equations

The next two examples illustrate approaches to solving some types of logarithmic equations.

EXAMPLE 23 Solve $\log (x + 3) + \log x = 1$ and check.

Solution

$\log (x + 3) + \log x = 1$ Combine left side using $\log M + \log N = \log MN$.
$\log [x(x + 3)] = 1$ Change to equivalent exponential form.
$x(x + 3) = 10^1$ Write in $ax^2 + bx + c = 0$ form.
$x^2 + 3x - 10 = 0$ Solve.
$(x + 5)(x - 2) = 0$
$x = -5, \; 2$

Check $x = -5$:

$\log(-5 + 3) + \log(-5)$ is not defined (Why?)

$x = 2$:

$\log (2 + 3) + \log 2 = \log 5 + \log 2$
$= \log (5 \cdot 2) = \log 10 = 1$

Remember, answers should be checked in the original equation to see whether any should be discarded.

4-5 Exponential and Logarithmic Equations

PROBLEM 23 Solve $\log(x - 15) = 2 - \log x$ and check.

EXAMPLE 24 Solve $(\ln x)^2 = \ln x^2$.

Solution

$(\ln x)^2 = \ln x^2$

$(\ln x)^2 = 2 \ln x$

$(\ln x)^2 - 2 \ln x = 0$

$(\ln x)(\ln x - 2) = 0$

$\ln x = 0$ or $\ln x - 2 = 0$

$x = e^0 \qquad \qquad \ln x = 2$

$x = 1 \qquad \qquad x = e^2$

This is a quadratic equation in ln x. Move all nonzero terms to the left and factor.

PROBLEM 24 Solve $\log x^2 = (\log x)^2$.

Answers to Matched Problems

19. 0.2263
20. More than double in 9 years, but not quite double in 8 years
21. 9.9 miles
22. 1.195
23. 20
24. 1, 100

Exercise 4-5

A scientific hand calculator will prove useful in many of the problems in this exercise.

A *Solve to three significant digits.*

1. $10^{-x} = 0.0347$
2. $10^x = 14.3$
3. $10^{3x+1} = 92$
4. $10^{5x-2} = 348$
5. $e^x = 3.65$
6. $e^{-x} = 0.0142$
7. $e^{2x-1} = 405$
8. $e^{3x+5} = 23.8$
9. $5^x = 18$
10. $3^x = 4$
11. $2^{-x} = 0.238$
12. $3^{-x} = 0.074$

Solve exactly.

13. $\log 5 + \log x = 2$
14. $\log x - \log 8 = 1$
15. $\log x + \log(x - 3) = 1$
16. $\log(x - 9) + \log 100x = 3$

B *Solve to three significant digits.*

17. $2 = 1.05^x$
18. $3 = 1.06^x$
19. $e^{-1.4x} = 13$
20. $e^{0.32x} = 632$
21. $123 = 500e^{-0.12x}$
22. $438 = 200e^{0.25x}$

Solve exactly.

23. $\log x - \log 5 = \log 2 - \log(x - 3)$
24. $\log(6x + 5) - \log 3 = \log 2 - \log x$

25. $(\ln x)^3 = \ln x^4$
26. $(\log x)^3 = \log x^4$
27. $\ln (\ln x) = 1$
28. $\log (\log x) = 1$
29. $x^{\log x} = 100x$
30. $3^{\log x} = 3x$

C In Problems 31–36 solve for the indicated letter in terms of all others using common or natural logs, whichever produces the simplest results.

31. $I = I_0 e^{-kx}$ for x (x-ray absorption)
32. $A = P(1 + i)^n$ for n (compound interest)
33. $N = 10 \log \left(\dfrac{I}{I_0} \right)$ for I (sound intensity—decibels)
34. $t = \dfrac{-1}{k} (\ln A - \ln A_0)$ for A (radioactive decay)
35. $I = \dfrac{E}{R} (1 - e^{-Rt/L})$ for t (electric circuits)
36. $S = R \dfrac{(1 + i)^n - 1}{i}$ for n (future value of an annuity)

37. Find the fallacy.

$$3 > 2$$
$$(\log \tfrac{1}{2})3 > (\log \tfrac{1}{2})2$$
$$3 \log \tfrac{1}{2} > 2 \log \tfrac{1}{2}$$
$$\log (\tfrac{1}{2})^3 > \log (\tfrac{1}{2})^2$$
$$(\tfrac{1}{2})^3 > (\tfrac{1}{2})^2$$
$$\tfrac{1}{8} > \tfrac{1}{4}$$

38. Find the fallacy.

$$-2 < -1$$
$$\ln e^{-2} < \ln e^{-1}$$
$$2 \ln e^{-1} < \ln e^{-1}$$
$$2 < 1$$

APPLICATIONS

39. *Compound interest.* How long will it take a sum of money to double if it is invested at 15% interest compounded annually (see Example 20)?

40. *Compound interest.* How long will it take money to quadruple if it is invested at 20% interest compounded annually (see Example 20)?

41. *Bacterial growth.* A single cholera bacterium divides every $\tfrac{1}{2}$ hour to produce two complete cholera bacteria. If we start with a colony

of 5,000 bacteria, then after t hours we will have

$$A = 5{,}000 \cdot 2^{2t}$$

bacteria. How long will it take for A to equal 1,000,000?

42. *Astronomy.* An optical instrument is required to observe stars beyond the sixth magnitude, the limit of ordinary vision. However, even optical instruments have their limitations. The limiting magnitude L of any optical telescope with lens diameter D in inches is given by

$$L = 8.8 + 5.1 \log D$$

(A) Find the limiting magnitude for a homemade 6-inch reflecting telescope.
(B) Find the diameter of a lens that would have a limiting magnitude of 20.6.

43. *World population.* A mathematical model for world population growth over short periods of time is given by

$$P = P_0 e^{rt}$$

where

$P_0 =$ Population at $t = 0$
$r =$ Rate compounded continuously
$t =$ Time in years
$P =$ Population at time t

How long will it take the earth's population to double if it continues to grow at its current rate of 2% per year (compounded continuously)? [*Hint:* Given $r = 0.02$, find t so that $P = 2P_0$.]

44. *World population.* If the world population is now 4 billion people and if it continues to grow at 2% per year (compounded continuously), how long will it be before there is only 1 square yard of land per person? Use the formula in Problem 43 and the fact that there is 1.7×10^{14} square yards of land on earth.

45. *Nuclear reactors—strontium-90.* Radioactive strontium-90 is used in nuclear reactors and decays according to

$$A = Pe^{-0.0248t}$$

where P is the amount present at $t = 0$, and A is the amount remaining after t years. Find the half-life of strontium-90; that is, find t so that $A = 0.5P$.

46. *Archaeology—carbon-14 dating.* Cosmic-ray bombardment of the atmosphere produces neutrons, which in turn react with nitrogen to produce radioactive carbon-14. Radioactive carbon-14 enters all

living tissues through carbon dioxide, which is first absorbed by plants. As long as a plant or animal is alive, carbon-14 is maintained in a constant amount in its tissues. Once dead, however, it ceases taking in carbon and, to the slow beat of time, the carbon-14 diminishes by radioactive decay according to the equation

$$A = A_0 e^{-0.000\,124t}$$

where t is time in years. Estimate the age of a skull uncovered in an archaeological site if 10% of the original amount of carbon-14 is still present. [Hint: Find t such that $A = 0.1 A_0$.]

47. *Sound intensity—decibels.* Because of the extraordinary range of sensitivity of the human ear (a range of over 1,000 million million to 1), it is helpful to use a logarithmic scale to measure sound intensity over this range rather than an absolute scale. The unit of measure is called the *decibel*, after the inventor of the telephone, Alexander Graham Bell. If we let N be the number of decibels, I the power of the sound in question in watts per cubic centimeter, and I_0 the power of sound just below the threshold of hearing (approximately 10^{-16} watt/square centimeter),

$$I = I_0 10^{N/10}$$

show that this formula can be written in the form

$$N = 10 \log \left(\frac{I}{I_0} \right)$$

48. *Sound intensity—decibels.* Use the formula in Problem 47 (with $I_0 = 10^{-16}$ watt/square centimeter) to find the decibel ratings of the following sounds:
    (A) Whisper (10^{-13} watt/square centimeter)
    (B) Normal conversation (3.16×10^{-10} watt/square centimeter)
    (C) Heavy traffic (10^{-8} watt/square centimeter)
    (D) Jet plane with afterburner (10^{-1} watt/square centimeter)

49. *Earth science.* For relatively clear bodies of freshwater or salt water, light intensity is reduced according to the exponential function

$$I = I_0 e^{-kd}$$

where I is the intensity at d feet below the surface, and I_0 is the intensity at the surface; k is called the coefficient of extinction. Two of the clearest bodies of water in the world are the freshwater Crystal Lake in Wisconsin ($k = 0.0485$) and the saltwater Sargasso Sea off the West Indies ($k = 0.009\,42$). Find the depths (to the nearest foot) in these two bodies of water at which the light is reduced to 1% of that at the surface.

50. *Psychology—learning.* In learning a particular task, such as typing or swimming, one progresses faster at the beginning and then levels off. If you plot the level of performance against time, you will obtain a curve of the type shown in the figure. This is called a *learning curve* and can be very closely approximated by an exponential equation of the form $y = a(1 - e^{-cx})$, where a and c are positive constants. Curves of this type have applications in psychology, education, and industry. Suppose a particular person's history of learning to type is given by the exponential equation $N = 80(1 - e^{-0.08n})$, where N is the number of words per minute typed after n weeks of instruction. Approximately how many weeks did it take the person to learn to type sixty words per minute?

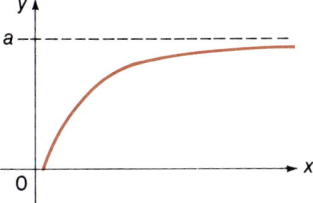

Section 4-6 Chapter Review

IMPORTANT TERMS AND SYMBOLS

4-1 Exponential Functions. Exponential function, base, domain, range, graphs, base e, properties, b^x ($b > 0$, $b \neq 1$)

4-2 Logarithmic Functions. Logarithmic function, base, domain, range, graphs, logarithmic–exponential identities, $\log_b x$ ($b > 0$, $b \neq 1$)

4-3 Properties of Logarithmic Functions. Basic properties, use of properties

4-4 Logarithms to Various Bases. Common logarithm, natural logarithm, change-of-base formula

4-5 Exponential and Logarithmic Equations. Exponential equations, logarithmic equations

Exercise 4-6 Chapter Review

Work through all the problems in this chapter review and check answers in the back of the book. (Answers to all problems are there, and following each answer is a number in italics indicating the section in which that type of problem is discussed.) Where weaknesses show up, review appropriate sections in the text. When you are satisfied that you know the material, take the practice test following this review.

4 Exponential and Logarithmic Functions

You will find a scientific hand calculator useful in many of the problems in this exercise.

A
1. Write $m = 10^n$ in logarithmic form with base 10.
2. Write $\log x = y$ in exponential form.

Solve for x exactly. Do not use a calculator or table.

3. $\log_2 x = 3$ 4. $\log_x 25 = 2$ 5. $\log_3 27 = x$

Solve for x to three significant digits.

6. $10^x = 17.5$ 7. $e^x = 143{,}000$

In Problems 8 and 9, solve for x exactly. Do not use a calculator or table.

8. $\log x - 2 \log 3 = 2$ 9. $\log x + \log(x - 3) = 1$

B
10. Write $\ln y = x$ in exponential form.
11. Write $x = e^y$ in logarithmic form with base e.

Solve for x exactly. Do not use a calculator or table.

12. $\log_{1/4} 16 = x$ 13. $\log_x 9 = -2$ 14. $\log_{16} x = \frac{3}{2}$
15. $\log_x e^5 = 5$ 16. $10^{\log_{10} x} = 33$ 17. $\ln x = 0$

Solve for x to three significant digits.

18. $25 = 5(2)^x$ 19. $4{,}000 = 2{,}500 e^{0.12x}$
20. $0.01 = e^{-0.05x}$

In Problems 21–24 solve for x exactly. Do not use a table or calculator.

21. $\log 3x^2 - \log 9x = 2$
22. $\log x - \log 3 = \log 4 - \log(x + 4)$
23. $(\log x)^3 = \log x^9$
24. $\ln(\log x) = 1$
25. Calculate $\log_5 23$ to three significant digits.

C
26. Write $\ln y = -5t + \ln c$ in an exponential form free of logarithms; then solve for y in terms of the other letters.
27. For $f = \{(x, y) | y = \log_2 x\}$, graph f and f^{-1} using the same coordinate system. What are the domains and ranges for f and f^{-1}?
28. Explain why 1 cannot be used as a logarithmic base.
29. Prove that $\log_b(M/N) = \log_b M - \log_b N$.

APPLICATIONS
30. *Population growth.* Many countries in the world have a population growth rate of 3% (or more) per year. At this rate how long, to the nearest year, will it take a population to double? Use the

population growth model

$$P = P_0(1.03)^t$$

which assumes annual compounding. Compute the answer to three significant digits.

31. *Population growth.* Repeat Problem 30 using the continuous population growth model

$$P = P_0 e^{0.03t}$$

which assumes continuous compounding. Compute the answer to three significant digits.

32. *Carbon-14 dating.* How long, to three significant digits, will it take for the carbon-14 to diminish to 1% of the original amount after the death of a plant or animal?

$$A = A_0 e^{-0.000\ 124t} \quad \text{where } t \text{ is time in years}$$

33. *x-ray absorption.* Solve $x = -(1/k) \ln (I/I_0)$ for I in terms of the other letters.

34. *Amortization—time payments.* Solve $r = P\{i/[1 - (1 + i)^{-n}]\}$ for n in terms of the other letters.

Practice Test Chapter 4

Take this practice test as if it were a graded test. Allow yourself up to 50 minutes. Work the problems without looking back in the chapter. Correct your work using the answers (keyed to appropriate sections) in the back of the book.

1. Write $\ln y = -x^2$ in equivalent exponential form.
2. Write $\frac{1}{2} \log_b x + 3 \log_b y - 2 \log_b z$ in terms of a single logarithm with a coefficient of 1.

Solve for x to three significant digits.

3. $10^x = 42.6$ **4.** $5^{2x-3} = 7.08$
5. $125 = 500 e^{-0.000\ 124x}$

In Problems 6–10 solve for x exactly. Do not use a calculator or table.

6. $\log_{1/5} 125 = x$ **7.** $\log_{16} x = \frac{3}{4}$
8. $\log_x 100 = -2$ **9.** $\log (x + 4) = 1 - \log (x - 5)$
10. $\log (\ln x) = 1$

11. Write $f^{-1}(x)$ if $f(x) = e^x$. What are the domain and range of f and f^{-1}?

12. Find $\log_{12} 8$ to three significant digits.

13. *Continuous compound interest.* How long, to three significant digits, would it take money to quadruple if invested at 15% interest compounded continuously? ($A = Pe^{rt}$)

14. *Continuous compound interest.* Solve $A = Pe^{rt}$ for r in terms of the other letters. Use either common or natural logs, whichever produces the simplest result.

15. *Present value of an annuity.* Solve $P = R\{[1 - (1 + i)^{-n}]/i\}$ for n in terms of the other letters.

Trigonometric Functions

5

5-1 Introduction
5-2 The Wrapping Function
5-3 Circular Functions
5-4 Angles
5-5 Trigonometric Functions
5-6 Graphs of Trigonometric Functions
5-7 Graphing $y = A \sin(Bx + C)$ and $y = A \cos(Bx + C)$
5-8 Inverse Trigonometric Functions
5-9 Chapter Review

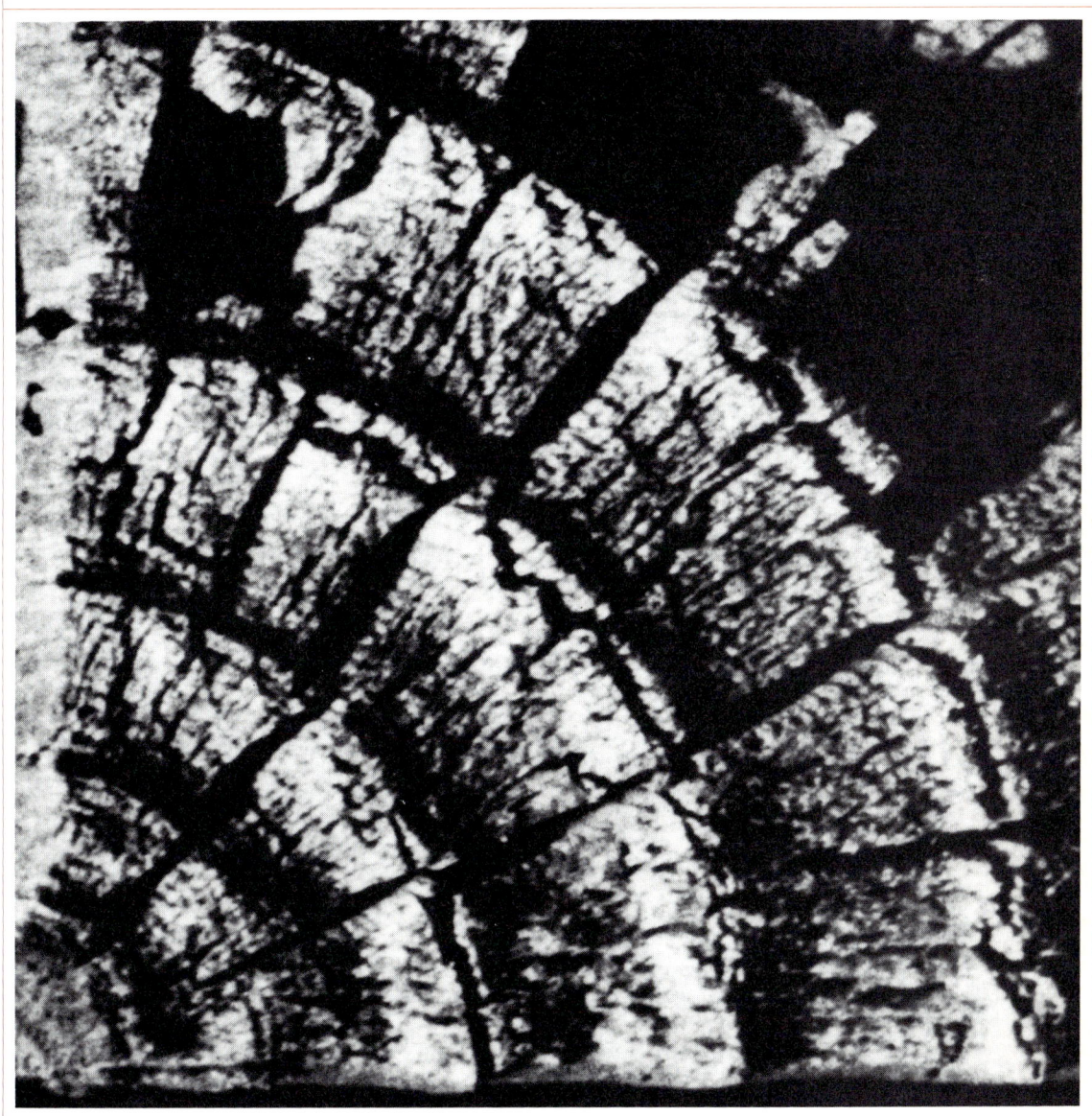

A natural design of mathematical interest. Can you guess the source? See the back of the book.

Chapter 5 ■ Trigonometric Functions

Section 5-1 Introduction

Trigonometric functions seem to have had their origins with the Greeks' investigation of the indirect measurement of distances and angles in the "celestial sphere" as well as their remeasurements of lands flooded by the Nile. The word *trigonometry*, based on the Greek words for "triangle measurement," was first used as the title for a text by the German mathematician Pitiscus in A.D. 1600.

Originally the trigonometric functions were restricted to angle domains and their applications to the indirect measurement of angles and distances. These functions gradually broke free of these restrictions and we now have trigonometric functions with real number domains. Modern applications range over many types of problems that have little or nothing to do with angles or triangles (e.g., periodic phenomena such as sound, light, and electrical waves; business cycles; and planetary motion).

Our approach to the subject will not follow historical lines in that we will first introduce trigonometric functions (circular functions) with real number domains; then we will define trigonometric functions with angle domains.

> Calculator versus Table Evaluation
>
> Calculator evaluation of trigonometric functions is emphasized. Table evaluation of trigonometric functions is included in Appendix A-8 for those who still desire that approach.

Section 5-2 The Wrapping Function

- Definition of the Wrapping Function
- Exact Values for Particular Real Numbers
- The Wrapping Function Is Not One-to-One

- Definition of the Wrapping Function

The important *circular function* definitions that will be presented in the

5-2 The Wrapping Function

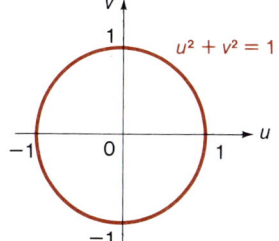

FIGURE 1 Unit circle

next section are based on a function W, called the *wrapping function*, whose domain is the set of real numbers and whose range is the set of points on a *unit circle*. By **unit circle** we mean a circle of radius 1 with center at the origin of a Cartesian coordinate system (see Fig. 1).*

In defining the **wrapping function**, we "wrap" a real number line with origin at (1, 0) around the unit circle—the positive real axis is wrapped counterclockwise and the negative real axis is wrapped clockwise. In this way each real number on the real line is paired with a unique point, called a **circular point**, on the unit circle (see Fig. 2).

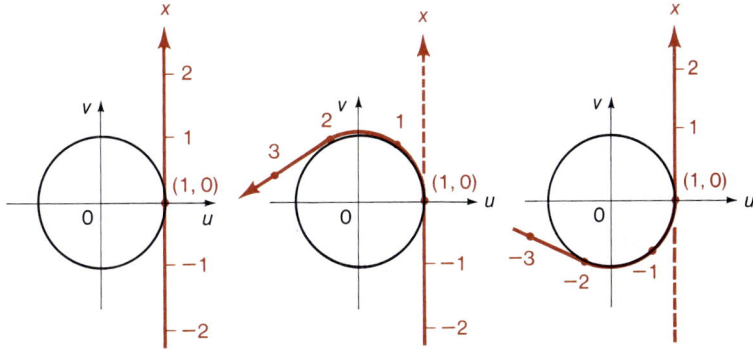

FIGURE 2 The wrapping function

To locate the circular point associated with a number such as 37 or −105, the number line will be wrapped many times around the circle.

An equivalent way of pairing real numbers with points on a unit circle is to think in terms of *arc length* (assuming we know what arc length is). To find the circular point associated with the real number x, we start at (1, 0) and move $|x|$ along the unit circle, counterclockwise if x is positive and clockwise if x is negative (see Fig. 3).

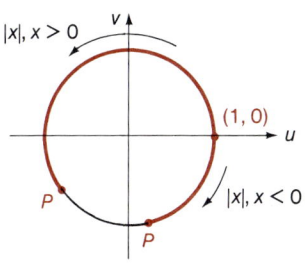

FIGURE 3

It is important to be able to find the coordinates (a, b) of the circular point P associated with a given real number x so that we can write $W(x) = (a, b)$. In general, this is difficult and requires the use of a calculator or table. However, for certain real numbers (integer multiples of $\pi/6$, $\pi/4$, $\pi/3$, and $\pi/2$) we can find the exact coordinates of the corresponding circular points using simple geometric properties of a circle. We will be doing this frequently in the following sections and chapters.

■ Exact Values for Particular Real Numbers

We start our investigation by finding the circumference of a unit circle. Since radius $R = 1$, the circumference is

$$2\pi R = 2\pi(1) = 2\pi \quad \text{Circumference of a unit circle}$$

* We use the variables u and v instead of x and y so that x can be used without ambiguity as an independent variable in defining the wrapping and circular functions. Both these functions use the unit circle in their definitions.

One-fourth, one-half, and three-fourths of the circumference are, respectively, $\pi/2$, π, and $3\pi/2$. The circular points corresponding to these real numbers are on the coordinate axes, and hence, their coordinates are easily determined (see Fig. 4).

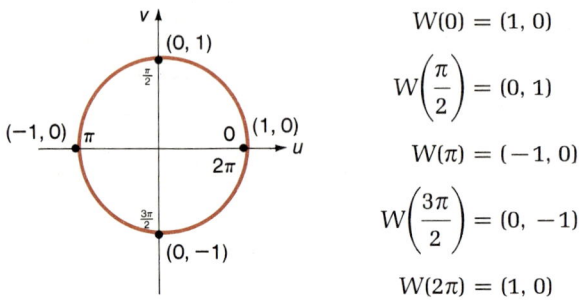

$W(0) = (1, 0)$

$W\left(\dfrac{\pi}{2}\right) = (0, 1)$

$W(\pi) = (-1, 0)$

$W\left(\dfrac{3\pi}{2}\right) = (0, -1)$

$W(2\pi) = (1, 0)$

FIGURE 4

Following the same procedure, we can find the coordinates of any circular point on a coordinate axis; that is, for any circular point corresponding to a real number that is an integer multiple of $\pi/2$.

EXAMPLE 1 Find the coordinates of the circular points:

(A) $W(-\pi/2)$ (B) $W(5\pi/2)$

Solution (A) Starting at (1, 0), we go one-fourth the way around the unit circle in a clockwise direction (see Fig. 4). Thus,

$$W\left(\dfrac{-\pi}{2}\right) = (0, -1)$$

(B) Starting at (1, 0) and proceeding counterclockwise, we count quarter-circle steps, $\pi/2$, $2\pi/2$, $3\pi/2$, $4\pi/2$, and ending at $5\pi/2$. Thus, the circular point is on the positive vertical axis, and we have

$$W\left(\dfrac{5\pi}{2}\right) = (0, 1)$$

PROBLEM 1 Find the coordinates of the circular points:

(A) $W(-\pi)$ (B) $W(3\pi)$

We now find the coordinates of the circular point $W(\pi/4)$. Since $\pi/4$ is one-half the arc joining (1, 0) and (0, 1), the circular point $W(\pi/4)$ must lie on the line $v = u$ (see Fig. 5). Since $W(\pi/4)$ is on the line $v = u$ and on the circle $u^2 + v^2 = 1$, its coordinates (a, b) must satisfy both equations. That is,

$a = b$ and $a^2 + b^2 = 1$

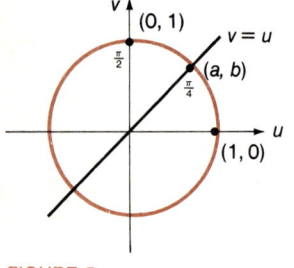

FIGURE 5

5-2 The Wrapping Function

Substituting a for b in the second equation, we have

$$a^2 + a^2 = 1$$
$$2a^2 = 1$$
$$a^2 = \frac{1}{2}$$
$$a = \frac{1}{\sqrt{2}} \quad \text{a is positive, since $W(\pi/4)$ is in the first quadrant}$$

Using the first equation, we see that

$$b = a = \frac{1}{\sqrt{2}}$$

Thus,

$$W\left(\frac{\pi}{4}\right) = \left(\frac{1}{\sqrt{2}}, \frac{1}{\sqrt{2}}\right) \tag{1}$$

Using symmetry properties of a circle (a unit circle is symmetric with respect to both axes and the origin) we can easily find the coordinates of any circular point that is reflected across the vertical axis, horizontal axis, or origin from $W(\pi/4)$.

EXAMPLE 2 Find the coordinates of the circular points:

(A) $W(5\pi/4)$ (B) $W(-\pi/4)$

Solution (A) Starting at (1, 0) and counting in one-eighth circle steps counterclockwise ($\pi/4, 2\pi/4, 3\pi/4, 4\pi/4, 5\pi/4$), we find ourselves in the third quadrant on the circle halfway between $(-1, 0)$ and $(0, -1)$, as indicated in the figure. Using symmetry with respect to the origin, we have

$$W\left(\frac{5\pi}{4}\right) = \left(\frac{-1}{\sqrt{2}}, \frac{-1}{\sqrt{2}}\right)$$

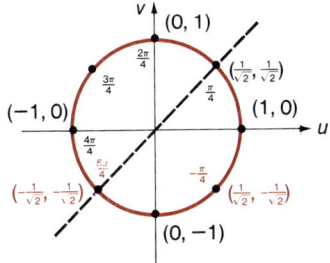

(B) Starting at (1, 0), we proceed one-eighth the way around the unit circle in a clockwise direction and end up in the fourth quadrant on the circle halfway between (0, −1) and (1, 0), as indicated in the figure. Using symmetry with respect to the horizontal axis, we see that

$$W\left(\frac{-\pi}{4}\right) = \left(\frac{1}{\sqrt{2}}, \frac{-1}{\sqrt{2}}\right)$$

PROBLEM 2 Find the coordinates of the circular points:

(A) $W(3\pi/4)$ (B) $W(-7\pi/4)$

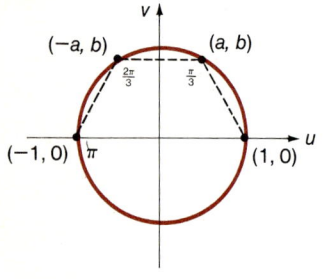

FIGURE 6

We continue our investigation by finding the coordinates of the circular point $W(\pi/3)$. Referring to Figure 6, we divide the upper semicircle from (1, 0) to (−1, 0) into thirds. The circular points $W(\pi/3)$ and $W(2\pi/3)$ are symmetric with respect to the v axis; hence, if $W(\pi/3)$ is given coordinates (a, b), then $W(2\pi/3)$ must have coordinates $(-a, b)$. The chord joining $W(2\pi/3)$ and $W(\pi/3)$ is thus $2a$ units long. Using the distance formula (see Section 2-1), we find the length of the chord joining $W(0)$ and $W(\pi/3)$ to be given by $\sqrt{(a-1)^2 + b^2}$. The two chords are equal, since equal arcs subtend equal chords on the same circle. Thus,

$$\sqrt{(a-1)^2 + b^2} = 2a$$

Squaring both sides, we obtain

$$(a-1)^2 + b^2 = 4a^2$$
$$a^2 - 2a + 1 + b^2 = 4a^2$$
$$a^2 + b^2 - 2a + 1 = 4a^2 \quad a^2 + b^2 = 1 \text{ (Why?)}$$
$$1 - 2a + 1 = 4a^2$$
$$4a^2 + 2a - 2 = 0$$
$$2a^2 + a - 1 = 0$$
$$(2a - 1)(a + 1) = 0$$
$$a = \tfrac{1}{2} \quad a = -1 \text{ must be discarded (Why?)}$$

Substitute $a = \tfrac{1}{2}$ into $a^2 + b^2 = 1$ and solve for b:

$$\left(\frac{1}{2}\right)^2 + b^2 = 1$$

$$b^2 = \frac{3}{4}$$

$$b = \frac{\sqrt{3}}{2} \quad b = -\frac{\sqrt{3}}{2} \text{ must be discarded (Why?)}$$

5-2 The Wrapping Function

Thus,

$$W\left(\frac{\pi}{3}\right) = \left(\frac{1}{2}, \frac{\sqrt{3}}{2}\right)$$

Proceeding in a similar manner, or using symmetry with respect to the line $v = u$, we can obtain

$$W\left(\frac{\pi}{6}\right) = \left(\frac{\sqrt{3}}{2}, \frac{1}{2}\right)$$

The key results from the above discussion for the first quadrant are summarized in Figure 7.

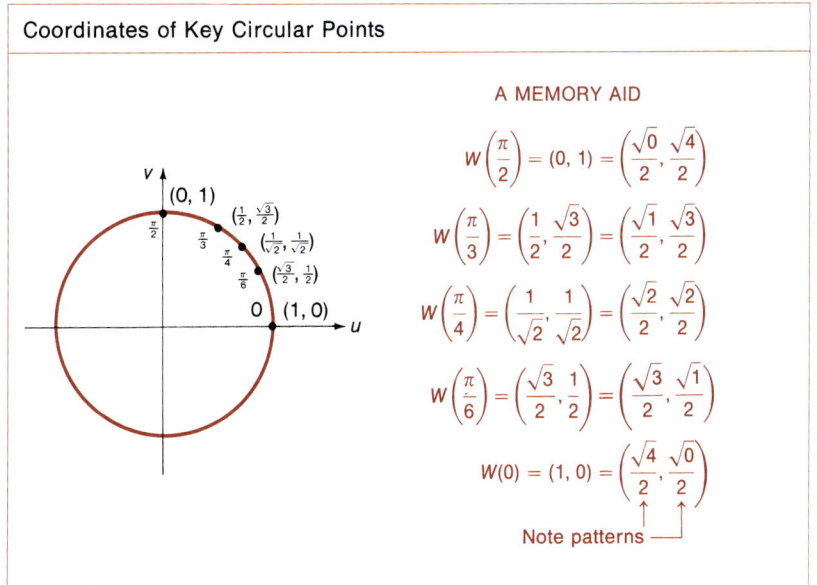

FIGURE 7

It is important that you memorize these first quadrant relationships. Using these and unit circle symmetry, we can find the coordinates of *any* circular point that corresponds to *any* integer multiple of $\pi/6$, $\pi/4$, $\pi/3$, and $\pi/2$.

EXAMPLE 3   Find the coordinates of the circular points:

(A) $W(5\pi/6)$ (B) $W(-2\pi/3)$

Solution (A) Note that $5\pi/6$ is $\pi/6$ less than $\pi = 6\pi/6$. Locate $5\pi/6$ in the second quadrant and use Figure 7 and symmetry with respect to the vertical axis to find $W(5\pi/6)$.

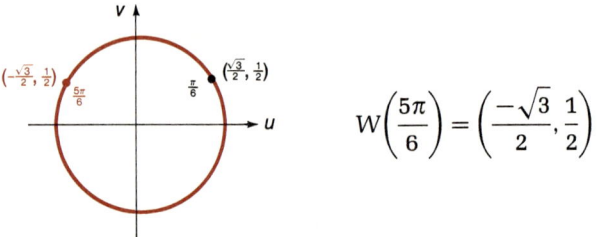

(B) Note that $-2\pi/3$ is $\pi/3$ more than $-\pi = -3\pi/3$. Locate $-2\pi/3$ in the third quadrant and use Figure 7 and symmetry with respect to the origin to find $W(-2\pi/3)$.

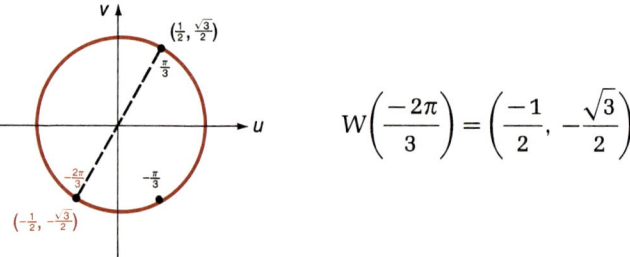

PROBLEM 3 Find the coordinates of the circular points:

(A) $W(5\pi/3)$ (B) $W(-7\pi/6)$

■ The Wrapping Function Is Not One-to-One

It is easy to see that the wrapping function is not a one-to-one function. Each domain value corresponds to exactly one range value, but each range value corresponds to infinitely many domain values. For example, we see that

$$W\left(\frac{\pi}{2}\right) = (0, 1)$$

That is, exactly one range value corresponds to the domain value $\pi/2$. But how many domain values correspond to the range value $(0, 1)$? Every time we go around the circle 2π units in either direction we will be back at the same circular point. Thus, if asked to solve

$$W(x) = (0, 1)$$

5-2 The Wrapping Function

we would have to write

$$x = \frac{\pi}{2} + 2k\pi \qquad k \text{ any integer}$$

and there are infinitely many domain values of W that correspond to the range value (0, 1). In general:

A Wrapping Function Property

For all real numbers x,

$$W(x) = W(x + 2k\pi) \qquad k \text{ any integer*}$$

(The wrapping function is not one-to-one.)

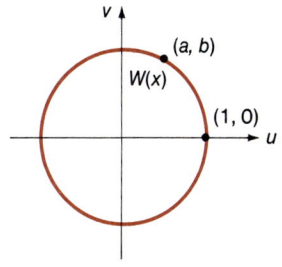

We will have more to say about the implications of this important property of the wrapping function in subsequent sections.

Answers to Matched Problems
1. (A) $(-1, 0)$ (B) $(-1, 0)$
2. (A) $(-1/\sqrt{2}, 1/\sqrt{2})$ (B) $(1/\sqrt{2}, 1/\sqrt{2})$
3. (A) $(1/2, -\sqrt{3}/2)$ (B) $(-\sqrt{3}/2, 1/2)$

Exercise 5-2

A Find the coordinates for each circular point. Sketch your own figures and do not look back in the text.

1. $W(\pi)$
2. $W(0)$
3. $W(6\pi)$
4. $W(3\pi)$
5. $W(-\pi)$
6. $W(-5\pi)$
7. $W\left(\dfrac{3\pi}{2}\right)$
8. $W\left(\dfrac{\pi}{2}\right)$
9. $W\left(\dfrac{-\pi}{2}\right)$
10. $W\left(\dfrac{-3\pi}{2}\right)$
11. $W\left(\dfrac{11\pi}{2}\right)$
12. $W\left(\dfrac{-15\pi}{2}\right)$

*Think of a point P moving around the unit circle in either direction. Every time P covers a distance of 2π (the circumference of the circle), it will be back at the point where it started.

B Find the coordinates for each circular point. Sketch your own figures and do not look back in the text.

13. $W\left(\dfrac{\pi}{4}\right)$ 14. $W\left(\dfrac{\pi}{3}\right)$ 15. $W\left(\dfrac{\pi}{6}\right)$

16. $W\left(\dfrac{-\pi}{6}\right)$ 17. $W\left(\dfrac{-\pi}{3}\right)$ 18. $W\left(\dfrac{-\pi}{4}\right)$

19. $W\left(\dfrac{2\pi}{3}\right)$ 20. $W\left(\dfrac{11\pi}{6}\right)$ 21. $W\left(\dfrac{-3\pi}{4}\right)$

22. $W\left(\dfrac{-7\pi}{6}\right)$ 23. $W\left(\dfrac{5\pi}{3}\right)$ 24. $W\left(\dfrac{7\pi}{4}\right)$

25. $W\left(\dfrac{13\pi}{4}\right)$ 26. $W\left(\dfrac{-10\pi}{3}\right)$ 27. $W\left(\dfrac{-25\pi}{6}\right)$

28. $W\left(\dfrac{-21\pi}{4}\right)$

If you multiply the coordinates of the indicated circular point, will you get a positive or negative number? Answer the question (without actually finding or multiplying the coordinates) by determining the quadrant in which each circular point lies. [Note: $\pi/2 \approx 1.57$, $\pi \approx 3.14$, $3\pi/2 \approx 4.71$, and $2\pi \approx 6.28$.]

29. $W(2)$ 30. $W(1)$ 31. $W(3)$ 32. $W(4)$
33. $W(5)$ 34. $W(7)$ 35. $W(-2.5)$ 36. $W(-4.5)$
37. $W(-6.1)$ 38. $W(-1.8)$ 39. $W(8.23)$ 40. $W(12.4)$

C If $W(x) = (a, b)$, indicate whether the following are true (T) or false (F). Sketching figures should help you decide.

41. $W(x + \pi) = (-a, -b)$ 42. $W(x + \pi) = (a, b)$
43. $W(-x) = (-a, b)$ 44. $W(-x) = (a, -b)$
45. $W(x + 2\pi) = (a, b)$ 46. $W(x + 2\pi) = (-a, -b)$

Find all solutions x, $-2\pi \leq x \leq 2\pi$, such that:

47. $W(x) = \left(\dfrac{1}{\sqrt{2}}, \dfrac{1}{\sqrt{2}}\right)$ 48. $W(x) = \left(\dfrac{\sqrt{3}}{2}, \dfrac{1}{2}\right)$

49. $W(x) = \left(\dfrac{-1}{2}, \dfrac{\sqrt{3}}{2}\right)$ 50. $W(x) = \left(\dfrac{1}{2}, \dfrac{-\sqrt{3}}{2}\right)$

51. $W(x) = \left(\dfrac{-\sqrt{3}}{2}, \dfrac{-1}{2}\right)$ 52. $W(x) = \left(\dfrac{-1}{\sqrt{2}}, \dfrac{1}{\sqrt{2}}\right)$

Find all solutions to each equation.

53. $W(x) = W\left(\dfrac{\pi}{4}\right)$

54. $W(x) = W\left(\dfrac{2\pi}{3}\right)$

Section 5-3 Circular Functions

- Definition of the Circular Functions
- Exact Values for Particular Real Numbers
- Sign Properties
- Basic Identities
- Calculator Evaluation

- **Definition of the Circular Functions**

In Section 5-2 we saw that the wrapping function W pairs each real number x with an ordered pair of real numbers (a, b), the coordinates of the circular point $W(x)$. We will use this association to construct the six **circular functions** (also called **trigonometric functions***): **sine, cosine, tangent, cotangent, secant,** and **cosecant**. The values of these functions for a real number x are denoted by **sin x, cos x, tan x, cot x, sec x,** and **csc x,** respectively. These values are expressed in terms of the coordinates of the circular point $W(x) = (a, b)$ as indicated in Definition 1.

DEFINITION 1

Circular Functions

If x is a real number and (a, b) are the coordinates of the circular point $W(x)$, then

$$\sin x = b \qquad \csc x = \dfrac{1}{b}, \quad b \neq 0$$

$$\cos x = a \qquad \sec x = \dfrac{1}{a}, \quad a \neq 0$$

$$\tan x = \dfrac{b}{a}, \quad a \neq 0 \qquad \cot x = \dfrac{a}{b}, \quad b \neq 0$$

* Strictly speaking, the term *trigonometric* is used when we are dealing with angle domains and *circular* is used when we are dealing with real number domains. We will not insist on this distinction and will often, as is the convention, use *trigonometric* for both.

Exact Values for Particular Real Numbers

Using the results in Section 5-2, we can evaluate any one of the six circular functions exactly (when it exists) for integer multiples of the real numbers $\pi/6$, $\pi/4$, $\pi/3$, and $\pi/2$. Figure 7 (which you should have memorized) and symmetry properties of the unit circle are central to the process. Later in this section we will show how a calculator can be used to evaluate the circular functions to eight or ten significant digits for arbitrary real numbers. You might ask why we don't go directly to the calculator? The answer is that there are many situations in which it is more desirable to work with exact forms, if available, than the corresponding decimal approximations that are produced by a calculator.

EXAMPLE 4 Evaluate each circular function for $x = \pi/3$.

Solution From Section 5-2 we know that

$$W\left(\frac{\pi}{3}\right) = \left(\frac{1}{2}, \frac{\sqrt{3}}{2}\right)$$

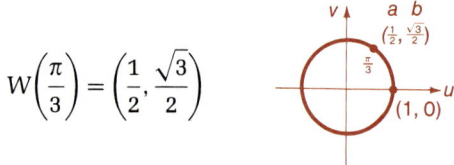

Thus,

$$\sin \frac{\pi}{3} = b = \frac{\sqrt{3}}{2} \qquad \csc \frac{\pi}{3} = \frac{1}{b} = \frac{1}{\sqrt{3}/2} = \frac{2}{\sqrt{3}}$$

$$\cos \frac{\pi}{3} = a = \frac{1}{2} \qquad \sec \frac{\pi}{3} = \frac{1}{a} = \frac{1}{1/2} = 2$$

$$\tan \frac{\pi}{3} = \frac{b}{a} = \frac{\sqrt{3}/2}{1/2} = \sqrt{3} \qquad \cot \frac{\pi}{3} = \frac{a}{b} = \frac{1/2}{\sqrt{3}/2} = \frac{1}{\sqrt{3}}$$

PROBLEM 4 Evaluate each circular function for $x = \pi/6$.

EXAMPLE 5 Evaluate exactly:

(A) $\sin(5\pi/6)$ (B) $\cot(-\pi)$ (C) $\sec(-2\pi/3)$ (D) $\tan(7\pi/4)$

Solution Sketch a figure for each part; then use Figure 7 and symmetry properties of a unit circle.

5-3 Circular Functions 281

(A)

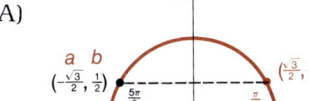

$$\sin \frac{5\pi}{6} = b = \frac{1}{2}$$

(B)

$$\cot(-\pi) = \frac{a}{b} = \frac{-1}{0}$$

Not defined

(C)

$$\sec\left(-\frac{2\pi}{3}\right) = \frac{1}{a} = \frac{1}{-1/2} = -2$$

(D)

$$\tan \frac{7\pi}{4} = \frac{b}{a} = \frac{-1/\sqrt{2}}{1/\sqrt{2}} = -1$$

PROBLEM 5   Evaluate exactly:

(A) $\cos(5\pi/6)$ (B) $\sin(-3\pi/4)$ (C) $\csc(3\pi)$ (D) $\tan(-\pi/3)$

5 Trigonometric Functions

■ **Sign Properties**

As a circular point $W(x)$ moves from quadrant to quadrant its coordinates (a, b) will undergo sign changes; hence, the circular functions will also undergo sign changes. It is useful to know the sign of each circular function in each quadrant. Table 1 shows the sign behavior for each function.

TABLE 1   SIGN PROPERTIES

Circular Function	Sign in Quadrant			
	I	II	III	IV
$\sin x = b$	+	+	−	−
$\csc x = 1/b$	+	+	−	−
$\cos x = a$	+	−	−	+
$\sec x = 1/a$	+	−	−	+
$\tan x = b/a$	+	−	+	−
$\cot x = a/b$	+	−	+	−

```
          v
   a b    |   a b
  (−, +)  |  (+, +)
    II    |    I
──────────┼────────── u
    a b   |   a b
  (−, −)  |  (+, −)
    III   |    IV
```

It is not necessary to memorize Table 1, since the sign of each function for each quadrant is easily determined from its definition (which *should* be memorized).

EXAMPLE 6   Indicate the quadrant in which (both) $\tan x < 0$ and $\sin x > 0$.

Solution $\tan x < 0$ in quadrants II and IV

$\sin x > 0$ in quadrants I and II

Thus, $\tan x < 0$ and $\sin x > 0$ in quadrant II.

PROBLEM 6   Indicate the quadrant in which (both) $\cos x > 0$ and $\cot x < 0$.

■ **Basic Identities**

Returning to the definitions of the circular functions and noting that

$$\sin x = b \quad \text{and} \quad \cos x = a$$

we can obtain the following useful relationships among the six circular functions:

$$\csc x = \frac{1}{b} = \frac{1}{\sin x} \tag{1}$$

$$\sec x = \frac{1}{a} = \frac{1}{\cos x} \tag{2}$$

$$\cot x = \frac{a}{b} = \frac{1}{b/a} = \frac{1}{\tan x} \tag{3}$$

$$\tan x = \frac{b}{a} = \frac{\sin x}{\cos x} \tag{4}$$

$$\cot x = \frac{a}{b} = \frac{\cos x}{\sin x} \tag{5}$$

Because the circular points $W(x)$ and $W(-x)$ are symmetric with respect to the horizontal axis (Fig. 8), we have the following sign properties:

$$\sin(-x) = -b = -\sin x \qquad \text{Sine is an odd function.} \tag{6}$$

$$\cos(-x) = a = \cos x \qquad \text{Cosine is an even function.} \tag{7}$$

$$\tan(-x) = \frac{-b}{a} = -\frac{b}{a} = -\tan x \qquad \text{Tangent is an odd function.} \tag{8}$$

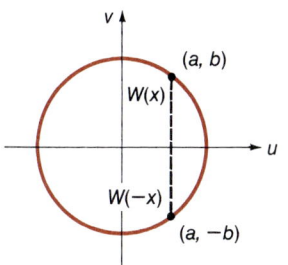

FIGURE 8

Finally, because $(a, b) = (\cos x, \sin x)$ is on the unit circle $u^2 + v^2 = 1$, it follows that

$$(\cos x)^2 + (\sin x)^2 = 1$$

which is usually written as

$$\sin^2 x + \cos^2 x = 1 \tag{9}$$

where $\sin^2 x$ and $\cos^2 x$ are concise ways of writing $(\sin x)^2$ and $(\cos x)^2$, respectively.

Equations (1) through (9) are called **basic identities**. They hold true for all replacements of x by real numbers for which both sides of an equation are defined. The basic identities (1) through (9) must be memorized along with the definitions of the six circular functions, since the material will be used extensively in developments that follow. (Note that most of Chapter 6 is devoted to trigonometric identities.)

EXAMPLE 7   Use the basic identities to find the values of the other five circular functions given $\sin x = -1/2$ and $\tan x > 0$.

Solution We first note that the circular point $W(x)$ is in quadrant III, since that is the only quadrant in which $\sin x < 0$ and $\tan x > 0$. We next find $\cos x$ using identity (9):

$$\sin^2 x + \cos^2 x = 1 \qquad \text{Identity (9)}$$

$$\left(\frac{-1}{2}\right)^2 + \cos^2 x = 1$$

$$\cos^2 x = \frac{3}{4}$$

$$\cos x = \frac{-\sqrt{3}}{2} \qquad \text{Since } W(x) \text{ is in quadrant III}$$

Now, since we have values for $\sin x$ and $\cos x$, we will be able to find values for the other four circular functions using identities (1), (2), (4), and (5).

$$\csc x = \frac{1}{\sin x} = \frac{1}{-1/2} = -2 \qquad \text{Use identity (1).}$$

$$\sec x = \frac{1}{\cos x} = \frac{1}{-\sqrt{3}/2} = \frac{-2}{\sqrt{3}} \qquad \text{Use identity (2).}$$

$$\tan x = \frac{\sin x}{\cos x} = \frac{-1/2}{-\sqrt{3}/2} = \frac{1}{\sqrt{3}} \qquad \text{Use identity (4).}$$

$$\cot x = \frac{\cos x}{\sin x} = \frac{-\sqrt{3}/2}{-1/2} = \sqrt{3} \qquad \text{Use identity (5).}$$
[*Note:* We could also use identity (3).]

In Example 7 it is important to note that we were able to find the values of the other five circular functions without finding x.

PROBLEM 7 Use the basic identities to find the values of the other five circular functions given $\cos x = 1/\sqrt{2}$ and $\cot x < 0$.

EXAMPLE 8 Given x in Example 7, find (using basic identities):

(A) $\sin(-x)$ (B) $\sec(-x)$ (C) $\tan(-x)$

Solution (A) $\sin(-x) = -\sin x$ Use identity (6).

$$= -\left(-\frac{1}{2}\right) = \frac{1}{2}$$

5-3 Circular Functions

(B) $\sec(-x) = \dfrac{1}{\cos(-x)}$ Use identity (2).

$ = \dfrac{1}{\cos x}$ Use identity (7).

$ = \dfrac{1}{-\sqrt{3}/2} = \dfrac{-2}{\sqrt{3}}$

(C) $\tan(-x) = -\tan x$ Use identity (8).

$ = -\dfrac{1}{\sqrt{3}} = \dfrac{-1}{\sqrt{3}}$

PROBLEM 8   Given x in Problem 7 above, find (using basic identities):

(A) $\sin(-x)$ (B) $\cos(-x)$ (C) $\cot(-x)$

■ Calculator Evaluation

Evaluating circular functions for real numbers other than integer multiples of $\pi/6$, $\pi/4$, $\pi/3$, and $\pi/2$ is difficult without the use of a calculator or table. Using advanced mathematics (calculus), scientific calculators are internally programmed to evaluate these functions automatically to an accuracy of eight or ten significant digits.

If you look at the function keys on your scientific calculator, you will find three buttons labeled

$\boxed{\text{SIN}} \quad \boxed{\text{COS}} \quad \boxed{\text{TAN}}$ (10)

These buttons are used to evaluate the sine, cosine, and tangent functions directly. A careful look at the function keys on your calculator will also reveal that there are no keys for cosecant, secant, and cotangent. Why is it not necessary to have these additional keys? Because of the **reciprocal identities** (1)–(3),

$\csc x = \dfrac{1}{\sin x}$ (1)

$\sec x = \dfrac{1}{\cos x}$ (2)

$\cot x = \dfrac{1}{\tan x}$ (3)

we can use the function keys for sine, cosine, and tangent, along with the reciprocal function key

$\boxed{1/x}$

to obtain csc x, sec x, and cot x. Some examples should make the process clear.

> Before commencing with the examples and exercises, read the instruction book accompanying your calculator to determine how to put it in radian (rad) mode. It is in this mode that we can evaluate the circular functions for real numbers. (This process is justified in Section 5-5 when we discuss trigonometric functions with angle domains.) Forgetting to set the right mode before starting calculations involving circular or trigonometric functions is a frequent cause of error when using a calculator.

EXAMPLE 9 Evaluate to four significant digits using a calculator:

(A) sin 2   (B) tan(−1.612)   (C) csc 3.2

Solution (A) sin 2 = 0.9093 $\boxed{2}\ \boxed{\text{SIN}}$

(B) tan(−1.612) = 24.26 $\boxed{1.612}\ \boxed{+/-}\ \boxed{\text{TAN}}$

(C) csc 3.2 = −17.13 $\boxed{3.2}\ \boxed{\text{SIN}}\ \boxed{1/x}$

PROBLEM 9 Evaluate to four significant digits using a calculator:

(A) cos 4   (B) sec 1.605   (C) cot(−3.133)

Answers to Matched Problems

4. $\sin(\pi/6) = 1/2$, $\cos(\pi/6) = \sqrt{3}/2$, $\tan(\pi/6) = 1/\sqrt{3}$, $\csc(\pi/6) = 2$, $\sec(\pi/6) = 2/\sqrt{3}$, $\cot(\pi/6) = \sqrt{3}$
5. (A) $-\sqrt{3}/2$ (B) $-1/\sqrt{2}$ (C) Not defined   (D) $-\sqrt{3}$
6. Quadrant IV
7. $\sin x = -1/\sqrt{2}$, $\csc x = -\sqrt{2}$, $\sec x = \sqrt{2}$, $\tan x = -1$, $\cot x = -1$
8. (A) $1/\sqrt{2}$ (B) $1/\sqrt{2}$ (C) 1
9. (A) −0.6536   (B) −29.24   (C) 116.4

Exercise 5-3 ■ *Figure 7 in Section 5-2, the definition of the circular functions, and the basic identities should now be memorized. Work the problems in this exercise without looking back in the text. Draw lots of pictures, if necessary.*

A 1. Write the value of each circular function in terms of the coordinates (a, b) of the circular point $W(x)$.

(A) cos x   (B) csc x   (C) cot x
(D) sec x   (E) tan x   (F) sin x

5-3 Circular Functions

2. Given $W(x) = (a, b)$, identify each quantity using one of the circular function values: sin x, cos x, and so on.
(A) b (B) $1/a$ (C) b/a
(D) $1/b$ (E) a (F) a/b

Find the exact value of each (if it exists) without the use of a calculator.

3. $\cos 0$ **4.** $\sin 0$ **5.** $\sin \dfrac{\pi}{6}$

6. $\cos \dfrac{\pi}{6}$ **7.** $\sin \dfrac{\pi}{2}$ **8.** $\cos \dfrac{\pi}{2}$

9. $\tan \dfrac{\pi}{3}$ **10.** $\cos \dfrac{\pi}{3}$ **11.** $\tan \dfrac{\pi}{2}$

12. $\cot 0$ **13.** $\sec 0$ **14.** $\cot \dfrac{\pi}{4}$

15. $\sec \dfrac{\pi}{4}$ **16.** $\csc \dfrac{\pi}{3}$ **17.** $\tan \dfrac{\pi}{4}$

18. $\tan 0$ **19.** $\csc 0$ **20.** $\cot \dfrac{\pi}{6}$

In which quadrant must $W(x)$ lie so that:

21. $\cos x < 0$ **22.** $\tan x > 0$ **23.** $\sin x > 0$
24. $\sec x > 0$ **25.** $\cot x < 0$ **26.** $\csc x < 0$

Evaluate to four significant digits using a scientific calculator.

27. $\cos 2.288$ **28.** $\sin 3.104$ **29.** $\tan(-4.644)$
30. $\sec(-1.555)$ **31.** $\csc 1.571$ **32.** $\cot 0.7854$

B *Find the exact value of each (if it exists) without the use of a calculator.*

33. $\cos \pi$ **34.** $\sin \dfrac{3\pi}{2}$ **35.** $\cos\left(-\dfrac{\pi}{2}\right)$

36. $\tan \pi$ **37.** $\sin \dfrac{3\pi}{4}$ **38.** $\cos \dfrac{2\pi}{3}$

39. $\cot 2\pi$ **40.** $\tan\left(-\dfrac{3\pi}{2}\right)$ **41.** $\tan\left(-\dfrac{\pi}{6}\right)$

42. $\cot\left(-\dfrac{\pi}{3}\right)$ **43.** $\cos\left(-\dfrac{\pi}{6}\right)$ **44.** $\sin\left(-\dfrac{\pi}{4}\right)$

45. $\sin \dfrac{5\pi}{4}$ **46.** $\cos \dfrac{3\pi}{4}$ **47.** $\sin \dfrac{7\pi}{6}$

48. $\cos \dfrac{11\pi}{6}$ 49. $\sec \dfrac{5\pi}{3}$ 50. $\csc \dfrac{4\pi}{3}$

51. $\cot\left(-\dfrac{3\pi}{4}\right)$ 52. $\tan \dfrac{5\pi}{4}$ 53. $\sec\left(-\dfrac{\pi}{4}\right)$

54. $\csc\left(-\dfrac{\pi}{3}\right)$ 55. $\cos\left(-\dfrac{5\pi}{6}\right)$ 56. $\tan\left(-\dfrac{4\pi}{3}\right)$

For which values of x, $0 \leq x \leq 2\pi$, is each of the following not defined?

57. cos x   58. sin x   59. tan x
60. cot x   61. sec x   62. csc x

How does the indicated functional value vary as x varies over the indicated intervals? [Hint: Draw a unit circle and note that W(x) = (a, b) = (cos x, sin x).]

63. sin x:
   (A) $[0, \pi/2]$ (B) $[\pi/2, \pi]$ (C) $[\pi, 3\pi/2]$ (D) $[3\pi/2, 2\pi]$
64. cos x:
   (A) $[0, \pi/2]$ (B) $[\pi/2, \pi]$ (C) $[\pi, 3\pi/2]$ (D) $[3\pi/2, 2\pi]$

Compute to four significant digits using a calculator.

65. sin(cos 0.3157)   66. cos(tan 5.183)
67. cos[csc(−1.408)]   68. sec[cot(−3.566)]

Use appropriate identities to solve Problems 69–74.

69. Find sin(−x) if sin x = $-\tfrac{1}{2}$.
70. Find cos(−x) if cos x = $-\tfrac{1}{2}$.
71. Find tan(−x) if tan x = $-\sqrt{3}$.
72. Find sec(−x) if sec x = 1.
73. Find cot(−x) if cot x = 5.
74. Find csc(−x) if csc x = −1.

C Use the basic identities to find the values of the other five circular functions given the indicated information.

75. cos x = $\dfrac{1}{2}$ and   tan x < 0

76. sin x = $\dfrac{\sqrt{3}}{2}$ and   cot x < 0

77. sin x = $-\dfrac{1}{\sqrt{2}}$ and   cos x < 0

78. sec x = 2   and   sin x < 0

5-3 Circular Functions

79. $\tan x = \sqrt{3}$ and $\sin x < 0$

80. $\cot x = -1$ and $\sin x > 0$

Find the least positive x (in terms of π) for which:

81. $\cos x = -1$

82. $\sin x = \dfrac{-\sqrt{3}}{2}$

83. $\cot x = -\sqrt{3}$

84. $\tan x = -1$

85. $\sec x = \dfrac{-2}{\sqrt{3}}$

86. $\csc x = -\sqrt{2}$

In Problems 87 and 88 fill in the blanks citing the appropriate identity (1)–(9).

87. STATEMENT REASON

$\cot^2 x + 1 = \left(\dfrac{\cos x}{\sin x}\right)^2 + 1$ (A) _____

$= \dfrac{\cos^2 x}{\sin^2 x} + 1$ Algebra

$= \dfrac{\cos^2 x + \sin^2 x}{\sin^2 x}$ Algebra

$= \dfrac{1}{\sin^2 x}$ (B) _____

$= \left(\dfrac{1}{\sin x}\right)^2$ Algebra

$= \csc^2 x$ (C) _____

88. STATEMENT REASON

$\tan^2 x + 1 = \left(\dfrac{\sin x}{\cos x}\right)^2 + 1$ (A) _____

$= \dfrac{\sin^2 x}{\cos^2 x} + 1$ Algebra

$= \dfrac{\sin^2 x + \cos^2 x}{\cos^2 x}$ Algebra

$= \dfrac{1}{\cos^2 x}$ (B) _____

$= \left(\dfrac{1}{\cos x}\right)^2$ Algebra

$= \sec^2 x$ (C) _____

APPLICATIONS *If an n-sided regular polygon is inscribed in a circle of radius R, then it can be shown that the area of the polygon is given by*

$$A = \frac{1}{2} nR^2 \sin \frac{2\pi}{n}$$

Compute each area exactly, and then to four significant digits using a calculator (if it is not an integer).

89. $n = 12$, $R = 5$ meters **90.** $n = 4$, $R = 3$ inches

91. $n = 3$, $R = 4$ inches **92.** $n = 8$, $R = 10$ centimeters

Section 5-4 Angles

- **Angles**
- **Degree and Radian Measure**
- **From Degrees to Radians and Vice Versa**

In Section 5-3 we defined the circular functions (with real number domains) using circular points and the wrapping function. Circular functions are widely used in advanced mathematics and certain applied areas such as electrical and aeronautical engineering. In calculus, these functions are utilized in a variety of significant ways.

Historically, as was mentioned earlier, trigonometric functions first came into being relative to triangle measurements, and in this context they had angle domains. Trigonometric functions with angle domains are still very important in surveying, navigation, and many other areas in technology, engineering, and science; hence, they should be considered along with the circular functions. To this end we will discuss angles and their measure in this section and trigonometric functions with angle domains in the next section. As you would probably guess, the circular functions and the trigonometric functions are very closely related.

- **Angles**

In plane geometry an angle is usually thought of as the set of points on two rays (or two line segments) that have a common end point O (Fig. 9). The two rays (or line segments) that form an angle are called the **sides** of the angle, and the common end point is called the **vertex**. We will label angles with single Greek letters, such as θ, α, β, and γ, or with three points on the angle (one on each side and the vertex).

5-4 Angles

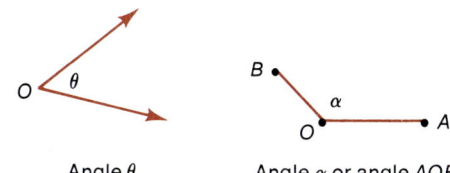

FIGURE 9   Angle θ Angle α or angle AOB

For defining trigonometric functions with angle domains, it is useful to think of an angle in more general terms as follows: To form an **angle** θ, we start with one side, called the **initial side**, in a fixed position. Then, starting with the second side, called the **terminal side**, in the same position as the initial side, we rotate the terminal side in a plane about O until it reaches its final position. A counterclockwise rotation produces a positive angle and a clockwise rotation a **negative** angle. The amount of rotation in either direction is not restricted [Fig. 10(a) and (b)]. Two different angles may have the same initial and terminal sides; such angles are said to be **coterminal** [Fig. 10(c)].

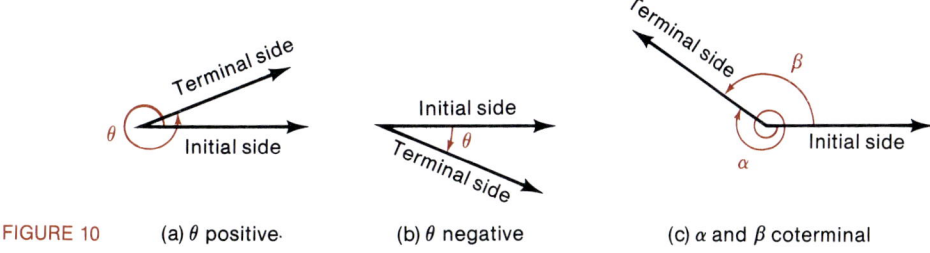

FIGURE 10   (a) θ positive   (b) θ negative   (c) α and β coterminal

An angle in a rectangular coordinate system is said to be in **standard position** if its vertex is at the origin and the initial side is along the positive x axis. If the terminal side of an angle in standard position lies along a coordinate axis, the angle is said to be a **quadrantal angle**. If the terminal side does not lie along a coordinate axis, then the angle is often referred to in terms of the quadrant in which the terminal side lies (Fig. 11).

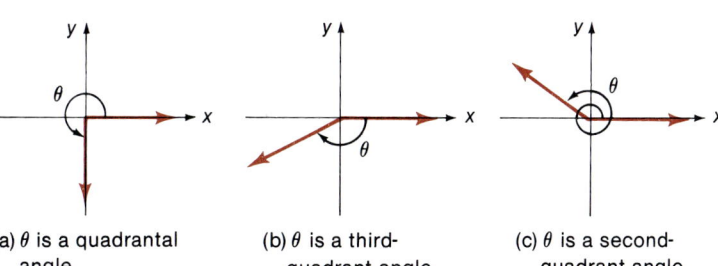

FIGURE 11   Angles in standard positions   (a) θ is a quadrantal angle   (b) θ is a third-quadrant angle   (c) θ is a second-quadrant angle

■ Degree and Radian Measure

Just as line segments are measured in centimeters, meters, inches, or miles, angles are measured in different units. The two most commonly used units for angle measure are degree and radian.

Degree Measure

An angle generated by one complete rotation is said to have a measure of 360 degrees (360°). An angle generated by $\frac{1}{360}$ of a complete rotation is said to have a measure of **1 degree** (1°). The symbol ° denotes degrees.

Certain angles have special names. Figure 12 shows a **straight angle**, a **right angle**, an **acute angle**, and an **obtuse angle**. (As was mentioned earlier, the sides of an angle can either be rays or line segments.)

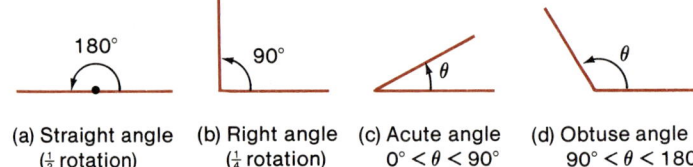

FIGURE 12
(a) Straight angle ($\frac{1}{2}$ rotation) (b) Right angle ($\frac{1}{4}$ rotation) (c) Acute angle $0° < \theta < 90°$ (d) Obtuse angle $90° < \theta < 180°$

Two positive angles are **complementary** if their sum is 90° and **supplementary** if their sum is 180°.

A degree can be divided further using either decimals or minutes and seconds. For the latter we divide a degree into minutes and seconds just as an hour is divided into minutes and seconds; that is, each degree is divided into sixty equal parts called **minutes**, and each minute is divided into sixty equal parts called **seconds**. Symbolically, minutes are represented by ′ and seconds are represented by ″. Thus,

12°23′14″

is a concise way of writing 12 degrees, 23 minutes, and 14 seconds.

Most scientific calculators compute with degrees in decimal form. (To facilitate these calculations, many calculators can directly convert a degree–minute–second representation into a decimal equivalent and vice versa.)

5-4 Angles   293

EXAMPLE 10 Convert $21°47'12''$ to decimal degrees.

Solution $21°47'12'' = (21 + \frac{47}{60} + \frac{12}{3,600})° = 21.787°$ To three decimal places*

Calculator operations

A: $\boxed{21}\boxed{+}\boxed{(}\boxed{(}\boxed{47}\boxed{\div}\boxed{60}\boxed{)}\boxed{+}\boxed{(}\boxed{(}\boxed{12}\boxed{\div}\boxed{3600}\boxed{)}\boxed{=}$

P: $\boxed{21}\boxed{\text{ENTER}}\boxed{47}\boxed{\text{ENTER}}\boxed{60}\boxed{\div}\boxed{+}\boxed{12}\boxed{\text{ENTER}}\boxed{3600}\boxed{\div}\boxed{+}$

If your calculator can perform this conversion directly (in one step), try it.

PROBLEM 10 Convert $193°17'34''$ to decimal degrees.

Degree measure of angles is used extensively in engineering, surveying, and navigation. Another unit of angle measure, called the *radian*, is better suited for scientific work and certain engineering applications.

Radian Measure

If the vertex of an angle θ is placed at the center of a circle with radius $R > 0$, and the length of the arc subtended on the circumference is s, then the **radian measure of θ** is given by

$\theta = \dfrac{s}{R}$ radians

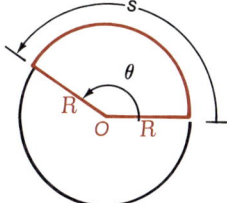

If $s = R$, then

$\theta = \dfrac{R}{R} = 1$ radian

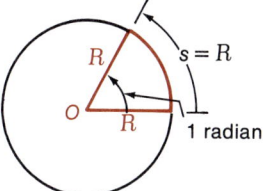

Note: s and R must be measured in the same units.

---

* If an angle is measured to the nearest second, then the converted decimal form should not go beyond three decimal places.

In plane geometry it is shown that if R_1 and R_2 are the radii of two concentric circles* with the same central angle θ, and if s_1 and s_2 are the respective arcs subtended by θ on each circle, then

$$\frac{s_1}{R_1} = \frac{s_2}{R_2}$$

Thus, we conclude that the radian measure of an angle is independent of the size of the circle.

EXAMPLE 11 What is the radian measure of a central angle subtended by an arc of 24 meters in a circle of radius 6 meters?

Solution $\theta = \dfrac{s}{R} = \dfrac{24 \text{ meters}}{6 \text{ meters}} = 4$ radians

Note: The units in which the arc length and radius are measured cancel; hence, we are left with a "unitless" or pure number. For this reason, the word *radian* is often omitted when we are dealing with the radian measure of angles unless a special emphasis is desired.

PROBLEM 11 What is the radian measure of a central angle subtended by an arc of 60 feet in a circle of radius 12 feet?

■ **From Degrees to Radians and Vice Versa**

What is the radian measure of an angle of one complete rotation (360°)? The arc length s subtended by an angle of one complete rotation is the circumference of the circle—that is, $2\pi R$. Thus,

$$\theta = \frac{s}{R} = \frac{2\pi R}{R} = 2\pi \text{ radians}$$

And we conclude that the two measures are related by:

360° = 2π radians

From this we see that an angle of one-half rotation (180°) is half as much; thus,

180° = π radians

This relationship is useful to remember because the radian measure of many special angles can be obtained by starting with it. For example,

$$90° = \frac{180°}{2} = \frac{\pi}{2} \text{ radians} \qquad \text{If we divide 180 by 2, we also divide } \pi \text{ by 2.}$$

* Two circles are concentric if they have the same center.

5-4 Angles

$$45° \;\middle|\; = \frac{180°}{4} \;\middle|\; = \frac{\pi}{4} \text{ radians}$$ If we divide 180 by 4, we also divide π by 4.

$$30° \;\middle|\; = \frac{180°}{6} \;\middle|\; = \frac{\pi}{6} \text{ radians}$$ If we divide 180 by 6, we also divide π by 6.

$$60° \;\middle|\; = \frac{180°}{3} \;\middle|\; = \frac{\pi}{3} \text{ radians}$$ If we divide 180 by 3, we also divide π by 3.

$$270° \;\middle|\; = \frac{3}{2}(180°) \;\middle|\; = \frac{3\pi}{2} \text{ radians}$$ If we multiply 180 by $\frac{3}{2}$, we also multiply π by $\frac{3}{2}$.

We summarize these results in Figure 13 for easy reference. These correspondences should be memorized, since they (and multiples of them) will be used extensively in work that follows.

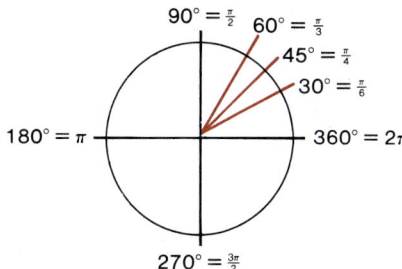

FIGURE 13

In general, to convert degree measure to radian measure and vice versa, we can use the following proportion or corresponding conversion formulas:

Radian–Degree Conversion

If $\theta°$ represents degree measure and θ represents radian measure, then

$$\frac{\theta°}{180°} = \frac{\theta}{\pi} \quad \text{or}$$

$$\theta° = \frac{180°}{\pi} \theta \quad \text{Radians to degrees}$$

$$\theta = \frac{\pi}{180°} \theta° \quad \text{Degrees to radians}$$

EXAMPLE 12 Find exactly:

(A) The radian measure of an angle of 75°
(B) The degree measure of an angle of $11\pi/12$ radians

Solution (A) $\theta = \dfrac{\theta}{180°} \theta°$

$= \dfrac{\pi}{180°}(75°) = \dfrac{5\pi}{12}$

(B) $\theta° = \dfrac{180°}{\pi} \theta$

$= \dfrac{180°}{\pi}\left(\dfrac{11\pi}{12}\right) = 165°$

PROBLEM 12 Find exactly:

(A) The radian measure of an angle of 240°
(B) The degree measure of an angle of $5\pi/3$ radians

EXAMPLE 13 Find to three decimal places:

(A) The radian measure of an angle of 41°12′
(B) The (decimal) degree measure of an angle of 5 radians

Solution (A) First convert 41°12′ to decimal degrees, then use $\theta = (\pi/180)\theta°$:

$41°12' = \left(41 + \dfrac{12}{60}\right)° = 41.2°$

$\theta = \dfrac{\pi}{180}(41.2)$ A: $\boxed{\pi}\boxed{\div}\boxed{180}\boxed{\times}\boxed{41.2}\boxed{=}$

$= 0.719$ P: $\boxed{\pi}\boxed{\text{ENTER}}\boxed{180}\boxed{\div}\boxed{41.2}\boxed{\times}$

(B) Use $\theta° = (180/\pi)\theta$:

$\theta° = \dfrac{180}{\pi}(5)$ A: $\boxed{180}\boxed{\div}\boxed{\pi}\boxed{\times}\boxed{5}\boxed{=}$

$= 286.479°$ P: $\boxed{180}\boxed{\text{ENTER}}\boxed{\pi}\boxed{\div}\boxed{5}\boxed{\times}$

PROBLEM 13 Find to two decimal places:

(A) The radian measure of 125°23′
(B) The (decimal) degree measure of 1 radian

EXAMPLE 14 A belt connects a 2-inch-radius pulley with a 5-inch-radius pulley. If the larger pulley turns through 10 radians, through how many radians will the smaller pulley turn?

5-4 Angles

Solution We first sketch the accompanying figure. When the larger pulley turns through 10 radians, the point P on its circumference will travel the same

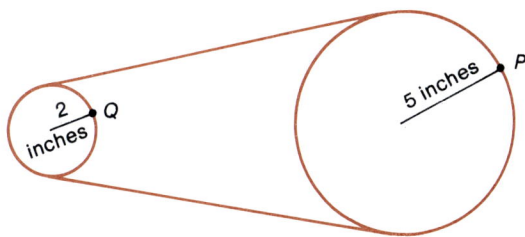

distance (arc length) that point Q on the smaller circle travels. For the larger pulley:

$$\theta = \frac{s}{R}$$

$$s = R\theta = (5)(10) = 50 \text{ inches}$$

For the smaller pulley:

$$\theta = \frac{s}{R} = \frac{50}{2} = 25 \text{ radians}$$

PROBLEM 14 In Example 14, through how many radians will the larger pulley turn if the smaller pulley turns through 5 radians?

Answers to Matched Problems

10. $193.293°$
11. 5
12. (A) $4\pi/3$ (B) $300°$
13. (A) 2.188 (B) $57.296°$
14. 2 radians

Exercise 5-4

In all problems if an angle is expressed by a number that is not in degrees, it is assumed to be in radians.

A *Find the degree measure of each of the following angles, keeping in mind that an angle of one complete rotation corresponds to $360°$.*

1. $\frac{1}{9}$ rotation
2. $\frac{1}{5}$ rotation
3. $\frac{3}{4}$ rotation
4. $\frac{3}{8}$ rotation
5. $\frac{9}{8}$ rotations
6. $\frac{7}{6}$ rotations

Find the radian measure of a central angle θ subtended by an arc s in a circle of radius R where R and s are as given.

7. $R = 4$ centimeters, $s = 24$ centimeters
8. $R = 8$ inches, $s = 16$ inches

9. $R = 12$ feet, $s = 30$ feet
10. $R = 18$ meters, $s = 27$ meters

Find the radian measure of each of the following angles, keeping in mind that an angle of one complete rotation corresponds to 2π radians.

11. $\frac{1}{8}$ rotation
12. $\frac{1}{6}$ rotation
13. $\frac{3}{4}$ rotation
14. $\frac{5}{12}$ rotation
15. $\frac{13}{12}$ rotations
16. $\frac{11}{8}$ rotations

B Find the exact radian measure (in terms of π) of each angle.

17. $30°, 60°, 90°, 120°, 150°, 180°$
18. $60°, 120°, 180°, 240°, 300°, 360°$
19. $-45°, -90°, -135°, -180°$
20. $-90°, -180°, -270°, -360°$

Find the exact degree measure of each angle.

21. $\dfrac{\pi}{3}, \dfrac{2\pi}{3}, \pi, \dfrac{4\pi}{3}, \dfrac{5\pi}{3}, 2\pi$
22. $\dfrac{\pi}{6}, \dfrac{\pi}{3}, \dfrac{\pi}{2}, \dfrac{2\pi}{3}, \dfrac{5\pi}{6}, \pi$
23. $-\dfrac{\pi}{2}, -\pi, -\dfrac{3\pi}{2}, -2\pi$
24. $-\dfrac{\pi}{4}, -\dfrac{\pi}{2}, -\dfrac{3\pi}{4}, -\pi$

Convert to decimal degrees to three decimal places.

25. $5°51'33''$
26. $14°18'37''$
27. $354°8'29''$
28. $184°31'7''$

Find the radian measure to three decimal places for each angle.

29. $64°$
30. $25°$
31. $108.413°$
32. $203.097°$
33. $13°25'14''$
34. $56°11'52''$

Find the (decimal) degree measure to two decimal places for each angle.

35. 0.93
36. 0.08
37. 1.13
38. 3.07
39. -2.35
40. -1.72

Indicate whether the angle is a I, II, III, or IV quadrant angle or a quadrantal angle. All angles are in standard position in a rectangular coordinate system. (A sketch may be of help in some problems.)

41. $187°$
42. $135°$
43. $-200°$
44. $-60°$
45. 4
46. 3
47. $270°$
48. $360°$
49. -1
50. -6
51. $\dfrac{5\pi}{3}$
52. $\dfrac{2\pi}{3}$
53. $-\dfrac{7\pi}{6}$
54. $-\dfrac{3\pi}{4}$
55. $-\pi$
56. $-\dfrac{3\pi}{2}$

57. 820°   **58.** −565°   **59.** $\dfrac{13\pi}{4}$ **60.** $\dfrac{23\pi}{3}$

C *Which angles are coterminal with 30°? All are in standard position in a rectangular coordinate system.*

61. 390°   **62.** −330°   **63.** $\dfrac{\pi}{6}$ **64.** $-\dfrac{11\pi}{6}$

65. −690°   **66.** 750°

Which angles are coterminal with $3\pi/4$? All are in standard position in a rectangular coordinate system.

67. $-\dfrac{3\pi}{4}$ **68.** $-\dfrac{7\pi}{4}$ **69.** 135°   **70.** −225°

71. $\dfrac{11\pi}{4}$ **72.** $-\dfrac{5\pi}{4}$

APPLICATIONS

73. *Engineering.* A bicycle has a front wheel with a diameter of 40 centimeters and a back wheel of diameter 60 centimeters. Through what angle in radians does the front wheel turn if the back wheel turns through 8 radians?

74. *Astronomy.* A line from the earth to the sun sweeps out an angle of how many radians in one week? Assume the earth's orbit is circular and express the answer in terms of π and as a decimal fraction to two decimal places.

75. *Engineering.* Through how many radians does a 10-centimeter-diameter pulley turn when 10 meters of rope have been pulled through it without slippage? [*Hint:* Use $\theta = s/R$.]

Section 5-5 Trigonometric Functions

- Definition of the Trigonometric Functions
- Calculator Evaluation of Trigonometric Functions
- Definition of the Trigonometric Functions—Alternate Form
- Exact Values for Special Angles and Real Numbers

- Definition of the Trigonometric Functions

We are now ready to define trigonometric functions with angle domains. Since we have already defined the circular functions with real number domains, we will take advantage of these results and define the trigonometric functions with angle domains in terms of the circular functions. To each of the six circular functions we associate a trigonometric function of the same name. If θ is an angle, we assign values to $\sin \theta$, $\cos \theta$, $\tan \theta$, $\csc \theta$, $\sec \theta$, and $\cot \theta$ as given in Definition 2.

DEFINITION 2 | **Trigonometric Functions with Angle Domains**

If θ is an angle with radian measure x, then the value of each **trigonometric function** at θ is given by its value at the real number x.

TRIGONOMETRIC FUNCTION		CIRCULAR FUNCTION
$\sin \theta$	=	$\sin x$
$\cos \theta$	=	$\cos x$
$\tan \theta$	=	$\tan x$
$\csc \theta$	=	$\csc x$
$\sec \theta$	=	$\sec x$
$\cot \theta$	=	$\cot x$

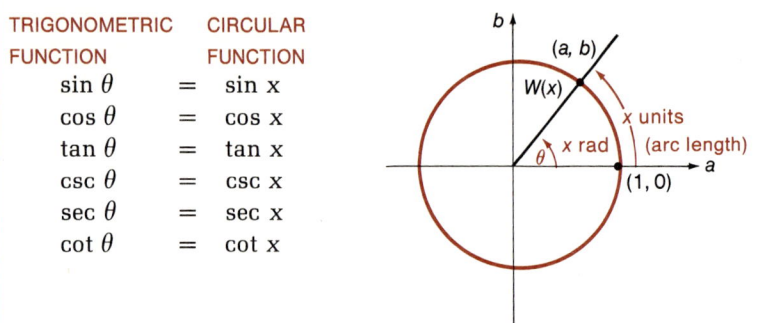

Note: To reduce the number of different symbols in certain figures, the u and v axes we started with will often be labeled as the a and b axes, respectively.

The figure in Definition 2 makes use of the important fact that in a unit circle the arc length subtended by an angle of x radians is x units long and vice versa:

$$s = R\theta = 1 \cdot x = x$$

EXAMPLE 15 Evaluate exactly without a calculator:

(A) $\sin 30°$ (B) $\cos 180°$ (C) $\csc(-150°)$

Solution (A) $\sin 30° = \sin\left(\dfrac{\pi}{6} \text{ rad}\right) = \sin \dfrac{\pi}{6} = \dfrac{1}{2}$

(B) $\cos 180° = \cos(\pi \text{ rad}) = \cos \pi = -1$

(C) $\csc(-150°) = \csc\left(\dfrac{-5\pi}{6} \text{ rad}\right) = \csc\left(\dfrac{-5\pi}{6}\right) = -2$

PROBLEM 15 Evaluate exactly without a calculator:

(A)   tan 45°   (B)   sin 90°   (C)   sec(−120°)

■ **Calculator Evaluation of Trigonometric Functions**

How do we evaluate trigonometric functions for arbitrary angles? Just as a calculator or table can be used to approximate circular functions for arbitrary real numbers, a calculator or table can be used to approximate trigonometric functions for arbitrary angles.

Most scientific calculators have a choice of three trigonometric modes: degree (decimal), radian, or grad.

$$\text{Right angle} = 90° = \frac{\pi}{2} \text{ radians} = 100 \text{ grads}$$

The **grad unit** is used in certain engineering applications; it will not be used in this book. We repeat a warning stated earlier:

> Read the instruction book accompanying your scientific calculator to determine how to put your calculator in degree or radian mode. Forgetting to set the right mode before starting calculations involving trigonometric functions is a frequent cause of error when using a calculator.

Using a calculator with degree and radian modes, we can evaluate trigonometric functions directly for angles in either degree or radian measure without having to convert degree measure to radian measure first. However, when working in degree mode, degree measure must be in decimal degrees. (Some calculators can convert degree-minute-second forms to decimal degrees and vice versa by the push of a button.)

EXAMPLE 16 Find to four significant digits using a scientific calculator:

(A)   sin 48°   (B)   cos(−103°)
(C)   tan 260.35°   (D)   sin(−338°32′43″)

Solution   Since all these problems involve degrees, set the calculator in degree mode.

(A)   Enter 48 and press $\boxed{\text{SIN}}$ to obtain

$$\sin 48° = 0.7431 \quad \boxed{48}\ \boxed{\text{SIN}}$$

(B) Enter -103 and press $\boxed{\text{COS}}$ to obtain
$$\cos(-103°) = -0.2250 \quad \boxed{103}\,\boxed{+/-}\,\boxed{\text{COS}}$$
[*Note:* Some calculators use $\boxed{\text{chs}}$ in place of $\boxed{+/-}$.]

(C) $\tan 260.35 = 5.881 \quad \boxed{260.35}\,\boxed{\text{TAN}}$

(D) $\sin(-338°32'43'') = \sin(-338.545\,2\ldots)$ Change to decimal degrees.
$$= -0.3658 \quad \boxed{338.5452\ldots}\,\boxed{+/-}\,\boxed{\text{SIN}}$$

PROBLEM 16 Find to four significant digits using a scientific calculator:

(A) $\cos 34°$ (B) $\sin(-257°)$
(C) $\tan 93.25°$ (D) $\cos(-12°13'55'')$

Note: Table evaluation of trigonometric functions is included in Appendix A-8 for those who still desire that approach.

How do we evaluate $\csc\theta$, $\sec\theta$, and $\cot\theta$? Because of Definition 2, the same basic identities hold for the trigonometric functions as for the circular functions. In particular.

$$\csc\theta = \frac{1}{\sin\theta} \qquad \sec\theta = \frac{1}{\cos\theta} \qquad \cot\theta = \frac{1}{\tan\theta}$$

Thus, for example, to find $\csc 28°$, we find $\sin 28°$ and then press the reciprocal button $\boxed{1/x}$.

EXAMPLE 17 Find to four significant digits using a scientific calculator:

(A) $\csc 28°$ (B) $\sec(-405.23°)$
(C) $\cot 83.451°$ (D) $\sec(47°11'23'')$

Solution Be sure the calculator is in degree mode.

(A) Enter 28, press $\boxed{\text{SIN}}$, and then press $\boxed{1/x}$ to obtain
$$\csc 28° = 2.130 \quad \boxed{28}\,\boxed{\text{SIN}}\,\boxed{1/x}$$

(B) Enter -405.23, press $\boxed{\text{COS}}$, and then press $\boxed{1/x}$ to obtain
$$\sec(-405.23°) = 1.420 \quad \boxed{405.23}\,\boxed{+/-}\,\boxed{\text{COS}}\,\boxed{1/x}$$

(C) $\cot 83.451° = 0.1148 \quad \boxed{83.451}\,\boxed{\text{TAN}}\,\boxed{1/x}$

(D) $\sec(47°11'23'') = \sec(47.189\,7\ldots)$ Change to decimal degrees.
$$= 1.472 \quad \boxed{147.1897\ldots}\,\boxed{\text{COS}}\,\boxed{1/x}$$

PROBLEM 17 Find to four significant digits using a scientific calculator:

(A) $\cot 75°$ (B) $\csc(-123.62°)$
(C) $\sec 178.33°$ (D) $\cot(-68°54'12'')$

5-5 Trigonometric Functions

For an angle in radian measure, we set the calculator in radian mode and evaluate exactly as we evaluated the corresponding circular function in Section 5-3. For example,

$$\sin(5.138 \text{ rad}) = \sin 5.138 = -0.9108$$

■ Definition of the Trigonometric Functions—Alternate Form

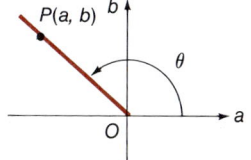

FIGURE 14

For many applications involving the use of trigonometric functions (including triangle applications) it is useful to write Definition 2 in an alternate form—a form that utilizes the coordinates of an arbitrary point, other than (0, 0), on the terminal side of an angle θ (see Fig. 14).

This alternate form of Definition 2 is easily found by inserting a unit circle in Figure 14, dropping perpendiculars from points P and Q to the horizontal axis (Fig. 15), and utilizing the fact that ratios of corresponding sides of similar triangles are proportional.

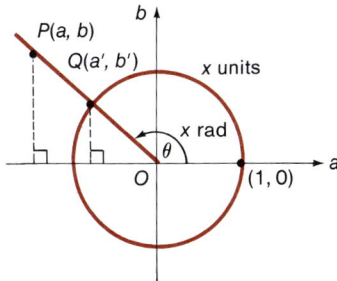

FIGURE 15

Letting $R = d(O, P)$ and noting that $d(O, Q) = 1$, we have

$$\sin \theta = \sin x = b' = \frac{b'}{1} = \frac{b}{R} \qquad \text{b and b' will always have the same sign}$$

$$\cos \theta = \cos x = a' = \frac{a'}{1} = \frac{a}{R} \qquad \text{a and a' will always have the same sign}$$

The values of the other four trigonometric functions can be obtained using basic identities. For example,

$$\tan \theta = \frac{\sin \theta}{\cos \theta} = \frac{b/R}{a/R} = \frac{b}{a}$$

We now have the very useful alternate form of Definition 2 given in the box at the top of the next page.

**DEFINITION 2
(ALTERNATE FORM)**

Trigonometric Functions with Angle Domains

If θ is an arbitrary angle in standard position in a rectangular coordinate system and $P(a, b)$ is a point R units from the origin ($R > 0$) on the terminal side of θ, then

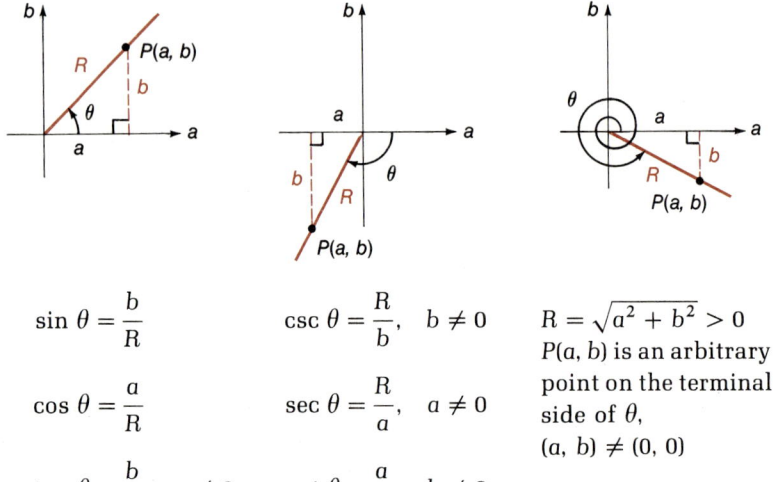

$\sin \theta = \dfrac{b}{R}$ $\csc \theta = \dfrac{R}{b}, \quad b \neq 0$ $R = \sqrt{a^2 + b^2} > 0$

$\cos \theta = \dfrac{a}{R}$ $\sec \theta = \dfrac{R}{a}, \quad a \neq 0$ $P(a, b)$ is an arbitrary point on the terminal side of θ,

$\tan \theta = \dfrac{b}{a}, \quad a \neq 0$ $\cot \theta = \dfrac{a}{b}, \quad b \neq 0$ $(a, b) \neq (0, 0)$

Domains: Sets of all possible angles for which the ratios are defined
Ranges: Subsets of the set of real numbers

Note: The right triangle formed by dropping a perpendicular from $P(a, b)$ to the horizontal axis is called the **reference triangle** associated with the angle θ. We will often refer to this triangle.

The alternate form of Definition 2 should be memorized. As a memory aid, note that when $R = 1$, then $P(a, b)$ is on a unit circle, and all function values correspond to the values in Definition 1 for circular functions. In fact, using the alternate form of Definition 2 in conjunction with the original statement of Definition 2, we now have an alternate way of evaluating circular functions:

If C is any circular function and T is the corresponding trigonometric function, then for any real number x,

$$C(x) = T(x \text{ rad}) \tag{1}$$

That is, sin x = sin(x rad), cos x = cos(x rad), and so on.

It is because of (1) that we are able to evaluate circular functions using a calculator set in radian mode (see Section 5-3). Generally, unless a certain emphasis is desired, we will not use "rad" after a real number. That is, we will interpret expressions such as "sin 5.73" as the "circular function value sin 5.73" or the "trigonometric function value sin(5.73 rad)" by the context in which the expression occurs or the form we wish to emphasize. We will remain flexible and will often switch back and forth between circular function emphasis and trigonometric function emphasis, depending on which approach provides the most enlightenment for a given situation.

It is useful to visualize the definition of the trigonometric functions in terms of a "function machine." Figure 16 illustrates a "cosine machine" with an angle domain.

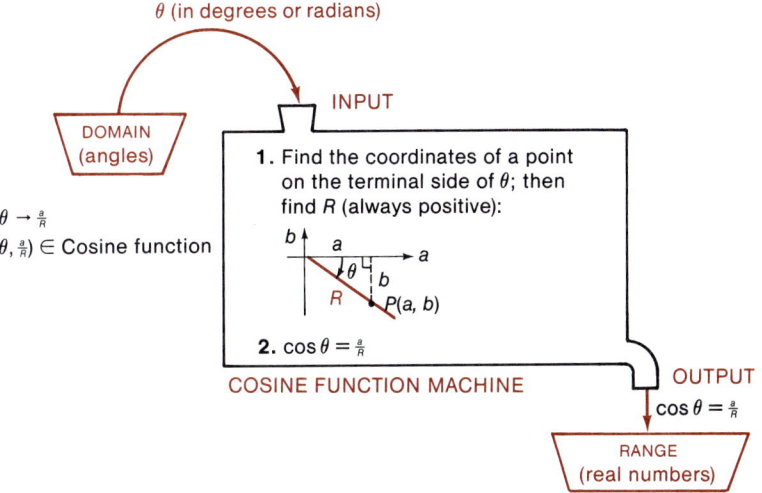

FIGURE 16 Function machine

EXAMPLE 18 Find the value of each of the six trigonometric functions for the illustrated angle θ with terminal side that contains $P(-3, -4)$.

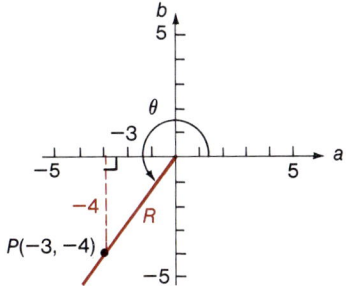

Solution $(a, b) = (-3, -4)$
$R = \sqrt{a^2 + b^2} = \sqrt{(-3)^2 + (-4)^2} = \sqrt{25} = 5$

$\sin \theta = \dfrac{b}{R} = \dfrac{-4}{5} = -\dfrac{4}{5}$ $\qquad \csc \theta = \dfrac{R}{b} = \dfrac{5}{-4} = -\dfrac{5}{4}$

$\cos \theta = \dfrac{a}{R} = \dfrac{-3}{5} = -\dfrac{3}{5}$ $\qquad \sec \theta = \dfrac{R}{a} = \dfrac{5}{-3} = -\dfrac{5}{3}$

$\tan \theta = \dfrac{b}{a} = \dfrac{-4}{-3} = \dfrac{4}{3}$ $\qquad \cot \theta = \dfrac{a}{b} = \dfrac{-3}{-4} = \dfrac{3}{4}$

PROBLEM 18 Find the value of each of the six trigonometric functions if the terminal side of θ contains the point $(-6, -8)$. [*Note:* This point lies on the terminal side of the angle in Example 18; hence, the final results should be the same as those obtained in Example 18.]

EXAMPLE 19 Find the value of each of the other five trigonometric functions for an angle θ (without finding θ) given that θ is a IV quadrant angle and $\sin \theta = -\frac{4}{5}$.

Solution The information given is sufficient for us to locate a reference triangle in quadrant IV for θ, even though we do not know what θ is. We sketch a reference triangle, label what we know, and then complete the problem as indicated.

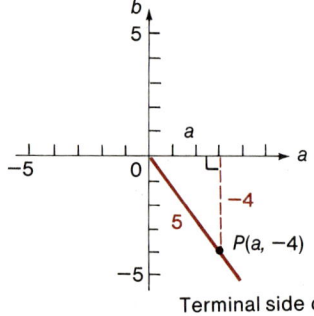

Since $\sin \theta = b/R = -\frac{4}{5}$, we know that $b = -4$ and $R = 5$ (R is never negative). If we can find a, then we can determine the value of the other five functions.

Use the Pythagorean theorem to find a:

$a^2 + (-4)^2 = 5^2$ $\qquad$ a cannot be negative because θ is a IV quadrant angle.

$\qquad a^2 = 9$

$\qquad a = 3$

5-5 Trigonometric Functions

Using $(a, b) = (3, -4)$ and $R = 5$, we have

$$\cos\theta = \frac{a}{R} = \frac{3}{5} \qquad \sec\theta = \frac{R}{a} = \frac{5}{3} \qquad \csc\theta = \frac{R}{b} = \frac{5}{-4} = -\frac{5}{4}$$

$$\tan\theta = \frac{b}{a} = \frac{-4}{3} = -\frac{4}{3} \qquad \cot\theta = \frac{a}{b} = \frac{3}{-4} = -\frac{3}{4}$$

PROBLEM 19   Find the value of the other five trigonometric functions for an angle θ (without finding θ) given that θ is a II quadrant angle and $\tan\theta = -\frac{3}{4}$.

■ **Exact Values for Special Angles and Real Numbers**

Assuming a trigonometric function is defined, it can be evaluated exactly without the use of a calculator or a table (which is different from finding approximate values using a calculator or a table) for any integer multiple of 30°, 45°, 60°, 90°, $\pi/6$, $\pi/4$, $\pi/3$, or $\pi/2$. With a little practice you will be able to determine these values mentally. Working with exact values has advantages over working with approximate values in many situations.

The easiest angles to deal with are **quadrantal angles**—that is, angles with their terminal side lying along a coordinate axis. These angles are integer multiples of 90° or $\pi/2$. It is easy to find the coordinates of a point on a coordinate axis. Since any nonorigin point will do, we shall, for convenience, choose points one unit from the origin (Fig. 17).

FIGURE 17

Points: $(0, 1)$, $(-1, 0)$, $(1, 0)$, $(0, -1)$.

In each case $R = \sqrt{a^2 + b^2} = 1$, a positive number.

EXAMPLE 20   Find:

(A) $\sin 90°$ (B) $\cos \pi$ (C) $\tan(-2\pi)$ (D) $\cot(-180°)$

Solution For each, visualize the location of the terminal side of the angle relative to Figure 17. With a little practice, you should be able to do most of the following mentally.

(A) $\sin 90° = \boxed{\dfrac{b}{R}} = \dfrac{1}{1} = 1$ $(a, b) = (0, 1)$, $R = 1$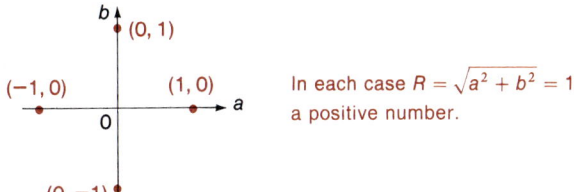

(B) $\cos \pi = \boxed{\dfrac{a}{R}} = \dfrac{-1}{1} = -1 \quad (a, b) = (-1, 0), \quad R = 1$

(C) $\tan(-2\pi) = \boxed{\dfrac{b}{a}} = \dfrac{0}{1} = 0 \quad (a, b) = (1, 0), \quad R = 1$

(D) $\cot(-180°) = \boxed{\dfrac{a}{b}} = \dfrac{-1}{0} \quad (a, b) = (-1, 0), \quad R = 1$

Not defined

PROBLEM 20 Find:

(A) $\sin(3\pi/2)$ (B) $\sec(-\pi)$ (C) $\tan 90°$ (D) $\cot(-270°)$

If a reference triangle of a given angle is a 30°–60° right triangle or a 45° right triangle, then we will be able to the find exact nonorigin coordinates on the terminal side of the given angle. Because the reference triangle plays a very important role in this process, we restate its definition as well as that of a **reference angle** in the box.

Reference Triangle and Angle

1. To form a **reference triangle** for θ, drop a perpendicular from a point $P(a, b)$ on the terminal side of θ to the horizontal axis.
2. The **reference angle** α is the acute angle (always taken positive) between the terminal side of θ and the horizontal axis.

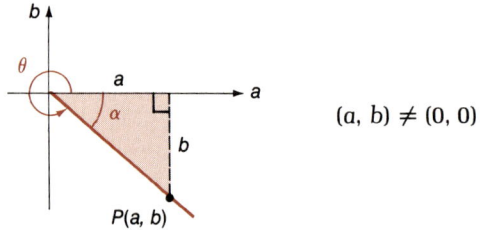

$(a, b) \neq (0, 0)$

A 30°–60° right triangle forms half of an equilateral triangle, as indicated in Figure 18. Because all sides are equal in an equilateral triangle, we can apply the Pythagorean theorem to obtain a useful relationship among the three sides of the original triangle.

5-5 Trigonometric Functions

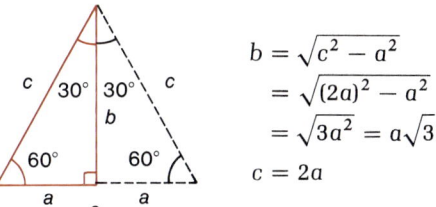

$$b = \sqrt{c^2 - a^2}$$
$$= \sqrt{(2a)^2 - a^2}$$
$$= \sqrt{3a^2} = a\sqrt{3}$$
$$c = 2a$$

FIGURE 18

Similarly, using the Pythagorean theorem on a 45° right triangle, we obtain the result shown in Figure 19.

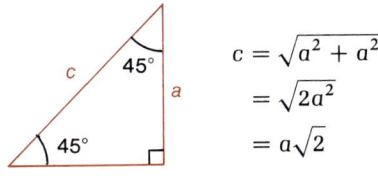

$$c = \sqrt{a^2 + a^2}$$
$$= \sqrt{2a^2}$$
$$= a\sqrt{2}$$

FIGURE 19

We summarize these results in Figure 20 along with two frequently used special cases for convenient reference. The sides of these special triangles should be memorized; they will be used often in this and subsequent sections.

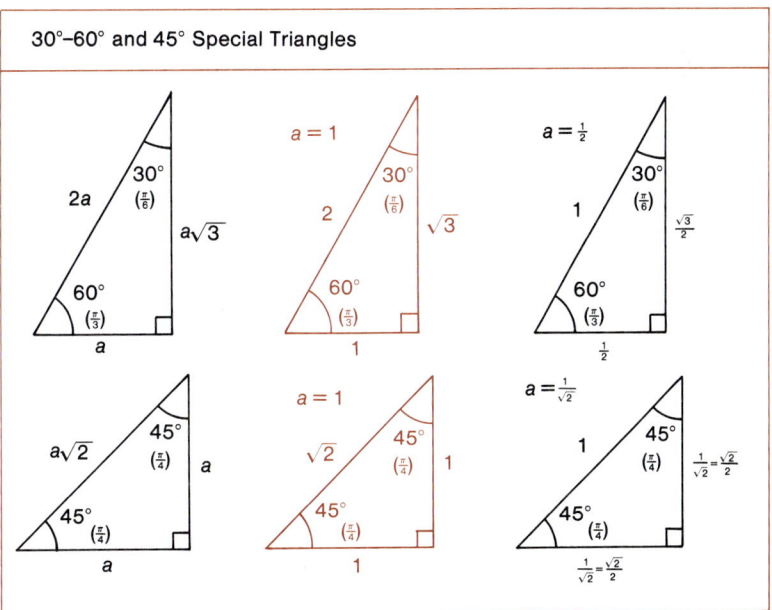

FIGURE 20

The two triangles shown in color in Figure 20 are the easiest to remember. The others can be obtained from these by multiplying or dividing the length of each side by the same nonzero quantity. If an angle (or a real number) has a 30°–60° or a 45° reference triangle, then we can use Figure 20 to find exact nonorigin coordinates on the terminal side of the angle. Using the alternate form of Definition 2, we can find the exact value of any of the six trigonometric functions for the indicated angle (or real number).

EXAMPLE 21 Evaluate exactly.

(A) $\cos 60°$, $\sin(\pi/3)$, $\tan(\pi/3)$ (B) $\sin 45°$, $\cot(\pi/4)$, $\sec(\pi/4)$

Solution (A) Use the special 30°–60° triangle with sides 1, 2, and $\sqrt{3}$ as the reference triangle and use 60° or $\pi/3$ as the reference angle. Use the sides of the reference triangle to determine $P(a, b)$ and R; then use the appropriate definitions.

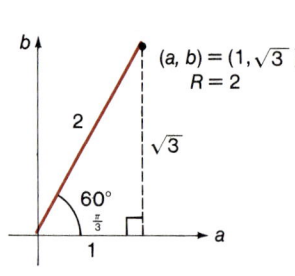

$$\cos 60° = \frac{a}{R} = \frac{1}{2}$$

$$\sin \frac{\pi}{3} = \frac{b}{R} = \frac{\sqrt{3}}{2}$$

$$\tan \frac{\pi}{3} = \frac{b}{a}$$
$$= \frac{\sqrt{3}}{1} = \sqrt{3}$$

(B) Use the special 45° triangle with sides 1, 1, and $\sqrt{2}$ as the reference triangle and use 45° or $\pi/4$ as the reference angle. Use the sides of the reference triangle to determine $P(a, b)$; then use the appropriate definitions.

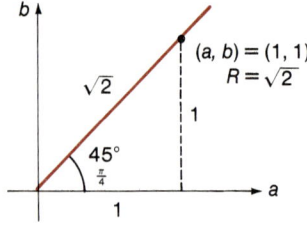

$$\sin 45° = \frac{b}{R} = \frac{1}{\sqrt{2}} \text{ or } \frac{\sqrt{2}}{2}$$

$$\cot \frac{\pi}{4} = \frac{a}{b} = \frac{1}{1} = 1$$

$$\sec \frac{\pi}{4} = \frac{R}{a} = \frac{\sqrt{2}}{1} = \sqrt{2}$$

5-5 Trigonometric Functions   311

PROBLEM 21   Evaluate exactly.

(A) $\cos 45°$, $\tan(\pi/4)$, $\csc(\pi/4)$ (B) $\sin 30°$, $\cos(\pi/6)$, $\cot(\pi/6)$

Before proceeding, it is useful to observe from a geometric point of view multiples of $\pi/3$ (60°), $\pi/6$ (30°), and $\pi/4$ (45°). These are illustrated in Figure 21.

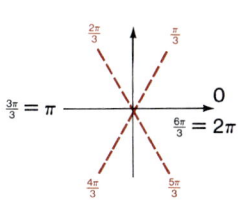
(a) Multiples of $\frac{\pi}{3}$ (60°)

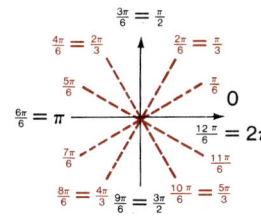
(b) Multiples of $\frac{\pi}{6}$ (30°)

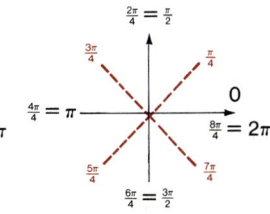
(c) Multiples of $\frac{\pi}{4}$ (45°)

FIGURE 21   Multiples of special angles

EXAMPLE 22   Evaluate exactly.

(A) $\cos(7\pi/4)$ (B) $\sin(2\pi/3)$ (C) $\tan 210°$
(D) $\sec(-240°)$ (E) $\csc(-5\pi/6)$ (F) $\cot 315°$

Solution   Each angle (or real number) has a 30°–60° or a 45° reference triangle. Locate it, determine (a, b) and R as in Example 21, and then evaluate.

(A) $\cos \dfrac{7\pi}{4} = \dfrac{1}{\sqrt{2}}$ or $\dfrac{\sqrt{2}}{2}$ (B) $\sin \dfrac{2\pi}{3} = \dfrac{\sqrt{3}}{2}$

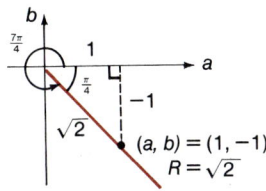   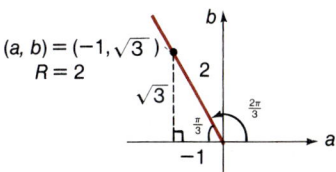

(C) $\tan 210° = \dfrac{-1}{-\sqrt{3}} = \dfrac{1}{\sqrt{3}}$ or $\dfrac{\sqrt{3}}{3}$ (D) $\sec(-240°) = \dfrac{2}{-1} = -2$

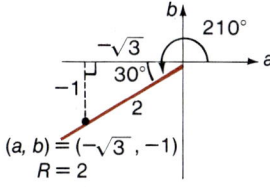   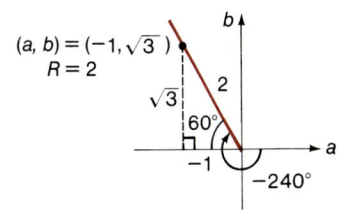

(E) $\csc \dfrac{-5\pi}{6} = \dfrac{2}{-1} = -2$ (F) $\cot 315° = \dfrac{1}{-1} = -1$

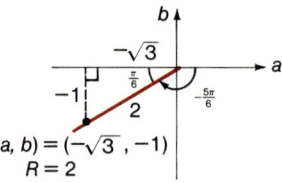 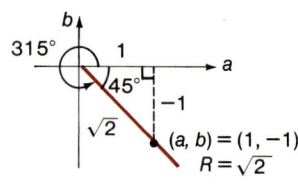

PROBLEM 22 Evaluate exactly.

(A) $\tan(-\pi/4)$ (B) $\sin 210°$ (C) $\cos(2\pi/3)$
(D) $\csc(-240°)$ (E) $\cot(-5\pi/6)$ (F) $\sec 420°$

Now let us reverse the problem; that is, let the exact value of one of the six trigonometric functions be given and assume this value corresponds to one of the special reference triangles. Can we find a least positive θ for which the trigonometric function has that value? Example 23 shows how.

EXAMPLE 23 Find the least positive θ in degree and radian measure for which each is true.

(A) $\tan \theta = 1/\sqrt{3}$ (B) $\sec \theta = -\sqrt{2}$

Solution (A) $\tan \pi = \dfrac{b}{a} = \dfrac{1}{\sqrt{3}}$

Thus, $(a, b) = (\sqrt{3}, 1)$ or $(-\sqrt{3}, -1)$. The least positive θ for which this is true is a I quadrant angle with reference triangle as drawn:

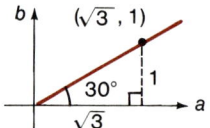

$\theta = 30°$ or $\dfrac{\pi}{6}$

(B) $\sec \theta = \dfrac{R}{a} = \dfrac{\sqrt{2}}{-1}$ Because $R > 0$

a is negative in the II and III quadrants. The smallest positive θ is associated with a 45° reference triangle in the II quadrant as drawn:

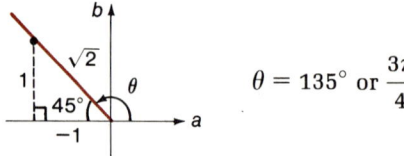

$\theta = 135°$ or $\dfrac{3\pi}{4}$

PROBLEM 23   Find the least positive θ in degree and radian measure for which each is true.

(A) $\sin \theta = \sqrt{3}/2$ (B) $\cos \theta = -1/\sqrt{2}$

Remark: After quite a bit of practice, the reference triangle figures in Examples 22 and 23 can be visualized mentally; however, when in doubt, draw a figure.

COMMENT   In evaluating trigonometric functions for integer multiples of 30°, 45°, 60°, $\pi/6$, $\pi/4$, or $\pi/3$, should one use Figure 7 in Section 5-2 with symmetry properties of unit circle or reference triangles and the special 30°–60° and 45° triangles? Many people find the triangle approach easier to remember and use; however, some prefer the other approach. Use the approach that is easiest for you.

Answers to Matched Problems

15. (A) 1   (B) 1   (C) -2
16. (A) 0.8290   (B) 0.9744   (C) -17.61 (D) 0.9773
17. (A) 0.2679   (B) -1.201 (C) -1.000 (D) -0.3858
18. $\sin \theta = -\frac{4}{5}$, $\cos \theta = -\frac{3}{5}$, $\tan \theta = \frac{4}{3}$, $\csc \theta = -\frac{5}{4}$, $\sec \theta = -\frac{5}{3}$, $\cot \theta = \frac{3}{4}$
19. $\sin \theta = \frac{3}{5}$, $\cos \theta = -\frac{4}{5}$, $\csc \theta = \frac{5}{3}$, $\sec \theta = -\frac{5}{4}$, $\cot \theta = -\frac{4}{3}$
20. (A) -1 (B) -1 (C) Not defined   (D) 0
21. (A) $\cos 45° = 1/\sqrt{2}$, $\tan(\pi/4) = 1$, $\csc(\pi/4) = \sqrt{2}$
     (B) $\sin 30° = \frac{1}{2}$, $\cos(\pi/6) = \sqrt{3}/2$, $\cot(\pi/6) = \sqrt{3}$
22. (A) -1 (B) $-\frac{1}{2}$ (C) $-\frac{1}{2}$ (D) $2/\sqrt{3}$
     (E) $\sqrt{3}$ (F) 2
23. (A) 60° or $\pi/3$ (B) 135° or $3\pi/4$

Exercise 5-5

A   *Find the value of each of the six trigonometric functions for an angle θ that has a terminal side containing the indicated point.*

1. (6, 8)   2. $(-3, 4)$ 3. $(-2, -1)$
4. $(-2, -2)$ 5. $(-1, \sqrt{3})$ 6. $(\sqrt{3}, 1)$

Evaluate to four significant digits using a scientific calculator. Make sure your calculator is in the correct mode (degree or radian) for each problem.

7. $\sin 25°$ 8. $\tan 89°$ 9. $\cot 12$
10. $\csc 13$ 11. $\sin 2.137$ 12. $\tan 4.327$
13. $\cot(-431.41°)$ 14. $\sec(-247.39°)$ 15. $\sin(113°27'13'')$
16. $\cos(235°12'47'')$ 17. $\sec(367°51'24'')$ 18. $\cot(105°44'18'')$

Evaluate exactly without using a calculator or table.

19. sin 0°
20. cos 0°
21. tan 60°
22. cos 30°
23. sin 45°
24. csc 60°
25. sec 45°
26. cot 45°
27. cot 0°
28. cot 90°
29. tan 90°
30. sec 0°

B *Evaluate exactly without using a calculator or table.*

31. cos 120°
32. sin 150°
33. $\cos \dfrac{3\pi}{2}$
34. $\sin \dfrac{\pi}{2}$
35. cot(−60°)
36. sec(−30°)
37. $\cos\left(-\dfrac{\pi}{6}\right)$
38. $\cot\left(-\dfrac{\pi}{4}\right)$
39. $\sin \dfrac{3\pi}{4}$
40. $\cos \dfrac{2\pi}{3}$
41. csc 150°
42. cot 225°
43. sec 300°
44. cot 330°
45. $\cot \dfrac{3\pi}{2}$
46. csc(−π)
47. $\tan\left(-\dfrac{4\pi}{3}\right)$
48. $\sec \dfrac{11\pi}{6}$
49. cos 510°
50. tan 690°
51. sec(−270°)
52. csc(−540°)
53. $\cos \dfrac{8\pi}{3}$
54. $\cot\left(-\dfrac{19\pi}{6}\right)$

Find the least positive θ in degree and radian measure for which:

55. $\cos \theta = \dfrac{-1}{2}$
56. $\sin \theta = \dfrac{-\sqrt{3}}{2}$
57. $\sin \theta = \dfrac{-1}{2}$
58. $\tan \theta = -\sqrt{3}$
59. $\csc \theta = \dfrac{-2}{\sqrt{3}}$
60. $\sec \theta = -\sqrt{2}$

C *Find the value of each of the other five trigonometric functions for an angle θ (without finding θ) given the indicated information. Sketching a reference triangle should prove helpful.*

61. $\sin \theta = \dfrac{3}{5}$ and $\cos \theta < 0$

62. $\tan \theta = -\dfrac{4}{3}$ and $\sin \theta < 0$

63. $\cos \theta = -\dfrac{\sqrt{5}}{3}$ and $\cot \theta > 0$

64. $\cos \theta = \dfrac{-\sqrt{5}}{3}$ and $\tan \theta > 0$

65. $\tan\theta = -\sqrt{2}$ and $\sin\theta < 0$

66. $\tan\theta = -\sqrt{2}$ and $\cos\theta > 0$

67. Which trigonometric functions are not defined when the terminal side of an angle lies along the positive or negative vertical axis?

68. Which trigonometric functions are not defined when the terminal side of an angle lies along the positive or negative horizontal axis?

APPLICATIONS

69. *Physics–engineering.* The figure illustrates a piston connected to a wheel that turns three revolutions per second; hence, the angle θ is being generated at $3(2\pi) = 6\pi$ radians per second, or $\theta = 6\pi t$, where t is time in seconds. If P is at $(1, 0)$ when $t = 0$, show that

$$y = b + \sqrt{4^2 - a^2}$$
$$= \sin 6\pi t + \sqrt{16 - (\cos 6\pi t)^2}$$

for $t \geq 0$.

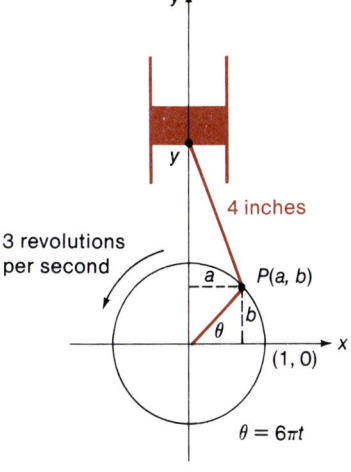

70. *Physics–engineering.* In Problem 69, find the position of the piston y when $t = 0.2$ second to three significant digits.

Section 5-6 Graphs of Trigonometric Functions

- Periodic Functions
- Graphs of $y = \sin x$ and $y = \cos x$
- Graphs of $y = \tan x$ and $y = \cot x$
- Graphs of $y = \csc x$ and $y = \sec x$

- Periodic Functions

Consider the graphs of sunrise times and sound waves shown in Figure 22 (page 316). A common feature appears to be that both types of phenomena are repetitive; that is, both appear to be periodic. Trigonometric functions are particularly well suited to describe phenomena of this nature.

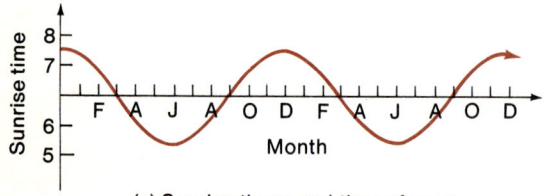
(a) Sunrise times and time of year

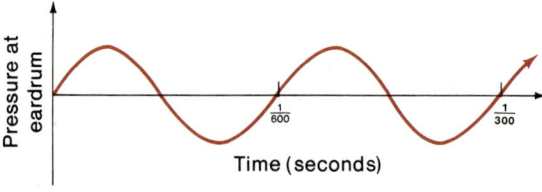
(b) Sound wave arriving at eardrum

FIGURE 22

Because a unit circle has a circumference of 2π, we find for a given value of x (see Fig. 23) that we will come back to the circular point $W(x) = (a, b)$ if we add any integer multiple of 2π to x.* Thus, for x any real number and for k any integer,

$$\sin(x + 2k\pi) = \sin x$$
$$\cos(x + 2k\pi) = \cos x$$

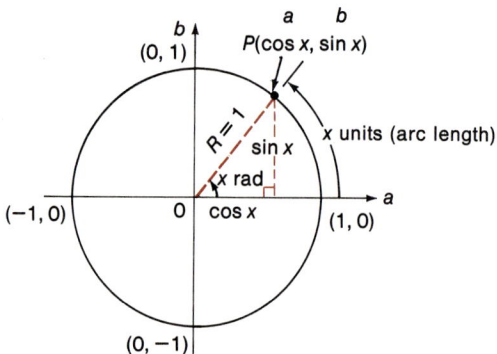

FIGURE 23

Functions with this kind of repetitive behavior are called **periodic functions**. In general:

Periodic Functions

A function f is **periodic** if there exists a positive real number p such that

$$f(x + p) = f(x)$$

for all x in the domain of f. The smallest such positive p, if it exists, is called the **fundamental period of f** (or often just the **period of f**).

Both the sine and cosine functions are periodic with period 2π.

---

* Think of a point P moving around the unit circle in either direction. Every time P covers a distance of 2π, the circumference of the circle, it will be back at the point where it started.

5-6 Graphs of Trigonometric Functions

■ **Graphs of $y = \sin x$ and $y = \cos x$**

We start by graphing

$y = \sin x \qquad x$ a real number

Before we start graphing, it is useful to look at the sine function (with real number domain) in terms of a function machine as shown in Figure 24.

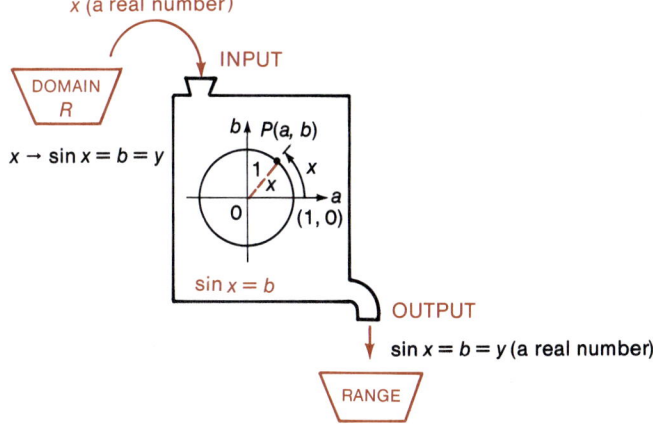

FIGURE 24 Sine function machine

We are interested in graphing all ordered pairs of real numbers (x, y) produced by the sine function machine. We could use a calculator or a table to find range values associated with domain values. However, we can speed up the process of sketching $y = \sin x$ by directly observing how $\sin x = b = y$ varies as $P(a, b)$ moves around the unit circle. This variation is illustrated in Figure 25.

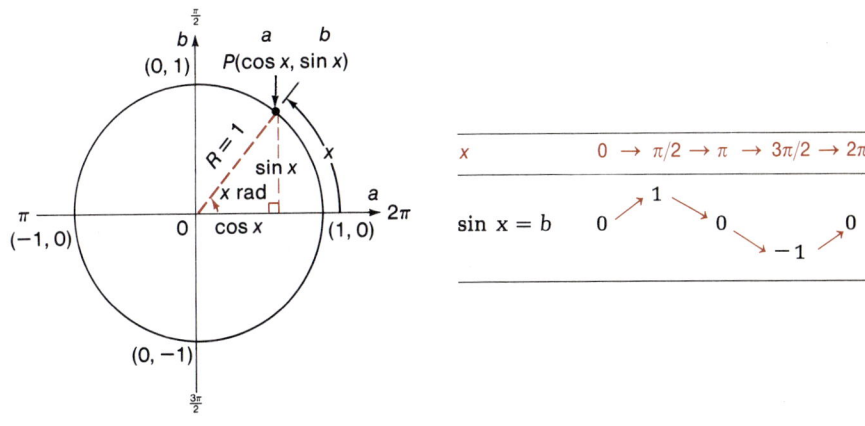

FIGURE 25

To sketch the graph of $y = \sin x$ for the interval $[0, 2\pi]$, we use the results in Figure 25 supplemented where necessary by special real values (integer multiples of $\pi/6$ or $\pi/4$) or calculator values. The final graph is shown in Figure 26.

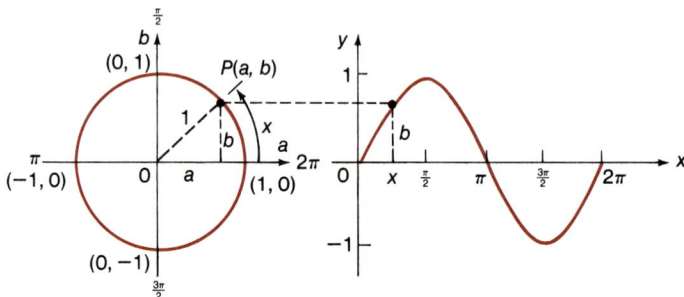

FIGURE 26

Earlier, we found that the sine function is periodic with period 2π. Thus, to complete the graph of $y = \sin x$, we need only to repeat the graph in Figure 26 to the left and right over intervals of 2π to produce as much of the general graph as we please (Fig. 27).

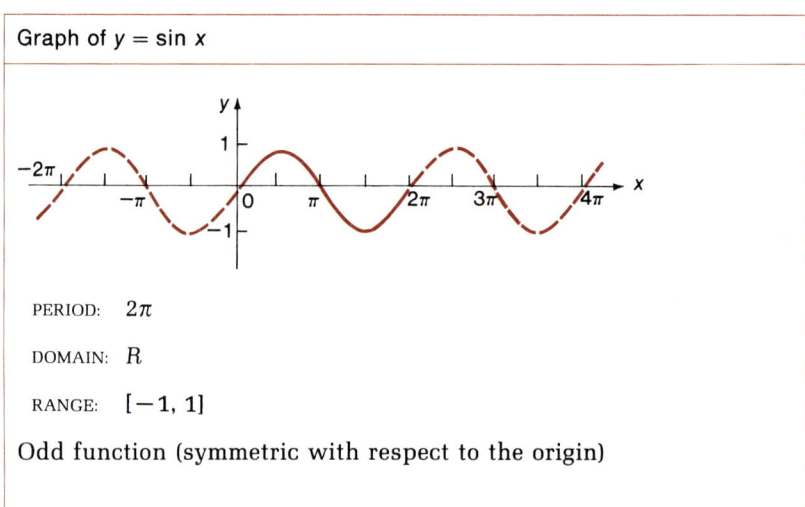

FIGURE 27

In Figure 27, we indicated multiples of π (special real numbers) on the x axis for convenience and clarity; all other real numbers are also associated with points on the x axis. This graph (as well as the others in this section) with its special characteristics should be learned so that it can be sketched from memory.

Proceeding in the same way for the cosine function, we can obtain its graph. Figure 28 shows how $\cos x = a = y$ varies as $P(a, b)$ moves around a unit circle.

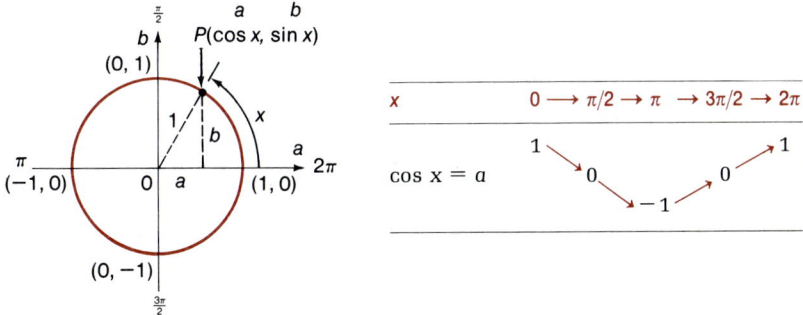

FIGURE 28

Using the results in Figure 28, the fact that the cosine function is periodic with period 2π, and special or calculator values where necessary, we obtain Figure 29.

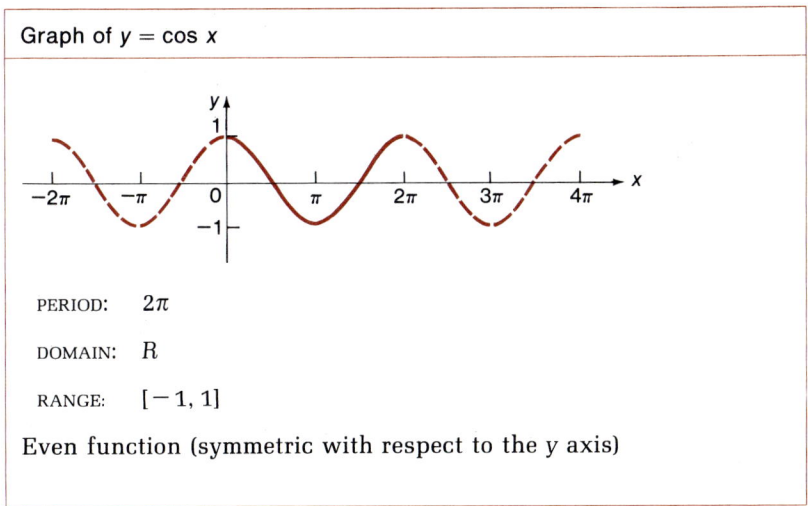

FIGURE 29

The key features of the graph of $y = \cos x$ should be learned so that it can be sketched from memory.

■ **Graphs of $y = \tan x$ and $y = \cot x$**

We first discuss the graph of $y = \tan x$; then from this graph, because $\cot x = 1/\tan x$, we will be able to get the graph of $y = \cot x$ using reciprocals of ordinates.

Referring to Figure 30, we see that whenever $P(a, b)$ is on the horizontal axis (that is, whenever $x = k\pi$, k an integer), then $a = \pm 1$, $b = 0$, and $\tan x = b/a = 0/(\pm 1) = 0$. Whenever $P(a, b)$ is on the vertical axis (that is, when $x = \pi/2 + k\pi$, k an integer), then $a = 0$, $b = \pm 1$, and $\tan x = b/a = (\pm 1)/0$ is not defined (the tangent function is discontinuous).

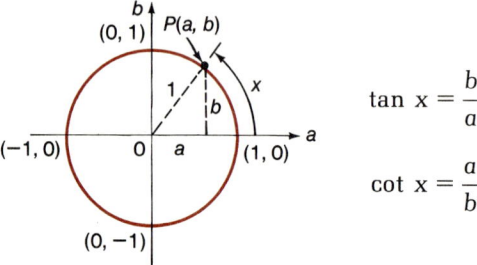

FIGURE 30

The values of x such that $P(a, b)$ is on the horizontal axis (in Fig. 30) are 0's for $\tan x$ or x intercepts for the graph of $y = \tan x$. The values of x such that $P(a, b)$ is on the vertical axis (in Fig. 30) are points through which vertical asymptotes (see Section 2-6) pass.

x intercepts: $x = k\pi$, k an integer

Vertical asymptotes: $x = \dfrac{\pi}{2} + k\pi$, k an integer

This information is summarized in Figure 31.

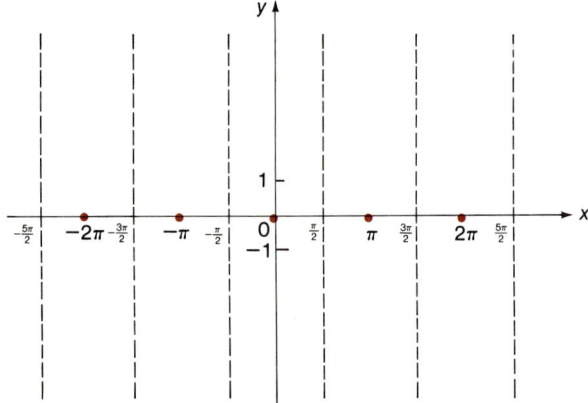

FIGURE 31

Let us investigate the behavior of $\tan x$ in more detail over the interval $(-\pi/2, \pi/2)$. Referring to Figure 30, we see that as x approaches $\pi/2$ from below $[x \to (\pi/2)^-]$, a approaches 0 through positive values $(a \to 0^+)$ and b approaches 1 $(b \to 1)$, and we conclude that $\tan x = b/a$ increases without bound (tan x → ∞). On the other hand, as x approaches $-\pi/2$ from above $[x \to (-\pi/2)^+]$, a approaches 0 through positive values ($a \to 0^+$) and b approaches -1 ($b \to -1$), and we see that tan x = b/a decreases without bound (tan x → $-\infty$). Noting that $\tan(-\pi/4) = -1$, tan 0 = 0, and $\tan(\pi/4) = 1$, we sketch the graph of y = tan x for the interval $(-\pi/2, \pi/2)$ in Figure 32.

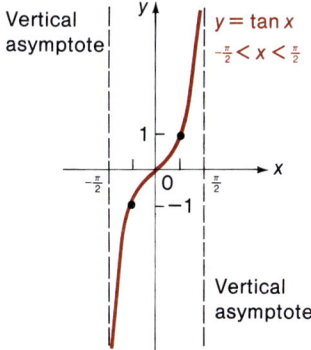

FIGURE 32

To reinforce the sketch in Figure 32, let us use a calculator to form a table of values of tan x with x approaching $\pi/2 \approx 1.570\ 7963$ from the left:

x	0	0.5	1.0	1.5	1.57	1.5707	1.570 7963
tan x	0	0.5	1.6	14.5	1,255	10,381	37,320,396

A similar table can be constructed with x approaching $-\pi/2 \approx -1.570\ 7963$ from the right:

x	0	−0.5	−1	−1.5	−1.57	−1.5707	−1.570 7963
tan x	0	−0.5	−1.6	−14.5	−1,255	−10,381	−37,320,396

Proceeding in the same way for the other intervals between asymptotes, we see that the tangent function is periodic with period π. Thus, to complete the graph of y = tan x, we need only repeat the graph in Figure 32 to the left and right over intervals of π to produce as much of the general graph as we need (see Fig. 33, page 322). The main characteristics of the graph of y = tan x should be learned so that it can be sketched from memory.

To graph y = cot x, we simply take reciprocals of the ordinate values in the graph of y = tan x in Figure 33. Note that the x intercepts and vertical asymptotes are interchanged. The graph of y = cot x is shown in Figure 34. The main characteristics of the graph should be learned so that it can be sketched from memory.

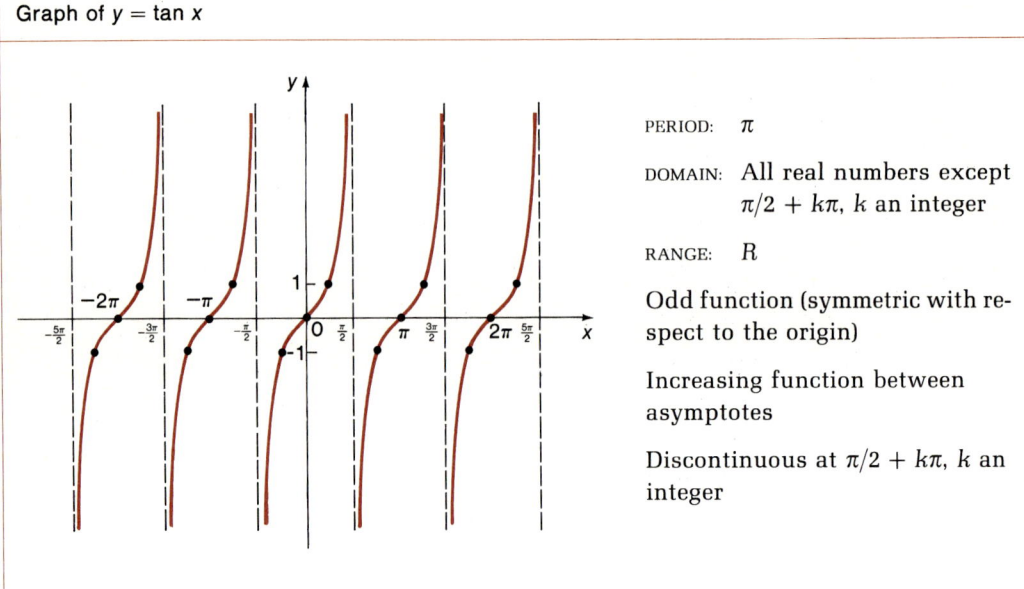

FIGURE 33

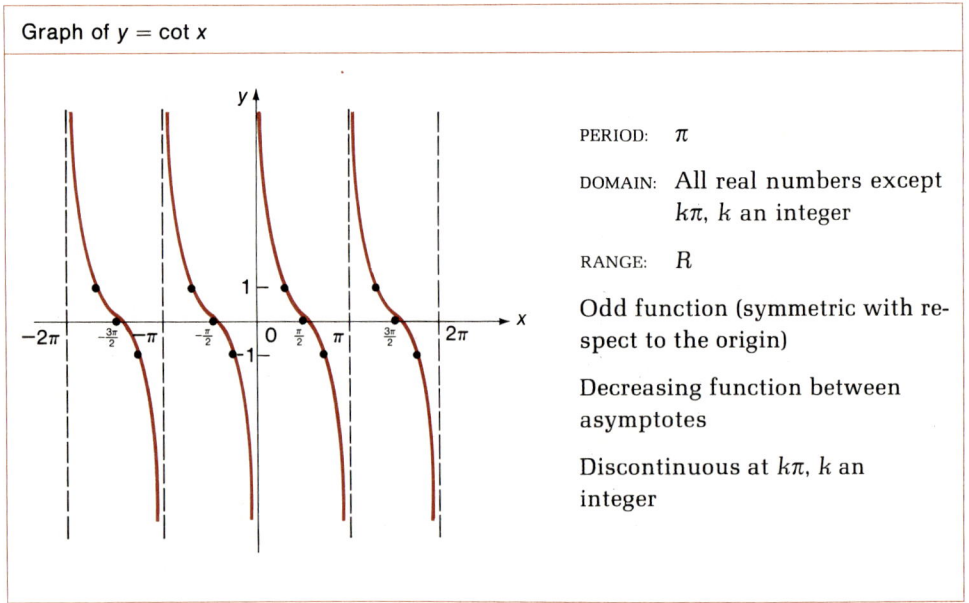

FIGURE 34

5-6 Graphs of Trigonometric Functions

■ **Graphs of $y = \csc x$ and $y = \sec x$**

Just as we obtained the graph of $y = \cot x$ by taking reciprocals of the ordinates in the graph of $y = \tan x$, since

$$\csc x = \frac{1}{\sin x} \quad \text{and} \quad \sec x = \frac{1}{\cos x}$$

we can obtain the graphs of $y = \csc x$ and $y = \sec x$ by taking reciprocals of the ordinates in the graphs of $y = \sin x$ and $y = \cos x$, respectively. Vertical asymptotes will occur at the x intercepts of the latter.

Graph of $y = \csc x$

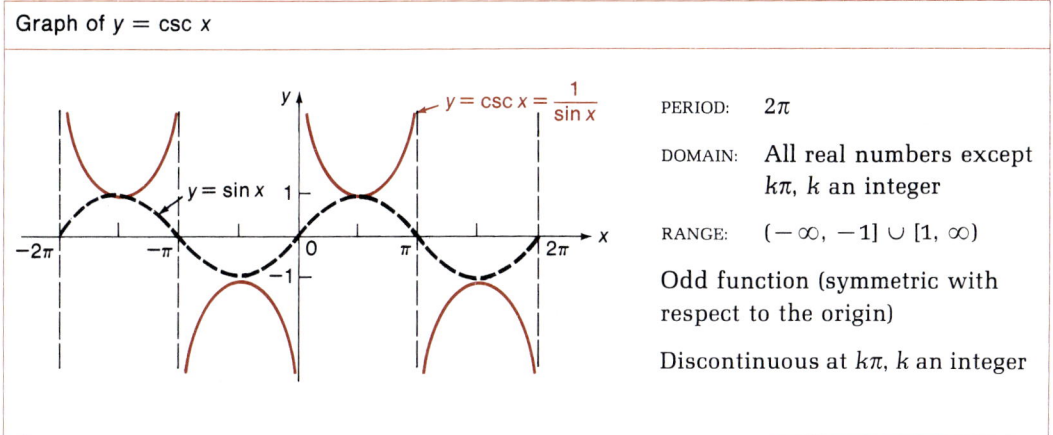

PERIOD: 2π

DOMAIN: All real numbers except $k\pi$, k an integer

RANGE: $(-\infty, -1] \cup [1, \infty)$

Odd function (symmetric with respect to the origin)

Discontinuous at $k\pi$, k an integer

FIGURE 35

Graph of $y = \sec x$

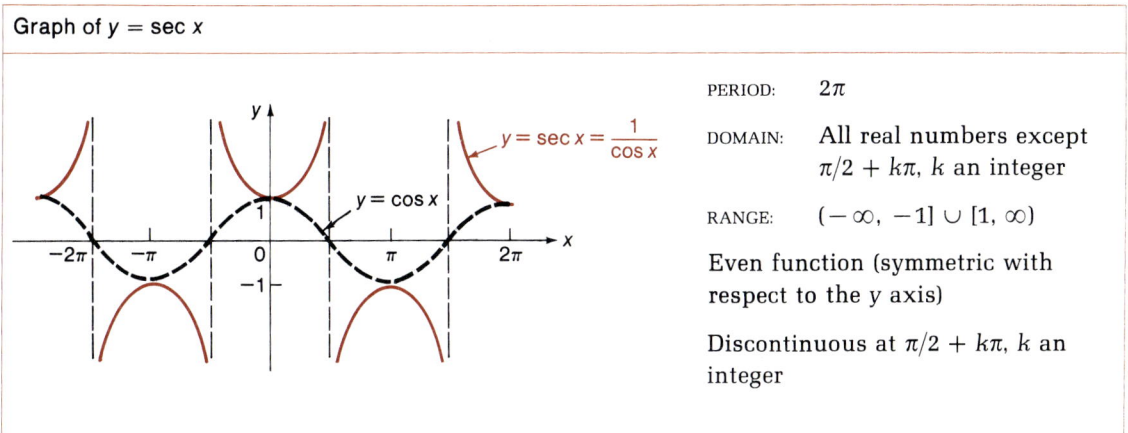

PERIOD: 2π

DOMAIN: All real numbers except $\pi/2 + k\pi$, k an integer

RANGE: $(-\infty, -1] \cup [1, \infty)$

Even function (symmetric with respect to the y axis)

Discontinuous at $\pi/2 + k\pi$, k an integer

FIGURE 36

The graphs of $y = \csc x$ and $y = \sec x$ are shown in Figures 35 and 36, respectively. As a graphing aid, we sketch in broken lines of $y = \sin x$ and $y = \cos x$ first and then draw vertical asymptotes through the x intercepts.

Exercise 5-6

This figure will be useful in many of the problems in this exercise.

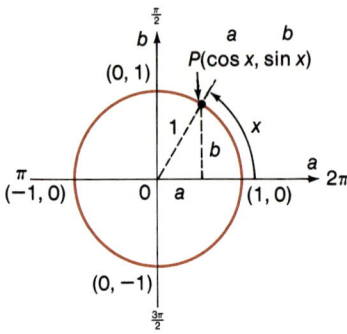

A 1. Complete the following table:

	VALUE AT x				
FUNCTION	0	$\pi/2$	π	$3\pi/2$	2π
sin x					0
cos x		0			
tan x	0				
cot x					
sec x		Not defined			
csc x				−1	

B 2. Use $P(a, b)$ on a unit circle (see the figure) to complete the following table (↗ means increasing and ↘ means decreasing):

	x VARIES FROM 0 TO $\pi/2$				x VARIES FROM $\pi/2$ TO π				x VARIES FROM π TO $3\pi/2$				x VARIES FROM $3\pi/2$ TO 2π			
FUNCTION																
$x \neq 0, \pi/2, \pi, 3\pi/2, 2\pi$	+	−	↗	↘	+	−	↗	↘	+	−	↗	↘	+	−	↗	↘
$y = \sin x = b$	✓		✓		✓			✓		✓		✓		✓	✓	
$y = \cos x = a$																
$y = \tan x = b/a$																
$y = \cot x = a/b$																

Make a rough sketch of each of the following trigonometric functions without looking at the text or using a calculator (or table). Label the points (in terms of π) where the curves cross the x axis.

3. $y = \cos x$, $-5\pi/2 \leq x \leq 7\pi/2$
4. $y = \sin x$, $-2\pi \leq x \leq 4\pi$
5. $y = \tan x$, $-3\pi/2 \leq x \leq 5\pi/2$
6. $y = \cot x$, $-\pi \leq x \leq 2\pi$

C 7. From a rough sketch of $y = \sin x$, $-\pi \leq x \leq 2\pi$, graph $y = \csc x$ by estimating the reciprocals of ordinate values on the sine curve.

8. From a rough sketch of $y = \cos x$, $-\pi/2 \leq x \leq 3\pi/2$, graph $y = \sec x$ by estimating the reciprocals of ordinate values on the cosine curve.

Section 5-7 Graphing $y = A \sin(Bx + C)$ and $y = A \cos(Bx + C)$

- $y = A \sin x$ and $y = A \cos x$
- $y = A \sin Bx$ and $y = A \cos Bx$
- $y = A \sin(Bx + C)$ and $y = A \cos(Bx + C)$

Now that we have discussed the graphs of $y = \sin x$ and $y = \cos x$, we are ready to consider graphs of the more general forms

$$y = A \sin(Bx + C) \quad \text{and} \quad y = A \cos(Bx + C) \qquad (1)$$

These equations are extremely important in pure and applied mathematics. They are often used in the analysis of sound waves, radio waves, x rays, gamma rays, visible light, infrared radiation, ultraviolet radiation, seismic waves, ocean waves, electrical circuits, electrical generators, vibrations, bridge and building construction, spring–mass systems, bow waves of boats, sonic booms, and so on. Phenomena that can be described by either of equations (1), where x represents time, are referred to as **simple harmonic motion**, and certain types of analysis involving these equations are called **harmonic analysis**.

Sketching graphs of equations (1) is not difficult if we attack the problem step by step. Essential to the process, however, is a sound knowledge of the graphs of $y = \sin x$ and $y = \cos x$ discussed in Section 5-6. Our approach will be to see what effect A, B, and C have on the basic graphs of $y = \sin x$ and $y = \cos x$. A brief review of Section 2-3 should prove helpful, since we will use some of that material in this section.

- $y = A \sin x$ and $y = A \cos x$

We first investigate the effect of A by comparing

$$y = \sin x \quad \text{and} \quad y = A \sin x$$

The graph of $y = A \sin x$ can be obtained from the graph of $y = \sin x$ by multiplying each ordinate value of the latter by A. The graph of $y = A \sin x$ will still cross the x axis where the graph of $y = \sin x$ crosses the x axis, because A times 0 is 0, but the maximum deviation of the graph of $y = \sin x$ from the x axis will change. Because $\sin x$ has period 2π,

$$A \sin(x + 2\pi) = A \sin x$$

Hence, $A \sin x$ also has period 2π. Compare the graphs of $y = \frac{1}{2} \sin x$ and $y = -2 \sin x$ with the graph of $y = \sin x$ in Figure 37.

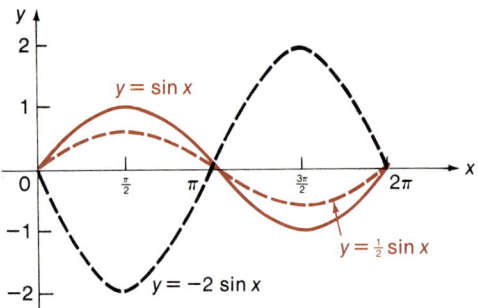

FIGURE 37

The constant $|A|$, the maximum deviation of the graph of $A \sin x$ from the x axis, is called the **amplitude** of $y = A \sin x$. Thus, $y = \frac{1}{2} \sin x$ has an amplitude of $\left|\frac{1}{2}\right| = \frac{1}{2}$, $y = -2 \sin x$ has an amplitude of $|-2| = 2$, and $y = \sin x$ has an amplitude of 1. The negative sign in $y = -2 \sin x$ turns the graph of $y = 2 \sin x$ upside down; that is, the graph of $y = -2 \sin x$ is the same as the graph of $y = 2 \sin x$ reflected across the x axis. The effect of A is to increase or decrease the ordinates (y values) of $y = \sin x$ without changing the abscissas (x values).

A similar analysis applies to

$$y = A \cos x$$

and we conclude that this function also has a period of 2π and an amplitude of $|A|$.

■ $y = A \sin Bx$ and $y = A \cos Bx$

We now investigate the effect of B by comparing

$$y = \sin x \quad \text{and} \quad y = \sin Bx \qquad B > 0$$

Both have the same amplitude, 1, but how do their periods compare? The period of $\sin x$ is 2π. To find the period of $f(x) = \sin Bx$, we seek the smallest positive number P such that

$$f(x + P) = f(x)$$

5-7 Graphing $y = A \sin(Bx + C)$ and $y = A \cos(Bx + C)$

To this end we see that

$$f(x + P) = \sin B(x + P) = \sin(Bx + BP) = \sin Bx = f(x)$$

if $BP = 2\pi$ or $P = 2\pi/B$. Thus,

$$\text{Period of } \sin Bx = P = \frac{2\pi}{B}$$

What is the period of $y = \sin 2x$?

$$P = \frac{2\pi}{B} = \frac{2\pi}{2} = \pi \quad \text{Half the period for sin } x$$

What is the period of $y = \sin(x/2)$?

$$P = \frac{2\pi}{B} = \frac{2\pi}{\frac{1}{2}} = 4 \quad \text{Double the period for sin } x$$

Thus, the effect of B is to compress or stretch the basic sine curve; that is, **B changes the period of sin x**.

Now compare the graphs of $y = \sin 2x$ and $y = \sin(x/2)$ with the graph of $y = \sin x$ in Figure 38. To graph these functions, we mark off one period (starting at the origin), divide it into four equal parts, and sketch in the modified sine curve. The graph over one period may then be extended to cover any desired interval.

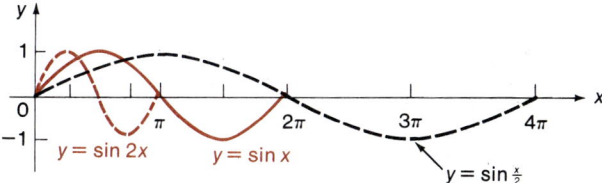

FIGURE 38

A similar analysis applies to $y = \cos Bx$, where $B > 0$. Combining the discussions on amplitude and period, we summarize the results as follows:

For $y = A \sin Bx$ or $y = A \cos Bx$, $B > 0$:

$$\text{Amplitude} = |A| \qquad \text{Period} = \frac{2\pi}{B}$$

If $0 < B < 1$, the basic sine or cosine curve is stretched.

If $B > 1$, the basic sine or cosine curve is compressed.

Let us now consider a couple of examples where we show how graphs of $y = A \sin Bx$ and $y = A \cos Bx$ can be sketched rather quickly.

EXAMPLE 24 State the amplitude and period for $y = 2 \sin 2x$, and graph the equation for $-\pi \le x \le 2\pi$.

Solution Amplitude $= |2| = 2$ Period $= \dfrac{2\pi}{2} = \pi$

To sketch the graph, divide the interval $[0, \pi]$ into four equal parts, locate the high and low points and x intercepts, and then sketch in one period and extend this sketch to cover the desired interval. Adjust the scales on the horizontal and vertical axes so that the figure is clear. Both scales do not have to be the same.

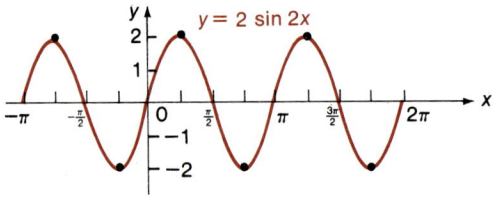

PROBLEM 24 State the amplitude and period for $y = -\frac{1}{2} \sin(x/2)$, and graph the equation for $-4\pi \le x \le 4\pi$.

EXAMPLE 25 State the amplitude and period for $y = -3 \cos(\pi x/2)$, and graph the equation for $-4 \le x \le 4$.

Solution Amplitude $= |-3| = 3$ Period $= \dfrac{2\pi}{\pi/2} = 4$

The graph of $y = -3 \cos(\pi x/2)$ is the same as the graph of $y = 3 \cos(\pi x/2)$ reflected across the x axis. Divide the interval $[0, 4]$ into four equal parts, locate the high and low points and x intercepts, and then sketch in one period and extend this sketch to cover the desired interval.

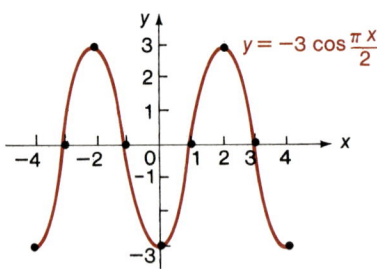

PROBLEM 25 State the amplitude and period for $y = \frac{1}{4} \cos 2\pi x$, and sketch the graph of the equation for $-1 \leq x \leq 1$.

■ **$y = A \sin(Bx + C)$ and $y = A \cos(Bx + C)$**

Finally, we are ready to consider the graphs of equations of the form

$$y = A \sin(Bx + C) \quad \text{and} \quad y = A \cos(Bx + C) \tag{2}$$

We will find that the graphs of these equations are simply the graphs of

$$y = A \sin Bx \quad \text{and} \quad y = A \cos Bx \tag{3}$$

shifted to the left or right. How much and in what direction? We discussed this type of shift in Section 2-3. Rewriting equations (2) in the form

$$y = A \sin B\left(x + \frac{C}{B}\right) \quad \text{and} \quad y = A \cos B\left(x + \frac{C}{B}\right) \tag{4}$$

we find that the graphs of equations (4) are the same as the corresponding graphs of equations (3) shifted to the *left* C/B units if C/B is positive or shifted to the *right* $|C/B|$ units if C/B is negative.

Compare the graphs of $y = \sin(x + \pi/2)$ and $y = \sin(x - \pi/2)$ with the graph of $y = \sin x$ in Figure 39. The graph of $y = \sin(x + \pi/2)$ is simply the graph of $y = \sin x$ shifted to the left $\pi/2$ units, and the graph of $y = \sin(x - \pi/2)$ is simply the graph of $y = \sin x$ shifted to the right $\pi/2$ units. This shift of the basic curve is referred to as a **phase shift**. And we say that the graph of $y = \sin(x + \pi/2)$ has a phase shift of $\pi/2$ units to the left and the graph of $y = \sin(x - \pi/2)$ has a phase shift of $\pi/2$ units to the right. The shift is just the opposite of what you might expect from the signs of $\pi/2$. The negative sign is associated with a shift to the right and the positive sign with a shift to the left.

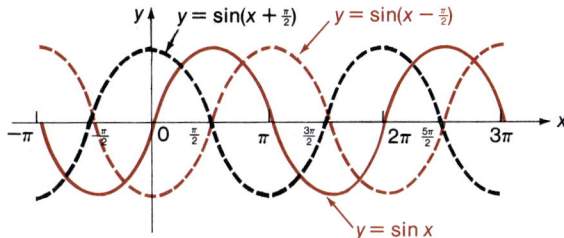

FIGURE 39

In general, for

$$y = A \sin(Bx + C) = A \sin B\left(x + \frac{C}{B}\right)$$

$$y = A \cos(Bx + C) = A \cos B\left(x + \frac{C}{B}\right)$$

the constant C/B is the phase shift (as described). The graphs of $y = A \sin Bx$ and $y = A \cos Bx$ are shifted to the left C/B units if C/B is positive or to the right $|C/B|$ units if C/B is negative.

We now summarize all the results obtained in this section in one convenient statement.

For $y = A \sin(Bx + C)$ or $y = A \cos(Bx + C)$, $B > 0$:

$$\text{Amplitude} = |A| \qquad \text{Period} = \frac{2\pi}{B}$$

$$\text{Phase shift} = \begin{cases} |C/B| \text{ units to the right if } C/B < 0 \\ C/B \text{ units to the left if } C/B > 0 \end{cases}$$

EXAMPLE 26 State the amplitude, period, and phase shift for $y = \frac{1}{2}\cos(4x - \pi)$. Graph the equation for $-\pi \le x \le \pi$.

Solution First write $y = \frac{1}{2}\cos(4x - \pi) = \frac{1}{2}\cos 4[x + (-\pi/4)]$.

$$\text{Amplitude} = \frac{1}{2} \qquad \text{Period} = \frac{2\pi}{4} = \frac{\pi}{2}$$

$$\text{Phase shift} = \frac{\pi}{4} \text{ unit to the right}$$

To graph the equation, first sketch the graph of $y = \frac{1}{2}\cos 4x$ over one period from 0 to $\pi/2$ on scratch paper; then shift the graph $\pi/4$ units to the right and extend the graph to fill out the interval $[-\pi, \pi]$.

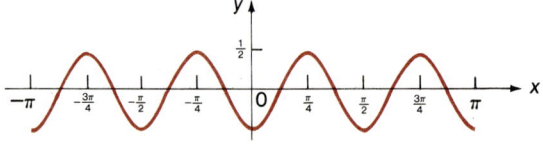

PROBLEM 26 State the amplitude, period, and phase shift for $y = \frac{3}{4}\sin(2x + \pi)$. Graph the equation for $-\pi \le x \le \pi$.

Proceeding as with the sine function, one can obtain similar results for the tangent and cotangent functions, with the exception that we do not define amplitude for the latter.

5-7 Graphing $y = A \sin(Bx + C)$ and $y = A \cos(Bx + C)$ 331

For $y = A \tan(Bx + C)$ and $y = A \cot(Bx + C)$, $B > 0$:

Period $= \pi/B$

Phase shift $= \begin{cases} |C/B| \text{ units to the right if } C/B < 0 \\ C/B \text{ units to the left if } C/B > 0 \end{cases}$

Answers to Matched Problems

24. Amplitude: $\tfrac{1}{2}$; period: 4π

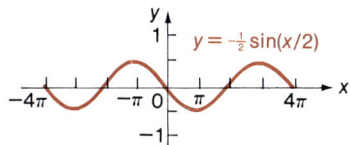

25. Amplitude: $\tfrac{1}{4}$; period: 1

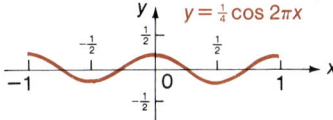

26. Amplitude: $\tfrac{3}{4}$; period: π; phase shift: $\pi/2$ units to the left

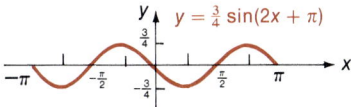

Exercise 5-7 ■ **A** State the amplitude and period of each function, and graph the function over the indicated interval.

1. $y = 3 \sin x$, $-2\pi \leq x \leq 2\pi$
2. $y = \tfrac{1}{4} \cos x$, $-2\pi \leq x \leq 2\pi$
3. $y = -\tfrac{1}{2} \cos x$, $-2\pi \leq x \leq 2\pi$
4. $y = -2 \sin x$, $-2\pi \leq x \leq 2\pi$
5. $y = \sin 3x$, $-\pi \leq x \leq 2\pi$
6. $y = \cos 2x$, $-\pi \leq x \leq \pi$

7. $y = \cos(x/2)$, $-4\pi \leq x \leq 4\pi$
8. $y = \sin(x/3)$, $-6\pi \leq x \leq 6\pi$
9. $y = \sin \pi x$, $-2 \leq x \leq 2$
10. $y = \cos \pi x$, $-2 \leq x \leq 2$
11. $y = 3 \cos 2x$, $-\pi \leq x \leq \pi$
12. $y = 2 \sin 4x$, $-\pi \leq x \leq \pi$

B 13. $y = \frac{1}{2} \sin 2\pi x$, $-2 \leq x \leq 2$
14. $y = \frac{1}{3} \cos 2\pi x$, $-2 \leq x \leq 2$
15. $y = -3 \cos(x/2)$, $-4\pi \leq x \leq 4\pi$
16. $y = -\frac{1}{4} \sin(x/2)$, $-4\pi \leq x \leq 4\pi$

State the amplitude, period, and phase shift of each function, and graph the function over the indicated interval.

17. $y = \sin(x + \pi/2)$, $-\pi \leq x \leq 2\pi$
18. $y = \sin(x - \pi/2)$, $-\pi \leq x \leq 3\pi$
19. $y = \frac{1}{2} \cos(x - \pi/4)$, $-\pi \leq x \leq 3\pi$
20. $y = 2 \sin(x + \pi/4)$, $-2\pi \leq x \leq 2\pi$
21. $y = \sin[\pi(x - 1)]$, $-2 \leq x \leq 3$
22. $y = \cos[2\pi(x - \frac{1}{2})]$, $-1 \leq x \leq 2$
23. $y = 3 \cos(\pi x + \pi/2)$, $-2 \leq x \leq 2$
24. $y = 2 \sin(\pi x - \pi/4)$, $-1 \leq x \leq 3$
25. $y = 4 \cos(2x - \pi)$, $-\pi \leq x \leq 3\pi$
26. $y = -2 \cos(4x + \pi)$, $-\pi \leq x \leq \pi$

C 27. $y = 3.5 \sin\left[\dfrac{\pi}{2}(t + 0.5)\right]$, $0 \leq t \leq 10$

28. $y = 5.4 \sin\left[\dfrac{\pi}{2.5}(t - 1)\right]$, $0 \leq t \leq 6$

29. $y = 50 \cos[2\pi(t - 0.25)]$, $0 \leq t \leq 2$
30. $y = 25 \cos[5\pi(t - 0.1)]$, $0 \leq t \leq 2$

Indicate the period and phase shift, and graph each function.

31. $y = 3 \tan 2x$, $-\pi \leq x \leq \pi$
32. $y = 2 \cot 4x$, $0 < x < \pi/2$
33. $y = 4 \tan(2x + \pi)$, $-\pi \leq x \leq \pi$
34. $y = -3 \cot(\pi x - \pi)$, $-1 \leq x \leq 2$

APPLICATIONS

35. *Physics–engineering.* A 6-pound weight hanging from the end of a spring is pulled $\frac{1}{3}$ foot below the equilibrium position and then released. If air resistance and friction are neglected, the distance x that the weight is from the equilibrium position relative to time t in seconds is given by

$$x = \tfrac{1}{3} \cos 8t$$

State the period and amplitude of this function, and graph it for $0 \le t \le \pi$.

36. *Physics–engineering.* An alternating current generator generates a current given by

$$I = 30 \sin 120t$$

where t is time in seconds. What are the amplitude and period of this function? What is the frequency of the current; that is, how many cycles will be completed in 1 second?

37. *Physics–engineering.* The thin, plastic disk in the accompanying figure is rotated at three revolutions per second, starting at $\theta = 0$ (thus at the end of t seconds, $\theta = 6\pi t$—why?). If the disk has a radius of 3, show that the position of the shadow on the y scale from the small steel ball B is given by

$$y = 3 \sin 6\pi t$$

Graph this equation for $0 \le t \le 1$.

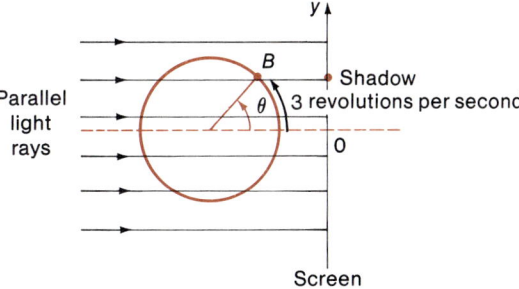

38. *Physics–engineering.* If in Problem 37 the disk started rotating at $\theta = \pi/2$, show that the position of the shadow at time t (in seconds) is given by

$$y = 3 \sin\left(6\pi t + \frac{\pi}{2}\right)$$

Graph this equation for $0 \le t \le 1$.

Section 5-8 Inverse Trigonometric Functions

- General Comments
- Inverse Sine Function
- Inverse Cosine Function
- Inverse Tangent Function
- Inverse Cotangent, Secant, and Cosecant Functions (Optional)

- **General Comments**

A brief review of the general concept of inverse relations and functions discussed in Section 2-7 should prove helpful before proceeding with this section. In the box we restate a few important facts about inverse functions from that section.

Facts about Inverse Functions

Let f^{-1} be the inverse function of the one-to-one function f. (A function f has an inverse that is a function if f is one to one over its domain.)

1. $f^{-1} = \{(b, a) \mid (a, b) \in f\}$
2. Domain of f^{-1} = Range of f;   Range of f^{-1} = Domain of f
3. $x = f^{-1}(y)$ if and only if $f(x) = y$ for y in the domain of f^{-1} and x in the domain of f
4. $f[f^{-1}(y)] = y$ for y in the domain of f^{-1}
   $f^{-1}[f(x)] = x$ for x in the domain of f
5.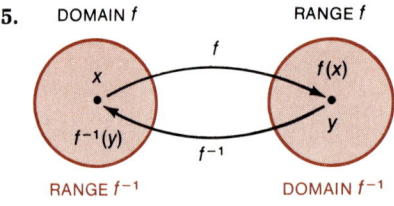

Because all the trigonometric functions are periodic, none are one to one; hence, none have inverses that are functions. To rectify this problem, we can restrict the domain of each so that it will become one to one over the restricted domain. Thus, for this restricted domain, a unique inverse function is guaranteed.

We now turn to the inverse sine, cosine, and tangent functions. These functions will be particularly useful to us when we solve trigonometric equations in Section 6-6.

■ **Inverse Sine Function**

How can we restrict the domain of the sine function so that it becomes one to one? This can be done in infinitely many ways. A fairly natural and generally accepted way is illustrated in Figure 40.

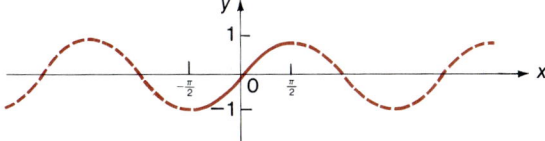

FIGURE 40 $y = \sin x$ is one to one over $[-\pi/2, \pi/2]$

If we restrict the domain of the sine function to the interval $[-\pi/2, \pi/2]$, we see that the sine function is increasing and thus is one to one. Note that each range value from -1 to 1 is assumed exactly once as x moves from $-\pi/2$ to $\pi/2$. We use this restricted sine function to define the **inverse sine function**.

Inverse Sine Function

The **inverse sine function**, denoted by **sin^{-1}** or **arcsin**, is defined as the inverse of the restricted sine function $y = \sin x$, $-\pi/2 \leq x \leq \pi/2$. Thus, $y = \sin^{-1} x = \arcsin x$ is equivalent to $\sin y = x$, where $-\pi/2 \leq y \leq \pi/2$ and $-1 \leq x \leq 1$. The inverse sine of x, or the arcsine of x, is the number or angle y, $-\pi/2 \leq y \leq \pi/2$, whose sine is x.

To graph $y = \sin^{-1} x$, we take the coordinates of each point on the graph of $y = \sin x$, $-\pi/2 \leq x \leq \pi/2$, and reverse the order. For example, $(\pi/2, 1)$, $(0, 0)$, and $(-\pi/2, -1)$ are on the graph of $y = \sin x$; hence, $(1, \pi/2)$, $(0, 0)$, and $(-1, -\pi/2)$ are on the graph of $y = \sin^{-1} x$. The graph of $y = \sin^{-1} x$ is the graph of $y = \sin x$, $-\pi/2 \leq x \leq \pi/2$, reflected across the line $y = x$. Compare the graphs in Figure 41 (page 336). The two graphs are drawn separately for increased clarity.

5 Trigonometric Functions

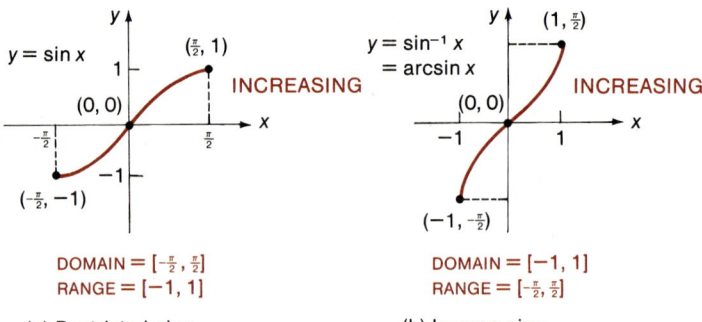

(a) Restricted sine function

(b) Inverse sine function

FIGURE 41

Finally, we state the important sine–inverse sine identities (which follow from the general properties of inverse functions). (See page 156.)

Sine–Inverse Sine Identities

$\sin(\sin^{-1} x) = x \quad -1 \leq x \leq 1$

$\sin^{-1}(\sin x) = x \quad -\pi/2 \leq x \leq \pi/2$

$\sin(\sin^{-1} 0.5) = 0.5 \quad \sin^{-1}[\sin(-1.3)] = -1.3 \quad \sin^{-1}(\sin 2) \neq 2$

Note: 2 is not in the restricted domain of the sine function.

EXAMPLE 27 Find exact values without using a calculator or a table.

(A) $\sin^{-1}(\sqrt{3}/2)$ (B) $\arcsin(-\frac{1}{2})$
(C) $\sin^{-1}(\sin 1.2)$ (D) $\cos(\sin^{-1} \frac{2}{3})$

Solution (A) $y = \sin^{-1}(\sqrt{3}/2)$ is equivalent to

$\sin y = \dfrac{\sqrt{3}}{2} \quad -\dfrac{\pi}{2} \leq y \leq \dfrac{\pi}{2}$

$y = \dfrac{\pi}{3} = \sin^{-1} \dfrac{\sqrt{3}}{2}$

Reference triangle associated with y

(B) $y = \arcsin(-\frac{1}{2})$ is equivalent to

$\sin y = -\dfrac{1}{2} \quad -\dfrac{\pi}{2} \leq y \leq \dfrac{\pi}{2}$

$y = -\dfrac{\pi}{6} = \arcsin(-\frac{1}{2})$

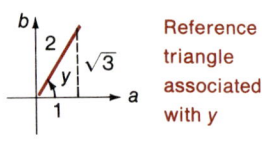

Reference triangle associated with y

[*Note:* $y \neq 11\pi/6$, even though $\sin 11\pi/6 = -\frac{1}{2}$. (Why?)]

(C) $\sin^{-1}(\sin 1.2) = 1.2$ Sine–inverse sine identity
(D) Let $y = \sin^{-1}\frac{2}{3}$, then $\sin y = \frac{2}{3}$, $-\pi/2 \le y \le \pi/2$.

Draw the reference triangle associated with y; then $\cos y = \cos(\sin^{-1}\frac{2}{3})$ can be determined directly from the triangle (after finding the third side) without actually finding y.

$$a^2 + b^2 = c^2$$
$$a = \sqrt{3^2 - 2^2}$$
$$= \sqrt{5}$$

Thus, $\cos(\sin^{-1}\frac{2}{3}) = \cos y = \sqrt{5}/3$.

PROBLEM 27   Find exact values without using a calculator or a table.

(A) $\arcsin(\sqrt{2}/2)$ (B) $\sin^{-1}(-1)$
(C) $\sin[\sin^{-1}(-0.4)]$ (D) $\tan[\sin^{-1}(-1/\sqrt{5})]$

EXAMPLE 28   Find to four significant digits using a scientific calculator.

(A) $\sin^{-1}(0.8432)$ (B) $\arcsin(-0.3042)$
(C) $\sin^{-1} 1.357$ (D) $\cot[\sin^{-1}(-0.1087)]$

Solution   *Note:* The buttons used to obtain $\sin^{-1}$ vary among different brands of calculators. (Read the user's manual for your calculator.) Two common designations are $\boxed{\text{SIN}^{-1}}$ and the combination $\boxed{\text{INV}}\boxed{\text{SIN}}$. We will use the latter, realizing that your calculator may have something else. For all these problems set your calculator in the radian mode.

(A) $\sin^{-1}(0.8432) = 1.003$ $\boxed{0.8432}\boxed{\text{INV}}\boxed{\text{SIN}}$

(B) $\arcsin(-0.3042) = -0.3091$ $\boxed{0.3042}\boxed{+/-}\boxed{\text{INV}}\boxed{\text{SIN}}$

(C) $\sin^{-1} 1.357 =$ Error   1.357 is not in the domain of $\sin^{-1}$

(D) $\cot[\sin^{-1}(-0.1087)] = -9.145$ $\boxed{0.1087}\boxed{+/-}\boxed{\text{INV}}\boxed{\text{SIN}}\boxed{\text{TAN}}\boxed{1/x}$

PROBLEM 28   Find to four significant digits using a scientific calculator.

(A) $\arcsin 0.2903$ (B) $\sin^{-1}(-0.7633)$
(C) $\arcsin(-2.305)$ (D) $\sec[\sin^{-1}(-0.3446)]$

■ **Inverse Cosine Function**

To restrict the cosine function so that it becomes one to one, we choose the interval $[0, \pi]$. Over this interval the cosine function is decreasing, and each range value is assumed exactly once as x moves from 0 to π (see Fig. 42, page 338). We use this restricted cosine function to define the **inverse cosine function.**

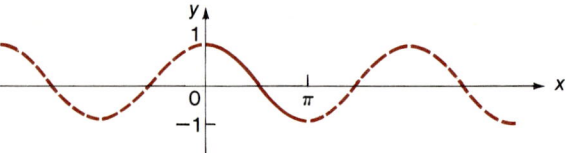

FIGURE 42 $y = \cos x$ is one to one over $[0, \pi]$

Inverse Cosine Function

The **inverse cosine function**, denoted by **cos^{-1}** or **arccos**, is defined as the inverse of the restricted cosine function, $y = \cos x$, $0 \leq x \leq \pi$. Thus, $y = \cos^{-1} x = \arccos x$ is equivalent to $\cos y = x$, where $0 \leq y \leq \pi$ and $-1 \leq x \leq 1$. The inverse cosine of x, or the arccosine of x, is the number or angle y, $0 \leq y \leq \pi$, whose cosine is x.

Figure 43 compares the graphs of the restricted cosine function and its inverse. Notice that $(0, 1)$, $(\pi/2, 0)$, and $(\pi, -1)$ are on the restricted cosine graph. Reversing the coordinates gives us three points on the graph of the inverse cosine function.

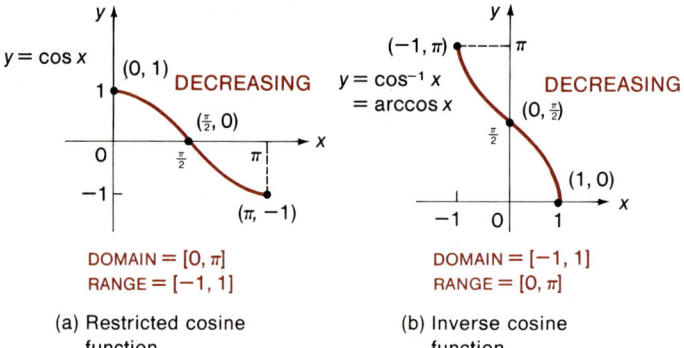

FIGURE 43

(a) Restricted cosine function

(b) Inverse cosine function

We complete the discussion by giving the cosine–inverse cosine identities:

Cosine–Inverse Cosine Identities

$\cos(\cos^{-1} x) = x$ $-1 \leq x \leq 1$

$\cos^{-1}(\cos x) = x$ $0 \leq x \leq \pi$

$\cos(\cos^{-1} 0.5) = 0.5$ $\cos^{-1}(\cos 2) = 2$ $\cos^{-1}[\cos(-1)] \neq -1$

EXAMPLE 29 Find exact values without using a calculator or a table.

(A) $\cos^{-1} \frac{1}{2}$
(B) $\arccos(-\sqrt{3}/2)$
(C) $\cos(\cos^{-1} 0.7)$
(D) $\sin[\cos^{-1}(-\frac{1}{3})]$

Solution (A) $y = \cos^{-1} \frac{1}{2}$ is equivalent to

$$\cos y = \frac{1}{2} \quad 0 \le y \le \pi$$

$$y = \frac{\pi}{3} = \cos^{-1}\frac{1}{2}$$

Reference triangle associated with y

(B) $y = \arccos(-\sqrt{3}/2)$ is equivalent to

$$\cos y = -\frac{\sqrt{3}}{2} \quad 0 \le y \le \pi$$

$$y = \frac{5\pi}{6} = \arccos\left(-\frac{\sqrt{3}}{2}\right)$$

Reference triangle associated with y

[Note: $y \ne -5\pi/6$, even though $\cos(-5\pi/6) = -\sqrt{3}/2$. (Why?)]

(C) $\cos(\cos^{-1} 0.7) = 0.7$ Cosine–inverse cosine identity

(D) Let $y = \cos^{-1}(-\frac{1}{3})$; then $\cos y = -\frac{1}{3}$, $0 \le y \le \pi$. Draw a reference triangle associated with y; then $\sin y = \sin[\cos^{-1}(-\frac{1}{3})]$ can be determined directly from the triangle (after finding the third side) without actually finding y.

$$a^2 + b^2 = c^2$$
$$b = \sqrt{3^2 - (-1)^2}$$
$$= \sqrt{8} = 2\sqrt{2}$$

Thus, $\sin[\cos^{-1}(-\frac{1}{3})] = \sin y = 2\sqrt{2}/3$.

PROBLEM 29 Find exact values without using a calculator or a table.

(A) $\arccos(\sqrt{2}/2)$ (B) $\cos^{-1}(-1)$
(C) $\cos^{-1}(\cos 3.05)$ (D) $\cot[\cos^{-1}(-1/\sqrt{5})]$

EXAMPLE 30 Find to four significant digits using a scientific calculator.

(A) $\cos^{-1} 0.4325$ (B) $\arccos(-0.8976)$
(C) $\cos^{-1} 2.137$ (D) $\csc[\cos^{-1}(-0.0349)]$

Solution Set your calculator in the radian mode.

(A) $\cos^{-1} 0.4325 = 1.124$ $\boxed{0.4325}\ \boxed{\text{INV}}\ \boxed{\text{COS}}$

(B) $\arccos(-0.8976) = 2.685$ $\boxed{0.8976}\ \boxed{+/-}\ \boxed{\text{INV}}\ \boxed{\text{COS}}$

(C) $\cos^{-1} 2.137 = $ Error 2.137 is not in the domain of $\cos^{-1}$

(D) $\csc[\cos^{-1}(-0.0349)] = 1.001$ $\boxed{0.0349}\ \boxed{+/-}\ \boxed{\text{INV}}\ \boxed{\text{COS}}\ \boxed{\text{SIN}}\ \boxed{1/x}$

PROBLEM 30 Find to four significant digits using a scientific calculator.

(A) $\arccos 0.6773$ (B) $\cos^{-1}(-0.8114)$
(C) $\arccos(-1.003)$ (D) $\cot[\cos^{-1}(-0.5036)]$

■ Inverse Tangent Function

To restrict the tangent function so that it becomes one to one, we choose the interval $(-\pi/2, \pi/2)$. Over this interval the tangent function is increasing, and each range value is assumed exactly once as x moves across this restricted domain (Fig. 44). We use this restricted tangent function to define the **inverse tangent function**.

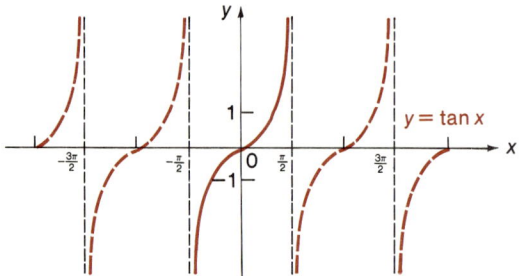

FIGURE 44 $y = \tan x$ is one to one over $(-\pi/2, \pi/2)$

Inverse Tangent Function

The **inverse tangent function**, denoted by **tan⁻¹** or **arctan**, is defined as the inverse of the restricted tangent function, $y = \tan x$, $-\pi/2 < x < \pi/2$. Thus, $y = \tan^{-1} x = \arctan x$ is equivalent to $\tan y = x$, where $-\pi/2 < y < \pi/2$ and x is any real number. The inverse tangent of x, or the arctangent of x, is the number or angle y, $-\pi/2 < y < \pi/2$, whose tangent is x.

Figure 45 compares the graphs of the restricted tangent function and its inverse. Notice that $(-\pi/4, -1)$, $(0, 0)$, and $(\pi/4, 1)$ are on the restricted

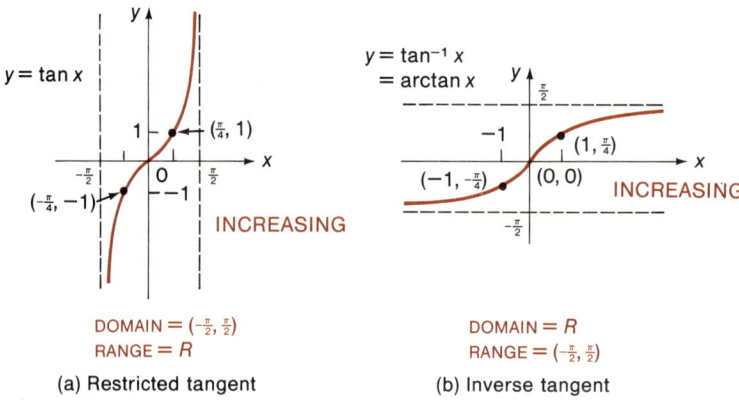

FIGURE 45

tangent graph. Reversing the coordinates gives us three points on the graph of the inverse tangent function. Also note that the vertical asymptotes become horizontal asymptotes.

We now state the tangent–inverse tangent identities:

Tangent–Inverse Tangent Identities

$\tan(\tan^{-1} x) = x$ for all x

$\tan^{-1}(\tan x) = x$ $-\pi/2 < x < \pi/2$

$\tan(\tan^{-1} 25) = 25$ $\tan^{-1}(\tan 1.2) = 1.2$ $\tan^{-1}[\tan(-\pi)] \neq -\pi$

■ **Inverse Cotangent, Secant, and Cosecant Functions (Optional)**

For completeness, we include the definitions and graphs of the inverse cotangent, secant, and cosecant functions.

Inverse Cotangent, Secant, and Cosecant Functions

$y = \cot^{-1} x$ is equivalent to $x = \cot y$ where $0 < y < \pi$ and $x \in R$

$y = \sec^{-1} x$ is equivalent to $x = \sec y$ where $0 \leq y \leq \pi$, $y \neq \pi/2$, and $|x| \geq 1$

$y = \csc^{-1} x$ is equivalent to $x = \csc y$ where $-\pi/2 \leq y \leq \pi/2$, $y \neq 0$, and $|x| \geq 1$

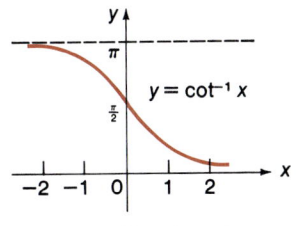

DOMAIN: All real numbers
RANGE: $0 < y < \pi$

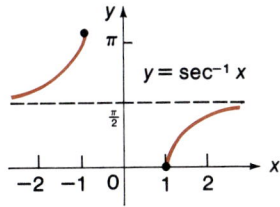

DOMAIN: $x \leq -1$ or $x \geq 1$
RANGE: $0 \leq y \leq \pi, y \neq \frac{\pi}{2}$

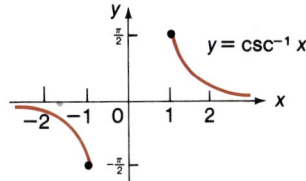

DOMAIN: $x \leq -1$ or $x \geq 1$
RANGE: $-\frac{\pi}{2} \leq y \leq \frac{\pi}{2}, y \neq 0$

[*Note:* The definitions of $\sec^{-1}$ and $\csc^{-1}$ are not universally agreed upon.]

Answers to Matched Problems

27. (A) $\pi/4$ (B) $-\pi/2$ (C) -0.4 (D) $-\frac{1}{2}$

28. (A) 0.2945 (B) -0.8684 (C) Not defined   (D) 1.065

29. (A) $\pi/4$ (B) π (C) 3.05 (D) $-\frac{1}{2}$

30. (A) 0.8267 (B) 2.517 (C) Not defined   (D) -0.5829

Exercise 5-8

A Find exact values without using a calculator or a table.

1. $\sin^{-1} 0$
2. $\cos^{-1} 0$
3. $\arccos \dfrac{\sqrt{3}}{2}$
4. $\arcsin \dfrac{\sqrt{3}}{2}$
5. $\tan^{-1} 1$
6. $\arctan \sqrt{3}$
7. $\cos^{-1} \dfrac{1}{2}$
8. $\sin^{-1} \dfrac{\sqrt{2}}{2}$
9. $\arctan \dfrac{1}{\sqrt{3}}$
10. $\arccos 1$
11. $\tan^{-1} 0$
12. $\sin^{-1} \dfrac{1}{2}$

Evaluate to four significant digits using a scientific calculator.

13. $\cos^{-1} 0.4038$
14. $\sin^{-1} 0.9103$
15. $\tan^{-1} 43.09$
16. $\arctan 103.7$
17. $\arcsin 1.131$
18. $\arccos 3.051$

B Find exact values without using a calculator or a table.

19. $\arccos\left(\dfrac{-1}{2}\right)$
20. $\arcsin\left(\dfrac{-\sqrt{2}}{2}\right)$
21. $\tan^{-1}(-1)$
22. $\arctan(-\sqrt{3})$
23. $\sin^{-1}\left(\dfrac{-\sqrt{3}}{2}\right)$
24. $\cos^{-1}(-1)$
25. $\cos^{-1}\left(\dfrac{-\sqrt{3}}{2}\right)$
26. $\sin^{-1}(-1)$
27. $\sin[\sin^{-1}(-0.6)]$
28. $\tan(\tan^{-1} 25)$
29. $\tan^{-1}[\tan(-1.5)]$
30. $\cos^{-1}(\cos 2.3)$
31. $\tan\left(\cos^{-1} \dfrac{1}{2}\right)$
32. $\sin\left(\cos^{-1} \dfrac{\sqrt{3}}{2}\right)$
33. $\cos\left[\sin^{-1}\left(\dfrac{-\sqrt{2}}{2}\right)\right]$
34. $\sec\left[\sin^{-1}\left(\dfrac{-\sqrt{3}}{2}\right)\right]$
35. $\cot\left[\cos^{-1}\left(\dfrac{-1}{2}\right)\right]$
36. $\csc[\tan^{-1}(-1)]$

Evaluate to four significant digits using a scientific calculator.

37. $\tan^{-1}(-4.038)$
38. $\arctan(-10.04)$
39. $\sec[\sin^{-1}(-0.0399)]$
40. $\cot[\cos^{-1}(-0.7003)]$
41. $\csc[\tan^{-1}(-4.118)]$
42. $\tan[\sin^{-1}(-0.4618)]$
43. $\sqrt{2} + \tan^{-1} \sqrt[3]{5}$
44. $\sqrt{5} + \cos^{-1}(1 - \sqrt{2})$

C Write as an algebraic expression in x free of trigonometric or inverse trigonometric functions.

45. $\sin(\cos^{-1} x)$
46. $\cos(\sin^{-1} x)$
47. $\tan(\arcsin x)$
48. $\cos(\arctan x)$

For $f(x)$ as given, find $f^{-1}(x)$. How must x be restricted in $f^{-1}(x)$?

49. $f(x) = 3 + 5 \sin(x - 1), \quad 1 - \dfrac{\pi}{2} \le x \le 1 + \dfrac{\pi}{2}$

50. $f(x) = 4 + 2 \cos(x - 3), \quad 3 \le x \le 3 + \pi$

Section 5-9 Chapter Review

IMPORTANT TERMS AND SYMBOLS

5-2 The Wrapping Function. Unit circle, wrapping function, circular point, coordinates of key circular points

5-3 Circular Functions. Circular functions, sine, cosine, tangent, cotangent, secant, cosecant, sin x, cos x, tan x, cot x, sec x, csc x, exact values, sign properties, basic identities, reciprocal identities, calculator evaluation

5-4 Angles. Angle, sides, vertex, initial side, terminal side, positive angle, negative angle, coterminal, standard position, quadrantal angle, degree measure, straight angle, right angle, acute angle, obtuse angle, complementary angles, supplementary angles, minutes, seconds, radian measure,

$$\dfrac{\theta°}{180°} = \dfrac{\theta}{\pi}$$

5-5 Trigonometric Functions. Trigonometric functions with angle domains, $\sin \theta$, $\cos \theta$, $\tan \theta$, $\csc \theta$, $\sec \theta$, $\cot \theta$, exact values, calculator evaluation, trigonometric functions with angle domains (alternate form), reference triangle, reference angle, 30°–60° and 45° special triangles

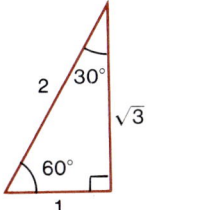 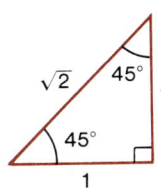

5-6 Graphs of Trigonometric Functions. Periodic functions; fundamental period of f; graphs of $y = \sin x$, $y = \cos x$, $y = \tan x$, $y = \cot x$, $y = \csc x$, and $y = \sec x$

5-7 Graphing $y = A \sin(Bx + C)$ and $y = A \cos(Bx + C)$. Simple harmonic motion, amplitude, period, phase shift; for $y = A \sin(Bx + C)$ or $y = A \cos(Bx + C)$, $B > 0$:

$$\text{Amplitude} = |A| \qquad \text{Period} = \frac{2\pi}{B}$$

$$\text{Phase shift} = \begin{cases} |C/B| \text{ units to the right if } C/B < 0 \\ C/B \text{ units to the left if } C/B > 0 \end{cases}$$

for $y = A \tan(Bx + C)$ or $y = A \cot(Bx + C)$, $B > 0$:

$$\text{Period} = \frac{\pi}{B}$$

$$\text{Phase shift} = \begin{cases} |C/B| \text{ units to the right if } C/B < 0 \\ C/B \text{ units to the left if } C/B > 0 \end{cases}$$

5-8 Inverse Trigonometric Functions. Restricted sine function, inverse sine function, $\sin^{-1}$, arcsine, sine–inverse sine identities, inverse cosine function, $\cos^{-1}$, arccosine, cosine–inverse cosine identities, inverse tangent function, $\tan^{-1}$, arctangent, tangent–inverse tangent identities

$$\sin^{-1} x \qquad \cos^{-1} x \qquad \tan^{-1} x$$
$$\text{arcsin } x \qquad \text{arccos } x \qquad \text{arctan } x$$

Exercise 5-9 Chapter Review

Work through all the problems in this chapter review and check answers in the back of the book. (Answers to all problems are there, and following each answer is a number in italics indicating the section in which that type of problem is discussed.) Where weaknesses show up, review appropriate sections in the text. When you are satisfied that you know the material, take the practice test following this review.

A 1. Find the coordinates of each circular point:
(A) $W(\pi/6)$ (B) $W(\pi/4)$

2. In a circle of radius 3 centimeters, find the length of an arc subtended by an angle of 2.5 radians.

3. In which quadrants must $W(x)$ lie for each of the following to be negative?
(A) $\sin x$ (B) $\cos x$ (C) $\tan x$

4. If $(4, -3)$ is on the terminal side of angle θ, find:
(A) $\sin \theta$ (B) $\sec \theta$ (C) $\cot \theta$

5. Complete the table using exact values (do not use a calculator or a table).

θ degrees	θ radians	sin θ	cos θ	tan θ	csc θ	sec θ	cot θ
0°					ND*		
30°							
45°	$\pi/4$		$1/\sqrt{2}$				
60°							
90°							
180°							
270°							
360°							

* ND = Not defined

6. What is the period of each of the following?
   (A) $y = \cos x$ (B) $y = \csc x$ (C) $y = \tan x$

7. Indicate the domain and range of each (x a real number).
   (A) $y = \sin x$ (B) $y = \tan x$

8. Sketch a graph of $y = \sin x$, $-2\pi \leq x \leq 2\pi$.

9. Sketch a graph of $y = \cot x$, $-\pi < x < \pi$.

B 10. Find the coordinates of the circular points:
   (A) $W(-2\pi/3)$ (B) $W(7\pi/2)$

11. Indicate whether the angle is a I, II, III, or IV quadrant angle or a quadrantal angle.
   (A) $-210°$ (B) $5\pi/2$ rad (C) 4.2 rad

12. Which of the following angles are coterminal with 120°?
   (A) $-240°$ (B) $-7\pi/6$ (C) 840°

13. Which of the following have the same value as cos 3?
   (A) $\cos 3°$ (B) $\cos(3 \text{ rad})$ (C) $\cos(3 + 2\pi)$

14. For which values of x, $0 \leq x < 2\pi$, is each of the following not defined?
   (A) tan x (B) cot x (C) csc x

Evaluate exactly without the use of a calculator or a table.

15. tan 0   16. sec 90°   17. $\cos^{-1} 1$

18. $\cos\left(-\dfrac{3\pi}{4}\right)$ 19. $\sin^{-1}\dfrac{\sqrt{2}}{2}$ 20. $\csc 300°$

21. $\arctan\sqrt{3}$ 22. $\sin 570°$ 23. $\tan^{-1}(-1)$

24. $\cot\left(-\dfrac{4\pi}{3}\right)$ 25. $\arcsin\left(-\dfrac{1}{2}\right)$ 26. $\cos^{-1}\left(-\dfrac{\sqrt{3}}{2}\right)$

27. $\cos(\cos^{-1} 0.33)$ 28. $\csc[\tan^{-1}(-1)]$

29. $\sin\left[\arccos\left(-\dfrac{1}{2}\right)\right]$

Evaluate Problems 30–37 to four significant digits using a scientific calculator.

30. $\cos 423.7°$ 31. $\tan 93°46'17''$ 32. $\sec(-2.073)$
33. $\sin^{-1}(-0.8277)$ 34. $\arccos(-1.3281)$ 35. $\tan^{-1} 75.14$
36. $\csc[\cos^{-1}(-0.4081)]$ 37. $\sin^{-1}(\tan 1.345)$

38. Solve for $-2\pi \leq x \leq 2\pi$.

   (A) $\sin x = 0$ (B) $\cot x = 0$

39. For what values of k does:

   (A) $\tan(x + k\pi) = \tan x$ (B) $\cos(x + 2k\pi) = \cos x$

40. Sketch a graph of $y = -2\cos \pi x$, $-1 \leq x \leq 3$. Indicate amplitude and period.

41. Sketch a graph of $y = 3\sin[(x/2) + (\pi/2)]$, $-4\pi \leq x \leq 4\pi$.

42. Indicate the amplitude, period, and phase shift for the graph of $y = -2\cos[(\pi/2)x - (\pi/4)]$. Do not graph.

43. Sketch a graph of $y = \cos^{-1} x$ and indicate the domain and range.

C 44. Find exactly the least positive real number for which:

   (A) $\cos x = -\dfrac{1}{2}$ (B) $\csc x = -\sqrt{2}$

45. Sketch a graph of $y = \sec x$, $-\pi/2 < x < 3\pi/2$.

46. Sketch a graph of $y = \tan^{-1} x$ and indicate the domain and range.

47. Indicate the period and phase shift for the graph of $y = -5\tan(\pi x + \pi/2)$. Do not graph.

48. Indicate whether an odd function, an even function, or neither.

   (A) Sine   (B) Cosine   (C) Tangent

49. Evaluate to four significant digits using a calculator:

   $\sqrt{3.204} - \sin^{-1}(-0.6443)$

50. Write as an algebraic expression in x free of trigonometric or inverse trigonometric functions: $\sec(\sin^{-1} x)$

51. Given $\cos x = -\dfrac{3}{5}$ and $\tan x < 0$, use basic identities to find:

   (A) $\sin x$ (B) $\cot x$ (C) $\sin(-x)$

Practice Test Chapter 5

Take this practice test as if it were a graded test. Allow yourself up to 50 minutes. Work the problems without looking back in the chapter. Correct your work using the answers (keyed to appropriate sections) in the back of the book.

1. Each of the following, except one, is a quadrantal angle. Which one?
   (A) $-3\pi/2$ (B) $300°$ (C) $-900°$ (D) 11π

2. Each of the following, except one, is a true statement for all values of x for which both sides are defined. Which one?
   (A) $\sin(x + 2\pi) = \sin x$ (B) $\tan(x + \pi) = \tan x$
   (C) $\sec(x + \pi) = \sec x$ (D) $\cot(x + 2\pi) = \cot x$

3. Given $\sin x = -\frac{4}{5}$ and $\cos x > 0$, use basic identities to find:
   (A) $\cos x$ (B) $\tan x$ (C) $\csc(-x)$

Evaluate exactly without using a calculator or a table.

4. $\sin\left(\dfrac{-\pi}{3}\right)$ 5. $\cos\dfrac{7\pi}{6}$ 6. $\cos^{-1}(-1)$

7. $\tan\left[\sin^{-1}\left(-\dfrac{1}{2}\right)\right]$ 8. $\csc 3\pi$ 9. $\cot(-120°)$

10. $\tan(\tan^{-1} 25)$ 11. $\arccos\left(-\dfrac{\sqrt{3}}{2}\right)$

Evaluate Problems 12–15 to four significant digits using a calculator.

12. $\cot(-0.0133)$ 13. $\sec[\arctan(-13.76)]$
14. $\cos 103°22'51''$ 15. $\csc(-21.78°)$

16. Sketch a graph of $y = 2 \sin \pi x$, $-2 \le x \le 2$. Indicate amplitude and period.

17. Sketch a graph of $y = -\cos(x/2 + \pi/2)$, $0 \le x \le 4\pi$. Indicate amplitude, period, and phase shift.

18. Sketch a graph of $y = \sin^{-1} x$ and indicate the domain and range.

19. Solve: $\sin\theta = -\dfrac{1}{2}$, $-180° \le \theta \le 180°$

20. Write as an algebraic expression in terms of x free of any trigonometric or inverse trigonometric functions: $\csc(\cos^{-1} x)$

Trigonometric Identities and Conditional Equations ▪ 6

6-1 Introduction
6-2 Basic Identities and Their Use
6-3 Addition, Subtraction, and Cofunction Identities
6-4 Double-Angle and Half-Angle Identities
6-5 Product and Factor Identities
6-6 Trigonometric Equations
6-7 Chapter Review

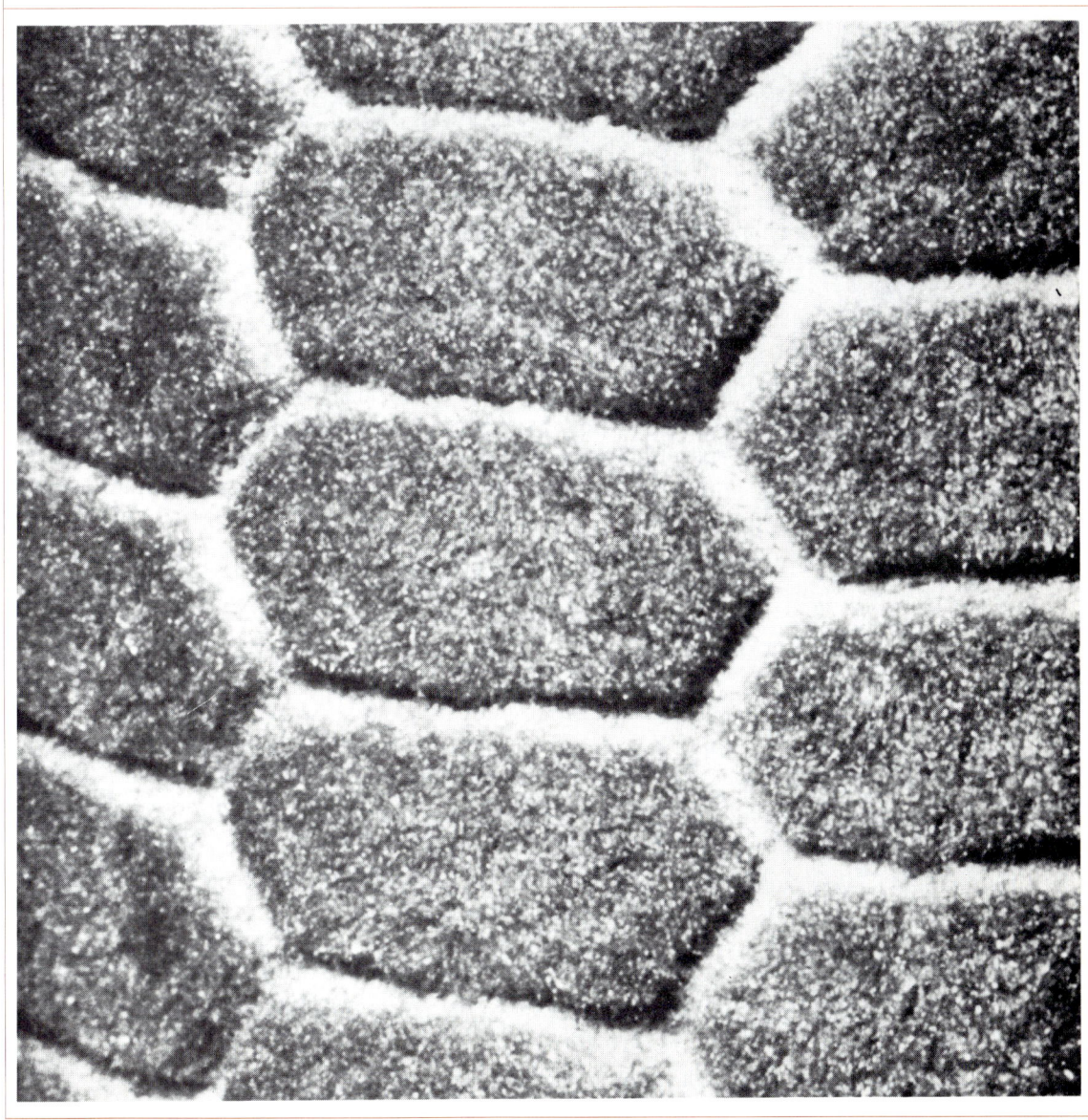

A natural design of mathematical interest. Can you guess the source? See the back of the book.

Chapter 6 ■ Trigonometric Identities and Conditional Equations

Section 6-1 Introduction

In this chapter we will consider two types of equations involving trigonometric functions: The first type is called an *identity* and the second type a *conditional equation*. In general, if

$$f(x) = g(x) \tag{1}$$

holds for all replacements of x by real numbers for which both f and g are defined, then equation (1) is called an **identity**. On the other hand, if equation (1) holds for some replacements of x by real numbers and does not hold for others (assuming f and g are defined for all replacements used), then equation (1) is called a **conditional equation**. For example, from algebra,

$$x^2 - x - 6 = (x - 3)(x + 2)$$

is an identity, whereas

$$x^2 - x - 6 = 0$$

is a conditional equation.

Trigonometric identities form an important part of any development of trigonometry. Familiarity with this subject will prove very useful in courses that follow this one. In almost any place that trigonometry is used you will find that identities play an important role. The next four sections deal with trigonometric identities and the last section with conditional trigonometric equations.

Section 6-2 Basic Identities and Their Use

- Basic Identities
- Establishing Other Identities

6-2 Basic Identities and Their Use

■ Basic Identities

In Section 5-3 we introduced nine basic trigonometric identities and used some of them when graphing the trigonometric functions in Section 5-6. We now add two more and list all eleven for convenient reference. These eleven basic identities will be used very frequently in the work that follows and should be memorized.

Basic Trigonometric Identities

RECIPROCAL IDENTITIES

$$\csc x = \frac{1}{\sin x} \qquad \sec x = \frac{1}{\cos x} \qquad \cot x = \frac{1}{\tan x}$$

QUOTIENT IDENTITIES

$$\tan x = \frac{\sin x}{\cos x} \qquad \cot x = \frac{\cos x}{\sin x}$$

IDENTITIES FOR NEGATIVES

$$\sin(-x) = -\sin x \qquad \cos(-x) = \cos x \qquad \tan(-x) = -\tan x$$

PYTHAGOREAN IDENTITIES

$$\sin^2 x + \cos^2 x = 1 \qquad \tan^2 x + 1 = \sec^2 x \qquad 1 + \cot^2 x = \csc^2 x$$

All of these identities were established in Section 5-3 (the second and third Pythagorean identities were established in Problems 87 and 88 in Exercise 5-3). An easy way to remember the second and third Pythagorean identities is to note that the second can be obtained from the first by dividing both sides of the first by $\cos^2 x$, and the third can be obtained from the first by dividing both sides of the first by $\sin^2 x$:

$$\sin^2 x + \cos^2 x = 1 \qquad \qquad \sin^2 x + \cos^2 x = 1$$

$$\frac{\sin^2 x}{\cos^2 x} + \frac{\cos^2 x}{\cos^2 x} = \frac{1}{\cos^2 x} \qquad \frac{\sin^2 x}{\sin^2 x} + \frac{\cos^2 x}{\sin^2 x} = \frac{1}{\sin^2 x}$$

$$\tan^2 x + 1 = \sec^2 x \qquad \qquad 1 + \cot^2 x = \csc^2 x$$

■ Establishing Other Identities

As indicated earlier, when working with trigonometric expressions, it is often desirable to convert one form to an equivalent form that may be more useful. This section is designed to give you experience in this process. The eleven basic identities listed in the box will be used frequently, so they should be learned before proceeding further. The following examples illustrate some of the techniques used to establish certain identities. To become proficient in the use of identities, it is important that you work out many problems on your own.

EXAMPLE 1 Establish the identity: $\cos x \tan x = \sin x$

Proof Generally, we proceed by starting with the more complicated of the two sides, and transform that side into the other side in one or more steps using basic identities, algebra, or other established identities. Thus,

$$\cos x \tan x = \cos x \frac{\sin x}{\cos x} \qquad \text{Quotient identity}$$

$$= \sin x \qquad \text{Algebra}$$

PROBLEM 1 Establish the identity: $\sin x \cot x = \cos x$

EXAMPLE 2 Establish the identity: $\sec(-x) = \sec x$

Proof $$\sec(-x) = \frac{1}{\cos(-x)} \qquad \text{Reciprocal identity}$$

$$= \frac{1}{\cos x} \qquad \text{Identity for negatives}$$

$$= \sec x \qquad \text{Reciprocal identity}$$

PROBLEM 2 Establish the identity: $\csc(-x) = -\csc x$

EXAMPLE 3 Establish the identity: $\cot x \cos x + \sin x = \csc x$

Proof $$\cot x \cos x + \sin x = \frac{\cos x}{\sin x} \cos x + \sin x \qquad \text{Quotient identity}$$

$$= \frac{\cos^2 x}{\sin x} + \sin x \qquad \text{Algebra}$$

$$= \frac{\cos^2 x + \sin^2 x}{\sin x} \qquad \text{Algebra}$$

$$= \frac{1}{\sin x} \qquad \text{Pythagorean identity}$$

$$= \csc x \qquad \text{Reciprocal identity}$$

PROBLEM 3 Establish the identity: $\tan x \sin x + \cos x = \sec x$

6-2 Basic Identities and Their Use

Even though there is no fixed method of proof that works for all identities, there are certain steps one can take that will help in many cases.

Suggested Steps in Proving Identities

1. Start with the more complicated side of the identity and transform it into the simpler side.
2. Try algebraic operations such as multiplying, factoring, combining fractions into single fractions, and splitting single fractions into two or more fractions.
3. If other steps fail, express each function in terms of sine and cosine functions, and then perform appropriate algebraic operations.
4. At each step keep the other side of the identity in mind. This often suggests what one should do to get there.

EXAMPLE 4 Establish the identity: $\dfrac{1 + \sin x}{\cos x} + \dfrac{\cos x}{1 + \sin x} = 2 \sec x$

Proof

$$\dfrac{1 + \sin x}{\cos x} + \dfrac{\cos x}{1 + \sin x} = \dfrac{(1 + \sin x)^2 + \cos^2 x}{\cos x (1 + \sin x)} \qquad \text{Algebra}$$

$$= \dfrac{1 + 2 \sin x + \sin^2 x + \cos^2 x}{\cos x (1 + \sin x)} \qquad \text{Algebra}$$

$$= \dfrac{1 + 2 \sin x + 1}{\cos x (1 + \sin x)} \qquad \text{Pythagorean identity}$$

$$= \dfrac{2 + 2 \sin x}{\cos x (1 + \sin x)} \qquad \text{Algebra}$$

$$= \dfrac{2(1 + \sin x)}{\cos x (1 + \sin x)} \qquad \text{Algebra}$$

$$= \dfrac{2}{\cos x} \qquad \text{Algebra}$$

$$= 2 \sec x \qquad \text{Reciprocal identity}$$

PROBLEM 4 Establish the identity: $\dfrac{1 + \cos x}{\sin x} + \dfrac{\sin x}{1 + \cos x} = 2 \csc x$

EXAMPLE 5 Establish the identity: $\dfrac{\sin^2 x + 2 \sin x + 1}{\cos^2 x} = \dfrac{1 + \sin x}{1 - \sin x}$

Proof
$$\frac{\sin^2 x + 2\sin x + 1}{\cos^2 x} = \frac{(\sin x + 1)^2}{\cos^2 x} \qquad \text{Algebra}$$

$$= \frac{(\sin x + 1)^2}{1 - \sin^2 x} \qquad \text{Pythagorean identity}$$

$$= \frac{(1 + \sin x)^2}{(1 - \sin x)(1 + \sin x)} \qquad \text{Algebra}$$

$$= \frac{1 + \sin x}{1 - \sin x} \qquad \text{Algebra}$$

PROBLEM 5 Establish the identity: $\sec^4 x - 2\sec^2 x \tan^2 x + \tan^4 x = 1$

EXAMPLE 6 Establish the identity: $\dfrac{\tan x - \cot x}{\tan x + \cot x} = 1 - 2\cos^2 x$

Proof
$$\frac{\tan x - \cot x}{\tan x + \cot x} = \frac{\dfrac{\sin x}{\cos x} - \dfrac{\cos x}{\sin x}}{\dfrac{\sin x}{\cos x} + \dfrac{\cos x}{\sin x}} \qquad \text{Change to sines and cosines (quotient identities).}$$

$$= \frac{(\sin x)(\cos x)\left(\dfrac{\sin x}{\cos x} - \dfrac{\cos x}{\sin x}\right)}{(\sin x)(\cos x)\left(\dfrac{\sin x}{\cos x} + \dfrac{\cos x}{\sin x}\right)} \qquad \text{Multiply numerator and denominator by } (\sin x)(\cos x) \text{ and use algebra to transform complex fraction into simple fraction.}$$

$$= \frac{\sin^2 x - \cos^2 x}{\sin^2 x + \cos^2 x}$$

$$= \frac{1 - \cos^2 x - \cos^2 x}{1} \qquad \text{Pythagorean identity}$$

$$= 1 - 2\cos^2 x \qquad \text{Algebra}$$

PROBLEM 6 Establish the identity: $\cot x - \tan x = \dfrac{2\cos^2 x - 1}{\sin x \cos x}$

> Just observing how others prove identities will not make you good at it. You must prove a large number on your own. With practice the process will seem less complicated.

Answers to Matched Problems

1. $\sin x \cot x = \sin x \dfrac{\cos x}{\sin x} = \cos x$

2. $\csc(-x) = \dfrac{1}{\sin(-x)} = \dfrac{1}{-\sin x} = -\csc x$

3. $\tan x \sin x + \cos x = \dfrac{\sin^2 x}{\cos x} + \cos x = \dfrac{\sin^2 x + \cos^2 x}{\cos x}$

$= \dfrac{1}{\cos x} = \sec x$

4. $\dfrac{1 + \cos x}{\sin x} + \dfrac{\sin x}{1 + \cos x} = \dfrac{(1 + \cos x)^2 + \sin^2 x}{\sin x(1 + \cos x)}$

$= \dfrac{1 + 2\cos x + \cos^2 x + \sin^2 x}{\sin x(1 + \cos x)}$

$= \dfrac{2(1 + \cos x)}{\sin x(1 + \cos x)} = 2 \csc x$

5. $\sec^4 x - 2\sec^2 x \tan^2 x + \tan^4 x = (\sec^2 x - \tan^2 x)^2 = 1^2 = 1$

6. $\cot x - \tan x = \dfrac{\cos x}{\sin x} - \dfrac{\sin x}{\cos x} = \dfrac{\cos^2 x - \sin^2 x}{\sin x \cos x}$

$= \dfrac{\cos^2 x - (1 - \cos^2 x)}{\sin x \cos x} = \dfrac{2\cos^2 x - 1}{\sin x \cos x}$

Exercise 6-2 ■ Prove that Problems 1–60 are identities.

A

1. $\sin x \csc x = 1$
2. $\cos x \sec x = 1$
3. $\sin \theta \sec \theta = \tan \theta$
4. $\cos \theta \csc \theta = \cot \theta$
5. $\cot u \sec u \sin u = 1$
6. $\tan \theta \csc \theta \cos \theta = 1$
7. $\tan(-x) = -\tan x$
8. $\cot(-x) = -\cot x$
9. $\dfrac{\sin^2 t}{\cos t} + \cos t = \sec t$
10. $\dfrac{\cos^2 t}{\sin t} + \sin t = \csc t$
11. $\sin^2 \theta = 1 - \cos^2 \theta$
12. $\sec^2 \theta (1 - \cos^2 \theta) = \tan^2 \theta$
13. $\sec^2 \theta (1 - \sin^2 \theta) = 1$
14. $\tan^2 x \cos^2 x = 1 - \cos^2 x$
15. $\cot^2 x \sin^2 x = 1 - \sin^2 x$
16. $\cos^2 m = 1 - \sin^2 m$
17. $\dfrac{\cos x}{1 - \sin^2 x} = \sec x$
18. $(1 - \sin t)(1 + \sin t) = \cos^2 t$
19. $(1 - \cos u)(1 + \cos u) = \sin^2 u$
20. $\dfrac{\sin u}{1 - \cos^2 u} = \csc u$
21. $\cos^2 x - \sin^2 x = 1 - 2\sin^2 x$
22. $(\sin x + \cos x)^2 = 1 + 2\sin x \cos x$
23. $\csc^2 x - \cot^2 x = 1$

24. $\sec^2 u - \tan^2 u = 1$

25. $\cot x + \sec x = \dfrac{\cos x + \tan x}{\sin x}$

26. $\sin m(\csc m - \sin m) = \cos^2 m$

27. $\dfrac{1}{\csc^2 x} + \dfrac{1}{\sec^2 x} = 1$

28. $\dfrac{\sin \theta}{\csc \theta} + \dfrac{\cos \theta}{\sec \theta} = 1$

B   29. $\dfrac{1 - (\sin x - \cos x)^2}{\sin x} = 2 \cos x$

30. $\dfrac{1 - \cos^2 y}{(1 - \sin y)(1 + \sin y)} = \tan^2 y$

31. $\tan^2 x - \sin^2 x = \tan^2 x \sin^2 x$

32. $\sec^2 x + \csc^2 x = \sec^2 x \csc^2 x$

33. $\dfrac{\csc \theta}{\cot \theta + \tan \theta} = \cos \theta$

34. $\dfrac{1 + \sec \theta}{\sin \theta + \tan \theta} = \csc \theta$

35. $\ln \tan x = \ln \sin x - \ln \cos x$

36. $\ln \cot x = \ln \cos x - \ln \sin x$

37. $\ln \cot x = -\ln \tan x$ 38. $\ln \csc x = -\ln \sin x$

39. $\dfrac{1 - \cos A}{1 + \cos A} = \dfrac{\sec A - 1}{\sec A + 1}$ 40. $\dfrac{1 - \csc y}{1 + \csc y} = \dfrac{\sin y - 1}{\sin y + 1}$

41. $\sin^4 w - \cos^4 w = 1 - 2 \cos^2 w$

42. $\sin^4 x + 2 \sin^2 x \cos^2 x + \cos^4 x = 1$

43. $\sec x - \dfrac{\cos x}{1 + \sin x} = \tan x$

44. $\csc n - \dfrac{\sin n}{1 + \cos n} = \cot n$

45. $\dfrac{\cos x}{\csc x + 1} + \dfrac{\cos x}{\csc x - 1} = 2 \tan x$

46. $\dfrac{\cos x}{1 - \sin x} + \dfrac{\cos x}{1 + \sin x} = 2 \sec x$

47. $\dfrac{\cos^2 z - 3 \cos z + 2}{\sin^2 z} = \dfrac{2 - \cos z}{1 + \cos z}$

48. $\dfrac{\sin^2 t + 4 \sin t + 3}{\cos^2 t} = \dfrac{3 + \sin t}{1 - \sin t}$

49. $\dfrac{\cos^3 \theta - \sin^3 \theta}{\cos \theta - \sin \theta} = 1 + \sin \theta \cos \theta$

50. $\dfrac{\cos^3 u + \sin^3 u}{\cos u + \sin u} = 1 - \sin u \cos u$

51. $\dfrac{\tan x}{\sin x - 2 \tan x} = \dfrac{1}{\cos x - 2}$

52. $\dfrac{1 - \cot^2 x}{\tan^2 x - 1} = \cot^2 x$

53. $(\sec x - \tan x)^2 = \dfrac{1 - \sin x}{1 + \sin x}$

54. $(\cot u - \csc u)^2 = \dfrac{1 - \cos u}{1 + \cos u}$

55. $\dfrac{\csc^4 x - 1}{\cot^2 x} = 2 + \cot^2 x$

56. $\dfrac{\sec^4 x - 1}{\tan^2 x} = 2 + \tan^2 x$

57. $\dfrac{1 + \sin v}{\cos v} = \dfrac{\cos v}{1 - \sin v}$

58. $\dfrac{\sin x}{1 - \cos x} = \dfrac{1 + \cos x}{\sin x}$

59. $\tan^2 \theta - \sin^2 \theta = \tan^2 \theta \sin^2 \theta$

60. $\sec^2 u + \csc^2 u = \sec^2 u \csc^2 u$

Show that each equation is not an identity by finding a value for which both sides are defined but are not equal to each other. For example, cos x = sin x is not an identity, since they are not equal at x = π/2.

61. $\sin^2 \theta - 2 \sin \theta + 1 = 0$
62. $\cos^2 x + 2 \sin x = 1$
63. $\cos^2 x - \sin^2 x = 1$
64. $\cot x = \tan x$
65. $1 - 2 \tan^2 x = \cot^2 x$
66. $\tan^2 x + \cot^2 x = -1$
67. $\sqrt{1 - \cos^2 x} = \sin x$
68. $\sqrt{1 - \sin^2 x} = \cos x$

C *Prove that Problems 69–72 are identities.*

69. $\dfrac{2 \sin^2 x + 3 \cos x - 3}{\sin^2 x} = \dfrac{2 \cos x - 1}{1 + \cos x}$

70. $\dfrac{3 \cos^2 z + 5 \sin z - 5}{\cos^2 z} = \dfrac{3 \sin z - 2}{1 + \sin z}$

71. $\dfrac{\tan u + \sin u}{\tan u - \sin u} - \dfrac{\sec u + 1}{\sec u - 1} = 0$

72. $\dfrac{\sin x \cos y + \cos x \sin y}{\cos x \cos y - \sin x \sin y} = \dfrac{\tan x + \tan y}{1 - \tan x \tan y}$

Each of the equations in Problems 73–80 is an identity in certain quadrants associated with x. Indicate which quadrants.

73. $\sqrt{1 - \cos^2 x} = -\sin x$
74. $\sqrt{1 - \sin^2 x} = \cos x$

75. $\sqrt{1 - \cos^2 x} = \sin x$ 76. $\sqrt{1 - \sin^2 x} = -\cos x$

77. $\sqrt{1 - \sin^2 x} = |\cos x|$ 78. $\sqrt{1 - \cos^2 x} = |\sin x|$

79. $\dfrac{\sin x}{\sqrt{1 - \sin^2 x}} = \tan x$ 80. $\dfrac{\sin x}{\sqrt{1 - \sin^2 x}} = -\tan x$

Section 6-3 Addition, Subtraction, and Cofunction Identities

- Addition and Subtraction Identities for Cosine
- Cofunction Identities
- Addition and Subtraction Identities for Sine and Tangent
- Summary and Use
- Calculator Exercise

- Addition and Subtraction Identities for Cosine

The basic identities discussed in Section 6-2 involved only one variable. We will now consider an important identity, called a *subtraction identity for cosine*, that involves two variables:

$$\cos(x - y) = \cos x \cos y + \sin x \sin y \qquad (1)$$

Many other useful identities can be readily established from this particular one.

We will sketch a proof of equation (1) assuming x and y are in the interval $(0, 2\pi)$ and $x > y > 0$. Identity (1) holds, however, for all real numbers and angles in radian or degree measure.

We associate x and y with arcs and angles on a unit circle as indicated in Figure 1(a). Using the definitions of the circular functions in terms of arc length given in Section 5-3, we label the terminal points of x and y as shown in Figure 1(a).

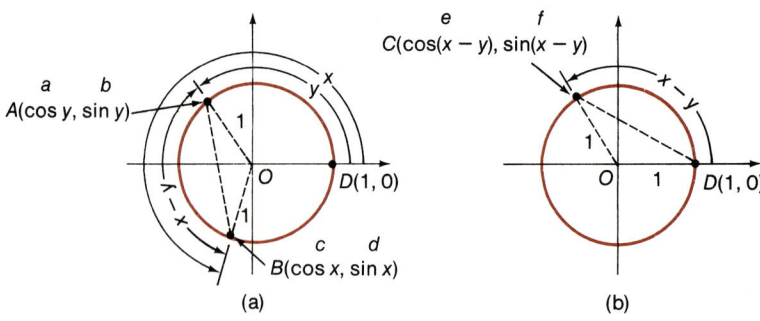

FIGURE 1   (a)   (b)

6-3 Addition, Subtraction, and Cofunction Identities

Now if we rotate the triangle AOB clockwise about the origin until the terminal point A coincides with $D(1, 0)$, then terminal point B will be at C [see Fig. 1(b)]. Thus, since rotation preserves lengths,

$$d(A, B) = d(C, D)$$
$$\sqrt{(c - a)^2 + (d - b)^2} = \sqrt{(1 - e)^2 + (0 - f)^2}$$
$$(c - a)^2 + (d - b)^2 = (1 - e)^2 + f^2$$
$$c^2 - 2ac + a^2 + d^2 - 2db + b^2 = 1 - 2e + e^2 + f^2$$
$$(c^2 + d^2) + (a^2 + b^2) - 2ac - 2db = 1 - 2e + (e^2 + f^2) \tag{2}$$

Since $c^2 + d^2 = 1$, $a^2 + b^2 = 1$, and $e^2 + f^2 = 1$ (Why?), equation (2) becomes

$$e = ac + bd \tag{3}$$

Replacing e, a, c, b, and d with $\cos(x - y)$, $\cos y$, $\cos x$, $\sin y$, and $\sin x$, respectively (see Fig. 1), we obtain:

$$\cos(x - y) = \cos y \cos x + \sin y \sin x$$
$$= \cos x \cos y + \sin x \sin y \tag{4}$$

If we replace y with $-y$ in equation (4) and use the identities for negatives, we obtain:

$$\mathbf{\cos(x + y) = \cos x \cos y - \sin x \sin y} \tag{5}$$

■ Cofunction Identities

To obtain addition and subtraction identities for the sine and tangent functions, we first derive *cofunction* identities directly from equation (1), the subtraction identity for cosine.

$$\cos(x - y) = \cos x \cos y + \sin x \sin y$$
$$\cos\left(\frac{\pi}{2} - y\right) = \cos\frac{\pi}{2}\cos y + \sin\frac{\pi}{2}\sin y$$
$$= (0)\cos y + (1)\sin y$$
$$= \sin y$$

Thus,

$$\mathbf{\cos\left(\frac{\pi}{2} - y\right) = \sin y} \tag{6}$$

for y any real number or angle in radian measure. If y is in degree measure, replace $\pi/2$ with $90°$.

Now, if in equation (6) we let $y = \pi/2 - x$, we have

$$\cos\left[\frac{\pi}{2} - \left(\frac{\pi}{2} - x\right)\right] = \sin\left(\frac{\pi}{2} - x\right)$$

$$\cos x = \sin\left(\frac{\pi}{2} - x\right)$$

or

$$\sin\left(\frac{\pi}{2} - x\right) = \cos x \qquad (7)$$

where x is any real number or angle in radian measure. If x is in degree measure, replace $\pi/2$ with $90°$.

Finally, we state the cofunction identity for tangent and leave its derivation to Problem 9 in Exercise 6-3.

$$\tan\left(\frac{\pi}{2} - x\right) = \cot x \qquad (8)$$

for x any real number or angle in radian measure. If x is in degree measure, replace $\pi/2$ with $90°$.

Remark If $0 < x < 90°$, then x and $90° - x$ are complementary angles. Originally, "cosine," "cotangent," and "cosecant" meant, respectively, "complements sine," "complements tangent," and "complements secant." Now we simply refer to cosine, cotangent, and cosecant as **cofunctions** of sine, tangent, and secant, respectively.

■ Addition and Subtraction Identities for Sine and Tangent

To derive a subtraction identity for sine, we use equations (7), (1), and (6) as follows:

$\sin(x - y) = \cos\left[\dfrac{\pi}{2} - (x - y)\right]$ Use equation (6).

$\qquad\qquad = \cos\left[\left(\dfrac{\pi}{2} - x\right) - (-y)\right]$ Algebra

$\qquad\qquad = \cos\left(\dfrac{\pi}{2} - x\right)\cos(-y) + \sin\left(\dfrac{\pi}{2} - x\right)\sin(-y)$ Use equation (1).

$\qquad\qquad = \sin x \cos y - \cos x \sin y$ Use equations (6) and (7) and identities for negatives.

The same result is obtained by replacing $\pi/2$ with $90°$. Thus,

$$\sin(x - y) = \sin x \cos y - \cos x \sin y \qquad (9)$$

Now, if we replace y with $-y$ (a good exercise for the reader), we obtain

$$\sin(x + y) = \sin x \cos y + \cos x \sin y \qquad (10)$$

6-3 Addition, Subtraction, and Cofunction Identities

It is not difficult to derive addition and subtraction identities for the tangent function. See if you can supply the reason for each step.

$$\tan(x - y) = \frac{\sin(x - y)}{\cos(x - y)}$$

$$= \frac{\sin x \cos y - \cos x \sin y}{\cos x \cos y + \sin x \sin y}$$

$$= \frac{\dfrac{\sin x \cos y}{\cos x \cos y} - \dfrac{\cos x \sin y}{\cos x \cos y}}{\dfrac{\cos x \cos y}{\cos x \cos y} + \dfrac{\sin x \sin y}{\cos x \cos y}}$$

$$= \frac{\tan x - \tan y}{1 + \tan x \tan y}$$

Thus, for all angles or real numbers x and y,

$$\tan(x - y) = \frac{\tan x - \tan y}{1 + \tan x \tan y} \tag{11}$$

And if we replace y in equation (11) with $-y$ (another good exercise for the reader), we obtain

$$\tan(x + y) = \frac{\tan x + \tan y}{1 - \tan x \tan y} \tag{12}$$

■ **Summary and Use**

Before proceeding with examples illustrating the use of these new identities, review the list given in the box on page 362.

EXAMPLE 7 Simplify $\sin(x - \pi)$ using a subtraction identity.

Solution Use the subtraction identity for sine, replacing y with π.

$$\sin(x - y) = \sin x \cos y - \cos x \sin y$$
$$\sin(x - \pi) = \sin x \cos \pi - \cos x \sin \pi$$
$$= (\sin x)(-1) - (\cos x)(0)$$
$$= -\sin x$$

PROBLEM 7 Simplify $\cos(x + 3\pi/2)$ using an addition identity.

EXAMPLE 8 Write $\sin 75°$ in the form $\cos \theta$, $0 \leq \theta \leq 90°$.

Solution Use $\sin x = \cos(90° - x)$. Thus,

$$\sin 75° = \cos(90° - 75°) = \cos 15°$$

> **Summary of Identities**
>
> ADDITION IDENTITIES
>
> $$\sin(x + y) = \sin x \cos y + \cos x \sin y$$
> $$\cos(x + y) = \cos x \cos y - \sin x \sin y$$
> $$\tan(x + y) = \frac{\tan x + \tan y}{1 - \tan x \tan y}$$
>
> SUBTRACTION IDENTITIES
>
> $$\sin(x - y) = \sin x \cos y - \cos x \sin y$$
> $$\cos(x - y) = \cos x \cos y + \sin x \sin y$$
> $$\tan(x - y) = \frac{\tan x - \tan y}{1 + \tan x \tan y}$$
>
> COFUNCTION IDENTITIES
>
> (replace $\pi/2$ with $90°$ if x is in degrees)
>
> $$\sin\left(\frac{\pi}{2} - x\right) = \cos x \qquad \tan\left(\frac{\pi}{2} - x\right) = \cot x \qquad \sec\left(\frac{\pi}{2} - x\right) = \csc x$$

PROBLEM 8 Write $\cos 37°$ in the form $\sin \theta$, $0 \leq \theta \leq 90°$.

EXAMPLE 9 Find the value of $\tan 75°$ in exact radical form.

Solution   Since we can write $75° = 45° + 30°$, the sum of two special angles, we can use the addition identity for tangents with $x = 45°$ and $y = 30°$.

$$\tan(x + y) = \frac{\tan x + \tan y}{1 - \tan x \tan y}$$

$$\tan(45° + 30°) = \frac{\tan 45° + \tan 30°}{1 - \tan 45° \tan 30°}$$

$$= \frac{1 + \dfrac{1}{\sqrt{3}}}{1 - 1 \cdot \dfrac{1}{\sqrt{3}}} \qquad \text{Multiply numerator and denominator by } \sqrt{3}.$$

$$= \frac{\sqrt{3} + 1}{\sqrt{3} - 1} \qquad \text{Rationalize denominator.}$$

$$= 2 + \sqrt{3}$$

PROBLEM 9 Find the value of cos 15° in exact radical form.

EXAMPLE 10 Find the exact value of cos(x + y), given sin x = $\frac{3}{5}$, cos y = $\frac{4}{5}$, x is in quadrant II, and y is in quadrant I. Do not use a calculator or a table.

Solution We start with the addition identity for cosine:

$$\cos(x + y) = \cos x \cos y - \sin x \sin y$$

We know sin x and cos y but not sin y and cos x. We find the latter two values using reference triangles and the Pythagorean theorem:

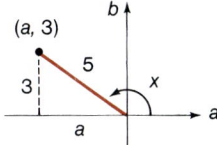 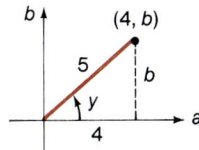

$a = -\sqrt{5^2 - 3^2} = -4$ $\qquad$ $b = \sqrt{5^2 - 4^2} = 3$

$\cos x = -\frac{4}{5}$ $\qquad\qquad\qquad$ $\sin y = \frac{3}{5}$

Thus,

$$\cos(x + y) = \cos x \cos y - \sin x \sin y$$

$$= \left(-\frac{4}{5}\right)\left(\frac{4}{5}\right) - \left(\frac{3}{5}\right)\left(\frac{3}{5}\right) = \frac{-25}{25} = -1$$

PROBLEM 10 Find the exact value of sin(x − y), given sin x = $-\frac{2}{3}$, cos y = $\sqrt{5}/3$, x is in quadrant III, and y is in quadrant IV.

EXAMPLE 11 Establish the identity: $\tan x + \cot y = \dfrac{\cos(x - y)}{\cos x \sin y}$

Proof

$$\dfrac{\cos(x - y)}{\cos x \sin y} = \dfrac{\cos x \cos y + \sin x \sin y}{\cos x \sin y} \qquad \text{Subtraction identity for cosine}$$

$$= \dfrac{\cos x \cos y}{\cos x \sin y} + \dfrac{\sin x \sin y}{\cos x \sin y} \qquad \text{Algebra}$$

$$= \cot y + \tan x \qquad \text{Quotient identities}$$

$$= \tan x + \cot y$$

PROBLEM 11 Establish the identity: $\cot y - \cot x = \dfrac{\sin(x - y)}{\sin x \sin y}$

■ **Calculator Exercise**

Let us use a calculator to evaluate, to four significant digits, both sides of the identity

$$\cos(x - y) = \cos x \cos y + \sin x \sin y$$

for $x = 2.317$ and $y = 1.583$. We start by setting the calculator in radian mode.

Evaluation of left side:

$$\cos(2.317 - 1.583) = 0.7425$$

A: [2.317] [−] [1.583] [=] [COS]

P: [2.317] [ENTER] [1.583] [−] [COS]

Evaluation of right side:

$$\cos 2.317 \cos 1.583 + \sin 2.317 \sin 1.583 = 0.7425$$

A: [2.317] [COS] [×] [1.583] [COS] [−] [+] [(] [2.317] [SIN] [×] [1.583] [SIN] [)] [=]

P: [2.317] [COS] [ENTER] [1.583] [COS] [×] [2.317] [SIN] [ENTER] [1.583] [SIN] [×] [+]

Both sides are the same to four significant digits. In similar evaluations, we would not always expect both sides to be exactly the same because of rounding errors.

Answers to Matched Problems

7. $\sin x$

8. $\sin 53°$

9. $(1 + \sqrt{3})/2\sqrt{2}$

10. $(-4\sqrt{5})/9$

11. $\dfrac{\sin(x - y)}{\sin x \sin y} = \dfrac{\sin x \cos y - \cos x \sin y}{\sin x \sin y}$

$= \dfrac{\cancel{\sin x} \cos y}{\cancel{\sin x} \sin y} - \dfrac{\cos x \cancel{\sin y}}{\sin x \cancel{\sin y}}$

$= \cot y - \cot x$

Exercise 6-3 ■ **A** We can use addition identities to establish periodic properties for the trigonometric functions. Establish the following identities using the addition identities.

1. $\sin(x + 2\pi) = \sin x$
2. $\cos(x + 2\pi) = \cos x$
3. $\tan(x + \pi) = \tan x$
4. $\cot(x + \pi) = \cot x$
5. $\cos(x + 2k\pi) = \cos x$, k an integer

6. $\sin(x + 2k\pi) = \sin x$, k an integer
7. $\cot(x + k\pi) = \cot x$, k an integer
8. $\tan(x + k\pi) = \tan x$, k an integer

Establish each identity using cofunction identities for sine and cosine and basic identities discussed in the last section.

9. $\tan\left(\dfrac{\pi}{2} - x\right) = \cot x$
10. $\cot\left(\dfrac{\pi}{2} - x\right) = \tan x$
11. $\sec\left(\dfrac{\pi}{2} - x\right) = \csc x$
12. $\csc\left(\dfrac{\pi}{2} - x\right) = \sec x$

Convert to forms involving sin x, cos x, and/or tan x using addition or subtraction identities.

13. $\sin(x - 45°)$
14. $\sin(30° - x)$
15. $\cos(x + 180°)$
16. $\sin(180° - x) = \sin x$
17. $\tan\left(\dfrac{\pi}{4} - x\right)$
18. $\tan\left(x + \dfrac{\pi}{3}\right)$

B *Find, using appropriate identities, exact values for each. Do not use a calculator or a table.*

19. $\sin 75°$
20. $\sec 75°$
21. $\cos \dfrac{\pi}{12}$
    [Hint: $(\pi/12) = (\pi/4) - (\pi/6)$]
22. $\sin \dfrac{7\pi}{12}$
    [Hint: $(7\pi/12) = (\pi/3) + (\pi/4)$]
23. $\sin 22° \cos 38° + \cos 22° \sin 38°$
24. $\cos 74° \cos 44° + \sin 74° \sin 44°$
25. $\dfrac{\tan 110° - \tan 50°}{1 + \tan 110° \tan 50°}$
26. $\dfrac{\tan 27° + \tan 18°}{1 - \tan 27° \tan 18°}$

Find $\sin(x - y)$ and $\tan(x + y)$ exactly without a calculator or a table using the information given and appropriate identities.

27. $\sin x = \tfrac{2}{3}$, $\cos y = -\tfrac{1}{4}$, x is in quadrant II, and y is in quadrant III
28. $\sin x = -\tfrac{3}{5}$, $\sin y = \sqrt{8}/3$, x is in quadrant IV, and y is in quadrant I
29. $\cos x = -\tfrac{1}{3}$, $\tan y = -\tfrac{1}{2}$, x is in quadrant II, and y is in quadrant III
30. $\tan x = \tfrac{3}{4}$, $\tan y = -\tfrac{1}{2}$, x is in quadrant III, and y is in quadrant IV

Establish each identity.

31. $\sin 2x = 2 \sin x \cos x$
32. $\cos 2x = \cos^2 x - \sin^2 x$

33. $\cot(x - y) = \dfrac{\cot x \cot y + 1}{\cot y - \cot x}$

34. $\cot(x + y) = \dfrac{\cot x \cot y - 1}{\cot x + \cot y}$

35. $\cot 2x = \dfrac{\cot^2 x - 1}{2 \cot x}$

36. $\tan 2x = \dfrac{2 \tan x}{1 - \tan^2 x}$

37. $\tan x - \tan y = \dfrac{\sin(x - y)}{\cos x \cos y}$

38. $\cot x - \tan y = \dfrac{\cos(x + y)}{\sin x \cos y}$

39. $\tan(x + y) = \dfrac{\cot x + \cot y}{\cot x \cot y - 1}$

40. $\tan(x - y) = \dfrac{\cot y - \cot x}{\cot x \cot y + 1}$

41. $\dfrac{\sin(x + h) - \sin x}{h} = \sin x \left(\dfrac{\cos h - 1}{h} \right) + \cos x \left(\dfrac{\sin h}{h} \right)$

42. $\dfrac{\cos(x + h) - \cos x}{h} = \cos x \left(\dfrac{\cos h - 1}{h} \right) - \sin x \left(\dfrac{\sin h}{h} \right)$

C In Problems 43–46 evaluate exactly as real numbers without the use of tables or a calculator.

43. $\sin [\cos^{-1}(-\tfrac{4}{5}) + \sin^{-1}(-\tfrac{3}{5})]$ 44. $\cos [\sin^{-1}(-\tfrac{3}{5}) + \cos^{-1} \tfrac{4}{5}]$

45. $\sin [\arccos \tfrac{1}{2} + \arcsin(-1)]$ 46. $\cos [\arccos(-\tfrac{3}{2}) - \arcsin(-\tfrac{1}{2})]$

47. Express $\sin(\sin^{-1} x + \cos^{-1} y)$ in an equivalent form free of trigonometric and inverse trigonometric functions.

48. Express $\cos(\sin^{-1} x - \cos^{-1} y)$ in an equivalent form free of trigonometric and inverse trigonometric functions.

Establish the identities in Problems 49 and 50. [Hint: $\sin(x + y + z) = \sin[(x + y) + z]$.]

49. $\sin(x + y + z) = \sin x \cos y \cos z + \cos x \sin y \cos z$
$\qquad\qquad\qquad + \cos x \cos y \sin z - \sin x \sin y \sin z$

50. $\cos(x + y + z) = \cos x \cos y \cos z - \sin x \sin y \cos z$
$\qquad\qquad\qquad - \sin x \cos y \sin z - \cos x \sin y \sin z$

51. Use the information in the figure to show that
$$\tan(\theta_2 - \theta_1) = \frac{m_2 - m_1}{1 + m_1 m_2}$$

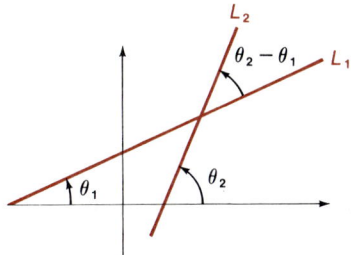

$\tan \theta_1 = $ Slope of $L_1 = m_1$
$\tan \theta_2 = $ Slope of $L_2 = m_2$

52. Find the acute angle of intersection between the two lines $y = 3x + 1$ and $y = \frac{1}{2}x - 1$. (Use the results of Problem 51.)

Evaluate both sides of the subtraction identity for sine and the addition identity for tangent for the indicated values of x and y. Evaluate to four significant digits using a scientific calculator.

53. $x = 3.042$, $y = 2.384$ **54.** $x = 5.288$, $y = 1.769$
55. $x = 128.3°$, $y = 25.62°$ **56.** $x = 42.08°$, $y = 68.37°$

Section 6-4 Double-Angle and Half-Angle Identities

- Double-Angle Identities
- Half-Angle Identities

In this section we will develop another important set of identities called *double-* and *half-angle identities*. We can derive these identities directly from the addition and subtraction identities given in Section 6-3. Even though the names use the word *angle*, the new identities hold for real numbers as well.

- **Double-Angle Identities**

We start with the addition identity for sine:

$\sin(x + y) = \sin x \cos y + \cos x \sin y$

and replace y with x to obtain

$\sin(x + x) = \sin x \cos x + \cos x \sin x$

or

$$\sin 2x = 2 \sin x \cos x \qquad (1)$$

If we start with the addition identity for cosine:

$$\cos(x + y) = \cos x \cos y - \sin x \sin y$$

and replace y with x, we obtain

$$\cos(x + x) = \cos x \cos x - \sin x \sin x$$

or

$$\cos 2x = \cos^2 x - \sin^2 x \qquad (2)$$

Now, using the Pythagorean identity

$$\sin^2 x + \cos^2 x = 1 \qquad (3)$$

in the form

$$\cos^2 x = 1 - \sin^2 x \qquad (4)$$

and substituting it into equation (2), we get

$$\cos 2x = 1 - \sin^2 x - \sin^2 x$$
$$\cos 2x = 1 - 2 \sin^2 x \qquad (5)$$

Or, if we use equation (3) in the form

$$\sin^2 x = 1 - \cos^2 x$$

and substitute it into equation (2), we get

$$\cos 2x = \cos^2 x - (1 - \cos^2 x)$$
$$\cos 2x = \cos^2 x - 1 \qquad (6)$$

Double-angle identities can be established for the tangent function in the same way by starting with the addition formula for tangent. This is left as exercises for the reader (Problems 19–21 in Exercise 6-4).

We list the double-angle identities for convenient reference.

Double-Angle Identities

$$\sin 2x = 2 \sin x \cos x$$
$$\cos 2x = \cos^2 x - \sin^2 x = 1 - 2 \sin^2 x = 2 \cos^2 x - 1$$
$$\tan 2x = \frac{2 \tan x}{1 - \tan^2 x} = \frac{2 \cot x}{\cot^2 x - 1} = \frac{2}{\cot x - \tan x}$$

6-4 Double-Angle and Half-Angle Identities

EXAMPLE 12 Establish the identity: $\cos 2x = \dfrac{1 - \tan^2 x}{1 + \tan^2 x}$

Proof We start with the right side:

$$\dfrac{1 - \tan^2 x}{1 + \tan^2 x} = \dfrac{1 - \dfrac{\sin^2 x}{\cos^2 x}}{1 + \dfrac{\sin^2 x}{\cos^2 x}} \qquad \text{Quotient identities}$$

$$= \dfrac{\cos^2 x - \sin^2 x}{\cos^2 x + \sin^2 x} \qquad \text{Algebra}$$

$$= \cos^2 x - \sin^2 x \qquad \text{Pythagorean identity}$$
$$= \cos 2x \qquad \text{Double-angle identity}$$

PROBLEM 12 Establish the identity: $\sin 2x = \dfrac{2 \tan x}{1 + \tan^2 x}$

EXAMPLE 13 Find the exact value, without using a calculator or a table, of sin 2x and cos 2x if $\tan x = -\tfrac{3}{4}$ and x is in quadrant IV.

Solution First draw the reference triangle for x and find any unknown sides:

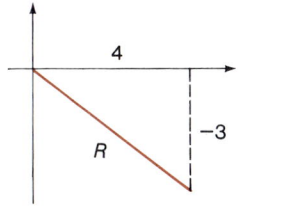

$R = \sqrt{(-3)^2 + 4^2} = 5$

$\sin x = -\dfrac{3}{5}$

$\cos x = \dfrac{4}{5}$

Now use double-angle identities for sine and cosine:

$\sin 2x = 2 \sin x \cos x = 2(-\tfrac{3}{5})(\tfrac{4}{5}) = -\tfrac{24}{25}$

$\cos 2x = 2 \cos^2 x - 1 = 2(\tfrac{4}{5})^2 - 1 = \tfrac{7}{25}$

PROBLEM 13 Find the exact value, without using a calculator or a table, of cos 2x and tan 2x if $\sin x = \tfrac{4}{5}$ and x is in quadrant II.

■ **Half-Angle Identities**

Half-angle identities are simply double-angle identities stated in an alternate form. Let us start with the double-angle identity for cosine in the form

$\cos 2m = 1 - 2 \sin^2 m$

Now replace m with $x/2$ and solve for $\sin(x/2)$ [if $2m$ is twice m, then m is half of $2m$—think about this]:

$$\cos x = 1 - 2 \sin^2 \frac{x}{2}$$

$$\sin^2 \frac{x}{2} = \frac{1 - \cos x}{2}$$

$$\sin \frac{x}{2} = \pm \sqrt{\frac{1 - \cos x}{2}} \tag{7}$$

where the choice of the sign is determined by the quadrant in which $x/2$ lies.

To obtain a half-angle identity for cosine, we start with the double-angle identity for cosine in the form

$$\cos 2m = 2 \cos^2 m - 1$$

and let $m = x/2$ to obtain

$$\cos \frac{x}{2} = \pm \sqrt{\frac{1 + \cos x}{2}} \tag{8}$$

where the sign is determined by the quadrant in which $x/2$ lies.

To obtain a half-angle identity for tangent, we can use the quotient identity and the half-angle formulas for sine and cosine:

$$\tan \frac{x}{2} = \frac{\sin \dfrac{x}{2}}{\cos \dfrac{x}{2}} = \frac{\pm \sqrt{\dfrac{1 - \cos x}{2}}}{\pm \sqrt{\dfrac{1 + \cos x}{2}}} = \pm \sqrt{\frac{1 - \cos x}{1 + \cos x}} \tag{9}$$

where the sign is determined by the quadrant in which $x/2$ lies.

Simpler versions of equation (9) can be obtained as follows:

$$\left| \tan \frac{x}{2} \right| = \sqrt{\frac{1 - \cos x}{1 + \cos x}} \tag{10}$$

$$= \sqrt{\frac{1 - \cos x}{1 + \cos x} \cdot \frac{1 + \cos x}{1 + \cos x}}$$

$$= \sqrt{\frac{1 - \cos^2 x}{(1 + \cos x)^2}}$$

$$= \sqrt{\frac{\sin^2 x}{(1 + \cos x)^2}}$$

$$= \frac{|\sin x|}{1 + \cos x}$$

6-4 Double-Angle and Half-Angle Identities

since $1 + \cos x$ is never negative. In fact, we can drop all absolute values, since it can be shown that $\tan x/2$ and $\sin x$ always have the same sign (a good exercise for the reader). Thus,

$$\tan \frac{x}{2} = \frac{\sin x}{1 + \cos x} \tag{11}$$

By multiplying the numerator and the denominator in the radicand in equation (10) by $1 - \cos x$ and reasoning as before, we can obtain

$$\tan \frac{x}{2} = \frac{1 - \cos x}{\sin x} \tag{12}$$

We now list all of the half-angle identities for convenient reference.

Half-Angle Identities

$$\sin \frac{x}{2} = \pm \sqrt{\frac{1 - \cos x}{2}}$$

$$\cos \frac{x}{2} = \pm \sqrt{\frac{1 + \cos x}{2}}$$

$$\tan \frac{x}{2} = \pm \sqrt{\frac{1 - \cos x}{1 + \cos x}} = \frac{\sin x}{1 + \cos x} = \frac{1 - \cos x}{\sin x}$$

where the sign is determined by the quadrant in which $x/2$ lies.

EXAMPLE 14 Compute the exact value of $\sin 165°$ without a calculator or a table using a half-angle identity.

Solution

$\sin 165° = \sin \dfrac{330°}{2}$ Use half-angle identity for sine.

$ = \sqrt{\dfrac{1 - \cos 330°}{2}}$ We use a positive radical, since $\sin 165°$ is positive.

$ = \sqrt{\dfrac{1 - \left(\dfrac{\sqrt{3}}{2}\right)}{2}}$

$ = \dfrac{\sqrt{2 - \sqrt{3}}}{2}$

PROBLEM 14 Compute the exact value of $\tan 105°$ without a calculator or a table using a half-angle identity.

EXAMPLE 15 Find the exact value of $\cos(x/2)$ and $\cot(x/2)$ without using a calculator or a table if $\sin x = -\frac{3}{5}$, $\pi < x < 3\pi/2$.

Solution Draw a reference triangle in the third quadrant and find cos x; then use appropriate half-angle identities.

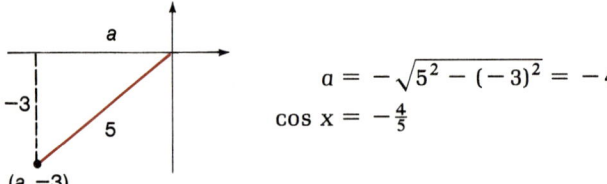

$a = -\sqrt{5^2 - (-3)^2} = -4$

$\cos x = -\frac{4}{5}$

If $\pi < x < 3\pi/2$, then

$\frac{\pi}{2} < \frac{x}{2} < \frac{3\pi}{4}$ Divide each member of $\pi < x < 3\pi/2$ by 2.

Thus, $x/2$ is in the second quadrant where cosine and cotangent are negative, and

$$\cos \frac{x}{2} = -\sqrt{\frac{1 + \cos x}{2}}$$

$$= -\sqrt{\frac{1 + (-\frac{4}{5})}{2}}$$

$$= -\sqrt{\frac{1}{10}} \quad \text{or} \quad \frac{-\sqrt{10}}{10}$$

$$\cot \frac{x}{2} = \frac{1}{\tan(x/2)} = \frac{\sin x}{1 - \cos x}$$

$$= \frac{-\frac{3}{5}}{1 - (-\frac{4}{5})} = -\frac{1}{3}$$

PROBLEM 15 Find the exact value of $\sin(x/2)$ and $\tan(x/2)$ if $\cot x = -\frac{4}{3}$, $\pi/2 < x < \pi$.

EXAMPLE 16 Establish the identity: $\sin^2 \frac{x}{2} = \frac{\tan x - \sin x}{2 \tan x}$

Proof $\sin \frac{x}{2} = \pm \sqrt{\frac{1 - \cos x}{2}}$ Half-angle identity for sine

$\sin^2 \frac{x}{2} = \frac{1 - \cos x}{2}$ Square both sides.

$= \frac{\tan x}{\tan x} \cdot \frac{1 - \cos x}{2}$ Algebra

$= \frac{\tan x - \tan x \cos x}{2 \tan x}$ Algebra

$= \frac{\tan x - \sin x}{2 \tan x}$ Quotient identity

PROBLEM 16 Establish the identity: $\cos^2 \dfrac{x}{2} = \dfrac{\tan x + \sin x}{2 \tan x}$

Answers to Matched Problems

12. $\dfrac{2 \tan x}{1 + \tan^2 x} = \dfrac{2\left(\dfrac{\sin x}{\cos x}\right)}{1 + \dfrac{\sin^2 x}{\cos^2 x}} = \dfrac{\cos^2 x \left[2\left(\dfrac{\sin x}{\cos x}\right)\right]}{\cos^2 x \left(1 + \dfrac{\sin^2 x}{\cos^2 x}\right)}$

$= \dfrac{2 \sin x \cos x}{\cos^2 x + \sin^2 x} = 2 \sin x \cos x = \sin 2x$

13. $\cos 2x = -\tfrac{7}{25}$, $\tan 2x = \tfrac{24}{7}$
14. $-\sqrt{3} - 2$
15. $\sin (x/2) = 3\sqrt{10}/10$, $\tan (x/2) = 3$
16. $\cos^2 \dfrac{x}{2} = \dfrac{1 + \cos x}{2} = \dfrac{\tan x}{\tan x} \cdot \dfrac{1 + \cos x}{2}$

$= \dfrac{\tan x + \tan x \cos x}{2 \tan x} = \dfrac{\tan x + \sin x}{2 \tan x}$

Exercise 6-4

A *Verify each identity for the indicated values.*

1. $\sin 2x = 2 \sin x \cos x$, $\quad x = 45°$
2. $\cos 2x = \cos^2 x - \sin^2 x$, $\quad x = 30°$
3. $\tan 2x = \dfrac{2 \tan x}{1 - \tan^2 x}$, $\quad x = \dfrac{\pi}{6}$
4. $\tan 2x = \dfrac{2}{\cot x - \tan x}$, $\quad x = \dfrac{\pi}{3}$
5. $\sin \dfrac{x}{2} = \pm \sqrt{\dfrac{1 - \cos x}{2}}$, $\quad x = \pi$ (Choose the right sign.)
6. $\cos \dfrac{x}{2} = \pm \sqrt{\dfrac{1 + \cos x}{2}}$, $\quad x = \dfrac{\pi}{2}$ (Choose the right sign.)

Find the exact value without a calculator or a table using double- and half-angle identities.

7. $\sin 22.5°$ 8. $\tan 75°$ 9. $\cos 67.5°$ 10. $\tan 15°$

B *Establish the following identities.*

11. $(\sin x + \cos x)^2 = 1 + \sin 2x$
12. $\sin 2x = (\tan x)(1 + \cos 2x)$
13. $\sin^2 x = \tfrac{1}{2}(1 - \cos 2x)$ 14. $\cos^2 x = \tfrac{1}{2}(\cos 2x + 1)$

15. $1 - \cos 2x = \tan x \sin 2x$

16. $\dfrac{2 \tan x}{\sin 2x} = \sec^2 x$

17. $2 \cos^2 \dfrac{x}{2} = \dfrac{\sin^2 x}{1 - \cos x}$

18. $2 \sin^2 \dfrac{x}{2} = \dfrac{\sin^2 x}{1 + \cos x}$

19. $\tan 2x = \dfrac{2 \tan x}{1 - \tan^2 x}$

20. $\tan 2x = \dfrac{2 \cot x}{\cot^2 x - 1}$

21. $\tan 2x = \dfrac{2}{\cot x - \tan x}$

22. $\cot 2x = \dfrac{\cot^2 x - 1}{2 \cot x}$

23. $\cot 2x = \dfrac{1 - \tan^2 x}{2 \tan x}$

24. $\cot 2x = \dfrac{\cot x - \tan x}{2}$

25. $2 \csc 2x = \dfrac{1 + \tan^2 x}{\tan x}$

26. $\sec^2 x = (\sec 2x)(2 - \sec^2 x)$

27. $\tan \dfrac{x}{2} = \dfrac{1 - \cos x}{\sin x}$

28. $\cot 2x = \dfrac{\cot x - \tan x}{\cot x + \tan x}$

Compute the exact values of sin 2x, cos 2x, and tan 2x using the information given and appropriate identities. Do not use a calculator or a table.

29. $\cos x = \tfrac{4}{5}, \quad 0° < x < 90°$
30. $\sin x = \tfrac{3}{5}, \quad 0° < x < 90°$
31. $\sin x = \tfrac{3}{5}, \quad \pi/2 < x < \pi$
32. $\cos x = -\tfrac{4}{5}, \quad \pi/2 < x < \pi$
33. $\tan x = -\tfrac{5}{12}, \quad -\pi/2 < x < 0$
34. $\cot x = -\tfrac{5}{12}, \quad -\pi/2 < x < 0$

In Problems 35–40, compute the exact values of sin(x/2), cos(x/2), and tan(x/2) using the information given and appropriate identities. Do not use a calculator or a table.

35. $\cos x = \tfrac{1}{3}, \quad 0° < x < 90°$
36. $\sin x = \tfrac{4}{5}, \quad 0° < x < 90°$
37. $\sin x = -\tfrac{1}{3}, \quad \pi < x < 3\pi/2$
38. $\cos x = -\tfrac{1}{4}, \quad \pi < x < 3\pi/2$

6-5 Product and Factor Identities

39. $\cot x = \frac{3}{4}$, $-\pi < x < -\pi/2$
40. $\tan x = \frac{3}{4}$, $-\pi < x < -\pi/2$
41. Find the exact values of $\sin \theta$ and $\cos \theta$, given $\tan 2\theta = -\frac{4}{3}$, $0 < \theta < 90°$. (See Problems 13 and 14.)
42. Find the exact values of $\sin \theta$ and $\cos \theta$, given $\tan 2\theta = -\frac{3}{4}$, $0 < \theta < 90°$.

C Establish the following identities.

43. $\cos 3x = 4 \cos^3 x - 3 \cos x$
44. $\sin 3x = 3 \sin x - 4 \sin^3 x$
45. $\cos 4x = 8 \cos^4 x - 8 \cos^2 x + 1$
46. $\sin 4x = (\cos x)(4 \sin x - 8 \sin^3 x)$

Find the exact value of each without using a calculator or a table.

47. $\cos(2 \cos^{-1} \frac{3}{5})$
48. $\sin(2 \cos^{-1} \frac{3}{5})$
49. $\tan[2 \cos^{-1}(-\frac{4}{5})]$
50. $\tan[2 \tan^{-1}(-\frac{3}{4})]$
51. $\cos[\frac{1}{2} \cos^{-1}(-\frac{3}{5})]$
52. $\sin[\frac{1}{2} \tan^{-1}(-\frac{4}{3})]$

CALCULATOR PROBLEMS

Verify each of the following identities for the indicated value of x. Compute values to five significant digits using a scientific calculator.

(A) $\tan 2x = \dfrac{2 \tan x}{1 - \tan^2 x}$
(B) $\cos \dfrac{x}{2} = \pm \sqrt{\dfrac{1 + \cos x}{2}}$ (Choose the right sign.)

53. $x = 252.06°$
54. $x = 72.358°$
55. $x = 0.934\ 57$
56. $x = 4$

Section 6-5 Product and Factor Identities

- Product Identities
- Factor Identities

We conclude our work on identities by developing the product and factor identities. These identities are easily derived from the addition and subtraction identities developed in Section 6-3.

- Product Identities

Let us add, left side to left side and right side to right side, the addition and subtraction identities for sine:

$$\sin(x + y) = \sin x \cos y + \cos x \sin y$$
$$\sin(x - y) = \sin x \cos y - \cos x \sin y$$
$$\overline{\sin(x + y) + \sin(x - y) = 2 \sin x \cos y}$$

or

$$\sin x \cos y = \tfrac{1}{2}[\sin(x + y) + \sin(x - y)]$$

Similarly, by adding or subtracting the appropriate addition and subtraction identities, we can obtain three other product identities. These identities are listed here for convenient reference.

Product Identities

$$\sin x \cos y = \tfrac{1}{2}[\sin(x + y) + \sin(x - y)]$$
$$\cos x \sin y = \tfrac{1}{2}[\sin(x + y) - \sin(x - y)]$$
$$\sin x \sin y = \tfrac{1}{2}[\cos(x - y) - \cos(x + y)]$$
$$\cos x \cos y = \tfrac{1}{2}[\cos(x + y) + \cos(x - y)]$$

EXAMPLE 17 Write the product $\cos 3t \sin t$ as a sum or difference.

Solution $\cos x \sin y = \tfrac{1}{2}[\sin(x + y) - \sin(x - y)]$ Let $x = 3t$ and $y = t$.
$\cos 3t \sin t = \tfrac{1}{2}[\sin(3t + t) - \sin(3t - t)]$
$= \tfrac{1}{2} \sin 4t - \tfrac{1}{2} \sin 2t$

PROBLEM 17 Write the product $\cos 5\theta \cos 2\theta$ as a sum or difference.

EXAMPLE 18 Evaluate $\sin 105° \sin 15°$ exactly using an appropriate product identity.

Solution $\sin x \sin y = \tfrac{1}{2}[\cos(x - y) - \cos(x + y)]$
$\sin 105° \sin 15° = \tfrac{1}{2}[\cos(105° - 15°) - \cos(105° + 15°)]$
$= \tfrac{1}{2}[\cos 90° - \cos 120°]$
$= \tfrac{1}{2}[0 - (-\tfrac{1}{2})] = \tfrac{1}{4}$ or 0.25

PROBLEM 18 Evaluate $\cos 165° \sin 75°$ exactly using an appropriate product identity.

■ **Factor Identities**

The product identities can be transformed into equivalent forms called *factor identities*. These identities are used to express sums and differences involving sines and cosines as products involving sines and cosines. We illustrate the transformation for one identity. The other three identities can be obtained by following similar procedures.

6-5 Product and Factor Identities

Let us start with the product identity:

$$\sin \alpha \cos \beta = \tfrac{1}{2}[\sin(\alpha + \beta) + \sin(\alpha - \beta)] \qquad (1)$$

We would like

$$\alpha + \beta = x \qquad \alpha - \beta = y$$

Solving this system, we have

$$\alpha = \frac{x + y}{2} \qquad \beta = \frac{x - y}{2} \qquad (2)$$

Substituting (2) into equation (1) and simplifying, we obtain

$$\sin x + \sin y = 2 \sin \frac{x + y}{2} \cos \frac{x - y}{2}$$

All four factor identities are listed here for convenient reference.

Factor Identities

$$\sin x + \sin y = 2 \sin \frac{x + y}{2} \cos \frac{x - y}{2}$$

$$\sin x - \sin y = 2 \cos \frac{x + y}{2} \sin \frac{x - y}{2}$$

$$\cos x + \cos y = 2 \cos \frac{x + y}{2} \cos \frac{x - y}{2}$$

$$\cos x - \cos y = -2 \sin \frac{x + y}{2} \sin \frac{x - y}{2}$$

EXAMPLE 19 Write the difference $\sin 7\theta - \sin 3\theta$ as a product.

Solution
$$\sin x - \sin y = 2 \cos \frac{x + y}{2} \sin \frac{x - y}{2}$$

$$\sin 7\theta - \sin 3\theta = 2 \cos \frac{7\theta + 3\theta}{2} \sin \frac{7\theta - 3\theta}{2}$$

$$= 2 \cos 5\theta \sin 2\theta$$

PROBLEM 19 Write the sum $\cos 3t + \cos t$ as a product.

EXAMPLE 20 Find the exact value of $\sin 105° - \sin 15°$ using an appropriate factor identity.

6 Trigonometric Identities and Conditional Equations

Solution

$$\sin x - \sin y = 2 \cos \frac{x+y}{2} \sin \frac{x-y}{2}$$

$$\sin 105° - \sin 15° = 2 \cos \frac{105° + 15°}{2} \sin \frac{105° - 15°}{2}$$

$$= 2 \cos 60° \sin 45°$$

$$= 2 \left(\frac{1}{2}\right)\left(\frac{\sqrt{2}}{2}\right) = \frac{\sqrt{2}}{2}$$

PROBLEM 20 Find the exact value of $\cos 165° - \cos 75°$ using an appropriate factor identity.

Answers to Matched Problems

17. $\cos 5\theta \cos 2\theta = \frac{1}{2} \cos 7\theta + \frac{1}{2} \cos 3\theta$
18. $(-\sqrt{3} - 2)/4$
19. $\cos 3t + \cos t = 2 \cos 2t \cos t$
20. $-\sqrt{6}/2$

Exercise 6-5

A Write each product as a sum or difference involving sine and cosine.

1. $\cos 7A \cos 5A$
2. $\sin 3m \cos m$
3. $\cos 2\theta \sin 3\theta$
4. $\sin u \sin 3u$

Write each difference or sum as a product involving sines and cosines.

5. $\cos 7\theta + \cos 5\theta$
6. $\sin 3t + \sin t$
7. $\sin u - \sin 5u$
8. $\cos 5w - \cos 9w$

B Evaluate each exactly using an appropriate identity.

9. $\cos 75° \sin 15°$
10. $\sin 195° \cos 75°$
11. $\sin 105° \sin 165°$
12. $\cos 15° \cos 75°$

Evaluate each exactly using an appropriate identity.

13. $\sin 195° + \sin 105°$
14. $\cos 285° + \cos 195°$
15. $\sin 75° - \sin 165°$
16. $\cos 15° - \cos 105°$

Use addition and subtraction identities to establish.

17. $\sin x \sin y = \frac{1}{2}[\cos(x - y) - \cos(x + y)]$
18. $\cos x \cos y = \frac{1}{2}[\cos(x + y) + \cos(x - y)]$

Use appropriate substitutions in the product identities to obtain the following.

19. $\sin x - \sin y = 2 \cos \frac{x+y}{2} \sin \frac{x-y}{2}$

6-6 Trigonometric Equations

20. $\cos x - \cos y = -2 \sin \dfrac{x+y}{2} \sin \dfrac{x-y}{2}$

Establish each identity.

21. $\dfrac{\cos t - \cos 3t}{\sin t + \sin 3t} = \tan t$

22. $\dfrac{\sin 2t + \sin 4t}{\cos 2t - \cos 4t} = \cot t$

23. $\dfrac{\sin x + \sin y}{\cos x + \cos y} = \tan \dfrac{x+y}{2}$

24. $\dfrac{\sin x - \sin y}{\cos x - \cos y} = -\cot \dfrac{x+y}{2}$

25. $\dfrac{\cos x - \cos y}{\sin x + \sin y} = -\tan \dfrac{x-y}{2}$

26. $\dfrac{\cos x + \cos y}{\sin x - \sin y} = \cot \dfrac{x-y}{2}$

27. $\dfrac{\sin x + \sin y}{\sin x - \sin y} = \dfrac{\tan \tfrac{1}{2}(x+y)}{\tan \tfrac{1}{2}(x-y)}$

28. $\dfrac{\cos x + \cos y}{\cos x - \cos y} = -\cot \dfrac{x+y}{2} \cot \dfrac{x-y}{2}$

C 29. $\sin x \sin y \sin z = \tfrac{1}{4}[\sin(x+y-z) + \sin(y+z-x)$
$+ \sin(z+x-y) - \sin(x+y+z)]$

30. $\cos x \cos y \cos z = \tfrac{1}{4}[\cos(x+y-z) + \cos(y+z-x)$
$+ \cos(z+x-y) + \cos(x+y+z)]$

CALCULATOR PROBLEMS

Verify each of the following identities for the indicated values of x and y. Evaluate each side to five significant digits.

(A) $\cos x \sin y = \tfrac{1}{2}[\sin(x+y) - \sin(x-y)]$

(B) $\cos x + \cos y = 2 \cos \dfrac{x+y}{2} \cos \dfrac{x-y}{2}$

31. $x = 50.137°, \quad y = 18.044°$
32. $x = 172.63°, \quad y = 20.177°$
33. $x = 0.039\ 17, \quad y = 0.610\ 52$
34. $x = 1.1255, \quad y = 3.6014$

Section 6-6 Trigonometric Equations

- Given *a*, *b*, *c*, Find All *x* So That sin *x* = *a*, cos *x* = *b*, tan *x* = *c*
- More General Trigonometric Equations

You have now had experience in solving algebraic equations, exponential equations, and logarithmic equations. In this section we will consider the solution of equations involving trigonometric functions. Earlier in this chapter we studied a certain type of trigonometric equation called an *identity*. Recall that a trigonometric identity is a trigonometric equation that is true for all replacements of the variable for which each member of the equation is defined. A trigonometric equation that is true for only certain values of the variable for which both members are defined (and is not true for others) is called a **conditional equation**. Conditional trigonometric equations are the subject of this section.

We will start our investigation by solving simple equations of the form $\sin x = a$, $\cos x = b$, and $\tan x = c$; then we will turn to more general equations. Because inverse trigonometric functions will play an important part in our approach to solving trigonometric equations, a brief review of Section 5-8 should prove helpful.

■ **Given a, b, c, Find All x So That sin x = a, cos x = b, tan x = c**

Given a, b, c, how do we write *all* the possible solutions to the equations

$$\sin x = a \qquad \cos x = b \qquad \tan x = c \qquad (1)$$

assuming solutions exist. Using the definitions of $\sin^{-1}$, $\cos^{-1}$, and $\tan^{-1}$ and the periodic properties of sine, cosine, and tangent, we can easily express all existing solutions to equations (1) in terms of appropriate inverse functions.

Study the material in Boxes 1–3 very carefully before proceeding. Inverses are associated with only the solid part of each graph. Using the periodic properties of the various trigonometric functions, all solutions of equations (1) can be written as indicated.

BOX 1

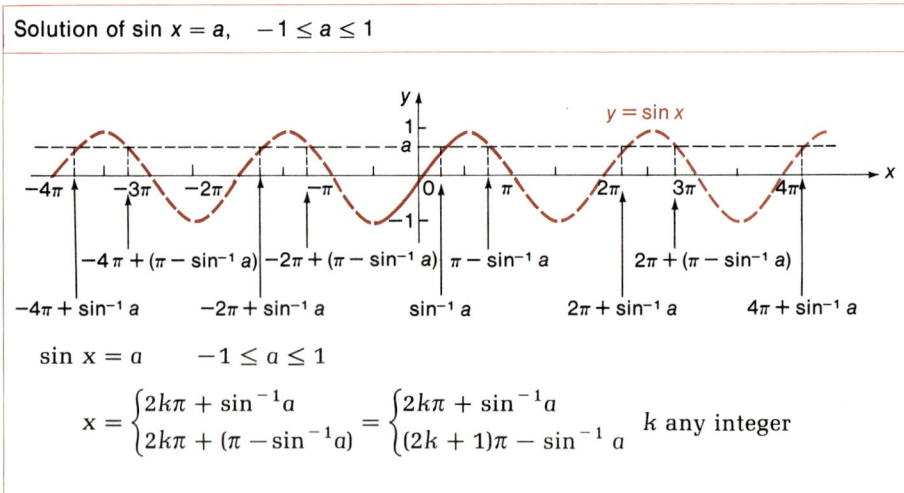

6-6 Trigonometric Equations

BOX 2

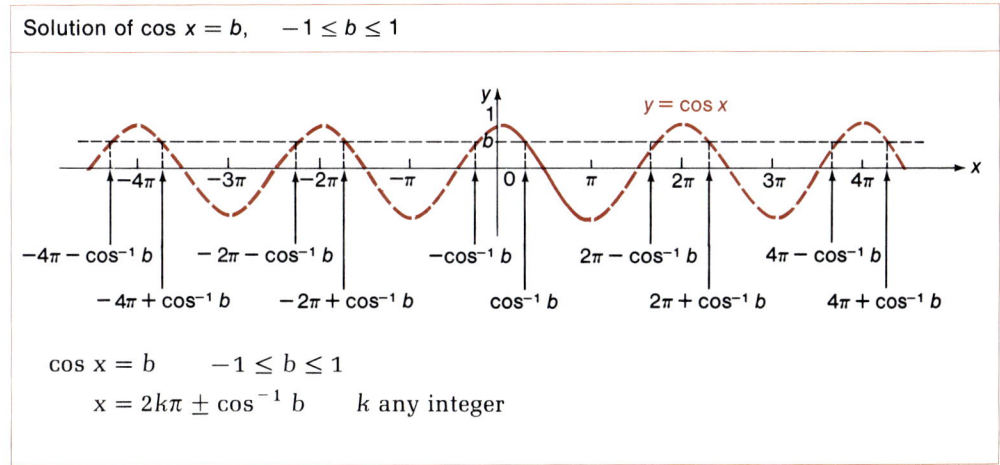

$$\cos x = b \quad -1 \le b \le 1$$
$$x = 2k\pi \pm \cos^{-1} b \quad k \text{ any integer}$$

BOX 3

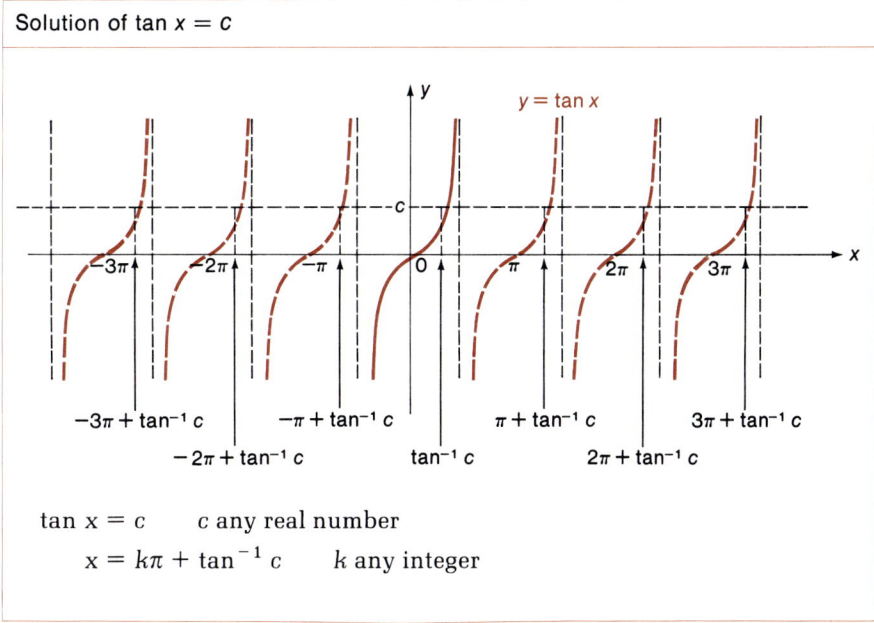

$$\tan x = c \quad c \text{ any real number}$$
$$x = k\pi + \tan^{-1} c \quad k \text{ any integer}$$

EXAMPLE 21 Solve, evaluating inverse functions to four significant digits.

(A) $\sin x = 0.8032$ (B) $\cos x = -0.3591$ (C) $\tan x = 12.32$

Solution (A) $\sin x = 0.8032$ Refer to Box 1.

$$x = \begin{cases} 2k\pi + \sin^{-1} 0.8032 \\ (2k+1)\pi - \sin^{-1} 0.8032 \end{cases}$$

Evaluate $\sin^{-1} 0.8032$ using a scientific calculator set in radian mode.

$$= \begin{cases} 2k\pi + 0.9326 \\ (2k+1)\pi - 0.9326 \end{cases} \quad k \text{ any integer}$$

(B) $\cos x = -0.3591$ Refer to Box 2.

$$x = 2k\pi \pm \cos^{-1}(-0.3591)$$
$$= 2k\pi \pm 1.938 \quad k \text{ any integer}$$

(C) $\tan x = 12.32$ Refer to Box 3.

$$x = k\pi + \tan^{-1} 12.32$$
$$= k\pi + 1.490 \quad k \text{ any integer}$$

PROBLEM 21   Solve, evaluating inverse functions to four significant digits.

(A) $\sin x = -0.2219$ (B) $\cos x = 0.4635$ (C) $\tan x = -23.08$

If a, b, or c in equations (1) can be associated with special reference triangles (30°–60° or 45° triangles) or quadrantal angles, then equations (1) can be solved exactly without using calculators or tables, if solutions exist. And if we are interested in solutions only over a small interval, say $[0, 2\pi)$, then the direct use of reference triangles will produce the desired results fairly easily.

EXAMPLE 22   Find the exact solutions of each over the interval $[0, 2\pi)$ without the use of a calculator or a table.

(A) $\sin x = \frac{1}{2}$ (B) $\cos x = -\sqrt{2}/2$

Solution   (A) $\sin x = \frac{1}{2}$: Where is $\sin x$ positive? In the first and second quadrants. What special reference triangle is associated with $\sin x = \frac{1}{2}$? A 30°–60° triangle. We sketch these triangles in the first and second quadrants and determine x.

$$x = \frac{\pi}{6}, \frac{5\pi}{6}$$

Referring to the material in Box 1 we could also write

$$x = \begin{cases} \sin^{-1} \frac{1}{2} \\ \pi - \sin^{-1} \frac{1}{2} \end{cases} = \begin{cases} \pi/6 \\ 5\pi/6 \end{cases}$$

(B) $\cos x = -\sqrt{2}/2 = -1/\sqrt{2}$: x is associated with the 45° reference triangle in the second and third quadrants. Thus,

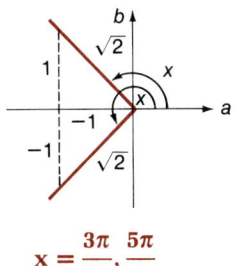

$$x = \frac{3\pi}{4}, \frac{5\pi}{4}$$

Referring to the material in Box 2, we could also write

$$x = \begin{cases} \cos^{-1}(-\sqrt{2}/2) \\ 2\pi - \cos^{-1}(-\sqrt{2}/2) \end{cases} = \begin{cases} 3\pi/4 \\ 5\pi/4 \end{cases}$$

PROBLEM 22 Find exact solutions over the interval $(-\pi, \pi)$ without the use of a calculator or a table.

(A) $\sin x = -\tfrac{1}{2}$ (B) $\tan x = 1$

EXAMPLE 23 Find all solutions over the interval $[0, 2\pi)$ for $\csc x = -14.07$. Compute answer(s) to four significant digits.

Solution $\csc x = -14.07$ Use reciprocal identity.

$\sin x = -\frac{1}{14.07}$ Refer to Box 1.

$= \begin{cases} 2\pi + \sin^{-1}(-\frac{1}{14.07}) \\ \pi - \sin^{-1}(-\frac{1}{14.07}) \end{cases}$ Calculate with a scientific calculator.

$= \begin{cases} 6.212 \\ 3.213 \end{cases}$

PROBLEM 23 Find all solutions over the interval $[0, 2\pi)$ for $\cot x = 0.0251$. Compute answer(s) to four significant digits.

■ **More General Trigonometric Equations**

We are now ready to solve trigonometric equations of a more general nature. No particular method or rule will always lead to the solution set of every trigonometric equation one is likely to encounter. Solving trigonometric equations often requires the use of identities, ingenuity, and perseverance. The following suggestions may help you get started.

> **Suggestions for Solving Trigonometric Equations**
>
> 1. If more than one trig function is present, use identities to try to write the equation in terms of one trig function.
> 2. Regard one particular trig function as the unknown and solve for it.
> 3. Algebraic manipulation such as factoring sometimes helps.
> 4. After solving for a trig function, regard the variable as an unknown and solve for it following the procedures in the first part of this section.

EXAMPLE 24 Solve $\sin 2x = \sin x$ exactly for x over the interval $[0, 2\pi)$.

Solution

$\sin 2x = \sin x$ Use a double-angle identity.

$2 \sin x \cos x = \sin x$

$2 \sin x \cos x - \sin x = 0$ Factor left side.

$\sin x (2 \cos x - 1) = 0$

Therefore, $\sin x = 0$ or $2 \cos x - 1 = 0$.

$\sin x = 0$ x is associated with a quadrantal angle

$x = 0, \pi$

$2 \cos x - 1 = 0$

$\cos x = \tfrac{1}{2}$

$x = \pi/3, 5\pi/3$

or

$x = \begin{cases} \cos^{-1} \tfrac{1}{2} \\ 2\pi - \cos^{-1} \tfrac{1}{2} \end{cases}$

$= \begin{cases} \pi/3 \\ 5\pi/3 \end{cases}$

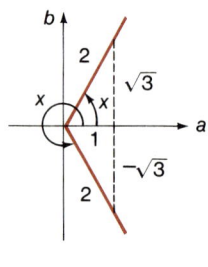

Thus, $x = 0, \pi, \pi/3, 5\pi/3$.

PROBLEM 24 Solve $\sin^2 x = \tfrac{1}{2} \sin 2x$ exactly for x over the interval $[0, 2\pi)$ without the use of a calculator or a table.

EXAMPLE 25 Solve $\cos^2 \theta = \cos \theta + \sin^2 \theta$, $0° \le \theta < 360°$, exactly without using a calculator or a table.

Solution

$\cos^2 \theta = \cos \theta + \sin^2 \theta$ Use a Pythagorean identity.

$\cos^2 \theta = \cos \theta + 1 - \cos^2 \theta$

$2 \cos^2 \theta - \cos \theta - 1 = 0$ A quadratic in $\cos \theta$.

$(\cos \theta - 1)(2 \cos \theta + 1) = 0$

6-6 Trigonometric Equations

Therefore,

$$\cos \theta - 1 = 0$$
$$\cos \theta = 1$$
$$\theta = 0°$$

or

$$2 \cos \theta + 1 = 0$$
$$\cos \theta = -\tfrac{1}{2}$$
$$\theta = 120°, 240°$$

or

$$\theta = \begin{cases} \cos^{-1}(-\tfrac{1}{2}) \\ 360° - \cos^{-1}(-\tfrac{1}{2}) \end{cases}$$
$$= \begin{cases} 120° \\ 240° \end{cases}$$

Thus, $\theta = 0°, 120°, 240°$.

PROBLEM 25 Solve $\sin^2 \theta = \cos^2 \theta - \sin \theta$, $0 \le \theta < 360°$, exactly without using a calculator or a table.

EXAMPLE 26 Solve $\cos 2x = 2(\sin x - 1)$ for all real x. Compute inverse functions to four significant digits.

Solution

$$\cos 2x = 2(\sin x - 1)$$ Use double-angle identity.
$$1 - 2 \sin^2 x = 2 \sin x - 2$$
$$2 \sin^2 x + 2 \sin x - 3 = 0$$ Quadratic in sin x

$$\sin x = \frac{-2 \pm \sqrt{4 - 4(2)(-3)}}{2(2)}$$

$$\sin x = -1.822\ 88 \quad \text{or} \quad 0.822\ 88$$

Thus,

$$\sin x = -1.822\ 88 \quad \text{No solution. Why?}$$

or

$$\sin x = 0.822\ 88$$

$$x = \begin{cases} 2k\pi + \sin^{-1} 0.822\ 88 \\ (2k+1)\pi - \sin^{-1} 0.822\ 88 \end{cases}$$

$$= \begin{cases} 2k\pi + 0.9665 \\ (2k+1)\pi - 0.9665 \end{cases} \quad k \text{ any integer}$$

PROBLEM 26 Solve $\cos 2x = 4 \cos x - 2$ for all real x. Compute inverse functions to four significant digits.

We finish with an example that involves a solution technique that can introduce extraneous solutions. The answers must be checked to see whether any must be discarded.

EXAMPLE 27 Solve $\csc x + \cot x = 1$ exactly over the interval $[0, 2\pi)$ without using a calculator or a table.

Solution To solve this equation, we square both sides after subtracting cot x from each side. We can then use Pythagorean identities to transform the equation into one involving only one type of trigonometric function. The squaring process can create extraneous solutions, so all solutions must be checked in the original equation to see whether any must be discarded.

$$\csc x = 1 - \cot x$$
$$\csc^2 x = 1 - 2 \cot x + \cot^2 x$$
$$1 + \cot^2 x = 1 - 2 \cot x + \cot^2 x$$
$$0 = -2 \cot x$$
$$\cot x = 0$$
$$x = \pi/2, 3\pi/2$$

Checking these in the original equation, we find that $3\pi/2$ must be discarded. Hence, $x = \pi/2$ is the only solution.

PROBLEM 27 Solve $1 + \cos x = \sin x$ exactly over the interval $[0, 2\pi)$ without using a calculator or a table.

Answers to Matched Problems
21. (A) $2k\pi - 0.2238, (2k + 1)\pi + 0.2238$, k any integer
    (B) $2k\pi \pm 1.089$, k any integer   (C) $k\pi - 1.528$, k any integer
22. (A) $-5\pi/6, -\pi/6$ (B) $-3\pi/4, \pi/4$
23. $1.546, 4.687$
24. $0, \pi/4, \pi, 5\pi/4$
25. $30°, 150°, 270°$
26. $2k\pi \pm 1.274$, k any integer
27. $\pi/2, \pi$

Exercise 6-6

In the following problems, "solving exactly" means finding exact solutions without using a calculator or a table.

A Solve exactly for the indicated variable over the specified interval.

1. $1 + \cos x = 0$, $[0, 2\pi)$
2. $1 - \sin x = 0$, $[0, 2\pi)$
3. $1 + \sqrt{2} \sin \theta = 0$, $[0°, 360°)$
4. $1 - \sqrt{2} \cos \theta = 0$, $[0°, 360°)$
5. $4 \cos^2 x - 3 = 0$, $[0, 2\pi)$
6. $2 \sin^2 x - 1 = 0$, $[0, 2\pi)$

6-6 Trigonometric Equations 387

7. $\sin^2 \theta = \sin \theta$, $[0°, 180°]$
8. $\cos^2 \theta = \cos \theta$, $[0°, 180°]$
9. $2 \sin x \cos x = \cos x$, $[0, 2\pi)$
10. $2 \sin x \cos x = \sin x$, $[0, 2\pi)$

Solve for all real x. Compute inverse functions to four significant digits using a scientific calculator.

11. $\sin x = 0.2977$
12. $\sin x = -0.6109$
13. $\cos x = -0.8861$
14. $\cos x = 0.8392$
15. $\tan x = 13.08$
16. $\tan x = -1.309$
17. $\sec x = -8.613$
18. $\cot x = 0.0496$
19. $\cot x = -3.478$
20. $\csc x = 42.29$

B *Solve exactly for the indicated variable over the specified interval.*

21. $2 \sin^2 x = 3 \sin x - 1$, $[0, 2\pi)$
22. $2 \cos^2 x + \cos x = 1$, $[0, 2\pi)$
23. $2 \cos 2x = 1$, $[0, 2\pi)$
24. $2 \sin 2x = \sqrt{3}$, $[0, 2\pi)$
25. $\sin x = \cos x$, $(-\infty, \infty)$
26. $\sqrt{3} \sin x - \cos x = 0$, $(-\infty, \infty)$
27. $4 \cos^2 2x - 4 \cos 2x + 1 = 0$, $[0, \pi]$
28. $2 \sin^2(x/2) - 3 \sin(x/2) + 1 = 0$, $[0, 2\pi)$
29. $\sin 2x = \sin x$, $(-\infty, \infty)$
30. $\sin 2x + \cos x = 0$, $(-\infty, \infty)$
31. $\sin^2 \theta + 2 \cos \theta = -2$, $[0°, 360°)$
32. $2 \cos^2 \theta + 3 \sin \theta = 0$, $[0°, 360°)$

Solve to four significant digits over the specified interval.

33. $4 \cos^2 \theta = 7 \cos \theta + 2$, $[0°, 180°]$
34. $6 \sin^2 \theta + 5 \sin \theta = 6$, $[0°, 90°]$
35. $\cos^2 x = 3 - 5 \cos x$, $[0, 2\pi)$
36. $2 \sin^2 x = 1 - 2 \sin x$, $[0, 2\pi)$

Solve for all real x. Compute inverse functions to four significant digits.

37. $\cos^2 x = 3 - 5 \cos x$
38. $2 \sin^2 x = 1 - 2 \sin x$

C *Solve exactly for the indicated variable over the specified interval. (Use factor identities for Problems 45–48.)*

39. $2 \sin^2 x + 3 \cos x = 3$, $(-\infty, \infty)$
40. $2 \cos^2 x = 3 \sin x + 3$, $(-\infty, \infty)$

41. $\sin x + \cos x = 1$, $[0, 2\pi)$
42. $\cos x - \sin x = 1$, $[0, 2\pi)$
43. $\sec x + \tan x = 1$, $[0, 2\pi)$
44. $\tan x - \sec x = 1$, $[0, 2\pi)$
45. $\sin 3x + \sin x = 0$, $[0, \pi]$
46. $\cos 5x + \cos 3x = 0$, $[0, \pi]$
47. $\sin 5x - \sin 3x = \sin x$, $[0, \pi/2]$
48. $\cos 3x - \cos x = \sin x$, $[0, \pi]$

APPLICATIONS

49. *Electrical current.* An alternating current generator produces a current given by the equation

$$I = 30 \sin 120\pi t$$

where t is time in seconds and I is amperes. Find the least positive t to four significant digits such that $I = 25$ amperes.

50. *Electrical current.* Find the least positive t in Problem 49, to four significant digits, such that $I = -10$ amperes.

51. *Astronomy.* The planet Mercury travels around the sun in an elliptical orbit given approximately by

$$r = \frac{3.44 \times 10^7}{1 - 0.206 \cos \theta}$$

(see the accompanying figure). Find the least positive θ (in decimal degrees to three significant digits) such that Mercury is 3.78×10^7 miles from the sun.

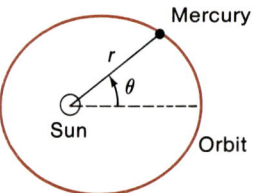

52. *Astronomy.* In Problem 51, find the least positive θ (in decimal degrees to three significant digits) such that Mercury is 3.09×10^7 miles from the sun.

Find simultaneous solutions for each system of equations over the interval $[0°, 360°]$. These are **polar equations**, which will be discussed in Chapter 7.

53. $r = 2 \sin \theta$
    $r = 2(1 - \sin \theta)$

54. $r = 2 \sin \theta$
    $r = \sin 2\theta$

Section 6-7 Chapter Review

IMPORTANT TERMS AND SYMBOLS

6-2 Basic Identities and Their Use. Reciprocal identities, quotient identities, identities for negatives, Pythagorean identities

6-3 Addition, Subtraction, and Cofunction Identities

6-4 Double-Angle and Half-Angle Identities

6-5 Product and Factor Identities

6-6 Trigonometric Equations

Exercise 6-7 Chapter Review

Work through all the problems in this chapter review and check answers in the back of the book. (Answers to all problems are there, and following each answer is a number in italics indicating the section in which that type of problem is discussed.) Where weaknesses show up, review appropriate sections in the text. When you are satisfied that you know the material, take the practice test following this review.

A *Establish each identity.*

1. $\tan x + \cot x = \sec x \csc x$
2. $\sec^4 x - 2 \sec^2 x \tan^2 x + \tan^4 x = 1$
3. $\dfrac{1}{1 - \sin x} + \dfrac{1}{1 + \sin x} = 2 \sec^2 x$
4. $\cos\left(x - \dfrac{3\pi}{2}\right) = -\sin x$

Solve exactly over the indicated interval.

5. $\sqrt{2} \cos \theta + 1 = 0$, $[0°, 360°)$
6. $\sin x \tan x - \sin x = 0$, $[0, 2)$

Solve for all real x. Compute inverse functions to four significant digits.

7. $\sin x = 0.7088$
8. $\cot x = -0.04473$

B *Establish each identity in Problems 9–14.*

9. $(1 - \cos x)(\csc x + \cot x) = \sin x$
10. $\cot x - \tan x = \dfrac{4 \cos^2 x - 2}{\sin 2x}$
11. $\left(\dfrac{1 - \cot x}{\csc x}\right)^2 = 1 - \sin 2x$

12. $\tan m + \tan n = \dfrac{\sin(m+n)}{\cos m \cos n}$

13. $\tan(x+y) = \dfrac{\cot x + \cot y}{\cot x \cot y - 1}$

14. $\cot \dfrac{x}{2} = \dfrac{\sin x}{1 - \cos x}$

15. Given $\sin x = -\tfrac{3}{5}$, $-\pi/2 \leq x \leq \pi/2$. Compute exact values for each.
    (A) $\cos \dfrac{x}{2}$ (B) $\sin 2x$ (C) $\tan 2x$

16. Write in a factored form: $\sin 3x - \sin x$

17. Write as a sum: $2 \cos 5x \cos 2x$

Solve Problems 18–20 exactly over the indicated interval.

18. $4 \sin^2 x - 3 = 0$, $[0, 2\pi)$

19. $2 \sin^2 \theta + \cos \theta = 1$, $[0°, 180°]$

20. $\sin 2x = \sqrt{3} \sin x$, $(-\infty, \infty)$

21. Solve $\tan^2 x = 2 \tan x + 1$ for x over the interval $[0, \pi)$. Compute the answer to four significant digits.

C 22. Find the exact value of $\sin[2 \tan^{-1}(-\tfrac{3}{4})]$.

23. Find the exact value of $\sin[\sin^{-1}(\tfrac{3}{5}) + \cos^{-1}(\tfrac{4}{5})]$.

Solve exactly over the indicated interval.

24. $\cos^2 2x = \cos 2x + \sin^2 2x$, $[0, \pi)$

25. $\sin 7x - \sin x = \sin 3x$, $[0, \pi/2]$

26. $\cos x = 1 - \sin x$, $[0, 2\pi)$

Practice Test Chapter 6

Take this practice test as if it were a graded test. Allow yourself up to 50 minutes. Work the problems without looking back in the chapter. Correct your work using the answers (keyed to appropriate sections) in the back of the book.

Establish each identity in Problems 1–3.

1. $\dfrac{1 - 2\cos x - 3\cos^2 x}{\sin^2 x} = \dfrac{1 - 3\cos x}{1 - \cos x}$

2. $\cos 2x = \dfrac{1 - \tan^2 x}{1 + \tan^2 x}$

Practice Test Chapter 6

3. $\dfrac{1 + \sin x}{\cos x} = \dfrac{\cos x}{1 - \sin x}$

4. Simplify: $\sin\left(x + \dfrac{9\pi}{2}\right)$

5. Given $\tan x = \frac{-3}{4}$, $\pi/2 \le x \le \pi$, find:
   (A) $\sin \dfrac{x}{2}$ (B) $\cos 2x$

6. Write $\sin 7x \cos 3x$ as a sum.

Solve exactly over the indicated interval.

7. $2 \sin^2 x - \sin x = 0$, $(-\infty, \infty)$

8. $2 \sin^2 \theta + 5 \cos \theta + 1 = 0$, $[0°, 360°)$

9. $1 - \cos x = \sqrt{3} \sin x$, $[0, 2\pi)$

 10. Solve $\sin^2 x + 2 = 4 \sin x$ for x over the interval $[0, 2\pi)$. Compute the answers to four significant digits.

Additional Topics in Trigonometry ■7

7-1 Solutions of Right Triangles
7-2 Law of Sines
7-3 Law of Cosines
7-4 Vectors
7-5 Polar and Rectangular Coordinates
7-6 Sketching Polar Graphs
7-7 Complex Numbers in Rectangular and Polar Forms
7-8 De Moivre's Theorem
7-9 Chapter Review

A natural design of mathematical interest. Can you guess the source? See the back of the book.

Chapter 7 ■ Additional Topics in Trigonometry

Section 7-1 Solutions of Right Triangles

We have applied trigonometric functions to problems involving periodic phenomena; in this and the next two sections we will apply trigonometric functions to the problem of solving triangles. We start by considering the right triangle.

We find that if in a right triangle we are given either one side and an acute angle or two sides, we will be able to find the other angles and sides of the triangle by using trigonometric functions and a scientific calculator or a table. To start, we locate a right triangle in the first quadrant of a rectangular coordinate system and observe, from the definitions of the trigonometric functions, six trigonometric ratios involving the sides of the triangle. (Note that the right triangle is the reference triangle for the angle θ.)

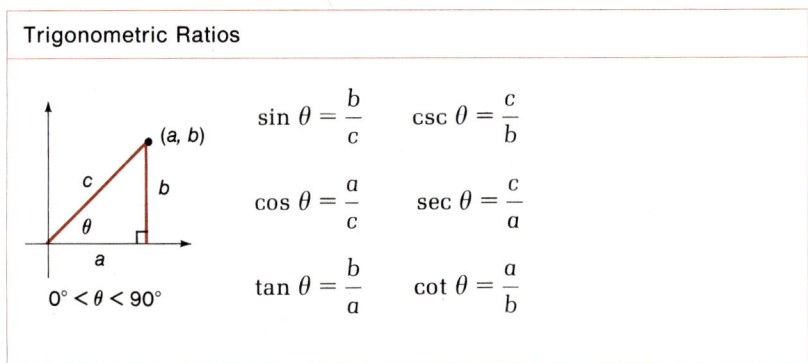

Trigonometric Ratios

$\sin \theta = \dfrac{b}{c}$ $\csc \theta = \dfrac{c}{b}$

$\cos \theta = \dfrac{a}{c}$ $\sec \theta = \dfrac{c}{a}$

$\tan \theta = \dfrac{b}{a}$ $\cot \theta = \dfrac{a}{b}$

$0° < \theta < 90°$

Side b is often referred to as the **side opposite** angle θ, a as the **side adjacent** to angle θ, and c as the **hypotenuse**. Using these designations for an arbitrary right triangle removed from a coordinate system, we have:

7-1 Solutions of Right Triangles

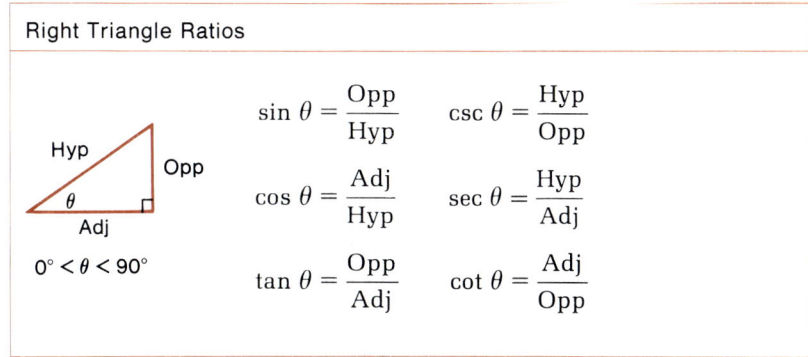

Right Triangle Ratios

$$\sin \theta = \frac{\text{Opp}}{\text{Hyp}} \qquad \csc \theta = \frac{\text{Hyp}}{\text{Opp}}$$

$$\cos \theta = \frac{\text{Adj}}{\text{Hyp}} \qquad \sec \theta = \frac{\text{Hyp}}{\text{Adj}}$$

$$0° < \theta < 90°$$

$$\tan \theta = \frac{\text{Opp}}{\text{Adj}} \qquad \cot \theta = \frac{\text{Adj}}{\text{Opp}}$$

The use of these relationships in solving right triangles is made clear in the following examples. (To **solve a right triangle** is to find all the unknown sides and angles of the triangle.) A scientific hand calculator will be used extensively in this chapter. Unless noted to the contrary, we will assume a labeling of a right triangle as indicated in Figure 1.

Regarding computational accuracy, we will be guided by the table below. Also, we will use = in many places, realizing the accuracy indicated in the table is all that is assumed.

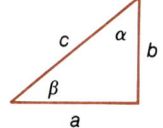

FIGURE 1 Standard labeling, unless stated to the contrary

ANGLE TO NEAREST	SIGNIFICANT DIGITS FOR SIDE MEASURE
1°	2
10′ or 0.1°	3
1′ or 0.01°	4
10″ or 0.001°	5

EXAMPLE 1   Solve a right triangle with $c = 6.25$ feet and $\beta = 32°10'$.

Solution   First draw a figure and label the parts.

Solve for α $\alpha = 90° - 32°10' = 57°50'$

Solve for b
$$\sin \beta = \frac{b}{c}$$

$$\sin 32°10' = \frac{b}{6.25}$$

$$b = 6.25 \sin 32°10'$$

$$= 3.33 \text{ feet}$$

Calculator operations: Convert 32°10' to decimal degrees and set calculator in degree mode.

A: [10] [÷] [60] [+] [32] [=] [SIN] [×] [6.25] [=]
P: [10] [ENTER] [60] [÷] [32] [+] [SIN] [6.25] [×]

Solve for a
$$\cos \beta = \frac{a}{c}$$

$$\cos 32°10' = \frac{a}{6.25}$$

$$a = 6.25 \cos 32°10'$$

$$= 5.29 \text{ feet} \qquad \text{Calculator operations similar to before}$$

PROBLEM 1 Solve a right triangle with $c = 27.3$ meters and $\alpha = 47°50'$.

EXAMPLE 2 Solve a right triangle with $a = 4.32$ centimeters and $b = 2.62$ centimeters.

Solution Draw a figure and label the known parts.

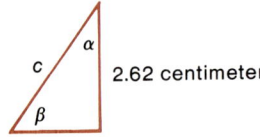

2.62 centimeters

4.32 centimeters

Solve for β
$$\tan \beta = \frac{2.62}{4.32}$$

$$\beta = \tan^{-1} \frac{2.62}{4.32}$$

A: [2.62] [÷] [4.32] [=] [INV] [TAN]
P: [2.62] [ENTER] [4.32] [÷] [INV] [TAN]

$$= 31.2° \quad \text{or} \quad 31°10' \qquad 0.2° = [(0.2)(60)]' = 12' \approx 10' \text{ to nearest } 10'$$

Solve for α $\quad \alpha = 90° - 31°10' = 58°50'$

Solve for c
$$\sin \beta = \frac{2.62}{c}$$

$$c = \frac{2.62}{\sin 31.2°} = 5.06 \text{ centimeters}$$

or, using the Pythagorean theorem:

$$c = \sqrt{4.32^2 + 2.62^2} = 5.05 \text{ centimeters}$$

PROBLEM 2 Solve a right triangle with $a = 1.38$ kilometers and $b = 6.73$ kilometers.

EXAMPLE 3 If a pentagon (a five-sided regular polygon) is inscribed in a circle of radius 5.35 centimeters, find the length of one side of the pentagon.

Solution Sketch a figure and insert triangle ACB with C at the center. Add the auxiliary line CD as indicated. We will find AD and double it.

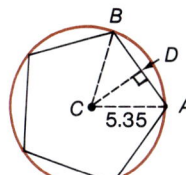

Angle $ACB = \dfrac{360°}{5} = 72°$

Angle $ACD = \dfrac{72°}{2} = 36°$

$$\sin(\text{angle } ACD) = \frac{AD}{AC}$$

$$AD = AC \sin(\text{angle } ACD)$$
$$= 5.35 \sin 36°$$
$$= 3.14 \text{ centimeters}$$
$$AB = 2AD = 6.28 \text{ centimeters}$$

PROBLEM 3 If a square of side 43.6 meters is inscribed in a circle, what is the radius of the circle?

Answers to Matched Problems
1. $\beta = 42°10'$, $a = 20.2$ meters, $b = 18.3$ meters
2. $\alpha = 11°40'$, $\beta = 78°20'$, $c = 6.87$ kilometers
3. 30.8 meters

Exercise 7-1

A scientific hand calculator will be useful in most problems.

A Solve each triangle in Problems 1–14 using the information given and the standard triangle labeling.

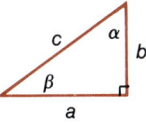

1. $\beta = 17°50'$, $c = 3.45$
2. $\beta = 33°40'$, $b = 22.4$
3. $\beta = 43°20'$, $a = 123$
4. $\beta = 62°30'$, $c = 42.5$
5. $\alpha = 23°0'$, $a = 54.0$
6. $\alpha = 54°$, $c = 4.3$

7. $\alpha = 53°12'$, $b = 23.82$
8. $\alpha = 35°44'$, $b = 6.482$
9. $a = 6.00$, $b = 8.46$
10. $a = 22.0$, $b = 46.2$
11. $b = 10.0$, $c = 12.6$
12. $b = 50.0$, $c = 165$
13. $a = 2.42$, $c = 3.22$
14. $a = 63.8$, $c = 134$

15. Find the height of a tree (growing on level ground) if at a point 100 feet from the base of the tree the angle of its top relative to the horizontal is found to be $65°20'$.

16. To measure the height of a cloud ceiling over an airport, a searchlight is directed straight upward to produce a lighted spot on the clouds. Five hundred meters away an observer reports the angle of the spot relative to the horizontal to be $32°10'$. How high (to the nearest meter) are the clouds above the airport?

B 17. If a train climbs at a constant angle of $1°23'$, how many vertical feet has it climbed after going 1 mile? (1 mile = 5,280 feet)

18. If a jet airliner climbs at an angle of $15°$ with a constant speed of 300 miles/hour, how long will it take (to the nearest minute) to reach an altitude of 8 miles? Assume there is no wind.

19. Find the distance d to the airplane shown in the figure (to the nearest foot) if θ on the range finder is $89°57'$.

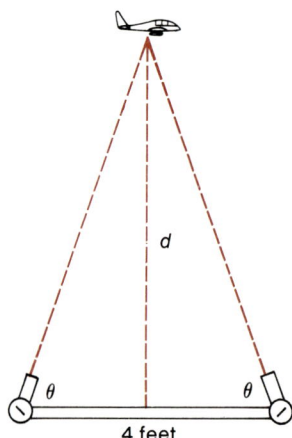

20. Find the distance d to the airplane shown in Problem 19 (to the nearest foot) if θ on the range finder is $89°59'$.

21. Find the diameter of the moon (to the nearest mile) if at 239,000 miles from the earth it subtends an angle of $32'$ relative to an observer on earth.

22. If the sun is 93,000,000 miles from earth and its diameter subtends an angle of $32'$ relative to an observer on earth, what is the diameter of the sun to two significant digits?

23. If a circle of radius 4 centimeters has a chord of length 3 centimeters, find the central angle subtended by this chord to the nearest degree.

24. Find the length of one side of a nine-sided regular polygon inscribed in a circle of radius 4.06 inches.

C 25. A river 1 mile wide flows at a rate of 1.5 miles/hour. If a man rows across the river, always heading straight for the other shore, how far downstream will he land if his rate of rowing in still water is 3 miles/hour?

26. In Problem 25, how far has the man traveled in crossing the river?

27. Find the area of a regular five-sided polygon inscribed in a circle of radius 6.02 centimeters.

28. Find the perimeter of a regular five-sided polygon inscribed in a circle of radius 6.02 inches.

29. Find R in the accompanying figure (to two significant digits) so that the circle is tangent to all three sides of the isosceles triangle. [*Hint:* The radius of a circle is perpendicular to a tangent line at the point of tangency.]

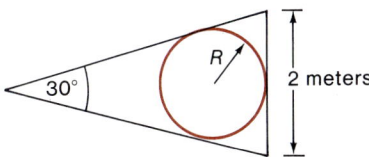

30. Find R in the accompanying figure (to two significant digits) so that the smaller circle is tangent to the larger circle and the two sides of the angle. (See the hint in Problem 29.)

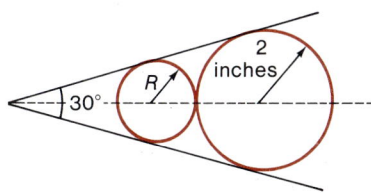

Section 7-2 Law of Sines

- Law of Sines
- Ambiguous Case

The law of sines and the law of cosines (to be developed in Section 7-3) play fundamental roles in solving **oblique triangles**—that is, triangles without a right angle. Every oblique triangle is either **acute** (all angles are between 0° and 90°) or **obtuse** (one angle is between 90° and 180°). Figure 2 illustrates the two cases.

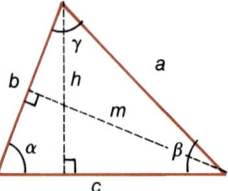

FIGURE 2   (a) Acute triangle   (b) Obtuse triangle

■ Law of Sines

The law of sines is relatively easy to prove using the right triangle properties studied in Section 7-1. We will also use the fact that

$$\sin(180° - x) = \sin x$$

which is easily obtained from a subtraction identity (a good exercise for the reader). Referring to the triangles in Figure 2, we proceed as follows: For each triangle,

$$\sin \alpha = \frac{h}{b} \quad \text{and} \quad \sin \beta = \frac{h}{a}$$

Therefore,

$$h = b \sin \alpha \quad \text{and} \quad h = a \sin \beta$$

Thus,

$$b \sin \alpha = a \sin \beta$$

and

$$\frac{\sin \alpha}{a} = \frac{\sin \beta}{b} \tag{1}$$

Similarly, for each triangle in Figure 2,

$$\sin \alpha = \frac{m}{c} \quad \text{and} \quad \sin \gamma = \sin(180° - \gamma) = \frac{m}{a}$$

Therefore,

$$m = c \sin \alpha \quad \text{and} \quad m = a \sin \gamma$$

Thus,

$$c \sin \alpha = a \sin \gamma$$

7-2 Law of Sines

and

$$\frac{\sin \alpha}{a} = \frac{\sin \gamma}{c} \qquad (2)$$

If we combine equations (1) and (2), we obtain the law of sines.

Law of Sines

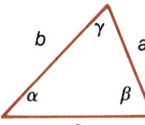

$$\frac{\sin \alpha}{a} = \frac{\sin \beta}{b} = \frac{\sin \gamma}{c}$$

This law is most useful when given:

Two sides and an angle opposite one of the sides

Two angles and a side opposite one of the angles

EXAMPLE 4 Solve the triangle:

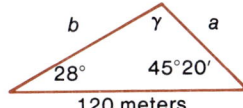

Solution

Solve for γ
$$\alpha + \beta + \gamma = 180°$$
$$\gamma = 180° - (\alpha + \beta)$$
$$= 180° - (28° + 45°20')$$
$$= 106°40'$$

Solve for a
$$\frac{\sin \alpha}{a} = \frac{\sin \gamma}{c}$$

$$a = \frac{c \sin \alpha}{\sin \gamma}$$

$$= \frac{120 \sin 28°}{\sin 106°40'}$$

$$= 58.8 \text{ meters}$$

A: [120] [×] [28] [SIN] [=] [÷] [(] [(] [106] [+] [40] [÷] [60] [)] [SIN] [)] [=]

P: [120] [ENTER] [28] [SIN] [×] [106] [ENTER] [40] [ENTER] [60] [÷] [+] [SIN] [÷]

Solve for b
$$\frac{\sin \beta}{b} = \frac{\sin \gamma}{c}$$

$$b = \frac{c \sin \beta}{\sin \gamma}$$

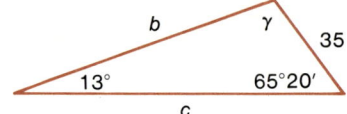

$$= 89.1 \text{ meters}$$

PROBLEM 4 Solve the triangle:

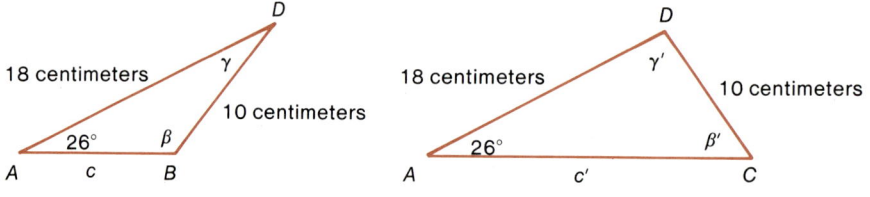

■ Ambiguous Case

One can specify parts of a triangle so that more than one triangle is possible. Consider Example 5.

EXAMPLE 5 Solve a triangle with $\alpha = 26°$, $a = 10$ centimeters, and $b = 18$ centimeters.

Solution If we try to draw a triangle with these values, we find that two triangles are possible:

Solve for β and β' Here we will see that there are two possible values for β; hence, there are two possible triangles.

$$\frac{\sin \beta}{b} = \frac{\sin \alpha}{a}$$

$$\sin \beta = \frac{b \sin \alpha}{a} = \frac{18 \sin 26°}{10} = 0.7891$$

Angle β can be either obtuse or acute:

$$\beta = 180° - \sin^{-1} 0.7891 \quad \text{or} \quad \beta' = \sin^{-1} 0.7891$$
$$= 180° - 52° = 128° \qquad\qquad\qquad = 52°$$

7-2 Law of Sines

Solve for γ and γ'
$$\gamma = 180° - (26° + 128°) = 26°$$
$$\gamma' = 180° - (26° + 52°) = 102°$$

Solve for c and c'
$$\frac{\sin \alpha}{a} = \frac{\sin \gamma}{c} \qquad \frac{\sin \alpha}{a} = \frac{\sin \gamma'}{c'}$$

$$c = \frac{a \sin \gamma}{\sin \alpha} \qquad c' = \frac{a \sin \gamma'}{\sin \alpha}$$

$$= \frac{10 \sin 26°}{\sin 26°} \qquad = \frac{10 \sin 102°}{\sin 26°}$$

$$= 10 \text{ centimeters} \qquad = 22 \text{ centimeters}$$

If an angle and its opposite side are specified, as well as an adjacent side, then there may be more than one triangle, or exactly one triangle, or even no triangle. Figure 3 illustrates four possibilities.

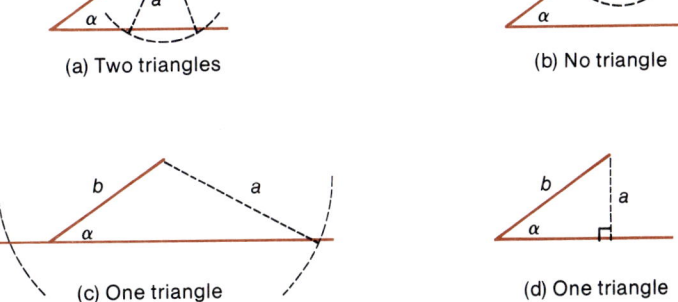

FIGURE 3 Given α, b, and a

(a) Two triangles (b) No triangle (c) One triangle (d) One triangle

Since the altitude h of the triangles in Figure 3 (if it exists) is $b \sin \alpha$, the possibilities illustrated in Figure 3 are summarized in Table 1. You need not memorize this table. Special cases become obvious in particular contexts.

TABLE 1

TRIANGLE IN FIGURE 3	CASE	NUMBER OF TRIANGLES DETERMINED
(a)	$b \sin \alpha < a < b$	2
(b)	$a < b \sin \alpha$	0
(c)	$a \geq b$	1
(d)	$a = b \sin \alpha$	1

PROBLEM 5 Solve the triangle(s) with $a = 8$ kilometers, $b = 10$ kilometers, and $\alpha = 35°$.

EXAMPLE 6 To measure the length d of a lake (see the figure), a base line AB (in the same plane as the lake) is established and measured to be 125 meters. Angles A and B are measured to be 41.6° and 124.3°, respectively. How long is the lake?

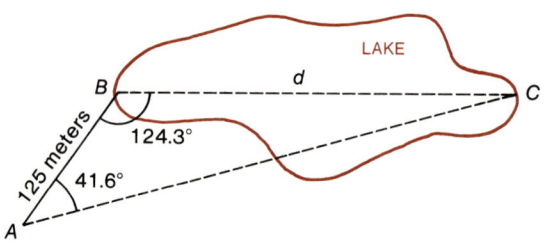

Solution Find angle C and use the law of sines.

$$\text{Angle } C = 180° - (124.3° + 41.6°)$$
$$= 14.1°$$

$$\frac{\sin 14.1°}{125} = \frac{\sin 41.6°}{d}$$

$$d = (125)\frac{\sin 41.6°}{\sin 14.1°}$$

$$= 341 \text{ meters}$$

PROBLEM 6 In Example 6, find the distance AC.

Answers to Matched Problems

4. $\gamma = 101°40'$, $b = 141$, $c = 152$
5. $\beta = 134°$, $\beta' = 46°$, $\gamma = 11°$, $\gamma' = 99°$, $c = 2.7$ kilometers, $c' = 14$ kilometers
6. 424 meters

Exercise 7-2

A scientific hand calculator will be useful in most problems.

Problems 1–12 refer to the labeling in the figure. Solve each triangle.

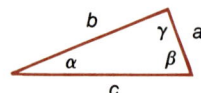

A
1. $\beta = 12°40'$, $\gamma = 100°$, $b = 13.1$
2. $\alpha = 41°$, $\beta = 77°$, $c = 100$

7-2 Law of Sines

3. $\beta = 27°30'$, $\gamma = 54°30'$, $a = 9.27$
4. $\beta = 105°$, $\gamma = 23°40'$, $a = 24.2$

B
5. $\alpha = 25°50'$, $a = 65.00$, $b = 105.0$, β obtuse
6. $\alpha = 33°20'$, $a = 30.0$, $b = 44.5$, β acute
7. $\beta = 31°40'$, $a = 12.0$, $b = 8.00$, α acute
8. $\alpha = 38°$, $a = 1.95$, $b = 2.43$, β obtuse
9. $a = 50$, $c = 40$, $\gamma = 30°$
10. $a = 23$, $b = 20$, $\beta = 37°$
11. $a = 14$, $b = 23$, $\alpha = 41°$
12. $\beta = 32°20'$, $a = 140$, $b = 60$

C
13. Mollweide's equation,

$$(a - b) \cos \frac{\gamma}{2} = c \sin \frac{\alpha - \beta}{2}$$

is often used to check the final solution of a triangle since all six parts of a triangle are involved in the equation. If, after substitution the left side does not equal the right side, then an error has been made in solving a triangle. Use this equation to check Problem 1.

14. Use Mollweide's equation from Problem 13 to check Problem 3.
15. Use the law of sines and suitable identities to show that for any triangle

$$\frac{a - b}{a + b} = \frac{\tan \frac{\alpha - \beta}{2}}{\tan \frac{\alpha + \beta}{2}}$$

16. Verify the formula in Problem 15 with values from Problem 1.

APPLICATIONS
17. *Coast guard.* Two lookout posts, A and B (10.0 miles apart), are established along a coast to watch for illegal ships coming within the 3-mile limit. If post A reports a ship S at angle $BAS = 37°30'$, and post B reports the same ship at angle $ABS = 20°0'$, how far is the ship from post A? How far is the ship from the shore (assuming the shore is along the line joining the two observation posts)?

18. *Fire lookout.* A fire at F is spotted from two fire lookout stations A and B, which are 10.0 miles apart. If station B reports the fire at angle $ABF = 53°0'$, and station A reports the fire at angle $BAF = 28°30'$, how far is the fire from station A? From station B? (Give the answer to three significant digits.)

19. *Natural science.* The tallest trees in the world grow in Redwood National Park in California; they are taller than a football field is long. Find the height of one of these trees, given the information in the figure.

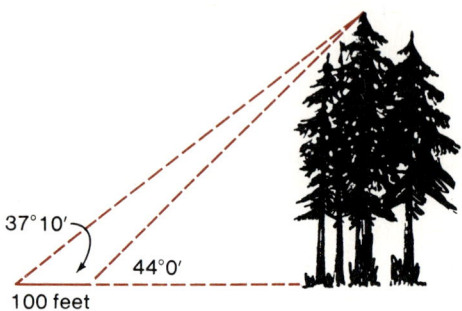

20. *Surveying.* To measure the height of Mt. Whitney in California, surveyors used the scheme shown in the figure in Problem 19. They set up a horizontal base line 2,000 feet long at the foot of the mountain and found the angle nearest the mountain to be 43°5′; the angle farthest from the mountain was found to be 38°0′. If the baseline was 5,000 feet above sea level, how high is Mt. Whitney above sea level?

21. *Astronomy.* The orbits of the Earth and Venus are approximately circular, with the sun at the center. A sighting of Venus is made from Earth, and the angle SEV is found to be 18°40′. If the radius of the orbit of the Earth is 1.495×10^8 kilometers, and the radius of the orbit of Venus is 1.085×10^8 kilometers, what are the possible distances from the Earth to Venus?

22. *Astronomy.* In Problem 21, find the maximum angle SEV. [*Hint:* The angle is maximum when a straight line joining the Earth and Venus is tangent to Venus's orbit.]

Section 7-3 Law of Cosines

If two sides and an included angle or three sides are given in a triangle, then the law of sines is not particularly helpful in solving the triangle. However, the Pythagorean theorem for right triangles can be generalized into another law, called the **law of cosines**, that will take care of these two situations.

7-3 Law of Cosines

Law of Cosines

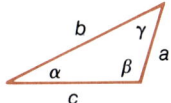

$a^2 = b^2 + c^2 - 2bc \cos \alpha$
$b^2 = a^2 + c^2 - 2ac \cos \beta$
$c^2 = a^2 + b^2 - 2ab \cos \gamma$

(All three equations say essentially the same thing.)

This law is most useful when given:

Three sides

Two sides and an included angle

We will establish $a^2 = b^2 + c^2 - 2bc \cos \alpha$. The other two equations can then be obtained from this one simply by relabeling the figure. We start by locating a triangle in a rectangular coordinate system. Figure 4 shows three typical triangles.

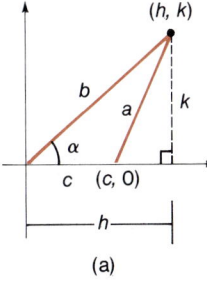 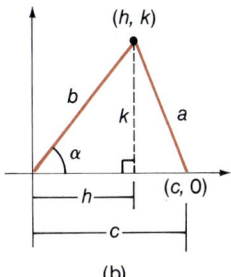 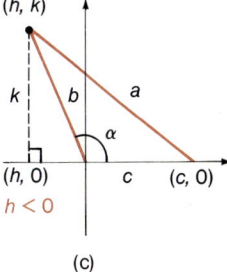

FIGURE 4   (a)   (b)   (c)

For an arbitrary triangle located as in Figure 4, we obtain, using the distance-between-two-points formula,

$$a = \sqrt{(h - c)^2 + (k - 0)^2}$$

or, squaring both sides,

$$a^2 = (h - c)^2 + k^2$$
$$= h^2 - 2hc + c^2 + k^2 \qquad (1)$$

From Figure 4 we note that

$$b^2 = h^2 + k^2$$

Substituting b^2 for $h^2 + k^2$ in equation (1), we obtain

$$a^2 = b^2 + c^2 - 2hc \qquad (2)$$

But

$$\cos \alpha = \frac{h}{b}$$

$$h = b \cos \alpha$$

Thus, by replacing h in equation (2) with $b \cos \alpha$, we reach our objective,

$$a^2 = b^2 + c^2 - 2bc \cos \alpha$$

[*Note:* If α is acute, then $\cos \alpha > 0$; if α is obtuse, then $\cos \alpha < 0$.]

EXAMPLE 7 Solve the triangle.

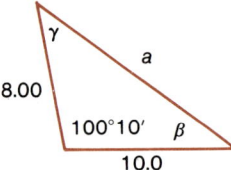

Solution

Solve for a

$$\begin{aligned}
a^2 &= b^2 + c^2 - 2bc \cos \alpha \\
&= (8.00)^2 + (10.0)^2 - 2(8.00)(10.0) \cos 100°10' \\
&= 192.241\ 9 \ldots \\
a &= \sqrt{192.241\ 9 \ldots} = 13.9 \quad \text{To three significant digits}
\end{aligned}$$

Calculator computation: First set calculator in degree mode and form decimal degrees:

A: [10] [÷] [60] [+] [100] [=] [COS] [×] [10] [×] [8] [×] [2] [+/−] [+] [10] [x^2] [+] [8] [x^2] [=] [√]

P: [10] [ENTER] [60] [÷] [100] [+] [COS] [10] [×] [8] [×] [2] [×] [+/−] [10] [x^2] [+] [8] [x^2] [+] [√]

Solve for β   We have a choice of using the law of sines or the law of cosines to find β. We choose the law of sines for simpler computation. We note first that both β and γ are acute angles. (Why?)

$$\frac{\sin \beta}{b} = \frac{\sin \alpha}{a} \qquad \text{Solve for } \sin \beta.$$

$$\sin \beta = \frac{b \sin \alpha}{a} = \frac{8.00 \sin 100°10'}{13.9} \qquad \text{Solve for } \beta; \text{ remember } \beta \text{ is acute.}$$

$$\beta = \sin^{-1}\left(\frac{8.00 \sin 100°10'}{13.9}\right)$$

$$= 34.5° \quad \text{or} \quad 34°30' \qquad \text{To the nearest } 0.1° \text{ or } 10'$$

Solve for γ
$$\gamma = 180° - (100°10' + 34°30')$$
$$= 45°20'$$

PROBLEM 7 Solve the triangle with $\alpha = 78°0'$, $b = 10.0$, and $c = 18.0$.

EXAMPLE 8 Solve a triangle with $a = 9.23$, $b = 5.04$, and $c = 10.6$.

Solution If we first draw the triangle roughly to scale, then two acute angles will generally be obvious. [A triangle must always have two acute angles and an obtuse angle, or three acute angles, or be a right triangle. (Why?)] We find the two acute angles first.

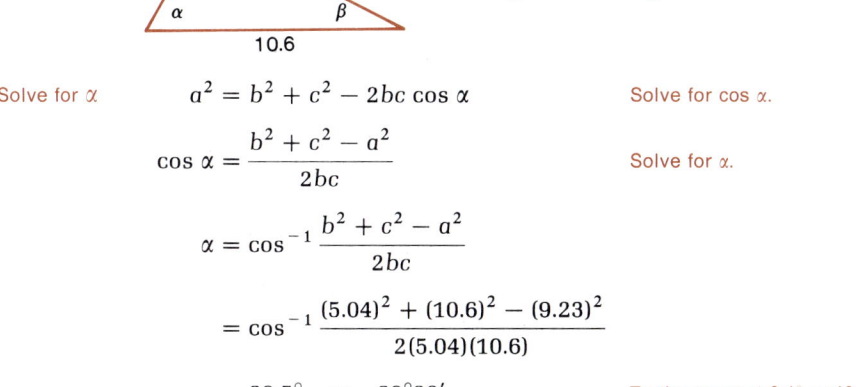

α and β are acute angles

Solve for α

$$a^2 = b^2 + c^2 - 2bc \cos \alpha \qquad \text{Solve for } \cos \alpha.$$

$$\cos \alpha = \frac{b^2 + c^2 - a^2}{2bc} \qquad \text{Solve for } \alpha.$$

$$\alpha = \cos^{-1} \frac{b^2 + c^2 - a^2}{2bc}$$

$$= \cos^{-1} \frac{(5.04)^2 + (10.6)^2 - (9.23)^2}{2(5.04)(10.6)}$$

$$= 60.5° \quad \text{or} \quad 60°30' \qquad \text{To the nearest } 0.1° \text{ or } 10'$$

Solve for β We can use either the law of cosines or the law of sines. We choose the latter because of simpler calculations.

$$\frac{\sin \alpha}{a} = \frac{\sin \beta}{b} \qquad \text{Solve for } \sin \beta.$$

$$\sin \beta = \frac{b \sin \alpha}{a} \qquad \text{Solve for } \beta; \text{ remember } \beta \text{ is acute.}$$

$$\beta = \sin^{-1} \frac{b \sin \alpha}{a}$$

$$= \sin^{-1} \frac{5.04 \sin 60.5°}{9.23}$$

$$= 28.4° \quad \text{or} \quad 28°20' \qquad \text{To the nearest } 0.1° \text{ or } 10'$$

Solve for γ $\gamma = 180° - (60°30' + 28°20') = 91°10'$

PROBLEM 8 Solve the triangle with $a = 1.20$, $b = 2.00$, and $c = 1.50$.

EXAMPLE 9 If a seven-sided regular polygon is inscribed in a circle of radius 22.8 centimeters, find the length of one side of the polygon.

Solution Sketch a figure and use the law of cosines:

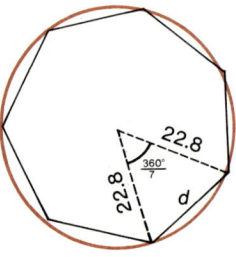
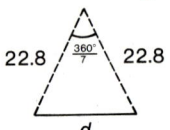

Actually, you only need to sketch the triangle:

$$d^2 = 22.8^2 + 22.8^2 - 2(22.8)(22.8) \cos \frac{360°}{7}$$

$$d = \sqrt{2(22.8)^2 - 2(22.8)^2 \cos \frac{360°}{7}}$$

$$= 19.8 \text{ centimeters}$$

PROBLEM 9 If an eleven-sided regular polygon is inscribed in a circle with radius 4.63 inches, find the length of one side of the polygon.

Answers to Matched Problems

7. $a = 18.7$, $\beta = 31°30'$, $\gamma = 70°30'$
8. $\alpha = 36°40'$, $\beta = 95°0'$, $\gamma = 48°20'$
9. 2.61 inches

Exercise 7-3

A scientific hand calculator will be useful in most problems.

Problems 1–8 refer to the labeling in the figure. Solve each triangle.

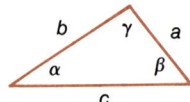

A
1. $\alpha = 50°40'$, $b = 7.03$, $c = 7.00$
2. $\beta = 57°0'$, $a = 6.00$, $c = 5.00$
3. $\gamma = 120°20'$, $a = 5.73$, $b = 10.2$
4. $\alpha = 135°50'$, $b = 8.44$, $c = 20.3$

B
5. $a = 4.00$, $b = 10.0$, $c = 9.00$
6. $a = 6.00$, $b = 5.00$, $c = 5.50$

7. $a = 10.5$ $b = 20.7$, $c = 12.2$
8. $a = 42.3$, $b = 76.8$, $c = 131$

C
9. Show, using the law of cosines, that if $\gamma = 90°$, then $c^2 = a^2 + b^2$ (the Pythagorean theorem).
10. Show, using the law of cosines, that if $c^2 = a^2 + b^2$, then $\cos \gamma = 90°$.
11. Check Problem 1 using Mollweide's equation:

$$(a - b) \cos \frac{\gamma}{2} = c \sin \frac{\alpha - \beta}{2}$$

12. Check Problem 3 using Mollweide's equation from Problem 11.

APPLICATIONS
13. *Geometry.* Two adjacent sides of a parallelogram meet at an angle of 35°10′ and have lengths of 3 and 8 feet. What is the length of the shortest diagonal of the parallelogram (to three significant digits)?
14. *Geometry.* What is the length of the longest diagonal of the parallelogram in Problem 13 (to three significant digits)?
15. *Navigation.* Los Angeles and Las Vegas are approximately 200 miles apart. A pilot 80 miles from Los Angeles finds that she is 6°20′ off course relative to her start in Los Angeles. How far is she from Las Vegas at this time? (Compute the answer to three significant digits.)
16. *Search and rescue.* At noon two search planes set out from San Francisco to find a downed plane in the ocean. Plane A travels due west at 400 miles/hour and plane B northwest at 500 miles/hour. At 2 PM plane A spots the survivors of the downed plane and radios plane B to come and assist in the rescue. How far is plane B from plane A at this time (to three significant digits)?
17. *Geometry.* Find the perimeter of a pentagon inscribed in a circle of radius 12.6 meters.
18. *Engineering.* Three circles of radius 2.03, 5.00, and 8.20 centimeters are tangent to one another (see the figure). Find to the nearest 10′ the three angles formed by the lines joining their centers.

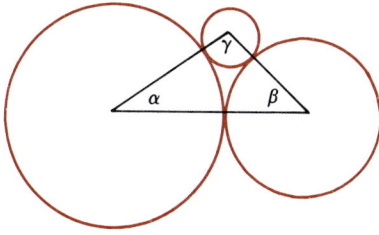

Section 7-4 Vectors

- Geometric Vectors and Their Addition
- Velocity Vectors
- Force Vectors
- Resolution of Vectors into Components
- Static Equilibrium

Many physical quantities, such as length, area, or volume, can be completely specified by a single real number. Other quantities, such as directed distances, velocities, and forces, require for their complete specification both a magnitude and a direction. The former are often called **scalar quantities** and the latter are called **vector quantities**.

In this book we limit our discussion to the intuitive idea of geometric vectors in a plane. In a more advanced treatment of the subject, geometric vectors become a special case of a more general type of vector. Vector ideas are very useful in many branches of science and mathematics.

- ### Geometric Vectors and Their Addition

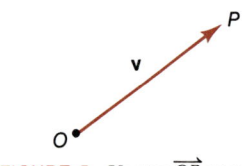

FIGURE 5 Vector $\overrightarrow{OP}$ or **v**

Formally, we define a **geometric vector in a plane** to be a directed line segment from a fixed point O to a point P in the plane. We denote the vector by $\overrightarrow{OP}$ or a single boldface letter, such as **v** (see Fig. 5). If you use a single letter for a vector when writing, place an arrow over it or a line under it, since it is difficult to write in boldface.

The **magnitude** of the vector $\overrightarrow{OP}$ is the length of the line segment from O to P and is denoted by $|\overrightarrow{OP}|$ or $|\mathbf{v}|$. The **direction** of the vector $\overrightarrow{OP}$ is the direction of the directed line segment from O to P. The **zero vector** is denoted by **0** and has magnitude zero and an arbitrary direction. Two vectors are **equal** if and only if they have the same magnitude and direction.*

The **sum of two vectors u and v** with different directions is the diagonal of the parallelogram formed using **u** and **v** as adjacent sides [see Fig. 6(a)]. The diagonal **u + v** is also called the **resultant** of the two vectors **u** and **v**, and **u** and **v** are called **components** of **u + v**.

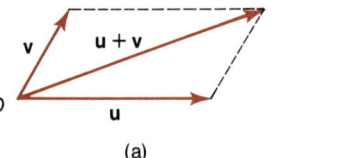
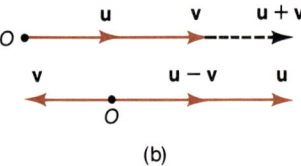

FIGURE 6 Vector addition   (a)   (b)

---

* Some prefer to say that two vectors are **equivalent** if they have the same magnitude and direction and to reserve equality only if they also have the same origin.

7-4 Vectors

If the two vectors **u** and **v** have the same direction, then their sum is a vector with this direction and magnitude $|\mathbf{u}| + |\mathbf{v}|$. If **u** and **v** have opposite directions, then their sum is a vector with magnitude $||\mathbf{u}| - |\mathbf{v}||$ and direction the same as the component with the largest magnitude. These last two cases are illustrated in Figure 6(b).

■ **Velocity Vectors**

A vector that represents the direction and speed of an object in motion is called a **velocity vector**.

EXAMPLE 10 An airplane has a compass heading of 90° (east) and an air speed of 120 miles/hour. The wind is blowing from the north (0°) to the south (180°) at 90 miles/hour. A pilot's velocity relative to the air is called **apparent velocity**, and his or her velocity relative to the ground is called **resultant** or **actual velocity**. The resultant velocity is the vector sum of the apparent velocity and the wind velocity. Find the resultant velocity; that is, find the actual speed and direction of the airplane relative to the ground.

Solution Let geometric vectors represent the various velocities as indicated in the figure. [*Note:* A navigational compass is marked clockwise in degrees starting at north.]

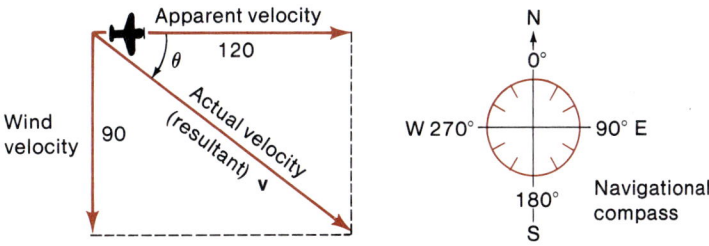

Using the Pythagorean theorem, we find the magnitude of the resultant vector to be

$$|\mathbf{v}| = \sqrt{120^2 + 90^2} = \sqrt{22{,}500} = 150 \text{ miles/hour}$$

To find θ, we see that

$$\tan \theta = \tfrac{90}{120}$$

Thus,

$$\theta = \tan^{-1} \tfrac{90}{120} \approx 37°$$

Actual heading $= 90° + 37° = 127°$

PROBLEM 10 A river is flowing east (90°) at 3 miles/hour. A boat crosses the river with a compass heading of 180° (south). If the speedometer on the boat reads 4 miles/hour, what are the boat's actual speed and direction (resultant velocity) relative to the river bottom?

■ Force Vectors

A vector that represents the direction and magnitude of an applied force is called a **force vector**. If an object is subjected to two forces, then the vector sum of these two forces produces a resultant force that affects the object in the same way as the two original forces taken together.

EXAMPLE 11 Two forces of 30 and 70 pounds act on a point in a plane. If the angle between the force vectors is 40°, what are the magnitude and direction (relative to the 70-pound force) of the resultant force?

Solution We start with a diagram, letting geometric vectors represent the various forces:

Because adjacent angles in a parallelogram are supplementary, $OAB = 180° - 40° = 140° = OCB$. We can now find the magnitude of the resultant vector **R** using the law of cosines.

$$|\mathbf{R}|^2 = 30^2 + 70^2 - 2(30)(70)\cos 140°$$
$$|\mathbf{R}| = \sqrt{30^2 + 70^2 - 2(30)(70)\cos 140°}$$
$$= 95 \text{ pounds} \quad \text{To two significant digits}$$

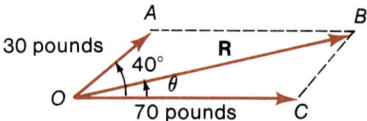

To find θ, the direction of **R**, we use the law of sines.

$$\frac{\sin \theta}{30} = \frac{\sin 140°}{95}$$
$$\sin \theta = \frac{30 \sin 140°}{95}$$

$$\theta = \sin^{-1} \frac{30 \sin 140°}{95} \approx 12° \quad \text{To the nearest degree}$$

PROBLEM 11 Repeat Example 11 using an angle of 100° between the two forces.

Resolution of Vectors into Components

Instead of adding vectors, many problems require the breaking down of vectors into components. Whenever a vector is expressed as a resultant of two vectors, these two vectors are called **components** of the given vector. For example, to find the **horizontal** and **vertical components of the vector v** in Figure 7, we find the magnitudes (the directions of the horizontal and vertical lines are already known) of these components using the sine and cosine functions.

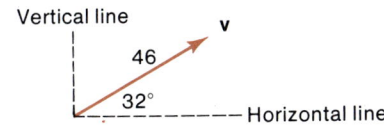

FIGURE 7

FIGURE 8

a is the horizontal component of **v**
b is the vertical component of **v**

The magnitude of the vertical component of **v** (see Figure 8) is:

$$\sin 32° = \frac{|\mathbf{h}|}{46}$$

$$|\mathbf{h}| = 46 \sin 32° = 24$$

Static Equilibrium

Now let us see how the component concept can be used to solve certain physics–engineering problems. We start with two basic ideas regarding forces and objects subjected to these forces:

1. A body at rest is said to be in **static equilibrium**.
2. For a body to remain in static equilibrium in a plane, it is necessary that the sum of the horizontal components and the sum of the vertical components of all forces acting on the body each be 0.

EXAMPLE 12 A weight of 1,000 pounds is suspended from two ropes as indicated in the figure. What is the tension on each rope?

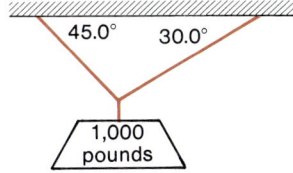

Solution   We introduce geometric vectors for each force involved. Let M_1 be the magnitude of the tension in the rope on the left and M_2 the magnitude of the tension in the rope on the right. We obtain the following corresponding force vector diagram:

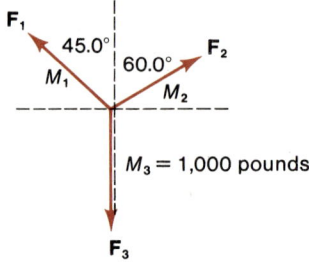

The magnitudes are:

Horizontal component of $\mathbf{F}_1 = M_1 \sin 45.0°$
Horizontal component of $\mathbf{F}_2 = M_2 \sin 60.0°$
Vertical component of $\mathbf{F}_1 = M_1 \cos 45.0°$
Vertical component of $\mathbf{F}_2 = M_2 \cos 60.0°$

For the system to be in static equilibrium, we must have

$-(\sin 45°)M_1 + (\sin 60°)M_2 = 0$ Horizontal
$(\cos 45°)M_1 + (\cos 60°)M_2 = 1{,}000$ Vertical

Solving this system of two equations and two unknowns (this process is reviewed in Section 9-1), we find that

$M_1 \approx 897$ pounds   and $M_2 = 732$ pounds

PROBLEM 12   A 400-pound sign is suspended as shown in the figure. Find the magnitudes of the forces in the rigid supporting members; that is, find $M_1 = |\mathbf{F}_1|$ and $M_2 = |\mathbf{F}_2|$ in the force diagram. Compute the answers to three significant digits.

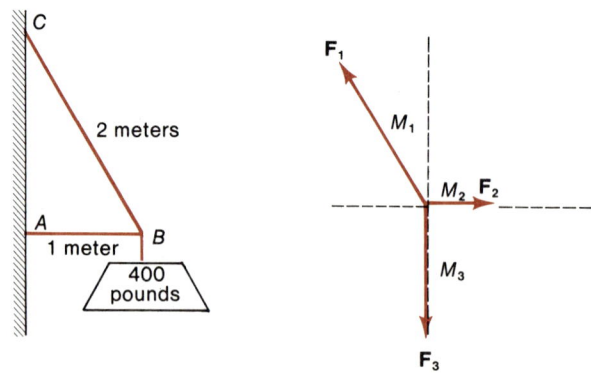

7-4 Vectors

EXAMPLE 13 A car weighing 3,000 pounds is on a driveway inclined 20°0′ to the horizontal. Neglecting friction, what force parallel to the driveway will keep the car from rolling downhill?

Solution As before, we draw a vector diagram:

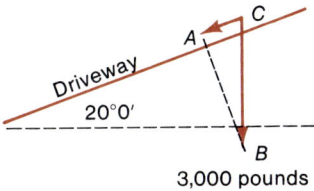

We find the component of $\vec{CB}$ parallel to the driveway—that is, the magnitude of $\vec{AC}$.

Angle $ABC = 20°0′$ (Why?)
$|\vec{CA}| = 3{,}000 \sin 20°$
$= 1{,}030$ pounds   To three significant digits

PROBLEM 13 What is the force (due to the weight of the car) perpendicular to the driveway?

Answers to Matched Problems
10. Resultant velocity: magnitude = 5 miles/hour, direction = 143°
11. $|\mathbf{R}| = 71$ pounds, $\theta = 25°$
12. Member AB has a compression force of 231 pounds; member BC has a tension force of 462 pounds.
13. 2,820 pounds

Exercise 7-4 ■ *Express all angles in decimal degrees. In any of the problems dealing with navigation, remember that a compass is divided clockwise into degrees starting at north.*

A In Problems 1–4, find the magnitudes, H and V, of the horizontal and vertical components, respectively, of the vector **v**, given $|\mathbf{v}|$ and θ.

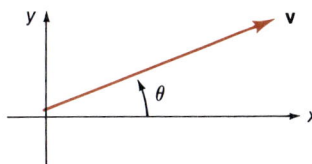

1. $|\mathbf{v}| = 30$ pounds, $\theta = 20°$
2. $|\mathbf{v}| = 250$ pounds, $\theta = 65°$
3. $|\mathbf{v}| = 20$ knots, $\theta = 70°$
4. $|\mathbf{v}| = 600$ miles/hour, $\theta = 43°$

5. A plane, after flying 2.5 hours at 800 miles/hour, with a true heading of 30°, is forced down. Determine how far north and how far east the plane is from its starting point; that is, resolve the displacement vector into horizontal and vertical components.

6. A force of 76 pounds acts at an angle of 60° on an object located at the origin of a rectangular coordinate system. What two forces in the x and y directions will have an equivalent effect on the object?

B In Problems 7–10, find M_3 and β, given M_1, M_2, and α.

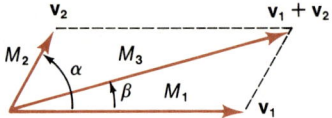

7. $M_1 = 60$ pounds, $M_2 = 20$ pounds, $\alpha = 70°$
8. $M_1 = 120$ pounds, $M_2 = 80$ pounds, $\alpha = 40°$
9. $M_1 = 20$ miles/hour, $M_2 = 3$ miles/hour, $\alpha = 53°$
10. $M_1 = 8$ knots, $M_2 = 2$ knots, $\alpha = 68°$

11. A boat capable of traveling 12 miles/hour on still water maintains a westward compass reading while crossing a river. If the river is flowing southward at 4.0 miles/hour, what is the velocity of the boat with respect to land? (Remember, velocity is a vector quantity requiring both a direction and a magnitude.)

12. A plane, after flying 3 hours at an air speed of 400 miles/hour with a compass heading of 300°, is forced down. If wind during the flight was blowing a steady 60 miles/hour from the south, find the position vector for the downed aircraft relative to the starting position.

13. An automobile weighing 4,000 pounds is standing on a smooth driveway that is inclined 5.0° with the horizontal. Find the force parallel to the driveway necessary to keep the car from rolling down the hill. Neglect all friction.

14. In Problem 13, what is the force perpendicular to the driveway?

15. A weight of 500 pounds is supported by two cables as illustrated. What is the tension in each cable?

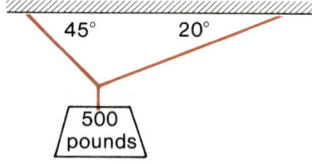

16. A weight of 300 pounds is supported by two poles as illustrated. What is the magnitude of the compression force in each pole?

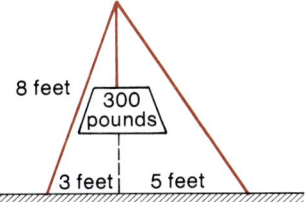

17. A weight of 5,000 pounds is supported as indicated in the figure. What are the magnitudes of the forces on the members AB and BC?

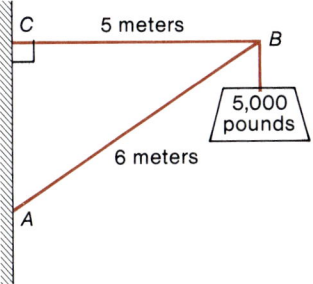

18. A weight of 1,000 pounds is supported as indicated in the figure. What are the magnitudes of the forces on the members AB and BC?

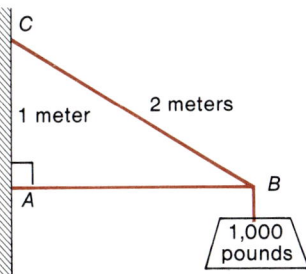

C 19. An airplane can cruise at 250 miles/hour in still air. If a steady wind of 40.0 miles/hour is blowing from the west, what compass heading should the pilot fly for his true course to be north (0°)? Compute the ground speed for this course.

20. Two docks are directly opposite each other on a southward flowing river. A boat pilot wishes to go in a straight line from the east dock to the west dock in a boat with a cruising speed of 8.0 knots on still water. If the river's current is 2.5 knots, what compass heading should be maintained while crossing the river? What is the actual speed of the boat relative to land?

21. A sailboat skipper is preparing for a race on San Francisco Bay. One leg of the race, between two buoys, has a compass bearing of 10°. If a 4-knot current is expected to flow in the direction of 270° and if the knotmeter on the boat is expected to read 6 knots (speed relative to water), what compass heading should the skipper sail to go from one buoy to the other in a straight line? What will be the actual (ground) speed of the boat for this course?

22. If two weights fastened together are placed on inclined planes as indicated in the figure, neglecting friction, which way will they slide?

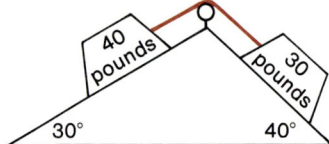

Section 7-5 Polar and Rectangular Coordinates

- Polar Coordinate System
- From Polar Form to Rectangular Form and Vice Versa

Until now we have had only one way of associating ordered pairs of numbers with points in a plane—namely, the rectangular coordinate system. On occasion, it is helpful to have available other coordinate systems. Of the many that are possible, the polar coordinate system ranks second in importance to the rectangular coordinate system.

- Polar Coordinate System

To form a polar coordinate system in a plane (see Fig. 9), we start with a fixed point O and call it the **pole** or **origin**. From this point we draw a half-line (usually horizontal and to the right) and call this line the **polar axis**.

If P is an arbitrary point in a plane, then we associate polar coordinates (r, θ) with it as follows: Starting with the polar axis as the initial side of an angle, we rotate the terminal side until it, or the extension of

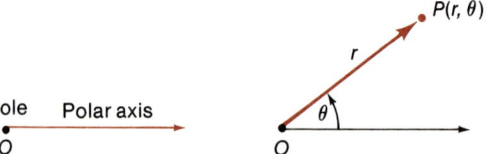

FIGURE 9 Polar coordinate system

it through the pole, passes through the point. The θ coordinate in (r, θ) is this angle, in degree or radian measure. The angle θ is positive if the rotation is counterclockwise and negative if the rotation is clockwise. The r coordinate in (r, θ) is the directed distance from the pole to the point P. It is positive if measured from the pole along the terminal side of θ and negative if measured along the terminal side extended through the pole. Figure 10 illustrates a point P with three different sets of polar coordinates. Study this figure carefully. The pole has polar coordinates $(0, \theta)$ for arbitrary θ. For example, $(0, 0°)$, $(0, \pi/3)$, and $(0, -371°)$ are all coordinates of the pole.

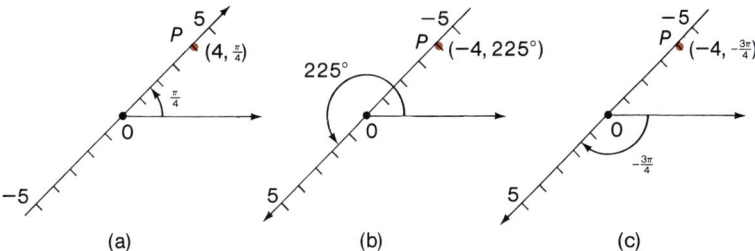

FIGURE 10

We now see a distinct difference between rectangular and polar coordinates for the same point. For a given point in a rectangular coordinate system, there exists exactly one set of rectangular coordinates. On the other hand, in a polar coordinate system a point has infinitely many sets of polar coordinates.

Just as graph paper with a rectangular grid is readily available for plotting rectangular coordinates, polar graph paper is available for plotting polar coordinates.

EXAMPLE 14 Plot the following points in a polar coordinate system:

(A) $A(3, 30°)$, $B(-8, 180°)$, $C(5, -135°)$, $D(-10, -45°)$
(B) $A(5, \pi/3)$, $B(-6, 5\pi/6)$, $C(7, -\pi/2)$, $D(-4, -\pi/6)$

Solution

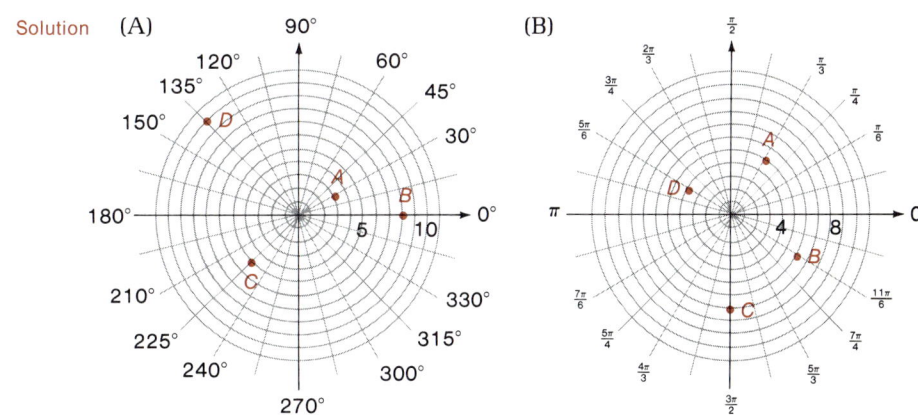

PROBLEM 14 Plot the following points in a polar coordinate system:

(A) $A(9, 45°)$, $B(-6, 150°)$, $C(5, -210°)$, $D(-7, -90°)$
(B) $A(10, \pi/6)$, $B(-4, -\pi)$, $C(-8, 7\pi/4)$, $D(6, -5\pi/6)$

EXAMPLE 15 For the point $(5, 30°)$, find three other sets of polar coordinates such that $-360° \leq \theta \leq 360°$.

Solution

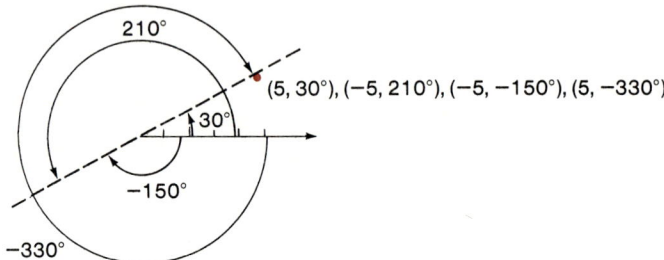

$(5, 30°), (-5, 210°), (-5, -150°), (5, -330°)$

PROBLEM 15 For the point $(8, \pi/4)$, find three other sets of polar coordinates such that $-2\pi \leq \theta \leq 2\pi$.

■ **From Polar Form to Rectangular Form and Vice Versa**

Often it is convenient to be able to transform coordinates or equations in rectangular form to polar form or vice versa. The following polar–rectangular relationships are useful in the process:

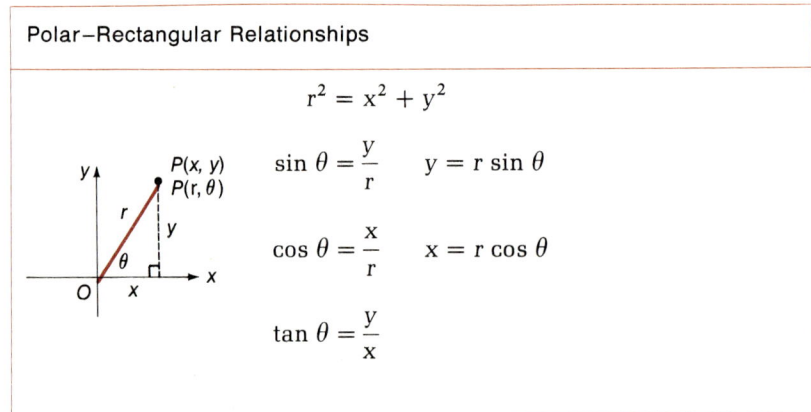

Polar–Rectangular Relationships

$$r^2 = x^2 + y^2$$

$$\sin \theta = \frac{y}{r} \qquad y = r \sin \theta$$

$$\cos \theta = \frac{x}{r} \qquad x = r \cos \theta$$

$$\tan \theta = \frac{y}{x}$$

EXAMPLE 16 Change $A(4, \pi/4)$, $B(-2, 5\pi/6)$, and $C(-6, -\pi/4)$ to rectangular coordinates.

7-5 Polar and Rectangular Coordinates

Solution   Use $x = r \cos \theta$ and $y = r \sin \theta$.

For A
$$x = 4 \cos \frac{\pi}{4} = 4 \cdot \frac{\sqrt{2}}{2} = 2\sqrt{2}$$

$$y = 4 \sin \frac{\pi}{4} = 4 \cdot \frac{\sqrt{2}}{2} = 2\sqrt{2}$$

The rectangular coordinates are $(2\sqrt{2}, 2\sqrt{2})$.

For B
$$x = -2 \cos \left(\frac{5\pi}{6}\right) = -2 \left(\frac{-\sqrt{3}}{2}\right) = \sqrt{3}$$

$$y = -2 \sin \left(\frac{5\pi}{6}\right) = -2 \left(\frac{1}{2}\right) = -1$$

The rectangular coordinates are $(\sqrt{3}, -1)$.

For C
$$x = -6 \cos \left(\frac{-\pi}{4}\right) = -6 \left(\frac{\sqrt{2}}{2}\right) = -3\sqrt{2}$$

$$y = -6 \sin \left(\frac{-\pi}{4}\right) = -6 \left(\frac{-\sqrt{2}}{2}\right) = 3\sqrt{2}$$

The rectangular coordinates are $(-3\sqrt{2}, 3\sqrt{2})$.

PROBLEM 16 Change $A(2, 60°)$, $B(-8, 210°)$, and $C(-6, -30°)$ to rectangular coordinates.

EXAMPLE 17 Change $(-\sqrt{3}, 1)$ to polar form with $r \geq 0$ and $0 \leq \theta < 2\pi$.

Solution

Find r
$$r^2 = x^2 + y^2$$
$$= (-\sqrt{3})^2 + 1^2 = 4$$
$$r = 2$$

Find θ
$$\cos \theta = -\frac{\sqrt{3}}{2}$$

$$\theta = \cos^{-1}\left(\frac{-\sqrt{3}}{2}\right) = \frac{5\pi}{6}$$

The polar coordinates are $(2, 5\pi/6)$.

PROBLEM 17 Change $(1, -\sqrt{3})$ to polar form with $r \geq 0$ and $-90° \leq \theta \leq 90°$.

EXAMPLE 18 Change $x^2 + y^2 - 4y = 0$ to polar form.

Solution   Use $r^2 = x^2 + y^2$ and $y = r \sin \theta$.

$$x^2 + y^2 - 4y = 0$$
$$r^2 - 4r \sin \theta = 0$$
$$r(r - 4 \sin \theta) = 0$$
$$r = 0 \quad \text{or} \quad r - 4 \sin \theta = 0$$

The graph of $r = 0$ is the pole. Because the pole is included in the graph of $r - 4 \sin \theta = 0$ (let $\theta = 0$), we can discard $r = 0$ and keep only

$$r - 4 \sin \theta = 0$$

or

$$r = 4 \sin \theta$$

PROBLEM 18   Change $x^2 + y^2 - 6x = 0$ to polar form.

EXAMPLE 19   Change $r = -3 \cos \theta$ to rectangular form.

Solution   The transformation of this equation (as it stands) into rectangular form is a little difficult. With a little trick, however, it becomes easy. We multiply both sides by r, which simply adds the pole to the graph. But the pole is already part of the graph of $r = -3 \cos \theta$ (let $\theta = \pi/2$), so we have not actually changed anything to do this.

$$r = -3 \cos \theta \quad \text{Multiply both sides by } r.$$
$$r^2 = -3r \cos \theta \quad r^2 = x^2 + y^2 \text{ and } r \cos \theta = x$$
$$x^2 + y^2 = -3x$$
$$x^2 + y^2 + 3x = 0$$

PROBLEM 19   Change $r + 2 \sin \theta = 0$ to rectangular form.

Answers to Matched Problems

14. (A) 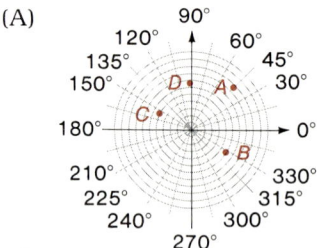 (B)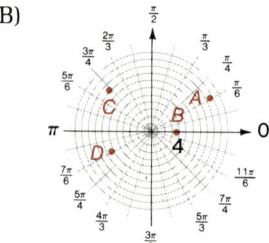

15. $(-8, 5\pi/4), (-8, -3\pi/4), (8, -7\pi/4)$
16. $(1, \sqrt{3}), (4\sqrt{3}, 4), (-3\sqrt{3}, 3)$ 17. $(2, -60°)$
18. $r = 6 \cos \theta$ 19. $x^2 + y^2 + 2y = 0$

7-5 Polar and Rectangular Coordinates

Exercise 7-5 ■ **A** *Plot in a polar coordinate system.*

1. $A(4, 0°)$, $B(7, 180°)$, $C(9, 45°)$
2. $A(8, 0°)$, $B(5, 90°)$, $C(6, 30°)$
3. $A(-4, 0°)$, $B(-7, 180°)$, $C(-9, 45°)$
4. $A(-8, 0°)$, $B(-5, 90°)$, $C(-6, 30°)$
5. $A(8, -45°)$, $B(6, -60°)$, $C(4, -30°)$
6. $A(5, -30°)$, $B(4, -45°)$, $C(9, -90°)$
7. $A(-8, -45°)$, $B(-6, -60°)$, $C(-4, -30°)$
8. $A(-5, -30°)$, $B(-4, -45°)$, $C(-9, -90°)$
9. $A(8, \pi/3)$, $B(4, \pi/4)$, $C(10, 0)$
10. $A(6, \pi/6)$, $B(5, \pi/2)$, $C(8, \pi/4)$
11. $A(-8, \pi/3)$, $B(-4, \pi/4)$, $C(-10, 0)$
12. $A(-6, \pi/6)$, $B(-5, \pi/2)$, $C(-8, \pi/4)$
13. $A(8, -\pi/3)$, $B(4, -\pi/4)$, $C(10, -\pi/6)$
14. $A(6, -\pi/6)$, $B(5, -\pi/2)$, $C(8, -\pi/4)$
15. $A(-6, -\pi/6)$, $B(-5, -\pi/2)$, $C(-8, -\pi/4)$
16. $A(-6, -\pi/2)$, $B(-5, -\pi/3)$, $C(-8, -\pi/4)$

Change to rectangular coordinates.

17. $(4, \pi/4)$
18. $(8, \pi/3)$
19. $(-8, \pi)$
20. $(-9, \pi/2)$
21. $(-6, \pi/6)$
22. $(-4, \pi/4)$
23. $(8, 7\pi/6)$
24. $(10, 5\pi/6)$
25. $(4, -7\pi/4)$
26. $(6, -7\pi/6)$
27. $(-5, -\pi/3)$
28. $(-4, -\pi/6)$

Change to polar coordinates with $r \geq 0$ and $0 \leq \theta \leq 2\pi$.

29. $(3, 3\sqrt{3})$
30. $(2\sqrt{3}, 2)$
31. $(-6\sqrt{3}, 6)$
32. $(-4\sqrt{2}, 4\sqrt{2})$
33. $(5\sqrt{2}, -5\sqrt{2})$
34. $(-4, -4\sqrt{3})$
35. $(-10, 0)$
36. $(0, -7)$

B *Plot in a polar coordinate system.*

37. $A(9, 120°)$, $B(-9, 120°)$, $C(-9, -120°)$
38. $A(5, 210°)$, $B(-5, 210°)$, $C(-5, -210°)$
39. $A(6, 4\pi/3)$, $B(-6, 4\pi/3)$, $C(-6, -4\pi/3)$
40. $A(7, 7\pi/4)$, $B(-7, 7\pi/4)$, $C(-7, -7\pi/4)$

Change to polar form.

41. $y^2 = 5y - x^2$
42. $6x - x^2 = y^2$

43. $3x - 5y = -2$ 44. $2x + 3y = 5$
45. $y = x$ 46. $x^2 + y^2 = 9$

Change to rectangular form.

47. $r(3 \cos \theta - 4 \sin \theta) = -1$ 48. $r(2 \cos \theta + \sin \theta) = 4$
49. $r = -2 \sin \theta$ 50. $r = 8 \cos \theta$
51. $\theta = \pi/4$ 52. $r = 4$

C 53. Change $(y - 3)^2 = 4(x^2 + y^2)$ into polar form.
54. Change $r = 3/(\sin \theta - 2)$ into rectangular form.

Section 7-6 Sketching Polar Graphs

- Point-by-Point Plotting
- Rapid Sketching
- Application

- **Point-by-Point Plotting**

To graph an equation such as

$r = 8 \cos \theta$

in a polar coordinate system, we locate all points with coordinates that satisfy the equation. An approximation of the graph is found (as in rectangular coordinates) by making a table of values that satisfy the equation, plotting these, and then joining the points with a smooth curve. We can use special angles, a calculator, or a table. In Example 20, we use special angles.

EXAMPLE 20 Graph $r = 8 \cos \theta$.

Solution Form a table using special angles in radian measure, and continue until the graph repeats.

θ	0	$\pi/6$	$\pi/4$	$\pi/3$	$\pi/2$	$2\pi/3$	$3\pi/4$	$5\pi/6$	π
r	8	$4\sqrt{3}$	$4\sqrt{2}$	4	0	-4	$-4\sqrt{2}$	$-4\sqrt{3}$	-8

7-6 Sketching Polar Graphs

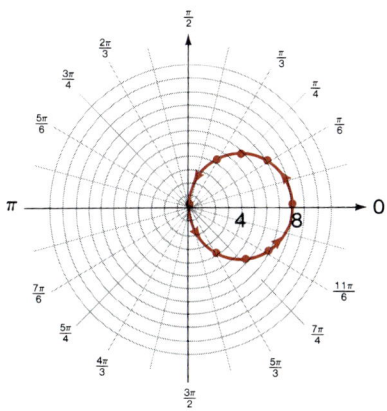

PROBLEM 20 Graph $r = 8 \sin \theta$.

EXAMPLE 21 Repeat Example 20, using a calculator and θ in degrees.

Solution

θ	0°	30°	45°	60°	90°	120°	135°	150°	180°
r	8	6.9	5.7	4	0	-4	-5.7	-6.9	-8

The graph is the same as that in Example 20, except degrees are marked around the polar coordinate system instead of radians. For accurate graphing, a calculator is of considerable help; where the graph is unclear, it is easy, using a calculator, to compute intermediate values to clarify regions of uncertainty.

PROBLEM 21 Complete the following table (to one decimal place accuracy) using a calculator for $r = 8 \sin \theta$.

θ	0°	15°	30°	45°	60°	75°	90°	105°	120°	135°	150°	165°	180°
r													

■ Rapid Sketching

If only a rough sketch of a polar equation involving $\sin \theta$ or $\cos \theta$ is desired, we can speed up the graphing process described earlier by taking advantage of the uniform variation of $\sin \theta$ and $\cos \theta$ as θ moves through each set of quadrant values. For reference, we can visualize the graphs of $y = \sin x$ and $y = \cos x$ in a rectangular coordinate system (Fig. 11). We can tell at a glance how each function behaves in each of the four quadrants, and we can use this information to sketch certain polar equations without the need for tedious point-by-point plotting. Examples should make the process clear.

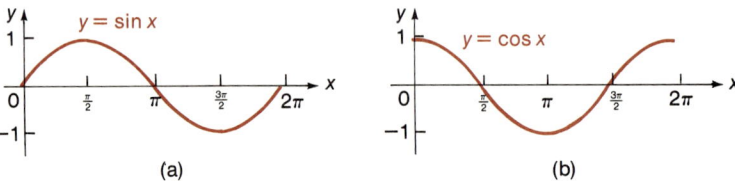

FIGURE 11   (a)   (b)

EXAMPLE 22   Sketch $r = 4 + 4 \cos \theta$ using rapid sketching techniques.

Solution   We set up a table that indicates how r varies as we let θ move through each set of quadrant values:

θ	$\cos \theta$		$4 \cos \theta$		$r = 4 + 4 \cos \theta$
0 to $\pi/2$	1 to	0	4 to	0	8 to 4
$\pi/2$ to π	0 to	-1	0 to	-4	4 to 0
π to $3\pi/2$	-1 to	0	-4 to	0	0 to 4
$3\pi/2$ to 2π	0 to	1	0 to	4	4 to 8

Notice that as θ increases from 0 to $\pi/2$, $\cos \theta$ decreases from 1 to 0, $4 \cos \theta$ decreases from 4 to 0, and $r = 4 + 4 \cos \theta$ decreases from 8 to 4, and so on. Sketching these results, we obtain:

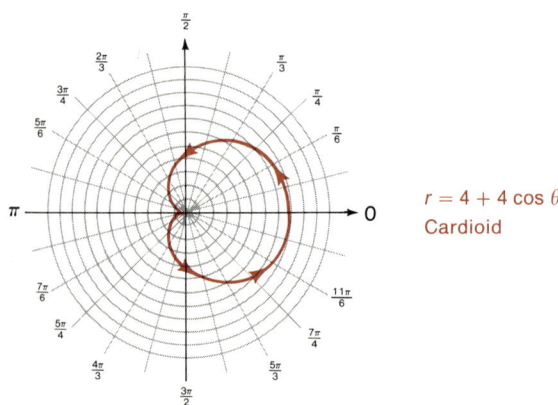

$r = 4 + 4 \cos \theta$
Cardioid

PROBLEM 22 Sketch $r = 5 + 5 \sin \theta$ using rapid sketching techniques.

EXAMPLE 23 Sketch $r = 8 \cos 2\theta$.

Solution We start by letting 2θ (instead of θ) range through each set of quadrant values; that is, we start with values for 2θ in the second column of the table, fill in the table to the right, and then fill in the first column for θ.

┌─── Start with the second column

θ	2θ	$\cos 2\theta$	$r = 8 \cos 2\theta$
0 to $\pi/4$	0 to $\pi/2$	1 to 0	8 to 0
$\pi/4$ to $\pi/2$	$\pi/2$ to π	0 to -1	0 to -8
$\pi/2$ to $3\pi/4$	π to $3\pi/2$	-1 to 0	-8 to 0
$3\pi/4$ to π	$3\pi/2$ to 2π	0 to 1	0 to 8
π to $5\pi/4$	2π to $5\pi/2$	1 to 0	8 to 0
$5\pi/4$ to $3\pi/2$	$5\pi/2$ to 3π	0 to -1	0 to -8
$3\pi/2$ to $7\pi/4$	3π to $7\pi/2$	-1 to 0	-8 to 0
$7\pi/4$ to 2π	$7\pi/2$ to 4π	0 to 1	0 to 8

As 2θ increases from 0 to $\pi/2$, θ increases from 0 to $\pi/4$, and r decreases from 8 to 0. As 2θ increases from $\pi/2$ to π, θ increases from $\pi/4$ to $\pi/2$, and r decreases from 0 to -8, and so on. We continue until the table starts to repeat. Plotting the results, we obtain the following graph:

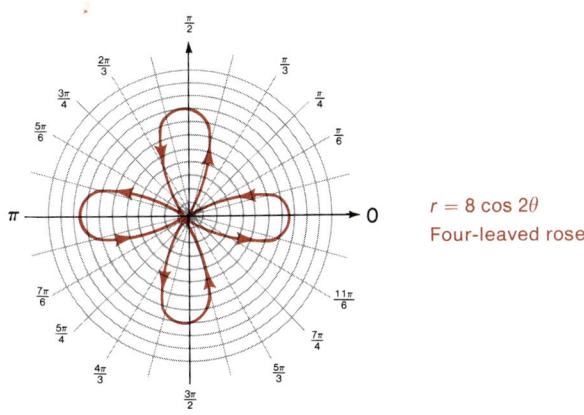

$r = 8 \cos 2\theta$
Four-leaved rose

PROBLEM 23 Sketch $r = 6 \sin 2\theta$.

430   7   Additional Topics in Trigonometry

Two of the simplest types of polar equations to graph in a polar coordinate system are $\theta =$ constant and $r =$ constant. Figure 12 illustrates the graphs of $\theta = \pi/4$ and $r = 5$.

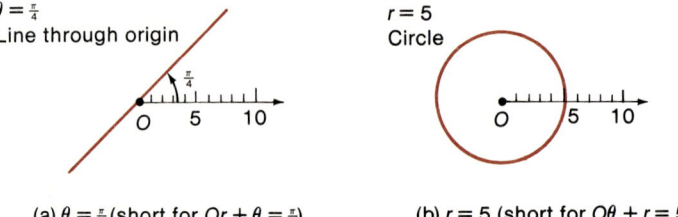

FIGURE 12   (a) $\theta = \frac{\pi}{4}$ (short for $Or + \theta = \frac{\pi}{4}$)   (b) $r = 5$ (short for $O\theta + r = 5$)

■ Application

Serious sailboat racers make polar plots of boat speed at various angles to the wind with various sail combinations at different wind speeds. With many polar plots for different sizes and types of sails at different wind speeds, they are able to accurately choose a sail for the optimum performance for different points of sail relative to any given wind strength. Figure 13 illustrates one such polar plot.

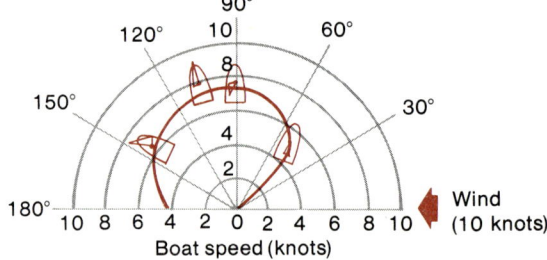

FIGURE 13   Polar diagram showing optimum sailing speed (in knots) at different sailing angles to the wind. The maximum speed appears to be about 7.5 knots at 105° off the wind (with spinnaker sail set).

Answers to Matched Problems   **20.** $r = 8 \sin \theta$

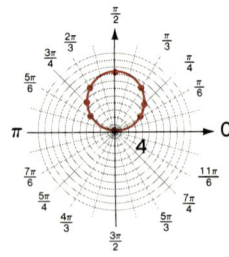

7-6 Sketching Polar Graphs   431

21. $r = 8 \sin \theta$

θ	0°	15°	30°	45°	60°	75°	90°	105°	120°	135°	150°	165°	180°
r	0	2.1	4	5.7	6.9	7.7	8	7.7	6.9	5.7	4	2.1	0

22. $r = 5 + 5 \sin \theta$
Cardioid

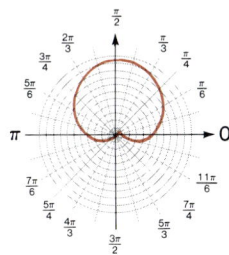

23. $r = 6 \sin 2\theta$
Four-leaved rose

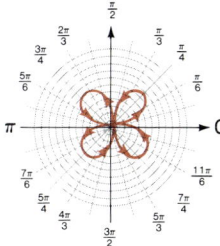

Exercise 7-6 ■ A

Graph Problems 1 and 2 using the special values 0, $\pi/6$, $\pi/4$, $\pi/3$, $\pi/2$, $2\pi/3$, $3\pi/4$, $5\pi/6$, and π. Join the resulting points with a smooth curve.

1. $r = 10 \sin \theta$ **2.** $r = 10 \cos \theta$

Graph Problems 3 and 4 for $0° \leq \theta \leq 360°$, using multiples of 30° starting at 0°. Use a calculator or special angle properties.

3. $r = 4 + 4 \sin \theta$ **4.** $r = 3 + 3 \cos \theta$

Graph Problems 5 and 6 for $0 \leq \theta \leq 2\pi$, using multiples of $\pi/2$ for 5 and $\pi/6$ for 6.

5. $r = \theta/2$ **6.** $r = \theta$

Graph each in a polar coordinate system.

7. $r = 8$ **8.** $r = 5$ **9.** $\theta = \pi/3$ **10.** $\theta = \pi/6$

B Use rapid sketching techniques to sketch each polar equation.

11. $r = 4 \sin \theta$
12. $r = 4 \cos \theta$
13. $r = 10 \sin 2\theta$
14. $r = 8 \cos 2\theta$
15. $r = 5 \cos 3\theta$
16. $r = 6 \sin 3\theta$
17. $r = 2 + 2 \sin \theta$
18. $r = 3 + 3 \cos \theta$
19. $r = 2 + 4 \sin \theta$
20. $r = 2 + 4 \cos \theta$
21. $r = 4 - 2 \cos \theta$
22. $r = 4 - 2 \sin \theta$
23. $r\theta = \pi, \quad \theta \geq 0$
24. $r = \theta/\pi, \quad \theta \geq 0$

C Use rapid sketching techniques to sketch the graph of each polar equation in Problems 25–28.

25. $r = 5 + 5 \sin(\theta/2)$
26. $r = 5 + 5 \cos(\theta/2)$
27. $r^2 = 64 \sin 2\theta$
28. $r^2 = 64 \cos 2\theta$

29. Find all ordered pairs of numbers (r, θ), $0 \leq \theta \leq \pi$, that satisfy the following system, and interpret geometrically:

$$r = 2 \cos \theta \quad \text{and} \quad r = 2 \sin \theta$$

[Note: (r_1, θ_1) must satisfy both equations; some points of intersection of the two graphs may have coordinates that do not satisfy both equations.]

30. Using a calculator, graph the equation

$$r = \frac{8}{1 - e \cos \theta}$$

for the values of e given.
(A) $e = \frac{1}{2}$ (B) $e = 1$ (C) $e = 2$

Section 7-7 Complex Numbers in Rectangular and Polar Forms

- Rectangular Form
- Polar Form
- Multiplication and Division in Polar Form

- Rectangular Form

Recall from Section 1-6 that a complex number is any number that can be written in the form

$a + bi$

where a and b are real numbers and i is the imaginary unit. Thus, associated with each complex number $a + bi$ is a unique ordered pair of real numbers

(a, b)

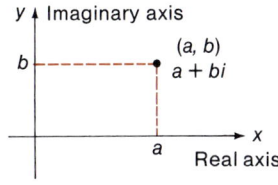

FIGURE 14 Complex plane

Conversely, associated with each ordered pair of real numbers (a, b) is a unique complex number $a + bi$. Associating these ordered pairs of real numbers with points in a rectangular coordinate system, we obtain a **complex plane** (see Fig. 14). When complex numbers are associated with points in a Cartesian coordinate system, we refer to the x axis as the **real axis** and the y axis as the **imaginary axis**. The complex number $a + bi$ is said to be in **rectangular form**.

EXAMPLE 24 Plot the following complex numbers in a complex plane:

$$A = 2 + 3i \qquad B = -3 + 5i \qquad C = -4 \qquad D = -3i$$

Solution

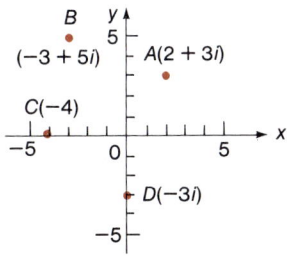

PROBLEM 24 Plot the following complex numbers in a complex plane:

$$A = 4 + 2i \qquad B = 2 - 3i \qquad C = -5 \qquad D = 4i$$

■ Polar Form

Complex numbers can also be written in **polar** (or **trigonometric**) **form**. Using the polar–rectangular relationships from Section 7-5:

$$x = r \cos \theta \quad \text{and} \quad y = r \sin \theta$$

we can write the complex number $z = x + iy$ in polar form as follows:

$$z = x + iy = r \cos \theta = ir \sin \theta = r(\cos \theta = i \sin \theta)$$

This rectangular–polar relationship is illustrated in Figure 15.

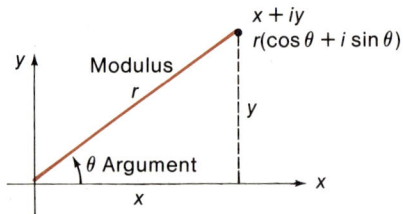

FIGURE 15

Because $\cos\theta$ and $\sin\theta$ are both periodic with period 2π, however, we have

$$\cos(\theta + 2k\pi) = \cos\theta$$
$$\sin(\theta + 2k\pi) = \sin\theta$$

k any integer

Thus, we can write a more general polar form for a complex number $z = x + iy$:

General Polar Form

$$z = x + iy = r[\cos(\theta + 2k\pi) + i\sin(\theta + 2k\pi)]$$
$$= r\,\text{cis}(\theta + 2k\pi) \qquad k \text{ any integer}$$

where

$$\text{cis}\,\theta = \cos\theta + i\sin\theta$$

and the quadrant for θ is determined by x and y.

The number r is called the **modulus** or **absolute value** of z, denoted by **mod z** or $|z|$. The polar angle that the line joining z to the origin makes with the positive x axis is called the **argument** of z and is denoted by **arg z**. From Figure 15 we can see the following relationships:

Modulus and Argument for $z = x + iy$

$$\text{mod } z = r = |z| = \sqrt{x^2 + y^2} \qquad \text{(always nonnegative)}$$
$$\arg z = \theta + 2k\pi \qquad k \text{ any integer}$$

where $\sin\theta = y/r$ and $\cos\theta = x/r$. (We will usually take the smallest positive angle for arg z when writing a complex number in polar form.)

EXAMPLE 25 Write in polar form, using the smallest positive angle for arg z.

(A) $z_1 = 1 + i$ (B) $z_2 = -\sqrt{3} + i$ (C) $z_3 = -5i$

Solution (A) Graph in a rectangular coordinate system first; then if x and y are associated with special angles, we can often determine r and θ by inspection.

$$r = \sqrt{2}$$
$$\theta = \frac{\pi}{4}$$
$$z_1 = \sqrt{2}\left(\cos\frac{\pi}{4} + i\sin\frac{\pi}{4}\right)$$
$$= \sqrt{2} \text{ cis } \frac{\pi}{4}$$

(B)
$$r = 2$$
$$\theta = \frac{5\pi}{6}$$
$$z_2 = 2\left(\cos\frac{5\pi}{6} + i\sin\frac{5\pi}{6}\right)$$
$$= 2 \text{ cis } \frac{5\pi}{6}$$

(C)
$$r = 5$$
$$\theta = \frac{3\pi}{2}$$
$$z_3 = 5\left(\cos\frac{3\pi}{2} + i\sin\frac{3\pi}{2}\right)$$
$$= 5 \text{ cis } \frac{3\pi}{2}$$

PROBLEM 25 Write in polar form:

(A) $z_1 = -1 + i$ (B) $z_2 = 1 - i\sqrt{3}$ (C) $z_3 = -6$

EXAMPLE 26 Write in rectangular form:

(A) $z_1 = 2 \text{ cis } \frac{\pi}{6}$ (B) $z_2 = 5 \text{ cis } \frac{\pi}{2}$ (C) $z_3 = 3 \text{ cis } 300°$

Solution (A) $z_1 = 2 \text{ cis } \dfrac{\pi}{6} = 2 \cos \dfrac{\pi}{6} + i2 \sin \dfrac{\pi}{6}$

$= 2\left(\dfrac{\sqrt{3}}{2}\right) + i2\left(\dfrac{1}{2}\right)$

$= \sqrt{3} + i$

(B) $z_2 = 5 \text{ cis } \dfrac{\pi}{2} = 5 \cos \dfrac{\pi}{2} + i5 \sin \dfrac{\pi}{2}$

$= 5(0) + i5 = 5i$

(C) $z_3 = 3 \text{ cis } 300° = 3 \cos 300° + i3 \sin 300°$

$= 3\left(\dfrac{1}{2}\right) + i3\left(-\dfrac{\sqrt{3}}{2}\right)$

$= \dfrac{3}{2} - \dfrac{3\sqrt{3}}{2}i$

PROBLEM 26 Write in rectangular form:

(A) $z_1 = 3 \text{ cis } \dfrac{2\pi}{3}$ (B) $z_2 = 2 \text{ cis } 210°$ (C) $z_3 = \sqrt{2} \text{ cis } \dfrac{7\pi}{4}$

■ **Multiplication and Division in Polar Form**

We will now see a particular advantage in representing complex numbers in polar form: Multiplication and division become easy. Let us start with multiplication.

$z_1 z_2 = (r_1 \text{ cis } \theta_1)(r_2 \text{ cis } \theta_2)$
$= r_1 r_2 (\cos \theta_1 + i \sin \theta_1)(\cos \theta_2 + i \sin \theta_2)$
$= r_1 r_2 (\cos \theta_1 \cos \theta_2 + i \cos \theta_1 \sin \theta_2$
$\qquad + i \sin \theta_1 \cos \theta_2 - \sin \theta_1 \sin \theta_2)$
$= r_1 r_2 [(\cos \theta_1 \cos \theta_2 - \sin \theta_1 \sin \theta_2)$
$\qquad + i(\cos \theta_1 \sin \theta_2 + \sin \theta_1 \cos \theta_2)]$
$= r_1 r_2 [\cos(\theta_1 + \theta_2) + i \sin(\theta_1 + \theta_2)]$ Addition identities
$= r_1 r_2 \text{ cis}(\theta_1 + \theta_2)$

Thus, to multiply two complex numbers in polar form, we multiply r_1 and r_2 and add θ_1 to θ_2. Similarly, one can show that

$$\dfrac{z_1}{z_2} = \dfrac{r_1 \text{ cis } \theta_1}{r_2 \text{ cis } \theta_2} = \dfrac{r_1}{r_2} \text{ cis}(\theta_1 - \theta_2)$$

7-7 Complex Numbers in Rectangular and Polar Forms

The proof of this quotient form is left to Problem 27 in Exercise 7-7. We summarize both forms for convenient reference.

Products and Quotients in Polar Form

$$z_1 z_2 = (r_1 \text{ cis } \theta_1)(r_2 \text{ cis } \theta_2) = r_1 r_2 \text{ cis}(\theta_1 + \theta_2)$$

$$\frac{z_1}{z_2} = \frac{r_1 \text{ cis } \theta_1}{r_2 \text{ cis } \theta_2} = \frac{r_1}{r_2} \text{ cis}(\theta_1 - \theta_2)$$

EXAMPLE 27 If $z_1 = 8 \text{ cis } 45°$ and $z_2 = 2 \text{ cis } 30°$, find: (A) $z_1 z_2$ (B) z_1/z_2

Solution (A) $z_1 z_2 = r_1 r_2 \text{ cis}(\theta_1 + \theta_2)$
$= 8 \cdot 2 \text{ cis}(45° + 30°)$
$= 16 \text{ cis } 75°$

(B) $\dfrac{z_1}{z_2} = \dfrac{r_1}{r_2} \text{ cis}(\theta_1 - \theta_2)$
$= \tfrac{8}{2} \text{ cis}(45° - 30°)$
$= 4 \text{ cis } 15°$

PROBLEM 27 If $z_1 = 9 \text{ cis } 85°$ and $z_2 = 3 \text{ cis } 25°$, find: (A) $z_1 z_2$ (B) z_1/z_2

Answers to Matched Problems

24.

25. (A) $z_1 = \sqrt{2} \text{ cis}(3\pi/4)$ (B) $z_2 = 2 \text{ cis}(5\pi/3)$
    (C) $z_3 = 6 \text{ cis } \pi$

26. (A) $z_1 = -\dfrac{3}{2} + \dfrac{3\sqrt{3}}{2} i$ (B) $z_2 = -\sqrt{3} - i$ (C) $z_3 = 1 - i$

27. (A) $z_1 z_2 = 27 \text{ cis } 110°$ (B) $z_1/z_2 = 3 \text{ cis } 60°$

Exercise 7-7   **A**   *Plot each set of complex numbers in a complex plane.*

1. $A = 3 + 4i$, $B = -2 - i$, $C = 2i$
2. $A = 4 + i$, $B = -3 + 2i$, $C = -3i$
3. $A = 3 - 3i$, $B = 4$, $C = -2 + 3i$
4. $A = -3$, $B = -2 - i$, $C = 4 + 4i$
5. $A = 2 \text{ cis } \dfrac{\pi}{3}$, $B = \sqrt{2} \text{ cis } \dfrac{\pi}{4}$, $C = 4 \text{ cis } \dfrac{\pi}{2}$

6. $A = 2 \text{ cis } \dfrac{\pi}{6}$, $\quad B = 4 \text{ cis } \pi$, $\quad C = \sqrt{2} \text{ cis } \dfrac{3\pi}{4}$

7. $A = 4 \text{ cis } 210°$, $\quad B = 3 \text{ cis } 20°$, $\quad C = 5 \text{ cis } 270°$

8. $A = 2 \text{ cis } 150°$, $\quad B = 3 \text{ cis } 310°$, $\quad C = 4 \text{ cis } 75°$

B Plot each set in a complex plane and change to polar form.

9. $z_1 = \sqrt{3} + i$, $\quad z_2 = -1 + i$, $\quad z_3 = 2i \quad$ (θ in radians)

10. $z_1 = -1 + i\sqrt{3}$, $\quad z_2 = -1 - i$, $\quad z_3 = -5 \quad$ (θ in radians)

11. $z_1 = 1 - i\sqrt{3}$, $\quad z_2 = \sqrt{3} + i$, $\quad z_3 = 4 \quad$ (θ in degrees)

12. $z_1 = 1 - i$, $\quad z_2 = -\sqrt{3} - i$, $\quad z_3 = -3i \quad$ (θ in degrees)

Plot each set in a complex plane and change to rectangular form.

13. $z_1 = 2 \text{ cis } \dfrac{\pi}{3}$, $\quad z_2 = \sqrt{2} \text{ cis } \dfrac{\pi}{4}$, $\quad z_3 = 3 \text{ cis } \dfrac{\pi}{2}$

14. $z_1 = 4 \text{ cis } \dfrac{\pi}{6}$, $\quad z_2 = 3 \text{ cis } \dfrac{5\pi}{4}$, $\quad z_3 = 4 \text{ cis } \pi$

15. $z_1 = 5 \text{ cis } 300°$, $\quad z_2 = \text{cis } 90°$, $\quad z_3 = 3 \text{ cis } 210°$

16. $z_1 = 4 \text{ cis } 330°$, $\quad z_2 = 7 \text{ cis } 180°$, $\quad z_3 = 2 \text{ cis } 60°$

17. $z_1 = 2.30 \text{ cis } 37°10'$, $\quad z_2 = 3.40 \text{ cis } 62°40'$

18. $z_1 = 1.70 \text{ cis } 41°20'$, $\quad z_2 = 2.60 \text{ cis } 28°30'$

Find $z_1 z_2$ and z_1/z_2 for each of the following pairs of complex numbers.

19. $z_1 = 7 \text{ cis } 82°$, $\quad z_2 = 2 \text{ cis } 31°$

20. $z_1 = 6 \text{ cis } 132°$, $\quad z_2 = 3 \text{ cis } 93°$

21. $z_1 = 5 \text{ cis } 52°$, $\quad z_2 = 4 \text{ cis } 83°$

22. $z_1 = 3 \text{ cis } 67°$, $\quad z_2 = 2 \text{ cis } 97°$

23. Find $(1 + i)^2$ directly and by converting $(1 + i)$ to polar form first.

24. Find $(1 + i)^3$ directly and by converting $(1 + i)$ to polar form first.

C **25.** Find $[\tfrac{1}{2} + i(\sqrt{3}/2)]^2$ in the form $a + bi$ by first converting $[\tfrac{1}{2} + i(\sqrt{3}/2)]$ to trigonometric form.

26. Find $[\tfrac{1}{2} - i(\sqrt{3}/2)]^3$ by following the same procedures as in Problem 25.

27. Prove: $\dfrac{z_1}{z_2} = \dfrac{r_1 \text{ cis } \theta_1}{r_2 \text{ cis } \theta_2} = \dfrac{r_1}{r_2} \text{cis}(\theta_1 - \theta_2)$

28. If $z = r \text{ cis } \theta$, show that $z^2 = r^2 \text{ cis } 2\theta$ and $z^3 = r^3 \text{ cis } 3\theta$. What do you think z^n is?

Section 7-8 De Moivre's Theorem

- De Moivre's Theorem, *n* a Natural Number
- *n*th Roots of *z*
- De Moivre's Theorem, *n* an Integer

- De Moivre's Theorem, *n* a Natural Number

By repeated application of the product formula discussed in Section 7-7, we can easily obtain natural number powers of complex numbers. For example,

$$(x + yi)^2 = (r \text{ cis } \theta)^2 = r^2 \text{ cis}(\theta + \theta) = r^2 \text{ cis } 2\theta$$
$$(x + yi)^3 = (r \text{ cis } \theta)^3 = (r \text{ cis } \theta)(r^2 \text{ cis } 2\theta) = r^3 \text{ cis } 3\theta$$

In general, we have De Moivre's theorem. This formula requires a method of proof, called *mathematical induction*, which is discussed in Section 11-2.

De Moivre's Theorem

$$z^n = (x + yi)^n = (r \text{ cis } \theta)^n = r^n \text{ cis } n\theta \qquad n \text{ a natural number}$$

EXAMPLE 28 Use De Moivre's theorem to find $(1 + i)^{10}$. Write the answer in the form $x + yi$.

Solution
$$(1 + i)^{10} = (\sqrt{2} \text{ cis } 45°)^{10}$$
$$= (\sqrt{2})^{10} \text{ cis } 10(45°)$$
$$= 32 \text{ cis } 450°$$
$$= 32 \text{ cis } 90° \quad \text{Why?}$$
$$= 32(\cos 90° + i \sin 90°)$$
$$= 32(0 + i)$$
$$= 32i$$

PROBLEM 28 Find $(1 + i\sqrt{3})^5$ and write the answer in the form $x + yi$. Use De Moivre's theorem.

nth Roots of z

Now let us take a look at roots of complex numbers. We say **w is an nth root of z**, n a natural number, if

$$w^n = z$$

For example, if $w^2 = z$, then w is a square root of z; if $w^3 = z$, then w is a cube root of z; and so on. Let us show that if

$$z = r \text{ cis } \theta \qquad (1)$$

then

$$r^{1/2} \text{ cis } \frac{\theta}{2} \qquad (2)$$

is a square root of z. We simply square expression (2), using De Moivre's theorem, to obtain (1):

$$\left(r^{1/2} \text{ cis } \frac{\theta}{2}\right)^2 = (r^{1/2})^2 \text{ cis } 2\left(\frac{\theta}{2}\right)$$

$$= r \text{ cis } \theta$$

We can proceed in the same way to show that $r^{1/n} \text{ cis}(\theta/n)$ is an nth root of $r \text{ cis } \theta$, n a natural number:

$$\left(r^{1/n} \text{ cis } \frac{\theta}{n}\right)^n = (r^{1/n})^n \text{ cis } n\left(\frac{\theta}{n}\right)$$

$$= r \text{ cis } \theta$$

But we can do even better than this. The nth-root theorem shows us how to find *all* the nth roots of a complex number.

---

nth-Root Theorem

$$r^{1/n} \text{ cis}\left(\frac{\theta}{n} + k\frac{360°}{n}\right) \qquad k = 0, 1, \ldots, (n-1)$$

are n distinct nth roots of $z = r \text{ cis } \theta$, and there are no others.

EXAMPLE 29 Find six distinct sixth roots of $z = -1 + i\sqrt{3}$ and graph them.

Solution First write $z = -1 + i\sqrt{3}$ in polar form:

$$z = -1 + i\sqrt{3} = 2 \text{ cis } 120°$$

All six roots are given by

$$2^{1/6} \text{ cis}\left(\frac{120°}{6} + k\frac{360°}{6}\right) = 2^{1/6} \text{ cis}(20° + k \, 60°) \qquad k = 0, 1, 2, 3, 4, 5$$

7-8 De Moivre's Theorem

All roots are easily graphed after the first root is located. The root points are equally spaced around a circle of radius $2^{1/6}$ at an angular increment of $60°$ from one root to the next.

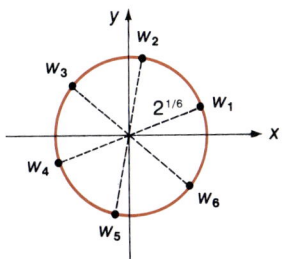

$$w_1 = 2^{1/6}\operatorname{cis}(20° + \mathbf{0}\cdot 60°) = 2^{1/6}\operatorname{cis} 20°$$
$$w_2 = 2^{1/6}\operatorname{cis}(20° + \mathbf{1}\cdot 60°) = 2^{1/6}\operatorname{cis} 80°$$
$$w_3 = 2^{1/6}\operatorname{cis}(20° + \mathbf{2}\cdot 60°) = 2^{1/6}\operatorname{cis} 140°$$
$$w_4 = 2^{1/6}\operatorname{cis}(20° + \mathbf{3}\cdot 60°) = 2^{1/6}\operatorname{cis} 200°$$
$$w_5 = 2^{1/6}\operatorname{cis}(20° + \mathbf{4}\cdot 60°) = 2^{1/6}\operatorname{cis} 260°$$
$$w_6 = 2^{1/6}\operatorname{cis}(20° + \mathbf{5}\cdot 60°) = 2^{1/6}\operatorname{cis} 320°$$

PROBLEM 29 Find five distinct fifth roots of $z = 1 + i$. Leave the answer in polar form.

■ **De Moivre's Theorem, *n* an Integer**

By defining z^0 and z^{-m}, m a natural number, as in real numbers—namely,

$$z^0 = 1 \quad \text{and} \quad z^{-m} = \frac{1}{z^m}$$

De Moivre's theorem can be extended to apply to *all* integers (negative and 0 as well as positive):

De Moivre's Theorem—Extended Form

$$z^n = (x + yi)^n = (r\operatorname{cis}\theta)^n = r^n\operatorname{cis} n\theta \qquad \text{n an integer}$$

The case where $n = 0$ is left to the reader. Let us consider the case in which n is a negative integer, say $n = -m$, where m is a positive integer. Then for $z = r\operatorname{cis}\theta$,

$$z^{-m} = \frac{1}{z^m} \qquad \text{Write 1 and } z \text{ in polar form.}$$

$$= \frac{1\operatorname{cis} 0°}{(r\operatorname{cis}\theta)^m} \qquad \text{Use De Moivre's theorem.}$$

$$= \frac{1\operatorname{cis} 0°}{r^m\operatorname{cis} m\theta} \qquad \text{Use the quotient formula.}$$

$$= r^{-m}\operatorname{cis}(0° - m\theta)$$

$$= r^{-m}\operatorname{cis}(-m\theta) \qquad \text{Simplify.}$$

EXAMPLE 30 Use De Moivre's theorem in the extended form to evaluate the expression $(\sqrt{3} + i)^4/(-1 + i\sqrt{3})^6$. Write the answer in the $x + yi$ form.

Solution
$$\frac{(\sqrt{3} + i)^4}{(-1 + i\sqrt{3})^6} = \frac{(2 \text{ cis } 30°)^4}{(2 \text{ cis } 120°)^6}$$
$$= (2 \text{ cis } 30°)^4 (2 \text{ cis } 120°)^{-6}$$
$$= (2^4 \text{ cis } 4 \cdot 30°)[2^{-6} \text{ cis}(-6 \cdot 120°)]$$
$$= 2^{-2} \text{ cis}(120° - 720°)$$
$$= \tfrac{1}{4} \text{ cis}(-600°)$$
$$= \tfrac{1}{4}[\cos(-600) + i \sin(-600)]$$
$$= \frac{1}{4}\left(-\frac{1}{2} + i\frac{\sqrt{3}}{2}\right)$$
$$= -\frac{1}{8} + i\frac{\sqrt{3}}{8}$$

PROBLEM 30 Use De Moivre's theorem in the extended form to evaluate the expression $(-\sqrt{3} + i)/(1 + i\sqrt{3})^5$. Write the answer in the form $x + yi$.

Answers to Matched Problems
28. $32 \text{ cis } 300° = 16 - i16\sqrt{3}$
29. $w_1 = 2^{1/10} \text{ cis } 9°$, $w_2 = 2^{1/10} \text{ cis } 81°$, $w_3 = 2^{1/10} \text{ cis } 153°$, $w_4 = 2^{1/10} \text{ cis } 225°$, $w_5 = 2^{1/10} \text{ cis } 297°$
30. $-\sqrt{3}/32 - (1/32)i$

Exercise 7-8

A Use De Moivre's theorem to find each of the following. Write the answers in the $x + yi$ form.

1. $(2 \text{ cis } 30°)^3$
2. $(4 \text{ cis } 15°)^3$
3. $(\sqrt{2} \text{ cis } 10°)^6$
4. $(\sqrt{2} \text{ cis } 15°)^8$
5. $(\sqrt{3} + i)^8$
6. $(1 + i\sqrt{3})^3$
7. $\left(-\frac{1}{2} + \frac{\sqrt{3}}{2}i\right)^3$
8. $\left(-\frac{1}{2} - \frac{\sqrt{3}}{2}i\right)^3$

B Use De Moivre's theorem to find each of the following. Write the answers in the $x + yi$ form.

9. $(-\sqrt{3} + i)^5$
10. $(-\sqrt{3} - i)^4$
11. $(1 + i)^{-5}$
12. $(1 + i\sqrt{3})^{-8}$
13. $(-\sqrt{3} + i)^4(1 + i\sqrt{3})^3$
14. $(1 + i)^6(\sqrt{3} - i)^3$
15. $\dfrac{(1 - i\sqrt{3})^3(1 - i)^{-2}}{(1 + i)^4(\sqrt{3} + i)^{-3}}$
16. $\dfrac{(-\sqrt{3} + i)^{-4}(-1 - i)^6}{(-1 + i)^{-2}(-\sqrt{3} - i)^3}$

For n and z as indicated, find all nth roots of z. Leave the answers in trigonometric form.

17. $z = 8 \text{ cis } 30°, \quad n = 3$
18. $z = 8 \text{ cis } 45°, \quad n = 3$
19. $z = 8 \text{ cis } 60°, \quad n = 6$
20. $z = 16 \text{ cis } 90°, \quad n = 4$
21. $z = -1 + i, \quad n = 5$
22. $z = 1 - i, \quad n = 5$
23. $z = 8, \quad n = 3$
24. $z = 1, \quad n = 6$
25. $z = i, \quad n = 6$
26. $z = -i, \quad n = 5$
27. $z = -\sqrt{3} - i, \quad n = 6$
28. $z = 1 - i\sqrt{3}, \quad n = 4$

C 29. Use De Moivre's theorem to show that $r^{1/n} \text{ cis}(\theta/n)$ is an nth root of $z = r \text{ cis } \theta$.

30. Use De Moivre's theorem to show that $r^{1/n} \text{ cis}[(\theta + 2k\pi)/n], k \in I$, are nth roots of n.

31. Find all complex zeros for $P(x) = x^5 - 32$.
32. Find all complex zeros for $P(x) = x^6 + 1$.
33. Solve $x^5 + 1 = 0$ in the set of complex numbers.
34. Solve $x^3 - i = 0$ in the set of complex numbers.
35. Write $P(x) = x^6 + 64$ as a product of linear factors.
36. Write $P(x) = x^6 - 1$ as a product of linear factors.

Section 7-9 Chapter Review

IMPORTANT TERMS AND SYMBOLS

7-1 Solutions of Right Triangles. Right triangle ratios, side opposite, side adjacent, hypotenuse, solve a right triangle

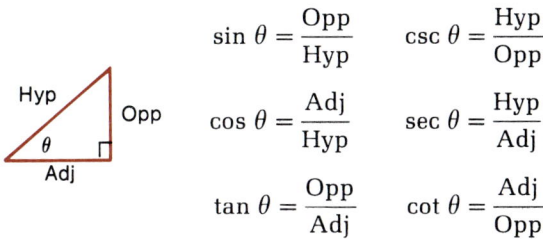

$$\sin \theta = \frac{\text{Opp}}{\text{Hyp}} \qquad \csc \theta = \frac{\text{Hyp}}{\text{Opp}}$$

$$\cos \theta = \frac{\text{Adj}}{\text{Hyp}} \qquad \sec \theta = \frac{\text{Hyp}}{\text{Adj}}$$

$$\tan \theta = \frac{\text{Opp}}{\text{Adj}} \qquad \cot \theta = \frac{\text{Adj}}{\text{Opp}}$$

7-2 Law of Sines. Oblique triangle, acute triangle, obtuse triangle, law of sines

$$\frac{\sin \alpha}{a} = \frac{\sin \beta}{b} = \frac{\sin \gamma}{c}$$

7-3 Law of Cosines. Law of cosines

$$a^2 = b^2 + c^2 - 2bc \cos \alpha$$
$$b^2 = a^2 + c^2 - 2ac \cos \beta$$
$$c^2 = a^2 + b^2 - 2ab \cos \gamma$$

7-4 Vectors. Scalar quantities, vector quantities, geometric vectors, direction, magnitude, zero vector, vector sum, resultant, components, velocity vector, force vector, static equilibrium

7-5 Polar and Rectangular Coordinates. Polar coordinate system, pole, polar axis, polar–rectangular relationships

7-6 Sketching Polar Graphs. Point-by-point plotting, rapid sketching

7-7 Complex Numbers in Rectangular and Polar Forms. Complex plane, real axis, imaginary axis, rectangular form, polar form, trigonometric form, general polar form, modulus, absolute value, argument, multiplication and division in polar form, $|z|$, mod z, arg z

7-8 De Moivre's Theorem. De Moivre's theorem (n a natural number), nth roots of z, De Moivre's theorem (n an integer)

Exercise 7-9 Chapter Review

Work through all the problems in this chapter review and check answers in the back of the book. (Answers to all problems are there, and following each answer is a number in italics indicating the section in which that type of problem is discussed.) Where weaknesses show up, review appropriate sections in the text. When you are satisfied that you know the material, take the practice test following this review.

Problems in this exercise use the following standard labeling of sides and angles:

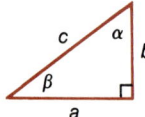 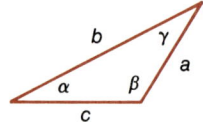

A
1. Solve a right triangle with $\alpha = 54°20'$ and $c = 2.00$.
2. Solve a triangle with $\gamma = 120°10'$, $c = 10.0$, and $b = 4.00$.
3. Find side a if $\alpha = 15°0'$, $b = 9.00$, and $c = 10.0$.
4. Plot in a polar coordinate system: $A(4, 60°)$, $B(9, -\pi/3)$, and $C(-6, 5\pi/6)$.
5. Plot in a rectangular complex plane: $A = 3 + 5i$, $B = -1 - i$, and $C = -3i$.

6. Plot in a polar complex plane: $A = 5 \text{ cis } 30°$, $B = 10 \text{ cis}(3\pi/2)$, and $C = 7 \text{ cis}(3\pi/4)$.
7. Find $[-\frac{1}{2} - (\sqrt{3}/2)i]^3$ using De Moivre's theorem. Write the final answer in the form $x + yi$.
8. Find $(2 \text{ cis } 15°)^4$. Write the final answer in the form $x + yi$.
9. Sketch a graph of $r = 6$ in a polar coordinate system.
10. Find the direction and magnitude of the vector sum $\mathbf{u} + \mathbf{v}$ relative to the horizontal vector.

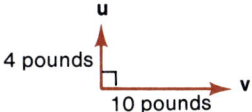

B 11. Find the perimeter of a square inscribed in a circle with radius 5.00.
12. Solve a triangle with $\alpha = 37°40'$, $a = 45.2$, and $c = 63.2$.
13. Solve a triangle with $\alpha = 120°30'$, $b = 4.00$, and $c = 8.00$.
14. Find the vector sum $\mathbf{u} + \mathbf{v}$ (direction and magnitude relative to the horizontal vector) for the figure.

15. Plot in a polar coordinate system: $A(7, 210°)$, $B(-10, 5\pi/3)$, and $C(-4, -7\pi/6)$.
16. Sketch a graph of $r = 8 + 8 \sin \theta$ in a polar coordinate system.
17. Change to polar form: $z_1 = -1 + i$, $z_2 = -1 + i\sqrt{3}$, and $z_3 = 5$.
18. Change to rectangular form: $z_1 = \sqrt{2} \text{ cis}(\pi/4)$, $z_2 = 3 \text{ cis } 210°$, and $z_3 = -2 \text{ cis}(2\pi/3)$.
19. If $z_1 = 6 \text{ cis } 92°$ and $z_2 = 2 \text{ cis } 32°$, find $z_1 z_2$ and z_1/z_2. Leave answers in polar form.
20. Write $(1 + i\sqrt{3})^{-4}$ in the form $a + bi$. Use De Moivre's theorem—extended form.

C 21. Find h in the figure.

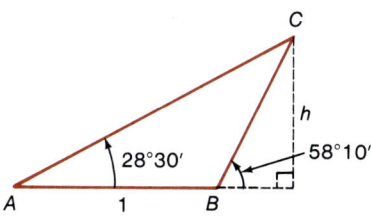

22. An airplane that can cruise at 500 miles/hour in still air is to fly due east. If the wind is blowing from the northeast at 50 miles/hour, what compass heading should the pilot choose? What will be the actual speed of the plane relative to the ground?
23. Sketch a graph of $r = 10 \cos 2\theta$ in a polar coordinate system.
24. Convert $x^2 + y^2 - 9y = 0$ to polar form.
25. Convert $r = 5 \sin \theta$ to rectangular form.
26. Find *all* solutions (real and complex) for $x^8 - 1 = 0$. Write roots in $x + yi$ form.

Practice Test Chapter 7

Take this practice test as if it were a graded test. Allow yourself up to 50 minutes. Work the problems without looking back in the chapter. Correct your work using the answers (keyed to appropriate sections) in the back of the book.

1. Solve the right triangle where $\alpha = 39°40'$ and $b = 7.00$.
2. Solve the triangle where $\beta = 55°0'$, $a = 5.00$, and $c = 7.00$.
3. Plot and label the following points in a polar coordinate system: $A(5, 240°)$, $B(8, -210°)$, $C(-4, 5\pi/3)$, and $D(-10, -5\pi/6)$.
4. (A) Change $1 - i\sqrt{3}$ to polar form, $r \geq 0$, $0° \leq \theta < 360°$.
   (B) Change 4 cis 330° to rectangular form.
5. If $z_1 = 8$ cis 25° and $z_2 = 4$ cis 19°, find:
   (A) $z_1 z_2$ (B) z_1/z_2
   Leave answers in polar form.
6. Use De Moivre's theorem to find all third roots of i. Write answers in rectangular form.
7. Find the vector sum $\mathbf{u} + \mathbf{v}$ (direction and magnitude relative to the horizontal vector) for the figure.

8. Sketch a graph of $\theta = \pi/6$ in a polar coordinate system.
9. Sketch a graph of $r = 6 + 4 \cos \theta$ in a polar coordinate system.
10. Transform $r = 8 \cos \theta$ to rectangular form.

Additional Topics in Analytic Geometry 8

8-1 Conic Sections
8-2 Parabola
8-3 Ellipse
8-4 Hyperbola
8-5 Translation of Axes
8-6 Rotation of Axes
8-7 Parametric Equations
8-8 Chapter Review

A natural design of mathematical interest. Can you guess the source? See the back of the book.

Chapter 8 ▪ Additional Topics in Analytic Geometry

Section 8-1 Conic Sections

Earlier (Section 2-4) we found that the graph of a first-degree equation in two variables

$$Ax + By = C \qquad (1)$$

not both A and B zero, is a straight line, and every straight line in a rectangular coordinate system has an equation of this form. If we increase the degree of equation (1) by 1, what kind of graphs will we have? That is, what kind of graphs will the relation specified by

$$Ax^2 + Bxy + Cy^2 + Dx + Ey + F = 0 \qquad (2)$$

yield for different sets of values of the coefficients (not all A, B, and C zero)? The graphs of equation (2) for various choices of the coefficients are plane curves obtainable by intersecting a cone* with a plane—thus, the name **conic section** (see Fig. 1).

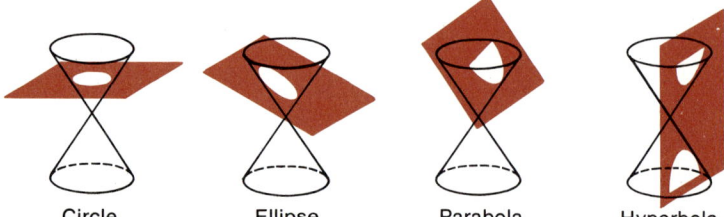

FIGURE 1 Conic sections   Circle   Ellipse   Parabola   Hyperbola

If a plane cuts clear through one nappe, then the intersection curve is called an **ellipse** (or if at right angles to the axis, a **circle**). If a plane cuts only one nappe, but does not cut clear through, then the intersection curve is called a **parabola**. Finally, if a plane cuts through both nappes, but not through the vertex, the resulting intersection curve is called a **hyperbola**. A plane passing through the vertex of the cone produces **degenerate conics**—a point, a line, or a pair of lines.

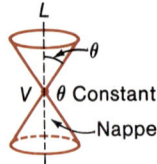

* Starting with a fixed line L and a fixed point V on L, then the surface formed by all straight lines through V making a constant angle with L is called a **right circular cone**. The fixed line L is called the **axis** of the cone and V its **vertex**. The two parts of the cone separated by the vertex are called **nappes**.

8-2 Parabola

In Section 2-1, we gave a coordinate-free definition of a circle and developed its general equation in a rectangular coordinate system. In this chapter we will give coordinate-free definitions of a parabola, ellipse, and hyperbola, and will develop standard equations for each of these conics in a rectangular coordinate system.

Conic sections have been of interest since the time of the Greeks, and they continue to be of interest and use. We will discuss some of their current uses in the context of the developments that follow.

Section 8-2 Parabola

- Definition of a Parabola
- Drawing a Parabola
- Standard Equations and Their Graphs
- Applications

In Chapter 2 we referred to the graphs of quadratic functions as parabolas (without actually defining a parabola) and said that we would have more to say about these curves later. It is now later. We start by giving a coordinate-free definition of a parabola; then, using the distance-between-two-points formula (Section 2-1), we will derive standard equations for parabolas in a rectangular coordinate system.

- Definition of a Parabola

The following is a definition of a parabola that does not depend on the coordinate of points in any coordinate system; thus, we refer to it as a **coordinate-free definition**.

Definition of a Parabola

A **parabola** is the set of all points in a plane equidistant from a fixed point F and a fixed line L in the plane. The fixed point F is called the **focus** and the fixed line L, the **directrix**. A line through the focus perpendicular to the directrix is called the **axis** and the point on the axis halfway between the directrix and focus is called the **vertex**.

Drawing a Parabola

Using the definition, we can draw a parabola with fairly simple equipment—a straightedge, a right-angle drawing triangle, a piece of string, a thumbtack, and a pencil. Referring to Figure 2, tape the straightedge along the line AB (the directrix) and place the thumbtack above the line AB (this is the focus). Place one leg of the triangle along the straightedge as indicated, then take a piece of string the same length as the other leg, tie one end to the thumbtack and fasten the other end with tape at C on the triangle. Now press the string to the edge of the triangle, and keeping the string taut, slide the triangle along the straightedge. The resulting curve will be part of a parabola, since DE will always equal DF.

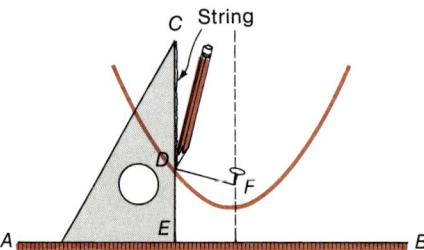

FIGURE 2   Drawing a parabola

Standard Equations and Their Graphs

Using the definition of a parabola and the distance-between-two-points formula

$$d = \sqrt{(x_2 - x_1)^2 + (y_2 - y_1)^2} \tag{1}$$

we will be able to derive simple standard equations for a parabola (located in a rectangular coordinate system) with its vertex at the origin and its axis along a coordinate axis. We start with the axis of the parabola along the x axis and the focus at $F(a, 0)$. We locate the parabola in a coordinate system as in Figure 3 and label key lines and points

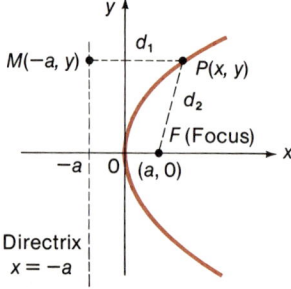 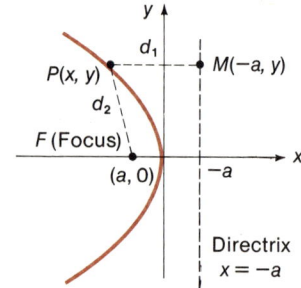

FIGURE 3   (a) $a > 0$, focus on positive x axis   (b) $a < 0$, focus on negative x axis

8-2 Parabola

(this is an important step in finding an equation of a geometric figure in a coordinate system). Note that the parabola opens to the right if $a > 0$ and to the left if $a < 0$. The vertex is at the origin, the directrix is $x = -a$, and the coordinates of M are $(-a, y)$.

The point $P(x, y)$ is a point on the parabola if and only if

$$d_1 = d_2$$
$$d(P, M) = d(P, F) \qquad \text{Use equation (1).}$$
$$\sqrt{(x + a)^2 + (y - y)^2} = \sqrt{(x - a)^2 + (y - 0)^2} \qquad \text{Square both sides.}$$
$$(x + a)^2 = (x - a)^2 + y^2 \qquad \text{Simplify.}$$
$$x^2 + 2ax + a^2 = x^2 - 2ax + a^2 + y^2$$
$$\mathbf{y^2 = 4ax} \qquad (2)$$

Equation (2) is the equation of a parabola with vertex at the origin and focus at $(a, 0)$.

Now let us locate the vertex at the origin and focus on the y axis at $(0, a)$. Looking at Figure 4, we note that the parabola opens upward if $a > 0$ and downward if $a < 0$. The directrix is $y = -a$ and the coordinates of N are $(x, -a)$. The point $P(x, y)$ is a point on the parabola if and only if

$$d_1 = d_2$$
$$d(P, N) = d(P, F) \qquad \text{Use equation (1).}$$
$$\sqrt{(x - x)^2 + (y + a)^2} = \sqrt{(x - 0)^2 + (y - a)^2} \qquad \text{Square both sides.}$$
$$(y + a)^2 = x^2 + (y - a)^2 \qquad \text{Simplify.}$$
$$y^2 + 2ay + a^2 = x^2 + y^2 - 2ay + a^2$$
$$\mathbf{x^2 = 4ay} \qquad (3)$$

Equation (3) is the equation of a parabola with vertex at the origin and focus at $(0, a)$.

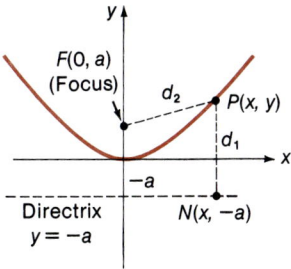 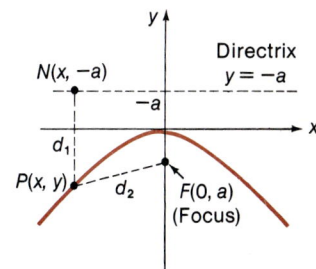

FIGURE 4   (a) $a > 0$, focus on positive y axis   (b) $a < 0$, focus on negative y axis

We summarize these results for easy reference:

Standard Equations of a Parabola

1. $y^2 = 4ax$
   Vertex: $(0, 0)$
   Focus: $(a, 0)$
   Directrix: $x = -a$

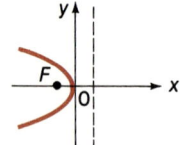

a < 0 (opens left)   a > 0 (opens right)

Symmetric with respect to the x axis.

2. $x^2 = 4ay$
   Vertex: $(0, 0)$
   Focus: $(0, a)$
   Directrix: $y = -a$

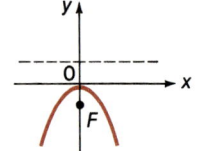

a < 0 (opens down)   a > 0 (opens up)

Symmetric with respect to the y axis.

[*Note:* In all cases the focus is inside the parabola.]

EXAMPLE 1 Graph $x^2 = -16y$, and locate the focus and directrix.

Solution To graph $x^2 = -16y$, assign y values that make the right side a perfect square (y must be nonpositive for x to be real), and solve for x. Since the coefficient of y is negative, a must be negative, and the parabola opens down.

x	0	±4	±8
y	0	−1	−4

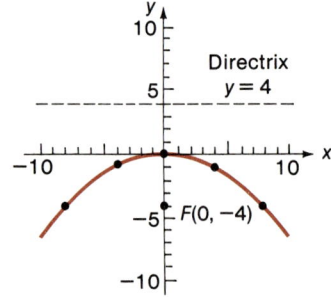

To find the focus and directrix, solve

$$4a = -16$$
$$a = -4$$

Focus: $(0, -4)$
Directrix: $y = -(-4) = 4$

PROBLEM 1   Graph $y^2 = -8x$, and locate the focus and directrix.

EXAMPLE 2   (A) Find the equation of a parabola having the origin as its vertex, the y axis as its axis, and $(-10, -5)$ on its graph.
(B) Find the coordinates of its focus and the equation of its directrix.

Solution   (A) The parabola is opening down and has an equation of the form $x^2 = 4ay$. Since $(-10, -5)$ is on the graph, we have:

$$x^2 = 4ay$$
$$(-10)^2 = 4a(-5)$$
$$100 = -20a$$
$$a = -5$$

Thus, the equation of the parabola is

$$x^2 = 4(-5)y$$
$$x^2 = -20y$$

(B) Focus: $(0, a) = (0, -5)$
Directrix: $y = -a$
$y = -(-5)$
$y = 5$

PROBLEM 2   (A) Find the equation of a parabola having the origin as its vertex, the x axis as its axis, and $(4, -8)$ on its graph.
(B) Find the coordinates of its focus and the equation of its directrix.

■ Applications

Parabolic forms are frequently encountered in the physical world. Suspension bridges, arch bridges, reflecting telescopes, radio telescopes, radar equipment, solar furnaces, and searchlights are only a few of the many real-world items that utilize parabolic forms in their design.

Figure 5(a) illustrates a parabolic reflector used in all reflecting telescopes—from the 3- to 6-inch home type to the 200-inch research instru-

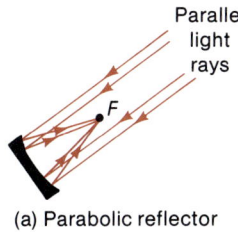
(a) Parabolic reflector

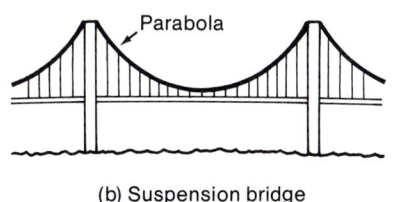
(b) Suspension bridge

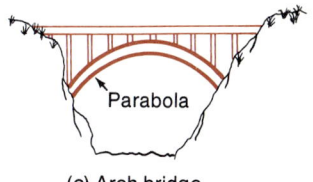
(c) Arch bridge

FIGURE 5   Uses of parabolic forms

ment on Mount Palomar in California. Parallel light rays from distant celestial bodies are reflected to the focus off a parabolic mirror. If the light source is the sun, then the parallel rays will be focused at F and we will have a solar furnace. Temperatures of over 6,000°C have been achieved by such furnaces. If we locate a light source at F, then the rays in Figure 5(a) will reverse and we will have a spotlight or a searchlight. Automobile headlights use parabolic reflectors with special lenses over the light to diffuse the rays into useful patterns.

Figure 5(b) shows a suspension bridge, such as the Golden Gate Bridge in San Francisco. The hanging cable is a parabola. It is interesting to note that a free-hanging cable, such as a telephone line, does not form a parabola. It forms another curve called a **catenary**.

Figure 5(c) shows a concrete arch bridge. If all of the loads on the arch are to be compression loads (concrete works very well under compression), then using physics and advanced mathematics (differential equations), one can show that the arch must be parabolic.

Answers to Matched Problems

1. Focus: $(-2, 0)$
   Directrix: $x = 2$

x	0	-2
y	0	± 4

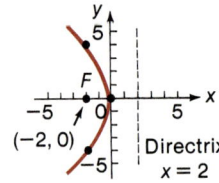

2. (A) $y^2 = 16x$
   (B) Focus: $(4, 0)$; Directrix: $x = -4$

Exercise 8-2

Graph each equation and locate the focus and directrix.

A
1. $y^2 = 4x$
2. $y^2 = 8x$
3. $x^2 = 8y$
4. $x^2 = 4y$
5. $y^2 = -12x$
6. $y^2 = -4x$
7. $x^2 = -4y$
8. $x^2 = -8y$

B
9. $y^2 = -20x$
10. $x^2 = -24y$
11. $x^2 = 10y$
12. $y^2 = 6x$

Find the coordinates to two decimal places of the focus for each parabola.

13. $y^2 = 39x$
14. $x^2 = 58y$
15. $x^2 = -105y$
16. $y^2 = -93x$
17. $y^2 = -77x$
18. $x^2 = -205y$

Find the equation of the parabola having its vertex at the origin, its axis as indicated, and passing through the indicated point.

19. y axis; (4, 2)
20. x axis; (4, 8)
21. x axis; (−3, 6)
22. y axis; (−5, 10)
23. y axis; (−6, −9)
24. x axis; (−6, −12)

C 25. Use the definition of a parabola and the distance formula to find the equation of a parabola with directrix $y = 4$ and focus at (2, 2).

26. Use the definition of a parabola and the distance formula to find the equation of a parabola with directrix $x = 2$ and focus at (6, 4).

APPLICATIONS 27. *Engineering.* The parabolic arch in the concrete bridge in the figure must have a clearance of 50 feet above the water and span a distance of 200 feet. Find the equation of the parabola after inserting a coordinate system with the origin at the vertex of the parabola and the vertical axis (pointing upward) along the axis of the parabola. Label the vertical axis the y axis and the horizontal axis (pointing to the right) the x axis.

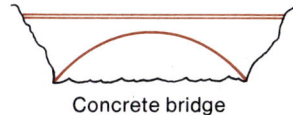
Concrete bridge

28. *Astronomy.* The cross section of a 6-inch diameter parabolic reflector is ground so that its vertex is 0.15 inch below the rim (see the figure).
    (A) Find the equation of the parabola after inserting an xy coordinate system with the vertex at the origin, the y axis pointing upward, the x axis pointing to the right, and the axis of the parabola the y axis.
    (B) How far is the focus from the vertex?

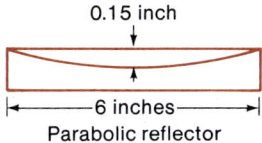
Parabolic reflector

Section 8-3 Ellipse

- Definition of an Ellipse
- Drawing an Ellipse
- Standard Equations and Their Graphs
- Applications

Definition of an Ellipse

We start with a coordinate-free definition of an ellipse. Using this definition, we will show how an ellipse can be drawn and we will derive simple standard equations for ellipses specially located in a Cartesian coordinate system.

Definition of an Ellipse

An ellipse is the set of all points P in a plane such that the sum of the distances of P from two fixed points in the plane is constant. The fixed points F' and F are called **foci** (each is a **focus**). Referring to the figure, the line segment $V'V$ through the foci is the **major axis**. The perpendicular bisector $B'B$ of the major axis is the **minor axis**. Each end of the major axis, V' and V, is called a **vertex**. The midpoint of the line segment $F'F$ is called the **center** of the ellipse.

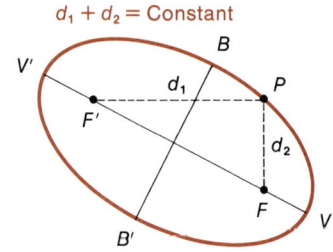

$d_1 + d_2 =$ Constant

Drawing an Ellipse

An ellipse is easy to draw. All you need is a piece of string, two thumbtacks, and a pencil or pen (see Fig. 6). Place the two thumbtacks in a piece of cardboard (these form the foci of the ellipse). Take a piece of string longer than the distance between the two thumbtacks (this represents the constant in the definition) and tie each end to a thumbtack. Finally, catch the tip of a pencil under the string and move it while keeping the string taut. The resulting figure is (by definition) an ellipse.

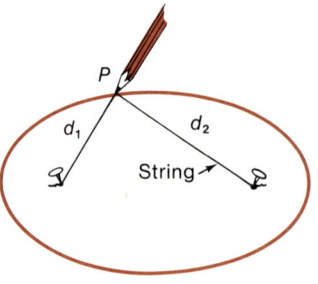

Note that $d_1 + d_2$ always adds up to the length of the string, which does not change.

FIGURE 6 Drawing an ellipse

8-3 Ellipse

Different shaped ellipses result, depending on the placement of thumbtacks and the length of the string joining them.

■ Standard Equations and Their Graphs

Using the definition of an ellipse and the distance-between-two-points formula, we now derive simple standard equations for an ellipse located in a rectangular coordinate system. We start by placing an ellipse in the coordinate system with the foci on the x axis equidistant from the origin at $F'(-c, 0)$ and $F(c, 0)$, as in Figure 7.

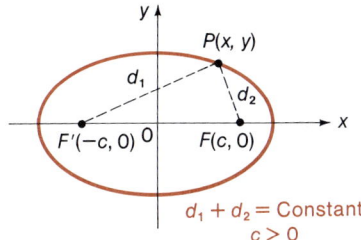

FIGURE 7

For reasons that will become clear soon, it is convenient to represent the constant sum $d_1 + d_2$ by $2a$, $a > 0$. Also, it is useful to note from Figure 7 and from the geometric fact that the sum of the lengths of any two sides of a triangle must be greater than the third side, that

$$d(F', P) + d(P, F) > d(F', F)$$
$$d_1 + d_2 > 2c$$
$$2a > 2c$$
$$a > c \qquad (1)$$

We will use this result in our derivation, which we now begin.

Referring to Figure 7, the point $P(x, y)$ is on the ellipse if and only if

$$d_1 + d_2 = 2a$$
$$d(P, F') + d(P, F) = 2a$$
$$\sqrt{(x + c)^2 + (y - 0)^2} + \sqrt{(x - c)^2 + (y - 0)^2} = 2a$$

After eliminating radicals and simplifying—a good exercise for the reader—we obtain

$$(a^2 - c^2)x^2 + a^2 y^2 = a^2(a^2 - c^2) \qquad (2)$$

or

$$\frac{x^2}{a^2} + \frac{y^2}{a^2 - c^2} = 1 \qquad (3)$$

Dividing both sides of equation (2) by $a^2(a^2 - c^2)$ is permitted, since neither a^2 nor $a^2 - c^2$ is zero. [From equation (1), $a > c$; thus, $a^2 > c^2$ and $a^2 - c^2 > 0$. In addition, since the foci were chosen so that $c > 0$, if $a > c$ and $c > 0$, then $a > 0$ and $a^2 > 0$.]

To simplify equation (3) further, we let

$$b^2 = a^2 - c^2 \qquad b > 0 \tag{4}$$

to obtain

$$\frac{x^2}{a^2} + \frac{y^2}{b^2} = 1 \tag{5}$$

From equation (5) we see that the x intercepts are $x = \pm a$ (also vertices) and the y intercepts are $y = \pm b$. Thus,

Major axis length = 2a

Minor axis length = 2b

To see that the major axis is longer than the minor axis, we show that $2a > 2b$. Returning to equation (4)

$$b^2 = a^2 - c^2 \quad a, b, c > 0$$
$$b^2 + c^2 = a^2$$
$$b^2 < a^2 \qquad \text{Definition of } <$$
$$b^2 - a^2 < 0$$
$$(b - a)(b + a) < 0 \qquad \text{Since } (b + a) \text{ is positive, } (b - a) \text{ must be negative.}$$
$$b - a < 0$$
$$b < a$$
$$2b < 2a$$
$$2a > 2b$$

$$\begin{pmatrix} \text{Length of} \\ \text{major axis} \end{pmatrix} > \begin{pmatrix} \text{Length of} \\ \text{minor axis} \end{pmatrix}$$

If we had started with the foci on the y axis at $F(0, c)$ and $F'(0, -c)$ as in Figure 8 instead of on the x axis as in Figure 7, then, following arguments similar to those used for the first derivation, we would obtain

$$\frac{x^2}{b^2} + \frac{y^2}{a^2} = 1 \qquad a > b \tag{6}$$

where the relationship among a, b, and c remains the same as before:

$$b^2 = a^2 - c^2 \tag{7}$$

The center is still at the origin, but the major axis is now along the y axis and the minor axis is along the x axis.

8-3 Ellipse

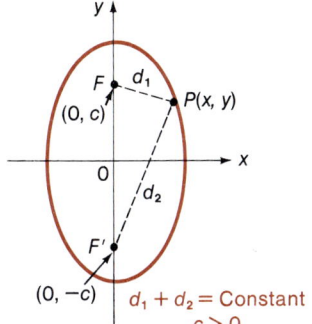

FIGURE 8

To sketch graphs of equations of the form (5) or (6) is a simple matter. We simply find the x and y intercepts and sketch in an appropriate ellipse. Since replacing x with $-x$, or y with $-y$ does not change either equation, we conclude that the graphs are symmetric with respect to the x axis, y axis, and the origin. If additional accuracy is required, additional points can be found with the aid of a hand calculator and the use of symmetry properties.

Given an equation of the form (5) or (6), how can we find the coordinates of the foci without memorizing or looking up the relation

$$b^2 = a^2 - c^2 \tag{8}$$

There is a simple geometric relationship in an ellipse that will enable us to get the same result using the Pythagorean theorem. To see this relationship refer to Figure 9(a). Then, using the definition of an ellipse and $2a$ for the constant sum (as we did in deriving the standard equations), we see that

$$d + d = 2a$$
$$2d = 2a$$
$$d = a$$

Thus, **the length of the line segment from the end of a minor axis to a focus is the same as half the length of a major axis**. This geometric

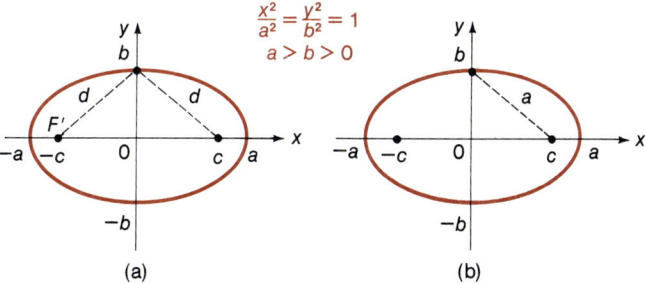

FIGURE 9   (a)   (b)

relationship is illustrated in Figure 9(b). Using the Pythagorean theorem for the triangle in Figure 9(b), we have

$$b^2 + c^2 = a^2$$

or

$$b^2 = a^2 - c^2 \quad \text{Equation (7) and (8) above}$$

or

$$c^2 = a^2 - b^2 \quad \text{Useful for finding the foci, given } a \text{ and } b$$

Thus, we can find the foci of an ellipse given the intercepts a and b simply by using the triangle alluded to in the boldface statement and the Pythagorean theorem.

We summarize all of these results for easy reference:

Standard Equations of an Ellipse

1. $\dfrac{x^2}{a^2} + \dfrac{y^2}{b^2} = 1 \qquad a > b > 0$

    x intercepts: $\pm a$ (vertices)
    y intercepts: $\pm b$
    Foci: $F'(-c, 0), F(c, 0)$
    $$c^2 = a^2 - b^2$$
    Major axis length $= 2a$
    Minor axis length $= 2b$

    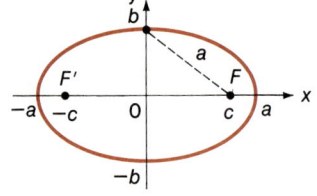

2. $\dfrac{x^2}{b^2} + \dfrac{y^2}{a^2} = 1 \qquad a > b > 0$

    x intercepts: $\pm b$
    y intercepts: $\pm a$ (vertices)
    Foci: $F'(0, -c), F(0, c)$
    $$c^2 = a^2 - b^2$$
    Major axis length $= 2a$
    Minor axis length $= 2b$

    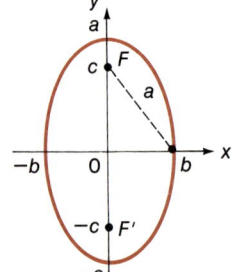

[Note: Both graphs are symmetric with respect to the x axis, y axis, and origin.]

8-3 Ellipse

Now let us consider some examples.

EXAMPLE 3 Sketch graphs of each equation, find the coordinates of the foci, and find the lengths of the major and minor axes.

(A) $9x^2 + 16y^2 = 144$ (B) $2x^2 + y^2 = 10$

Solution (A) First, write the equation in standard form by dividing both sides by 144.

$$9x^2 + 16y^2 = 144$$

$$\frac{9x^2}{144} + \frac{16y^2}{144} = \frac{144}{144}$$

$$\frac{x^2}{16} + \frac{y^2}{9} = 1$$

Locate the intercepts:

x intercepts: ± 4

y intercepts: ± 3

and sketch in the ellipse, as shown.

Foci: $c^2 = 4^2 - 3^2$ $F'(-\sqrt{7}, 0)$

$c^2 = 7$ $F(\sqrt{7}, 0)$

$c = \sqrt{7}$

Major axis length $= 2(4) = 8$

Minor axis length $= 2(3) = 6$

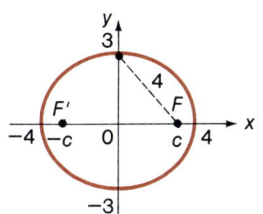

(B) Write the equation in standard form by dividing both sides by 10.

$$2x^2 + y^2 = 10$$

$$\frac{2x^2}{10} + \frac{y^2}{10} = \frac{10}{10}$$

$$\frac{x^2}{5} + \frac{y^2}{10} = 1$$

Locate the intercepts:

x intercepts: $\pm\sqrt{5} \approx 2.24$

y intercepts: $\pm\sqrt{10} \approx 3.16$

and sketch in the ellipse, as shown on the next page.

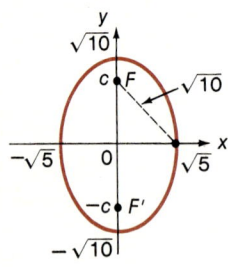

Foci: $c^2 = (\sqrt{10})^2 - (\sqrt{5})^2$ $F'(0, -\sqrt{5})$
$c^2 = 10 - 5$ $F(0, \sqrt{5})$
$c^2 = 5$
$c = \sqrt{5}$

Major axis length $= 2\sqrt{10} \approx 6.32$
Minor axis length $= 2\sqrt{5} \approx 4.47$

PROBLEM 3 Sketch graphs of each equation, find the coordinates of the foci, and find the lengths of the major and minor axes:

(A) $x^2 + 4y^2 = 4$ (B) $3x^2 + y^2 = 18$

EXAMPLE 4 Find an equation of an ellipse in the form

$$\frac{x^2}{M} + \frac{y^2}{N} = 1 \quad M, N > 0$$

if the center is at the origin, the major axis is along the y axis, and:

(A) Length of major axis = 20   (B) Length of major axis = 10
     Length of minor axis = 12   Distance of foci from center = 4

Solution   (A) Make a rough sketch of the ellipse and compute x and y intercepts.

$$\frac{x^2}{b^2} + \frac{y^2}{a^2} = 1$$

$$a = \frac{20}{2} = 10, \quad b = \frac{12}{2} = 6$$

$$\frac{x^2}{36} + \frac{y^2}{100} = 1$$

(B) Make a rough sketch of the ellipse; locate the focus and y intercepts, then determine the x intercepts using the special triangle relationship discussed earlier.

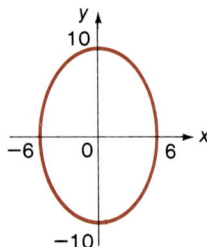

$$\frac{x^2}{b^2} + \frac{y^2}{a^2} = 1$$

$$a = \frac{10}{2} = 5$$

$$b^2 = 5^2 - 4^2 = 25 - 16 = 9$$

$$b = 3$$

$$\frac{x^2}{9} + \frac{y^2}{25} = 1$$

PROBLEM 4 Find an equation of an ellipse in the form

$$\frac{x^2}{M} + \frac{y^2}{N} = 1 \qquad M, N > 0$$

if the center is at the origin, the major axis is along the x axis, and:

(A) Length of major axis = 50
Length of minor axis = 30

(B) Length of minor axis = 16
Distance of foci from center = 6

■ Applications

You are no doubt aware of many uses of elliptical forms: orbits of satellites, planets, and comets; gears and cams; tabletops and domes in buildings are but a few examples.

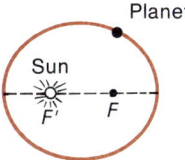
(a) Planetary motion

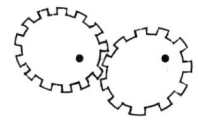
(b) Elliptical gears

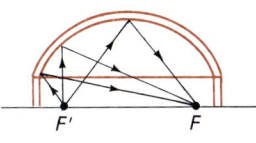
(c) Elliptical dome

FIGURE 10 Uses of elliptical forms

Johannes Kepler (1571–1630), a German astronomer, discovered that planets move in elliptical orbits, with the sun at a focus, and not in circular orbits as had been thought before [Fig. 10(a)]. Figure 10(b) shows a pair of elliptical gears with pivot points at foci. Such gears transfer constant rotational speed to variable rotational speed, and vice versa. Figure 10(c) shows an elliptical dome. An interesting property of such a dome is that a sound or light source at one focus will reflect off the dome and pass through the other focus. One of the chambers in the Capitol Building in Washington, D.C., has such a dome, and is referred to as a whispering room because a whispered sound at one focus can easily be heard at the other focus.

Answers to Matched Problems   3. (A)

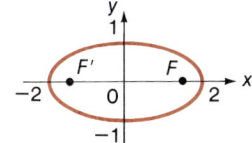

Foci: $F'(-\sqrt{3}, 0)$, $F(\sqrt{3}, 0)$
Major axis length = 4
Minor axis length = 2

3. (B)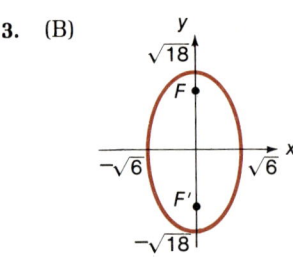

Foci: $F'(0, -\sqrt{12})$, $F(0, \sqrt{12})$
Major axis length $= 2\sqrt{18} \approx 8.49$
Minor axis length $= 2\sqrt{6} \approx 4.90$

4. (A) $\dfrac{x^2}{625} + \dfrac{y^2}{225} = 1$ (B) $\dfrac{x^2}{100} + \dfrac{y^2}{64} = 1$

Exercise 8-3 ■ *Sketch a graph of each equation, find the coordinates of the foci, and find the lengths of the major and minor axes.*

A
1. $\dfrac{x^2}{25} + \dfrac{y^2}{4} = 1$
2. $\dfrac{x^2}{9} + \dfrac{y^2}{4} = 1$
3. $\dfrac{x^2}{4} + \dfrac{y^2}{25} = 1$
4. $\dfrac{x^2}{4} + \dfrac{y^2}{9} = 1$
5. $x^2 + 9y^2 = 9$
6. $4x^2 + y^2 = 4$

B
7. $25x^2 + 9y^2 = 225$
8. $16x^2 + 25y^2 = 400$
9. $2x^2 + y^2 = 12$
10. $4x^2 + 3y^2 = 24$
11. $4x^2 + 7y^2 = 28$
12. $3x^2 + 2y^2 = 24$

Find an equation of an ellipse in the form

$$\dfrac{x^2}{M} + \dfrac{y^2}{N} = 1 \qquad M, N > 0$$

if the center is at the origin, and:

13. Major axis on x axis
    Major axis length $= 8$
    Minor axis length $= 6$

14. Major axis on x axis
    Major axis length $= 14$
    Minor axis length $= 10$

15. Major axis on y axis
    Major axis length $= 22$
    Minor axis length $= 16$

16. Major axis on y axis
    Major axis length $= 24$
    Minor axis length $= 18$

17. Major axis on x axis
    Major axis length $= 16$
    Distance of foci from center $= 6$

18. Major axis on x axis
    Major axis length $= 24$
    Distance of foci from center $= 10$

8-4 Hyperbola   465

19. Major axis on y axis
    Minor axis length = 20
    Distance of foci from center = $\sqrt{70}$

20. Major axis on y axis
    Minor axis length = 14
    Distance of foci from center = $\sqrt{200}$

C 21. Find an equation of the set of points in a plane, each of whose distance from (2, 0) is $\frac{1}{2}$ its distance from the line x = 8. Identify the geometric figure.

22. Find an equation of the set of points in a plane, each of whose distance from (0, 9) is $\frac{3}{4}$ its distance from the line y = 16. Identify the geometric figure.

APPLICATIONS 23. *Engineering.* The semielliptical arch in the concrete bridge in the figure must have a clearance of 12 feet above the water and span a distance of 40 feet. Find the equation of the ellipse after inserting a coordinate system with the center of the ellipse at the origin and the major axis on the x axis. The y axis points up and the x axis points to the right. How much clearance above the water is there 5 feet from the bank?

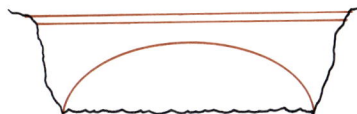

24. *Design.* A 4 × 8 foot elliptical tabletop is to be cut out of a 4 × 8 foot rectangular sheet of teak plywood (see figure). To draw the ellipse on the plywood, how far should the foci be located from each edge and how long a piece of string must be fastened to each focus to produce the ellipse (see Figure 6 in this section)? Compute answer to two decimal places.

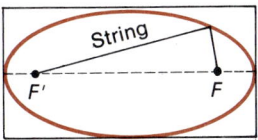

Section 8-4 Hyperbola

- Definition of a Hyperbola
- Drawing a Hyperbola
- Standard Equations and Their Graphs
- Applications

■ Definition of a Hyperbola

As before, we start with a coordinate-free definition of a hyperbola. Using this definition, we will show how a hyperbola can be drawn and we will derive simple standard equations for hyperbolas specially located in a Cartesian coordinate system.

Definition of a Hyperbola

A **hyperbola** is the set of all points P in a plane such that the absolute value of the difference of the distance of P to two fixed points in the plane is a positive constant. The fixed points F' and F are called **foci** (each a **focus**). The intersection points V' and V of the line through the foci and the two branches of the hyperbola are called **vertices** (each a **vertex**), and the line segment $V'V$ is called the **transverse axis**. The midpoint of the transverse axis is the **center** of the hyperbola.

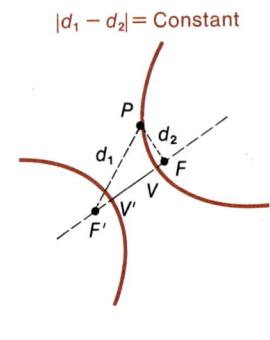

$|d_1 - d_2|$ = Constant

■ Drawing a Hyperbola

A hyperbola is not too difficult to draw. Thumbtacks, a straightedge, string, and a pencil are all that are needed (see Fig. 11). Place two thumbtacks in a piece of cardboard (these form the foci of the hyperbola).

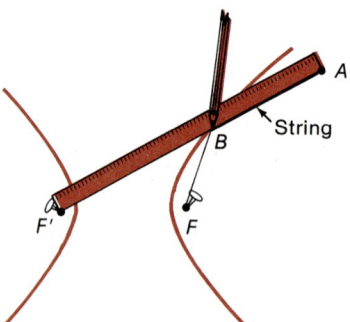

FIGURE 11 Drawing a hyperbola

8-4 Hyperbola

Rest one corner of the straightedge at the focus F' so that it is free to rotate about this point. Cut a piece of string shorter than the length of the straightedge, and fasten one end to the straightedge corner A and the other end to the thumbtack at F. Now push the string with a pencil up against the straightedge at B. Keeping the string taut, rotate the straightedge about F', keeping the corner at F'. The resulting curve will be part of a hyperbola. (Other parts of the hyperbola can be drawn by changing the position of the straightedge and string.) To see that the resulting curve meets the conditions of the definition, note that the difference of the distances BF' and BF is

$$\begin{aligned} BF' - BF &= BF' + BA - BF - BA \\ &= AF' - (BF + BA) \\ &= \begin{pmatrix} \text{Straightedge} \\ \text{length} \end{pmatrix} - \begin{pmatrix} \text{String} \\ \text{length} \end{pmatrix} \\ &= \text{Constant} \end{aligned}$$

■ Standard Equations and Their Graphs

Using the definition of a hyperbola and the distance-between-two-points formula, we now derive simple standard equations for a hyperbola located in a rectangular coordinate system. We start by placing a hyperbola in the coordinate system with the foci on the x axis equidistant from the origin at $F'(-c, 0)$ and $F(c, 0)$, $c > 0$, as in Figure 12.

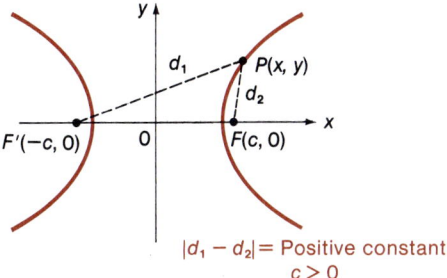

FIGURE 12

Just as for the ellipse (for reasons that will become clear soon), it is convenient to represent the constant difference by $2a$, $a > 0$. Also, it is useful to note from Figure 12 and from the geometric fact that the difference of two sides of a triangle is always less than the third side, that

$$|d_1 - d_2| < 2c$$
$$2a < 2c$$
$$a < c \tag{1}$$

We will use this result in our derivation, which we now begin.

Referring to Figure 12, the point P(x, y) is on the hyperbola if and only if

$$|d_1 - d_2| = 2a$$
$$|d(P, F') - d(P, F)| = 2a$$
$$\left|\sqrt{(x + c)^2 + y^2} - \sqrt{(x - c)^2 + y^2}\right| = 2a$$

After eliminating radicals and absolute value signs by appropriate use of squaring and simplifying (another good exercise for the reader), we have

$$(c^2 - a^2)x^2 - a^2 y^2 = a^2(c^2 - a^2) \tag{2}$$

or

$$\frac{x^2}{a^2} - \frac{y^2}{c^2 - a^2} = 1 \tag{3}$$

Dividing both sides of equation (2) by $a^2(c^2 - a^2)$ is permitted, since neither a^2 nor $(c^2 - a^2)$ is zero. [From equation (1), $a < c$; thus, $a^2 < c^2$ and $c^2 - a^2 > 0$. The constant a was chosen positive at the beginning.]

To simplify equation (3) further, we let

$$b^2 = c^2 - a^2 \qquad b > 0 \tag{4}$$

to obtain

$$\frac{x^2}{a^2} - \frac{y^2}{b^2} = 1 \tag{5}$$

From equation (5) we see that the x intercepts are $x = \pm a$ (also vertices) and there are no y intercepts. To see why there are no y intercepts, let $x = 0$ and solve for y:

$$\frac{0^2}{a^2} - \frac{y^2}{b^2} = 1$$

$$y^2 = -b^2$$
$$y = \pm\sqrt{-b^2} \quad \text{A complex number}$$

If we had started with the foci on the y axis at $F'(0, -c)$ and $F(0, c)$ as in Figure 13 instead of on the x axis as in Figure 12, then, following

8-4 Hyperbola

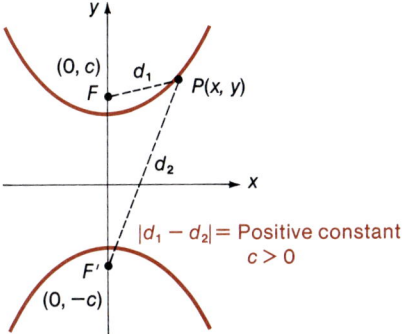

FIGURE 13

arguments similar to those used for the first derivation, we would obtain

$$\frac{y^2}{a^2} - \frac{x^2}{b^2} = 1 \qquad (6)$$

where the relationship among a, b, and c remains the same as before:

$$b^2 = c^2 - a^2 \qquad (7)$$

the center is still at the origin, but the transverse axis is now on the y axis.

Of the four conics—circles, ellipses, parabolas, and hyperbolas—it is interesting to find that the hyperbolas are the only ones that have asymptotes. Locating intercepts and asymptotes is a significant aid to sketching graphs of either equation (5) or (6).

To find the asymptotes for the graph of equation (5), we proceed as follows:

$$\frac{x^2}{a^2} - \frac{y^2}{b^2} = 1 \qquad \text{Solve equation (5) for } y \text{ in terms of } x.$$

$$\frac{y^2}{b^2} = \frac{x^2}{a^2} - 1$$

$$y^2 = \frac{b^2}{a^2}(x^2 - a^2)$$

$$y^2 = \frac{b^2 x^2}{a^2}\left(1 - \frac{a^2}{x^2}\right)$$

$$y = \pm \frac{b}{a} x \sqrt{1 - \frac{a^2}{x^2}}$$

As x gets larger, the radical approaches 1; hence, for large x, the graph of equation (5) behaves very much like the lines

$$y = \pm \frac{b}{a} x \quad \text{Asymptotes for } \frac{x^2}{a^2} - \frac{y^2}{b^2} = 1 \tag{8}$$

These lines are asymptotes for the graph of equation (5). That is, the graph approaches these lines, as indicated in Figure 14, as a point P(x, y) on the graph moves away from the origin. An easy way to draw the asymptotes is to first draw the rectangle as in Figure 14, then extend the diagonals. We will refer to this rectangle as the **asymptote rectangle**.

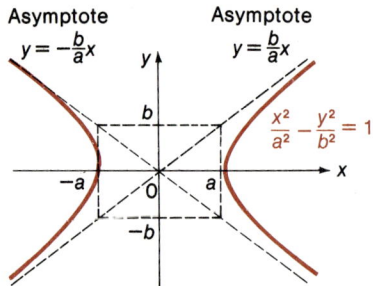

FIGURE 14

Starting with equation (6) and proceeding as we did for equation (5), we obtain the asymptotes

$$y = \pm \frac{a}{b} x \quad \text{Asymptotes for } \frac{y^2}{a^2} - \frac{x^2}{b^2} = 1 \tag{9}$$

The perpendicular bisector of the transverse axis, extending from one side of the asymptote rectangle to the other, is called the **conjugate axis** of the hyperbola.

Given an equation of the form (5) or (6), how can we find the coordinates of the foci without memorizing or looking up the relation

$$b^2 = c^2 - a^2 \tag{10}$$

Just as with the ellipse, there is a simple geometric relationship in a hyperbola that will enable us to get the same result using the Pythagorean theorem. To see this relationship, we write equation (10) in the form

$$c^2 = a^2 + b^2 \tag{11}$$

and note in the figures in the following box that the distance from the center to a focus is the same as the distance from the center to a corner of the asymptote rectangle. Stated in another way, **a circle (with center at the origin) that passes through all four corners of the asymptote rectangle**

8-4 Hyperbola

will also pass through all foci of hyperbolas with asymptotes determined by the diagonals of the rectangle.

We summarize all of the preceding results for convenient reference.

Standard Equations of a Hyperbola

1. $\dfrac{x^2}{a^2} - \dfrac{y^2}{b^2} = 1$

    x intercepts: $\pm a$ (vertices)
    y intercepts: none
    Foci: $F'(-c, 0)$, $F(c, 0)$

    $c^2 = a^2 + b^2$

    Transverse axis length $= 2a$
    Conjugate axis length $= 2b$

    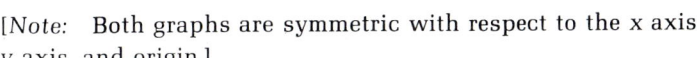

2. $\dfrac{y^2}{a^2} - \dfrac{x^2}{b^2} = 1$

    x intercepts: none
    y intercepts: $\pm a$ (vertices)
    Foci: $F'(0, -c)$, $F(0, c)$

    $c^2 = a^2 + b^2$

    Transverse axis length $= 2a$
    Conjugate axis length $= 2b$

[Note: Both graphs are symmetric with respect to the x axis, y axis, and origin.]

We are now ready to consider some examples.

EXAMPLE 5 Sketch graphs of each equation, find the coordinates of the foci, and find the lengths of the transverse and conjugate axes.

(A) $9x^2 - 16y^2 = 144$ (B) $16y^2 - 9x^2 = 144$ (C) $2x^2 - y^2 = 10$

Solution (A) First, write the equation in standard form by dividing both sides by 144.

$$9x^2 - 16y^2 = 144$$

$$\frac{x^2}{16} - \frac{y^2}{9} = 1$$

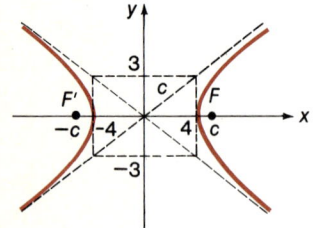

Locate x intercepts, $x = \pm 4$ (there are no y intercepts), sketch the asymptote using the asymptote rectangle, then sketch in the hyperbola.

Foci: $c^2 = 4^2 + 3^2$ $F'(-5, 0), F(5, 0)$
$c^2 = 25$
$c = 5$

Transverse axis length $= 2(4) = 8$
Conjugate axis length $= 2(3) = 6$

(B) $16y^2 - 9x^2 = 144$ Write equation in standard form.

$$\frac{y^2}{9} - \frac{x^2}{16} = 1$$

Locate y intercepts, $y = \pm 3$ (there are no x intercepts), sketch the asymptotes using the asymptote rectangle, then sketch in the hyperbola. (It is important to note that the transverse axis and the foci are on the y axis.)

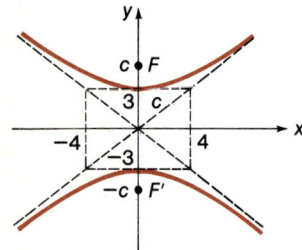

Foci: $c^2 = 4^2 + 3^2$ $F'(0, -5), F(0, 5)$
$c^2 = 25$
$c = 5$

Transverse axis length $= 2(3) = 6$
Conjugate axis length $= 2(4) = 8$

(C) $2x^2 - y^2 = 10$ Write equation in standard form.

$$\frac{x^2}{5} - \frac{y^2}{10} = 1$$

Locate x intercepts, $x = \pm\sqrt{5}$ (there are no y intercepts), sketch the asymptotes using the asymptote rectangle, then sketch in the hyperbola.

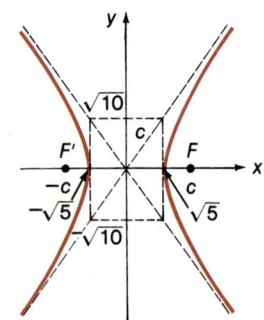

Foci: $c^2 = (\sqrt{5})^2 + (\sqrt{10})^2$ $F'(-\sqrt{15}, 0), F(\sqrt{15}, 0)$
$c^2 = 15$
$c = \sqrt{15}$

Transverse axis length $= 2\sqrt{5} \approx 4.47$
Conjugate axis length $= 2\sqrt{10} \approx 6.32$

Hyperbolas of the form

$$\frac{x^2}{M} - \frac{y^2}{N} = 1 \qquad \frac{y^2}{N} - \frac{x^2}{M} = 1 \qquad M, N > 0$$

are called **conjugate hyperbolas**. In Example 5, the hyperbolas in parts (A) and (B) are conjugate hyperbolas (they share the same asymptotes).

8-4 Hyperbola

PROBLEM 5 Sketch graphs of each equation, find the coordinates of the foci, and find the lengths of the transverse and conjugate axes.

(A) $16x^2 - 25y^2 = 400$ (B) $25y^2 - 16x^2 = 400$
(C) $y^2 - 3x^2 = 12$

EXAMPLE 6 Find an equation of a hyperbola in the form

$$\frac{y^2}{M} - \frac{x^2}{N} = 1 \qquad M, N > 0$$

if the center is at the origin, and:

(A) Length of transverse axis is 12
Length of conjugate axis is 20
(B) Length of transverse axis is 6
Distance of foci from center is 5

Solution (A) Start with

$$\frac{y^2}{a^2} - \frac{x^2}{b^2} = 1$$

and find a and b:

$$a = \frac{12}{2} = 6 \quad \text{and} \quad b = \frac{20}{2} = 10$$

Thus, the equation is

$$\frac{y^2}{36} - \frac{x^2}{100} = 1$$

(B) Start with

$$\frac{y^2}{a^2} - \frac{x^2}{b^2} = 1$$

and find a and b:

$$a = \frac{6}{2} = 3$$

To find b, sketch the asymptote rectangle, label known parts, and use the Pythagorean theorem.

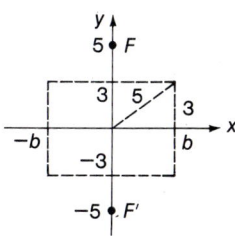

$b^2 = 5^2 - 3^2$ Note: A common error is to place the foci and transverse axes
$b^2 = 16$ on the wrong coordinate axis. Check this carefully at the
$b^2 = 4$ beginning.

Thus, the equation is

$$\frac{y^2}{9} - \frac{x^2}{16} = 1$$

PROBLEM 6 Find an equation of a hyperbola in the form

$$\frac{x^2}{M} - \frac{y^2}{N} = 1 \qquad M, N > 0$$

if the center is at the origin, and:
(A) Length of transverse axis is 50
    Length of conjugate axis is 30
(B) Length of conjugate axis is 12
    Distance of foci from center is 9

■ **Applications**

There are important uses of hyperbolic forms that you may not be aware of. They are encountered in the study of comets; the loran system of navigation for pleasure boats, ships, and aircraft; optics; and in contemporary architectural structures (the TWA building at Kennedy Airport is a hyperbolic paraboloid); to name a few of many examples.

Some comets from outer space will enter the sun's gravitational field, follow a hyperbolic path around the sun (with the sun at a focus), then leave, never to be seen again [Fig. 15(a)]. In the loran system of navigation, transmitting stations in three locations, S_1, S_2, and S_3 [see Fig. 15(b)], send out signals simultaneously. A ship with a receiver records the difference in the arrival times of the signals from S_1 and S_2 and also records the difference in arrival times of the signals from S_2 and S_3. The difference in arrival times can be transformed into differences of the distances that the ship is to S_1 and S_2 and to S_2 and S_3. Plotting all points so that these differences in distances remain constant produces two branches, p_1 and p_2, of a hyperbola with foci S_1 and S_2 and two branches, q_1 and q_2, of a hyperbola with foci S_2 and S_3. It is easy to tell which branches the ship is on by noting the arrival times of the signals from each station. The intersection of a branch from each hyperbola locates the ship. Most of these calculations are now done by

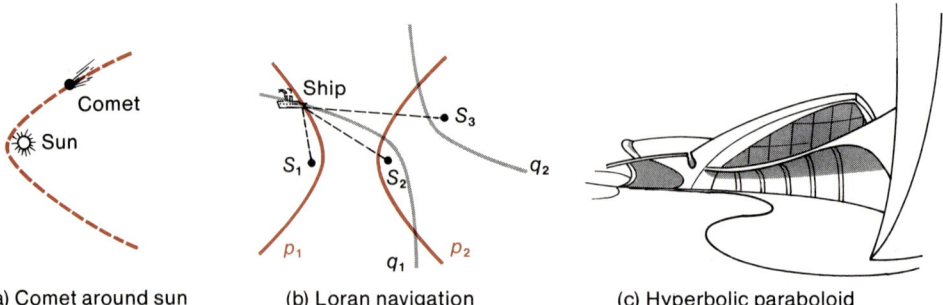

(a) Comet around sun   (b) Loran navigation   (c) Hyperbolic paraboloid

FIGURE 15 Uses of hyperbolic forms

shipboard computers, and positions in longitude and latitude are given. This system of navigation is widely used for coastal navigation. Inexpensive loran units are now found on many smaller pleasure boats. Figure 15(c) illustrates a hyperbolic paraboloid used architecturally. With such structures, thin concrete shells can span large spaces.

Answers to Matched Problems

5. (A) 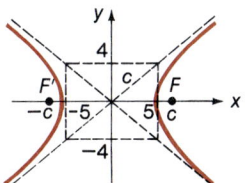 $\dfrac{x^2}{25} - \dfrac{y^2}{16} = 1$

Foci: $F'(-\sqrt{41}, 0), F(\sqrt{41}, 0)$
Transverse axis length = 10
Conjugate axis length = 8

(B) 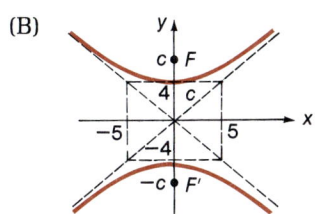 $\dfrac{y^2}{16} - \dfrac{x^2}{25} = 1$

Foci: $F'(0, -\sqrt{41}), F(0, \sqrt{41})$
Transverse axis length = 8
Conjugate axis length = 10

(C) 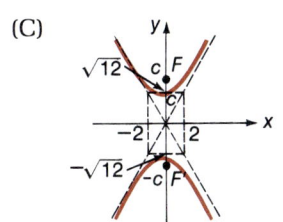 $\dfrac{y^2}{12} - \dfrac{x^2}{4} = 1$

Foci: $F'(0, -4), F(0, 4)$
Transverse axis length = $2\sqrt{12}$
≈ 6.93
Conjugate axis length = 4

6. (A) $\dfrac{x^2}{625} - \dfrac{y^2}{225} = 1$ (B) $\dfrac{x^2}{45} - \dfrac{y^2}{36} = 1$

Exercise 8-4

Sketch a graph of each equation, find the coordinates of the foci, and find the lengths of the transverse and conjugate axes.

A
1. $\dfrac{x^2}{9} - \dfrac{y^2}{4} = 1$ 2. $\dfrac{x^2}{9} - \dfrac{y^2}{25} = 1$

3. $\dfrac{y^2}{4} - \dfrac{x^2}{9} = 1$ 4. $\dfrac{y^2}{25} - \dfrac{x^2}{9} = 1$

5. $4x^2 - y^2 = 16$ 6. $x^2 - 9y^2 = 9$

7. $9y^2 - 16x^2 = 144$ 8. $4y^2 - 25x^2 = 100$

B
9. $3x^2 - 2y^2 = 12$ 10. $3x^2 - 4y^2 = 24$

11. $7y^2 - 4x^2 = 28$ 12. $3y^2 - 2x^2 = 24$

Find an equation of a hyperbola in the form

$$\frac{x^2}{M} - \frac{y^2}{N} = 1 \quad \text{or} \quad \frac{y^2}{N} - \frac{x^2}{M} = 1 \qquad M, N > 0$$

if the center is at the origin, and:

13. Transverse axis is on x axis
    Transverse axis length is 14
    Conjugate axis length is 10

14. Transverse axis is on x axis
    Transverse axis length is 8
    Conjugate axis length is 6

15. Transverse axis is on y axis
    Transverse axis length is 24
    Conjugate axis length is 18

16. Transverse axis is on y axis
    Transverse axis length is 16
    Conjugate axis length is 22

17. Transverse axis is on x axis
    Transverse axis length is 18
    Distance of foci from center is 11

18. Transverse axis is on x axis
    Transverse axis length is 16
    Distance of foci from center is 10

19. Conjugate axis is on x axis
    Conjugate axis length is 14
    Distance of foci from center is $\sqrt{200}$

20. Conjugate axis is on x axis
    Conjugate axis length is 10
    Distance of foci from center is $\sqrt{70}$

C *Eccentricity (Remark):* Problems 21 and 22 in this exercise and Problems 21 and 22 in Exercise 8-3 are related to a property of conics called **eccentricity**, which is denoted by a real number $e > 0$. Parabolas, ellipses, and hyperbolas can all be defined in terms of e, a fixed point (called a focus), and a fixed line not containing the point (called a directrix) as follows: "The set of points in a plane each of whose distance from a fixed point is e ($e > 0$) times its distance from a fixed line is either an ellipse ($0 < e < 1$), a parabola ($e = 1$), or a hyperbola ($e > 1$)."

21. Find an equation of the set of points in a plane each of whose distance from (3, 0) is $\frac{3}{2}$ its distance from the line $x = \frac{4}{3}$. Identify the geometric figure.

22. Find an equation of the set of points in a plane each of whose distance from (0, 4) is $\frac{4}{3}$ its distance from the line $y = \frac{9}{4}$. Identify the geometric figure.

APPLICATIONS

23. *Architecture.* An architect is interested in designing a thin-shelled dome in the shape of a hyperbolic paraboloid [Fig. (a)]. Find the equation of the hyperbola located in a coordinate system [Fig. (b)] satisfying the indicated conditions. How far is the hyper-

bola above the vertex 6 feet to the right of the vertex? Compute the answer to two decimal places.

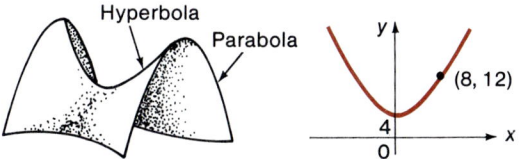

(a) Hyperbolic paraboloid   (b) Hyperbola part of dome

Section 8-5 Translation of Axes

- Transformation of Coordinates
- Translation of Axes
- Translation Used in Graphing
- Translation Used in Finding Equations of Conics

Recall that analytic geometry in a plane is a wedding of algebra and plane geometry. By introducing a coordinate system, we are able to form equations of certain plane curves and to graph certain equations. Starting with coordinate-free definitions of circles, parabolas, ellipses, and hyperbolas, we placed these figures in special places in a rectangular coordinate system and found standard equations for each. All equations that we found were second-degree equations and were special cases of the general second-degree equation:

$$Ax^2 + Bxy + Cy^2 + Dx + Ey + F = 0 \qquad (1)$$

where A, B, C, D, E, and F are real numbers. For example, the circle $x^2 + y^2 - 25 = 0$ has $A = 1$, $B = 0$, $C = 1$, $D = 0$, $E = 0$, and $F = -25$; and the parabola $y^2 - 16x = 0$ has $A = 0$, $B = 0$, $C = 1$, $D = -16$, $E = 0$, and $F = 0$; and so on.

The placing of the various conics in special places in a coordinate system resulted in particularly simple standard equations for these curves. What happens if the conics are moved away from the origin? It can be shown that they will still have second-degree equations of the form of equation (1) with most of the constants A, B, C, D, E, and F not zero. Our problems now are to consider graphing equation (1) in its more general form and to find equations of conic curves placed arbitrarily in a coordinate system. These problems form the subject matter for this and the next section.

- **Transformation of Coordinates**

Suppose we are given a coordinate system, a curve, and its equation. Let us consider the problem of finding the equation of the same curve with respect to another set of coordinate axes. The process of changing from one set of coordinate axes to another is called a **transformation of coordinates**. By suitable use of transformation of coordinates we can often transform equation (1) into one of the standard forms considered earlier. In this book we will limit our investigation to two types of transformations called *translation* and *rotation*. Translations will be considered in this section and rotations will be considered in the next section.

■ Translation of Axes

A **translation of coordinate axes** occurs when the new coordinate axes have the same sense as and are parallel to the old coordinate axes. To see how coordinates in the original system are changed when moving to the translated system, and vice versa, refer to Figure 16.

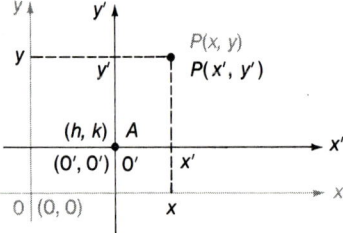

FIGURE 16   Translation of coordinates

A point P in the plane has two sets of coordinates: (x, y) in the original system and (x', y') in the translated system. If the coordinates of the origin of the translated system are (h, k) relative to the original system, then the old and new coordinates are related by:

Translation Formulas
1. $x = x' + h \qquad y = y' + k$
2. $x' = x - h \qquad y' = y - k$

It can be shown that these formulas hold for $A(h, k)$ located anywhere in the original coordinate system.

EXAMPLE 7   A curve has the equation

$$(x - 4)^2 + (y + 1)^2 = 36$$

If the origin is translated to $(4, -1)$, find the equation of the curve in the translated system and identify the curve.

Solution Since $(h, k) = (4, -1)$, use translation formulas

$$x' = x - h = x - 4$$
$$y' = y - k = y + 1$$

to obtain, after substitution,

$$x'^2 + y'^2 = 36$$

This is the equation of a circle of radius 6 with center at the new origin $(4, -1)$. Note that this result agrees with our general treatment of the circle in Section 2-1.

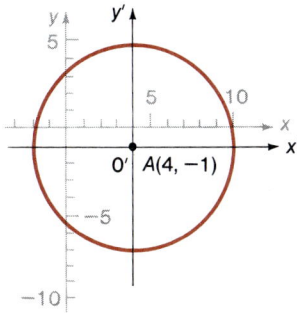

PROBLEM 7 A curve has the equation $(y + 2)^2 = 8(x - 3)$. If the origin is translated to $(3, -2)$, find an equation of the curve in the translated system. Identify the curve.

■ Translation Used in Graphing

Example 7 and Problem 7 suggest a process that could be very useful: A translation of coordinates might transform a complicated equation into a simpler standard form whose graph is easily recognized and drawn in the translated system. We will apply this technique to the second-degree equation

$$Ax^2 + Cy^2 + Dx + Ey + F = 0 \qquad (2)$$

that is, to equation (1) with $B = 0$. It can be shown that the graph of this equation is a conic, a degenerate conic, or that there is no graph. The main problem in using a translation to simplify equation (2) is in locating the origin of the translated system so that a simplification does, in fact, take place. The technique of "completing the square" will be the central tool used in locating the origin of the translated system. An example should make the process clear.

EXAMPLE 8 Use a translation of coordinates to transform

$$y^2 - 6y - 4x + 1 = 0$$

into one of the standard equations of a nondegenerate conic. Identify the curve and graph it.

Solution The equation has a second-degree term in y and first-degree terms in x and y. This suggests completing the square relative to y to obtain an equation that looks like

$$(y - k)^2 = M(x - h)$$

Once in this form, a translation that leads to a standard equation is easily identified.

$$y^2 - 6y - 4x + 1 = 0$$
$$y^2 - 6y = 4x - 1 \quad \text{\textcolor{red}{Complete the square by adding 9 to both sides.}}$$
$$y^2 - 6y + 9 = 4x + 8$$
$$(y - 3)^2 = 4(x + 2)$$

The greatest simplification takes place if we use the translation

$$x' = x + 2$$
$$y' = y - 3$$

that is, if we translate the origin to $(h, k) = (-2, 3)$. The equation in the translated system is

$$y'^2 = 4x'$$

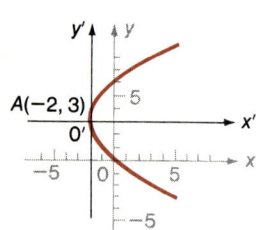

and we recognize this as the equation of a parabola opening to the right. Graph the curve relative to the translated coordinate axes, and note that if the translated axes are erased, we then have the graph of the original equation relative to the original coordinate system.

PROBLEM 8 Use translation of coordinates to transform

$$x^2 + y^2 - 8x - 4y + 4 = 0$$

into one of the standard equations of a nondegenerate conic. Identify the curve and graph it.

EXAMPLE 9 Use translation of coordinates to transform

$$9x^2 - 4y^2 - 36x - 24y - 36 = 0$$

into one of the standard equations of a nondegenerate conic. Identify the curve and graph it. Find the coordinates of any foci relative to the original system.

8-5 Translation of Axes

Solution

$$9x^2 - 4y^2 - 36x - 24y - 36 = 0$$

$$9x^2 - 36x \quad - 4y^2 - 24y \quad = 36 \quad \text{Complete the square relative to } x \text{ and to } y.$$

$$9(x^2 - 4x \quad) - 4(y^2 + 6y \quad) = 36$$

$$9(x^2 - 4x + 4) - 4(y^2 + 6y + 9) = 36 + 36 - 36$$

$$9(x - 2)^2 - 4(y + 3)^2 = 36$$

$$\frac{(x - 2)^2}{4} - \frac{(y + 3)^2}{9} = 1$$

The greatest simplification takes place if we use the translation

$$x' = x - 2$$
$$y' = y + 3$$

to obtain

$$\frac{x'^2}{4} - \frac{y'^2}{9} = 1$$

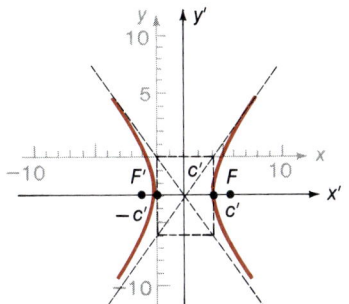

where $A(h, k) = A(2, -3)$ is the new origin $0'$. This is a hyperbola with transverse axis on the x' axis. Sketch the asymptote rectangle and vertices in the translated system, then sketch the graph.

To find the coordinates of the foci in the original system, first find the coordinates in the translated system.

$$c'^2 = 2^2 + 3^2 = 13$$
$$c' = \sqrt{13}$$
$$-c' = -\sqrt{13}$$

Thus, the coordinates in the translated system are

$$F'(-\sqrt{13}, 0) \quad \text{and} \quad F(\sqrt{13}, 0)$$

Now, use

$$x = x' + h = x' + 2$$
$$y = y' + k = y' - 3$$

to obtain

$$F'(-\sqrt{13} + 2, -3) \quad \text{and} \quad F(\sqrt{13} + 2, -3)$$

as the coordinates of the foci in the original system.

PROBLEM 9 Repeat Example 9 for $9x^2 + 16y^2 + 36x - 32y - 92 = 0$.

Translation Used in Finding Equations of Conics

We now reverse the problem: Given certain information about a conic in a rectangular coordinate system, find its equation.

EXAMPLE 10 Find the equation of a hyperbola with vertices on the line $x = -4$, conjugate axis on the line $y = 3$, length of the transverse axis 4, and length of the conjugate axis 6.

Solution Sketch the hyperbola in the coordinate system, translate the origin to the center of the hyperbola, write its equation in the translated system (a standard form), then translate the variables in the equation back to the original system.

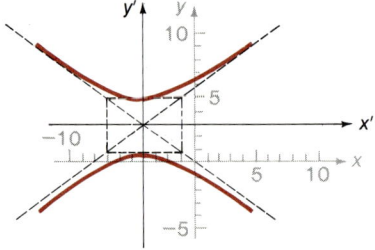

$$\frac{y'^2}{4} - \frac{x'^2}{9} = 1$$

$$(h, k) = (-4, 3)$$

Thus,

$$x' = x + 4$$
$$y' = y - 3$$

Translation formulas

and the equation in the original system is

$$\frac{(y-3)^2}{4} - \frac{(x+4)^2}{9} = 1$$

or written in the form of equation (2)

$$4x^2 - 9y^2 + 32x + 54y + 19 = 0$$

PROBLEM 10 Find the equation of an ellipse with foci on the line $x = 4$, minor axis on the line $y = -3$, length of the major axis 8, and length of the minor axis 4.

Answers to Matched Problems

7. $y'^2 = 8x'$; a parabola
8. $x'^2 + y'^2 = 16$; circle

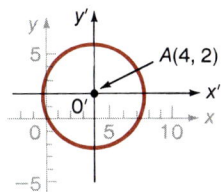

9. $\dfrac{x'^2}{16} + \dfrac{y'^2}{9} = 1$; ellipse

   Foci: $F'(-\sqrt{7} - 2, 1)$, $F(\sqrt{7} - 2, 1)$

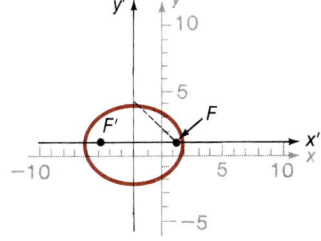

10. $\dfrac{(x-4)^2}{4} + \dfrac{(y+3)^2}{16} = 1$ or $4x^2 + y^2 - 32x + 6y + 57 = 0$

Exercise 8-5

A In Problems 1–8:
(A) Find the translation formulas that reduce each equation to a standard form for a nondegenerate conic. Write (h, k).
(B) Write the equation of the curve for the translated system.
(C) Identify the curve.

1. $(x - 3)^2 + (y - 5)^2 = 81$
2. $(x - 3)^2 = 8(y + 2)$
3. $\dfrac{(x + 7)^2}{9} + \dfrac{(y - 4)^2}{16} = 1$
4. $(x + 2)^2 + (y + 6)^2 = 36$
5. $(y + 9)^2 = 16(x - 4)$
6. $\dfrac{(y - 9)^2}{10} - \dfrac{(x + 5)^2}{6} = 1$
7. $\dfrac{(x + 8)^2}{12} + \dfrac{(y + 3)^2}{8} = 1$
8. $\dfrac{(x + 7)^2}{25} - \dfrac{(y - 8)^2}{50} = 1$

In Problems 9–14:
(A) Transform each equation into a standard form for a nondegenerate conic by a suitable translation of coordinates. Identify the conic.
(B) Indicate the translation formulas used and indicate (h, k).

9. $16(x - 3)^2 - 9(y + 2)^2 = 144$
10. $(y + 2)^2 - 12(x - 3) = 0$
11. $6(x + 5)^2 + 5(y + 7)^2 = 30$
12. $12(y - 5)^2 - 8(x - 3)^2 = 24$
13. $(x + 6)^2 + 24(y - 4) = 0$
14. $4(x - 7)^2 + 7(x - 3)^2 = 28$

B Use a translation of coordinates to transform each equation into an equation for a nondegenerate conic. Identify the curve and graph it.

15. $4x^2 + 9y^2 - 16x - 36y + 16 = 0$
16. $16x^2 + 9y^2 + 64x + 54y + 1 = 0$
17. $x^2 + 8x + 8y = 0$
18. $y^2 + 12x + 4y - 32 = 0$
19. $x^2 + y^2 + 12x + 10y + 45 = 0$
20. $x^2 + y^2 - 8x - 6y = 0$
21. $-9x^2 + 16y^2 - 72x - 96y - 144 = 0$
22. $16x^2 - 25y^2 - 160x = 0$

C Find the coordinates of any foci relative to the original coordinate system in:

23. Problem 15 **24.** Problem 16 **25.** Problem 17
26. Problem 18 **27.** Problem 21 **28.** Problem 22

Section 8-6 Rotation of Axes

- Rotation of Axes
- Rotation Used in Graphing

In the preceding section we found that when $B = 0$ in

$$Ax^2 + Bxy + Cy^2 + Dx + Ey + F = 0 \quad (A, B, C \text{ not all zero}) \quad (1)$$

we could, by a suitable translation of axes (except for degenerate cases), transform equation (1) into one of the standard equations for a conic. What happens if $B \neq 0$ in equation (1)? In this case we will show that a suitable *rotation of axes* will transform equation (1) into a new equation with the $x'y'$ term missing. Then, if necessary, we can proceed with a translation of axes to obtain a standard equation of a conic (except for degenerate cases). From this type of argument it can be shown that the graph of a general second-degree equation in two variables [equation (1)] is a conic (circle, parabola, ellipse, hyperbola), a degenerate conic (pair of lines, a line, a point), or no graph.

- **Rotation of Axes**

We now introduce a transformation of coordinates from an xy system to an $x'y'$ system that is accomplished by a rotation of axes. The origin is

8-6 Rotation of Axes

kept fixed and the x' and y' axes are obtained by rotating the x and y axes counterclockwise as shown in Figure 17.

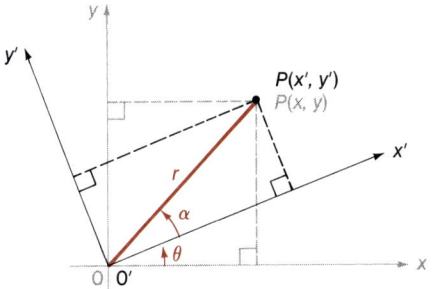

FIGURE 17 Rotation of axes

Referring to Figure 17 and using trigonometry we have

$$x' = r \cos \alpha \qquad y' = r \sin \alpha \qquad (2)$$

and

$$x = r \cos(\theta + \alpha) \qquad y = r \sin(\theta + \alpha) \qquad (3)$$

Using addition identities from trigonometry for equation (3), we obtain

$$\begin{aligned}
x &= r \cos(\theta + \alpha) \\
&= r(\cos \theta \cos \alpha - \sin \theta \sin \alpha) \\
&= r \cos \theta \cos \alpha - r \sin \theta \sin \alpha \\
&= (r \cos \alpha) \cos \theta - (r \sin \alpha) \sin \theta \quad \text{Use equations (2)} \\
&= x' \cos \theta - y' \sin \theta \qquad (4)
\end{aligned}$$

$$\begin{aligned}
y &= r \sin(\theta + \alpha) \\
&= r(\sin \theta \cos \alpha + \cos \theta \sin \alpha) \\
&= r \sin \theta \cos \alpha + r \cos \theta \sin \alpha \\
&= (r \cos \alpha)\sin \theta + (r \sin \alpha)\cos \theta \quad \text{Use equations (2)} \\
&= x' \sin \theta + y' \cos \theta \qquad (5)
\end{aligned}$$

Solving equations (4) and (5) for x' and y' in terms of x and y, we obtain transformation formulas for the reverse direction.

$$\begin{aligned}
x' &= x \cos \theta + y \sin \theta \\
y' &= -x \sin \theta + y \cos \theta
\end{aligned} \qquad (6)$$

Summarizing these results for convenient reference, we have:

Rotation Formulas

If the xy coordinate axes are rotated counterclockwise through an angle of θ, then the xy and x'y' coordinates are related by:

1. $x = x' \cos\theta - y' \sin\theta$
   $y = x' \sin\theta + y' \cos\theta$

2. $x' = x \cos\theta + y \sin\theta$
   $y' = -x \sin\theta + y \cos\theta$

These formulas hold for P any point in the original coordinate system and θ any counterclockwise rotation.

■ **Rotation Used in Graphing**

We now investigate how rotation formulas are used in graphing.

EXAMPLE 11 Transform the equation

$$xy = -2$$

using a rotation of axes through 45°. Graph the new equation and identify the curve.

Solution   Use the rotation formulas:

$$x = x' \cos 45° - y' \sin 45° = \frac{\sqrt{2}}{2}(x' - y')$$

$$y = x' \sin 45° + y' \cos 45° = \frac{\sqrt{2}}{2}(x' + y')$$

$$xy = -2$$

$$\frac{\sqrt{2}}{2}(x' - y') \frac{\sqrt{2}}{2}(x' + y') = -2$$

$$\frac{1}{2}(x'^2 - y'^2) = -2$$

$$\frac{x'^2}{2} - \frac{y'^2}{2} = -2$$

$$\frac{y'^2}{4} - \frac{x'^2}{4} = 1$$

This is a standard equation for a hyperbola.

$xy = -2$

$\dfrac{y'^2}{4} - \dfrac{x'^2}{4} = 1$

Hyperbola

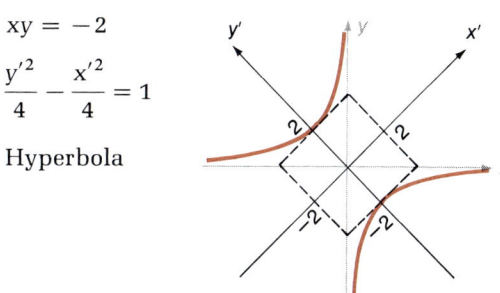

Notice that the asymptotes in the rotated system are the x and y axes in the original system.

PROBLEM 11 Transform the equation $2xy = 1$ using a rotation of axes through $45°$. Graph the new equation and identify the curve.

How do we find how much rotation is necessary so that the $x'y'$ term drops out from a transformed second-degree equation? To find out, we substitute

$$x = x' \cos \theta - y' \sin \theta$$
$$y = x' \sin \theta + y' \cos \theta$$

into equation (1) to obtain

$$A(x' \cos \theta - y' \sin \theta)^2 + B(x' \cos \theta - y' \sin \theta)(x' \sin \theta + y' \cos \theta) + C(x' \sin \theta + y' \cos \theta)^2 + D(x' \cos \theta - y' \sin \theta) + E(x' \sin \theta + y' \cos \theta) + F = 0$$

After multiplying and collecting terms we have

$$A'x'^2 + B'x'y' + C'y'^2 + D'x' + E'y' + F = 0 \qquad (7)$$

where

$$B' = 2(C - A)\sin \theta \cos \theta + B(\cos^2 \theta - \sin^2 \theta) \qquad (8)$$

In order for the $x'y'$ term in equation (7) to drop out, B' must be 0. (We do not worry about A', C', D', and E' at this point. All will automatically be determined once we find θ so that $B' = 0$.) We set the right side of equation (8) equal to 0 and solve for θ.

$$2(C - A)\sin \theta \cos \theta + B(\cos^2 \theta - \sin^2 \theta) = 0$$

Using the double-angle identities from trigonometry ($\sin 2\theta = 2 \sin \theta \cos \theta$

and $\cos 2\theta = \cos^2 \theta - \sin^2 \theta$), we obtain

$$(C - A)\sin 2\theta + B\cos 2\theta = 0$$

$$B\cos 2\theta = (A - C)\sin 2\theta$$

$$\frac{\cos 2\theta}{\sin 2\theta} = \frac{A - C}{B}$$

$$\cot 2\theta = \frac{A - C}{B} \qquad (9)$$

Thus, if we choose θ so that $\cot 2\theta = (A - C)/B$, then $B' = 0$ and the $x'y'$ term in equation (7) will drop out. There is always an angle 2θ between 0 and 180° that solves equation (9); thus, there is always an angle θ between 0 and 90° that solves equation (9).

Angle of Rotation to Eliminate the $x'y'$ Term

Find θ, $0 < \theta < 90°$, so that

$$\cot 2\theta = \frac{A - C}{B}$$

where A, B, and C are the coefficients in

$$Ax^2 + Bxy + Cy^2 + Dx + Ey + F = 0$$

EXAMPLE 12 Given the equation $17x^2 - 6xy + 9y^2 = 72$, find the angle of rotation so that the transformed equation will have no $x'y'$ term. Sketch and identify the graph.

Solution $\quad 17x^2 - 6xy + 9y^2 = 72 \qquad (10)$

$$\cot 2\theta = \frac{A - C}{B} = \frac{17 - 9}{-6} = -\frac{4}{3}$$

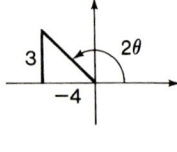

Thus, 2θ is a second-quadrant angle and using the reference triangle in the margin, we see that $\cos 2\theta = -\frac{4}{5}$. We can find the rotation formulas exactly by the use of the half-angle identities

$$\sin \theta = \sqrt{\frac{1 - \cos 2\theta}{2}} \qquad \text{and} \qquad \cos \theta = \sqrt{\frac{1 + \cos 2\theta}{2}}$$

Using these identities, we obtain

$$\sin \theta = \sqrt{\frac{1 - (-\frac{4}{5})}{2}} = \frac{3}{\sqrt{10}} \qquad \text{and} \qquad \cos \theta = \sqrt{\frac{1 + (-\frac{4}{5})}{2}} = \frac{1}{\sqrt{10}}$$

8-6 Rotation of Axes

Hence, the rotation formulas are

$$x = \frac{3}{\sqrt{10}} x' - \frac{3}{\sqrt{10}} y' \quad \text{and} \quad y = \frac{3}{\sqrt{10}} x' + \frac{1}{\sqrt{10}} y' \tag{11}$$

Substituting equation (11) into equation (10), we have

$$17 \left(\frac{1}{\sqrt{10}} x' - \frac{3}{\sqrt{10}} y' \right)^2 - 6 \left(\frac{1}{\sqrt{10}} x' - \frac{3}{\sqrt{10}} y' \right) \left(\frac{3}{\sqrt{10}} x' + \frac{1}{\sqrt{10}} y' \right) + 9 \left(\frac{3}{\sqrt{10}} x' + \frac{1}{\sqrt{10}} y' \right)^2 = 72$$

$$\frac{17}{10} (x' - 3y')^2 - \frac{6}{10} (x' - 3y')(3x' + y') + \frac{9}{10} (3x' + y')^2 = 72$$

Further simplification leads to

$$\frac{x'^2}{9} + \frac{y'^2}{4} = 1$$

which is a standard equation for an ellipse. To graph, we rotate the original axes through an angle of θ determined as follows:

$$\cot 2\theta = -\tfrac{4}{3}$$

$$2\theta \approx 143.1301°$$

$$\theta \approx 71.57°$$

We could also use either

$$\sin \theta = \frac{3}{\sqrt{10}} \quad \text{or} \quad \cos \theta = \frac{1}{\sqrt{10}}$$

to determine the angle of rotation. Summarizing these results, we have

$$17x^2 - 6xy + 9y^2 = 72$$

$$\frac{x'^2}{9} + \frac{y'^2}{4} = 1$$

$$\theta \approx 71.57°$$

PROBLEM 12 Given the equation $3x^2 + 26\sqrt{3}\, xy - 23y^2 = 144$, find the angle of rotation so that the transformed equation will have no $x'y'$ term. Sketch and identify the graph.

Answers to Matched Problems   **11.** $x'^2 - y'^2 = 1$; hyperbola   **12.** $\dfrac{x'^2}{9} - \dfrac{y'^2}{4} = 1$; $\theta = 30°$

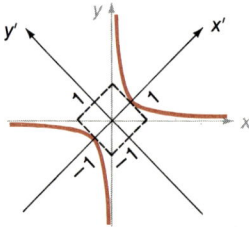 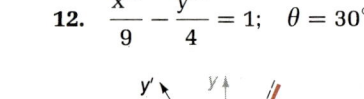

Exercise 8-6

A Find the transformed equation when the axes are rotated through the indicated angle. Sketch and identify the graph.

1. $x^2 + y^2 = 49$, $\theta = 45°$
2. $x^2 + y^2 = 25$, $\theta = 60°$
3. $2x^2 + \sqrt{3}xy + y^2 - 10 = 0$, $\theta = 30°$
4. $x^2 + 8xy + y^2 - 75 = 0$, $\theta = 45°$

B Given the indicated equation, find the angle of rotation so that the transformed equation will have no $x'y'$ term. Sketch and identify the graph.

5. $x^2 - 4xy + y^2 = 12$
6. $x^2 + xy + y^2 = 6$
7. $8x^2 - 4xy + 5y^2 = 36$
8. $5x^2 - 4xy + 8y^2 = 36$
9. $x^2 - 2\sqrt{3}xy + 3y^2 - 16\sqrt{3}x - 16y = 0$
10. $x^2 + 2\sqrt{3}xy + 3y^2 + 8\sqrt{3}x - 8y = 0$

C By use of a rotation followed by a translation, transform each equation into a standard equation. Sketch and identify the curve.

11. $x^2 + 2\sqrt{3}\,xy + 3y^2 - 8\sqrt{3}x - 8y - 4 = 0$
12. $73x^2 + 72xy + 52y^2 - 260x - 320y + 400 = 0$
13. Show that for any rotation of axes transformation, the equation of a circle $x^2 + y^2 = R^2$ transforms into $x'^2 + y'^2 = R^2$.
14. Show that for the general second-degree equation (1), $B'^2 - 4A'C' = B^2 - 4AC$ under a rotation of axes. This is one step in the proof of the following identification theorem.

THEOREM

$B^2 - 4AC$	Type of Curve
Negative	Ellipse, circle, point, no curve
Zero	Parabola, two lines, one line, no curve
Positive	Hyperbola, two lines

Section 8-7 Parametric Equations

- Parametric Equations and Plane Curves
- Projectile Motion
- Cycloid

■ **Parametric Equations and Plane Curves**

Consider the two equations

$$x = t + 1$$
$$y = t^2 - 2t \quad t \in (-\infty, \infty) \tag{1}$$

Each value of t determines a value of x, a value of y, and hence, an ordered pair (x, y). The set of all such ordered pairs (x, y) determined by letting t assume all values on some interval I constitutes a relation. To graph the relation we set up a table (see Table 1) involving t, x, and y, then plot the ordered pairs (x, y) as in Figure 18. The variable t, called a **parameter**, does not appear on the graph. Equations (1) are called **parametric equations** because both x and y are expressed in terms of the parameter t.

TABLE 1

t	0	1	2	3	4	5	−1	−2	−3
x	1	2	3	4	5	6	0	−1	−2
y	0	−1	0	3	8	15	3	8	15

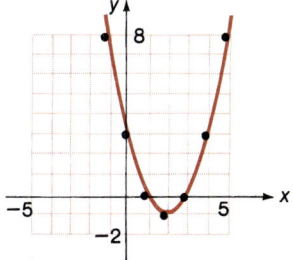

FIGURE 18

In some cases it is possible to eliminate the parameter by solving one of the equations for t and substituting into the other. In the example just considered, solving the first equation for t in terms of x, we have

$$t = x - 1$$

then, substituting the result into the second equation, we obtain

$$y = (x - 1)^2 - 2(x - 1)$$
$$= x^2 - 4x + 3$$

We recognize this as the equation of a parabola, as we would guess from Figure 18.

In other cases, it may not be easy or possible to eliminate the parameter to obtain an equation in just x and y. For example, for

$$x = t + \log t$$
$$y = t - e^t \qquad t \in (0, \infty)$$

you will not find it possible to solve either equation for t in terms of functions we have considered.

Is there more than one parametric representation for a plane curve? The answer is yes. In fact, there is an unlimited number of parametric representations for the same plane curve. The following are two additional (of infinitely many) representations of the parabola in Figure 18.

$$x = t + 3$$
$$y = t^2 + 2t \qquad t \in (-\infty, \infty) \qquad (2)$$

$$x = t$$
$$y = t^2 - 4t + 3 \qquad t \in (-\infty, \infty) \qquad (3)$$

Why are we interested in parametric representations of plane curves? It turns out that this approach is more general than using two-variable equations as we have been doing. In addition, the approach generalizes to curves in three- and higher-dimensional spaces. For this purpose we introduce the following general definition of a plane curve in a rectangular coordinate system.

Definition of a Plane Curve

A **plane curve** is the set of all points (x, y) such that

$$x = f(t)$$
$$y = g(t) \qquad t \in I$$

where f and g are both defined on the interval I.

There are other important reasons for using parametric representations of plane curves; these will be brought out in the discussion and examples that follow.

EXAMPLE 13 Graph the plane curve given parametrically by

$$x = 8 \cos \theta$$
$$y = 4 \sin \theta \qquad \theta \in (-\infty, \infty) \qquad (4)$$

Identify the curve by eliminating the parameter.

8-7 Parametric Equations

Solution Construct a table and graph.

θ	0	$\pi/6$	$\pi/3$	$\pi/2$	$2\pi/3$	$5\pi/6$	π	$7\pi/6$	$4\pi/3$	$3\pi/2$	$5\pi/3$	$11\pi/6$	2π
x	8	$4\sqrt{3}$	4	0	-4	$-4\sqrt{3}$	-8	$-4\sqrt{3}$	-4	0	4	$4\sqrt{3}$	8
y	0	2	$2\sqrt{3}$	4	$2\sqrt{3}$	2	0	-2	$-2\sqrt{3}$	-4	$-2\sqrt{3}$	-2	0

We eliminate the parameter θ as follows:

$\left(\dfrac{x}{8}\right)^2 = \cos^2\theta$ Divide both sides of equations (4) by 8 and 4, respectively, then square each side of each equation and add left side to left side and right side to right side. Recall $\sin^2\theta + \cos^2\theta = 1$.

$\left(\dfrac{y}{4}\right)^2 = \sin^2\theta$

$\dfrac{x^2}{64} + \dfrac{y^2}{16} = \cos^2\theta + \sin^2\theta$

$\dfrac{x^2}{64} + \dfrac{y^2}{16} = 1$

The graph is an ellipse.

PROBLEM 13 Graph the plane curve given parametrically by $x = 4\cos\theta$, $y = 4\sin\theta$, $\theta \in [0, \infty)$. Identify the curve by eliminating the parameter.

■ **Projectile Motion**

Using Newton's laws of motion and advanced mathematics, it can be shown that the motion of a projectile is given (neglecting air resistance) by

$$x = (v_0 \cos \alpha)t$$
$$y = (v_0 \sin \alpha)t - 16t^2 \qquad t \in [0, b] \qquad (5)$$

where the constant v_0 is the initial speed of the projectile in the direction of α with the horizontal (see Fig. 19). The parameter t represents time in seconds, and x and y are in feet. Solving the first equation in (5)

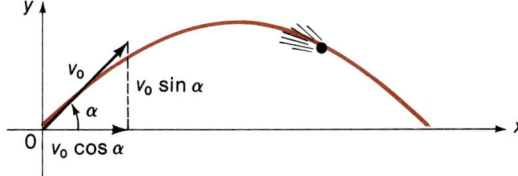

FIGURE 19   Projectile motion

for t in terms of x and substituting into the second equation, we obtain after simplification (a good exercise for the reader):

$$y = (\tan \alpha)x - \frac{16}{v_0^2 \cos^2 \alpha} x^2 \tag{6}$$

We recognize equation (6) as an equation of a parabola.

We now see another advantage of parametric representations of plane curves: The parameter t in many applications represents time, and from equations such as (5) we can not only determine the path that an object takes, but we can tell where the object is at any time t. In addition (using physics and calculus), we can perform operations on equations (5) that will yield new equations that can be used to determine the velocity and the acceleration of the projectile at any time t. From equation (6), an equation in two variables, we recognize that the projectile follows a particular parabolic path, but it tells us little else.

By keeping the initial speed v_0 of the projectile constant and varying the angle α in Figure 19, we obtain different parabolic paths described by the projectile and different ranges. The maximum range is obtained when $\alpha = 45°$. Irrespective of the angle α used, $0° \leq \alpha \leq 180°$, for a given initial speed v_0 there are places in the air and on the ground (staying in the same vertical plane) that the projectile cannot reach. It is an interesting fact (that can be proved using more advanced mathematics) that the reachable region is separated from the nonreachable region by a parabola called an **envelope** of the other parabolas (see Fig. 20).

FIGURE 20 Reachable region of a projectile

■ Cycloid

We now consider an unusual curve called a **cycloid**, which has a fairly simple parametric representation and a very complicated representation in terms of x and y only. The path traced by a point on the rim of a circle that rolls along a line is called a **cycloid**. To derive parametric equations for a cycloid we roll a circle of radius a along the x axis with the tracing point P on the rim starting at the origin (see Fig. 21).

Since the circle rolls along the x axis without slipping (referring to Fig. 21) we see that

$$\overline{OS} = \text{arc } PS$$
$$= a\theta \quad \theta \text{ in radians} \tag{7}$$

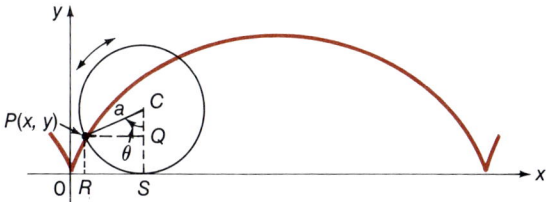

FIGURE 21 Cycloid

where S is the point of contact between the circle and the x axis. Referring to triangle CPQ, we see that

$$\overline{PQ} = a \sin \theta \qquad 0 \leq \theta \leq \pi/2 \qquad (8)$$

$$\overline{QC} = a \cos \theta \qquad 0 \leq \theta \leq \pi/2 \qquad (9)$$

Using these results, we have

$$\begin{aligned}
x &= \overline{OR} \\
&= \overline{OS} - \overline{RS} \\
&= (\text{arc } PS) - \overline{PQ} \qquad \text{Use equations (7) and (8)} \\
&= a\theta - a \sin \theta
\end{aligned}$$

$$\begin{aligned}
y &= \overline{RP} \\
&= \overline{SC} - \overline{QC} \qquad \text{Use equation (9) and the fact that } \overline{SC} = a \\
&= a - a \cos \theta
\end{aligned}$$

Even though θ in equations (8) and (9) was restricted so that $0 \leq \theta \leq \pi/2$, it can be shown that the derived parametric equations generate the whole cycloid for $\theta \in (-\infty, \infty)$. (The graph specifies a periodic function with period $2\pi a$.) Thus, in general:

Parametric Equations for a Cycloid

For a circle of radius a rolled along the x axis, the resulting cycloid generated by a point on the rim starting at the origin is given by

$$\begin{aligned}
x &= a\theta - a \sin \theta \\
y &= a - a \cos \theta
\end{aligned} \qquad \theta \in (-\infty, \infty)$$

The cycloid is a very good example of a curve that would have been very difficult to represent without the use of a parameter.

A cycloid has a very interesting physical property. An object sliding without friction from a point P to a point Q lower than P, but not on the same vertical line as P, will arrive at Q in a shorter time traveling along a cycloid than on any other path (see Fig. 22).

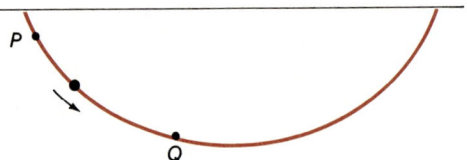

FIGURE 22 Cycloid path

Answers to Matched Problem   13. $x^2 + y^2 = 16$; circle

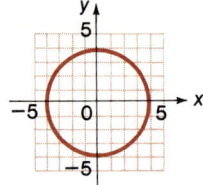

Exercise 8-7

A In Problems 1–10 plot each plane curve by use of a table of values (see Example 13). Obtain an equation in x and y by eliminating the parameter and identify the curve. The interval for the parameter is the whole real line, unless stated to the contrary.

1. $x = -t$, $\quad y = 2t - 2$ 2. $x = t$, $\quad y = t + 1$
3. $x = -t^2$, $\quad y = 2t^2 - 2$ 4. $x = t^2$, $\quad y = t^2 + 1$
5. $x = 3t$, $\quad y = -2t$ 6. $x = 2t$, $\quad y = t$
7. $x = \frac{1}{4}t^2$, $\quad y = t$ 8. $x = 2t$, $\quad y = t^2$
9. $x = \frac{1}{4}t^4$, $\quad y = t^2$ 10. $x = 2t^2$, $\quad y = t^4$

B In Problems 11–24 obtain an equation in x and y by eliminating the parameter. Use the simpler of the two forms to plot the curve. Identify the curve if it is a curve we have identified by name.

11. $x = 3 \sin \theta$, $\quad y = 4 \cos \theta$
12. $x = 3 \sin \theta$, $\quad y = 3 \cos \theta$
13. $x = 2 + 2 \sin \theta$, $\quad y = 3 + 2 \cos \theta$
14. $x = 3 + 4 \sin \theta$, $\quad y = 2 + 2 \cos \theta$
15. $x = t - 2$, $\quad y = \dfrac{2}{2 - t}$; $\quad t \neq 2$
16. $x = t - 1$, $\quad y = \dfrac{2}{t - 1}$; $\quad t \neq 1$

17. $x = t - 1, \quad y = \sqrt{t}; \quad t \geq 0$

18. $x = t^3, \quad y = t^2 + 1$

C **19.** $x = t^2, \quad y = t^{-2}; \quad t \neq 0$

20. $x = e^t, \quad y = e^{-t}$

21. $x = \cos 2\theta, \quad y = 4 \sin \theta$

22. $x = 3 \sec^2 \theta, \quad y = 2 \tan^2 \theta$

23. $x = \dfrac{8}{t^2 + 4}, \quad y = \dfrac{4t}{t^2 + 4}$

24. $x = \dfrac{4t}{t^2 + 1}, \quad y = \dfrac{4t^2}{t^2 + 1}$

Graph, using a hand calculator, one period $(0 \leq \theta \leq 2\pi)$ of each cycloid in Problems 25–26.

25. $x = \theta - \sin \theta, \quad y = 1 - \cos \theta$

26. $x = 2\theta - 2 \sin \theta, \quad y = 2 - 2 \cos \theta$

APPLICATIONS

27. *Plane motion.* An object follows a path as given by

$$x = 5 \sin(6\pi t)$$
$$y = 5 \cos(6\pi t) \quad t \geq 0$$

where t is time in seconds and x and y are in feet.
(A) What are the coordinates of the object when $t = 0.1$ second? (Compute answers to one decimal place.)
(B) Eliminate the parameter and graph the resulting equation in x and y. Identify the path.

28. *Plane motion.* Repeat the preceding problem for

$$x = 4 \sin(\pi t)$$
$$y = 2 \cos(\pi t) \quad t \geq 0$$

29. *Projectile motion.* A projectile is fired with an initial speed of 1,000 feet per second at an angle of 45° to the horizontal. Neglecting air resistance, find:
(A) The time of impact.
(B) The horizontal distance covered (range) in feet and miles at time of impact.
(C) Maximum height in feet of projectile.
(Compute all answers to three decimal places using a hand calculator.)

30. *Projectile motion.* Repeat the preceding problem if the same projectile is fired at 40° to the horizontal instead of 45°.

Section 8-8 Chapter Review

IMPORTANT TERMS AND SYMBOLS

8-1 Conic Sections. Right circular cone, conic section, circle, ellipse, parabola, hyperbola, degenerate conics (point, line, pair of lines)

8-2 Parabola. Definition, focus, directrix, axis, vertex, drawing a parabola, standard equations, $y^2 = 4ax$, $x^2 = 4ay$

8-3 Ellipse. Definition, foci, major axis, minor axis, vertices, center, drawing an ellipse, standard equations,

$$\frac{x^2}{a^2} + \frac{y^2}{b^2} = 1 \quad (a > b > 0), \qquad \frac{x^2}{b^2} + \frac{y^2}{a^2} = 1 \quad (a > b > 0)$$

8-4 Hyperbola. Definition, foci, vertices, transverse axis, conjugate axis, center, drawing a hyperbola, standard equations, asymptotes,

$$\frac{x^2}{a^2} - \frac{y^2}{b^2} = 1, \qquad \frac{y^2}{a^2} - \frac{x^2}{b^2} = 1$$

8-5 Translation of Axes. Transformation of coordinates, translation of axes, translation formulas, translation used in graphing, translation used in finding equations of conics, $x = x' + h$, $y = y' + k$, $x' = x - h$, $y' = y - k$

8-6 Rotation of Axes. Rotation of axes, rotation formulas, rotation used in graphing, $x = x' \cos \theta - y' \sin \theta$, $y = x' \sin \theta + y' \cos \theta$, $x' = x \cos \theta + y \sin \theta$, $y' = -x \sin \theta + y \cos \theta$

8-7 Parametric Equations. Parametric equations, parameter, defintion of a plane curve, graphing parametric equations

Exercise 8-8 Chapter Review

Work through all the problems in this chapter review and check answers in the back of the book. (Answers to all problems are there, and following each answer is a number in italics indicating the section in which that type of problem is discussed.) Where weaknesses show up, review appropriate sections in the text. When you are satisfied that you know the material, take the practice test following this review.

A Graph each equation and locate foci. Locate the directrix for any parabolas. Find the lengths of major, minor, transverse, and conjugate axes where applicable.

1. $9x^2 + 25y^2 = 225$
2. $x^2 = -12y$
3. $25y^2 - 9x^2 = 225$

In Problems 4–6:
(A) Transform each equation into a standard form for a nondegenerate conic by a suitable translation of coordinates. Identify the conic.
(B) Indicate the translation formulas used and indicate (h, k).

4. $4(y + 2)^2 - 25(x - 4)^2 = 100$
5. $(x + 5)^2 = -12(y + 4)$
6. $16(x - 6)^2 + 9(y - 4)^2 = 144$
7. Given the equation $x^2 - \sqrt{3}xy + 2y^2 - 10 = 0$, find the transformed equation when the axes are rotated through 30°. Sketch and identify the graph.
8. Plot the curve given parametrically by

$$x = -t^2$$
$$y = -\tfrac{1}{2}t^2 + 1$$

Obtain an equation in x and y by eliminating the parameter, and identify the curve.

B

9. Find the equation of the parabola having its vertex at the origin, its axis the x axis, and (−4, −2) on its graph.
10. Find an equation of an ellipse in the form

$$\frac{x^2}{M} + \frac{y^2}{N} = 1 \quad M, N > 0$$

if the center is at the origin, the major axis is on the y axis, the minor axis length is 6, and the distance of the foci from the center is 4.

11. Find an equation of a hyperbola in the form

$$\frac{y^2}{M} - \frac{x^2}{N} = 1 \quad M, N > 0$$

if the center is at the origin, the conjugate axis length is 8, and the foci are 5 units from the center.

In Problems 12–14 use a translation of coordinates to transform each equation into a standard equation for a nondegenerate conic. Identify the curve and graph it.

12. $16x^2 + 4y^2 + 96x - 16y + 96 = 0$
13. $x^2 - 4x - 8y - 20 = 0$
14. $4x^2 - 9y^2 + 24x - 36y - 36 = 0$
15. Given the equation $5x^2 + 26xy + 5y^2 + 72 = 0$, find the angle of rotation so that the transformed equation will have no x'y' term. Sketch and identify the graph.

16. Given the parametric equations of a plane curve, $x = -2 + 2\sin\theta$ and $y = 3 + 4\cos\theta$, obtain an equation in x and y by eliminating the parameter. Use the simpler of the two forms to plot the curve. Identify the curve.

C **17.** Use the definition of a parabola and the distance formula to find the equation of a parabola with directrix $x = 6$ and focus at (2, 4).

18. Find an equation of the set of points in a plane each of whose distance from (4, 0) is twice its distance from the line $x = 1$. Identify the geometric figure.

19. Find an equation of the set of points in a plane each of whose distance from (4, 0) is $\frac{2}{3}$ its distance from the line $x = 9$. Identify the geometric figure.

For Problems 20–22 find the coordinates of any foci relative to the original coordinate system.

20. Problem 12   **21.** Problem 13   **22.** Problem 14

23. Given the parametric equations of a plane curve

$$x = 2^t$$
$$y = 2^{-t}$$

obtain an equation in x and y by eliminating the parameter. Use the simpler of the two forms to graph the curve. Identify the curve.

Practice Test Chapter 8

Take this practice test as if it were a graded test. Allow yourself up to 50 minutes. Work the problems without looking back in the chapter. Correct your work using the answers (keyed to appropriate sections) in the back of the book.

1. Graph $y^2 = -4x$ and identify the curve.

2. Find the coordinates of the focus and the equation of the directrix for the graph in Problem 1.

3. Graph $4y^2 - 9x^2 = 36$ and identify the curve.

4. For the graph in Problem 3, find the coordinates of the foci and the length of the conjugate axis.

5. Use a translation of coordinates to transform

$$x^2 + 4y^2 - 4x + 24y + 24 = 0$$

into a standard equation for a nondegenerate conic. Identify the curve and graph it.

6. Write the coordinates of the foci of the conic in Problem 5 in terms of the original coordinate system and find the length of the major axis.

7. Plot the curve given parametrically by

$$x = t - 2$$
$$y = -2t + 6$$

Eliminate the parameter and identify the curve.

8. Given the parametric equations of a plane curve, $x = 4 \sin \theta$ and $y = 3 + 3 \cos \theta$, obtain an equation in x and y by eliminating the parameter. Use the simpler of the two forms to plot the curve. Identify the curve.

9. Find the equation of a hyperbola in the form

$$\frac{x^2}{M} - \frac{y^2}{N} = 1 \qquad M, N > 0$$

if the center is at the origin, the transverse axis length is 8, and the foci are $\sqrt{20}$ units from the center.

10. Write the rotation formulas that will eliminate the $x'y'$ term when used on the equation

$$7x^2 - 6\sqrt{3}xy + 13y^2 - 16 = 0$$

Systems of Equations and Inequalities   ▪9

9-1 Systems of Linear Equations—A Review
9-2 Systems and Augmented Matrices—An Introduction
9-3 Gauss–Jordan Elimination
9-4 Systems Involving Second-Degree Equations
9-5 Systems of Linear Inequalities
9-6 Chapter Review

A natural design of mathematical interest. Can you guess the source? See the back of the book.

Chapter 9 ▪ Systems of Equations and Inequalities

In this chapter we will first review how systems of equations are solved using techniques learned in elementary algebra. These techniques are suitable for systems involving two or three variables, but they are not suitable for systems involving larger numbers of variables. After this review, we will introduce techniques that are more suitable for solving systems with larger numbers of variables. These new techniques form the basis for computer solutions of large-scale systems.

Section 9-1 Systems of Linear Equations—A Review

- Systems in Two Variables
- Systems in Three Variables
- Application

▪ Systems in Two Variables

To establish basic concepts, consider the following simple example: If two children have a combined weight of 80 kilograms and one weighs 20 kilograms more than the other, what is the weight of each?

Let x = Weight of heavier child
y = Weight of lighter child

Then $x + y = 80$
$x - y = 20$

We now have a system of two equations and two unknowns. To solve this system we find all ordered pairs of real numbers that satisfy both equations. In general, we are interested in solving linear systems of the type

$ax + by = h$
$cx + dy = k$

where a, b, c, d, h, and k are real constants. The **solution set** of this system is the set of all ordered pairs of numbers such that each ordered pair

9-1 Systems of Linear Equations—A Review

satisfies each equation in the system. We will consider three methods of solving such systems, each with certain advantages, depending on the situation.

SOLUTION BY GRAPHING To solve the weight problem by graphing, we graph both equations in the same coordinate system. Then the coordinates of any points that the graphs have in common must be solutions to the system, since they must satisfy both equations.

EXAMPLE 1 Solve the weight problem by graphing:

$$x + y = 80$$
$$x - y = 20$$

Solution

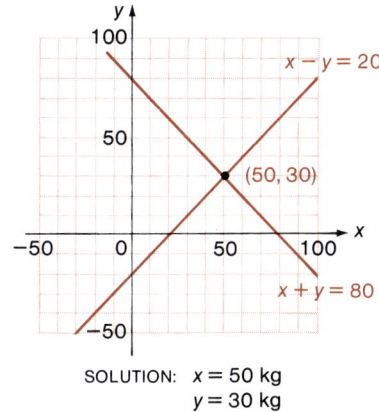

SOLUTION: $x = 50$ kg
$y = 30$ kg

PROBLEM 1 Solve by graphing:

$$x + y = 12$$
$$x - y = 4$$

It is clear that the preceding example (and problem) has exactly one solution, since the lines have exactly one point of intersection. In general, lines in a rectangular coordinate system are related to each other in one of the three ways illustrated in the next example.

EXAMPLE 2 Solve each of the following systems by graphing.

(A) $2x - 3y = 2$
     $x + 2y = 8$

(B) $4x + 6y = 12$
     $2x + 3y = -6$

(C) $2x - 3y = -6$
     $-x + \frac{3}{2}y = 3$

Solution (A) 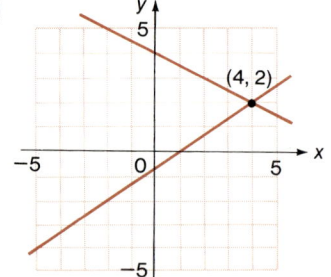 (B)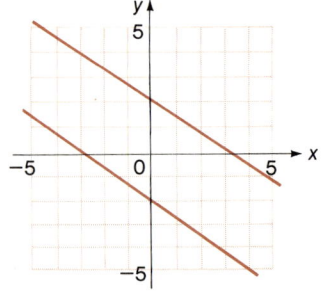

Lines intersect at
one point only.
Exactly one solution:
$x = 4, \quad y = 2$

Lines are parallel
(each has slope $-\frac{2}{3}$).
No solution.

(C)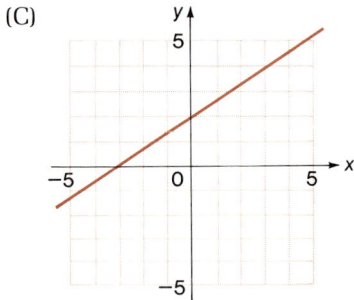

Lines coincide.
Infinitely many
solutions.

PROBLEM 2 Solve each of the following systems by graphing.

(A) $2x + 3y = 12$ (B) $\quad x - 3y = -3$ (C) $\quad 2x - 3y = 12$
$\quad\;\; x - 3y = -3$ $-2x + 6y = 12$ $-x + \frac{3}{2}y = -6$

By interpreting a system of two linear equations in two unknowns geometrically, we gain useful information about what to expect in the way of solutions to the system. Since two lines in a coordinate system must intersect in exactly one point, be parallel, or coincide, we conclude that the system has (1) exactly one solution, (2) no solution, or (3) infinitely many solutions. In addition, graphs frequently reveal relationships in problems that might otherwise be hidden. Generally, however, graphic methods give us only rough approximations of solutions. The methods of elimination by substitution and elimination by addition to be considered next will yield solutions to any decimal accuracy desired—assuming solutions exist.

9-1 Systems of Linear Equations—A Review

SOLUTION BY ELIMINATION USING SUBSTITUTION

Choose one of the two equations in a system and solve for one variable in terms of the other (make a choice that avoids fractions, if possible). Then substitute the result into the other equation and solve the resulting linear equation in one variable. Now substitute this result back into either of the original equations to find the second variable. An example should make the process clear.

EXAMPLE 3 Solve by elimination using substitution.

$$2x - 3y = 7$$
$$3x - y = 7$$

Solution Solve either equation for one variable in terms of the other; then substitute into the remaining equation. In this problem we can avoid fractions if we solve for y in terms of x in the second equation.

$3x - y = 7$ Solve the second equation for y in terms of x.
$-y = -3x + 7$
$y = 3x - 7$ Substitute into the first equation to eliminate y.

$2x - 3y = 7$ First equation
$2x - 3(3x - 7) = 7$
$2x - 9x + 21 = 7$
$-7x = -14$
$\mathbf{x = 2}$

Now replace x with 2 in $y = 3x - 7$ to find y:

$y = 3x - 7$
$= 3(2) - 7$
$\mathbf{y = -1}$

Thus, $(2, -1)$ is the unique solution to the original system.

Check

$2x - 3y = 7$ $3x - y = 7$
$2(2) - 3(-1) \stackrel{?}{=} 7$ $3(2) - (-1) \stackrel{?}{=} 7$
$7 \stackrel{\checkmark}{=} 7$ $7 \stackrel{\checkmark}{=} 7$

PROBLEM 3 Solve by elimination using substitution.

$$3x - 4y = 18$$
$$2x + y = 1$$

SOLUTION BY ELIMINATION USING ADDITION

Now we turn to **elimination using addition**. This is probably the most important method of solution, since it is readily generalized to higher-order systems. The method involves the replacement of systems of equations with simpler *equivalent systems* (by performing appropriate operations) until we obtain a system with an obvious solution. **Equivalent systems** of equations are, as you would expect, systems that have exactly the same solution set. Theorem 1 lists operations that produce equivalent systems.

THEOREM 1 | Producing Equivalent Systems

Equivalent systems of equations result if:

1. Two equations are interchanged.
2. An equation is multiplied by a nonzero constant.
3. A constant multiple of another equation is added to a given equation.

EXAMPLE 4 Solve by elimination using addition.

$$3x - 2y = 8$$
$$2x + 5y = -1$$

Solution We use Theorem 1 to eliminate one of the variables and thus obtain a system with an obvious solution.

$$3x - 2y = 8$$
$$2x + 5y = -1$$

If we multiply the top equation by 5, the bottom by 2, and then add, we can eliminate y.

$$15x - 10y = 40$$
$$\underline{4x + 10y = -2}$$
$$19x = 38$$
$$\mathbf{x = 2}$$

Now substitute $x = 2$ back into either of the original equations, say the second equation, and solve for y ($x = 2$ paired with either of the two original equations produces an equivalent system).

$$2(2) + 5y = -1$$
$$5y = -5$$
$$\mathbf{y = -1}$$

Check

$$3x - 2y = 8 \qquad\qquad 2x + 5y = -1$$
$$3(2) - 2(-1) \stackrel{?}{=} 8 \qquad 2(2) + 5(-1) \stackrel{?}{=} -1$$
$$8 \stackrel{\checkmark}{=} 8 \qquad\qquad -1 \stackrel{\checkmark}{=} -1$$

9-1 Systems of Linear Equations—A Review

PROBLEM 4 Solve by elimination using addition.
$$6x + 3y = 3$$
$$5x + 4y = 7$$

Let us see what happens in the elimination process when a system either has no solution or has infinitely many solutions. Consider the following system:
$$2x + 6y = -3$$
$$x + 3y = 2$$

Multiplying the second equation by -2 and adding, we obtain
$$2x + 6y = -3$$
$$\underline{-2x - 6y = -4}$$
$$0 = -7$$

We have obtained a contradiction. An assumption that the original system has solutions must be false (otherwise, we have proved that $0 = -7$). Thus, the system has no solutions. The graphs of the equations are parallel. Systems with no solutions are said to be **inconsistent**. Systems of equations that have solutions are said to be **consistent**.

Now consider the system
$$x - \tfrac{1}{2}y = 4$$
$$-2x + y = -8$$

If we multiply the top equation by 2 and add the result to the bottom equation, we get
$$2x - y = 8$$
$$\underline{-2x + y = -8}$$
$$0 = 0$$

Obtaining $0 = 0$ by addition implies that the equations are equivalent. (Why?) Hence, the two equations have the same solution set, and the system has infinitely many solutions. If $x = k$, then $y = 2k - 8$; that is, $(k, 2k - 8)$ is a solution for any real number k. Such a system is said to be **dependent**. The variable k is called a **parameter**; replacing it with any real number produces a particular solution to the system.

■ Systems in Three Variables

Now that we know how to solve systems of linear equations in two variables, there is no reason to stop there. Systems of the form

$$a_1x + b_1y + c_1z = k_1$$
$$a_2x + b_2y + c_2z = k_2 \qquad (1)$$
$$a_3x + b_3y + c_3z = k_3$$

as well as higher-order systems are encountered frequently. In fact, systems of equations are so important in solving real-world problems that there are whole courses devoted to this one topic. A triplet of numbers $x = x_0$, $y = y_0$, and $z = z_0$ [also written as an ordered triplet (x_0, y_0, z_0)] is a **solution** of system (1) if each equation is satisfied by this triplet. The set of all such ordered triplets of numbers is called the **solution set** of the system. Two systems are said to be **equivalent** if they have the same solution set. Linear equations in three variables represent planes in a three-dimensional space. Trying to visualize how three planes can intersect will give you insight as to what kind of solution sets are possible for system (1).

In this section we will use an extension of the method of elimination to solve systems in the form of (1). In the next section we will consider techniques for solving linear systems that are more compatible with a computer approach to solving such systems. In practice, most linear systems involving more than three variables are usually solved with the aid of a computer.

Steps in Solving Systems of Form (1)

1. Choose two equations from the system and eliminate one of the three variables using elimination by addition. The result is generally one equation in two unknowns.
2. Now eliminate the same variable from the unused equation and one of those used in step 1. We (generally) obtain another equation in two variables.
3. The two equations from steps 1 and 2 form a system of two equations and two unknowns. Solve as described in the first part of this section.
4. Substitute the solution from step 3 into any of the three original equations and solve for the third variable to complete the solution of the original system.

EXAMPLE 5 Solve.

$$3x - 2y + 4z = 6 \quad (2)$$
$$2x + 3y - 5z = -8 \quad (3)$$
$$5x - 4y + 3z = 7 \quad (4)$$

Solution *Step 1.* We look at the coefficients of the variables and choose to eliminate y from equations (2) and (4) because of the convenient coefficients

9-1 Systems of Linear Equations—A Review

-2 and -4. Multiply equation (2) by -2 and add to equation (4):

$$
\begin{array}{rl}
-6x + 4y - 8z = -12 & \quad -2[\text{Equation (2)}] \\
\underline{5x - 4y + 3z = 7} & \quad \text{Equation (4)} \\
-x \phantom{{}+4y} - 5z = -5 &
\end{array}
\tag{5}
$$

Step 2. Now we eliminate y (the same variable) from equations (2) and (3):

$$
\begin{array}{rl}
9x - 6y + 12z = 18 & \quad 3[\text{Equation (2)}] \\
\underline{4x + 6y - 10z = -16} & \quad 2[\text{Equation (3)}] \\
13x \phantom{{}+6y} + 2z = 2 &
\end{array}
\tag{6}
$$

Step 3. From steps 1 and 2 we obtain the system

$$-x - 5z = -5 \quad \text{These equations, along with (2), (3), or (4), form} \tag{5}$$
$$13x + 2z = 2 \quad \text{a system equivalent to the original system.} \tag{6}$$

We solve this system as in the first part of this section:

$$
\begin{array}{rl}
-13x - 65z = -65 & \quad 13[\text{Equation (5)}] \\
\underline{13x + 2z = 2} & \quad \text{Equation (6)} \\
 -63z = -63 &
\end{array}
$$

$$\mathbf{z = 1}$$

Substitute $z = 1$ back into either equation (5) or (6) [we choose equation (5)] to find x:

$$
\begin{aligned}
-x - 5z &= -5 \\
-x - 5(1) &= -5 \\
-x &= 0
\end{aligned}
\tag{5}
$$

$$\mathbf{x = 0}$$

Step 4. Substitute $x = 0$ and $z = 1$ back into any of the three original equations [we choose equation (2)] to find y:

$$
\begin{aligned}
3x - 2y + 4z &= 6 \\
3(0) - 2y + 4(1) &= 6 \\
-2y + 4 &= 6 \\
-2y &= 2
\end{aligned}
\tag{2}
$$

$$\mathbf{y = -1}$$

Thus, the solution to the original system is $(0, -1, 1)$, or $x = 0$, $y = -1$, $z = 1$.

Check To check the solution, we must check *each* equation in the original system:

$$3x - 2y + 4z = 6 \qquad 2x + 3y - 5z = -8$$
$$3(0) - 2(-1) + 4(1) \stackrel{?}{=} 6 \qquad 2(0) + 3(-1) - 5(1) \stackrel{?}{=} -8$$
$$6 \stackrel{\checkmark}{=} 6 \qquad -8 \stackrel{\checkmark}{=} -8$$

$$5x - 4y + 3z = 7$$
$$5(0) - 4(-1) + 3(1) \stackrel{?}{=} 7$$
$$7 \stackrel{\checkmark}{=} 7$$

PROBLEM 5 Solve.

$$3x - 4y + 2z = -9$$
$$2x + 5y - 3z = 5$$
$$4x - 2y - 4z = -12$$

In the process just described, if we encounter an equation that states a contradiction, such as $0 = -2$, then we must conclude that the system has no solution (that is, the system is inconsistent). If, on the other hand, one of the equations turns out to be $0 = 0$, either the system has infinitely many solutions or it has none. We must proceed further to determine which. Notice how this last result differs from the two-equation–two-unknown case. There, when we obtained $0 = 0$, we *knew* that there were infinitely many solutions. We will have more to say about this in the following sections.

■ Application

We now consider a real-world problem that leads to a system of equations.

EXAMPLE 6 *Production scheduling.* A small manufacturing plant makes three types of inflatable boats: one-person, two-person, and four-person models. Each boat requires the services of three departments, as listed in the table. The cutting, assembly, and packaging departments have available a maximum of 380, 330, and 120 work-hours per week, respectively. How many boats of each type must be produced each week for the plant to operate at full capacity?

	ONE-PERSON BOAT	TWO-PERSON BOAT	FOUR-PERSON BOAT
Cutting department	0.6 hr	1.0 hr	1.5 hr
Assembly department	0.6 hr	0.9 hr	1.2 hr
Packaging department	0.2 hr	0.3 hr	0.5 hr

9-1 Systems of Linear Equations—A Review

Solution   Let x = Number of one-person boats produced per week
y = Number of two-person boats produced per week
z = Number of four-person boats produced per week

Then $0.6x + 1.0y + 1.5z = 380$ Cutting department
$0.6x + 0.9y + 1.2z = 330$ Assembly department
$0.2x + 0.3y + 0.5z = 120$ Packaging department

We can clear the system of decimals, if desired, by multiplying each side of each equation by 10. Thus,

$6x + 10y + 15z = 3,800$ (7)
$6x + 9y + 12z = 3,300$ (8)
$2x + 3y + 5z = 1,200$ (9)

Let us start by eliminating x from equations (7) and (8):

$$\text{Add}\begin{cases} 6x + 10y + 15z = 3,800 & \text{Equation (7)} \\ -6x - 9y - 12z = -3,300 & -1[\text{Equation (8)}] \end{cases}$$
$$y + 3z = 500$$

Now we eliminate x from equations (7) and (9):

$$\text{Add}\begin{cases} 6x + 10y + 15z = 3,800 & \text{Equation (7)} \\ -6x - 9y - 15z = -3,600 & -3[\text{Equation (9)}] \end{cases}$$
$$\mathbf{y = 200}$$

Substituting $y = 200$ into $y + 3z = 500$, we can solve for z:

$200 + 3z = 500$
$3z = 300$
$\mathbf{z = 100}$

Now use equation (7), (8), or (9) to find x [we use (9)]:

$2x + 3y + 5z = 1,200$
$2x + 3(200) + 5(100) = 1,200$
$2x = 100$
$\mathbf{x = 50}$

Thus, each week, the company should produce 50 one-person boats, 200 two-person boats, and 100 four-person boats to operate at full capacity. The check of the solution is left to the reader.

PROBLEM 6   Repeat Example 6 assuming the cutting, assembly, and packaging departments have available a maximum of 260, 234, and 82 work-hours per week, respectively.

Answers to Matched Problems

1. $(8, 4)$, or $x = 8$ and $y = 4$
2. (A) $(3, 2)$, or $x = 3$ and $y = 2$ (B) No solution
   (C) Infinite number of solutions
3. $(2, -3)$, or $x = 2$ and $y = -3$
4. $(-1, 3)$, or $x = -1$ and $y = 3$
5. $(-1, 2, 1)$, or $x = -1$, $y = 2$, and $z = 1$
6. 100 one-person boats, 140 two-person boats, and 40 four-person boats

Exercise 9-1

A *Solve by graphing.*

1. $3x - 2y = 12$
   $7x + 2y = 8$
2. $3x - y = 2$
   $x + 2y = 10$
3. $3u + 5v = 15$
   $6u + 10v = -30$
4. $m + 2n = 4$
   $2m + 4n = -8$

Solve by elimination using substitution.

5. $x - y = 4$
   $x + 3y = 12$
6. $2x - y = 3$
   $x + 2y = 14$
7. $3x - y = 7$
   $2x + 3y = 1$
8. $2x + y = 6$
   $x - y = -3$

Solve by elimination using addition.

9. $2x + 3y = 1$
   $3x - y = 7$
10. $2m - n = 10$
    $m - 2n = -4$
11. $4x + 3y = 26$
    $3x - 11y = -7$
12. $9x - 3y = 24$
    $11x + 2y = 1$

Solve by elimination using either substitution or addition.

13. $3x - 6y = -9$
    $-2x + 4y = 6$
14. $2x - 3y = -2$
    $-4x + 6y = 7$
15. $7m + 12n = -1$
    $5m - 3n = 7$
16. $3x + 8y = 4$
    $15x + 10y = -10$
17. $2x + 4y = -8$
    $x + 2y = 4$
18. $-6x + 10y = -30$
    $3x - 5y = 15$
19. $y = 0.08x$
    $y = 100 + 0.04x$
20. $y = 0.07x$
    $y = 80 + 0.05x$

B *Solve.*

21. $0.3u - 0.6v = 0.18$
    $0.5u + 0.2v = 0.54$
22. $0.2x - 0.5y = 0.07$
    $0.8x - 0.3y = 0.79$

9-1 Systems of Linear Equations—A Review

23. $x - 3y + z = 4$
    $-x + 4y - 4z = 1$
    $2x - y + 5z = -3$

24. $2x + y - z = 5$
    $x - 2y - 2z = 4$
    $3x + 4y + 3z = 3$

25. $3u - 2v + 3w = 11$
    $2u + 3v - 2w = -5$
    $u + 4v - w = -5$

26. $2a + 4b + 3c = 6$
    $a - 3b + 2c = -7$
    $-a + 2b - c = 5$

C 27. $3x - 2y - 4z = -8$
    $4x + 3y - 5z = -5$
    $6x - 5y + 2z = -17$

28. $2x - 3y + 3z = -15$
    $3x + 2y - 5z = 19$
    $5x - 4y - 2z = -2$

29. $-x + 2y - z = -4$
    $2x + 5y - 4z = -16$
    $x + y - z = -4$

30. $x - 8y + 2z = -1$
    $x - 3y + z = 1$
    $2x - 11y + 3z = 2$

APPLICATIONS

31. *Puzzle.* A friend of yours came out of the post office having spent $8.80 on 20¢ and 15¢ stamps. If she bought forty-seven stamps in all, how many of each type did she buy?

32. *Puzzle.* A parking meter contains only nickels and dimes worth $6.05. If there are eighty-nine coins in all, how many of each type are there?

33. *Chemistry.* A chemist has two concentrations of hydrochloric acid in stock: a 50% solution and an 80% solution. How much of each should she mix to obtain 100 milliliters of a 68% solution?

34. *Chemistry.* Repeat Problem 33 assuming the 50% stock solution is replaced with a 60% stock solution.

35. *Business.* A jeweler has two bars of gold alloy in stock, one 12 carat and the other 18 carat (24-carat gold is pure gold, 12-carat gold is 12/24 pure, 18-carat gold is 18/24 pure, and so on). How many grams of each alloy must be mixed to obtain 10 grams of 14-carat gold?

36. *Business.* Repeat Problem 35 assuming the jeweler has only 10-carat and pure gold in stock.

37. *Nutrition.* Animals in an experiment are to be kept on a strict diet. Each animal is to receive, among other things, 20 grams of protein and 6 grams of fat. The laboratory technician is able to purchase two food mixes with the compositions shown in the table. How many grams of each mix should be used to obtain the right diet for a single animal?

MIX	PROTEIN (%)	FAT (%)
A	10	6
B	20	2

38. *Nutrition.* A biologist in a nutrition experiment wants to prepare a special diet for her experimental animals. She requires a food mixture that contains, among other things, 20 ounces of protein and 6 ounces of fat. Food mixes are available with the compositions shown in the table. How many ounces of each mix should be used to prepare the diet mix?

MIX	PROTEIN (%)	FAT (%)
A	20	2
B	10	6

39. *Earth science.* An earthquake emits a primary wave and a secondary wave. Near the surface of the earth the primary wave travels at about 5 miles/second, and the secondary wave at about 3 miles/second. From the time lag between the two waves arriving at a given station, it is possible to estimate the distance to the quake. (The *epicenter* can be located by obtaining distance bearings at three or more stations.) Suppose a station measured a time difference of 16 seconds between the arrival of the two waves. How long did each wave travel, and how far was the earthquake from the station?

40. *Earth science.* A ship using sound-sensing devices above and below water recorded a surface explosion 6 seconds sooner by its underwater device than its above-water device. Sound travels in air at about 1,100 feet/second and in seawater at about 5,000 feet/second.
    (A) How long did it take each sound wave to reach the ship?
    (B) How far was the explosion from the ship?

41. *Production scheduling.* A garment industry manufactures three shirt styles. Each style of shirt requires the services of three departments, as listed in the table. The cutting, sewing, and packaging departments have available a maximum of 1,160, 1,560, and 480 work-hours per week, respectively. How many of each style shirt must be produced each week for the plant to operate at full capacity?

	STYLE A	STYLE B	STYLE C
Cutting department	0.2 hr	0.4 hr	0.3 hr
Sewing department	0.3 hr	0.5 hr	0.4 hr
Packaging department	0.1 hr	0.2 hr	0.1 hr

42. *Production scheduling.* Repeat Problem 41 with the cutting, sewing, and packaging departments having available a maximum of 1,180, 1,560, and 510 work-hours per week, respectively.

43. *Diet.* In an experiment involving mice, a zoologist finds she needs a food mix that contains, among other things, 23 grams of protein, 6.2 grams of fat, and 16 grams of moisture. She has on hand mixes with the compositions shown in the table. How many grams of each mix should she use to get the desired diet mix?

MIX	PROTEIN (%)	FAT (%)	MOISTURE (%)
A	20	2	15
B	10	6	10
C	15	5	5

44. *Diet.* Repeat Problem 43 assuming the diet mix is to contain 18.5 grams of protein, 4.9 grams of fat, and 13 grams of moisture.

Section 9-2 Systems and Augmented Matrices—An Introduction

- Introduction
- Augmented Matrices
- Solving Linear Systems Using Augmented Matrix Methods

- Introduction

Most linear systems of any consequence involve large numbers of equations and unknowns. These systems are solved using computers, since hand methods would be impractical (try solving even a five-equation–five-unknown problem and you will understand why). However, even if you have a computer facility to help solve your problem, it is still important for you to know how to formulate the problem so that it can be solved by a computer. In addition, it is helpful to have at least a general idea of how computers solve these problems. And, finally, it is important for you to know how to interpret the results.

Even though the procedures and notation introduced in this and the next section are more involved than those used in the preceding section, it is important to keep in mind that our objective is not to find an efficient hand method for solving large-scale systems (there is none), but rather to find a process that generalizes readily for computer use. It turns out that you will receive an added bonus for your efforts, since several of the processes developed in this and the next section will be of additional use in Section 10-3.

■ Augmented Matrices

In solving systems of equations by elimination in the preceding section the coefficients of the variables and constant terms played a central role. The process can be made more efficient for generalization and computer work by the introduction of a mathematical form called a *matrix*. A **matrix** is a rectangular array of numbers written within brackets. Some examples are

$$\begin{bmatrix} 3 & 5 \\ 0 & -2 \end{bmatrix} \quad \begin{bmatrix} 2 \\ -3 \\ 0 \end{bmatrix} \quad \begin{bmatrix} 1 & -1 & 0 & 5 \end{bmatrix}$$

$$\begin{bmatrix} -1 & 2 & -5 & 0 \\ 0 & 3 & 2 & 1 \end{bmatrix} \quad \begin{bmatrix} 1 & 0 & 0 \\ 0 & 1 & 0 \\ 0 & 0 & 1 \end{bmatrix}$$

Each number in a matrix is called an **element** of the matrix.

Associated with each linear system of the form*

$$\begin{aligned} a_1 x_1 + b_1 x_2 &= k_1 \\ a_2 x_1 + b_2 x_2 &= k_2 \end{aligned} \tag{1}$$

where x_1 and x_2 are variables, is a matrix called the **augmented matrix** of the system:

$$\begin{bmatrix} a_1 & b_1 & | & k_1 \\ a_2 & b_2 & | & k_2 \end{bmatrix} \quad \begin{matrix} \leftarrow \text{Row 1 } (R_1) \\ \leftarrow \text{Row 2 } (R_2) \end{matrix} \tag{2}$$

with Column 1 (C_1), Column 2 (C_2), Column 3 (C_3).

This matrix contains the essential parts of system (1). The vertical bar is included only to separate the coefficients of the variables from the constant terms. Our objective is to learn how to manipulate augmented matrices in such a way that a solution to system (1) will result, if a solution exists. The manipulative process is a direct outgrowth of the elimination process discussed in Section 9-1.

---

* We are gradually shifting notation for variables and constants to **subscript notation**. Subscript notation is more convenient for the generalization of these concepts, since in large systems one soon runs out of letters.

Recall that two linear systems are said to be **equivalent** if they have exactly the same solution set. How did we transform linear systems into equivalent linear systems? We used Theorem 1, which we restate here for convenient reference.

THEOREM 1

Producing Equivalent Systems

A system of linear equations is transformed into an equivalent system if:

1. Two equations are interchanged.
2. An equation is multiplied by a nonzero constant.
3. A constant multiple of another equation is added to a given equation.

Paralleling the previous discussion, we say that two augmented matrices are **row-equivalent**, denoted by the symbol $\sim$ between the two matrices, if they are augmented matrices of equivalent systems of equations. (Think about this.) How do we transform augmented matrices into row-equivalent matrices? We use Theorem 2, which is a direct consequence of Theorem 1.

THEOREM 2

Producing Row-Equivalent Matrices

An augmented matrix is transformed into a row-equivalent matrix if:

1. Two rows are interchanged ($R_i \leftrightarrow R_j$).
2. A row is multiplied by a nonzero constant ($kR_i \to R_i$).
3. A constant multiple of another row is added to a given row ($R_i + kR_j \to R_i$).

[*Note:* The arrow $\to$ means "replaces."]

■ Solving Linear Systems Using Augmented Matrix Methods

The use of Theorem 2 in solving systems in the form of (1) is best illustrated by examples.

EXAMPLE 7 Solve, using augmented matrix methods.

$$3x_1 + 4x_2 = 1$$
$$x_1 - 2x_2 = 7$$

(3)

Solution We start by writing the augmented matrix corresponding to (3).

$$\begin{bmatrix} 3 & 4 & | & 1 \\ 1 & -2 & | & 7 \end{bmatrix} \qquad (4)$$

Our objective is to use row operations from Theorem 2 to try to transform (4) into the form

$$\begin{bmatrix} 1 & 0 & | & m \\ 0 & 1 & | & n \end{bmatrix} \qquad (5)$$

where m and n are real numbers. The solution to system (3) will then be obvious, since matrix (5) will be the augmented matrix of the following system:

$$x_1 = m$$
$$x_2 = n$$

We now proceed to use row operations to transform (4) into form (5).

Step 1. To get a 1 in the upper left corner, we interchange rows 1 and 2 (Theorem 2-1).

$$\begin{bmatrix} 3 & 4 & | & 1 \\ 1 & -2 & | & 7 \end{bmatrix} \xrightarrow{R_1 \leftrightarrow R_2} \begin{bmatrix} 1 & -2 & | & 7 \\ 3 & 4 & | & 1 \end{bmatrix} \quad \text{Now you see why we wanted Theorem 1-1.}$$

Step 2. To get a 0 in the lower left corner, we multiply R_1 by (-3) and add to R_2 (Theorem 2-3)—this changes R_2 but not R_1. Some people find it useful to write $(-3)R_1$ outside the matrix to help reduce errors in arithmetic, as shown.

$$\begin{array}{ccc} -3 & 6 & -21 \leftarrow\!-\!-\!-\!-\!\rceil \\ \end{array}$$
$$\begin{bmatrix} 1 & -2 & | & 7 \\ 3 & 4 & | & 1 \end{bmatrix} \xrightarrow{R_2 + (-3)R_1 \to R_2} \begin{bmatrix} 1 & -2 & | & 7 \\ 0 & 10 & | & -20 \end{bmatrix}$$

Step 3. To get a 1 in the second row, second column, we multiply R_2 by $\frac{1}{10}$ (Theorem 2-2).

$$\begin{bmatrix} 1 & -2 & | & 7 \\ 0 & 10 & | & -20 \end{bmatrix} \xrightarrow{\frac{1}{10}R_2 \to R_2} \begin{bmatrix} 1 & -2 & | & 7 \\ 0 & 1 & | & -2 \end{bmatrix}$$

Step 4. To get a 0 in the first row, second column, we multiply R_2 by 2 and add the result to R_1 (Theorem 2-3)—this changes R_1 but not R_2.

$$\begin{array}{ccc} 0 & 2 & -4 \leftarrow\!-\!-\!-\!-\!\rceil \\ \end{array}$$
$$\begin{bmatrix} 1 & -2 & | & 7 \\ 0 & 1 & | & -2 \end{bmatrix} \xrightarrow{R_1 + 2R_2 \to R_1} \begin{bmatrix} 1 & 0 & | & 3 \\ 0 & 1 & | & -2 \end{bmatrix}$$

9-2 Systems and Augmented Matrices—An Introduction

We have accomplished our objective! The last matrix is the augmented matrix for the system

$$x_1 = 3$$
$$x_2 = -2$$
(6)

Since system (6) is equivalent to system (3), our starting system, we have solved (3); that is, $x_1 = 3$ and $x_2 = -2$.

Check
$$3x_1 + 4x_2 = 1 \qquad x_1 - 2x_2 = 7$$
$$3(3) + 4(-2) \stackrel{?}{=} 1 \qquad 3 - 2(-2) \stackrel{?}{=} 7$$
$$9 - 8 \stackrel{\checkmark}{=} 1 \qquad 3 + 4 \stackrel{\checkmark}{=} 7$$

This process is written more compactly as follows:

Step 1: Need a 1 here
$$\begin{bmatrix} 3 & 4 & | & 1 \\ 1 & -2 & | & 7 \end{bmatrix} \quad R_1 \leftrightarrow R_2$$

Step 2: Need a 0 here
$$\sim \begin{bmatrix} 1 & -2 & | & 7 \\ 3 & 4 & | & 1 \end{bmatrix} \quad R_2 + (-3)R_1 \to R_2$$
$$ \quad -3 \quad\; 6 \quad -21$$

Step 3: Need a 1 here
$$\sim \begin{bmatrix} 1 & -2 & | & 7 \\ 0 & 10 & | & -20 \end{bmatrix} \quad \tfrac{1}{10}R_2 \to R_2$$

Step 4: Need a 0 here
$$\sim \begin{bmatrix} 1 & -2 & | & 7 \\ 0 & 1 & | & -2 \end{bmatrix} \quad R_1 + 2R_2 \to R_1$$
$$ \quad 0 \quad\; 2 \quad -4$$

$$\sim \begin{bmatrix} 1 & 0 & | & 3 \\ 0 & 1 & | & -2 \end{bmatrix}$$

Therefore, $x_1 = 3$ and $x_2 = -2$.

PROBLEM 7 Solve, using augmented matrix methods.
$$2x_1 - x_2 = -7$$
$$x_1 + 2x_2 = 4$$

EXAMPLE 8 Solve, using augmented matrix methods:
$$2x_1 - 3x_2 = 7$$
$$3x_1 + 4x_2 = 2$$

Solution

Step 1: Need a 1 here
$$\begin{bmatrix} 2 & -3 & | & 7 \\ 3 & 4 & | & 2 \end{bmatrix} \quad \tfrac{1}{2}R_1 \to R_1$$

Step 2: Need a 0 here
$$\sim \begin{bmatrix} 1 & -\tfrac{3}{2} & | & \tfrac{7}{2} \\ 3 & 4 & | & 2 \end{bmatrix} \quad R_2 + (-3)R_1 \to R_2$$

$$-3 \quad \tfrac{9}{2} \quad -\tfrac{21}{2}$$

Step 3: Need a 1 here
$$\sim \begin{bmatrix} 1 & -\tfrac{3}{2} & | & \tfrac{7}{2} \\ 0 & \tfrac{17}{2} & | & -\tfrac{17}{2} \end{bmatrix} \quad \tfrac{2}{17}R_2 \to R_2$$

Step 4: Need a 0 here
$$\sim \begin{bmatrix} 1 & -\tfrac{3}{2} & | & \tfrac{7}{2} \\ 0 & 1 & | & -1 \end{bmatrix} \quad R_1 + \tfrac{3}{2}R_2 \to R_1$$

$$0 \quad -\tfrac{3}{2} \quad -\tfrac{3}{2}$$

$$\sim \begin{bmatrix} 1 & 0 & | & 2 \\ 0 & 1 & | & -1 \end{bmatrix}$$

Thus, $x_1 = 2$ and $x_2 = -1$.

PROBLEM 8 Solve, using augmented matrix methods.

$$5x_1 - 2x_2 = 12$$
$$2x_1 + 3x_2 = 1$$

EXAMPLE 9 Solve, using augmented matrix methods.

$$2x_1 - x_2 = 4$$
$$-6x_1 + 3x_2 = -12$$

Solution
$$\begin{bmatrix} 2 & -1 & | & 4 \\ -6 & 3 & | & -12 \end{bmatrix} \quad \begin{array}{l} \tfrac{1}{2}R_1 \to R_1 \text{ (this produces a 1 in the upper left corner)} \\ \tfrac{1}{3}R_2 \to R_2 \text{ (this simplifies } R_2\text{)} \end{array}$$

$$\sim \begin{bmatrix} 1 & -\tfrac{1}{2} & | & 2 \\ -2 & 1 & | & -4 \end{bmatrix} \quad R_2 + 2R_1 \to R_2 \text{ (this produces a 0 in the lower left corner)}$$

$$2 \quad -1 \quad 4$$

$$\sim \begin{bmatrix} 1 & -\tfrac{1}{2} & | & 2 \\ 0 & 0 & | & 0 \end{bmatrix}$$

The last matrix corresponds to the system

$$x_1 - \tfrac{1}{2}x_2 = 2$$
$$0x_1 + 0x_2 = 0$$

Thus, $x_1 = \tfrac{1}{2}x_2 + 2$. Hence, for any real number t,

$$x_2 = t \quad \text{and} \quad x_1 = \tfrac{1}{2}t + 2$$

that is, $(\frac{1}{2}t + 2, t)$ is a solution. For example, if $t = 6$, then $(5, 6)$ is a solution; if $t = -2$, then $(1, -2)$ is a solution; and so on. Geometrically, the graphs of the two original equations coincide and there are infinitely many solutions. In general, if we end up with a row of 0's in an augmented matrix for a two-equation–two-unknown system, the system is dependent and there are infinitely many solutions.

PROBLEM 9 Solve, using augmented matrix methods.

$$-2x_1 + 6x_2 = 6$$
$$3x_1 - 9x_2 = -9$$

EXAMPLE 10 Solve, using augmented matrix methods.

$$2x_1 + 6x_2 = -3$$
$$x_1 + 3x_2 = 2$$

Solution

$$\begin{bmatrix} 2 & 6 & | & -3 \\ 1 & 3 & | & 2 \end{bmatrix} \quad R_1 \leftrightarrow R_2$$

$$\sim \begin{bmatrix} 1 & 3 & | & 2 \\ 2 & 6 & | & -3 \end{bmatrix} \quad R_2 + (-2)R_1 \to R_2$$

$$ \quad -2 \; -6 \quad\quad -4$$

$$\sim \begin{bmatrix} 1 & 3 & | & 2 \\ 0 & 0 & | & -7 \end{bmatrix} \quad R_2 \text{ implies the contradiction: } 0 = -7$$

The system is inconsistent and has no solution—otherwise, we have proved that $0 = -7$! Thus, if in a row of an augmented matrix we obtain all 0's to the left of the vertical bar and a nonzero number to the right of the bar, then the system is inconsistent and there are no solutions.

PROBLEM 10 Solve, using augmented matrix methods.

$$2x_1 - x_2 = 3$$
$$4x_1 - 2x_2 = -1$$

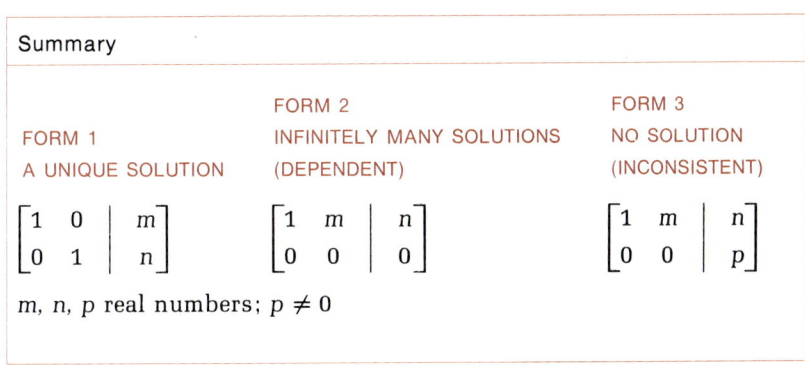

Summary

	FORM 2	FORM 3
FORM 1	INFINITELY MANY SOLUTIONS	NO SOLUTION
A UNIQUE SOLUTION	(DEPENDENT)	(INCONSISTENT)
$\begin{bmatrix} 1 & 0 & \| & m \\ 0 & 1 & \| & n \end{bmatrix}$	$\begin{bmatrix} 1 & m & \| & n \\ 0 & 0 & \| & 0 \end{bmatrix}$	$\begin{bmatrix} 1 & m & \| & n \\ 0 & 0 & \| & p \end{bmatrix}$

m, n, p real numbers; $p \neq 0$

The process of solving systems of equations described in this section is referred to as **Gauss–Jordan elimination**. We will use this method to solve larger-scale systems in the next section, including systems where the number of equations and the number of variables are not the same.

Answers to Matched Problems

7. $x_1 = -2, x_2 = 3$
8. $x_1 = 2, x_2 = -1$
9. The system is dependent. For t any real number, $x_2 = t, x_1 = 3t - 3$ is a solution.
10. Inconsistent—no solution

Exercise 9-2

A Perform each of the indicated row operations on the following matrix.

$$\begin{bmatrix} 1 & -3 & | & 2 \\ 4 & -6 & | & -8 \end{bmatrix}$$

1. $R_1 \leftrightarrow R_2$
2. $\frac{1}{2}R_2 \to R_2$
3. $-4R_1 \to R_1$
4. $-2R_1 \to R_1$
5. $2R_2 \to R_2$
6. $-1R_2 \to R_2$
7. $R_2 + (-4)R_1 \to R_2$
8. $R_1 + (-\frac{1}{2})R_2 \to R_1$
9. $R_2 + (-2)R_1 \to R_2$
10. $R_2 + (-3)R_1 \to R_2$
11. $R_2 + (-1)R_1 \to R_2$
12. $R_2 + (1)R_1 \to R_2$

Solve, using augmented matrix methods.

13. $x_1 + x_2 = 5$
    $x_1 - x_2 = 1$
14. $x_1 - x_2 = 2$
    $x_1 + x_2 = 6$

B Solve, using augmented matrix methods.

15. $x_1 - 2x_2 = 1$
    $2x_1 - x_2 = 5$
16. $x_1 + 3x_2 = 1$
    $3x_1 - 2x_2 = 14$
17. $x_1 - 4x_2 = -2$
    $-2x_1 + x_2 = -3$
18. $x_1 - 3x_2 = -5$
    $-3x_1 - x_2 = 5$
19. $3x_1 - x_2 = 2$
    $x_1 + 2x_2 = 10$
20. $2x_1 + x_2 = 0$
    $x_1 - 2x_2 = -5$
21. $x_1 + 2x_2 = 4$
    $2x_1 + 4x_2 = -8$
22. $2x_1 - 3x_2 = -2$
    $-4x_1 + 6x_2 = 7$
23. $2x_1 + x_2 = 6$
    $x_1 - x_2 = -3$
24. $3x_1 - x_2 = -5$
    $x_1 + 3x_2 = 5$
25. $3x_1 - 6x_2 = -9$
    $-2x_1 + 4x_2 = 6$
26. $2x_1 - 4x_2 = -2$
    $-3x_1 + 6x_2 = 3$
27. $4x_1 - 2x_2 = 2$
    $-6x_1 + 3x_2 = -3$
28. $-6x_1 + 2x_2 = 4$
    $3x_1 - x_2 = -2$

C *Solve, using augmented matrix methods.*

29. $3x_1 - x_2 = 7$
$2x_1 + 3x_2 = 1$

30. $2x_1 - 3x_2 = -8$
$5x_1 + 3x_2 = 1$

31. $3x_1 + 2x_2 = 4$
$2x_1 - x_2 = 5$

32. $4x_1 + 3x_2 = 26$
$3x_1 - 11x_2 = -7$

33. $0.2x_1 - 0.5x_2 = 0.07$
$0.8x_1 - 0.3x_2 = 0.79$

34. $0.3x_1 - 0.6x_2 = 0.18$
$0.5x_1 - 0.2x_2 = 0.54$

Section 9-3 Gauss–Jordan Elimination

- Reduced Matrices
- Solving Systems by Gauss–Jordan Elimination

Now that you have had some experience with row operations on simple augmented matrices, we will consider systems involving more than two variables. In addition, we will not require that a system have the same number of equations as variables.

- **Reduced Matrices**

Our objective is to start with the augmented matrix of a linear system and transform it, using row operations from Theorem 2 in the preceding section, into a simple form where the solution can be read by inspection. The simple form we will obtain is called the *reduced form*, and we define it as follows:

Reduced Matrix

A matrix is in **reduced form** if:

1. Each row consisting entirely of 0's is below any row having at least one nonzero element.
2. The leftmost nonzero element in each row is 1.
3. The column containing the leftmost 1 of a given row has 0's above and below the 1.
4. The leftmost 1 in any row is to the right of the leftmost 1 in the preceding row.

EXAMPLE 11 The following matrices are in reduced form. Check each one carefully to convince yourself that the conditions in the definition are met.

$$\begin{bmatrix} 1 & 0 & | & 2 \\ 0 & 1 & | & -3 \end{bmatrix} \quad \begin{bmatrix} 1 & 0 & 0 & | & 2 \\ 0 & 1 & 0 & | & -1 \\ 0 & 0 & 1 & | & 3 \end{bmatrix} \quad \begin{bmatrix} 1 & 0 & | & 3 \\ 0 & 1 & | & -1 \\ 0 & 0 & | & 0 \end{bmatrix}$$

$$\begin{bmatrix} 1 & 4 & 0 & 0 & | & -3 \\ 0 & 0 & 1 & 0 & | & 2 \\ 0 & 0 & 0 & 1 & | & 6 \end{bmatrix} \quad \begin{bmatrix} 1 & 0 & 4 & | & 0 \\ 0 & 1 & 3 & | & 0 \\ 0 & 0 & 0 & | & 1 \end{bmatrix}$$

PROBLEM 11 The matrices below are not in reduced form. Indicate which condition in the definition is violated for each matrix.

(A) $\begin{bmatrix} 1 & 0 & | & 2 \\ 0 & 3 & | & -6 \end{bmatrix}$ (B) $\begin{bmatrix} 1 & 5 & 4 & | & 3 \\ 0 & 1 & 2 & | & -1 \\ 0 & 0 & 0 & | & 0 \end{bmatrix}$

(C) $\begin{bmatrix} 0 & 1 & 2 & | & -3 \\ 1 & -2 & 3 & | & 0 \\ 0 & 0 & 1 & | & 2 \end{bmatrix}$ (D) $\begin{bmatrix} 1 & 2 & 0 & | & 3 \\ 0 & 0 & 0 & | & 0 \\ 0 & 0 & 1 & | & 4 \end{bmatrix}$

EXAMPLE 12 Write the linear system corresponding to each reduced augmented matrix and solve.

(A) $\begin{bmatrix} 1 & 0 & 0 & | & 2 \\ 0 & 1 & 0 & | & -1 \\ 0 & 0 & 1 & | & 3 \end{bmatrix}$ (B) $\begin{bmatrix} 1 & 0 & 4 & | & 0 \\ 0 & 1 & 3 & | & 0 \\ 0 & 0 & 0 & | & 1 \end{bmatrix}$

(C) $\begin{bmatrix} 1 & 0 & 2 & | & -3 \\ 0 & 1 & -1 & | & 8 \\ 0 & 0 & 0 & | & 0 \end{bmatrix}$ (D) $\begin{bmatrix} 1 & 4 & 0 & 0 & 3 & | & -2 \\ 0 & 0 & 1 & 0 & -2 & | & 0 \\ 0 & 0 & 0 & 1 & 2 & | & 4 \end{bmatrix}$

Solution (A) $x_1 = 2$
$ x_2 = -1$
$ x_3 = 3$

The solution is obvious: $x_1 = 2$, $x_2 = -1$, $x_3 = 3$.

(B) $x_1 + 4x_3 = 0$
$ x_2 + 3x_3 = 0$
$0x_1 + 0x_2 + 0x_3 = 1$

The last equation implies $0 = 1$, which is a contradiction. Hence, the system is inconsistent and has no solution.

(C) $x_1 + 2x_3 = -3$ We disregard the equation corresponding to the third
$ x_2 - x_3 = 8$ row in the matrix, since it is satisfied by all values of x_1, x_2, and x_3.

When a reduced system (a system corresponding to a reduced augmented matrix) has more variables than equations, the system is dependent and has infinitely many solutions. To represent these solutions, it is useful to divide the variables into two types: **basic variables** and **nonbasic variables**. To represent the infinitely many solutions to the system, we solve for the basic variables in terms of the nonbasic variables. This can be accomplished very easily if we *choose as basic variables the first variable (with a nonzero coefficient) in each equation of the reduced system*. Since each of these variables occurs in exactly one equation, it is easy to solve for each in terms of the other variables, the nonbasic variables. Returning to our original system, we choose x_1 and x_2 (the first variable in each equation) as basic variables and x_3 as a nonbasic variable. We then solve for the basic variables x_1 and x_2 in terms of the nonbasic variable x_3:

$$x_1 = -2x_3 - 3$$
$$x_2 = x_3 + 8$$

If we let $x_3 = t$, then for any real number t,

$$x_1 = -2t - 3$$
$$x_2 = t + 8$$
$$x_3 = t$$

is a solution. For example,

If $t = 0$, then

$$x_1 = -2(0) - 3 = -3$$
$$x_2 = 0 + 8 = 8$$
$$x_3 = 0$$

is a solution.

If $t = -2$, then

$$x_1 = -2(-2) - 3 = 1$$
$$x_2 = -2 + 8 = 6$$
$$x_3 = -2$$

is a solution.

(D) $x_1 + 4x_2 \qquad\quad + 3x_5 = -2$
$\qquad\qquad\; x_3 \quad - 2x_5 = 0$
$\qquad\qquad\qquad\; x_4 + 2x_5 = 4$

Solve for x_1, x_3, and x_4 (basic variables) in terms of x_2 and x_5 (nonbasic variables).

$$x_1 = -4x_2 - 3x_5 - 2$$
$$x_3 = 2x_5$$
$$x_4 = -2x_5 + 4$$

If we let $x_2 = s$ and $x_5 = t$, then for any real numbers s and t,

$$x_1 = -4s - 3t - 2$$
$$x_2 = s$$
$$x_3 = 2t$$
$$x_4 = -2t + 4$$
$$x_5 = t$$

is a solution. The system is dependent and has infinitely many solutions. Can you find two?

PROBLEM 12 Write the linear system corresponding to each reduced augmented matrix and solve.

(A) $\begin{bmatrix} 1 & 0 & 0 & | & -5 \\ 0 & 1 & 0 & | & 3 \\ 0 & 0 & 1 & | & 6 \end{bmatrix}$ (B) $\begin{bmatrix} 1 & 2 & -3 & | & 0 \\ 0 & 0 & 0 & | & 1 \\ 0 & 0 & 0 & | & 0 \end{bmatrix}$

(C) $\begin{bmatrix} 1 & 0 & -2 & | & 4 \\ 0 & 1 & 3 & | & -2 \\ 0 & 0 & 0 & | & 0 \end{bmatrix}$ (D) $\begin{bmatrix} 1 & 0 & 3 & 2 & | & 5 \\ 0 & 1 & -2 & -1 & | & 3 \\ 0 & 0 & 0 & 0 & | & 0 \end{bmatrix}$

■ **Solving Systems by Gauss–Jordan Elimination**

We are now ready to outline a step-by-step procedure for solving systems of linear equations called the *Gauss–Jordan elimination* method. The method provides us with a systematic way of transforming augmented matrices into a reduced form from which we can write the solution to the original system by inspection, if a solution exists. The method will also reveal when a solution fails to exist.

EXAMPLE 13 Solve by Gauss–Jordan elimination.

$$2x_1 - 2x_2 + x_3 = 3$$
$$3x_1 + x_2 - x_3 = 7$$
$$x_1 - 3x_2 + 2x_3 = 0$$

Solution Write the augmented matrix and follow the steps indicated at the right.

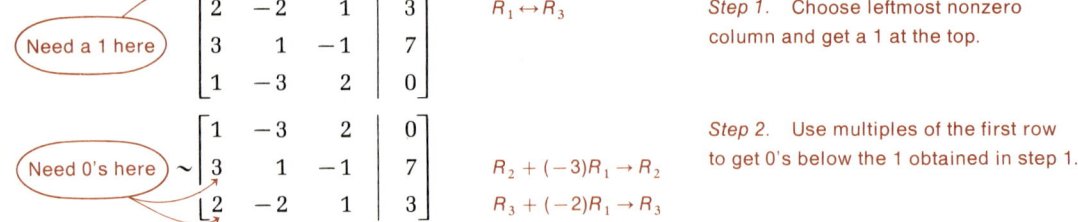

9-3 Gauss–Jordan Elimination

$\text{(Need a 1 here)} \sim \begin{bmatrix} 1 & -3 & 2 & | & 0 \\ 0 & 10 & -7 & | & 7 \\ 0 & 4 & -3 & | & 3 \end{bmatrix} \quad \frac{1}{10}R_2 \to R_2$

$\text{(Need a 0 here)} \sim \begin{bmatrix} 1 & -3 & 2 & | & 0 \\ 0 & 1 & -\frac{7}{10} & | & \frac{7}{10} \\ 0 & 4 & -3 & | & 3 \end{bmatrix} \quad R_3 + (-4)R_2 \to R_3$

$\text{(Need a 1 here)} \sim \begin{bmatrix} 1 & -3 & 2 & | & 0 \\ 0 & 1 & -\frac{7}{10} & | & \frac{7}{10} \\ 0 & 0 & -\frac{1}{5} & | & \frac{1}{5} \end{bmatrix} \quad (-5)R_3 \to R_3$

$\text{(Need 0's here)} \sim \begin{bmatrix} 1 & -3 & 2 & | & 0 \\ 0 & 1 & -\frac{7}{10} & | & \frac{7}{10} \\ 0 & 0 & 1 & | & -1 \end{bmatrix} \quad \begin{matrix} R_1 + (-2)R_3 \to R_1 \\ R_2 + \frac{7}{10}R_3 \to R_2 \end{matrix}$

$\text{(Need a 0 here)} \sim \begin{bmatrix} 1 & -3 & 0 & | & 2 \\ 0 & 1 & 0 & | & 0 \\ 0 & 0 & 1 & | & -1 \end{bmatrix} \quad R_1 + 3R_2 \to R_1$

$\sim \begin{bmatrix} 1 & 0 & 0 & | & 2 \\ 0 & 1 & 0 & | & 0 \\ 0 & 0 & 1 & | & -1 \end{bmatrix}$

Step 3. Mentally delete R_1 and repeat steps 1 and 2 with the **submatrix** (the matrix that remains after deleting the top row and the first column). Continue the above process (steps 1–3) until it is not possible to go further; then proceed with step 4.

Mentally delete R_1 and R_2.

Since steps 1–3 cannot be carried further, proceed to step 4.

Step 4. Return deleted rows. Begin with the bottom nonzero row and use appropriate multiples of it to get 0's above the leftmost 1. Continue the process, moving up row by row, until the matrix is in reduced form.

The matrix is in reduced form, and we can write the solution to the original system by inspection.

Solution: $x_1 = 2$, $x_2 = 0$, $x_3 = -1$. It is left to the reader to check this solution.

Steps 1–4 outlined in the solution of Example 13 are referred to as **Gauss–Jordan elimination**. The steps are summarized in the box for easy reference.

Gauss–Jordan Elimination

1. Choose the leftmost nonzero column and use appropriate row operations to get a 1 at the top.
2. Use multiples of the first row to get 0's in all places below the 1 obtained in step 1.
3. Delete (mentally) the top row and first column of the matrix. Repeat steps 1 and 2 with the **submatrix** (the matrix that remains after deleting the top row and first column). Continue this process (steps 1–3) until it is not possible to go further.

4. Consider the whole matrix obtained after mentally returning all the rows to the matrix. Begin with the bottom nonzero row and use appropriate multiples of it to get 0's above the leftmost 1. Continue this process, moving up row by row, until the matrix is finally in reduced form.

[Note: If at any point in the above process we obtain a row with all 0's to the left of the vertical line and a nonzero number to the right, we can stop, since we will have a contradiction ($0 = n$, $n \neq 0$). We can then conclude that the system has no solution.]

PROBLEM 13 Solve by Gauss–Jordan elimination.

$$3x_1 + x_2 - 2x_3 = 2$$
$$x_1 - 2x_2 + x_3 = 3$$
$$2x_1 - x_2 - 3x_3 = 3$$

EXAMPLE 14 Solve by Gauss–Jordan elimination.

$$2x_1 - x_2 + 4x_3 = -2$$
$$3x_1 + 2x_2 - x_3 = 1$$

Solution

$$\begin{bmatrix} 2 & -1 & 4 & | & -2 \\ 3 & 2 & -1 & | & 1 \end{bmatrix} \quad \tfrac{1}{2}R_1 \to R_1$$

(Need a 1 here)

$$\sim \begin{bmatrix} 1 & -\tfrac{1}{2} & 2 & | & -1 \\ 3 & 2 & -1 & | & 1 \end{bmatrix} \quad R_2 + (-3)R_1 \to R_2$$

(Need a 0 here)

$$\sim \begin{bmatrix} 1 & -\tfrac{1}{2} & 2 & | & -1 \\ 0 & \tfrac{7}{2} & -7 & | & 4 \end{bmatrix} \quad \tfrac{2}{7}R_2 \to R_2$$

(Need a 1 here)

$$\sim \begin{bmatrix} 1 & -\tfrac{1}{2} & 2 & | & -1 \\ 0 & 1 & -2 & | & \tfrac{8}{7} \end{bmatrix} \quad R_1 + \tfrac{1}{2}R_2 \to R_1$$

(Need a 0 here)

$$\sim \begin{bmatrix} 1 & 0 & 1 & | & -\tfrac{3}{7} \\ 0 & 1 & -2 & | & \tfrac{8}{7} \end{bmatrix}$$

The matrix is now in reduced form. Write the corresponding system and the solution.

$$x_1 + x_3 = -\tfrac{3}{7}$$
$$x_2 - 2x_3 = \tfrac{8}{7}$$

Solve for the basic variables x_1 and x_2 in terms of the nonbasic variable x_3.

$$x_1 = -x_3 - \tfrac{3}{7}$$
$$x_2 = 2x_3 + \tfrac{8}{7}$$

If $x_3 = t$, then for t any real number,

$x_1 = -t - \frac{3}{7}$
$x_2 = 2t + \frac{8}{7}$
$x_3 = t$

is a solution.

Remark: In general, it can be proved that a system with more variables than equations cannot have a unique solution.

PROBLEM 14 Solve by Gauss–Jordan elimination.

$3x_1 + 6x_2 - 3x_3 = 2$
$2x_1 - x_2 + 2x_3 = -1$

EXAMPLE 15 Solve by Gauss–Jordan elimination.

$2x_1 - x_2 = -4$
$x_1 + 2x_2 = 3$
$3x_1 - x_2 = -1$

Solution

$$\begin{bmatrix} 2 & -1 & | & -4 \\ 1 & 2 & | & 3 \\ 3 & -1 & | & -1 \end{bmatrix} \quad R_1 \leftrightarrow R_2$$

$$\sim \begin{bmatrix} 1 & 2 & | & 3 \\ 2 & -1 & | & -4 \\ 3 & -1 & | & -1 \end{bmatrix} \quad \begin{array}{l} R_2 + (-2)R_1 \to R_2 \\ R_3 + (-3)R_1 \to R_3 \end{array}$$

$$\sim \begin{bmatrix} 1 & 2 & | & 3 \\ 0 & -5 & | & -10 \\ 0 & -7 & | & -10 \end{bmatrix} \quad -\tfrac{1}{5}R_2 \to R_2$$

$$\sim \begin{bmatrix} 1 & 2 & | & 3 \\ 0 & 1 & | & 2 \\ 0 & -7 & | & -10 \end{bmatrix} \quad R_3 + 7R_2 \to R_3$$

$$\sim \begin{bmatrix} 1 & 2 & | & 3 \\ 0 & 1 & | & 2 \\ 0 & 0 & | & 4 \end{bmatrix} \quad \begin{array}{l} \text{We stop the Gauss–Jordan elimination even though} \\ \text{the matrix is not in reduced form, since the last row} \\ \text{produces a contradiction.} \end{array}$$

The last row implies $0 = 4$, which is a contradiction; therefore, the system has no solution.

PROBLEM 15 Solve by Gauss–Jordan elimination.

$3x_1 + x_2 = 5$
$2x_1 + 3x_2 = 1$
$x_1 - x_2 = 3$

Answers to Matched Problems

11. (A) Condition 2 is violated: The 3 in the second row should be a 1.
    (B) Condition 3 is violated: In the second column, the 5 should be a 0.
    (C) Condition 4 is violated: The leftmost 1 in the second row is not to the right of the leftmost 1 in the first row.
    (D) Condition 1 is violated: The all-zero second row should be at the bottom.

12. (A) $x_1 = -5$
    $ x_2 = 3$
    $ x_3 = 6$
    Solution:
    $x_1 = -5, x_2 = 3, x_3 = 6$

    (B) $x_1 + 2x_2 - 3x_3 = 0$
    $0x_1 + 0x_2 + 0x_3 = 1$
    $0x_1 + 0x_2 + 0x_3 = 0$
    Inconsistent; no solution.

    (C) $x_1 - 2x_3 = 4$
    $ x_2 + 3x_3 = -2$
    Dependent: let $x_3 = t$.
    Then for any real t,
    $x_1 = 2t + 4$
    $x_2 = -3t - 2$
    $x_3 = t$
    is a solution.

    (D) $x_1 + 3x_3 + 2x_4 = 5$
    $ x_2 - 2x_3 - x_4 = 3$
    Dependent: let $x_3 = s$ and $x_4 = t$. Then for any real s and t,
    $x_1 = -3s - 2t + 5$
    $x_2 = 2s + t + 3$
    $x_3 = s$
    $x_4 = t$
    is a solution.

13. $x_1 = 1, \quad x_2 = -1, \quad x_3 = 0$

14. $x_1 = -\frac{3}{5}t - \frac{4}{15}, \quad x_2 = \frac{4}{5}t + \frac{7}{15}, \quad x_3 = t, \quad t$ any real number

15. $x_1 = 2, \quad x_2 = -1$

Exercise 9-3

A *Indicate whether each matrix is in reduced form.*

1. $\left[\begin{array}{cc|c} 1 & 0 & 2 \\ 0 & 1 & -1 \end{array}\right]$

2. $\left[\begin{array}{cc|c} 0 & 1 & 2 \\ 1 & 0 & -1 \end{array}\right]$

3. $\left[\begin{array}{ccc|c} 1 & 0 & 2 & 3 \\ 0 & 0 & 0 & 0 \\ 0 & 1 & -1 & 4 \end{array}\right]$

4. $\left[\begin{array}{ccc|c} 1 & 0 & 0 & -2 \\ 0 & 1 & 0 & 0 \\ 0 & 0 & 1 & 1 \end{array}\right]$

5. $\left[\begin{array}{ccc|c} 0 & 1 & 0 & 2 \\ 0 & 0 & 3 & -1 \\ 0 & 0 & 0 & 0 \end{array}\right]$

6. $\left[\begin{array}{ccc|c} 1 & 3 & 0 & 0 \\ 0 & 0 & 1 & 0 \\ 0 & 0 & 0 & 1 \end{array}\right]$

7. $\left[\begin{array}{ccc|c} 1 & 2 & 0 & 3 & 2 \\ 0 & 0 & 1 & -1 & 0 \end{array}\right]$

8. $\left[\begin{array}{ccc|c} 0 & 1 & 2 & 1 \\ 1 & 0 & -3 & 2 \end{array}\right]$

9-3 Gauss–Jordan Elimination

Write the linear system corresponding to each reduced augmented matrix and solve.

9. $\begin{bmatrix} 1 & 0 & 0 & | & -2 \\ 0 & 1 & 0 & | & 3 \\ 0 & 0 & 1 & | & 0 \end{bmatrix}$

10. $\begin{bmatrix} 1 & 0 & 0 & 0 & | & -2 \\ 0 & 1 & 0 & 0 & | & 0 \\ 0 & 0 & 1 & 0 & | & 1 \\ 0 & 0 & 0 & 1 & | & 3 \end{bmatrix}$

11. $\begin{bmatrix} 1 & 0 & -2 & | & 3 \\ 0 & 1 & 1 & | & -5 \\ 0 & 0 & 0 & | & 0 \end{bmatrix}$

12. $\begin{bmatrix} 1 & -2 & 0 & | & -3 \\ 0 & 0 & 1 & | & 5 \\ 0 & 0 & 0 & | & 0 \end{bmatrix}$

13. $\begin{bmatrix} 1 & 0 & | & 0 \\ 0 & 1 & | & 0 \\ 0 & 0 & | & 1 \end{bmatrix}$

14. $\begin{bmatrix} 1 & 0 & | & 5 \\ 0 & 1 & | & -3 \\ 0 & 0 & | & 0 \end{bmatrix}$

15. $\begin{bmatrix} 1 & -2 & 0 & -3 & | & -5 \\ 0 & 0 & 1 & 3 & | & 2 \end{bmatrix}$

16. $\begin{bmatrix} 1 & 0 & -2 & 3 & | & 4 \\ 0 & 1 & -1 & 2 & | & -1 \end{bmatrix}$

B *Use row operations to change each matrix to reduced form.*

17. $\begin{bmatrix} 1 & 2 & | & -1 \\ 0 & 1 & | & 3 \end{bmatrix}$

18. $\begin{bmatrix} 1 & 3 & | & 1 \\ 0 & 2 & | & -4 \end{bmatrix}$

19. $\begin{bmatrix} 1 & 0 & -3 & | & 1 \\ 0 & 1 & 2 & | & 0 \\ 0 & 0 & 3 & | & -6 \end{bmatrix}$

20. $\begin{bmatrix} 1 & 0 & 4 & | & 0 \\ 0 & 1 & -3 & | & -1 \\ 0 & 0 & -2 & | & 2 \end{bmatrix}$

21. $\begin{bmatrix} 1 & 2 & -2 & | & -1 \\ 0 & 3 & -6 & | & 1 \\ 0 & -1 & 2 & | & -\frac{1}{3} \end{bmatrix}$

22. $\begin{bmatrix} 0 & -2 & 8 & | & 1 \\ 2 & -2 & 6 & | & -4 \\ 0 & -1 & 4 & | & \frac{1}{2} \end{bmatrix}$

Solve, using Gauss–Jordan elimination.

23. $2x_1 + 4x_2 - 10x_3 = -2$
$3x_1 + 9x_2 - 21x_3 = 0$
$x_1 + 5x_2 - 12x_3 = 1$

24. $3x_1 + 5x_2 - x_3 = -7$
$x_1 + x_2 + x_3 = -1$
$2x_1 + 11x_3 = 7$

25. $3x_1 + 8x_2 - x_3 = -18$
$2x_1 + x_2 + 5x_3 = 8$
$2x_1 + 4x_2 + 2x_3 = -4$

26. $2x_1 + 7x_2 + 15x_3 = -12$
$4x_1 + 7x_2 + 13x_3 = -10$
$3x_1 + 6x_2 + 12x_3 = -9$

27. $2x_1 - x_2 - 3x_3 = 8$
$x_1 - 2x_2 = 7$

28. $2x_1 + 4x_2 - 6x_3 = 10$
$3x_1 + 3x_2 - 3x_3 = 6$

29. $2x_1 + 3x_2 - x_3 = 1$
$x_1 - 2x_2 + 2x_3 = -2$

30. $x_1 - 3x_2 + 2x_3 = -1$
$3x_1 + 2x_2 - x_3 = 2$

31. $2x_1 + 2x_2 = 2$
$x_1 + 2x_2 = 3$
$ -3x_2 = -6$

32. $2x_1 - x_2 = 0$
$3x_1 + 2x_2 = 7$
$x_1 - x_2 = -1$

33. $2x_1 - x_2 = 0$
    $3x_1 + 2x_2 = 7$
    $x_1 - x_2 = -2$

34. $x_1 - 3x_2 = 5$
    $2x_1 + x_2 = 3$
    $x_1 - 2x_2 = 5$

35. $3x_1 - 4x_2 - x_3 = 1$
    $2x_1 - 3x_2 + x_3 = 1$
    $x_1 - 2x_2 + 3x_3 = 2$

36. $3x_1 + 7x_2 - x_3 = 11$
    $x_1 + 2x_2 - x_3 = 3$
    $2x_1 + 4x_2 - 2x_3 = 10$

C Solve, using Gauss–Jordan elimination.

37. $2x_1 - 3x_2 + 3x_3 = -15$
    $3x_1 + 2x_2 - 5x_3 = 19$
    $5x_1 - 4x_2 - 2x_3 = -2$

38. $3x_1 - 2x_2 - 4x_3 = -8$
    $4x_1 + 3x_2 - 5x_3 = -5$
    $6x_1 - 5x_2 + 2x_3 = -17$

39. $5x_1 - 3x_2 + 2x_3 = 13$
    $2x_1 + 4x_2 - 3x_3 = -9$
    $4x_1 - 2x_2 + 5x_3 = 13$

40. $4x_1 - 2x_2 + 3x_3 = 0$
    $3x_1 - 5x_2 - 2x_3 = -12$
    $2x_1 + 4x_2 - 3x_3 = -4$

41. $x_1 + 2x_2 - 4x_3 - x_4 = 7$
    $2x_1 + 5x_2 - 9x_3 - 4x_4 = 16$
    $x_1 + 5x_2 - 7x_3 - 7x_4 = 13$

42. $2x_1 + 4x_2 + 5x_3 + 4x_4 = 8$
    $x_1 + 2x_2 + 2x_3 + x_4 = 3$

APPLICATIONS Solve all of the following problems using Gauss–Jordan elimination.

43. *Production scheduling.* A small manufacturing plant makes three types of inflatable boats: one-person, two-person, and four-person models. Each boat requires the services of three departments, as listed in the table. The cutting, assembly, and packaging departments have available a maximum of 380, 330, and 120 work-hours per week, respectively. How many boats of each type must be produced each week for the plant to operate at full capacity?

	ONE-PERSON BOAT	TWO-PERSON BOAT	FOUR-PERSON BOAT
Cutting department	0.5 hr	1.0 hr	1.5 hr
Assembly department	0.6 hr	0.9 hr	1.2 hr
Packaging department	0.2 hr	0.3 hr	0.5 hr

44. *Production scheduling.* Repeat Problem 43 assuming the cutting, assembly, and packaging departments have available a maximum of 350, 330, and 115 work-hours per week, respectively.

45. *Production scheduling.* Work Problem 43 assuming the packaging department is no longer used.

46. *Production scheduling.* Work Problem 44 assuming the packaging department is no longer used.

47. *Production scheduling.* Work Problem 43 assuming the four-person boat is no longer produced.

48. *Production scheduling.* Work Problem 44 assuming the four-person boat is no longer produced.

49. *Nutrition.* A dietitian in a hospital is to arrange a special diet using three basic foods. The diet is to include exactly 340 units of calcium, 180 units of iron, and 220 units of vitamin A. The number of units per ounce of each special ingredient for each of the foods is indicated in the table. How many ounces of each food must be used to meet the diet requirements?

	UNITS PER OUNCE		
	Food A	Food B	Food C
Calcium	30	10	20
Iron	10	10	20
Vitamin A	10	30	20

50. *Nutrition.* Repeat Problem 49 if the diet is to include exactly 400 units of calcium, 160 units of iron, and 240 units of vitamin A.

51. *Nutrition.* Solve Problem 49 with the assumption that food C is no longer available.

52. *Nutrition.* Solve Problem 50 with the assumption that food C is no longer available.

53. *Nutrition.* Solve Problem 49 assuming the vitamin A requirement is deleted.

54. *Nutrition.* Solve Problem 50 assuming the vitamin A requirement is deleted.

55. *Sociology.* Two sociologists have grant money to study school busing in a particular city. They wish to conduct an opinion survey using 600 telephone contacts and 400 house contacts. Survey company A has personnel to do thirty telephone and ten house contacts per hour; survey company B can handle twenty telephone and twenty house contacts per hour. How many hours should be scheduled for each firm to produce exactly the number of contacts needed?

56. *Sociology.* Repeat Problem 55 if 650 telephone contacts and 350 house contacts are needed.

Section 9-4 Systems Involving Second-Degree Equations

- Systems with First-Degree and Second-Degree Equations
- Systems with Second-Degree Equations Only

In this section we will investigate systems of the form

$$4x^2 + y^2 = 25 \qquad x^2 - y^2 = 5 \qquad x^2 + 3xy + y^2 = 20$$
$$2x + y = 7 \qquad x^2 + 2y^2 = 17 \qquad xy - y^2 = 0$$

It can be shown that such systems have at most four solutions.

- **Systems with First-Degree and Second-Degree Equations**

If a system involves a first-degree and a second-degree equation, the method of elimination using substitution is effective. We solve the first-degree equation for one variable in terms of the other, and substitute into the second-degree equation to obtain a quadratic in one variable. An example should make the process clear.

EXAMPLE 16 Solve the system.

$$4x^2 + y^2 = 25$$
$$2x + y = 7$$

Solution

$2x + y = 7$	Solve the first-degree equation for y in terms of x; then substitute into the second-degree equation.
$y = 7 - 2x$	
$4x^2 + y^2 = 25$	Second-degree equation.
$4x^2 + (7 - 2x)^2 = 25$	Simplify and write in standard quadratic form.
$8x^2 - 28x + 24 = 0$	Divide through by 4 to simplify further.
$2x^2 - 7x + 6 = 0$	Solve.
$(2x - 3)(x - 2) = 0$	
$x = \frac{3}{2}, 2$	

These values are substituted back into the linear equation to find the corresponding values for y. (Note that if we substitute these values back into the second-degree equations, we may obtain "extraneous" roots; try it and see why.)

For $x = \frac{3}{2}$,

$$2(\tfrac{3}{2}) + y = 7$$
$$y = 4$$

9-4 Systems Involving Second-Degree Equations

For $x = 2$,
$$2(2) + y = 7$$
$$y = 3$$

Thus $(\frac{3}{2}, 4)$ and $(2, 3)$ are solutions to the system, as can easily be checked.

PROBLEM 16 Solve.
$$2x^2 - y^2 = 1$$
$$3x + y = 2$$

■ Systems with Second-Degree Equations Only

We now look at a couple of systems where both equations are second degree.

EXAMPLE 17 Solve.
$$x^2 - y^2 = 5$$
$$x^2 + 2y^2 = 17$$

Solution This type of system can be solved by elimination using addition. Multiply the second equation by -1 and add.

$$x^2 - y^2 = 5$$
$$\underline{-x^2 - 2y^2 = -17}$$
$$-3y^2 = -12$$
$$y^2 = 4$$
$$y = \pm 2 \qquad \text{Now substitute } y = 2 \text{ and } y = -2 \text{ back into either original equation to find } x.$$

For $y = 2$,
$$x^2 - (2)^2 = 5$$
$$x = \pm 3$$

For $y = -2$,
$$x^2 - (-2)^2 = 5$$
$$x = \pm 3$$

Thus, $(3, -2)$, $(3, 2)$, $(-3, -2)$, and $(-3, 2)$ are the four solutions to the system. The check of the solutions is left to the reader.

PROBLEM 17 Solve.
$$2x^2 - 3y^2 = 5$$
$$3x^2 + 4y^2 = 16$$

EXAMPLE 18   Solve.
$$x^2 + 3xy + y^2 = 20$$
$$xy - y^2 = 0$$

Solution   Factor the left side of the equation that has a 0 constant term.
$$xy - y^2 = 0$$
$$y(x - y) = 0$$
$$y = 0 \quad \text{or} \quad y = x$$

Thus, the original system is equivalent to the two systems:

$$\begin{array}{ccc} y = 0 & & y = x \\ x^2 + 3xy + y^2 = 20 & \text{or} & x^2 + 3xy + y^2 = 20 \end{array}$$

These systems are solved as in Example 16 by substitution.

First system:

$$y = 0$$
$$x^2 + 3xy + y^2 = 20$$
$$x^2 + 3x(0) + (0)^2 = 20$$
$$x^2 = 20$$
$$x = \pm\sqrt{20} = \pm 2\sqrt{5}$$

Substitute $y = 0$ in the second equation and solve for x.

Second system:

$$y = x$$
$$x^2 + 3xy + y^2 = 20$$
$$x^2 + 3xx + x^2 = 20$$
$$5x^2 = 20$$
$$x^2 = 4$$
$$x = \pm 2$$

Substitute $y = x$ in the second equation and solve for x; then substitute these values back into $y = x$ to find y.

For $x = 2$, $y = 2$. For $x = -2$, $y = -2$.
The solutions for the original system are $(2\sqrt{5}, 0)$, $(-2\sqrt{5}, 0)$, $(2, 2)$, and $(-2, -2)$. The check of the solutions is left to the reader.

PROBLEM 18   Solve.
$$x^2 + xy - y^2 = 4$$
$$2x^2 - xy = 0$$

Example 18 is somewhat specialized; however, it suggests a procedure that is effective for some problems.

9-4 Systems Involving Second-Degree Equations

Answers to Matched Problems
16. $(1, -1), (\frac{5}{7}, -\frac{1}{7})$
17. $(2, 1), (2, -1), (-2, 1), (-2, -1)$
18. $(0, 2i), (0, -2i), (2i, 4i), (-2i, -4i)$

Exercise 9-4

A *Solve each system.*

1. $x^2 + y^2 = 169$
   $x = -12$

2. $x^2 + y^2 = 25$
   $y = -4$

3. $8x^2 - y^2 = 16$
   $y = 2x$

4. $y^2 = 2x$
   $x = y - \frac{1}{2}$

5. $2x^2 - 3y^2 = 25$
   $x + y = 0$

6. $x^2 + 4y^2 = 32$
   $x + 2y = 0$

7. $y^2 = -x$
   $x - 2y = 5$

8. $x^2 = 2y$
   $3x = y + 5$

9. $2x^2 + y^2 = 24$
   $x^2 - y^2 = -12$

10. $x^2 - y^2 = 3$
    $x^2 + y^2 = 5$

11. $x^2 + y^2 = 10$
    $16x^2 + y^2 = 25$

12. $x^2 - 2y^2 = 1$
    $x^2 + 4y^2 = 25$

B
13. $xy = -4$
    $y - x = 2$

14. $xy - 6 = 0$
    $x - y = 4$

15. $x^2 - 2xy + y^2 = 1$
    $x - 2y = 2$

16. $x^2 + xy - y^2 = -5$
    $y - x = 3$

17. $2x^2 + 3y^2 = -4$
    $4x^2 + 2y^2 = 8$

18. $2x^2 - 3y^2 = 10$
    $x^2 + 4y^2 = -17$

19. $x^2 - y^2 = 2$
    $y^2 = x$

20. $x^2 + y^2 = 20$
    $x^2 = y$

21. $x^2 + y^2 = 5$
    $x^2 = 4(2 - y)$

22. $x^2 + y^2 = 16$
    $y^2 = 4 - x$

23. $x^2 - y^2 = 3$
    $xy = 2$

24. $2x^2 + y^2 = 18$
    $xy = 4$

C
25. $2x^2 - xy + y^2 = 8$
    $x^2 - y^2 = 0$

26. $x^2 + 2xy + y^2 = 36$
    $x^2 - xy = 0$

27. $x^2 + xy - 3y^2 = 3$
    $x^2 + 4xy + 3y^2 = 0$

28. $x^2 - 2xy + 2y^2 = 16$
    $x^2 - y^2 = 0$

APPLICATIONS

29. *Numbers.* Find two numbers such that their sum is 1 and their product is 1.

30. *Numbers.* Find two numbers such that their difference is 1 and their product is 1. (Let x be the larger number and y the smaller number.)

31. Geometry. Find the dimensions of a rectangle with an area of 60 square inches if its diagonal is 13 inches long.

32. Geometry. Find the dimensions of a rectangle with an area of 32 square meters if its perimeter is 36 meters long.

33. Supply and demand. The daily demand equation for a certain brand of ball-point pen in a given city is $dp = 1{,}000$, and the supply equation is $s = 5p - 50$, where d is the number of pen shoppers who are willing to buy at p¢ each, and s is the number of suppliers who are willing to sell at p¢ each. At what price will supply equal demand; that is, at what price will $s = d$?

Section 9-5 Systems of Linear Inequalities

- Single-Inequality Statements
- Systems of Inequality Statements
- Application

■ Single-Inequality Statements

We know how to graph first-degree equations such as

$$y = 2x - 3 \quad \text{and} \quad 2x - 3y = 5$$

but how do we graph first-degree inequalities such as

$$y \leq 2x - 3 \quad \text{and} \quad 2x - 3y > 5$$

We will find that graphing these inequalities is almost as easy as graphing the equalities. The following discussion leads to a simple solution of the problem.

A vertical line divides a plane into left and right **half-planes**; a nonvertical line divides a plane into upper and lower half-planes as indicated in Figure 1.

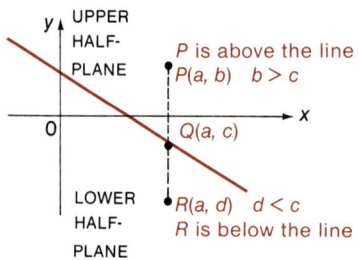

FIGURE 1

Now let us compare the graphs of the following:

$$y < 2x - 3 \qquad y = 2x - 3 \qquad y > 2x - 3$$

9-5 Systems of Linear Inequalities

We start by graphing $y = 2x - 3$. For a fixed x, equality holds if a point is on the line. For the same x, if a point is below the line, then $y < 2x - 3$, and if a point is above the line, then $y > 2x - 3$. See Figure 2. Since the same results are obtained for each point on the x axis, we conclude that the graph of $y > 2x - 3$ is the upper half-plane determined by the line $y = 2x - 3$, and the graph of $y < 2x - 3$ is the lower half-plane determined by the same line.

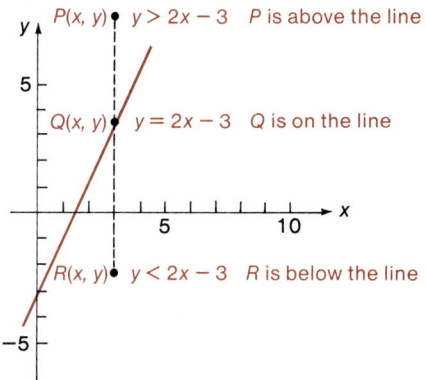

FIGURE 2

In graphing $y > 2x - 3$, we show the line $y = 2x - 3$ as a broken line, indicating that it is not part of the graph; in graphing $y \geq 2x - 3$, we show the line $y = 2x - 3$ as a solid line, indicating that it is part of the graph. Figure 3 illustrates four typical cases.

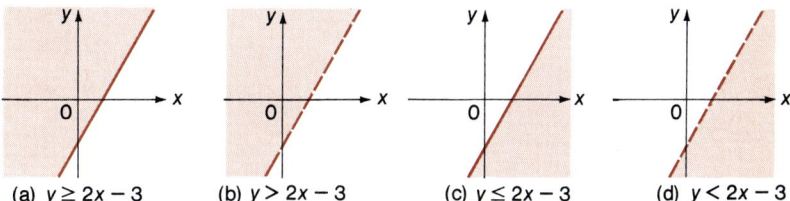

FIGURE 3 (a) $y \geq 2x - 3$ (b) $y > 2x - 3$ (c) $y \leq 2x - 3$ (d) $y < 2x - 3$

THEOREM 3

Graphs of Linear Inequalities

The graph of a linear inequality

$$Ax + By < C \quad \text{or} \quad Ax + By > C$$

with $B \neq 0$, is either the upper half-plane or the lower half-plane (but not both) determined by the line $Ax + By = C$. If $B = 0$, then the graph of

$$Ax < C \quad \text{or} \quad Ax > C$$

is either the left half-plane or the right half-plane (but not both) determined by the line $Ax = C$.

As a consequence of Theorem 3, we state a simple and fast mechanical procedure for graphing linear inequalities.

> **Procedure For Graphing Linear Inequalities**
>
> *Step 1.* Graph $Ax + By = C$ as a broken line if equality is not included in the original statements, or as a solid line if equality is included.
>
> *Step 2.* Choose a test point anywhere in the plane not on the line [the origin (0, 0) often requires the least computation] and substitute the coordinates into the inequality.
>
> *Step 3.* The graph of the original inequality includes the half-plane containing the test point if the inequality is satisfied by that point, or the half-plane not containing that point if the inequality is not satisfied by that point.

EXAMPLE 19 Graph: $3x - 4y \leq 12$

Solution *Step 1.* Graph $3x - 4y = 12$ as a solid line, since equality is included in the original statement.

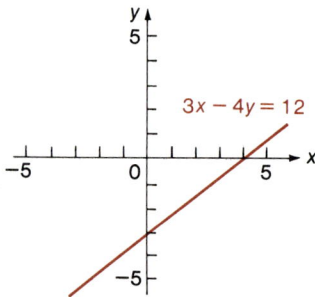

Step 2. Pick a convenient test point above or below the line. The origin (0, 0) requires the least computation. Substituting (0, 0) into the inequality

$$3x - 4y \leq 12$$

$$3(0) - 4(0) = 0 \leq 12$$

produces a true statement; therefore, (0, 0) is in the solution set.

9-5 Systems of Linear Inequalities

Step 3. The line $3x - 4y = 12$ and the half-plane containing the origin form the graph of $3x - 4y \leq 12$.

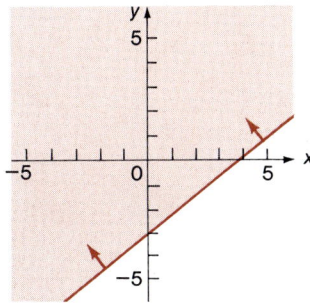

PROBLEM 19   Graph: $2x + 3y < 6$

EXAMPLE 20   Graph: (A) $y > -3$ (B) $2x \leq 5$

Solution   (A) 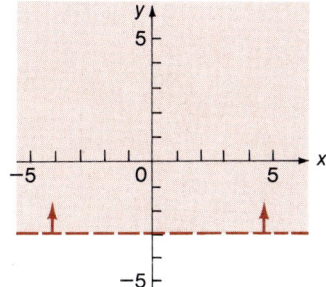   (B)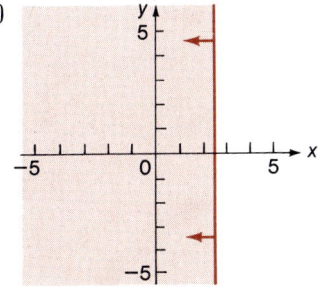

PROBLEM 20   Graph: (A) $y \leq 2$ (B) $3x > -8$

■ Systems of Inequality Statements

As in systems of linear equations in two variables, we say that the ordered pair of numbers (x_0, y_0) is a solution of a system of linear inequalities in two variables if the ordered pair satisfies each inequality in the system. Thus, **the graph of a system of linear inequalities is the intersection of the graphs of each inequality in the system**. In this book we will limit our investigation of solutions of systems of inequalities to graphical methods. An example will illustrate the process.

EXAMPLE 21   Solve the following linear system graphically.

$0 \leq x \leq 8$

$0 \leq y \leq 4$

Solution   This system is actually equivalent to the system

$$\left.\begin{array}{l} x \geq 0 \\ x \leq 8 \\ y \geq 0 \\ y \leq 4 \end{array}\right\}$$ The solution to the system is the intersection of all four solution sets.

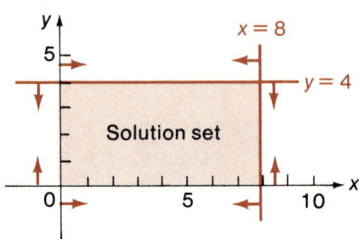

PROBLEM 21   Solve graphically.

$$2 \leq x \leq 6$$
$$1 \leq y \leq 3$$

EXAMPLE 22   Solve graphically.

$$3x + 5y \leq 60$$
$$4x + 2y \leq 40$$
$$x \geq 0$$
$$y \geq 0$$

Solution

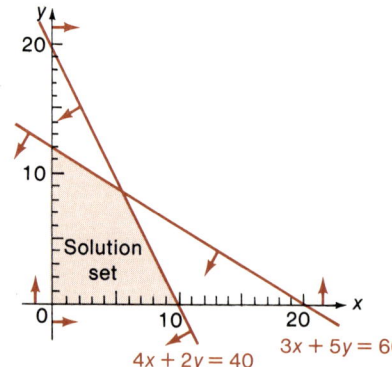

PROBLEM 22   Solve graphically.

$$x + 2y \geq 12$$
$$3x + 2y \geq 24$$
$$x \geq 0$$
$$y \geq 0$$

■ Application

EXAMPLE 23 *Production scheduling.* A manufacturer of surfboards makes a standard model and a competition model. The relevant manufacturing data are shown in the table. What combinations of boards can be produced each week so as not to exceed the number of work-hours available in each department per week?

	STANDARD MODEL (WORK-HOURS PER BOARD)	COMPETITION MODEL (WORK-HOURS PER BOARD)	MAXIMUM WORK-HOURS AVAILABLE PER WEEK
Fabricating	6	8	120
Finishing	1	3	30

Solution Let x and y be the respective number of standard and competition boards produced per week. These variables are restricted as follows:

$$6x + 8y \leq 120 \quad \text{Fabricating}$$
$$x + 3y \leq 30 \quad \text{Finishing}$$
$$x \geq 0$$
$$y \geq 0$$

The solution set of this system of inequalities is the shaded area in the figure and is referred to as the **feasible region**. Any point within the shaded area would represent a possible production schedule. Any point outside the shaded area would represent an impossible schedule. For example, it would be possible to produce 10 standard boards and five competition boards per week, but it would not be possible to produce thirteen standard boards and six competition boards per week.

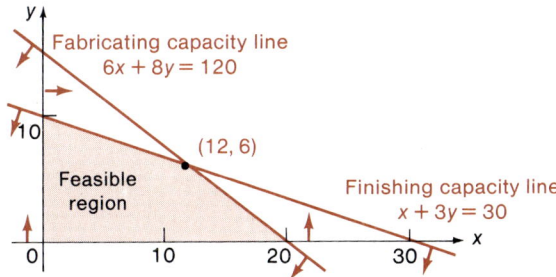

PROBLEM 23 Repeat Example 23 using 5 hours for fabricating a standard board in place of 6 hours, and a maximum of 27 work-hours for the finishing department.

Answers to Matched Problems

19.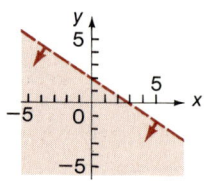

20. (A) 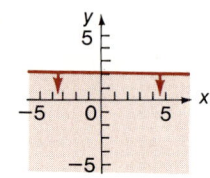   (B)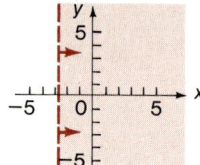

21. 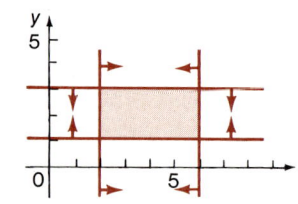   22.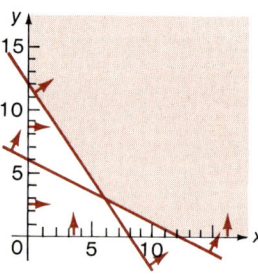

23.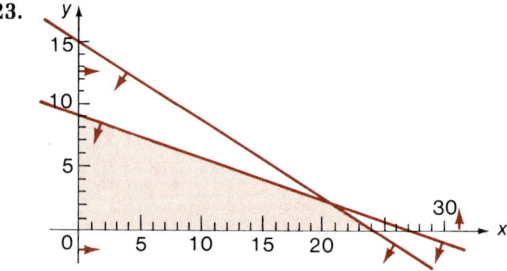

Exercise 9-5 ■ A   Graph each inequality.

1. $2x - 3y < 6$
2. $3x + 4y < 12$
3. $3x + 2y \geq 18$
4. $3y - 2x \geq 24$
5. $y \leq \frac{2}{3}x + 5$
6. $y \geq \frac{x}{3} - 2$
7. $y < 8$
8. $x > -5$
9. $-3 \leq y < 2$
10. $-1 < x \leq 3$

B *Find the solution set of each system graphically.*

11. $-2 \leq x < 2$
    $-1 < y \leq 6$

12. $-4 \leq x < -1$
    $-2 < y \leq 5$

13. $2x + y \leq 8$
    $0 \leq x \leq 3$
    $0 \leq y \leq 5$

14. $x + 3y \leq 12$
    $0 \leq x \leq 8$
    $0 \leq y \leq 3$

15. $2x + y \leq 8$
    $x + 3y \leq 12$
    $x \geq 0$
    $y \geq 0$

16. $x + 2y \leq 10$
    $3x + y \leq 15$
    $x \geq 0$
    $y \geq 0$

17. $6x + 3y \leq 24$
    $3x + 6y \leq 30$
    $x \geq 0$
    $y \geq 0$

18. $2x + y \leq 10$
    $x + 2y \leq 8$
    $x \geq 0$
    $y \geq 0$

19. $3x + 4y \geq 8$
    $4x + 3y \geq 24$
    $x \geq 0$
    $y \geq 0$

20. $x + 2y \geq 8$
    $2x + y \geq 10$
    $x \geq 0$
    $y \geq 0$

C 21. $3x + 5y \geq 60$
    $4x + 2y \geq 40$
    $2 \leq x \leq 14$
    $6 \leq y \leq 18$

22. $2x + y \geq 8$
    $x + 2y \geq 10$
    $1 \leq x \leq 7$
    $3 \leq y \leq 9$

APPLICATIONS

23. *Manufacturing—resource allocation.* A manufacturing company makes two types of water skis: a trick ski and a slalom ski. The trick ski requires 6 work-hours for fabricating and 1 work-hour for finishing. The slalom ski requires 4 work-hours for fabricating and 1 work-hour for finishing. The maximum work-hours available per day for fabricating and finishing are 108 and 24, respectively. If x is the number of trick skis and y is the number of slalom skis produced per day, write a system of inequalities that indicates appropriate restraints on x and y. Find the set of feasible solutions graphically for the number of each type of ski that can be produced.

24. *Nutrition.* A dietitian in a hospital is to arrange a special diet using two foods. Each ounce of food M contains 30 units of calcium, 10 units of iron, and 10 units of vitamin A. Each ounce of food N contains 10 units of calcium, 10 units of iron, and 30 units of vitamin A. The minimum requirements in the diet are 360 units of calcium, 160 units of iron, and 240 units of vitamin A. If x is the number of ounces of food M used and y is the number of ounces of food N used, write a system of linear inequalities that reflects the conditions indicated. Find the set of feasible solutions graphically for the amount of each kind of food that can be used.

Section 9-6 Chapter Review

IMPORTANT TERMS AND SYMBOLS

9-1 Systems of Linear Equations—A Review. Systems in two variables, solutions, solution set, equivalent systems, solution by graphing, solution by elimination using substitution, solution by elimination using addition, inconsistent systems, dependent systems, parameter, systems in three variables, solution, solution set, equivalent systems

$$a_1x + b_1y = k_1 \qquad a_1x + b_1y + c_1z = k_1$$
$$a_2x + b_2y = k_2 \qquad a_2x + b_2y + c_2z = k_2$$
$$a_3x + b_3y + c_3z = k_3$$

9-2 Systems and Augmented Matrices—An Introduction. Matrix, element, column, row, augmented matrix, row-equivalent matrices, row operations, Gauss–Jordan elimination

$$R_i \leftrightarrow R_j \qquad kR_i \to R_i \qquad R_i + kR_j \to R_i$$

9-3 Gauss–Jordan Elimination. Reduced form of a matrix, basic variables, nonbasic variables, submatrix, Gauss–Jordan elimination

9-4 Systems Involving Second-Degree Equations. First- and second-degree systems, solution by elimination using substitution, second-degree systems, solution by elimination using addition, solution by elimination using substitution

9-5 Systems of Linear Inequalities. Half-planes, left and right half-planes, upper and lower half-planes, solving single-inequality statements geometrically, solving systems of inequality statements geometrically

Exercise 9-6 Chapter Review

Work through all the problems in this chapter review and check answers in the back of the book. (Answers to all problems are there, and following each answer is a number in italics indicating the section in which that type of problem is discussed.) Where weaknesses show up, review appropriate sections in the text. When you are satisfied that you know the material, take the practice test following this review.

A Solve by elimination using substitution or addition.

1. $2x + y = 7$
   $3x - 2y = 0$

2. $3x - 6y = 5$
   $-2x + 4y = 1$

3. $4x - 3y = -8$
   $-2x + \frac{3}{2}y = 4$

4. $x + 2y + z = 3$
   $2x + 3y + 4z = 3$
   $x + 2y + 3z = 1$

5. $x^2 + y^2 = 2$
   $2x - y = 3$

6. $3x^2 - y^2 = -6$
   $2x^2 + 3y^2 = 29$

Solve by graphing.

7. $3x - 2y = 8$
   $x + 3y = -1$

8. $2x + y \leq 8$
   $2x + 3y \leq 12$
   $x \geq 0$
   $y \geq 0$

B *Solve using Gauss–Jordan elimination.*

9. $3x_1 + 2x_2 = 3$
   $x_1 + 3x_2 = 8$

10. $x_1 + x_2 = 1$
    $x_1 - x_3 = -2$
    $x_2 + 2x_3 = 4$

11. $x_1 + 2x_2 + 3x_3 = 1$
    $2x_1 + 3x_2 + 4x_3 = 3$
    $x_1 + 2x_2 + x_3 = 3$

12. $x_1 + 2x_2 - x_3 = 2$
    $2x_1 + 3x_2 + x_3 = -3$
    $3x_1 + 5x_2 = -1$

13. $x_1 - 2x_2 = 1$
    $2x_1 - x_2 = 0$
    $x_1 - 3x_2 = -2$

14. $x_1 + 2x_2 - x_3 = 2$
    $3x_1 - x_2 + 2x_3 = -3$

Solve Problems 15 and 16.

15. $x^2 - y^2 = 2$
    $y^2 = x$

16. $2x^2 + xy + y^2 = 8$
    $x^2 - y^2 = 0$

17. Solve graphically.

    $2x + y \geq 8$
    $x + 3y \geq 12$
    $x \geq 0$
    $y \geq 0$

C 18. Solve using Gauss–Jordan elimination.

$x_1 + x_2 + x_3 = 7{,}000$
$0.04x_1 + 0.05x_2 + 0.06x_3 = 360$
$0.04x_1 + 0.05x_2 - 0.06x_3 = 120$

19. Solve.

$x^2 - xy + y^2 = 4$
$x^2 + xy - 2y^2 = 0$

APPLICATIONS

20. *Business.* A container contains 120 packages. Some of the packages weigh $\frac{1}{2}$ pound each, and the rest weigh $\frac{1}{3}$ pound each. If the total contents of the container weigh 48 pounds, how many are there of each type of package? Solve using two-equations–two-unknowns methods.

21. *Geometry.* Find the dimensions of a rectangle with an area of 48 square meters and a perimeter of 28 meters. Solve using two-equations–two-unknowns methods.

22. *Diet.* A lab assistant wishes to obtain a food mix that contains, among other things, 27 grams of protein, 5.4 grams of fat, and 19 grams of moisture. He has available mixes of the compositions as listed in the table. How many grams of each mix should be used to get the desired diet mix? Set up a system of equations and solve using Gauss–Jordan elimination.

MIX	PROTEIN (%)	FAT (%)	MOISTURE (%)
A	30	3	10
B	20	5	20
C	10	4	10

Practice Test Chapter 9

Take this practice test as if it were a graded test. Allow yourself up to 50 minutes. Work the problems without looking back in the chapter. Correct your work using the answers (keyed to appropriate sections) in the back of the book.

Solve by graphing.

1. $2x + y = 6$
   $x + 4y = -4$

2. $3x - 4y \geq 24$

3. $x + 2y \leq 8$
   $2x + y \leq 10$
   $x \geq 0$
   $y \geq 0$

Solve by elimination using substitution or addition.

4. $x^2 = y$
   $y = 2x - 2$

5. $4x^2 - y^2 = -5$
   $3x^2 + 2y^2 = 21$

6. $x^2 + 2xy + y^2 = 8$
   $x^2 - xy = 0$

Solve using Gauss–Jordan elimination.

7. $2x_1 + 4x_2 = 2$
   $3x_1 + 8x_2 = 1$

8. $2x_1 + x_2 + x_3 = 3$
   $x_1 + 2x_2 - 4x_3 = -6$
   $2x_1 + x_2 - x_3 = 1$

9. $4x_1 - 8x_2 = 8$
   $3x_1 - 4x_2 = 9$
   $2x_1 - 4x_2 = 2$

10. $x_1 + 2x_2 + 8x_3 = -1$
    $3x_1 + 7x_2 + 29x_3 = -2$

Matrices and Determinants

10

10-1 Matrix Addition; Multiplication of a Matrix by a Number
10-2 Matrix Multiplication
10-3 Inverse of a Square Matrix; Matrix Equations
10-4 Determinant Functions
10-5 Properties of Determinants
10-6 Cramer's Rule
10-7 Chapter Review

A natural design of mathematical interest. Can you guess the source? See the back of the book.

Chapter 10 ■ Matrices and Determinants

Section 10-1 Matrix Addition; Multiplication of a Matrix by a Number

- Dimension of a Matrix
- Matrix Addition
- Multiplication of a Matrix by a Number
- Application

In the last chapter we introduced the important idea of matrix. In this and the following two sections, we will develop this concept further.

■ Dimension of a Matrix

Recall that we defined a **matrix** as any rectangular array of numbers enclosed within brackets. The **size** or **dimension of a matrix** is important relative to operations on matrices. We define an **m × n matrix** (read "m by n matrix") to be one with m rows and n columns. It is important to note that the number of rows is always given first. If a matrix has the same number of rows and columns, it is called a **square matrix**. A matrix with only one column is called a **column matrix**, and one with only one row is called a **row matrix**. These definitions are illustrated by the following:

$$
\begin{array}{cccc}
3 \times 2 & 3 \times 3 & 4 \times 1 & 1 \times 4 \\
\begin{bmatrix} -2 & 5 \\ 0 & -2 \\ 3 & 6 \end{bmatrix} &
\begin{bmatrix} 0.5 & 0.2 & 1.0 \\ 0.0 & 0.3 & 0.5 \\ 0.7 & 0.0 & 0.2 \end{bmatrix} &
\begin{bmatrix} 3 \\ -2 \\ 1 \\ 0 \end{bmatrix} &
[2 \;\; \tfrac{1}{2} \;\; 0 \;\; -\tfrac{2}{3}]
\end{array}
$$

Square matrix   Column matrix   Row matrix

Two matrices are **equal** if they have the same dimension and their corresponding elements are equal. For example,

$$
\begin{bmatrix} a & b & c \\ d & e & f \end{bmatrix} = \begin{bmatrix} u & v & w \\ x & y & z \end{bmatrix} \quad \text{if and only if} \quad \begin{array}{ccc} a=u & b=v & c=w \\ d=x & e=y & f=z \end{array}
$$

(both 2 × 3)

10-1 Matrix Addition; Multiplication of a Matrix by a Number

■ **Matrix Addition**

The **sum of two matrices** of the same dimension is a matrix with elements that are the sums of the corresponding elements of the two given matrices. Addition is not defined for matrices with different dimensions.

EXAMPLE 1 (A) $\begin{bmatrix} a & b \\ c & d \end{bmatrix} + \begin{bmatrix} w & x \\ y & z \end{bmatrix} = \begin{bmatrix} (a+w) & (b+x) \\ (c+y) & (d+z) \end{bmatrix}$

(B) $\begin{bmatrix} 2 & -3 & 0 \\ 1 & 2 & -5 \end{bmatrix} + \begin{bmatrix} 3 & 1 & 2 \\ -3 & 2 & 5 \end{bmatrix} = \begin{bmatrix} 5 & -2 & 2 \\ -2 & 4 & 0 \end{bmatrix}$

PROBLEM 1 Add.

$\begin{bmatrix} 3 & 2 \\ -1 & -1 \\ 0 & 3 \end{bmatrix} + \begin{bmatrix} -2 & 3 \\ 1 & -1 \\ 2 & -2 \end{bmatrix}$

Because we add two matrices by adding their corresponding elements, it follows from the properties of real numbers that matrices of the same dimension are commutative and associative relative to addition. That is, if A, B, and C are matrices of the same dimension, then

$A + B = B + A$ Commutative

$(A + B) + C = A + (B + C)$ Associative

A matrix with elements that are all 0's is called a **zero matrix**. For example,

$[0 \ 0 \ 0]$ $\begin{bmatrix} 0 & 0 \\ 0 & 0 \end{bmatrix}$ $\begin{bmatrix} 0 \\ 0 \\ 0 \\ 0 \end{bmatrix}$ $\begin{bmatrix} 0 & 0 & 0 & 0 \\ 0 & 0 & 0 & 0 \\ 0 & 0 & 0 & 0 \end{bmatrix}$

are zero matrices of different dimensions. [*Note:* "0" may be used to denote the zero matrix of any dimension.] The **negative of a matrix M**, denoted by $-M$, is a matrix with elements that are the negative of the elements in M. Thus, if

$M = \begin{bmatrix} a & b \\ c & d \end{bmatrix}$

then

$-M = \begin{bmatrix} -a & -b \\ -c & -d \end{bmatrix}$

Note that $M + (-M) = 0$ (a zero matrix).

If A and B are matrices of the same dimension, then we define **subtraction** as follows:

$$A - B = A + (-B)$$

Thus, to subtract matrix B from matrix A, we simply subtract corresponding elements.

EXAMPLE 2
$$\begin{bmatrix} 3 & -2 \\ 5 & 0 \end{bmatrix} - \begin{bmatrix} -2 & 2 \\ 3 & 4 \end{bmatrix} = \begin{bmatrix} 3 & -2 \\ 5 & 0 \end{bmatrix} + \begin{bmatrix} 2 & -2 \\ -3 & -4 \end{bmatrix} = \begin{bmatrix} 5 & -4 \\ 2 & -4 \end{bmatrix}$$

PROBLEM 2 Subtract: $[2 \quad -3 \quad 5] - [3 \quad -2 \quad 1]$

■ Multiplication of a Matrix by a Number

Finally, the **product of a number k and a matrix M**, denoted by kM, is a matrix formed by multiplying each element of M by k. This definition is partly motivated by the fact that if M is a matrix, then we would like $M + M$ to equal $2M$.

EXAMPLE 3
$$-2 \begin{bmatrix} 3 & -1 & 0 \\ -2 & 1 & 3 \\ 0 & -1 & -2 \end{bmatrix} = \begin{bmatrix} -6 & 2 & 0 \\ 4 & -2 & -6 \\ 0 & 2 & 4 \end{bmatrix}$$

PROBLEM 3 Find: $10 \begin{bmatrix} 1.3 \\ 0.2 \\ 3.5 \end{bmatrix}$

We now consider an application that uses various operations.

■ Application

EXAMPLE 4 Ms. Smith and Mr. Jones are salespeople in a new-car agency that sells only two models. August was the last month for this year's models, and next year's models were introduced in September. Gross dollar sales for each month are given in the following matrices:

AUGUST SALES
	Compact	Luxury
Ms. Smith	$18,000	$36,000
Mr. Jones	$36,000	0

$= A$

SEPTEMBER SALES
	Compact	Luxury
Ms. Smith	$72,000	$144,000
Mr. Jones	$90,000	$108,000

$= B$

(For example, Ms. Smith had $18,000 in compact sales in August, and Mr. Jones had $108,000 in luxury car sales in September.)

(A) What was the combined dollar sales in August and September for each person and each model?

10-1 Matrix Addition; Multiplication of a Matrix by a Number

(B) What was the increase in dollar sales from August to September?

(C) If both salespeople receive 5% commissions on gross dollar sales, compute the commission for each person for each model sold in September.

Solution (A) $A + B = \begin{bmatrix} \$90{,}000 & \$180{,}000 \\ \$126{,}000 & \$108{,}000 \end{bmatrix}$ Ms. Smith / Mr. Jones
(Compact   Luxury)

(B) $B - A = \begin{bmatrix} \$54{,}000 & \$108{,}000 \\ \$54{,}000 & \$108{,}000 \end{bmatrix}$ Ms. Smith / Mr. Jones
(Compact   Luxury)

(C) $0.05B = \begin{bmatrix} (0.05)(\$72{,}000) & (0.05)(\$144{,}000) \\ (0.05)(\$90{,}000) & (0.05)(\$108{,}000) \end{bmatrix}$

$= \begin{bmatrix} \$3{,}600 & \$7{,}200 \\ \$4{,}500 & \$5{,}400 \end{bmatrix}$ Ms. Smith / Mr. Jones
(Compact   Luxury)

In Example 4 we chose a relatively simple example involving an agency with only two salespeople and two models. Consider the more realistic problem of an agency with nine models and perhaps seven salespeople—then you can begin to see the value of matrix methods.

PROBLEM 4 Repeat Example 4 with

$$A = \begin{bmatrix} \$36{,}000 & \$36{,}000 \\ \$18{,}000 & \$36{,}000 \end{bmatrix} \text{ and } B = \begin{bmatrix} \$90{,}000 & \$108{,}000 \\ \$72{,}000 & \$108{,}000 \end{bmatrix}$$

Answers to Matched Problems

1. $\begin{bmatrix} 1 & 5 \\ 0 & -2 \\ 2 & 1 \end{bmatrix}$ 2. $[-1 \quad -1 \quad 4]$ 3. $\begin{bmatrix} 13 \\ 2 \\ 35 \end{bmatrix}$

4. (A) $\begin{bmatrix} \$126{,}000 & \$144{,}000 \\ \$90{,}000 & \$144{,}000 \end{bmatrix}$ (B) $\begin{bmatrix} \$54{,}000 & \$72{,}000 \\ \$54{,}000 & \$72{,}000 \end{bmatrix}$

(C) $\begin{bmatrix} \$4{,}500 & \$5{,}400 \\ \$3{,}600 & \$5{,}400 \end{bmatrix}$

Exercise 10-1

A Problems 1–18 refer to the following matrices:

$$A = \begin{bmatrix} 2 & -1 \\ 3 & 0 \end{bmatrix} \quad B = \begin{bmatrix} -3 & 1 \\ 2 & -3 \end{bmatrix} \quad C = \begin{bmatrix} 2 \\ -3 \\ 0 \end{bmatrix}$$

$$D = \begin{bmatrix} 1 \\ 3 \\ 5 \end{bmatrix} \qquad E = \begin{bmatrix} -4 & 1 & 0 & -2 \end{bmatrix} \qquad F = \begin{bmatrix} 2 & -3 \\ -2 & 0 \\ 1 & 2 \\ 3 & 5 \end{bmatrix}$$

1. What are the dimensions of B? Of E?
2. What are the dimensions of F? Of D?
3. What element is in the third row and second column of matrix F?
4. What element is in the second row and first column of matrix F?
5. Write a zero matrix of the same dimension as B.
6. Write a zero matrix of the same dimension as E.
7. Identify all column matrices.
8. Identify all row matrices.
9. Identify all square matrices.
10. How many additional columns would F have to have to be a square matrix?
11. Find $A + B$.
12. Find $C + D$.
13. Write the negative of matrix C.
14. Write the negative of matrix B.
15. Find $D - C$.
16. Find $A - A$.
17. Find $5B$.
18. Find $-2E$.

B In Problems 19–24 perform the indicated operations.

19. $\begin{bmatrix} -2 & 3 & 0 \end{bmatrix} + 2\begin{bmatrix} 1 & -1 & 2 \end{bmatrix}$

20. $\begin{bmatrix} 230 \\ 120 \end{bmatrix} + 3\begin{bmatrix} 20 \\ 60 \end{bmatrix}$

21. $1{,}000 \begin{bmatrix} 0.25 & 0.36 \\ 0.04 & 0.35 \end{bmatrix}$

22. $100 \begin{bmatrix} 0.32 & 0.05 & 0.17 \\ 0.22 & 0.03 & 0.21 \end{bmatrix}$

23. $2\begin{bmatrix} 1 & 2 \\ -1 & 3 \\ 0 & -2 \end{bmatrix} - \begin{bmatrix} 3 & 4 \\ -2 & 0 \\ 1 & -3 \end{bmatrix}$

24. $-2\begin{bmatrix} 1 & 3 & 0 \\ -2 & -1 & 1 \end{bmatrix} - \begin{bmatrix} -3 & -1 & 1 \\ 0 & 2 & -1 \end{bmatrix}$

C 25. Find a, b, c, and d so that

$$\begin{bmatrix} a & b \\ c & d \end{bmatrix} + \begin{bmatrix} 2 & -3 \\ 0 & 1 \end{bmatrix} = \begin{bmatrix} 1 & -2 \\ 3 & -4 \end{bmatrix}$$

26. Find w, x, y, and z so that

$$\begin{bmatrix} 4 & -2 \\ -3 & 0 \end{bmatrix} + \begin{bmatrix} w & x \\ y & z \end{bmatrix} = \begin{bmatrix} 2 & -3 \\ 0 & 5 \end{bmatrix}$$

APPLICATIONS

27. *Cost analysis.* A company with two different plants manufactures guitars and banjos. Its production costs for each instrument are given in the following matrices:

$$\begin{array}{c} \text{PLANT X} \\ \text{Guitar \quad Banjo} \end{array}$$
$$\begin{array}{c} \text{Materials} \\ \text{Labor} \end{array} \begin{bmatrix} \$30 & \$25 \\ \$60 & \$80 \end{bmatrix} = A \qquad \begin{array}{c} \text{PLANT Y} \\ \text{Guitar \quad Banjo} \end{array} \begin{bmatrix} \$36 & \$27 \\ \$54 & \$74 \end{bmatrix} = B$$

Find $\frac{1}{2}(A + B)$, the average cost of production for the two plants.

28. *Heredity.* Gregor Mendel (1822–1884), a Bavarian monk and botanist, made discoveries that revolutionized the science of heredity. In one experiment he crossed dihybrid yellow round peas (yellow and round are dominant characteristics; the peas also contained green and wrinkled as recessive genes) and obtained 560 peas of the types indicated in the matrix:

$$\begin{array}{c} \quad \text{Round} \quad \text{Wrinkled} \end{array}$$
$$\begin{array}{c} \text{Yellow} \\ \text{Green} \end{array} \begin{bmatrix} 319 & 101 \\ 108 & 32 \end{bmatrix} = M$$

Suppose he carried out a second experiment of the same type and obtained 640 peas of the types indicated in this matrix:

$$\begin{array}{c} \quad \text{Round} \quad \text{Wrinkled} \end{array}$$
$$\begin{array}{c} \text{Yellow} \\ \text{Green} \end{array} \begin{bmatrix} 370 & 124 \\ 110 & 36 \end{bmatrix} = N$$

If the results of the two experiments are combined, write the resulting matrix $M + N$. Compute the decimal fraction of the total number of peas (1,200) in each category of the combined results. [*Hint:* Compute $(1/1{,}200)(M + N)$.]

29. *Psychology.* Two psychologists independently carried out studies on the relationship between height and aggressive behavior in women over 18 years of age. The results of the studies are summarized in the following matrices:

$$\begin{array}{c} \text{PROFESSOR ALDQUIST} \\ \text{Under 5 ft} \quad 5\text{–}5\tfrac{1}{2}\text{ ft} \quad \text{Over } 5\tfrac{1}{2}\text{ ft} \end{array}$$
$$\begin{array}{c} \text{Passive} \\ \text{Aggressive} \end{array} \begin{bmatrix} 70 & 122 & 20 \\ 30 & 118 & 80 \end{bmatrix} = A$$

$$\begin{array}{c} \text{PROFESSOR KELLEY} \\ \text{Under 5 ft} \quad 5\text{–}5\tfrac{1}{2}\text{ ft} \quad \text{Over } 5\tfrac{1}{2}\text{ ft} \end{array}$$
$$\begin{array}{c} \text{Passive} \\ \text{Aggressive} \end{array} \begin{bmatrix} 65 & 160 & 30 \\ 25 & 140 & 75 \end{bmatrix} = B$$

The two psychologists decided to combine their results and publish a joint paper. Write the matrix $A + B$ illustrating their combined results. If you have a hand calculator, compute the decimal fraction of the total sample in each category of the combined study. [*Hint:* Compute $(1/935)(A + B)$.]

Section 10-2 Matrix Multiplication

- Dot Product
- Matrix Product
- Multiplication Properties
- Application

In this section we are going to introduce two types of matrix multiplication that will at first seem rather strange. In spite of this apparent strangeness, these operations are well founded in the general theory of matrices and, as we will see, are extremely useful in practical problems.

- Dot Product

We start by defining the dot product of two special matrices.

Dot Product

The **dot product** of a $1 \times n$ row matrix and an $n \times 1$ column matrix is a real number given by:

$$\underset{1 \times n}{[a_1 \ a_2 \cdots a_n]} \cdot \underset{n \times 1}{\begin{bmatrix} b_1 \\ b_2 \\ \vdots \\ b_n \end{bmatrix}} = a_1 b_1 + a_2 b_2 + \cdots + a_n b_n \quad \text{A real number}$$

The dot between the two matrices is important. If the dot is omitted, the multiplication is of another type, which we will consider later.

EXAMPLE 5
$$[2 \ -3 \ 0] \cdot \begin{bmatrix} -5 \\ 2 \\ -2 \end{bmatrix} = (2)(-5) + (-3)(2) + (0)(-2)$$
$$= -10 - 6 + 0 = -16$$

10-2 Matrix Multiplication

PROBLEM 5
$$[-1 \quad 0 \quad 3 \quad 2] \cdot \begin{bmatrix} 2 \\ 3 \\ 4 \\ -1 \end{bmatrix} = ?$$

EXAMPLE 6 A factory produces a slalom water ski that requires 4 work-hours in the fabricating department and 1 work-hour in the finishing department. Fabricating personnel receive \$8 per hour and finishing personnel receive \$6 per hour. Total labor cost per ski is given by the dot product:

$$[4 \quad 1] \cdot \begin{bmatrix} 8 \\ 6 \end{bmatrix} = (4)(8) + (1)(6) = 32 + 6 = \$38 \text{ per ski}$$

PROBLEM 6 If the factory in Example 6 also produces a trick water ski that requires 6 work-hours in the fabricating department and 1.5 work-hours in the finishing department, write a dot product between appropriate row and column matrices that will give the total labor cost for this ski. Compute the cost.

■ **Matrix Product**

It is important to remember that the dot product of a row matrix and a column matrix is a real number and not a matrix. We now define a matrix product for certain matrices.

Matrix Product

The **product of two matrices** A and B is defined only on the assumption that the number of columns in A is equal to the number of rows in B. If A is an $m \times p$ matrix and B is a $p \times n$ matrix, then the matrix product of A and B, denoted by AB, is an $m \times n$ matrix whose element in the ith row and jth column is the dot product of the ith row matrix of A and the jth column matrix of B.

It is important to check dimensions before starting the multiplication process. If matrix A has dimension $a \times b$ and matrix B has dimension $c \times d$, then if $b = c$, the product AB will exist and will have dimension

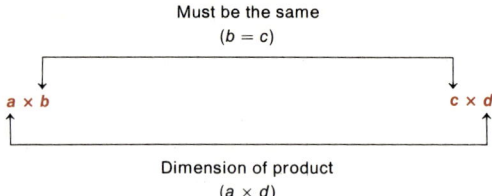

FIGURE 1

$a \times d$. This is shown schematically in Figure 1. The definition is not as complicated as it might first seem. An example should help to clarify the process. For

$$A = \begin{bmatrix} 2 & 3 & -1 \\ -2 & 1 & 2 \end{bmatrix} \quad \text{and} \quad B = \begin{bmatrix} 1 & 3 \\ 2 & 0 \\ -1 & 2 \end{bmatrix}$$

A is 2×3, B is 3×2, and AB will be 2×2. The four dot products used to produce the four elements in AB (usually calculated mentally for small whole numbers) are shown in the following large matrix.

$$\overset{2 \times 3}{\begin{bmatrix} 2 & 3 & -1 \\ -2 & 1 & 2 \end{bmatrix}} \overset{3 \times 2}{\begin{bmatrix} 1 & 3 \\ 2 & 0 \\ -1 & 2 \end{bmatrix}} = \begin{bmatrix} [2\ 3\ -1] \cdot \begin{bmatrix} 1 \\ 2 \\ -1 \end{bmatrix} & [2\ 3\ -1] \cdot \begin{bmatrix} 3 \\ 0 \\ 2 \end{bmatrix} \\ [-2\ 1\ 2] \cdot \begin{bmatrix} 1 \\ 2 \\ -1 \end{bmatrix} & [-2\ 1\ 2] \cdot \begin{bmatrix} 3 \\ 0 \\ 2 \end{bmatrix} \end{bmatrix} = \overset{2 \times 2}{\begin{bmatrix} 9 & 4 \\ -2 & -2 \end{bmatrix}}$$

EXAMPLE 7

$$\overset{3 \times 2}{\begin{bmatrix} 2 & 1 \\ 1 & 0 \\ -1 & 2 \end{bmatrix}} \overset{2 \times 4}{\begin{bmatrix} 1 & -1 & 0 & 1 \\ 2 & 1 & 2 & 0 \end{bmatrix}} = \overset{3 \times 4}{\begin{bmatrix} 4 & -1 & 2 & 2 \\ 1 & -1 & 0 & 1 \\ 3 & 3 & 4 & -1 \end{bmatrix}}$$

PROBLEM 7   Find the product.

$$\begin{bmatrix} 1 & -1 & 0 \\ 2 & 1 & 2 \end{bmatrix} \begin{bmatrix} 2 & -1 \\ 1 & 0 \\ -1 & 2 \end{bmatrix}$$

■ **Multiplication Properties**

In the arithmetic of real numbers it does not matter in which order we multiply; for example, $5 \times 7 = 7 \times 5$. In matrix multiplication it does make a difference; that is, MN does not always equal NM, even if both multiplications are defined (see Problems 11, 12, 25, 27, and 28 in Exercise 10-2). Also, MN may be zero with neither M nor N zero.

Matrices do, however, have other general properties. We state three important properties without proof. Assuming all products and sums are defined for the indicated matrices A, B, and C, then for k a real number:

1. $(AB)C = A(BC)$ Associative property
2. $A(B + C) = AB + AC$ Left-hand distributive property
3. $(B + C)A = BA + CA$ Right-hand distributive property
4. $k(AB) = (kA)B = A(kB)$

Since matrix multiplication is not commutative, properties 2 and 3 must be listed as distinct properties (see Problems 27 and 28 in Exercise 10-2).

■ **Application**

The next example illustrates the use of the dot and matrix product in a business application.

EXAMPLE 8   *Production scheduling.* Let us combine the time requirements discussed in Example 6 and Problem 6 into one matrix:

$$\begin{array}{c} \\ \text{Trick ski} \\ \text{Slalom ski} \end{array} \begin{array}{cc} \text{Fabricating} & \text{Finishing} \\ \text{department} & \text{department} \\ \begin{bmatrix} 6 \text{ hr} & 1.5 \text{ hr} \\ 4 \text{ hr} & 1 \text{ hr} \end{bmatrix} \end{array} = A$$

Now suppose the company has two manufacturing plants X and Y in different parts of the country and that their hourly rates for each department are given in the following matrix:

$$\begin{array}{c} \\ \text{Fabricating department} \\ \text{Finishing department} \end{array} \begin{array}{cc} \text{Plant X} & \text{Plant Y} \\ \begin{bmatrix} \$8 & \$7 \\ \$6 & \$4 \end{bmatrix} \end{array} = B$$

To find the total labor costs for each ski at each factory, we multiply A and B:

$$AB = \overset{2 \times 2}{\begin{bmatrix} 6 & 1.5 \\ 4 & 1 \end{bmatrix}} \overset{2 \times 2}{\begin{bmatrix} 8 & 7 \\ 6 & 4 \end{bmatrix}} = \overset{\text{X} \quad \text{Y}}{\begin{bmatrix} \$57 & \$48 \\ \$38 & \$32 \end{bmatrix}} \begin{array}{l} \text{Trick ski} \\ \text{Slalom ski} \end{array}$$

Notice that the dot product of the first row matrix of A and the first column matrix of B gives us the labor costs, $57, for a trick ski manufactured at plant X; the dot product of the second row matrix of A and the second column matrix of B gives us the labor costs, $32, for manufacturing a slalom ski at plant Y; and so on.

Example 8 is, of course, oversimplified. Companies that manufacture many different items in many different plants deal with matrices that have very large numbers of rows and columns.

PROBLEM 8 Repeat Example 8 with

$$A = \begin{bmatrix} 7 \text{ hr} & 2 \text{ hr} \\ 5 \text{ hr} & 1.5 \text{ hr} \end{bmatrix} \quad \text{and} \quad B = \begin{bmatrix} \$10 & \$8 \\ \$6 & \$4 \end{bmatrix}$$

Answers to Matched Problems

5. 8

6. $[6 \quad 1.5] \cdot \begin{bmatrix} 8 \\ 6 \end{bmatrix} = \$57$

7. $\begin{bmatrix} 1 & -1 \\ 3 & 2 \end{bmatrix}$

8. $\begin{array}{cc} \text{X} & \text{Y} \end{array}$
$\begin{bmatrix} \$82 & \$64 \\ \$59 & \$46 \end{bmatrix} \begin{array}{l} \text{Trick} \\ \text{Slalom} \end{array}$

Exercise 10-2

A *Find the dot products.*

1. $[2 \quad 4] \cdot \begin{bmatrix} 3 \\ 1 \end{bmatrix}$

2. $[3 \quad 1] \cdot \begin{bmatrix} 2 \\ 4 \end{bmatrix}$

3. $[-3 \quad 2] \cdot \begin{bmatrix} -1 \\ -2 \end{bmatrix}$

4. $[3 \quad -2] \cdot \begin{bmatrix} -4 \\ -1 \end{bmatrix}$

Find the matrix products.

5. $[2 \quad 5] \begin{bmatrix} 1 & -1 \\ 2 & 3 \end{bmatrix}$

6. $[1 \quad 3] \begin{bmatrix} 2 & 3 \\ 1 & -4 \end{bmatrix}$

7. $\begin{bmatrix} 3 & 4 \\ -1 & -2 \end{bmatrix} \begin{bmatrix} -1 \\ 2 \end{bmatrix}$

8. $\begin{bmatrix} -1 & 1 \\ 2 & -3 \end{bmatrix} \begin{bmatrix} 4 \\ -2 \end{bmatrix}$

9. $\begin{bmatrix} 2 & -3 \\ 1 & 2 \end{bmatrix} \begin{bmatrix} 1 & -1 \\ 0 & -2 \end{bmatrix}$

10. $\begin{bmatrix} -3 & 2 \\ 4 & -1 \end{bmatrix} \begin{bmatrix} -2 & 5 \\ -1 & 3 \end{bmatrix}$

11. $\begin{bmatrix} -5 & -2 \\ 1 & -3 \end{bmatrix} \begin{bmatrix} -2 & 1 \\ 0 & -3 \end{bmatrix}$

12. $\begin{bmatrix} -2 & 1 \\ 0 & -3 \end{bmatrix} \begin{bmatrix} -5 & -2 \\ 1 & -3 \end{bmatrix}$

B *Find the dot products.*

13. $[-1 \quad -2 \quad 2] \cdot \begin{bmatrix} 2 \\ -1 \\ 3 \end{bmatrix}$

14. $[-2 \quad 4 \quad 0] \cdot \begin{bmatrix} -1 \\ -3 \\ 2 \end{bmatrix}$

15. $[-1 \quad -3 \quad 0 \quad 5] \cdot \begin{bmatrix} 4 \\ -3 \\ -1 \\ 2 \end{bmatrix}$

16. $[-1 \quad 2 \quad 3 \quad -2] \cdot \begin{bmatrix} 3 \\ -2 \\ 0 \\ 4 \end{bmatrix}$

Find the matrix products.

17. $\begin{bmatrix} 2 & -1 & 1 \\ 1 & 3 & -2 \end{bmatrix} \begin{bmatrix} 1 & 3 \\ 0 & -1 \\ -2 & 2 \end{bmatrix}$

18. $\begin{bmatrix} -1 & -4 & 3 \\ 2 & 0 & 1 \end{bmatrix} \begin{bmatrix} 2 & -3 \\ 1 & 2 \\ 0 & -1 \end{bmatrix}$

19. $\begin{bmatrix} 1 & 3 \\ 0 & -1 \\ -2 & 2 \end{bmatrix} \begin{bmatrix} 2 & -1 & 1 \\ 1 & 3 & -2 \end{bmatrix}$

20. $\begin{bmatrix} 2 & -3 \\ 1 & 2 \\ 0 & -1 \end{bmatrix} \begin{bmatrix} -1 & -4 & 3 \\ 2 & 0 & 1 \end{bmatrix}$

21. $[3 \quad -2 \quad -4] \begin{bmatrix} 1 \\ 2 \\ -3 \end{bmatrix}$

22. $[1 \quad -2 \quad 2] \begin{bmatrix} 2 \\ -1 \\ 1 \end{bmatrix}$

23. $\begin{bmatrix} 1 \\ 2 \\ -3 \end{bmatrix} [3 \quad -2 \quad -4]$

24. $\begin{bmatrix} 2 \\ -1 \\ 1 \end{bmatrix} [1 \quad -2 \quad 2]$

C *In Problems 25–28 verify each statement using the following matrices:*

$$A = \begin{bmatrix} 1 & 2 \\ 0 & 1 \end{bmatrix} \quad B = \begin{bmatrix} 1 & 1 \\ 2 & 3 \end{bmatrix} \quad C = \begin{bmatrix} -3 & 1 \\ -1 & 2 \end{bmatrix}$$

25. $AB \neq BA$

26. $(AB)C = A(BC)$

27. $A(B + C) = AB + AC$

28. $(B + C)A = BA + CA$

APPLICATIONS

29. *Labor costs.* A company with manufacturing plants located in different parts of the country has work-hour and wage requirements for the manufacturing of three types of inflatable boats as given in the following two matrices:

WORK-HOURS PER BOAT

	Cutting department	Assembly department	Packaging department	
$M =$	0.6 hr	0.6 hr	0.2 hr	One-person boat
	1.0 hr	0.9 hr	0.3 hr	Two-person boat
	1.5 hr	1.2 hr	0.4 hr	Four-person boat

HOURLY WAGES

	Plant I	Plant II	
$N =$	\$6	\$7	Cutting department
	\$8	\$10	Assembly department
	\$3	\$4	Packaging department

(A) Find the labor costs for a one-person boat manufactured at plant I; that is, find the dot product

$$[0.6 \quad 0.6 \quad 0.2] \cdot \begin{bmatrix} 6 \\ 8 \\ 3 \end{bmatrix}$$

(B) Find the labor costs for a four-person boat manufactured at plant II. Set up a dot product as in part (A) and multiply.
(C) What is the dimension of MN?
(D) Find MN and interpret.

30. *Nutrition.* A nutritionist for a cereal company blends two cereals in different mixes. The amounts of protein, carbohydrate, and fat (in grams per ounce) in each cereal are given by matrix M. The amounts of each cereal used in the three mixes is given by matrix N.

$$M = \begin{matrix} & \text{Cereal A} & \text{Cereal B} & \\ & \begin{bmatrix} 4 \text{ grams} & 2 \text{ grams} \\ 20 \text{ grams} & 16 \text{ grams} \\ 3 \text{ grams} & 1 \text{ gram} \end{bmatrix} & \begin{matrix} \text{Protein} \\ \text{Carbohydrate} \\ \text{Fat} \end{matrix} \end{matrix}$$

$$N = \begin{matrix} & \text{Mix X} & \text{Mix Y} & \text{Mix Z} & \\ & \begin{bmatrix} 15 \text{ ounces} & 10 \text{ ounces} & 5 \text{ ounces} \\ 5 \text{ ounces} & 10 \text{ ounces} & 15 \text{ ounces} \end{bmatrix} & \begin{matrix} \text{Cereal A} \\ \text{Cereal B} \end{matrix} \end{matrix}$$

(A) Find the amount of protein in mix X by computing the dot product

$$[4 \quad 2] \cdot \begin{bmatrix} 15 \\ 5 \end{bmatrix}$$

(B) Find the amount of fat in mix Z. Set up a dot product as in part (A) and multiply.
(C) What is the dimension of MN?
(D) Find MN and interpret.
(E) Find $\frac{1}{20}$ MN and interpret.

31. *Politics.* In a local election a group hired a public relations firm to promote its candidate in three ways: telephone, house calls, and letters. The cost per contact is given in matrix M:

$$M = \begin{matrix} \text{Cost per} \\ \text{contact} \\ \begin{bmatrix} \$0.40 \\ \$0.75 \\ \$0.25 \end{bmatrix} \begin{matrix} \text{Telephone} \\ \text{House call} \\ \text{Letter} \end{matrix} \end{matrix}$$

The number of contacts of each type made in two adjacent cities is given in matrix N:

$$N = \begin{bmatrix} \text{Telephone} & \text{House call} & \text{Letter} \\ 1{,}000 & 500 & 5{,}000 \\ 2{,}000 & 800 & 8{,}000 \end{bmatrix} \begin{matrix} \text{Berkeley} \\ \text{Oakland} \end{matrix}$$

(A) Find the total amount spent in Berkeley by computing the dot product

$$[1{,}000 \quad 500 \quad 5{,}000] \cdot \begin{bmatrix} \$0.40 \\ \$0.75 \\ \$0.25 \end{bmatrix}$$

(B) Find the total amount spent in Oakland by computing the dot product of appropriate matrices.
(C) Compute NM and interpret.
(D) Multiply N by the matrix [1 1] and interpret.

Section 10-3 Inverse of a Square Matrix; Matrix Equations

- Identity Matrix for Multiplication
- Inverse of a Square Matrix
- Matrix Equations
- Application

- Identity Matrix for Multiplication

We know that

$$1a = a1 = a$$

for all real numbers a. The number 1 is called the **identity** for real number multiplication. Does the set of all matrices of a given dimension have an identity element for multiplication? The answer, in general, is no. However, the set of all **square matrices of order n** (dimension $n \times n$) does have an identity, and it is given as follows: The **identity element for multiplication** for the set of all square matrices of order n is the square matrix of order n, denoted by I, with 1's along the **main diagonal** (from upper left corner to lower right corner) and 0's elsewhere. For example,

$$\begin{bmatrix} 1 & 0 \\ 0 & 1 \end{bmatrix} \quad \text{and} \quad \begin{bmatrix} 1 & 0 & 0 \\ 0 & 1 & 0 \\ 0 & 0 & 1 \end{bmatrix}$$

are the identity matrices for square matrices of order 2 and 3, respectively.

EXAMPLE 9
$$\begin{bmatrix} 1 & 0 & 0 \\ 0 & 1 & 0 \\ 0 & 0 & 1 \end{bmatrix} \begin{bmatrix} a & b & c \\ d & e & f \\ g & h & i \end{bmatrix} = \begin{bmatrix} a & b & c \\ d & e & f \\ g & h & i \end{bmatrix}$$

$$= \begin{bmatrix} a & b & c \\ d & e & f \\ g & h & i \end{bmatrix} \begin{bmatrix} 1 & 0 & 0 \\ 0 & 1 & 0 \\ 0 & 0 & 1 \end{bmatrix}$$

PROBLEM 9 Multiply.

$$\begin{bmatrix} 1 & 0 \\ 0 & 1 \end{bmatrix} \begin{bmatrix} 2 & -3 \\ 5 & 7 \end{bmatrix} \quad \text{and} \quad \begin{bmatrix} 2 & -3 \\ 5 & 7 \end{bmatrix} \begin{bmatrix} 1 & 0 \\ 0 & 1 \end{bmatrix}$$

In general, we can show that if M is a square matrix of order n and I is the identity matrix of order n, then

IM = MI = M

■ Inverse of a Square Matrix

In the set of real numbers we know that for each real number a (except 0) there exists a real number a^{-1} such that

$$a^{-1}a = 1$$

The number a^{-1} is called the **inverse** of the number a relative to multiplication, or the **multiplicative inverse** of a. For example, 2^{-1} is the multiplicative inverse of 2, since $2^{-1} \cdot 2 = 1$. For each square matrix M does there exist an inverse matrix M^{-1} such that the following relation is true?

$$M^{-1}M = MM^{-1} = I$$

If M^{-1} exists for a given matrix M, then M^{-1} is called the **inverse of M relative to multiplication**. Let us use this definition to find M^{-1} for

$$M = \begin{bmatrix} 2 & 3 \\ 1 & 2 \end{bmatrix}$$

We are looking for

$$M^{-1} = \begin{bmatrix} a & c \\ b & d \end{bmatrix}$$

such that

$$MM^{-1} = M^{-1}M = I$$

Thus, we write

$$\overset{M}{\begin{bmatrix} 2 & 3 \\ 1 & 2 \end{bmatrix}} \overset{M^{-1}}{\begin{bmatrix} a & c \\ b & d \end{bmatrix}} = \overset{I}{\begin{bmatrix} 1 & 0 \\ 0 & 1 \end{bmatrix}}$$

and try to find a, b, c, and d so that the product of M and M^{-1} is the identity matrix I. Multiplying M and M^{-1} on the left side, we obtain

$$\begin{bmatrix} (2a+3b) & (2c+3d) \\ (a+2b) & (c+2d) \end{bmatrix} = \begin{bmatrix} 1 & 0 \\ 0 & 1 \end{bmatrix}$$

which is true only if

$$2a + 3b = 1 \qquad 2c + 3d = 0$$
$$a + 2b = 0 \qquad c + 2d = 1$$

Solving these two systems, we find that $a = 2$, $b = -1$, $c = -3$, and $d = 2$. Thus,

$$M^{-1} = \begin{bmatrix} 2 & -3 \\ -1 & 2 \end{bmatrix}$$

as is easily checked:

$$\overset{M}{\begin{bmatrix} 2 & 3 \\ 1 & 2 \end{bmatrix}} \overset{M^{-1}}{\begin{bmatrix} 2 & -3 \\ -1 & 2 \end{bmatrix}} = \overset{I}{\begin{bmatrix} 1 & 0 \\ 0 & 1 \end{bmatrix}} = \overset{M^{-1}}{\begin{bmatrix} 2 & -3 \\ -1 & 2 \end{bmatrix}} \overset{M}{\begin{bmatrix} 2 & 3 \\ 1 & 2 \end{bmatrix}}$$

Inverses do not always exist for square matrices. For example, if

$$M = \begin{bmatrix} 2 & 1 \\ 4 & 2 \end{bmatrix}$$

then, proceeding as before, we are led to the systems

$$2a + b = 1 \qquad 2c + d = 0$$
$$4a + 2b = 0 \qquad 4c + 2d = 1$$

These are both inconsistent and have no solution. Hence, M^{-1} does not exist.

Being able to find inverses, when they exist, leads to direct and simple solutions to many practical problems. At the end of this section, for example, we will show how inverses can be used to solve systems of linear equations.

The method outlined for finding M^{-1}, if it exists, gets very involved for matrices of order larger than 2. Now that we know what we are looking for, we can introduce the idea of the augmented matrix (considered in Sections 9-2 and 9-3) to make the process more efficient. For example, to find the inverse (if it exists) of

$$M = \begin{bmatrix} 1 & -1 & 1 \\ 0 & 2 & -1 \\ 2 & 3 & 0 \end{bmatrix}$$

we start as before and write

$$\overset{M}{\begin{bmatrix} 1 & -1 & 1 \\ 0 & 2 & -1 \\ 2 & 3 & 0 \end{bmatrix}} \overset{M^{-1}}{\begin{bmatrix} a & d & g \\ b & e & h \\ c & f & i \end{bmatrix}} = \overset{I}{\begin{bmatrix} 1 & 0 & 0 \\ 0 & 1 & 0 \\ 0 & 0 & 1 \end{bmatrix}}$$

which is true only if

$$\begin{array}{lll} a - b + c = 1 & d - e + f = 0 & g - h + i = 0 \\ 2b - c = 0 & 2e - f = 1 & 2h - i = 0 \\ 2a + 3b = 0 & 2d + 3e = 0 & 2g + 3h = 1 \end{array}$$

Now we write augmented matrices for each of the three systems:

$$\overset{\text{First}}{\begin{bmatrix} 1 & -1 & 1 & | & 1 \\ 0 & 2 & -1 & | & 0 \\ 2 & 3 & 0 & | & 0 \end{bmatrix}} \quad \overset{\text{Second}}{\begin{bmatrix} 1 & -1 & 1 & | & 0 \\ 0 & 2 & -1 & | & 1 \\ 2 & 3 & 0 & | & 0 \end{bmatrix}} \quad \overset{\text{Third}}{\begin{bmatrix} 1 & -1 & 1 & | & 0 \\ 0 & 2 & -1 & | & 0 \\ 2 & 3 & 0 & | & 1 \end{bmatrix}}$$

Since each matrix to the left of the vertical bar is the same, exactly the same row operations can be used on each total matrix to transform it into a reduced form. We can speed up the process substantially by combining all three augmented matrices into the single augmented matrix form

$$\begin{bmatrix} 1 & -1 & 1 & | & 1 & 0 & 0 \\ 0 & 2 & -1 & | & 0 & 1 & 0 \\ 2 & 3 & 0 & | & 0 & 0 & 1 \end{bmatrix} = [M|I] \tag{1}$$

We now try to perform row operations on matrix (1) until we obtain a row-equivalent matrix that looks like matrix (2):

$$\begin{bmatrix} \overset{I}{1} & 0 & 0 & | & \overset{B}{a} & d & g \\ 0 & 1 & 0 & | & b & e & h \\ 0 & 0 & 1 & | & c & f & i \end{bmatrix} \tag{2}$$

If this can be done, then the new matrix to the right of the vertical bar will be M^{-1}! Now let us try to transform (1) into a form like (2).

10-3 Inverse of a Square Matrix; Matrix Equations

$$\begin{array}{c c}
\overset{M}{\begin{bmatrix} 1 & -1 & 1 \\ 0 & 2 & -1 \\ 2 & 3 & 0 \end{bmatrix}} \overset{I}{\begin{bmatrix} 1 & 0 & 0 \\ 0 & 1 & 0 \\ 0 & 0 & 1 \end{bmatrix}} & R_3 + (-2)R_1 \to R_3 \\[2em]
\sim \begin{bmatrix} 1 & -1 & 1 & | & 1 & 0 & 0 \\ 0 & 2 & -1 & | & 0 & 1 & 0 \\ 0 & 5 & -2 & | & -2 & 0 & 1 \end{bmatrix} & \tfrac{1}{2}R_2 \to R_2 \\[2em]
\sim \begin{bmatrix} 1 & -1 & 1 & | & 1 & 0 & 0 \\ 0 & 1 & -\tfrac{1}{2} & | & 0 & \tfrac{1}{2} & 0 \\ 0 & 5 & -2 & | & -2 & 0 & 1 \end{bmatrix} & R_3 + (-5)R_2 \to R_3 \\[2em]
\sim \begin{bmatrix} 1 & -1 & 1 & | & 1 & 0 & 0 \\ 0 & 1 & -\tfrac{1}{2} & | & 0 & \tfrac{1}{2} & 0 \\ 0 & 0 & \tfrac{1}{2} & | & -2 & -\tfrac{5}{2} & 1 \end{bmatrix} & 2R_3 \to R_3 \\[2em]
\sim \begin{bmatrix} 1 & -1 & 1 & | & 1 & 0 & 0 \\ 0 & 1 & -\tfrac{1}{2} & | & 0 & \tfrac{1}{2} & 0 \\ 0 & 0 & 1 & | & -4 & -5 & 2 \end{bmatrix} & \begin{array}{l} R_1 + (-1)R_3 \to R_1 \\ R_2 + \tfrac{1}{2}R_3 \to R_2 \end{array} \\[2em]
\sim \begin{bmatrix} 1 & -1 & 0 & | & 5 & 5 & -2 \\ 0 & 1 & 0 & | & -2 & -2 & 1 \\ 0 & 0 & 1 & | & -4 & -5 & 2 \end{bmatrix} & R_1 + R_2 \to R_1 \\[2em]
\overset{I}{\sim \begin{bmatrix} 1 & 0 & 0 \\ 0 & 1 & 0 \\ 0 & 0 & 1 \end{bmatrix}} \overset{B}{\begin{bmatrix} 3 & 3 & -1 \\ -2 & -2 & 1 \\ -4 & -5 & 2 \end{bmatrix}} &
\end{array}$$

Converting back to systems of equations equivalent to our three original systems, we have

$$\begin{array}{lll}
a = 3 & d = 3 & g = -1 \\
b = -2 & e = -2 & h = 1 \\
c = -4 & f = -5 & i = 2
\end{array}$$

And these are just the elements of M^{-1} that we are looking for! Hence,

$$M^{-1} = \begin{bmatrix} 3 & 3 & -1 \\ -2 & -2 & 1 \\ -4 & -5 & 2 \end{bmatrix}$$

Note that this is the matrix to the right of the vertical line in the last augmented matrix. (You should check that $MM^{-1} = I$.)

10 Matrices and Determinants

> **Inverse of a Square Matrix M**
>
> If $[M|I]$ is transformed by row operations into $[I|B]$, then the resulting matrix B is M^{-1}. If, however, we obtain all 0's in one or more rows to the left of the vertical line, then M^{-1} will not exist.

EXAMPLE 10 Find M^{-1}, given

$$M = \begin{bmatrix} 3 & -1 \\ -4 & 2 \end{bmatrix}$$

Solution Can you identify the row operations used in the transformations?

$$\begin{bmatrix} 3 & -1 & | & 1 & 0 \\ -4 & 2 & | & 0 & 1 \end{bmatrix} \sim \begin{bmatrix} 1 & -\frac{1}{3} & | & \frac{1}{3} & 0 \\ -4 & 2 & | & 0 & 1 \end{bmatrix}$$

$$\sim \begin{bmatrix} 1 & -\frac{1}{3} & | & \frac{1}{3} & 0 \\ 0 & \frac{2}{3} & | & \frac{4}{3} & 1 \end{bmatrix}$$

$$\sim \begin{bmatrix} 1 & -\frac{1}{3} & | & \frac{1}{3} & 0 \\ 0 & 1 & | & 2 & \frac{3}{2} \end{bmatrix}$$

$$\sim \begin{bmatrix} 1 & 0 & | & 1 & \frac{1}{2} \\ 0 & 1 & | & 2 & \frac{3}{2} \end{bmatrix}$$

Thus,

$$M^{-1} = \begin{bmatrix} 1 & \frac{1}{2} \\ 2 & \frac{3}{2} \end{bmatrix}$$ Check by showing that $M^{-1}M = I$.

PROBLEM 10 Find M^{-1}, given

$$M = \begin{bmatrix} 2 & -6 \\ 1 & -2 \end{bmatrix}$$

■ **Matrix Equations**

We will now show how systems of equations can be solved using inverses of square matrices.

EXAMPLE 11 Solve the system

$$\begin{align} 3x_1 - x_2 &= k_1 \\ -4x_1 + 2x_2 &= k_2 \end{align} \quad (3)$$

10-3 Inverse of a Square Matrix; Matrix Equations

for:

(A) $k_1 = -5, k_2 = 8$ (B) $k_1 = 4, k_2 = -2$ (C) $k_1 = 0, k_2 = -4$

Solution Once we obtain the inverse of the coefficient matrix,

$$A = \begin{bmatrix} 3 & -1 \\ -4 & 2 \end{bmatrix}$$

we will be able to solve parts (A)–(C) very easily. To see why, we convert system (3) into the following equivalent **matrix equation**:

$$\overset{A}{\begin{bmatrix} 3 & -1 \\ -4 & 2 \end{bmatrix}} \overset{X}{\begin{bmatrix} x_1 \\ x_2 \end{bmatrix}} = \overset{B}{\begin{bmatrix} k_1 \\ k_2 \end{bmatrix}} \quad (4)$$

You should check that matrix equation (4) is equivalent to system (3) by multiplying the left side and then equating corresponding elements on the left with those on the right.

We are now interested in finding a column matrix X that will satisfy the matrix equation

$$AX = B$$

To solve this equation we multiply both sides by A^{-1} (if it exists) to isolate X on the left side:

$$\begin{aligned} AX &= B & &\text{Multiply both sides by } A^{-1}. \\ A^{-1}(AX) &= A^{-1}B & &\text{Use the associative property.} \\ (A^{-1}A)X &= A^{-1}B & &A^{-1}A = I \\ IX &= A^{-1}B & &IX = X \\ X &= A^{-1}B \end{aligned}$$

The inverse of A was found in Example 10 to be

$$A^{-1} = \begin{bmatrix} 1 & \frac{1}{2} \\ 2 & \frac{3}{2} \end{bmatrix}$$

Thus,

$$\overset{X}{\begin{bmatrix} x_1 \\ x_2 \end{bmatrix}} = \overset{A^{-1}}{\begin{bmatrix} 1 & \frac{1}{2} \\ 2 & \frac{3}{2} \end{bmatrix}} \overset{B}{\begin{bmatrix} k_1 \\ k_2 \end{bmatrix}}$$

To solve parts (A)–(C) we simply replace k_1 and k_2 with appropriate values and multiply.

(A) $\begin{bmatrix} x_1 \\ x_2 \end{bmatrix} = \begin{bmatrix} 1 & \frac{1}{2} \\ 2 & \frac{3}{2} \end{bmatrix} \begin{bmatrix} -5 \\ 8 \end{bmatrix} = \begin{bmatrix} -1 \\ 2 \end{bmatrix}$

Thus, $x_1 = -1$ and $x_2 = 2$.

(B) $\begin{bmatrix} x_1 \\ x_2 \end{bmatrix} = \begin{bmatrix} 1 & \frac{1}{2} \\ 2 & \frac{3}{2} \end{bmatrix} \begin{bmatrix} 4 \\ -2 \end{bmatrix} = \begin{bmatrix} 3 \\ 5 \end{bmatrix}$

Thus, $x_1 = 3$ and $x_2 = 5$.

(C) $\begin{bmatrix} x_1 \\ x_2 \end{bmatrix} = \begin{bmatrix} 1 & \frac{1}{2} \\ 2 & \frac{3}{2} \end{bmatrix} \begin{bmatrix} 0 \\ -4 \end{bmatrix} = \begin{bmatrix} -2 \\ -6 \end{bmatrix}$

Thus, $x_1 = -2$ and $x_2 = -6$.

PROBLEM 11 Solve the system

$$2x_1 - 3x_2 = k_1$$
$$-x_1 + 2x_2 = k_2$$

using the inverse of the coefficient matrix, for:
(A) $k_1 = 10$, $k_2 = -6$ (B) $k_1 = 1$, $k_2 = 0$ (C) $k_1 = -8$, $k_2 = 5$

A great advantage of using an inverse to solve a system of linear equations is that once the inverse is found, it can be used to solve any new system formed through a change in the constant terms. This method is not suited, however, for cases where the numbers of equations and unknowns are not the same. (Why?)

■ **Application**

The following application will illustrate the usefulness of the inverse method.

EXAMPLE 12 An investment adviser currently has two types of investments available for clients: a conservative investment A that pays 10% per year and an investment B of higher risk that pays 20% per year. Clients may divide their investments between the two to achieve any total return desired between 10% and 20%. However, the higher the desired return, the higher the risk. How should each client listed in the table invest to achieve the indicated return?

	CLIENT			
	1	2	3	k
Total investment	$20,000	$50,000	$10,000	k_1
Annual return desired	$ 2,400 (12%)	$ 7,500 (15%)	$ 1,300 (13%)	k_2

10-3 Inverse of a Square Matrix; Matrix Equations

Solution We will solve the problem for an arbitrary client k, using inverses, and then apply the result to the three specific clients.

Let $x_1 = $ Amount invested in A
$x_2 = $ Amount invested in B

Then $x_1 + x_2 = k_1$ Total invested
$0.1x_1 + 0.2x_2 = k_2$ Total annual return

Write as a matrix equation:

$$\underset{A}{\begin{bmatrix} 1 & 1 \\ 0.1 & 0.2 \end{bmatrix}} \underset{X}{\begin{bmatrix} x_1 \\ x_2 \end{bmatrix}} = \underset{B}{\begin{bmatrix} k_1 \\ k_2 \end{bmatrix}}$$

If A^{-1} exists, then

$$X = A^{-1}B$$

We now find A^{-1} by starting with $[A|I]$ and proceeding as discussed earlier in this section.

$$\begin{bmatrix} 1 & 1 & | & 1 & 0 \\ 0.1 & 0.2 & | & 0 & 1 \end{bmatrix} \quad 10R_2 \to R_2$$

$$\sim \begin{bmatrix} 1 & 1 & | & 1 & 0 \\ 1 & 2 & | & 0 & 10 \end{bmatrix} \quad R_2 + (-1)R_1 \to R_2$$

$$\sim \begin{bmatrix} 1 & 1 & | & 1 & 0 \\ 0 & 1 & | & -1 & 10 \end{bmatrix} \quad R_1 + (-1)R_2 \to R_1$$

$$\sim \begin{bmatrix} 1 & 0 & | & 2 & -10 \\ 0 & 1 & | & -1 & 10 \end{bmatrix}$$

Thus,

$$A^{-1} = \begin{bmatrix} 2 & -10 \\ -1 & 10 \end{bmatrix}$$

Check

$$\underset{A^{-1}}{\begin{bmatrix} 2 & -10 \\ -1 & 10 \end{bmatrix}} \underset{A}{\begin{bmatrix} 1 & 1 \\ 0.1 & 0.2 \end{bmatrix}} = \underset{I}{\begin{bmatrix} 1 & 0 \\ 0 & 1 \end{bmatrix}}$$

and

$$\underset{X}{\begin{bmatrix} x_1 \\ x_2 \end{bmatrix}} = \underset{A^{-1}}{\begin{bmatrix} 2 & -10 \\ -1 & 10 \end{bmatrix}} \underset{B}{\begin{bmatrix} k_1 \\ k_2 \end{bmatrix}}$$

To solve each client's investment problem, we replace k_1 and k_2 with appropriate values from the table and multiply by A^{-1}:

Client 1
$$\begin{bmatrix} x_1 \\ x_2 \end{bmatrix} = \begin{bmatrix} 2 & -10 \\ -1 & 10 \end{bmatrix} \begin{bmatrix} 20{,}000 \\ 2{,}400 \end{bmatrix} = \begin{bmatrix} 16{,}000 \\ 4{,}000 \end{bmatrix}$$

Solution: $x_1 = \$16{,}000$ in A, $x_2 = \$4{,}000$ in B

Client 2
$$\begin{bmatrix} x_1 \\ x_2 \end{bmatrix} = \begin{bmatrix} 2 & -10 \\ -1 & 10 \end{bmatrix} \begin{bmatrix} 50{,}000 \\ 7{,}500 \end{bmatrix} = \begin{bmatrix} 25{,}000 \\ 25{,}000 \end{bmatrix}$$

Solution: $x_1 = \$25{,}000$ in A, $x_2 = \$25{,}000$ in B

Client 3
$$\begin{bmatrix} x_1 \\ x_2 \end{bmatrix} = \begin{bmatrix} 2 & -10 \\ -1 & 10 \end{bmatrix} \begin{bmatrix} 10{,}000 \\ 1{,}300 \end{bmatrix} = \begin{bmatrix} 7{,}000 \\ 3{,}000 \end{bmatrix}$$

Solution: $x_1 = \$7{,}000$ in A, $x_2 = \$3{,}000$ in B

PROBLEM 12 Repeat Example 12 with investment A paying 8% and investment B paying 24%.

Answers to Matched Problems

9. $\begin{bmatrix} 2 & -3 \\ 5 & 7 \end{bmatrix}$ 10. $\begin{bmatrix} -1 & 3 \\ -\frac{1}{2} & 1 \end{bmatrix}$

11. (A) $x_1 = 2$, $x_2 = -2$ (B) $x_1 = 2$, $x_2 = 1$
    (C) $x_1 = -1$, $x_2 = 2$

12. $A^{-1} = \begin{bmatrix} 1.5 & -6.25 \\ -0.5 & 6.25 \end{bmatrix}$;

Client 1: $15,000 in A and $5,000 in B; Client 2: $28,125 in A and $21,875 in B; Client 3: $6,875 in A and $3,125 in B

Exercise 10-3

A *Perform the indicated operations.*

1. $\begin{bmatrix} 1 & 0 \\ 0 & 1 \end{bmatrix} \begin{bmatrix} 2 & -3 \\ 4 & 5 \end{bmatrix}$ 2. $\begin{bmatrix} 2 & -3 \\ 4 & 5 \end{bmatrix} \begin{bmatrix} 1 & 0 \\ 0 & 1 \end{bmatrix}$

3. $\begin{bmatrix} 1 & 0 & 0 \\ 0 & 1 & 0 \\ 0 & 0 & 1 \end{bmatrix} \begin{bmatrix} -2 & 1 & 3 \\ 2 & 4 & -2 \\ 5 & 1 & 0 \end{bmatrix}$ 4. $\begin{bmatrix} -2 & 1 & 3 \\ 2 & 4 & -2 \\ 5 & 1 & 0 \end{bmatrix} \begin{bmatrix} 1 & 0 & 0 \\ 0 & 1 & 0 \\ 0 & 0 & 1 \end{bmatrix}$

For each problem show that the two matrices are inverses of each other by showing that their product is the identity matrix I.

5. $\begin{bmatrix} 3 & -4 \\ -2 & 3 \end{bmatrix} \begin{bmatrix} 3 & 4 \\ 2 & 3 \end{bmatrix}$ 6. $\begin{bmatrix} 5 & -7 \\ -2 & 3 \end{bmatrix} \begin{bmatrix} 3 & 7 \\ 2 & 5 \end{bmatrix}$

10-3 Inverse of a Square Matrix; Matrix Equations

7. $\begin{bmatrix} 1 & -1 & 1 \\ 0 & 2 & -1 \\ 2 & 3 & 0 \end{bmatrix} \begin{bmatrix} 3 & 3 & -1 \\ -2 & -2 & 1 \\ -4 & -5 & 2 \end{bmatrix}$

8. $\begin{bmatrix} 3 & 3 & -1 \\ -2 & -2 & 1 \\ -4 & -5 & 2 \end{bmatrix} \begin{bmatrix} 1 & -1 & 1 \\ 0 & 2 & -1 \\ 2 & 3 & 0 \end{bmatrix}$

Find x_1 and x_2.

9. $\begin{bmatrix} x_1 \\ x_2 \end{bmatrix} = \begin{bmatrix} 3 & -2 \\ 1 & 4 \end{bmatrix} \begin{bmatrix} -2 \\ 1 \end{bmatrix}$

10. $\begin{bmatrix} x_1 \\ x_2 \end{bmatrix} = \begin{bmatrix} -2 & 1 \\ -1 & 2 \end{bmatrix} \begin{bmatrix} 3 \\ -2 \end{bmatrix}$

11. $\begin{bmatrix} x_1 \\ x_2 \end{bmatrix} = \begin{bmatrix} -2 & 3 \\ 2 & -1 \end{bmatrix} \begin{bmatrix} 3 \\ 2 \end{bmatrix}$

12. $\begin{bmatrix} x_1 \\ x_2 \end{bmatrix} = \begin{bmatrix} 3 & -1 \\ 0 & 2 \end{bmatrix} \begin{bmatrix} -2 \\ 1 \end{bmatrix}$

B *Given M as indicated, find M^{-1} and show that $M^{-1}M = I$.*

13. $\begin{bmatrix} 1 & 2 \\ 1 & 3 \end{bmatrix}$

14. $\begin{bmatrix} 2 & 1 \\ 5 & 3 \end{bmatrix}$

15. $\begin{bmatrix} 1 & 3 \\ 2 & 7 \end{bmatrix}$

16. $\begin{bmatrix} 2 & 1 \\ 1 & 1 \end{bmatrix}$

17. $\begin{bmatrix} 1 & -3 & 0 \\ 0 & 3 & 1 \\ 2 & -1 & 2 \end{bmatrix}$

18. $\begin{bmatrix} 2 & 9 & 0 \\ 1 & 2 & 3 \\ 0 & -1 & 1 \end{bmatrix}$

19. $\begin{bmatrix} 1 & 1 & 0 \\ 0 & 3 & -1 \\ 1 & 0 & 1 \end{bmatrix}$

20. $\begin{bmatrix} 1 & 0 & -1 \\ 2 & -1 & 0 \\ 1 & 1 & 1 \end{bmatrix}$

Write each system as a matrix equation and solve using inverses. [Note: The inverses were found in Problems 13–18.]

21. $x_1 + 2x_2 = k_1$
    $x_1 + 3x_2 = k_2$

    (A) $k_1 = 1$, $k_2 = 3$
    (B) $k_1 = 3$, $k_2 = 5$
    (C) $k_1 = -2$, $k_2 = 1$

22. $2x_1 + x_2 = k_1$
    $5x_1 + 3x_2 = k_2$

    (A) $k_1 = 2$, $k_2 = 13$
    (B) $k_1 = -2$, $k_2 = 4$
    (C) $k_1 = 1$, $k_2 = -3$

23. $x_1 + 3x_2 = k_1$
    $2x_1 + 7x_2 = k_2$

    (A) $k_1 = 2$, $k_2 = -1$
    (B) $k_1 = 1$, $k_2 = 0$
    (C) $k_1 = 3$, $k_2 = -1$

24. $2x_1 + x_2 = k_1$
    $x_1 + x_2 = k_2$

    (A) $k_1 = -1$, $k_2 = -2$
    (B) $k_1 = 2$, $k_2 = 3$
    (C) $k_1 = 2$, $k_2 = 0$

25. $x_1 - 3x_2 = k_1$
$ 3x_2 + x_3 = k_2$
$2x_1 - x_2 + 2x_3 = k_3$

(A) $k_1 = 1$, $k_2 = 0$, $k_3 = 2$
(B) $k_1 = -1$, $k_2 = 1$, $k_3 = 0$
(C) $k_1 = 2$, $k_2 = -2$, $k_3 = 1$

26. $2x_1 + 9x_2 = k_1$
$x_1 + 2x_2 + 3x_3 = k_2$
$ - x_2 + x_3 = k_3$

(A) $k_1 = 0$, $k_2 = 2$, $k_3 = 1$
(B) $k_1 = -2$, $k_2 = 0$, $k_3 = 1$
(C) $k_1 = 3$, $k_2 = 1$, $k_3 = 0$

C Write each system as a matrix equation and solve using inverses. [Note: The inverses were found in Problems 19 and 20.]

27. $x_1 + x_2 = k_1$
$ 3x_2 - x_3 = k_2$
$x_1 + x_3 = k_3$

(A) $k_1 = 2$, $k_2 = 0$, $k_3 = 4$
(B) $k_1 = 0$, $k_2 = 4$, $k_3 = -2$
(C) $k_1 = 4$, $k_2 = 2$, $k_3 = 0$

28. $x_1 - x_3 = k_1$
$2x_1 - x_2 = k_2$
$x_1 + x_2 + x_3 = k_3$

(A) $k_1 = 4$, $k_2 = 8$, $k_3 = 0$
(B) $k_1 = 4$, $k_2 = 0$, $k_3 = -4$
(C) $k_1 = 0$, $k_2 = 8$, $k_3 = -8$

Show that the inverses of the following matrices do not exist.

29. $\begin{bmatrix} 3 & 9 \\ 2 & 6 \end{bmatrix}$

30. $\begin{bmatrix} 2 & -4 \\ -3 & 6 \end{bmatrix}$

31. $\begin{bmatrix} 2 & 1 & 1 \\ 1 & 1 & 0 \\ -1 & -1 & 0 \end{bmatrix}$

32. $\begin{bmatrix} 1 & -1 & 0 \\ 2 & -1 & 1 \\ 0 & 1 & 1 \end{bmatrix}$

33. Show that $(A^{-1})^{-1} = A$ for

$$A = \begin{bmatrix} 3 & 4 \\ 2 & 3 \end{bmatrix}$$

34. Show that $(AB)^{-1} = B^{-1}A^{-1}$ for

$$A = \begin{bmatrix} 3 & 4 \\ 2 & 3 \end{bmatrix} \text{ and } B = \begin{bmatrix} 3 & 7 \\ 2 & 5 \end{bmatrix}$$

10-3 Inverse of a Square Matrix; Matrix Equations

APPLICATIONS *Solve using systems of equations and inverses.*

35. *Resource allocation.* A concert hall has 10,000 seats. If tickets are $4 and $8, how many of each type of ticket should be sold (assuming all seats can be sold) to bring in each of the returns indicated in the table? Use decimals in computing the inverse.

	CONCERT		
	1	2	3
Tickets sold	10,000	10,000	10,000
Return required	$56,000	$60,000	$68,000

36. *Production scheduling.* Labor and material costs for manufacturing two guitar models are given in the table below.

GUITAR MODEL	LABOR COST	MATERIAL COST
A	$30	$20
B	$40	$30

If a total of $3,000 a week is allowed for labor and material, how many of each model should be produced each week to exactly use each of the allocations of the $3,000 indicated in the following table? Use decimals in computing the inverse.

	WEEKLY ALLOCATION		
	1	2	3
Labor	$1,800	$1,750	$1,720
Material	$1,200	$1,250	$1,280

37. *Diets.* A biologist has available two commercial food mixes with the following percentages of protein and fat:

MIX	PROTEIN (%)	FAT (%)
A	20	2
B	10	6

How many ounces of each mix should be used to prepare each of the diets listed in the following table?

	DIET		
	1	2	3
Protein	20 ounces	10 ounces	10 ounces
Fat	6 ounces	4 ounces	6 ounces

Section 10-4 Determinant Functions

- Determinant Functions
- Second-Order Determinants
- Third-Order Determinants
- Remarks

■ Determinant Functions

In this section we are going to introduce a new function, called a **determinant function**. Its domain is the set of all square matrices with real elements, and its range is the set of all real numbers. If A is a square matrix, then the determinant of A is denoted by **det A** or simply by writing the array of elements in A using vertical lines in place of square brackets. For example,

$$\det \begin{bmatrix} 2 & -3 \\ 5 & 1 \end{bmatrix} = \begin{vmatrix} 2 & -3 \\ 5 & 1 \end{vmatrix}$$

$$\det \begin{bmatrix} 1 & -2 & 3 \\ 0 & 5 & -7 \\ -2 & 1 & 6 \end{bmatrix} = \begin{vmatrix} 1 & -2 & 3 \\ 0 & 5 & -7 \\ -2 & 1 & 6 \end{vmatrix}$$

The expressions on the right are often referred to simply as **determinants**.

A determinant of **order n** is one with n rows and n columns. In this section we will concentrate most of our attention on determining the value of determinants of orders 2 and 3. Many of the results and procedures we will discuss generalize completely to determinants of order n.

■ Second-Order Determinants

In general, we can symbolize a **second-order determinant** as follows:

$$\begin{vmatrix} a_{11} & a_{12} \\ a_{21} & a_{22} \end{vmatrix}$$

where we use a single letter with a **double subscript** to facilitate generalization to higher-order determinants. The first subscript number indicates the row in which the element lies, and the second subscript number indicates the column. Thus a_{21} is the element in the second row and first column, and a_{12} is the element in the first row and second column. Each second-order determinant represents a real number given by the following formula:

Value of a Second-Order Determinant

$$\begin{vmatrix} a_{11} & a_{12} \\ a_{21} & a_{22} \end{vmatrix} = a_{11}a_{22} - a_{21}a_{12} \tag{1}$$

Formula (1) is easily remembered if you notice that the expression on the right is the product of the principal diagonal (from upper left to lower right) minus the product of the secondary diagonal (from lower left to upper right).

EXAMPLE 13
$$\begin{vmatrix} -1 & 2 \\ -3 & 4 \end{vmatrix} = (-1)(-4) - (-3)(2) = 4 - (-6) = 10$$

PROBLEM 13 Find: $\begin{vmatrix} 3 & -5 \\ 4 & -2 \end{vmatrix}$

■ **Third-Order Determinants**

A determinant of order 3 is a square array of nine elements and represents a real number given by the following formula:

Value of a Third-Order Determinant

$$\begin{vmatrix} a_{11} & a_{12} & a_{13} \\ a_{21} & a_{22} & a_{23} \\ a_{31} & a_{32} & a_{33} \end{vmatrix} = a_{11}a_{22}a_{33} - a_{11}a_{32}a_{23} + a_{21}a_{32}a_{13} - a_{21}a_{12}a_{33} + a_{31}a_{12}a_{23} - a_{31}a_{22}a_{13} \tag{2}$$

Note that each term in the expansion on the right of equation (2) contains exactly one element from each row and each column. Don't panic! You do not need to memorize formula (2). After we introduce the ideas of "minor" and "cofactor," we will state a theorem that can be used to obtain the same result with much less memory strain.

The **minor of an element** in a third-order determinant is a second-order determinant obtained by deleting the row and column that contains the element. For example, in the determinant in formula (2),

$$\text{Minor of } a_{23} = \begin{vmatrix} a_{11} & a_{12} & a_{13} \\ a_{21} & a_{22} & a_{23} \\ a_{31} & a_{32} & a_{33} \end{vmatrix} = \begin{vmatrix} a_{11} & a_{12} \\ a_{31} & a_{32} \end{vmatrix}$$

$$\text{Minor of } a_{32} = \begin{vmatrix} a_{11} & a_{12} & a_{13} \\ a_{21} & a_{22} & a_{23} \\ a_{31} & a_{32} & a_{33} \end{vmatrix} = \begin{vmatrix} a_{11} & a_{13} \\ a_{21} & a_{23} \end{vmatrix}$$

A quantity closely associated with the minor of an element is the cofactor of an element. The **cofactor of an element** a_{ij} (from the ith row and jth column) is the product of the minor of a_{ij} and $(-1)^{i+j}$.

Cofactor

$$\text{Cofactor of } a_{ij} = (-1)^{i+j}(\text{Minor of } a_{ij})$$

Thus, a cofactor of an element is nothing more than a signed minor. The sign is determined by raising -1 to a power that is the sum of the numbers indicating the row and column in which the element lies. Note that $(-1)^{i+j}$ is -1 if $i+j$ is odd and 1 if $i+j$ is even. Thus, if we are given the determinant

$$\begin{vmatrix} a_{11} & a_{12} & a_{13} \\ a_{21} & a_{22} & a_{23} \\ a_{31} & a_{32} & a_{33} \end{vmatrix}$$

then

$$\text{Cofactor of } a_{23} = (-1)^{2+3} \begin{vmatrix} a_{11} & a_{12} \\ a_{31} & a_{32} \end{vmatrix} = - \begin{vmatrix} a_{11} & a_{12} \\ a_{31} & a_{32} \end{vmatrix}$$

$$\text{Cofactor of } a_{11} = (-1)^{1+1} \begin{vmatrix} a_{22} & a_{23} \\ a_{32} & a_{33} \end{vmatrix} = \begin{vmatrix} a_{22} & a_{23} \\ a_{32} & a_{33} \end{vmatrix}$$

EXAMPLE 14 Find the cofactors of -2 and 5 in the determinant

$$\begin{vmatrix} -2 & 0 & 3 \\ 1 & -6 & 5 \\ -1 & 2 & 0 \end{vmatrix}$$

10-4 Determinant Functions

Solution

$$\text{Cofactor of } -2 = (-1)^{1+1} \begin{vmatrix} -6 & 5 \\ 2 & 0 \end{vmatrix} = \begin{vmatrix} -6 & 5 \\ 2 & 0 \end{vmatrix}$$

$$= (-6)(0) - (2)(5) = -10$$

$$\text{Cofactor of } 5 = (-1)^{2+3} \begin{vmatrix} -2 & 0 \\ -1 & 2 \end{vmatrix} = - \begin{vmatrix} -2 & 0 \\ -1 & 2 \end{vmatrix}$$

$$= -[(-2)(2) - (-1)(0)] = 4$$

PROBLEM 14 Find the cofactors of 2 and 3 in the determinant in Example 14.

[Note: The sign in front of the minor, $(-1)^{i+j}$, can be determined rather mechanically by using a checkerboard pattern of $+$ and $-$ signs over the determinant, starting with $+$ in the upper left-hand corner:

$$\begin{matrix} + & - & + \\ - & + & - \\ + & - & + \end{matrix}$$

Use either the checkerboard or the exponent method, whichever is easier for you, to determine the sign in front of the minor.]

Now we are ready for the central theorem of this section, Theorem 1. It will provide us with an efficient means of evaluating third-order determinants.

THEOREM 1 | Value of a Third-Order Determinant

The value of a determinant of order 3 is the sum of three products obtained by multiplying each element of any one row (or each element of any one column) by its cofactor.

To prove this theorem we must show that the expansions indicated by the theorem for any row or any column (six cases) produce the expression on the right of formula (2). Proofs of special cases of this theorem are left to the C problems in Exercise 10-4.

EXAMPLE 15 Evaluate by expanding by (A) the first row, and (B) the second column.

$$\begin{vmatrix} 2 & -2 & 0 \\ -3 & 1 & 2 \\ 1 & -3 & -1 \end{vmatrix}$$

10 Matrices and Determinants

Solution (A) $\begin{vmatrix} 2 & -2 & 0 \\ -3 & 1 & 2 \\ 1 & -3 & -1 \end{vmatrix}$

$= a_{11} \begin{pmatrix} \text{Cofactor} \\ \text{of } a_{11} \end{pmatrix} + a_{12} \begin{pmatrix} \text{Cofactor} \\ \text{of } a_{12} \end{pmatrix} + a_{13} \begin{pmatrix} \text{Cofactor} \\ \text{of } a_{13} \end{pmatrix}$

$= 2 \left[(-1)^{1+1} \begin{vmatrix} 1 & 2 \\ -3 & -1 \end{vmatrix} \right] + (-2) \left[(-1)^{1+2} \begin{vmatrix} -3 & 2 \\ 1 & -1 \end{vmatrix} \right] + 0$

$= (2)(1)[(1)(-1) - (-3)(2)] + (-2)(-1)[(-3)(-1) - (1)(2)]$

$= (2)(5) + (2)(1) = 12$

(B) $\begin{vmatrix} 2 & -2 & 0 \\ -3 & 1 & 2 \\ 1 & -3 & -1 \end{vmatrix}$

$= a_{12} \begin{pmatrix} \text{Cofactor} \\ \text{of } a_{12} \end{pmatrix} + a_{22} \begin{pmatrix} \text{Cofactor} \\ \text{of } a_{22} \end{pmatrix} + a_{32} \begin{pmatrix} \text{Cofactor} \\ \text{of } a_{32} \end{pmatrix}$

$= (-2) \left[(-1)^{1+2} \begin{vmatrix} -3 & 2 \\ 1 & -1 \end{vmatrix} \right] + (1) \left[(-1)^{2+2} \begin{vmatrix} 2 & 0 \\ 1 & -1 \end{vmatrix} \right]$

$+ (-3) \left[(-1)^{3+2} \begin{vmatrix} 2 & 0 \\ -3 & 2 \end{vmatrix} \right]$

$= (-2)(-1)[(-3)(-1) - (1)(2)] + (1)(1)[(2)(-1) - (1)(0)]$

$+ (-3)(-1)[(2)(2) - (-3)(0)]$

$= (2)(1) + (1)(-2) + (3)(4)$

$= 12$

PROBLEM 15 Evaluate by expanding by (A) the first row, and (B) the third column.

$\begin{vmatrix} 2 & 1 & -1 \\ -2 & -3 & 0 \\ -1 & 2 & 1 \end{vmatrix}$

■ **Remarks**

1. It should now be apparent that we can greatly reduce the work involved in evaluating a determinant by choosing to expand (using Theorem 1) by the row or column with the greatest number of 0's.
2. Theorem 1 and the definitions of minors and cofactors generalize completely for determinants of arbitrary order.

3. Where are determinants used? Many equations and formulas have particularly simple and compact representations in determinant form that are easily remembered (see Problems 44–48 in Exercise 10-5). In addition, determinants are involved in theoretical work. For example, one can show that the inverse of a square matrix exists if and only if its determinant is not 0.

Answers to Matched Problems   **13.** 14   **14.** 13, −4
15. (A) 3   (B) 3

Exercise 10-4

A Evaluate each second-order determinant.

1. $\begin{vmatrix} 2 & 2 \\ -3 & 1 \end{vmatrix}$ **2.** $\begin{vmatrix} 2 & 4 \\ 3 & -1 \end{vmatrix}$ **3.** $\begin{vmatrix} 6 & -2 \\ -1 & -3 \end{vmatrix}$

4. $\begin{vmatrix} 5 & -4 \\ -2 & 2 \end{vmatrix}$ **5.** $\begin{vmatrix} 1.8 & -1.6 \\ -1.9 & 1.2 \end{vmatrix}$ **6.** $\begin{vmatrix} 0.5 & -3.2 \\ 1.4 & -6.7 \end{vmatrix}$

Given the determinant

$$\begin{vmatrix} a_{11} & a_{12} & a_{13} \\ a_{21} & a_{22} & a_{23} \\ a_{31} & a_{32} & a_{33} \end{vmatrix}$$

write the minor of each of the following elements.

7. a_{11} **8.** a_{33} **9.** a_{23} **10.** a_{22}

Write the cofactor of each of the following elements.

11. a_{11} **12.** a_{33} **13.** a_{23} **14.** a_{22}

Problems 15–22 pertain to the determinant.

$$\begin{vmatrix} -2 & 3 & 0 \\ 5 & 1 & -2 \\ 7 & -4 & 8 \end{vmatrix}$$

Write the minor of each of the following elements. (Leave the answer in determinant form.)

15. a_{11} **16.** a_{22} **17.** a_{32} **18.** a_{21}

Write the cofactor of each of the following elements and evaluate each.

19. a_{11} **20.** a_{22} **21.** a_{32} **22.** a_{21}

Evaluate Problems 23–28 using cofactors.

23. $\begin{vmatrix} 1 & 0 & 0 \\ -2 & 4 & 3 \\ 5 & -2 & 1 \end{vmatrix}$
24. $\begin{vmatrix} 2 & -3 & 5 \\ 0 & -3 & 1 \\ 0 & 6 & 2 \end{vmatrix}$

25. $\begin{vmatrix} 0 & 1 & 5 \\ 3 & -7 & 6 \\ 0 & -2 & -3 \end{vmatrix}$
26. $\begin{vmatrix} 4 & -2 & 0 \\ 9 & 5 & 4 \\ 1 & 2 & 0 \end{vmatrix}$

27. $\begin{vmatrix} -1 & 2 & -3 \\ -2 & 0 & -6 \\ 4 & -3 & 2 \end{vmatrix}$
28. $\begin{vmatrix} 0 & 2 & -1 \\ -6 & 3 & 1 \\ 7 & -9 & -2 \end{vmatrix}$

B Given the determinant

$$\begin{vmatrix} a_{11} & a_{12} & a_{13} & a_{14} \\ a_{21} & a_{22} & a_{23} & a_{24} \\ a_{31} & a_{32} & a_{33} & a_{34} \\ a_{41} & a_{42} & a_{43} & a_{44} \end{vmatrix}$$

write the cofactor (in determinant form) of each of the following elements.

29. a_{11} 30. a_{44} 31. a_{43} 32. a_{23}

Evaluate each of the following determinants using cofactors.

33. $\begin{vmatrix} 3 & -2 & -8 \\ -2 & 0 & -3 \\ 1 & 0 & -4 \end{vmatrix}$
34. $\begin{vmatrix} 4 & -4 & 6 \\ 2 & 8 & -3 \\ 0 & -5 & 0 \end{vmatrix}$

35. $\begin{vmatrix} 1 & 4 & 1 \\ 1 & 1 & -2 \\ 2 & 1 & -1 \end{vmatrix}$
36. $\begin{vmatrix} 3 & 2 & 1 \\ -1 & 5 & 1 \\ 2 & 3 & 1 \end{vmatrix}$

37. $\begin{vmatrix} 1 & 4 & 3 \\ 2 & 1 & 6 \\ 3 & -2 & 9 \end{vmatrix}$
38. $\begin{vmatrix} 4 & -6 & 3 \\ -1 & 4 & 1 \\ 5 & -6 & 3 \end{vmatrix}$

39. $\begin{vmatrix} 2 & 6 & 1 & 7 \\ 0 & 3 & 0 & 0 \\ 3 & 4 & 2 & 5 \\ 0 & 9 & 0 & 2 \end{vmatrix}$
40. $\begin{vmatrix} 0 & 1 & 0 & 1 \\ 2 & 4 & 7 & 6 \\ 0 & 3 & 0 & 1 \\ 0 & 6 & 2 & 5 \end{vmatrix}$

C 41. $\begin{vmatrix} -2 & 0 & 0 & 0 & 0 \\ 9 & -1 & 0 & 0 & 0 \\ 2 & 1 & 3 & 0 & 0 \\ -1 & 4 & 2 & 2 & 0 \\ 7 & -2 & 3 & 5 & 5 \end{vmatrix}$
42. $\begin{vmatrix} 2 & 0 & 0 & 0 & 0 \\ 0 & 3 & 0 & 0 & 0 \\ 0 & 0 & 2 & 0 & 0 \\ 0 & 0 & 0 & 1 & 0 \\ 0 & 0 & 0 & 0 & 4 \end{vmatrix}$

If all the letters in Problems 43–46 represent real numbers, show that each statement is true.

43. $\begin{vmatrix} a & b \\ ka & kb \end{vmatrix} = 0$

44. $\begin{vmatrix} a & b \\ c & d \end{vmatrix} = - \begin{vmatrix} b & a \\ d & c \end{vmatrix}$

45. $\begin{vmatrix} a & b \\ c & d \end{vmatrix} = \begin{vmatrix} a & c \\ b & d \end{vmatrix}$

46. $\begin{vmatrix} ka & kb \\ c & d \end{vmatrix} = k \begin{vmatrix} a & b \\ c & d \end{vmatrix}$

47. Show that the expansion of the determinant

$$\begin{vmatrix} a_{11} & a_{12} & a_{13} \\ a_{21} & a_{22} & a_{23} \\ a_{31} & a_{32} & a_{33} \end{vmatrix}$$

by the first column is the same as its expansion by the third row.

48. Repeat Problem 47, using the second row and the third column.

49. If

$$A = \begin{bmatrix} 2 & 3 \\ 1 & -2 \end{bmatrix} \quad \text{and} \quad B = \begin{bmatrix} -1 & 3 \\ 2 & 1 \end{bmatrix}$$

show that $\det(AB) = \det A \cdot \det B$.

50. If

$$A = \begin{bmatrix} a & b \\ c & d \end{bmatrix} \quad \text{and} \quad B = \begin{bmatrix} w & x \\ y & z \end{bmatrix}$$

show that $\det(AB) = \det A \cdot \det B$.

Section 10-5 Properties of Determinants

The following theorems greatly facilitate the task of evaluating determinants of order 3 or greater. Because the proofs for the general case are involved and notationally difficult, we will sketch only informal proofs for determinants of order 3.

THEOREM 2 | If each element of any row (or column) of a determinant is multiplied by a constant k, the new determinant is k times the original.

Partial Proof Let C_{ij} be the cofactor of a_{ij}. Then

$$\begin{vmatrix} ka_{11} & ka_{12} & ka_{13} \\ a_{21} & a_{22} & a_{23} \\ a_{31} & a_{32} & a_{33} \end{vmatrix} = ka_{11}C_{11} + ka_{12}C_{12} + ka_{13}C_{13}$$

$$= k(a_{11}C_{11} + a_{12}C_{12} + a_{13}C_{13})$$

$$= k\begin{vmatrix} a_{11} & a_{12} & a_{13} \\ a_{21} & a_{22} & a_{23} \\ a_{31} & a_{32} & a_{33} \end{vmatrix}$$

Theorem 2 also states that a factor common to all elements of a row (or column) can be taken out as a factor of the determinant.

EXAMPLE 16

$$\begin{vmatrix} 6 & 1 & 3 \\ -2 & 7 & -2 \\ 4 & 5 & 0 \end{vmatrix} = 2 \begin{vmatrix} 3 & 1 & 3 \\ -1 & 7 & -2 \\ 2 & 5 & 0 \end{vmatrix}$$

where 2 is a common factor of the first column.

PROBLEM 16 Take out factors common to any row or any column.

$$\begin{vmatrix} 3 & 2 & 1 \\ 6 & 3 & -9 \\ 1 & 0 & -5 \end{vmatrix}$$

THEOREM 3 | If every element in a row (or column) is 0, the value of the determinant is zero.

Theorem 3 is an immediate consequence of Theorem 2, and its proof is left as an exercise. It is illustrated in the following example:

$$\begin{vmatrix} 3 & -2 & 5 \\ 0 & 0 & 0 \\ -1 & 4 & 9 \end{vmatrix} = 0$$

THEOREM 4 | If two rows (or two columns) of a determinant are interchanged, the new determinant is the negative of the old.

A proof of Theorem 4 even for a determinant of order 3 is notationally involved. We suggest that you partially prove the theorem by direct expansion of the determinants before and after the interchange of two rows (or columns). The theorem is illustrated by the following example where the second and third columns are interchanged:

$$\begin{vmatrix} 1 & 0 & 9 \\ -2 & 1 & 5 \\ 3 & 0 & 7 \end{vmatrix} = - \begin{vmatrix} 1 & 9 & 0 \\ -2 & 5 & 1 \\ 3 & 7 & 0 \end{vmatrix}$$

THEOREM 5   If the corresponding elements are equal in two rows (or columns), the value of the determinant is zero.

Proof   The general proof of Theorem 5 is easy, making direct use of Theorem 4. If we start with a determinant D that has two rows (or columns) equal and we interchange the equal rows (or columns), the new determinant will be the same as the old. But by Theorem 4,

$$D = -D$$

hence,

$$2D = 0$$

and

$$D = 0$$

THEOREM 6   If a multiple of any row (or column) of a determinant is added to any other row (or column), the value of the determinant is not changed.

Partial Proof   If, in a general third-order determinant, we add a k multiple of the second column to the first, we obtain (where C_{ij} is the cofactor of a_{ij} in the original determinant)

$$\begin{vmatrix} a_{11} + ka_{12} & a_{12} & a_{13} \\ a_{21} + ka_{22} & a_{22} & a_{23} \\ a_{31} + ka_{32} & a_{32} & a_{33} \end{vmatrix} = (a_{11} + ka_{12})C_{11} + (a_{21} + ka_{22})C_{21} + (a_{31} + ka_{32})C_{31}$$

$$= (a_{11}C_{11} + a_{21}C_{21} + a_{31}C_{31}) + k(a_{12}C_{11} + a_{22}C_{21} + a_{32}C_{31})$$

$$= \begin{vmatrix} a_{11} & a_{12} & a_{13} \\ a_{21} & a_{22} & a_{23} \\ a_{31} & a_{32} & a_{33} \end{vmatrix} + k \begin{vmatrix} a_{12} & a_{12} & a_{13} \\ a_{22} & a_{22} & a_{23} \\ a_{32} & a_{32} & a_{33} \end{vmatrix} = \begin{vmatrix} a_{11} & a_{12} & a_{13} \\ a_{21} & a_{22} & a_{23} \\ a_{31} & a_{32} & a_{33} \end{vmatrix}$$

since the determinant following k is zero. (Why?)

10 Matrices and Determinants

Note the similarity in the process described in Theorem 6 to that used to obtain row-equivalent matrices. We use this theorem to transform a determinant without 0 elements into one that contains a row or column with all elements 0 but one. The determinant can then be easily expanded by this row (or column). An example best illustrates the process.

EXAMPLE 17 Evaluate the determinant.

$$\begin{vmatrix} 3 & -1 & 2 \\ -2 & 4 & -3 \\ 4 & -2 & 5 \end{vmatrix}$$

We use Theorem 6 to obtain two 0's in the first row, and then expand the determinant by this row. To start, we replace the third column with the sum of it and 2 times the second column to obtain a 0 in the a_{13} position.

$$\begin{vmatrix} 3 & -1 & 2 \\ -2 & 4 & -3 \\ 4 & -2 & 5 \end{vmatrix} = \begin{vmatrix} 3 & -1 & 0 \\ -2 & 4 & 5 \\ 4 & -2 & 1 \end{vmatrix} \qquad C_3 + 2C_2 \to C_3$$

Next, to obtain a 0 in the a_{11} position, we replace the first column with the sum of it and 3 times the second column.

$$\begin{vmatrix} 3 & -1 & 0 \\ -2 & 4 & 5 \\ 4 & -2 & 1 \end{vmatrix} = \begin{vmatrix} 0 & -1 & 0 \\ 10 & 4 & 5 \\ -2 & -2 & 1 \end{vmatrix} \qquad C_1 + 3C_2 \to C_1$$

Now it is an easy matter to expand this last determinant by the first row to obtain

$$0 + (-1)\left[(-1)^{1+2}\begin{vmatrix} 10 & 5 \\ -2 & 1 \end{vmatrix}\right] + 0 = 20$$

PROBLEM 17 Evaluate the following determinant by first using Theorem 6 to obtain 0's in the a_{11} and a_{31} positions, and then expand by the first column.

$$\begin{vmatrix} 3 & 10 & -5 \\ 1 & 6 & -3 \\ 2 & 3 & 4 \end{vmatrix}$$

Answers to Matched Problems   **16.** $3\begin{vmatrix} 3 & 2 & 1 \\ 2 & 1 & -3 \\ 1 & 0 & -5 \end{vmatrix}$ **17.** 44

* C_1, C_2, and C_3 represent columns 1, 2, and 3, respectively.

Exercise 10-5

A *For each statement, identify the theorem from this section that justifies it. Do not evaluate.*

1. $\begin{vmatrix} 16 & 8 \\ 0 & -1 \end{vmatrix} = 8 \begin{vmatrix} 2 & 1 \\ 0 & -1 \end{vmatrix}$

2. $\begin{vmatrix} 1 & -9 \\ 0 & -6 \end{vmatrix} = -3 \begin{vmatrix} 1 & 3 \\ 0 & 2 \end{vmatrix}$

3. $-2 \begin{vmatrix} 2 & 1 \\ -3 & 4 \end{vmatrix} = \begin{vmatrix} -4 & 1 \\ 12 & 4 \end{vmatrix}$

4. $4 \begin{vmatrix} -1 & 3 \\ 2 & 1 \end{vmatrix} = \begin{vmatrix} -4 & 12 \\ 2 & 1 \end{vmatrix}$

5. $\begin{vmatrix} 3 & 0 \\ -2 & 0 \end{vmatrix} = 0$

6. $\begin{vmatrix} 5 & -7 \\ 0 & 0 \end{vmatrix} = 0$

7. $\begin{vmatrix} 5 & -1 \\ 8 & 0 \end{vmatrix} = - \begin{vmatrix} -1 & 5 \\ 0 & 8 \end{vmatrix}$

8. $\begin{vmatrix} 6 & 9 \\ 0 & 1 \end{vmatrix} = - \begin{vmatrix} 0 & 1 \\ 6 & 9 \end{vmatrix}$

9. $\begin{vmatrix} 4 & 3 \\ 1 & 2 \end{vmatrix} = \begin{vmatrix} 4-4 & 3-8 \\ 1 & 2 \end{vmatrix}$

10. $\begin{vmatrix} 3 & 2 \\ 5 & 1 \end{vmatrix} = \begin{vmatrix} 3+4 & 2 \\ 5+2 & 1 \end{vmatrix}$

Theorem 6 was used to transform the determinant on the left to that on the right. Replace each letter x with an appropriate numeral to complete the transformation.

11. $\begin{vmatrix} -1 & 3 \\ 2 & -4 \end{vmatrix} = \begin{vmatrix} -1 & x \\ 2 & 2 \end{vmatrix}$

12. $\begin{vmatrix} -1 & 3 \\ 5 & -2 \end{vmatrix} = \begin{vmatrix} -1 & 3 \\ x & 13 \end{vmatrix}$

13. $\begin{vmatrix} -1 & 2 & 3 \\ 2 & 1 & 4 \\ 1 & 3 & 2 \end{vmatrix} = \begin{vmatrix} -1 & 2 & 0 \\ 2 & 1 & 10 \\ 1 & 3 & x \end{vmatrix}$

14. $\begin{vmatrix} -1 & 2 & 3 \\ 2 & 1 & 4 \\ 1 & 3 & 2 \end{vmatrix} = \begin{vmatrix} -1 & 0 & 3 \\ 2 & x & 4 \\ 1 & 5 & 2 \end{vmatrix}$

Use Theorem 6 to transform each determinant into one that contains a row (or column) with all elements 0 but one (if possible); then expand the transformed determinant by this row (or column).

15. $\begin{vmatrix} -1 & 0 & 3 \\ 2 & 5 & 4 \\ 1 & 5 & 2 \end{vmatrix}$

16. $\begin{vmatrix} -1 & 2 & 0 \\ 2 & 1 & 10 \\ 1 & 3 & 5 \end{vmatrix}$

17. $\begin{vmatrix} 3 & 5 & 0 \\ 1 & 1 & -2 \\ 2 & 1 & -1 \end{vmatrix}$

18. $\begin{vmatrix} 2 & 0 & 1 \\ -1 & -3 & 4 \\ 1 & 2 & 3 \end{vmatrix}$

B *For each statement, identify the theorem from this section that justifies it.*

19. $-2 \begin{vmatrix} 1 & 0 & 2 \\ 3 & -2 & 4 \\ 0 & 1 & 1 \end{vmatrix} = \begin{vmatrix} 1 & 0 & 2 \\ -6 & 4 & -8 \\ 0 & 1 & 1 \end{vmatrix}$

20. $\begin{vmatrix} 8 & 0 & 1 \\ 12 & -1 & 0 \\ 4 & 3 & 2 \end{vmatrix} = 4 \begin{vmatrix} 2 & 0 & 1 \\ 3 & -1 & 0 \\ 1 & 3 & 2 \end{vmatrix}$

21. $\begin{vmatrix} 1 & 2 & 0 \\ -1 & 3 & 0 \\ 0 & 1 & 0 \end{vmatrix} = 0$

22. $\begin{vmatrix} -2 & 5 & 13 \\ 1 & 7 & 12 \\ 0 & 8 & 15 \end{vmatrix} = - \begin{vmatrix} 5 & -2 & 13 \\ 7 & 1 & 12 \\ 8 & 0 & 15 \end{vmatrix}$

23. $\begin{vmatrix} 4 & 2 & -1 \\ 2 & 0 & 2 \\ -3 & 5 & -2 \end{vmatrix} = \begin{vmatrix} 4-4 & 2 & -1 \\ 2+8 & 0 & 2 \\ -3-8 & 5 & -2 \end{vmatrix}$

24. $\begin{vmatrix} 7 & 7 & 1 \\ -3 & -3 & 11 \\ 2 & 2 & 0 \end{vmatrix} = 0$

Theorem 6 was used to transform the determinant on the left to that on the right. Replace each letter with an appropriate numeral to complete the transformation.

25. $\begin{vmatrix} 2 & 1 & -1 \\ 3 & 4 & 1 \\ 1 & 2 & -2 \end{vmatrix} = \begin{vmatrix} 0 & 0 & -1 \\ x & 5 & 1 \\ -3 & y & -2 \end{vmatrix}$

26. $\begin{vmatrix} 3 & -1 & 1 \\ -2 & 4 & 3 \\ 1 & 5 & 2 \end{vmatrix} = \begin{vmatrix} 0 & -1 & 0 \\ 10 & 4 & 7 \\ x & 5 & y \end{vmatrix}$

27. $\begin{vmatrix} 7 & 9 & 4 \\ 2 & 3 & 1 \\ 3 & 4 & -2 \end{vmatrix} = \begin{vmatrix} -1 & x & 0 \\ 2 & 3 & 1 \\ 7 & y & 0 \end{vmatrix}$

28. $\begin{vmatrix} 5 & 2 & 3 \\ 3 & 1 & 2 \\ -4 & -3 & 5 \end{vmatrix} = \begin{vmatrix} x & 0 & -1 \\ 3 & 1 & 2 \\ 5 & 0 & y \end{vmatrix}$

Use Theorem 6 to transform each determinant into one that contains a row (or column) with all elements 0 but one (if possible); then expand the transformed determinant by this row (or column).

29. $\begin{vmatrix} 1 & 5 & 3 \\ 4 & 2 & 1 \\ 3 & 1 & 2 \end{vmatrix}$
30. $\begin{vmatrix} -1 & 5 & 1 \\ 2 & 3 & 1 \\ 3 & 2 & 1 \end{vmatrix}$

31. $\begin{vmatrix} 5 & 2 & -3 \\ -2 & 4 & 4 \\ 1 & -1 & 3 \end{vmatrix}$
32. $\begin{vmatrix} 5 & 3 & -6 \\ -1 & 1 & 4 \\ 4 & 3 & -6 \end{vmatrix}$

33. $\begin{vmatrix} 3 & -4 & 1 \\ 6 & -1 & 2 \\ 9 & 2 & 3 \end{vmatrix}$

34. $\begin{vmatrix} 2 & 3 & -1 \\ 5 & 4 & 7 \\ -4 & -6 & 2 \end{vmatrix}$

35. $\begin{vmatrix} 0 & 1 & 0 & 1 \\ 1 & -2 & 4 & 3 \\ 2 & 1 & 5 & 4 \\ 1 & 2 & 1 & 2 \end{vmatrix}$

36. $\begin{vmatrix} 2 & 3 & 1 & -1 \\ 3 & 1 & 2 & 1 \\ 0 & 5 & 4 & 0 \\ -1 & 2 & 3 & 0 \end{vmatrix}$

C 37. $\begin{vmatrix} 3 & 2 & 3 & 1 \\ 3 & -2 & 8 & 5 \\ 2 & 1 & 3 & 1 \\ 4 & 5 & 4 & -3 \end{vmatrix}$

38. $\begin{vmatrix} -1 & 4 & 2 & 1 \\ 5 & -1 & -3 & -1 \\ 2 & -1 & -2 & 3 \\ -3 & 3 & 3 & 3 \end{vmatrix}$

Prove each of the following statements.

39. $\begin{vmatrix} a & b & a \\ d & e & d \\ g & h & g \end{vmatrix} = 0$

40. $\begin{vmatrix} a & b & c \\ kd & ke & kf \\ g & h & i \end{vmatrix} = k \begin{vmatrix} a & b & c \\ d & e & f \\ g & h & i \end{vmatrix}$

41. $\begin{vmatrix} a_1 & b_1 & c_1 \\ a_2 & b_2 & c_2 \\ a_3 & b_3 & c_3 \end{vmatrix} = - \begin{vmatrix} b_1 & a_1 & c_1 \\ b_2 & a_2 & c_2 \\ b_3 & a_3 & c_3 \end{vmatrix}$

42. $\begin{vmatrix} a_1 & b_1 & c_1 \\ a_2 & b_2 & c_2 \\ a_3 & b_3 & c_3 \end{vmatrix} = \begin{vmatrix} a_1 + kc_1 & b_1 & c_1 \\ a_2 + kc_2 & b_2 & c_2 \\ a_3 + kc_3 & b_3 & c_3 \end{vmatrix}$

43. Show, without expanding, that $(2, 5)$ and $(-3, 4)$ satisfy the equation

$$\begin{vmatrix} x & y & 1 \\ 2 & 5 & 1 \\ -3 & 4 & 1 \end{vmatrix} = 0$$

44. Show that

$$\begin{vmatrix} x & y & 1 \\ 2 & 3 & 1 \\ -1 & 2 & 1 \end{vmatrix} = 0$$

is the equation of a line that passes through $(2, 3)$ and $(-1, 2)$.

45. Show that

$$\begin{vmatrix} x & y & 1 \\ x_1 & y_1 & 1 \\ x_2 & y_2 & 1 \end{vmatrix} = 0$$

is the equation of a line that passes through (x_1, y_1) and (x_2, y_2).

46. In analytic geometry it is shown that the area of a triangle with vertices (x_1, y_1), (x_2, y_2), and (x_3, y_3) is the absolute value of

$$\frac{1}{2} \begin{vmatrix} x_1 & y_1 & 1 \\ x_2 & y_2 & 1 \\ x_3 & y_3 & 1 \end{vmatrix}$$

Use this result to find the area of a triangle with vertices $(-1, 4)$, $(4, 8)$, and $(1, 1)$.

47. What can we say about the three points (x_1, y_1), (x_2, y_2), and (x_3, y_3) if

$$\begin{vmatrix} x_1 & y_1 & 1 \\ x_2 & y_2 & 1 \\ x_3 & y_3 & 1 \end{vmatrix} = 0$$

[*Hint:* See Problem 46.]

48. If the three points (x_1, y_1), (x_2, y_2), and (x_3, y_3) are all on the same line, what can we say about the value of the determinant

$$\begin{vmatrix} x_1 & y_1 & 1 \\ x_2 & y_2 & 1 \\ x_3 & y_3 & 1 \end{vmatrix}$$

Section 10-6 Cramer's Rule

- Two-Equations–Two-Unknowns
- Three-Equations–Three-Unknowns

Now let us see how determinants arise rather naturally in the process of solving systems of linear equations. We will start by investigating two equations and two unknowns, and then extend any results to three equations and three unknowns.

10-6 Cramer's Rule

■ Two-Equations–Two-Unknowns

Instead of thinking of each system of linear equations in two unknowns as a different problem, let us see what happens when we attempt to solve the general system

$$a_{11}x + a_{12}y = k_1 \tag{1A}$$
$$a_{21}x + a_{22}y = k_2 \tag{1B}$$

once and for all in terms of the unspecified real constants a_{11}, a_{12}, a_{21}, a_{22}, k_1, and k_2.

We proceed by multiplying equations (1A) and (1B) by suitable constants so that when the resulting equations are added, left side to left side and right side to right side, one of the variables drops out. Suppose we choose to eliminate y; what constant should we use to make the coefficients of y the same except for the signs? Multiply equation (1A) by a_{22} and (1B) by $-a_{12}$; then add.

$$\begin{aligned} a_{22}(1A): \quad & a_{11}a_{22}x + a_{12}a_{22}y = k_1 a_{22} \\ -a_{12}(1B): \quad & \underline{-a_{21}a_{12}x - a_{12}a_{22}y = -k_2 a_{12}} \\ & a_{11}a_{22}x - a_{21}a_{12}x + 0y = k_1 a_{22} - k_2 a_{12} \\ & (a_{11}a_{22} - a_{21}a_{12})x = k_1 a_{22} - k_2 a_{12} \\ & x = \frac{k_1 a_{22} - k_2 a_{12}}{a_{11}a_{22} - a_{21}a_{12}} \end{aligned}$$

$$a_{11}a_{22} - a_{21}a_{12} \neq 0$$

What do the numerator and denominator remind you of? From your experience with determinants in the last two sections, you should recognize these expressions as

$$x = \frac{\begin{vmatrix} k_1 & a_{12} \\ k_2 & a_{22} \end{vmatrix}}{\begin{vmatrix} a_{11} & a_{12} \\ a_{21} & a_{22} \end{vmatrix}}$$

Similarly, starting with system (1) and eliminating x (this is left as an exercise), we obtain

$$y = \frac{\begin{vmatrix} a_{11} & k_1 \\ a_{21} & k_2 \end{vmatrix}}{\begin{vmatrix} a_{11} & a_{12} \\ a_{21} & a_{22} \end{vmatrix}}$$

These results are summarized in Theorem 7, which is named after the Swiss mathematician, G. Cramer (1704–1752).

> **THEOREM 7** **Cramer's Rule for Two Equations and Two Unknowns**
>
> Given the system
> $$a_{11}x + a_{12}y = k_1$$
> $$a_{21}x + a_{22}y = k_2$$
> with
> $$D = \begin{vmatrix} a_{11} & a_{12} \\ a_{21} & a_{22} \end{vmatrix} \neq 0$$
> then
> $$x = \frac{\begin{vmatrix} k_1 & a_{12} \\ k_2 & a_{22} \end{vmatrix}}{D} \quad \text{and} \quad y = \frac{\begin{vmatrix} a_{11} & k_1 \\ a_{21} & k_2 \end{vmatrix}}{D}$$

The determinant D is called the **coefficient determinant**. If $D \neq 0$, then the system has exactly one solution, which is given by Cramer's rule. If, on the other hand, $D = 0$, then it can be shown that the system is either inconsistent or dependent; that is, the system either has no solutions or has an infinite number of solutions.

EXAMPLE 18 Solve using Cramer's rule.
$$2x - 3y = 7$$
$$-3x + y = -7$$

Solution
$$D = \begin{vmatrix} 2 & -3 \\ -3 & 1 \end{vmatrix} = -7$$

$$x = \frac{\begin{vmatrix} 7 & -3 \\ -7 & 1 \end{vmatrix}}{-7} = \frac{-14}{-7} = 2$$

$$y = \frac{\begin{vmatrix} 2 & 7 \\ -3 & -7 \end{vmatrix}}{-7} = \frac{7}{-7} = -1$$

PROBLEM 18 Solve using Cramer's rule.
$$3x + 2y = -3$$
$$-4x + 3y = -13$$

10-6 Cramer's Rule

■ **Three-Equations–Three-Unknowns**

Cramer's rule generalizes completely for any size linear system that has the same number of unknowns as equations. We state without proof in Theorem 8 the rule for three equations and three unknowns. (Augmented matrix methods can be used to prove this rule.)

THEOREM 8 | **Cramer's Rule for Three Equations and Three Unknowns**

Given the system

$$a_{11}x + a_{12}y + a_{13}z = k_1$$
$$a_{21}x + a_{22}y + a_{23}z = k_2$$
$$a_{31}x + a_{32}y + a_{33}z = k_3$$

with

$$D = \begin{vmatrix} a_{11} & a_{12} & a_{13} \\ a_{21} & a_{22} & a_{23} \\ a_{31} & a_{32} & a_{33} \end{vmatrix} \neq 0$$

then

$$x = \frac{\begin{vmatrix} k_1 & a_{12} & a_{13} \\ k_2 & a_{22} & a_{23} \\ k_3 & a_{32} & a_{33} \end{vmatrix}}{D} \qquad y = \frac{\begin{vmatrix} a_{11} & k_1 & a_{13} \\ a_{21} & k_2 & a_{23} \\ a_{31} & k_3 & a_{33} \end{vmatrix}}{D} \qquad z = \frac{\begin{vmatrix} a_{11} & a_{12} & k_1 \\ a_{21} & a_{22} & k_2 \\ a_{31} & a_{32} & k_3 \end{vmatrix}}{D}$$

It is easy to remember these determinant formulas for x, y, and z if one observes the following:

1. Determinant D is formed from the coefficients of x, y, and z, keeping the same relative position in the determinant as found in the system.
2. Determinant D appears in the denominators for x, y, and z.
3. The numerator for x can be obtained from D by replacing the coefficients of x—a_{11}, a_{21}, and a_{31}—with the constants k_1, k_2, and k_3, respectively. Similar statements can be made for the numerators for y and z.

EXAMPLE 19 Use Cramer's rule to solve.

$$x + y = 1$$
$$ 3y - z = -4$$
$$x + z = 3$$

Solution

$$D = \begin{vmatrix} 1 & 1 & 0 \\ 0 & 3 & -1 \\ 1 & 0 & 1 \end{vmatrix} = 2$$

$$x = \dfrac{\begin{vmatrix} 1 & 1 & 0 \\ -4 & 3 & -1 \\ 3 & 0 & 1 \end{vmatrix}}{2} = \dfrac{4}{2} = 2 \qquad y = \dfrac{\begin{vmatrix} 1 & 1 & 0 \\ 0 & -4 & -1 \\ 1 & 3 & 1 \end{vmatrix}}{2} = \dfrac{-2}{2} = -1$$

$$z = \dfrac{\begin{vmatrix} 1 & 1 & 1 \\ 0 & 3 & -4 \\ 1 & 0 & 3 \end{vmatrix}}{2} = \dfrac{2}{2} = 1$$

PROBLEM 19 Use Cramer's rule to solve.

$$3x - z = 5$$
$$x - y + z = 0$$
$$x + y = 0$$

In practice, Cramer's rule is rarely used to solve systems of order higher than 2 or 3; more efficient methods are available, including the methods discussed in Chapter 9. Cramer's rule is, however, a valuable tool in theoretical mathematics.

Answers to Matched Problems   **18.** $x = 1, \ y = -3$ **19.** $x = 1, \ y = -1, \ z = -2$

Exercise 10-6 ■ *Solve, using Cramer's rule.*

A
1. $x + 2y = 1$
   $x + 3y = -1$

2. $x + 2y = 3$
   $x + 3y = 5$

3. $2x + y = 1$
   $5x + 3y = 2$

4. $x + 3y = 1$
   $2x + 8y = 0$

5. $2x - y = -3$
   $-x + 3y = 4$

6. $2x + y = 1$
   $5x + 3y = 2$

B
7. $x + y = 0$
   $ 2y + z = -5$
   $-x + z = -3$

8. $x + y = -4$
   $ 2y + z = 0$
   $-x + z = 5$

9. $x + y = 1$
   $ 2y + z = 0$
   $-x + z = 0$

10. $x + y = -4$
    $ 2y + z = 3$
    $-x + z = 7$

11. $\begin{aligned} y + z &= -4 \\ x + 2z &= 0 \\ x - y &= 5 \end{aligned}$

12. $\begin{aligned} x - z &= 2 \\ 2x - y &= 8 \\ x + y + z &= 2 \end{aligned}$

13. $\begin{aligned} 2y - z &= -4 \\ x - y - z &= 0 \\ x - y + 2z &= 6 \end{aligned}$

14. $\begin{aligned} 2x + y &= 2 \\ x - y + z &= -1 \\ x + y + z &= -1 \end{aligned}$

C It is clear that $x = 0$, $y = 0$, $z = 0$ is a solution to each of the following systems. Use Cramer's rule to determine whether this solution is unique. [Hint: If $D \neq 0$, what can you conclude? If $D = 0$, what can you conclude?]

15. $\begin{aligned} x - 4y + 9z &= 0 \\ 4x - y + 6z &= 0 \\ x - y + 3z &= 0 \end{aligned}$

16. $\begin{aligned} 3x - y + 3z &= 0 \\ 5x + 5y - 9z &= 0 \\ -2x + y - 3z &= 0 \end{aligned}$

17. Prove Theorem 7 for y.

Section 10-7 Chapter Review

IMPORTANT TERMS AND SYMBOLS

10-1 Matrix Addition; Multiplication of a Matrix by a Number. Dimension of a matrix, $m \times n$ matrix, square matrix, column matrix, row matrix, equal matrices, matrix addition, addition properties, zero matrix, negative of a matrix, matrix subtraction, multiplication of a matrix by a number, $A + B$, $-B$, $A - B$, kA

10-2 Matrix Multiplication. Dot product, matrix product, multiplication properties, $A \cdot B$, AB

10-3 Inverse of a Square Matrix; Matrix Equations. Identity matrix for multiplication, main diagonal, inverse of a matrix relative to multiplication, matrix equation, I, M^{-1}, $AX = B$

10-4 Determinant Functions. Determinant function, determinant, second-order determinant, third-order determinant, det M

10-5 Properties of Determinants. Row operations, column operations

10-6 Cramer's Rule. Two-equations–two-unknowns, three-equations–three-unknowns, coefficient determinant

Exercise 10-7 Chapter Review

Work through all the problems in this chapter review and check answers in the back of the book. (Answers to all problems are there, and following each answer is a number in italics indicating the section in which that type of problem is discussed.) Where weaknesses show up, review appropriate sections in the text. When you are satisfied that you know the material, take the practice test following this review.

10 Matrices and Determinants

A In Problems 1–9 perform the operations that are defined, given the following matrices:

$$A = \begin{bmatrix} 1 & 2 \\ 3 & 1 \end{bmatrix} \quad B = \begin{bmatrix} 2 & 1 \\ 1 & 1 \end{bmatrix} \quad C = [2 \ \ 3] \quad D = \begin{bmatrix} 1 \\ 2 \end{bmatrix}$$

1. $A + B$
2. $B + D$
3. $A - 2B$
4. AB
5. AC
6. AD
7. DC
8. $C \cdot D$
9. $C + D$

10. Find the inverse of

$$A = \begin{bmatrix} 3 & 2 \\ 4 & 3 \end{bmatrix}$$

by appropriate row operations on $[A|I]$. Show that $A^{-1}A = I$.

11. Write the system

$$3x_1 + 2x_2 = k_1$$
$$4x_1 + 3x_2 = k_2$$

as a matrix equation and solve using the inverse found in Problem 10 for

(A) $k_1 = 3, \ k_2 = 5$ (B) $k_1 = 7, \ k_2 = 10$
(C) $k_1 = 4, \ k_2 = 2$

Evaluate Problems 12 and 13.

12. $\begin{vmatrix} 2 & -3 \\ -5 & -1 \end{vmatrix}$ 13. $\begin{vmatrix} 2 & 3 & -4 \\ 0 & 5 & 0 \\ 1 & -4 & -2 \end{vmatrix}$

14. Solve the system using Cramer's rule.

$$3x - 2y = 8$$
$$x + 3y = -1$$

B In Problems 15–20 perform the operations that are defined, given the following matrices:

$$A = \begin{bmatrix} 2 & -2 \\ 1 & 0 \\ 3 & 2 \end{bmatrix} \quad B = \begin{bmatrix} -1 \\ 2 \\ 3 \end{bmatrix} \quad C = [2 \ \ 1 \ \ 3]$$

$$D = \begin{bmatrix} 3 & -2 & 1 \\ -1 & 1 & 2 \end{bmatrix} \quad E = \begin{bmatrix} 3 & -4 \\ -1 & 0 \end{bmatrix}$$

15. $A + D$
16. $E + DA$
17. $DA - 3E$
18. $C \cdot B$
19. CB
20. $AD - BC$

21. Find the inverse of
$$A = \begin{bmatrix} 1 & 2 & 3 \\ 2 & 3 & 4 \\ 1 & 2 & 1 \end{bmatrix}$$
by appropriate row operations on $[A|I]$. Show that $A^{-1}A = I$.

22. Write the system
$$x_1 + 2x_2 + 3x_3 = k_1$$
$$2x_1 + 3x_2 + 4x_3 = k_2$$
$$x_1 + 2x_2 + x_3 = k_3$$
as a matrix equation and solve using the inverse found in Problem 21 for
(A) $k_1 = 1, \; k_2 = 3, \; k_3 = 3$ (B) $k_1 = 0, \; k_2 = 0, \; k_3 = -2$
(C) $k_1 = -3, \; k_2 = -4, \; k_3 = 1$

Evaluate Problems 23 and 24.

23. $\begin{vmatrix} -\frac{1}{4} & \frac{3}{2} \\ \frac{1}{2} & \frac{2}{3} \end{vmatrix}$
24. $\begin{vmatrix} 2 & -1 & 1 \\ -3 & 5 & 2 \\ 1 & -2 & 4 \end{vmatrix}$

25. Solve for y only using Cramer's rule.
$$x - 2y + z = -6$$
$$y - z = 4$$
$$2x + 2y + z = 2$$
Find the numerator and denominator first; then reduce.

C 26. Find the inverse of
$$A = \begin{bmatrix} 4 & 5 & 6 \\ 4 & 5 & -6 \\ 1 & 1 & 1 \end{bmatrix}$$
Show that $A^{-1}A = I$.

27. Clear the decimals in the system
$$0.04x_1 + 0.05x_2 + 0.06x_3 = 360$$
$$0.04x_1 + 0.05x_2 - 0.06x_3 = 120$$
$$x_1 + x_2 + x_3 = 7{,}000$$
by multiplying the first two equations by 100; then write the resulting system as a matrix equation and solve using the inverse found in Problem 26.

28. $\begin{vmatrix} -1 & 4 & 1 & 1 \\ 5 & -1 & 2 & -1 \\ 2 & -1 & 0 & 3 \\ -3 & 3 & 0 & 3 \end{vmatrix} = ?$

29. Show that
$$\begin{vmatrix} u & v \\ w & x \end{vmatrix} = \begin{vmatrix} u + kv & v \\ w + kx & x \end{vmatrix}$$

Practice Test   Chapter 10

Take this practice test as if it were a graded test. Allow yourself up to 50 minutes. Work the problems without looking back in the chapter. Correct your work using the answers (keyed to appropriate sections) in the back of the book.

In Problems 1–5 perform the indicated operations (if possible) given the following matrices:

$$A = \begin{bmatrix} 2 & -1 & 3 \\ -1 & 2 & 0 \end{bmatrix} \quad B = [1 \quad -2 \quad -3] \quad C = \begin{bmatrix} 4 \\ -1 \\ 2 \end{bmatrix}$$

$$D = \begin{bmatrix} 2 & -3 \\ -1 & 2 \end{bmatrix} \quad E = \begin{bmatrix} 1 & -2 & 0 \\ 3 & 1 & 2 \\ -1 & 0 & 1 \end{bmatrix}$$

1. $B \cdot C$ 2. DA 3. AD
4. $CB + 2E$ 5. $A + E$

6. Find the inverse for
$$A = \begin{bmatrix} 2 & -3 \\ 3 & -4 \end{bmatrix}$$
using row operations on $[A|I]$. Check by showing that $A^{-1}A = I$.

7. Write the system
$$2x_1 - 3x_2 = k_1$$
$$3x_1 - 4x_2 = k_2$$
as a matrix equation and solve using the inverse in Problem 6 for
 (A) $k_1 = 2, \quad k_2 = -3$ (B) $k_1 = -2, \quad k_2 = 1$

8. Evaluate.

$$\begin{vmatrix} 3 & -1 & 4 \\ -1 & 2 & -3 \\ 1 & 3 & 2 \end{vmatrix}$$

9. Solve the system for z only, using Cramer's rule.

$$2x - y + z = 9$$
$$x - 2z = -8$$
$$-x + 3y + z = 3$$

10. It is clear that (0, 0, 0) is a solution to the system

$$x_1 - x_2 + x_3 = 0$$
$$-x_1 - 3x_2 - x_3 = 0$$
$$x_1 + x_2 + x_3 = 0$$

Evaluate the coefficient determinant and determine whether (0, 0, 0) is the only solution or whether there are infinitely many others.

Sequences and Series .11

11-1 Sequences and Series
11-2 Mathematical Induction
11-3 Arithmetic Sequences and Series
11-4 Geometric Sequences and Series
11-5 Additional Applications
11-6 Binomial Formula
11-7 Chapter Review

A natural design of mathematical interest. Can you guess the source? See the back of the book.

Chapter 11 ■ Sequences and Series

In this chapter we are going to consider functions whose domains are special subsets of the set of integers; that is, subsets whose members are successive integers. These special functions, called **sequences**, are encountered with increased frequency as one progresses in mathematics.

Section 11-1 Sequences and Series

- Sequences
- Series

■ Sequences

Consider the function f given by

$$f(n) = 2n - 1 \qquad (1)$$

where the domain of f is the set of natural numbers N. The function f is an example of a sequence; however, one hardly ever sees sequences represented in this way. A special notation for sequences has evolved, which we now discuss.

To start, the range value $f(n)$ is usually symbolized more compactly with a symbol such as a_n. Thus, in place of equation (1) we would write

$$a_n = 2n - 1$$

and the domain would be understood to be the set of natural numbers N unless something was said to the contrary or the context indicated otherwise. The elements in the range are called **terms of the sequence**; a_1 is the first term, a_2 the second term, and a_n the nth term.

$a_1 = 2(1) - 1 = 1$ First term
$a_2 = 2(2) - 1 = 3$ Second term
$a_3 = 2(3) - 1 = 5$ Third term
$\quad\vdots \qquad\qquad \vdots$

11-1 Sequences and Series

When the terms in a sequence are written in their natural order with respect to domain values

$$a_1, a_2, a_3, \ldots, a_n, \ldots$$

or

$$1, 3, 5, \ldots, 2n - 1, \ldots$$

this ordered list of elements is often informally referred to as a sequence. A sequence is also represented in the abbreviated form $\{a_n\}$ where a symbol for the nth term is placed between braces. For example, we could refer to the sequence

$$1, 3, 5, \ldots, 2n - 1, \ldots$$

as the sequence $\{2n - 1\}$.

If the domain of a function is a finite set of successive integers, then the sequence is called a **finite sequence**; if the domain is an infinite set of successive integers, then the sequence is called an **infinite sequence**. The sequence $\{2n - 1\}$ is an infinite sequence. We now illustrate a finite sequence and another general way of specifying a sequence by use of a *recursion formula*.

EXAMPLE 1 List the terms of the sequence specified by

$$a_1 = 5$$
$$a_n = a_{n-1} + 2 \qquad n \in \{2, 3, 4\}$$

Solution $a_1 = 5$

$a_2 = a_{2-1} + 2 = a_1 + 2 = 5 + 2 = 7$
$a_3 = a_{3-1} + 2 = a_2 + 2 = 7 + 2 = 9$
$a_4 = a_{4-1} + 2 = a_3 + 2 = 9 + 2 = 11$

The formula $a_n = a_{n-1} + 2$ is called a **recursion formula** and is used to generate the terms of a sequence in terms of preceding terms. Of course, a starting term must be provided in order to use the formula. Recursion formulas are particularly suitable for use with calculators and computers (see Problems 55 and 56 in Exercise 11-1).

PROBLEM 1 Find the first five terms of a sequence specified by

$$a_1 = 4$$
$$a_n = \tfrac{1}{2}a_{n-1} \qquad n \geq 2$$

Now let us look at the problem in reverse; that is, given the first few terms of a sequence (assuming the terms of the sequence continue in the indicated pattern), find a_n in terms of n.

EXAMPLE 2 Find a_n in terms of n for the sequences whose first four terms are

(A) $5, 6, 7, 8, \ldots$ (B) $2, -4, 8, -16, \ldots$

Solution (A) $a_n = n + 4$ (B) $a_n = (-1)^{n+1} 2^n$

[Note: These representations are not unique. Also, since it is not stated to the contrary, the domain of each sequence is assumed to be the set of natural numbers N.]

PROBLEM 2 Find a_n in terms of n for

(A) $2, 4, 6, 8, \ldots$ (B) $1, -\frac{1}{2}, \frac{1}{4}, -\frac{1}{8}, \ldots$

■ Series

The indicated sum of the terms of a sequence is called a **series**. If the sequence is finite, the corresponding series is a **finite series**; if the sequence is infinite, the corresponding series is an **infinite series**. We will restrict our discussion to finite series in this section. For example,

$1, 2, 4, 8, 16$ Finite sequence

$1 + 2 + 4 + 8 + 16$ Finite series

Series are often represented in a compact form using summation notation. Consider the following examples:

$$\sum_{k=1}^{4} a_k = a_1 + a_2 + a_3 + a_4$$

$$\sum_{k=3}^{7} b_k = b_3 + b_4 + b_5 + b_6 + b_7$$

$$\sum_{k=0}^{n} c_k = c_0 + c_1 + c_2 + \cdots + c_n$$

The terms on the right are obtained from the left expression by successively replacing the **summing index** k with integers, starting with the first number indicated below $\sum$ and ending with the number that appears above $\sum$. Thus, for example, if we are given the sequence

$$\frac{1}{2}, \frac{1}{4}, \frac{1}{8}, \ldots, \frac{1}{2^n}$$

the corresponding series is

$$\frac{1}{2} + \frac{1}{4} + \frac{1}{8} + \cdots + \frac{1}{2^n}$$

or, more compactly,

$$\sum_{k=1}^{n} \frac{1}{2^k}$$

EXAMPLE 3 Write $\displaystyle\sum_{k=1}^{5} \frac{k-1}{k}$ without summation notation.

Solution
$$\sum_{k=1}^{5} \frac{k-1}{k} = \frac{1-1}{1} + \frac{2-1}{2} + \frac{3-1}{3} + \frac{4-1}{4} + \frac{5-1}{5}$$

$$= 0 + \frac{1}{2} + \frac{2}{3} + \frac{3}{4} + \frac{4}{5}$$

PROBLEM 3 Write $\displaystyle\sum_{k=0}^{5} \frac{(-1)^k}{2k+1}$ without summation notation.

EXAMPLE 4 Write the following series using summation notation.

$$1 - \frac{1}{2} + \frac{1}{3} - \frac{1}{4} + \frac{1}{5} - \frac{1}{6}$$

(A) Start the summing index at $k = 1$.
(B) Start the summing index at $k = 0$.

Solution (A) $(-1)^{k+1}$ provides the alternation of sign.
$1/k$ provides the other part of each term.
Thus, we can write

$$\sum_{k=1}^{6} \frac{(-1)^{k+1}}{k}$$

as can be easily checked.

(B) $(-1)^k$ provides the alternation of sign.
$1/(k+1)$ provides the other part of each term.
Thus, we can write

$$\sum_{k=0}^{5} \frac{(-1)^k}{k+1}$$

as can be checked.

PROBLEM 4 Write the following series using summation notation.

$$1 - \frac{2}{3} + \frac{4}{9} - \frac{8}{27} + \frac{16}{81}$$

(A) Start with $k = 1$.   (B) Start with $k = 0$.

Answers to Matched Problems

1. $4, 2, 1, \frac{1}{2}, \frac{1}{4}$
2. (A) $a_n = 2n$ (B) $a_n = (-1)^{n+1} 2^{1-n}$
3. $1 - \frac{1}{3} + \frac{1}{5} - \frac{1}{7} + \frac{1}{9} - \frac{1}{11}$
4. (A) $\sum_{k=1}^{5} \left(-\frac{2}{3}\right)^{k-1}$ (B) $\sum_{k=0}^{4} \left(-\frac{2}{3}\right)^{k}$

Exercise 11-1

A Write the first four terms for each sequence.

1. $a_n = n - 2$
2. $a_n = n + 3$
3. $a_n = \dfrac{n-1}{n+1}$
4. $a_n = \left(1 + \dfrac{1}{n}\right)^n$
5. $a_n = (-2)^{n+1}$
6. $a_n = \dfrac{(-1)^{n+1}}{n^2}$

7. Write the eighth term in the sequence in Problem 1.
8. Write the tenth term in the sequence in Problem 2.
9. Write the one-hundredth term in the sequence in Problem 3.
10. Write the two-hundredth term in the sequence in Problem 4.

Write each series in expanded form without summation notation.

11. $\sum_{k=1}^{5} k$
12. $\sum_{k=1}^{4} k^2$
13. $\sum_{k=1}^{3} \dfrac{1}{10^k}$
14. $\sum_{k=1}^{5} \left(\dfrac{1}{3}\right)^k$
15. $\sum_{k=1}^{4} (-1)^k$
16. $\sum_{k=1}^{6} (-1)^{k+1} k$

B Write the first five terms of each sequence.

17. $a_n = (-1)^{n+1} n^2$
18. $a_n = (-1)^{n+1} \left(\dfrac{1}{2^n}\right)$
19. $a_n = \dfrac{1}{3}\left(1 - \dfrac{1}{10^n}\right)$
20. $a_n = n[1 - (-1)^n]$
21. $a_n = \left(-\dfrac{1}{2}\right)^{n-1}$
22. $a_n = \left(-\dfrac{3}{2}\right)^{n-1}$

23. $a_1 = 7;\ a_n = a_{n-1} - 4,\ n \geq 2$
24. $a_1 = a_2 = 1;\ a_n = a_{n-1} + a_{n-2},\ n \geq 3$
25. $a_1 = 4;\ a_n = \frac{1}{4} a_{n-1},\ n \geq 2$
26. $a_1 = 2;\ a_n = 2 a_{n-1},\ n \geq 2$

Find a_n in terms of n.

27. $4, 5, 6, 7, \ldots$
28. $-2, -1, 0, 1, \ldots$
29. $3, 6, 9, 12, \ldots$
30. $-2, -4, -6, -8, \ldots$

11-1 Sequences and Series

31. $\frac{1}{2}, \frac{2}{3}, \frac{3}{4}, \frac{4}{5}, \ldots$ **32.** $\frac{1}{2}, \frac{3}{4}, \frac{5}{6}, \frac{7}{8}, \ldots$

33. $1, -1, 1, -1, \ldots$ **34.** $1, -2, 3, -4, \ldots$

35. $-2, 4, -8, 16, \ldots$ **36.** $1, -3, 5, -7, \ldots$

37. $x, \dfrac{x^2}{2}, \dfrac{x^3}{3}, \dfrac{x^4}{4}, \ldots$ **38.** $x, -x^3, x^5, -x^7, \ldots$

Write each series in expanded form without summation notation.

39. $\displaystyle\sum_{k=1}^{4} \dfrac{(-2)^{k+1}}{k}$ **40.** $\displaystyle\sum_{k=1}^{5} (-1)^{k+1}(2k-1)^2$

41. $\displaystyle\sum_{k=1}^{3} \dfrac{1}{k} x^{k+1}$ **42.** $\displaystyle\sum_{k=1}^{5} x^{k-1}$

43. $\displaystyle\sum_{k=1}^{5} \dfrac{(-1)^{k+1}}{k} x^{k}$ **44.** $\displaystyle\sum_{k=0}^{4} \dfrac{(-1)^{k} x^{2k+1}}{2k+1}$

Write each series using summation notation.

45. $S_4 = 1^2 + 2^2 + 3^2 + 4^2$ **46.** $S_5 = 2 + 3 + 4 + 5 + 6$

47. $S_5 = \dfrac{1}{2} + \dfrac{1}{2^2} + \dfrac{1}{2^3} + \dfrac{1}{2^4} + \dfrac{1}{2^5}$ **48.** $S_4 = 1 - \tfrac{1}{2} + \tfrac{1}{3} - \tfrac{1}{4}$

49. $S_n = 1 + \dfrac{1}{2^2} + \dfrac{1}{3^2} + \cdots + \dfrac{1}{n^2}$

50. $S_n = 2 + \tfrac{2}{3} + \tfrac{4}{3} + \cdots + \dfrac{n+1}{n}$

51. $S_n = 1 - 4 + 9 - \cdots + (-1)^{n+1} n^2$

52. $S_n = \tfrac{1}{2} + \tfrac{1}{4} + \tfrac{1}{8} + \cdots + \dfrac{(-1)^{n+1}}{2^n}$

C **53.** Show that: $\displaystyle\sum_{k=1}^{n} c a_k = c \sum_{k=1}^{n} a_k$

54. Show that: $\displaystyle\sum_{k=1}^{n} (a_k + b_k) = \sum_{k=1}^{n} a_k + \sum_{k=1}^{n} b_k$

CALCULATOR PROBLEMS

The sequence

$$a_n = \dfrac{a_{n-1}^2 + M}{2 a_{n-1}} \qquad n \geq 2; \quad M \text{ a positive real number}$$

can be used to find $\sqrt{M}$ to any decimal-place accuracy desired. To start the sequence, choose a_1 arbitrarily from the positive real numbers.

55. (A) Find the first four terms of the sequence

$$a_1 = 3 \qquad a_n = \dfrac{a_{n-1}^2 + 2}{2 a_{n-1}} \qquad n \geq 2$$

(B) Compare the terms with $\sqrt{2}$ from a calculator or a table.
(C) Repeat parts (A) and (B) by letting a_1 be any other positive number, say 1.

56. (A) Find the first four terms of the sequence
$$a_1 = 2 \qquad a_n = \frac{a_{n-1}^2 + 5}{2a_{n-1}} \qquad n \geq 2$$

(B) Find $\sqrt{5}$ in a table and compare with part (A).
(C) Repeat parts (A) and (B) by letting a_1 be any other positive number, say 3.

In calculus, it can be shown that

$$e^x = \sum_{k=0}^{\infty} \frac{x^k}{k!} \approx 1 + \frac{x}{1!} + \frac{x^2}{2!} + \frac{x^3}{3!} + \cdots + \frac{x^n}{n!}$$

the larger n, the better the approximation. Note that $0! = 1$ *and* $n! = 1 \cdot 2 \cdot 3 \cdots \cdot n$ *for* $n \in N$. *Problems 57 and 58 refer to this series.*

57. Approximate $e^{0.2}$ using the first five terms of the series. Compare this approximation with your calculator evaluation of $e^{0.2}$.

58. Approximate $e^{-0.5}$ using the first five terms of the series. Compare this approximation with your calculator evaluation of $e^{-0.5}$.

Section 11-2 Mathematical Induction

- Introduction
- Mathematical Induction
- Three Famous Problems

- Introduction

In common usage the word **induction** means the generalization from particular cases or facts. The ability to formulate general hypotheses from a limited number of facts is a distinguishing characteristic of a creative mathematician. The creative process does not stop here, however; these hypotheses must then be proved or disproved. In mathematics, we have a special method of proof called **mathematical induction** that ranks among the most important basic tools in a mathematician's tool box. This method of proof, using deductive reasoning, enters frequently into the second part of the process described above.

We illustrate the first part of the process by an example. Suppose we write the sums of consecutive odd integers as follows:

$$1 = 1$$
$$1 + 3 = 4$$
$$1 + 3 + 5 = 9$$
$$1 + 3 + 5 + 7 = 16$$
$$1 + 3 + 5 + 7 + 9 = 25$$

Is there something very regular about 1, 4, 9, 16, and 25? You no doubt guessed that each is a perfect square and, perhaps, even guessed that each is the square of the number of terms being added. Have we "discovered" a general property of integers? What does this property appear to be?

Conjecture P: The sum of the first n odd integers is n^2 for all positive integers n [that is, $1 + 3 + 5 + \cdots + (2n - 1) = n^2$ for $n \in N$].*

Thus far we have used ordinary induction to arrive at Conjecture P. But how do we prove that Conjecture P is true for all positive integers? Continuing by one-by-one testing will never accomplish a general proof—not in your lifetime or all of your descendants' lifetimes. Mathematical induction is the answer to this dilemma. Before we discuss this method of proof, let us consider another conjecture.

Conjecture Q: For each positive integer n, the number $n^2 - n + 41$ is a prime number.

It is important to recognize that a conjecture can be proved false if it fails for only one case, called a **counterexample**. Let us check the conjecture for a few particular cases:

n	$n^2 - n + 41$	Prime?
1	41	Yes
2	43	Yes
3	47	Yes
4	53	Yes
5	61	Yes

* The equation $1 + 3 + 5 + \cdots + (2n - 1) = n^2$ is a symbolic way of representing the statement, "The sum of the first n odd integers is n^2." On the left side of the equation, we start at 1 and add successive odd integers until we reach $2n - 1$, for a given n. If $n = 1$, then $2n - 1 = 2(1) - 1 = 1$, and we start at 1 and stop there! Thus, for $n = 1$, the equation becomes $1 = 1^2$. If $n = 2$, then $2n - 1 = 2(2) - 1 = 3$, and we start at 1 and stop at 3. Thus, for $n = 2$, the equation becomes $1 + 3 = 2^2$. For $n = 3$, then $2n - 1 = 2(3) - 1 = 5$, and we start at 1 and stop at 5. Thus, for $n = 3$, the equation becomes $1 + 3 + 5 = 3^2$. And so on.

It certainly appears that Conjecture Q has a good chance of being true. The reader may want to check a few more cases, and if she or he persists, it will be found that Conjecture Q is true for n up to 41. What happens at $n = 41$?

$$41^2 - 41 + 41 = 41^2$$

which is not prime. Thus, Conjecture Q is false; $n = 41$ provides a counterexample. Here we see the danger of generalizing without proof from a few special cases. This example was discovered by Euler (1707–1783).

■ **Mathematical Induction**

Now to discuss mathematical induction. To start, we state a rather obvious property of the integers as an axiom.

AXIOM

Axiom—Well-Ordering Principle

Let S be any set that contains one or more positive integers; then there must be a positive integer in S that is smaller than each of the others.

Sets in which we have a particular interest are sets of integers that are closed under the addition of 1; that is, if k is in the set, then $k + 1$ is in the set. We will refer to such sets as **inductive sets**. Now we state the important theorem of this section.

THEOREM 1

Principle of Mathematical Induction

If p is a positive integer and S is a set of integers such that

1. $p \in S$ Hypothesis 1
2. S is inductive   Hypothesis 2

then S contains all integers greater than or equal to p.

Theorem 1 certainly seems reasonable, since if $p \in S$, then by Hypothesis 2, $p + 1 \in S$; if $p + 1 \in S$, then by Hypothesis 2, $p + 2 \in S$; and so on. Clearly, all integers greater than or equal to p are in S. The only catch to this "proof" is in the use of "and so on" and "clearly." We proceed now to a rigorous proof of this important theorem.

Proof (by Contradiction) Assume, under the hypothesis of the theorem, S does not contain all integers greater than or equal to p. Let G be the set of all integers greater than p not in S. From the well-ordering axiom, G has a least element, say r, that is not p, since, by Hypothesis 1, $p \in S$. Thus $r - 1$, the integer preceding r, is in S. But by Hypothesis 2, if $r - 1 \in S$, then $(r - 1) + 1 = r \in S$, which is a contradiction, since r is in G. Our assumption must be false, and we conclude that S contains all integers from p on.

Let us now use Theorem 1 to prove that Conjecture P is true, as well as several other conjectures. To facilitate the writing of induction proofs in a more concise way, we introduce two special symbols:

SYMBOL	MEANING
$\therefore$	Therefore
$\Rightarrow$	Implies ("$p \Rightarrow q$" is read "p implies q" or, equivalently, "If p then q.")

EXAMPLE 5 Prove: $1 + 3 + 5 + \cdots + (2n - 1) = n^2, \quad n \in N$

Proof Write

$P_n: \quad 1 + 3 + 5 + \cdots + (2n - 1) = n^2$

$S = \{n \in N | P_n \text{ is true}\}$

where S is the truth set for the open statement P_n. To show that $S = N$, we must establish both parts of Theorem 1.

Part 1. Show that $1 \in S$.

$1 = 1^2$

$\therefore 1 \in S$

Part 2. Show that S is inductive (that is, prove generally that $k \in S \Rightarrow k + 1 \in S$).

We want to show that if P_n is true for $n = k$, it follows logically that P_n is true for $n = k + 1$. We write P_k and P_{k+1} first to obtain an idea of where we must start and where we must finish:

$P_k: \quad 1 + 3 + 5 + \cdots + (2k - 1) = k^2$

$P_{k+1}: \quad 1 + 3 + 5 + \cdots + (2k - 1) + (2k + 1) = (k + 1)^2$

Starting with P_k, we can add $2k + 1$ to both members, and after simplifying the right member we note that we have obtained P_{k+1} as a logical consequence of P_k:

$$1 + 3 + 5 + \cdots + (2k - 1) = k^2 \qquad P_k$$
$$1 + 3 + 5 + \cdots + (2k - 1) + (2k + 1) = k^2 + (2k + 1) \qquad \text{Equality property}$$
$$= (k + 1)^2 \qquad P_{k+1}$$

Thus, $k \in S \Rightarrow k + 1 \in S$, and S is inductive.

Conclusion. $S = N$; that is, P_n is true for all natural numbers n.

PROBLEM 5 Prove: $1 + 2 + 3 + \cdots + n = \dfrac{n(n + 1)}{2}$, $n \in N$

We are now in a position to prove the laws of exponents for natural numbers n. First, we redefine a^n, $n \in N$, using a recursion formula:

Definition of a^n

$a^1 = a$
$a^{n+1} = a^n a \qquad n \in N$

Thus,

$$a^4 = a^3 a = (a^2 a)a = [(a^1 a)a]a = [(aa)a]a$$

EXAMPLE 6 Prove that $(xy)^n = x^n y^n$ for all positive integers n.

Proof Write

P_n: $(xy)^n = x^n y^n$ and $S = \{n \in N \mid P_n \text{ is true}\}$

Part 1. Show that $1 \in S$.

$(xy)^1 = xy \qquad \text{Definition}$
$\qquad = x^1 y^1 \qquad \text{Definition}$

$\therefore 1 \in S$

Part 2. Show that S is inductive.

P_k: $(xy)^k = x^k y^k$
P_{k+1}: $(xy)^{k+1} = x^{k+1} y^{k+1}$

Here we start with the left member of P_{k+1} and use P_k to find the right member of P_{k+1}.

$$(xy)^{k+1} = (xy)^k(xy) \quad \text{Definition}$$
$$= x^k y^k xy \quad \text{Use of } P_k$$
$$= (x^k x)(y^k y) \quad \text{Property of real numbers}$$
$$= x^{k+1} y^{k+1} \quad \text{Definition}$$

Thus, $k \in S \Rightarrow k + 1 \in S$, and S is inductive.

Conclusion. $S = N$

PROBLEM 6 Prove that $(x/y)^n = x^n/y^n$ for all positive integers n.

We consider one last example. (Before we start, recall that integer p is **divisible** by integer q if $p = qr$ for some integer r.)

EXAMPLE 7 Prove that $4^{2n} - 1$ is divisible by 5 for all positive integers n.

Proof Write

P_n: $4^{2n} - 1$ is divisible by 5 and $S = \{n \in N \mid P_n \text{ is true}\}$

Part 1. Show that $1 \in S$.

$$4^{2 \cdot 1} - 1 = 15 \quad \text{Divisible by 5}$$
$$\therefore 1 \in S$$

Part 2. Show that S is inductive.

P_k: $\quad 4^{2k} - 1 = 5r \quad$ for some integer r

P_{k+1}: $\quad 4^{2(k+1)} - 1 = 5s \quad$ for some integer s

$$4^{2k} - 1 = 5r \quad P_k$$
$$4^2(4^{2k} - 1) = 4^2(5r) \quad \text{Property of equality (multiply each side by } 4^2\text{)}$$
$$4^{2k+2} - 4^2 = 4^2(5r) \quad \text{Property of a real number}$$
$$4^{2(k+1)} - 1 = 15 + 4^2(5r) \quad \text{Property of equality (add 15 to each side)}$$
$$= 5(3 + 16r) \quad \text{Property of a real number (factor out 5)}$$
$$= 5s \quad \text{where } s = (3 + 16r), \text{ an integer}$$

Thus, $k \in S \Rightarrow k + 1 \in S$, and S is inductive.

Conclusion. $S = N$

PROBLEM 7 Prove that $8^n - 1$ is divisible by 7 for all positive integers n.

Three Famous Problems

We conclude this section by stating three famous problems. Instant worldwide fame awaits anyone who can prove or disprove either of the first two; neither has been proved or disproved to date.

1. Goldbach's problem, 1742: Every positive even integer greater than 2 is the sum of two prime numbers.
2. Fermat's last theorem, 1637: For $n > 2$, $x^n + y^n = z^n$ does not have solutions in the natural numbers.
3. Each positive integer can be expressed as the sum of four or fewer squares of positive integers. (Considered by the early Greeks and finally proved in 1772 by Lagrange.)

5. Sketch of proof: Write

$$P_n: \quad 1 + 2 + 3 + \cdots + n = \frac{n(n + 1)}{2}$$

and

$$S = \{n \in N \mid P_n \text{ is true}\}$$

Part 1. Show that $1 \in S$.

$$1 = \frac{1(1 + 1)}{2}$$

$$= 1$$

$$\therefore 1 \in S$$

Part 2. Show that S is inductive. (Supply reasons.)

$$1 + 2 + 3 + \cdots + k = \frac{k(k + 1)}{2} \qquad P_k$$

$$1 + 2 + 3 + \cdots + k + (k + 1) = \frac{k(k + 1)}{2} + (k + 1)$$

$$= \frac{(k + 1)(k + 2)}{2} \qquad P_{k+1}$$

Thus, $k \in S \Rightarrow k + 1 \in S$, and S is inductive.

Conclusion. $S = N$

6. Sketch of proof: Write

$$P_n: \quad \left(\frac{x}{y}\right)^n = \frac{x^n}{y^n} \quad \text{and} \quad S = \{n \in N \mid P_n \text{ is true}\}$$

Part 1. Show that $1 \in S$. (Supply reasons.)

$$\left(\frac{x}{y}\right)^1 = \frac{x}{y}$$

$$= \frac{x^1}{y^1}$$

$$\therefore 1 \in S$$

Part 2. Show that S is inductive. (Supply reasons.)

$$\left(\frac{x}{y}\right)^{k+1} = \left(\frac{x}{y}\right)^k \frac{x}{y}$$

$$= \frac{x^k}{y^k} \frac{x}{y}$$

$$= \frac{x^k x}{y^k y}$$

$$= \frac{x^{k+1}}{y^{k+1}}$$

Thus, $k \in S \Rightarrow k + 1 \in S$, and S is inductive.

Conclusion. $S = N$

7. Sketch of proof: Write

P_n: $8^n - 1$ is divisible by 7   and $S = \{n \in N \mid P_n \text{ is true}\}$

Part 1. Show that $1 \in S$.

$8^1 - 1 = 7$

$\therefore 1 \in S$

Part 2. Show that S is inductive. (Supply reasons.)

$8^k - 1 = 7r$

$8(8^k - 1) = 8(7r)$

$8^{k+1} - 1 = 7 + 8(7r)$

$\qquad = 7(1 + 8r)$

Thus, $k \in S \Rightarrow k + 1 \in S$, and S is inductive.

Conclusion. $S = N$

Exercise 11-2 ■ A

Find the first positive integer n that causes the statement to fail.

1. $(3 + 5)^n = 3^n + 5^n$
2. $n < 10$
3. $n^2 = 3n - 2$
4. $n^3 + 11n = 6n^2 + 6$

Verify each open statement P_n for $n = 1, 2,$ and 3.

5. P_n: $2 + 6 + 10 + \cdots + (4n - 2) = 2n^2$
6. P_n: $4 + 8 + 12 + \cdots + 4n = 2n(n + 1)$
7. P_n: $a^5 a^n = a^{5+n}$
8. P_n: $(a^5)^n = a^{5n}$

9. P_n: $9^n - 1$ is divisible by 4
10. P_n: $4^n - 1$ is divisible by 3

Write P_k and P_{k+1} for each of the following.

11. P_n in Problem 5
12. P_n in Problem 6
13. P_n in Problem 7
14. P_n in Problem 8
15. P_n in Problem 9
16. P_n in Problem 10

Use mathematical induction to prove that each P_n holds for all positive integers n.

17. P_n in Problem 5
18. P_n in Problem 6
19. P_n in Problem 7
20. P_n in Problem 8
21. P_n in Problem 9
22. P_n in Problem 10

B Use mathematical induction to prove each of the following propositions for all positive integers n, unless restricted otherwise.

23. $2 + 2^2 + 2^3 + \cdots + 2^n = 2^{n+1} - 2$
24. $\dfrac{1}{2} + \dfrac{1}{4} + \dfrac{1}{8} + \cdots + \dfrac{1}{2^n} = 1 - \left(\dfrac{1}{2}\right)^n$
25. $1^2 + 3^2 + 5^2 + \cdots + (2n-1)^2 = \frac{1}{3}(4n^3 - n)$
26. $1 + 8 + 16 + \cdots + 8(n-1) = (2n-1)^2$
27. $1^2 + 2^2 + 3^2 + \cdots + n^2 = \dfrac{n(n+1)(2n+1)}{6}$
28. $1 \cdot 2 + 2 \cdot 3 + 3 \cdot 4 + \cdots + n(n+1) = \dfrac{n(n+1)(n+2)}{3}$
29. $\dfrac{a^n}{a^3} = a^{n-3}$; $n > 3$
30. $\dfrac{a^5}{a^n} = \dfrac{1}{a^{n-5}}$; $n > 5$
31. $a^m a^n = a^{m+n}$; $m, n \in N$
    [Hint: Choose m as an arbitrary element of N, and then use induction on n.]
32. $(a^n)^m = a^{mn}$; $m, n \in N$
33. $x^n - 1$ is divisible by $x - 1$, $x \neq 1$
    [Hint: Divisible means that $x^n - 1 = (x - 1)Q(x)$ for some polynomial $Q(x)$.]
34. $x^n - y^n$ is divisible by $x - y$, $x \neq y$
35. $x^{2n} - 1$ is divisible by $x - 1$, $x \neq 1$
36. $x^{2n} - 1$ is divisible by $x + 1$, $x \neq -1$

37. $1^3 + 2^3 + 3^3 + \cdots + n^3 = (1 + 2 + 3 + \cdots + n)^2$
    [*Hint:* See Problem 5 following Example 5.]

38. $\dfrac{1}{1 \cdot 2 \cdot 3} + \dfrac{1}{2 \cdot 3 \cdot 4} + \dfrac{1}{3 \cdot 4 \cdot 5} + \cdots + \dfrac{1}{n(n+1)(n+2)}$
    $= \dfrac{n(n+3)}{4(n+1)(n+2)}$

C *Discover a formula for each of the following, and prove your hypothesis using mathematical induction, $n \in N$.*

39. $2 + 4 + 6 + \cdots + 2n$

40. $\dfrac{1}{1 \cdot 2} + \dfrac{1}{2 \cdot 3} + \dfrac{1}{3 \cdot 4} + \cdots + \dfrac{1}{n(n+1)}$

41. The number of lines determined by n points in a plane, no three of which are collinear

42. The number of diagonals in a polygon with n sides

Prove Problems 43–46 true for all integers n as specified.

43. $a > 1 \Rightarrow a^n > 1, \quad n \in N$
44. $0 < a < 1 \Rightarrow 0 < a^n < 1, \quad n \in N$
45. $n^2 > 2n, \quad n \geq 3$ 46. $2^n > n^2, \quad n \geq 5$

47. Prove or disprove the generalization of the following two facts:
$$3^2 + 4^2 = 5^2$$
$$3^3 + 4^3 + 5^3 = 6^3$$

48. Prove or disprove: $n^2 + 21n + 1$ is a prime number for all natural numbers n.

If $\{a_n\}$ and $\{b_n\}$ are two sequences, we write $\{a_n\} = \{b_n\}$ if and only if $a_n = b_n$, $n \in N$. Use mathematical induction to show that $\{a_n\} = \{b_n\}$ where

49. $a_1 = 1, \quad a_n = a_{n-1} + 2; \quad b_n = 2n - 1$
50. $a_1 = 2, \quad a_n = a_{n-1} + 2; \quad b_n = 2n$
51. $a_1 = 2, \quad a_n = 2^2 a_{n-1}; \quad b_n = 2^{2n-1}$
52. $a_1 = 2, \quad a_n = 3 a_{n-1}; \quad b_n = 2 \cdot 3^{n-1}$

Section 11-3 Arithmetic Sequences and Series

- Arithmetic Sequences
- *n*th-Term Formula
- Finite Arithmetic Series

Arithmetic Sequences

Consider the sequence

5, 9, 13, 17, . . .

Can you guess what the fifth term is? If you guessed 21, you have observed that each term after the first can be obtained from the preceding one by adding 4 to it. This is an example of an arithmetic sequence.

Arithmetic Sequence

A sequence

$$a_1, a_2, a_3, \ldots, a_n, \ldots$$

is called an **arithmetic sequence** (or **arithmetic progression**) if there exists a constant d, called the **common difference**, such that

$$a_n - a_{n-1} = d$$

That is,

$$a_n = a_{n-1} + d \quad \text{for every } n > 1$$

EXAMPLE 8 Which sequence is an arithmetic sequence and what is its common difference?

(A) 1, 2, 3, 5, . . .   (B) 3, 5, 7, 9, . . .

Solution Sequence (B) is an arithmetic sequence with $d = 2$.

PROBLEM 8 Repeat Example 8 with (A) $-4, -1, 2, 5, \ldots$, and (B) 2, 4, 8, 16, ...

nth-Term Formula

Arithmetic sequences have several convenient properties. For example, one can derive formulas for the nth term in terms of n and the sum of any number of consecutive terms. To obtain an nth-term formula, we note that if $\{a_n\}$ is an arithmetic sequence, then

$$a_2 = a_1 + d$$
$$a_3 = a_2 + d = a_1 + 2d$$
$$a_4 = a_3 + d = a_1 + 3d$$

which suggests

$$a_n = a_1 + (n-1)d \quad \text{for every } n > 1$$

We have arrived at this formula by ordinary induction; its proof requires mathematical induction, which we leave as an exercise (see Problem 31 in Exercise 11-3).

EXAMPLE 9   If the first and tenth terms of an arithmetic sequence are 3 and 30, respectively, find the fiftieth term of the sequence.

Solution   First find d:

$$a_n = a_1 + (n-1)d$$
$$a_{10} = a_1 + (10-1)d$$
$$30 = 3 + 9d$$
$$d = 3$$

Now find a_{50}:

$$a_{50} = a_1 + (50-1)3$$
$$= 3 + 49 \cdot 3$$
$$= 150$$

PROBLEM 9   If the first and fifteenth terms of an arithmetic sequence are -5 and 23, respectively, find the seventy-third term of the sequence.

■ Finite Arithmetic Series

The sum of the terms of an arithmetic sequence is called an **arithmetic series**. We will derive two simple and very useful formulas for finding the sum of an arithmetic series. Let

$$S_n = a_1 + (a_1 + d) + \cdots + [a_1 + (n-2)d] + [a_1 + (n-1)d]$$

which is the sum of the first n terms of an arithmetic sequence. Reversing the order of the sum, we obtain

$$S_n = [a_1 + (n-1)d] + [a_1 + (n-2)d] + \cdots + (a_1 + d) + a_1$$

Adding left members and corresponding elements of the right members of the two equations, we see that

$$2S_n = [2a_1 + (n-1)d] + [2a_1 + (n-1)d] + \cdots + [2a_1 + (n-1)d]$$
$$= n[2a_1 + (n-1)d]$$

or

$$S_n = \frac{n}{2}[2a_1 + (n-1)d]$$

By replacing $a_1 + (n-1)d$ with a_n, we obtain a second useful formula for the sum:

$$S_n = \frac{n}{2}(a_1 + a_n)$$

The proof of the first sum formula by mathematical induction is left as an exercise (see Problem 32 in Exercise 11-3).

EXAMPLE 10 Find the sum of the first twenty-six terms of an arithmetic series if the first term is -7 and $d = 3$.

Solution
$$S_n = \frac{n}{2}[2a_1 + (n-1)d]$$
$$S_{26} = \frac{26}{2}[2(-7) + (26-1)3]$$
$$= 793$$

PROBLEM 10 Find the sum of the first fifty-two terms of an arithmetic series if the first term is 23 and $d = -2$.

EXAMPLE 11 Find the sum of all the odd numbers between 51 and 99, inclusive.

Solution First find n:
$$a_n = a_1 + (n-1)d$$
$$99 = 51 + (n-1)2$$
$$n = 25$$

Now find S_{25}:
$$S_n = \frac{n}{2}(a_1 + a_n)$$
$$S_{25} = \frac{25}{2}(51 + 99)$$
$$= 1{,}875$$

11-3 Arithmetic Sequences and Series

PROBLEM 11 Find the sum of all the even numbers between -22 and 52, inclusive.

Answers to Matched Problems
8. Sequence (A) with $d = 3$
9. 139
10. $-1,456$
11. 570

Exercise 11-3

A
1. Determine which of the following are arithmetic sequences. Find d and the next two terms for those that are.
   (A) 2, 4, 8, ...
   (B) 7, 6.5, 6, ...
   (C) $-11, -16, -21, \ldots$
   (D) $\frac{1}{2}, \frac{1}{6}, \frac{1}{18}, \ldots$
2. Repeat Problem 1 for
   (A) 5, $-1, -7, \ldots$
   (B) 12, 4, $\frac{4}{3}, \ldots$
   (C) $\frac{1}{2}, \frac{2}{3}, \frac{3}{4}, \ldots$
   (D) 16, 48, 80, ...

Let $a_1, a_2, a_3, \ldots, a_n, \ldots$ be an arithmetic sequence. In Problems 3–18 find the indicated quantities.

3. $a_1 = -5$, $d = 4$; $a_2 = ?$, $a_3 = ?$, $a_4 = ?$
4. $a_1 = -18$, $d = 3$; $a_2 = ?$, $a_3 = ?$, $a_4 = ?$
5. $a_1 = -3$, $d = 5$; $a_{15} = ?$, $S_{11} = ?$
6. $a_1 = 3$, $d = 4$; $a_{22} = ?$, $S_{21} = ?$
7. $a_1 = 1$, $a_2 = 5$; $S_{21} = ?$
8. $a_1 = 5$, $a_2 = 11$; $S_{11} = ?$
9. $a_1 = 7$, $a_2 = 5$; $a_{15} = ?$
10. $a_1 = -3$, $d = -4$; $a_{10} = ?$

B
11. $a_1 = 3$, $a_{20} = 117$; $d = ?$, $a_{101} = ?$
12. $a_1 = 7$, $a_8 = 28$; $d = ?$, $a_{25} = ?$
13. $a_1 = -12$, $a_{40} = 22$; $S_{40} = ?$
14. $a_1 = 24$, $a_{24} = -28$; $S_{24} = ?$
15. $a_1 = \frac{1}{3}$, $a_2 = \frac{1}{2}$; $a_{11} = ?$, $S_{11} = ?$
16. $a_1 = \frac{1}{6}$, $a_2 = \frac{1}{4}$; $a_{19} = ?$, $S_{19} = ?$
17. $a_3 = 13$, $a_{10} = 55$; $a_1 = ?$
18. $a_9 = -12$, $a_{13} = 3$; $a_1 = ?$

19. $S_{51} = \sum_{k=1}^{51} (3k + 3) = ?$
20. $S_{40} = \sum_{k=1}^{40} (2k - 3) = ?$

21. Find $g(1) + g(2) + g(3) + \cdots + g(51)$ if $g(t) = 5 - t$.
22. Find $f(1) + f(2) + f(3) + \cdots + f(20)$ if $f(x) = 2x - 5$.
23. Find the sum of all the even integers between 21 and 135.
24. Find the sum of all the odd integers between 100 and 500.

25. Show that the sum of the first n odd natural numbers is n^2, using appropriate formulas from this section.

26. Show that the sum of the first n even natural numbers is $n + n^2$, using appropriate formulas from this section.

27. For a given sequence in which $a_1 = -3$ and $a_n = a_{n-1} + 3$, $n > 1$, find a_n in terms of n.

28. For the sequence in Problem 27 find $S_n = \sum_{k=1}^{n} a_k$ in terms of n.

29. An object falling from rest in a vacuum near the surface of the earth falls 16 feet during the first second, 48 feet during the second second, 80 feet during the third second, and so on.
  (A) How far will the object fall during the eleventh second?
  (B) How far will the object fall in 11 seconds?
  (C) How far will the object fall in t seconds?

30. In investigating different job opportunities, you find that firm A will start you at $25,000 per year and guarantee you a raise of $1,200 each year, while firm B will start you at $28,000 per year but will guarantee you a raise of only $800 each year. Over a 15-year period how much would you receive from each firm?

C

31. Prove, using mathematical induction, that if $\{a_n\}$ is an arithmetic sequence, then
$$a_n = a_1 + (n-1)d$$

32. Prove, using mathematical induction, that if $\{a_n\}$ is an arithmetic sequence, then
$$S_n = \frac{n}{2}[2a_1 + (n-1)d]$$

33. Show that $(x^2 + xy + y^2)$, $(z^2 + xz + x^2)$, and $(y^2 + yz + z^2)$ are consecutive terms of an arithmetic progression if x, y, and z form an arithmetic progression. (From USSR Mathematical Olympiads, 1955–1956, Grade 9.)

34. Take 121 terms of each arithmetic progression 2, 7, 12, ... and 2, 5, 8, ... How many numbers will there be in common? (From USSR Mathematical Olympiads, 1955–1956, Grade 9.)

35. Given the system of equations
$$ax + by = c$$
$$dx + ey = f$$
where a, b, c, d, e, f is any arithmetic progression with a nonzero constant difference, show that the system has a unique solution.

Section 11-4 Geometric Sequences and Series

- Geometric Sequences
- *n*th-Term Formula
- Finite Geometric Series
- Infinite Geometric Series

■ **Geometric Sequences**

Consider the sequence

$$2, -4, 8, -16, \ldots$$

Can you guess what the fifth and sixth terms are? If you guessed 32 and -64, respectively, you have observed that each term after the first can be obtained from the preceding one by multiplying it by -2. This is an example of a *geometric sequence*.

Geometric Sequence

A sequence

$$a_1, a_2, a_3, \ldots, a_n, \ldots$$

is called a **geometric sequence** (or a **geometric progression**) if there exists a nonzero constant r, called the **common ratio**, such that

$$\frac{a_n}{a_{n-1}} = r$$

That is,

$$a_n = r a_{n-1} \quad \text{for every } n > 1$$

EXAMPLE 12   Which sequence is a geometric sequence and what is its common ratio?

(A) $2, 6, 8, 10, \ldots$ (B) $-1, 3, -9, 27, \ldots$

Solution   Sequence (B) is a geometric sequence with $r = -3$.

PROBLEM 12   Repeat Example 12 with (A) $\frac{1}{4}, \frac{1}{2}, 1, 2, \ldots$, and (B) $\frac{1}{2}, \frac{1}{4}, \frac{1}{16}, \frac{1}{256}, \ldots$.

nth-Term Formula

Just as with arithmetic sequences, geometric sequences have several convenient properties. It is easy to derive formulas for the nth term in terms of n and the sum of any number of consecutive terms. To obtain an nth-term formula, we note that if $\{a_n\}$ is a geometric sequence, then

$$a_2 = ra_1$$
$$a_3 = ra_2 = r^2 a_1$$
$$a_4 = ra_3 = r^3 a_1$$

which suggests that

$$a_n = a_1 r^{n-1} \qquad \text{for every } n > 1$$

We have arrived at this formula using ordinary induction; its proof requires mathematical induction, which we leave as an exercise.

EXAMPLE 13 Find the seventh term of the geometric sequence $1, \frac{1}{2}, \frac{1}{4}, \ldots$.

Solution $r = \frac{1}{2}$
$a_n = a_1 r^{n-1}$
$a_7 = 1(\frac{1}{2})^{7-1} = \frac{1}{64}$

PROBLEM 13 Find the eighth term of the geometric sequence $\frac{1}{64}, -\frac{1}{32}, \frac{1}{16}, \ldots$.

EXAMPLE 14 If the first and tenth terms of a geometric sequence are 1 and 2, respectively, find the common ratio r to two decimal places.

Solution $a_n = a_1 r^{n-1}$
$2 = 1 r^{10-1}$
$r = 2^{1/9} = 1.08$ Calculation by calculator using y^x button.

PROBLEM 14 If the first and eighth terms of a geometric sequence are 2 and 16, respectively, find the common ratio r to three decimal places.

Finite Geometric Series

The sum of the terms of a geometric sequence is called a **geometric series**. As was the case with an arithmetic series, we can derive two simple and very useful formulas for finding the **sum of a geometric series**.

11-4 Geometric Sequences and Series

Let
$$S_n = a_1 + a_1r + a_1r^2 + a_1r^3 + \cdots + a_1r^{n-2} + a_1r^{n-1}$$

which is the sum of the first n terms of a geometric sequence. Multiply both members by r to obtain

$$rS_n = a_1r + a_1r^2 + a_1r^3 + \cdots + a_1r^{n-1} + a_1r^n$$

Now subtract the left member of the second equation from the left member of the first, and the right member of the second equation from the right member of the first to obtain

$$S_n - rS_n = a_1 - a_1r^n$$
$$S_n(1 - r) = a_1 - a_1r^n$$

Thus,

$$S_n = \frac{a_1 - a_1r^n}{1 - r} \qquad r \neq 1$$

Since $a_n = a_1r^{n-1}$, or $ra_n = a_1r^n$, the sum formula can also be written in the form

$$S_n = \frac{a_1 - ra_n}{1 - r} \qquad r \neq 1$$

The proof of the first sum formula by mathematical induction is left as an exercise.

If $r = 1$, then
$$S_n = a_1 + a_1 1 + a_1 1^2 + \cdots + a_1 1^{n-1} = na_1$$

EXAMPLE 15 Find the sum of the first twenty terms of a geometric series if the first term is 1 and $r = 2$.

Solution
$$S_n = \frac{a_1 - a_1r^n}{1 - r}$$

$$= \frac{1 - 1 \cdot 2^{20}}{1 - 2} \approx 1{,}050{,}000 \quad \text{Calculation using a calculator}$$

PROBLEM 15 Find the sum (to two decimal places) of the first fourteen terms of a geometric series if the first term is $\frac{1}{64}$ and $r = -2$.

Infinite Geometric Series

Consider a geometric series with $a_1 = 5$ and $r = \frac{1}{2}$. What happens to the sum S_n as n increases? To answer this question, we first write the sum formula in the more convenient form

$$S_n = \frac{a_1 - a_1 r^n}{1 - r} = \frac{a_1}{1 - r} - \frac{a_1 r^n}{1 - r} \tag{1}$$

For $a_1 = 5$ and $r = \frac{1}{2}$,

$$S_n = 10 - 10\left(\frac{1}{2}\right)^n$$

Thus,

$$S_2 = 10 - 10\left(\frac{1}{4}\right)$$

$$S_4 = 10 - 10\left(\frac{1}{16}\right)$$

$$S_{10} = 10 - 10\left(\frac{1}{1,024}\right)$$

$$S_{20} = 10 - 10\left(\frac{1}{1,048,576}\right)$$

It appears that $(\frac{1}{2})^n$ becomes smaller and smaller as n increases, and that the sum gets closer and closer to 10.

In general, it is possible to show that, if $|r| < 1$ (that is, $-1 < r < 1$), then r^n will tend to 0 as n increases. Thus,

$$\frac{a_1 r^n}{1 - r}$$

in equation (1) will tend to 0 as n increases, and S_n will tend to

$$\frac{a_1}{1 - r}$$

In other words, if $|r| < 1$, then S_n can be made as close to

$$\frac{a_1}{1 - r}$$

as we wish by taking n sufficiently large. Thus, we define

$$S_\infty = \frac{a_1}{1 - r} \qquad |r| < 1$$

11-4 Geometric Sequences and Series

and call this the **sum of an infinite geometric series**. If $|r| \geq 1$, an infinite geometric series has no sum.

EXAMPLE 16 Represent the repeating decimal $0.45\overline{45}$ as the quotient of two integers. (Recall that a repeating decimal names a rational number, and that any rational number can be represented as the quotient of two integers.)

Solution

$$0.45\overline{45} = 0.45 + 0.0045 + 0.000045 + \cdots$$

The right member of the equation is an infinite geometric series with $a_1 = 0.45$ and $r = 0.01$. Thus,

$$S_\infty = \frac{a_1}{1 - r} = \frac{0.45}{1 - 0.01} = \frac{0.45}{0.99} = \frac{5}{11}$$

Hence, $0.45\overline{45}$ and $\frac{5}{11}$ name the same rational number. Check the result by dividing 5 by 11.

PROBLEM 16 Repeat Example 16 for $0.818\overline{181}$.

Answers to Matched Problems
12. Sequence (A) with $r = 2$ 13. -2
14. $r = 1.346$ 15. -85.33
16. $\frac{9}{11}$

Exercise 11-4

A calculator will be useful in some problems.

A 1. Determine which of the following are geometric sequences. Find r and the next two terms for those that are.
(A) $2, -4, 8, \ldots$ (B) $7, 6.5, 6, \ldots$
(C) $-11, -16, -21, \ldots$ (D) $\frac{1}{2}, \frac{1}{6}, \frac{1}{18}, \ldots$

2. Repeat Problem 1 for
(A) $5, -1, -7, \ldots$ (B) $12, 4, \frac{4}{3}, \ldots$
(C) $\frac{1}{2}, \frac{2}{3}, \frac{3}{4}, \ldots$ (D) $16, 48, 80, \ldots$

Let $a_1, a_2, a_3, \ldots, a_n, \ldots$ be a geometric sequence. Find each of the indicated quantities in Problems 3–16.

3. $a_1 = -6, \quad r = -\frac{1}{2}; \quad a_2 = ?, \quad a_3 = ?, \quad a_4 = ?$
4. $a_1 = 12, \quad r = \frac{2}{3}; \quad a_2 = ?, \quad a_3 = ?, \quad a_4 = ?$
5. $a_1 = 81, \quad r = \frac{1}{3}; \quad a_{10} = ?$
6. $a_1 = 64, \quad r = \frac{1}{2}; \quad a_{13} = ?$
7. $a_1 = 3, \quad a_7 = 2{,}187, \quad r = 3; \quad S_7 = ?$
8. $a_1 = 1, \quad a_7 = 729, \quad r = -3; \quad S_7 = ?$

B
9. $a_1 = 100$, $a_6 = 1$; $r = ?$
10. $a_1 = 10$, $a_{10} = 30$; $r = ?$
11. $a_1 = 5$, $r = -2$; $S_{10} = ?$
12. $a_1 = 3$, $r = 2$; $S_{10} = ?$
13. $a_1 = 9$, $a_4 = \frac{8}{3}$; $a_2 = ?$, $a_3 = ?$
14. $a_1 = 12$, $a_4 = -\frac{4}{9}$; $a_2 = ?$, $a_3 = ?$
15. $S_7 = \sum_{k=1}^{7} (-3)^{k-1} = ?$
16. $S_7 = \sum_{k=1}^{7} 3^k = ?$
17. Find $g(1) + g(2) + \cdots + g(10)$ if $g(x) = (\frac{1}{2})^x$.
18. Find $f(1) + f(2) + \cdots + f(10)$ if $f(x) = 2^x$.
19. Find a positive number x so that $-2 + x - 6$ is a three-term geometric series.
20. Find a positive number x so that $6 + x + 8$ is a three-term geometric series.

Find the sum of each infinite geometric series that has a sum.

21. $3 + 1 + \frac{1}{3} + \cdots$
22. $16 + 4 + 1 + \cdots$
23. $2 + 4 + 8 + \cdots$
24. $4 + 6 + 9 + \cdots$
25. $2 - \frac{1}{2} + \frac{1}{8} - \cdots$
26. $21 - 3 + \frac{3}{7} - \cdots$

Represent each repeating decimal fraction as the quotient of two integers.

27. $0.7777\overline{7}$
28. $0.555\overline{5}$
29. $0.5454\overline{54}$
30. $0.2727\overline{27}$
31. $3.216216\overline{216}$
32. $5.6363\overline{63}$

33. *Business.* If $P is invested at r% compounded annually, the amount A present after n years forms a geometric progression with a common ratio $(1 + r)$. Write a formula for the amount present after n years. How long will it take a sum of money P to double if invested at 6% interest compounded annually?

34. *Population growth.* If a population of A_0 people grows at the constant rate of r% per year, the population after t years forms a geometric progression with a common ratio $(1 + r)$. Write a formula for the total population after t years. If the world's population is increasing at the rate of 2% per year, how long will it take to double?

35. *Engineering.* A rotating flywheel coming to rest rotates 300 revolutions the first minute. If in each subsequent minute it rotates two-thirds as many times as in the preceding minute, how many revolutions will the wheel make before coming to rest?

36. *Physics.* The first swing of a bob on a pendulum is 10 inches. If on each subsequent swing it travels 0.9 as far as on the preceding swing, how far will the bob travel before coming to rest?

C 37. If in a given sequence, $a_1 = -2$ and $a_n = -3a_{n-1}$, $n > 1$, find a_n in terms of n.

38. For the sequence in Problem 37 find $S_n = \sum_{k=1}^{n} a_k$ in terms of n.

39. Prove, using mathematical induction, that if $\{a_n\}$ is a geometric sequence, then
$$a_n = a_1 r^{n-1} \qquad n \in N$$

40. Prove, using mathematical induction, that if $\{a_n\}$ is a geometric sequence, then
$$S_n = \frac{a_1 - a_1 r^n}{1 - r} \qquad n \in N$$

41. *Economics.* The government, through a subsidy program, distributes $1,000,000. If we assume that each individual or agency spends 0.8 of what is received, and 0.8 of this is spent, and so on, how much total increase in spending results from this government action? (Let $a_1 = \$800,000$.)

42. *Zeno's paradox.* Visualize a hypothetical 440-yard oval racetrack that has tapes stretched across the track at the halfway point and at each point that marks the halfway point of each remaining distance thereafter. A runner running around the track has to break the first tape before the second, the second before the third, and so on. From this point of view it appears that he will never finish the race. (This famous paradox is attributed to the Greek philosopher, Zeno, 495–435 B.C.) If we assume the runner runs at 440 yards/minute, the times between tape breakings form an infinite geometric progression. What is the sum of this progression?

Section 11-5 Additional Applications

This section includes additional applications involving progressions (mainly geometric) from many different fields. The problems are self-contained and require no previous knowledge of the subjects concerned.

Exercise 11-5

APPLICATIONS

Difficult problems are double-starred (★★), moderately difficult problems are single-starred (★), and the easier problems are not marked.

A calculator will be helpful in solving some of these problems.

Business and Economics

1. If you received $7,000 a year 11 years ago and now receive $14,000 a year, and if your salary has been increased the same amount

each year, what is that yearly increase and how much money have you received from the company over the 11 years?

2. Let us suppose the government has reduced taxes so that you have $600 more in spendable income. What is the net effect of this extra $600 on the economy? According to the "multiplier" doctrine in economics, the effect of the $600 is multiplied. Let us assume that you spend 0.7 of the $600 on consumer goods, that the producers of these goods in turn spend 0.7 of what they receive on consumer goods, and that this chain continues indefinitely, forming a geometric progression. What is the total amount spent on consumer goods if the process continues indefinitely? (Let $a_1 = \$420$.)

Earth Sciences

★3. If atmospheric pressure decreases (roughly) by a factor of 10 for each 10-mile increase in altitude up to 60 miles, and if the pressure is 15 pounds/square inch at sea level, what will the pressure be 40 miles up?

4. As dry air moves upward it expands, and in so doing cools at the rate of about 5°F for each 1,000-foot rise. This is known as the **adiabatic process**.

   (A) Temperatures at altitudes that are multiples of 1,000 form what kind of a sequence?

   (B) If the ground temperature is 80°F, write a formula for the temperature T_n in terms of n, if n is in thousands of feet.

Life Sciences—Ecology

5. A plant is eaten by an insect, an insect by a trout, a trout by a salmon, a salmon by a bear, and the bear is eaten by you. If only 20% of the energy is transformed from one stage to the next, how many calories must be supplied by plant food to provide you with 2,000 calories from the bear meat?

★6. If there are 30 years in a generation, how many direct ancestors did each of us have 600 years ago? (By *direct* ancestors we mean parents, grandparents, great-grandparents, and so on.)

★7. A single cholera bacterium divides every $\frac{1}{2}$ hour to produce two complete cholera bacteria. If we start with a colony of A_0 bacteria, in t hours (assuming adequate food supply) how many bacteria will we have?

★8. One leukemic cell injected into a healthy mouse will divide into two cells in about $\frac{1}{2}$ day; at the end of the day these two cells will divide again, with the doubling process continuing each half-day until there are 1 billion cells, at which time the mouse dies. On which day after the experiment is started does this happen?

11-5 Additional Applications

Astronomy ★★**9.** Ever since the time of the Greek astronomer Hipparchus (second century B.C.), the brightness of stars has been measured in terms of magnitude. The brightest stars (excluding the sun) are classed as magnitude 1, and the dimmest visible to the eye are classed as magnitude 6. In 1856, the English astronomer N. R. Pogson showed that first-magnitude stars are 100 times brighter than sixth-magnitude stars. If the ratio of brightness between consecutive magnitudes is constant, find this ratio. [*Hint:* If b_n is the brightness of an *n*th-magnitude star, find r for the geometric progression $b_1, b_2, b_3, \ldots$, given $b_1 = 100 b_6$.]

Music ★**10.** The notes on a piano, as measured in cycles per second, form a geometric progression.

(A) If A is 400 cycles per second and A′, 12 notes higher, is 800 cycles per second, find the constant ratio r.

(B) Find the cycles per second for C, three notes higher than A.

Geometry **11.** If the midpoints of the sides of an equilateral triangle are joined by straight lines, the new figure will be an equilateral triangle with a perimeter half the old. If we start with an equilateral triangle with perimeter 1, and form a sequence of "nested" equilateral triangles proceeding as above, what will be the total perimeter of all the triangles that can be formed in this way?

Photography **12.** The shutter speeds and f-stops on a camera are given as follows:

Shutter speeds: $1, \frac{1}{2}, \frac{1}{4}, \frac{1}{8}, \frac{1}{15}, \frac{1}{30}, \frac{1}{60}, \frac{1}{125}, \frac{1}{250}, \frac{1}{500}$

f-stops: 1.4, 2, 2.8, 4, 5.6, 8, 10.3, 16

These are very close to being geometric progressions. Estimate their common ratios.

Puzzles **13.** If you place 1¢ on the first square of a chessboard, 2¢ on the second square, 4¢ on the third, and so on, continuing to double the amount until all sixty-four squares are covered, how much money will be on the sixty-fourth square? How much money will there be on the whole board?

★**14.** If a sheet of very thin paper 0.001 inch thick is torn in half, and each half is again torn in half, and this process is repeated for a total of thirty-two times, how high will the stack of paper be if the pieces are placed one on top of the other? Give the answer to the nearest mile. (5,280 feet = 1 mile)

Section 11-6 Binomial Formula

- Factorial
- Binomial Formula

The binomial form

$$(a + b)^n$$

where n is a natural number, appears more frequently than you might expect. The coefficients in the expansion play an important role in probability studies. The binomial formula, which we will derive below, enables us to expand $(a + b)^n$ directly for n any natural number. Since the formula involves factorials, we digress for a moment to introduce this important concept.

- Factorial

For n a natural number, **n factorial**—denoted by $n!$—is the product of the first n natural numbers. **Zero factorial** is defined to be 1. Symbolically,

---

n Factorial

$n! = n(n - 1) \cdot \ldots \cdot 2 \cdot 1$

$1! = 1$

$0! = 1$

It is also useful to note that

$n! = n \cdot (n - 1)!$

EXAMPLE 17 (A) $4! = 4 \cdot 3! = 4 \cdot 3 \cdot 2! = 4 \cdot 3 \cdot 2 \cdot 1! = 4 \cdot 3 \cdot 2 \cdot 1 = 24$

(B) $5! = 5 \cdot 4 \cdot 3 \cdot 2 \cdot 1 = 120$

(C) $\dfrac{7!}{6!} = \dfrac{7 \cdot 6!}{6!} = 7$

(D) $\dfrac{8!}{5!} = \dfrac{8 \cdot 7 \cdot 6 \cdot 5!}{5!} = 336$

PROBLEM 17 Find: (A) 6! (B) 6!/5! (C) 9!/6!

The symbol $\binom{n}{r}$ is frequently used in probability studies and will be used by us shortly. It is called the **combinatorial symbol** and is defined for nonnegative r and n, as follows:

Combinatorial Symbol

For nonnegative integers r and n, $0 \leq r \leq n$,

$$\binom{n}{r} = \dfrac{n!}{r!(n-r)!}$$

$$= \dfrac{n(n-1)(n-2) \cdots (n-r+1)}{r(r-1) \cdots 2 \cdot 1}$$

EXAMPLE 18 (A) $\binom{8}{3} = \dfrac{8!}{3!(8-3)!} = \dfrac{8!}{3!5!} = \dfrac{8 \cdot 7 \cdot 6 \cdot 5!}{3 \cdot 2 \cdot 1 \cdot 5!} = 56$

(B) $\binom{7}{0} = \dfrac{7!}{0!(7-0)!} = \dfrac{7!}{7!} = 1$

PROBLEM 18 Find: (A) $\binom{9}{2}$ (B) $\binom{5}{5}$

■ **Binomial Formula**

We are now ready to try to discover a formula for the expansion of $(a + b)^n$ using ordinary induction; that is, we will look at a few special cases and try to postulate a general formula from them. If successful, we will try to prove that the formula holds for all natural numbers, using mathematical induction. To start, let us calculate directly the first five natural number powers of $(a + b)^n$:

$(a + b)^1 = a + b$
$(a + b)^2 = a^2 + 2ab + b^2$
$(a + b)^3 = a^3 + 3a^2b + 3ab^2 + b^3$
$(a + b)^4 = a^4 + 4a^3b + 6a^2b^2 + 4ab^3 + b^4$
$(a + b)^5 = a^5 + 5a^4b + 10a^3b^2 + 10a^2b^3 + 5ab^4 + b^5$

Observations

1. The expansion of $(a + b)^n$ has $(n + 1)$ terms.
2. The power of a decreases by 1 for each term as we move from left to right.
3. The power of b increases by 1 for each term as we move from left to right.
4. In each term the sum of the powers of a and b always adds up to n.
5. Starting with a given term, we can get the coefficient of the next term by multiplying the coefficient of the given term by the exponent of a and dividing by the number that represents the position of the term in the series of terms. For example, in the expansion of $(a + b)^4$, the coefficient of the third term is found from the second term by multiplying 4 and 3 and then dividing by 2 [that is, the coefficient of the third term = $(4 \cdot 3)/2 = 6$].

We now postulate the properties for the general case:

$$(a + b)^n = a^n + \frac{n}{1} a^{n-1}b + \frac{n(n-1)}{1 \cdot 2} a^{n-2}b^2$$

$$+ \frac{n(n-1)(n-2)}{1 \cdot 2 \cdot 3} a^{n-3}b^3 + \cdots + b^n$$

$$= \frac{n!}{0!(n-0)!} a^n + \frac{n!}{1!(n-1)!} a^{n-1}b + \frac{n!}{2!(n-2)!} a^{n-2}b^2$$

$$+ \frac{n!}{3!(n-3)!} a^{n-3}b^3 + \cdots + \frac{n!}{n!(n-n)!} b^n$$

$$= \binom{n}{0} a_n + \binom{n}{1} a^{n-1}b + \binom{n}{2} a^{n-2}b^2$$

$$+ \binom{n}{3} a^{n-3}b^3 + \cdots + \binom{n}{n} b^n$$

11-6 Binomial Formula

Thus, it appears that

Binomial Formula

$$(a + b)^n = \sum_{k=0}^{n} \binom{n}{k} a^{n-k} b^k \qquad n \geq 1$$

This result is known as the **binomial formula**, and we now proceed to prove that it holds for all natural numbers n.

Proof Write

$$P_n: \quad (a + b)^n = \sum_{j=0}^{n} \binom{n}{j} a^{n-j} b^j$$

$$S = \{n \in N \mid P_n \text{ is true}\}$$

Part 1. Show that $1 \in S$.

$$\sum_{j=0}^{1} \binom{1}{j} a^{1-j} b^j = \binom{1}{0} a + \binom{1}{1} b = a + b = (a + b)^1$$

$$\therefore 1 \in S$$

Part 2. Show that $k \in S \Rightarrow k + 1 \in S$.

$$P_k: \quad (a + b)^k = \sum_{j=0}^{k} \binom{k}{j} a^{k-j} b^j$$

$$P_{k+1}: \quad (a + b)^{k+1} = \sum_{j=0}^{k+1} \binom{k+1}{j} a^{k+1-j} b^j$$

Starting with P_k, we multiply both members by $a + b$ and try to obtain P_{k+1}:

$$(a + b)^k (a + b) = \left[\sum_{j=0}^{k} \binom{k}{j} a^{k-j} b^j \right] (a + b)$$

$$= \left[\binom{k}{0} a^k + \binom{k}{1} a^{k-1} b + \binom{k}{2} a^{k-2} b^2 + \cdots + \binom{k}{k} b^k \right] (a + b)$$

$$= \left[\binom{k}{0} a^{k+1} + \binom{k}{1} a^k b + \binom{k}{2} a^{k-1} b^2 + \cdots + \binom{k}{k} ab^k \right]$$

$$+ \left[\binom{k}{0} a^k b + \binom{k}{1} a^{k-1} b^2 + \cdots + \binom{k}{k-1} ab^k + \binom{k}{k} b^{k+1} \right]$$

$$(a+b)^k(a+b) = \binom{k}{0}a^{k+1} + \left[\binom{k}{0} + \binom{k}{1}\right]a^k b + \left[\binom{k}{1} + \binom{k}{2}\right]a^{k-1}b^2 + \cdots$$
$$+ \left[\binom{k}{k-1} + \binom{k}{k}\right]ab^k + \binom{k}{k}b^{k+1}$$

We now use the facts (the proofs left as exercises) that

$$\binom{k}{r-1} + \binom{k}{r} = \binom{k+1}{r} \qquad \binom{k}{0} = \binom{k+1}{0} \qquad \binom{k}{k} = \binom{k+1}{k+1}$$

to rewrite the right side as

$$\binom{k+1}{0}a^{k+1} + \binom{k+1}{1}a^k b + \binom{k+1}{2}a^{k-1}b^2 + \cdots + \binom{k+1}{k}ab^k + \binom{k+1}{k+1}b^{k+1}$$

$$= \sum_{j=0}^{k+1} \binom{k+1}{j} a^{k+1-j} b^j$$

Thus, $k \in S \Rightarrow k+1 \in S$, and S is inductive.

Conclusion. $S = N$

EXAMPLE 19 Use the binomial formula to expand $(x+y)^6$.

Solution
$$(x+y)^6 = \sum_{k=0}^{6} \binom{6}{k} x^{6-k} y^k$$
$$= \binom{6}{0}x^6 + \binom{6}{1}x^5 y + \binom{6}{2}x^4 y^2 + \binom{6}{3}x^3 y^3$$
$$+ \binom{6}{4}x^2 y^4 + \binom{6}{5}xy^5 + \binom{6}{6}y^6$$
$$= x^6 + 6x^5 y + 15x^4 y^2 + 20x^3 y^3 + 15x^2 y^4 + 6xy^5 + y^6$$

PROBLEM 19 Use the binomial formula to expand $(x+1)^5$.

EXAMPLE 20 Use the binomial formula to find the fourth term in the expansion of $(x-2)^{20}$.

Solution Fourth term $= \binom{20}{3} x^{17}(-2)^3$ In the expansion of $(a+b)^n$, the exponent of b in the rth term is $r-1$ and the exponent of a is $n-(r-1)$.

$$= \frac{20 \cdot 19 \cdot 18}{3 \cdot 2 \cdot 1} x^{17}(-8)$$
$$= -9,120 x^{17}$$

PROBLEM 20 Use the binomial formula to find the fifth term in the expansion of $(u-1)^{18}$.

11-6 Binomial Formula

Answers to Matched Problems

17. (A) 720 (B) 6 (C) 504
18. (A) 36 (B) 1
19. $x^5 + 5x^4 + 10x^3 + 10x^2 + 5x + 1$
20. $3{,}060u^{14}$

Exercise 11-6

A *Evaluate.*

1. $6!$
2. $4!$
3. $\dfrac{20!}{19!}$
4. $\dfrac{5!}{4!}$
5. $\dfrac{10!}{7!}$
6. $\dfrac{9!}{6!}$
7. $\dfrac{6!}{4!\,2!}$
8. $\dfrac{5!}{2!\,3!}$
9. $\dfrac{9!}{0!(9-0)!}$
10. $\dfrac{8!}{8!(8-8)!}$
11. $\dfrac{8!}{2!(8-2)!}$
12. $\dfrac{7!}{3!(7-3)!}$

Write as the quotient of two factorials.

13. 9
14. 12
15. $6 \cdot 7 \cdot 8$
16. $9 \cdot 10 \cdot 11 \cdot 12$

B *Evaluate.*

17. $\binom{9}{5}$
18. $\binom{5}{2}$
19. $\binom{6}{5}$
20. $\binom{7}{1}$
21. $\binom{9}{9}$
22. $\binom{5}{0}$
23. $\binom{17}{13}$
24. $\binom{20}{16}$

Expand, using the binomial formula.

25. $(m + n)^3$
26. $(x + 2)^3$
27. $(2x - 3y)^3$
28. $(3u + 2v)^3$
29. $(x - 2)^4$
30. $(x - y)^4$
31. $(m + 3n)^4$
32. $(3p - q)^4$
33. $(2x - y)^5$
34. $(2x - 1)^5$
35. $(m + 2n)^6$
36. $(2x - y)^6$

Find the indicated term in each expansion.

37. $(u + v)^{15}$; seventh term
38. $(a + b)^{12}$; fifth term
39. $(2m + n)^{12}$; eleventh term
40. $(x + 2y)^{20}$; third term
41. $[(w/2) - 2]^{12}$; seventh term
42. $(x - 3)^{10}$; fourth term
43. $(3x - 2y)^8$; sixth term
44. $(2p - 3q)^7$; fourth term

C 45. Evaluate $(1.01)^{10}$ to four decimal places, using the binomial formula.
[*Hint:* Let $1.01 = 1 + 0.01$.]

46. Evaluate $(0.99)^6$ to four decimal places, using the binomial formula.

47. Show that
$$\binom{n}{r} = \binom{n}{n-r}$$

48. Show that
$$\binom{n}{0} = \binom{n}{n}$$

49. Show that
$$\binom{k}{r-1} + \binom{k}{r} = \binom{k+1}{r}$$

50. Show that
$$\binom{k}{0} = \binom{k+1}{0}$$

51. Show that
$$\binom{k}{k} = \binom{k+1}{k+1}$$

52. Show that
$$\binom{n}{r}$$
is given by the recursion formula
$$\binom{n}{r} = \frac{n-r+1}{r}\binom{n}{r-1}$$
where $\binom{n}{0} = 1$

53. Write $2^n = (1+1)^n$ and expand, using the binomial formula, to obtain
$$2^n = \binom{n}{0} + \binom{n}{1} + \binom{n}{2} + \cdots + \binom{n}{n}$$

54. Can you guess what the next two rows in **Pascal's triangle** are? Compare the numbers in the triangle with the binomial coefficients obtained with the binomial formula.

$$\begin{array}{c}1\\1\ 1\\1\ 2\ 1\\1\ 3\ 3\ 1\\1\ 4\ 6\ 4\ 1\end{array}$$

Section 11-7 Chapter Review

IMPORTANT TERMS AND SYMBOLS

11-1 Sequences and Series. Sequence, terms of a sequence, finite sequence, infinite sequence, recursion formula, series, finite series, infinite series, summation notation, summing index

$$a_1, a_2, \ldots, a_n, \ldots \qquad \{a_n\}$$

$$a_1 + a_2 + \cdots + a_n \qquad \sum_{k=1}^{n} a_k$$

11-2 Mathematical Induction. Well-ordering principle, inductive sets, principle of mathematical induction

11-3 Arithmetic Sequences and Series. Arithmetic sequence, arithmetic progression, common difference, nth-term formula, finite arithmetic series, sum formulas

$$a_n = a_1 + (n-1)d$$

$$S_n = \frac{n}{2}[2a_1 + (n-1)d]$$

$$S_n = \frac{n}{2}(a_1 + a_n)$$

11-4 Geometric Sequences and Series. Geometric sequence, geometric progression, common ratio, nth-term formula, finite geometric series, infinite geometric series, sum formulas

$$a_n = a_1 r^{n-1}$$

$$S_n = \frac{a_1 - a_1 r^n}{1 - r} \qquad r \neq 1$$

$$S_n = \frac{a_1 - r a_n}{1 - r} \qquad r \neq 1$$

$$S_\infty = \frac{a_1}{1 - r} \qquad |r| < 1$$

11-5 Additional Applications

11-6 Binomial Formula. Factorial, combinatorial symbol, binomial formula

$$n! = n(n-1)(n-2) \cdots 1 \qquad \binom{n}{r} = \frac{n!}{r!(n-r)!}$$

$$(a+b)^n = \sum_{k=0}^{n} \binom{n}{k} a^{n-k} b^k$$

Exercise 11-7 Chapter Review

Work through all the problems in this chapter review and check answers in the back of the book. (Answers to all problems are there, and following each answer is a number in italics indicating the section in which that type of problem is discussed.) Where weaknesses show up, review appropriate sections in the text. When you are satisfied that you know the material, take the practice test following this review.

A 1. Determine whether the sequence is geometric, arithmetic, or neither.
   (A) $16, -8, 4, \ldots$ (B) $5, 7, 9, \ldots$
   (C) $-8, -5, -2, \ldots$ (D) $2, 3, 5, 8, \ldots$
   (E) $-1, 2, -4, \ldots$

(A) Write the first four terms of each sequence, (B) find a_{10}, and (C) find S_{10}.

2. $a_n = 2n + 3$
3. $a_n = 32(\tfrac{1}{2})^n$
4. $a_1 = -8; \quad a_n = a_{n-1} + 3, \quad n \geq 2$
5. $a_1 = -1; \quad a_n = (-2)a_{n-1}, \quad n \geq 2$
6. Find S_∞ in Problem 3.

Evaluate.

7. $6!$
8. $\dfrac{22!}{19!}$
9. $\dfrac{7!}{2!(7-2)!}$

Verify for $n = 1, 2,$ and 3.

10. $P_n: \quad 5 + 7 + 9 + \cdots + (2n + 3) = n^2 + 4n$
11. $P_n: \quad 2 + 4 + 8 + \cdots + 2^n = 2^{n+1} - 2$
12. $P_n: \quad 49^n - 1$ is divisible by 6

Write P_k and P_{k+1}.

13. For P_n in Problem 10
14. For P_n in Problem 11
15. For P_n in Problem 12

B Write without summation notation and find the sum.

16. $S_{10} = \sum_{k=1}^{10} (2k - 8)$
17. $S_7 = \sum_{k=1}^{7} \dfrac{16}{2^k}$

18. $S_\infty = 27 - 18 + 12 + \cdots = ?$

19. Write $S_n = \tfrac{1}{3} - \tfrac{1}{9} + \tfrac{1}{27} + \cdots + \dfrac{(-1)^{n+1}}{3^n}$ using summation notation, and find S_∞.

20. If in an arithmetic sequence $a_1 = 13$ and $a_7 = 31$, find the common difference d and the fifth term a_5.

21. Write $0.727\overline{272}$ as the quotient of two integers.

Evaluate.

22. $\dfrac{20!}{18!(20-18)!}$ **23.** $\binom{16}{12}$ **24.** $\binom{11}{11}$

25. Expand $(x - y)^5$ using the binomial formula.

26. Find the tenth term in the expansion of $(2x - y)^{12}$.

Establish each statement for all natural numbers, using mathematical induction.

27. P_n in Problem 10 **28.** P_n in Problem 11

29. P_n in Problem 12

C **30.** A free-falling body travels $g/2$ feet in the first second, $3g/2$ feet during the next second, $5g/2$ feet the next, and so on. Find the distance fallen during the twenty-fifth second, and the total distance fallen from the start to the end of the twenty-fifth second.

31. Expand $(x + i)^6$, i the complex unit, using the binomial formula.

Prove that each of the following statements holds for all positive integers, using mathematical induction.

32. $\sum_{k=1}^{n} k^3 = \left(\sum_{k=1}^{n} k \right)^2$

33. $x^{2n} - y^{2n}$ is divisible by $x - y$, $x \neq y$

34. $\dfrac{a^n}{a^m} = a^{n-m}$; $n > m$, $n, m \in N$

35. $\{a_n\} = \{b_n\}$ where $a_n = a_{n-1} + 2$, $a_1 = -3$, $b_n = -5 + 2n$

36. $(1!)1 + (2!)2 + (3!)3 + \cdots + (n!)n = (n + 1)! - 1$ (From USSR Mathematical Olympiad, 1955–1956, Grade 10.)

Practice Test Chapter 11

Take this practice test as if it were a graded test. Allow yourself up to 50 minutes. Work the problems without looking back in the chapter. Correct your work using the answers (keyed to appropriate sections) in the back of the book.

1. Determine whether the sequence is geometric, arithmetic, or neither. If geometric, find the common ratio r; if arithmetic, find the common difference d.

(A) $2, 5, 7, \ldots$ (B) $8, -2, \frac{1}{2}, \ldots$ (C) $7, 3, -1, \ldots$

2. Given the sequence $a_1 = 64$, $a_n = a_{n-1} - 4$, $n \geq 2$, find the first four terms, a_{51}, and S_{31}.

3. Given the sequence $\{64/2^n\}$, find the first four terms, a_{10}, and S_{10}.

4. Write $\sum_{k=1}^{4} k^2 x^{3k-1}$ without summation notation.

5. Write $\frac{2}{3} - \frac{4}{9} + \frac{8}{27} - \frac{16}{81}$ using summation notation starting the summing index at $k = 1$.

6. Write $0.018\overline{018}$ as the quotient of integers using the sum of an appropriate geometric series.

7. Use mathematical induction to prove the following proposition for all positive integers:

$$3 + 3^2 + 3^3 + \cdots + 3^n = \frac{3^{n+1} - 3}{2}$$

8. Evaluate. (A) $\dfrac{15!}{3!12!}$ (B) $\binom{23}{21}$

9. Expand using the binomial formula: $(3x - y)^4$

10. Write the sixth term in the expansion of $(x - 2)^9$.

Appendixes

Appendix A-1 Significant Digits

Most calculations involving problems of the real world deal with figures that are only approximate. It would therefore seem reasonable to assume that a final answer could not be any more accurate than the least accurate figure used in the calculation. This is an important point, since calculators tend to give the impression that greater accuracy is achieved than is warranted.

Suppose we wish to compute the length of the diagonal of a rectangular field from measurements of its sides of 237.8 meters and 61.3 meters. Using the Pythagorean theorem and a calculator, we find

$$d = \sqrt{237.8^2 + 61.3^2}$$
$$= 245.573\ 878\ldots$$

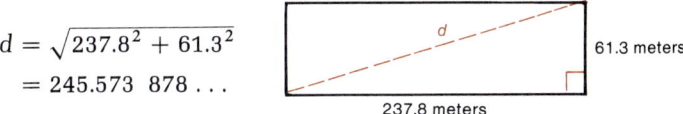

The calculator answer suggests an accuracy that is not justified. What accuracy is justified? To answer this question, we introduce the idea of **significant digits**.

The measurement 61.3 meters indicates that the measurement was made to the nearest tenth of a meter; that is, the actual width is between 61.25 and 61.35 meters. The number 61.3 has three significant digits. If we had written, instead, 61.30 meters as the width, then the actual width would be between 61.295 and 61.305 meters, and our measurement, 61.30 meters, would have four significant digits.

> The number of significant digits in a number is found by counting the digits from left to right, starting with the first nonzero digit and ending with the last digit present.

The significant digits in the following numbers are underlined:

<u>719.37</u>   <u>82,395</u>   <u>5.600</u>   0.000 <u>830</u>   0.000 0<u>8</u>

The definition takes care of all cases except one. Consider, for example, the number 7,800. It is not clear whether the number has been rounded to the hundreds place, the tens place, or the units place. This ambiguity can be resolved by writing this type of number in scientific notation.

A1

Thus,

$$7.8 \times 10^3 \quad \text{has two significant digits}$$
$$7.80 \times 10^3 \quad \text{has three significant digits}$$
$$7.800 \times 10^3 \quad \text{has four significant digits}$$

All three are equal to 7,800 when written without powers of 10.

In calculations involving multiplication, division, powers, and roots, we adopt the following convention:

We will round off the answer to match the number of significant digits in the number with the least number of significant digits used in the calculation.

Thus, in computing the length of the diagonal of the field, we would write the answer to three significant digits because the width, the least accurate of the two numbers involved, has three significant digits:

$$d = 246 \text{ meters} \quad \text{Three significant digits}$$

One final note: In rounding a number that is exactly halfway between a larger and a smaller number, we will use the convention of making the final result even.

EXAMPLE 1 Round each number to three significant digits.

(A) 43.0690   (B) 48.05   (C) 48.15   (D) $8.017\ 632 \times 10^{-3}$

Solution
(A) 43.1
(B) 48.0 ⎫ Use the convention of making the digit before the
(C) 48.2 ⎭ 5 even if it is odd, or leaving it alone if it is even.
(D) 8.02×10^{-3}

PROBLEM 1 Round each number to three significant digits:

(A) 3.1495   (B) 0.004 135   (C) 32,450   (D) $4.314\ 764\ 09 \times 10^{12}$

Answers to Matched Problem   1. (A) 3.15   (B) 0.004 14   (C) 32,400   (D) 4.31×10^{12}

Appendix A-2 Integer Exponents

- Integer Exponents
- Scientific Notation

- **Integer Exponents**

Table 1 summarizes the definition of the exponent form a^n, where the **exponent** n is an integer and the **base** a is a real number. Table 2 summarizes the properties of exponents.

Appendix A-2 Integer Exponents

TABLE 1 Definition of a^n, n an Integer and a Real

1. For n a positive integer:

$$a^n = a \cdot a \cdot \ldots \cdot a \qquad\qquad 3^5 = 3 \cdot 3 \cdot 3 \cdot 3 \cdot 3$$

n factors of a

2. For $n = 0$:

$$a^0 = 1 \qquad a \neq 0$$
$$0^0 \text{ is not defined} \qquad\qquad 132^0 = 1$$

3. For n a negative integer:

$$a^n = \frac{1}{a^{-n}} \qquad a \neq 0 \qquad\qquad 7^{-3} \boxed{= \frac{1}{7^{-(-3)}}} = \frac{1}{7^3}$$

[*Note*: In general, it can be shown that

$$a^{-n} = \frac{1}{a^n} \qquad\qquad a^{-5} = \frac{1}{a^5}; \quad a^{-(-3)} = \frac{1}{a^{-3}}$$

for *all* integers n.]

TABLE 2 Properties of Exponents

For n and m integers and a and b real numbers:

1. $a^m a^n = a^{m+n}$ $\qquad\qquad a^5 a^{-7} \boxed{= a^{5+(-7)}} = a^{-2}$

2. $(a^n)^m = a^{mn}$ $\qquad\qquad (a^3)^{-2} \boxed{= a^{(-2)3}} = a^{-6}$

3. $(ab)^m = a^m b^m$ $\qquad\qquad (ab)^3 = a^3 b^3$

4. $\left(\dfrac{a}{b}\right)^m = \dfrac{a^m}{b^m} \qquad b \neq 0 \qquad \left(\dfrac{a}{b}\right)^4 = \dfrac{a^4}{b^4}$

5. $\dfrac{a^m}{a^n} = \begin{cases} a^{m-n} \\ \dfrac{1}{a^{n-m}} \end{cases} \qquad a \neq 0$

$\dfrac{a^3}{a^{-2}} = a^{3-(-2)} = a^5$

$\dfrac{a^3}{a^{-2}} = \dfrac{1}{a^{-2-3}} = \dfrac{1}{a^{-5}}$

Many students are reasonably satisfied when we define 2^4 to be $2 \cdot 2 \cdot 2 \cdot 2$ but find it harder to accept 2^0 defined as 1 and 2^{-3} defined as $1/2^3$. If we want all five laws of exponents to continue to hold even if some of the exponents are 0 or negative integers, we do not have much choice as to how 2^0 or 2^{-3} must be defined. For example, if the exponent laws are to hold for *all* integer exponents, then, in particular,

$$2^0 \cdot 2^3 = 2^{0+3} = 2^3$$

Thus, 2^0 must be 1, the multiplicative identity. (The number 1 is the only real number that has the property $1 \cdot a = a$ for all real numbers a.)

What about 0^0? Proceeding as above,

$$0^0 \cdot 0^2 = 0^{0+2} = 0^2 = 0 \cdot 0 = 0$$

Thus, 0^0 could be any real number (since $0^2 = 0$ and the product of 0 and *any* number is 0); hence, 0^0 is not uniquely determined. For this reason we choose not to define the symbol 0^0.

Now let us turn to 2^{-3}. Again assuming the exponent laws hold for *all* integer exponents, then, in particular,

$$2^{-3} \cdot 2^3 = 2^{-3+3} = 2^0 = 1$$

Thus, 2^{-3} must be the multiplicative inverse (reciprocal) of 2^3; that is, 2^{-3} must be $1/2^3$.

This discussion is included to show you that the definitions of integer exponents on page A3 are reasonable. We have not given proofs of the definitions—definitions do not require proofs.

In all examples and discussion that follows we assume without statement that all variables are restricted to avoid division by 0.

EXAMPLE 1 (A) $(u^3 v^2)^0 = 1 \quad u \neq 0, \quad v \neq 0$ (B) $10^{-3} = \dfrac{1}{10^3} = \dfrac{1}{1{,}000} = 0.001$

(C) $x^{-8} = \dfrac{1}{x^8}$ (D) $\dfrac{x^{-3}}{y^{-5}} = \dfrac{x^{-3}}{1} \cdot \dfrac{1}{y^{-5}} = \dfrac{1}{x^3} \cdot \dfrac{y^5}{1} = \dfrac{y^5}{x^3}$

PROBLEM 1 Write (A) to (D) as decimal fractions and (E) and (F) with positive exponents.

(A) 635^0 (B) $(x^2)^0, \quad x \neq 0$ (C) 10^{-5}
(D) $1/10^{-3}$ (E) $1/x^{-4}$ (F) u^{-7}/v^{-3}

From the definition of negative exponents and the five laws of exponents, we can easily establish the following properties that are used very frequently when dealing with exponent forms.

Further Exponent Properties

For a and b any real numbers and m, n, and p any integers, then (division by 0 excluded)

1. $(a^m b^n)^p = a^{pm} b^{pn}$
2. $\left(\dfrac{a^m}{b^n}\right)^p = \dfrac{a^{pm}}{b^{pn}}$

3. $\dfrac{a^{-n}}{b^{-m}} = \dfrac{b^m}{a^n}$
4. $\left(\dfrac{a}{b}\right)^{-n} = \left(\dfrac{b}{a}\right)^n$

Appendix A-2 Integer Exponents

We prove properties 1 and 4 and leave 2 and 3 to the reader.

1. $(a^m b^n)^p = (a^m)^p (b^n)^p$ Exponent law 3
$ = a^{pm} b^{pn}$ Exponent law 2

4. $\left(\dfrac{a}{b}\right)^{-n} = \dfrac{a^{-n}}{b^{-n}}$ Exponent law 4

$\phantom{\left(\dfrac{a}{b}\right)^{-n}} = \dfrac{b^n}{a^n}$ Property 3

$\phantom{\left(\dfrac{a}{b}\right)^{-n}} = \left(\dfrac{b}{a}\right)^n$ Exponent law 4

EXAMPLE 2 Simplify using exponent properties, and express answers using positive exponents only.*

(A) $(3a^5)(2a^{-3}) \;\vert\; = (3 \cdot 2)(a^5 a^{-3}) \;\vert\; = 6a^2$

(B) $\dfrac{6x^{-2}}{8x^{-5}} = \dfrac{3x^{-2-(-5)}}{4} = \dfrac{3x^3}{4}$

(C) $(2a^{-3}b^2)^{-2} = 2^{-2} a^6 b^{-4} = \dfrac{a^6}{4b^4}$

(D) $\left(\dfrac{a^3}{b^5}\right)^{-2} = \left(\dfrac{b^5}{a^3}\right)^2 = \dfrac{b^{10}}{a^6}$

[Note: In simplifying exponent forms there is often more than one sequence of steps that will lead to the same result. For example, in Example 2D we can also proceed as follows:

$$\left(\dfrac{a^3}{b^5}\right)^{-2} = \dfrac{a^{-6}}{b^{-10}} = \dfrac{b^{10}}{a^6}$$

Use whichever sequence of steps makes sense to you; however, do not create new rules unless they can be justified.]

---

* By "simplify" we mean eliminate common factors from numerators and denominators and reduce to a minimum the number of times a given constant or variable appears in an expression. We ask that answers be expressed using positive exponents only in order to have a definite form for an answer. Elsewhere we will encounter situations where we will want negative exponents in a final answer.

PROBLEM 2 Simplify using exponent properties, and express answers using positive exponents only.

(A) $(5x^{-3})(3x^4)$ (B) $\dfrac{9y^{-7}}{6y^{-4}}$

(C) $(3x^4y^{-3})^{-2}$ (D) $\left(\dfrac{x^2}{y^4}\right)^{-3}$

EXAMPLE 3 Simplify using exponent properties, and express answers using positive exponents only.

(A) $\dfrac{4x^{-3}y^{-5}}{6x^{-4}y^3} = \dfrac{2x^{-3-(-4)}}{3y^{3-(-5)}} = \dfrac{2x}{3y^8}$

(B) $\left(\dfrac{m^{-3}m^3}{n^{-2}}\right)^{-2} = \left(\dfrac{m^{-3+3}}{n^{-2}}\right)^{-2} = \left(\dfrac{m^0}{n^{-2}}\right)^{-2} = \left(\dfrac{1}{n^{-2}}\right)^{-2} = \dfrac{1}{n^4}$

(C) $\left[\left(\dfrac{12x^{-3}}{8y^4}\right)^{-1}\right]^{-2} = \left(\dfrac{3x^{-3}}{2y^4}\right)^2 = \dfrac{9x^{-6}}{4y^8} = \dfrac{9}{4x^6y^8}$

(D) $(x+y)^{-3} = \dfrac{1}{(x+y)^3}$ Common error:
$(x+y)^{-3} = x^{-3} + y^{-3}$
$= \dfrac{1}{x^3} + \dfrac{1}{y^3}$

[Note: Exponent laws deal primarily with products and quotients. Some students cannot resist making up their own new "laws" (usually invalid) for sums and differences (see Example 3D).]

PROBLEM 3 Simplify using exponent properties, and express answers using positive exponents only.

(A) $\dfrac{6m^{-2}n^3}{15m^{-1}n^{-2}}$ (B) $\left(\dfrac{x^{-3}}{y^4y^{-4}}\right)^{-3}$

(C) $\left[\left(\dfrac{4a^2}{2b^5}\right)^2\right]^{-1}$ (D) $\dfrac{1}{(a-b)^{-2}}$

■ **Scientific Notation**

Scientific work often involves the use of very large numbers or very small numbers. For example, the average cell contains about 200,000,000,000,000 molecules, and the diameter of an electron is about 0.000 000 000 0004 centimeter. It is generally troublesome to write and work with numbers of this type in standard decimal form. The

Appendix A-2 Integer Exponents

two numbers written here cannot even be entered into most hand calculators as they are written. With exponents now defined for all integers, however, it is possible to express any decimal fraction as the product of a number between 1 and 10 and an integer power of 10— that is, in the form

$$a \times 10^n \qquad 1 \leq a < 10, \quad n \text{ an integer}$$

A number expressed in this form is said to be in **scientific notation**.

EXAMPLE 4

$$7 = 7 \times 10^0 \qquad\qquad 0.5 = 5 \times 10^{-1}$$
$$720 = 7.2 \times 10^2 \qquad\qquad 0.08 = 8 \times 10^{-2}$$
$$6,430 = 6.43 \times 10^3 \qquad\qquad 0.000\ 32 = 3.2 \times 10^{-4}$$
$$5,350,000 = 5.35 \times 10^6 \qquad 0.000\ 000\ 0738 = 7.38 \times 10^{-8}$$

Can you discover a rule relating the number of decimal places that the decimal is moved to the power of 10 that is used?

$$7{,}320{,}000 \quad = 7{,}320\ 000. \times 10^6 \quad = 7.32 \times 10^6$$
6 places left
positive exponent

$$0.000\ 000\ 54 \quad = 0.000\ 000\ 5.4 \times 10^{-7} \quad = 5.4 \times 10^{-7}$$
7 places right
negative exponent

PROBLEM 4 (A) Write each number in scientific notation: 430; 23,000; 345,000,000; 0.3; 0.0031; 0.000 000 683
(B) Write as decimal fractions: 4×10^3, 5.3×10^5, 2.53×10^{-2}, 7.42×10^{-6}

Most scientific and business hand calculators express very large and very small numbers in scientific notation. Read the instruction manual for your calculator to see how numbers in scientific notation are entered into your calculator. Numbers in scientific notation are displayed in most calculators as follows.

CALCULATOR DISPLAY   NUMBER REPRESENTED

| 5.427493 -17 | $5.427\ 493 \times 10^{-17}$

| 2.359779   12 | $2.359\ 779 \times 10^{12}$

Appendixes

EXAMPLE 5 Write each number that cannot be entered directly into your calculator in decimal form in scientific notation; then carry out the computation using your calculator. (Refer to the user's manual accompanying your calculator for the procedure.) Express the answer to three significant digits* in scientific notation.

$$\frac{(25.32)(325{,}100{,}000{,}000)}{(0.08)(0.000\ 000\ 000\ 000\ 0871)} = \frac{(25.32)(3.251 \times 10^{11})}{(0.08)(8.71 \times 10^{-14})}$$

$$= \boxed{1.181333\ \ 27} \quad \text{Calculator display}$$

$$= 1.18 \times 10^{27} \quad \text{To three significant digits}$$

CALCULATOR OPERATIONS

A: $\boxed{25.32}\ \boxed{\times}\ \boxed{3.251}\ \boxed{\text{EE}}\ \boxed{11}\ \boxed{=}\ \boxed{\div}\ \boxed{0.08}\ \boxed{\div}\ \boxed{8.71}\ \boxed{\text{EE}}\ \boxed{14}\ \boxed{+/-}\ \boxed{=}$

P: $\boxed{25.32}\ \boxed{\text{ENTER}}\ \boxed{3.251}\ \boxed{\text{EE}}\ \boxed{11}\ \boxed{\times}\ \boxed{0.08}\ \boxed{\div}\ \boxed{8.71}\ \boxed{\text{EE}}\ \boxed{14}\ \boxed{+/-}\ \boxed{\div}$

PROBLEM 5 Repeat Example 5 for

$$\frac{(0.371)(0.000\ 000\ 006\ 932)}{(532)(62{,}600{,}000{,}000)}$$

Answers to Matched Problems
1. (A) 1 (B) 1 (C) 0.000 01 (D) 1,000 (E) x^4 (F) v^3/u^7
2. (A) $15x$ (B) $3/2y^3$ (C) $y^6/9x^8$ (D) y^{12}/x^6
3. (A) $2n^5/5m$ (B) x^9 (C) $b^{10}/4a^4$ (D) $(a-b)^2$
4. (A) 4.3×10^2, 2.3×10^4, 3.45×10^8, 3×10^{-1}, 3.1×10^{-3}, 6.83×10^{-7}
   (B) 4,000; 530,000; 0.0253; 0.000 007 42
5. 7.72×10^{-23}

Exercise A-2 ■ A

Simplify Problems 1–16 and write the answers using positive exponents only.

1. $y^{-5}y^5$
2. $x^3 x^{-3}$
3. $(2x^2)(3x^3)(x^4)$
4. $(2x^5)(3x^7)(4x^2)$
5. $(3x^3 y^{-2})^2$
6. $(2cd^2)^{-3}$
7. $\left(\dfrac{ab^3}{c^2 d}\right)^4$
8. $\left(\dfrac{x^2 y}{2w^2}\right)^3$
9. $\dfrac{10^{23} \cdot 10^{-11}}{10^{-3} \cdot 10^{-2}}$

* For those not familiar with the meaning of "significant digits," see Appendix A-1 for a brief discussion of this concept.

Appendix A-2 Integer Exponents

10. $\dfrac{10^{-13} \cdot 10^{-4}}{10^{-21} \cdot 10^{3}}$

11. $\dfrac{4x^{-2}y^{-3}}{2x^{-3}y^{-1}}$

12. $\dfrac{2a^6 b^{-2}}{16a^{-3}b^2}$

13. $\left(\dfrac{n^{-3}}{n^{-2}}\right)^{-2}$

14. $\left(\dfrac{x^{-1}}{x^{-8}}\right)^{-1}$

15. $\dfrac{8 \times 10^3}{2 \times 10^{-5}}$

16. $\dfrac{18 \times 10^{12}}{6 \times 10^{-4}}$

Write the numbers in Problems 17–22 in scientific notation.

17. 32,250,000
18. 4,930
19. 0.085
20. 0.017
21. 0.000 000 0729
22. 0.000 592

In Problems 23–28 write each number as a decimal fraction.

23. 5×10^{-3}
24. 4×10^{-4}
25. 2.69×10^{7}
26. 6.5×10^{9}
27. 5.9×10^{-10}
28. 6.3×10^{-6}

B Simplify Problems 29–44 and write answers using positive exponents only.

29. $\dfrac{27x^{-5}x^5}{18y^{-6}y^2}$

30. $\dfrac{32n^5 n^{-8}}{24m^{-7}m^7}$

31. $\left(\dfrac{x^4 y^{-1}}{x^{-2}y^3}\right)^2$

32. $\left(\dfrac{m^{-2}n^3}{m^4 n^{-1}}\right)^2$

33. $\left(\dfrac{2x^{-3}y^2}{4xy^{-1}}\right)^{-2}$

34. $\left(\dfrac{6mn^{-2}}{3m^{-1}n^2}\right)^{-3}$

35. $\left[\left(\dfrac{u^3 v^{-1} w^{-2}}{u^{-2}v^{-2}w}\right)^{-2}\right]^2$

36. $\left[\left(\dfrac{x^{-2}y^3 t}{x^{-3}y^{-2}t^2}\right)^2\right]^{-1}$

37. $\left(\dfrac{3^3 x^0 y^{-2}}{2^3 x^3 y^{-5}}\right)^{-1}\left(\dfrac{3^3 x^{-1} y}{2^2 x^2 y^{-2}}\right)^2$

38. $\left(\dfrac{2^2 x^2 y^0}{8x^{-1}}\right)^{-2}\left(\dfrac{x^{-3}}{x^{-5}}\right)^3$

39. $(x + y)^{-2}$

40. $(a^2 - b^2)^{-1}$

C 41. $\dfrac{12(a + 2b)^{-3}}{6(a + 2b)^{-8}}$

42. $\dfrac{4(x - 3)^{-4}}{8(x - 3)^{-2}}$

43. $\dfrac{5(u - v + w)^8}{(u - v + w)^{11}}$

44. $\dfrac{3(x + y)^3 (x - y)^4}{(x - y)^2 6(x + y)^5}$

APPLICATIONS

45. *Earth science.* If the mass of the earth is approximately 6.1×10^{27} grams and each gram is 2.2×10^{-3} pound, what is the mass of the earth in pounds?

46. *Biology.* In 1929 Vernadsky, a biologist, estimated that all the free oxygen of the earth weighs 1.5×10^{21} grams and that it is

produced by life alone. If 1 gram is approximately 2.2×10^{-3} pound, what is the weight of the free oxygen in pounds?

47. *Computer science.* Today's fastest computers can perform a single addition in 10^{-8} second, and the next generation of computers working with tiny superconducting devices are expected to be able to perform a single addition in 10^{-10} second. How many additions will both types of computers be able to perform in 1 second? In 1 minute?

48. *Computer science.* If electricity travels in a computer circuit at the speed of light (1.86×10^5 miles/second), how far will electricity travel in the superconducting computer (see Problem 47) in the time it takes it to perform one addition? (Size of circuits is a critical problem in computer design.) Give the answer in miles, feet, and inches (1 mile = 5,280 feet). Compute answers to two significant digits.

CALCULATOR PROBLEMS

Evaluate Problems 49–52 to three significant digits using scientific notation where appropriate and a calculator.

49. $\dfrac{(32.7)(0.000\ 000\ 008\ 42)}{(0.0513)(80,700,000,000)}$

50. $\dfrac{(4,320)(0.000\ 000\ 000\ 704)}{(835)(635,000,000,000)}$

51. $\dfrac{(5,760,000,000)}{(527)(0.000\ 007\ 09)}$

52. $\dfrac{0.000\ 000\ 007\ 23}{(0.0933)(43,700,000,000)}$

Use y^x or a comparable key on your calculator to evaluate each of the following problems to five significant digits. (Read the instruction book accompanying your calculator.)

53. $(23.8)^8$

54. $(-302)^7$

55. $(-302)^{-7}$

56. $(23.8)^{-8}$

57. $(9,820,000,000)^3$

58. $(0.000\ 000\ 000\ 482)^{-4}$

Appendix A-3 Rational Exponents

- Roots of Real Numbers
- Rational Exponents

- Roots of Real Numbers

A **square root** of a number b is a number r such that $r^2 = b$, and a **cube root** of a number b is a number r such that $r^3 = b$.

Appendix A-3 Rational Exponents

2 is a square root of 4, since $2^2 = 4$.

-2 is a square root of 4, since $(-2)^2 = 4$.

2 is a cube root of 8, since $2^3 = 8$.

-2 is a cube root of -8, since $(-2)^3 = 8$.

In general, for n a natural number,

> r is an **nth root** of b if $r^n = b$

How many real square roots of 16 exist? Of 7? Of -4? How many real fourth roots of 7 exist? Of -7? How many real cube roots of -8 exist? Of 11? Theorem 1 (which we state without proof) answers these questions completely.

THEOREM 1

Number of Real nth Roots of a Real Number b

	n even	n odd
b positive	Two real nth roots -2 and 2 are both 4th roots of 16	One real nth root 2 is the only real cube root of 8
b negative	No real nth root -4 has no real square roots	One real nth root -2 is the only real cube root of -8

Thus:

7 has two real square roots, two real 4th roots, and so on.

10 has one real cube root, one real 5th root, and so on.

-13 has one real cube root, one real 5th root, and so on.

-8 has no real square roots, no real 4th roots, and so on.

What symbols do we use to represent the various kinds of real nth roots? We turn to this question now.

■ **Rational Exponents**

If all exponent laws are to continue to hold even if some of the exponents are not integers, then, in particular,

$$(6^{1/2})^2 = 6^{2/2} = 6 \quad \text{and} \quad (5^{1/3})^3 = 5^{3/3} = 5$$

Hence, $6^{1/2}$ must name a square root of 6, since $(6^{1/2})^2 = 6$. Similarly, $5^{1/3}$ must name a cube root of 5, since $(5^{1/3})^3 = 5$.

In general, for n a positive integer and b not negative when n is even,

$$(b^{1/n})^n = b^{n/n} = b$$

Thus, $b^{1/n}$ must name an nth root of b. Which of the two real nth roots of b does $b^{1/n}$ represent if n is even and b is positive? We answer this question in the definition of $b^{1/n}$.

Definition of $b^{1/n}$

For n a positive integer,

$b^{1/n}$ is an nth root of b

If n is even and b is positive, then $b^{1/n}$ represents the positive real nth root of b (sometimes called the **principal nth root** of b), $-b^{1/n}$ represents the negative real nth root of b, and $(-b)^{1/n}$ does not represent a real number. If n is odd and b is either positive or negative, then $b^{1/n}$ represents the real nth root of b. $0^{1/n} = 0$ for all positive integers n.

$25^{1/2} = 5$ $-25^{1/2} = -5$ $(-25)^{1/2}$ is not real
           $[-25^{1/2}$ and $(-25)^{1/2}$
           are not the same$]$
$32^{1/5} = 2$ $(-32)^{1/5} = -2$ $0^{1/4} = 0$

EXAMPLE 1   (A) $4^{1/2} = 2$ (B) $-4^{1/2} = -2$ (C) $(-4)^{1/2}$ is not a real number
(D) $(-8)^{1/3} = -2$ (E) $0^{1/5} = 0$

PROBLEM 1   Find integer representations of each of the following, if they exist.

(A) $9^{1/2}$ (B) $-9^{1/2}$ (C) $(-9)^{1/2}$
(D) $27^{1/3}$ (E) $(-27)^{1/3}$ (F) $0^{1/4}$

How should a symbol such as $7^{2/3}$ be defined? If the properties of exponents are to hold for rational exponents, then $7^{2/3} = (7^{1/3})^2$; that is, $7^{2/3}$ must represent the square of the cube root of 7. We are thus led to the general definition of $b^{m/n}$ and $b^{-m/n}$.

Appendix A-3 Rational Exponents

> **Definition of $b^{m/n}$ and $b^{-m/n}$**
>
> For m and n positive integers and b any real number, except b cannot be negative when n is even,
>
> $$b^{m/n} = (b^{1/n})^m \quad \text{and} \quad b^{-m/n} = \frac{1}{b^{m/n}}$$
>
> $4^{3/2} = (4^{1/2})^3 = 2^3 = 8 \qquad (-32)^{3/5} = [(-32)^{1/5}]^3 = (-2)^3 = -8$
>
> $4^{-3/2} = \frac{1}{4^{3/2}} = \frac{1}{8}$
>
> $(-4)^{3/2}$ is not a real number

We have discussed $b^{m/n}$ for all rational numbers m/n and real numbers b. It can be shown that all five laws of exponents discussed in Appendix A-2 continue to hold for rational number exponents as long as we avoid even roots of negative numbers. With the latter restriction in effect, the following useful relationship is an immediate consequence of the exponent properties:

> $$b^{m/n} = (b^{1/n})^m = (b^m)^{1/n}$$

To see why we require b to be positive when n is even, consider what happens when we relax that restriction. Can you resolve the following contradiction?

$$-1 = (-1)^{2/2} = [(-1)^2]^{1/2} = 1^{1/2} = 1$$

The second member of the equality chain, $(-1)^{2/2}$, involves an even root of a negative number, which is not real. Thus, we see that the properties of exponents do not necessarily hold when we are dealing with nonreal quantities unless further restrictions are imposed. One such restriction is to require all rational exponents be reduced to lowest terms.

EXAMPLE 2 Assume all letters represent positive real numbers.

(A) $8^{2/3} = (8^{1/3})^2 = 2^2 = 4 \quad \text{or} \quad 8^{2/3} = (8^2)^{1/3} = 64^{1/3} = 4$

(B) $(-8)^{5/3} = [(-8)^{1/3}]^5 = (-2)^5 = -32$

(C) $(3x^{1/3})(2x^{1/2}) = 6x^{1/3 + 1/2} = 6x^{5/6}$

(D) $\left(\dfrac{4x^{1/3}}{x^{1/2}}\right)^{1/2} = \dfrac{4^{1/2}x^{1/6}}{x^{1/4}} = \dfrac{2}{x^{1/4-1/6}} = \dfrac{2}{x^{1/12}}$

(E) $(x+2)^{3/4}(x+2)^{2/3} = (x+2)^{3/4+2/3} = (x+2)^{17/12}$

PROBLEM 2 Simplify, and express answers using positive exponents only. All letters represent positive real numbers.

(A) $9^{3/2}$ (B) $(-27)^{4/3}$
(C) $(5y^{3/4})(2y^{1/3})$ (D) $(2x^{-3/4}y^{1/4})^4$
(E) $\left(\dfrac{8x^{1/2}}{x^{2/3}}\right)^{1/3}$ (F) $\dfrac{6(a+b)^{2/3}}{8(a+b)^{1/2}}$

EXAMPLE 3 Evaluate to four significant digits using a hand calculator. (Refer to the instruction book for your particular calculator to see how exponential forms are evaluated.)

(A) $11^{3/4}$ (B) $3.1046^{-2/3}$ (C) $(0.000\ 000\ 008\ 437)^{3/11}$

Solution (A) First change $\tfrac{3}{4}$ to the decimal fraction 0.75; then evaluate $11^{0.75}$ using y^x (or a comparable key) on your calculator.

$11^{3/4} = 6.040$

A: $\boxed{11}\,\boxed{y^x}\,\boxed{(}\,\boxed{3}\,\boxed{\div}\,\boxed{4}\,\boxed{)}\,\boxed{=}$

P: $\boxed{11}\,\boxed{\text{ENTER}}\,\boxed{3}\,\boxed{\text{ENTER}}\,\boxed{4}\,\boxed{\div}\,\boxed{y^x}$

(B) $3.1046^{-2/3} = 0.4699$

A: $\boxed{3.1046}\,\boxed{y^x}\,\boxed{(}\,\boxed{2}\,\boxed{\div}\,\boxed{3}\,\boxed{+/-}\,\boxed{)}\,\boxed{=}$

P: $\boxed{3.1046}\,\boxed{\text{ENTER}}\,\boxed{2}\,\boxed{\text{ENTER}}\,\boxed{3}\,\boxed{\div}\,\boxed{+/-}\,\boxed{y^x}$

(C) $(0.000\ 000\ 008\ 437)^{3/11} = (8.437 \times 10^{-9})^{3/11}$
$= 0.006\ 281$

PROBLEM 3 Repeat Example 3 for:
(A) $2^{3/8}$ (B) $57.28^{-5/6}$ (C) $(83{,}240{,}000{,}000)^{5/3}$

Answers to Matched Problems

1. (A) 3 (B) -3 (C) Not a real number (D) 3
   (E) -3 (F) 0
2. (A) 27 (B) 81 (C) $10y^{13/12}$ (D) $16y/x^3$
   (E) $2/x^{1/18}$ (F) $[3(a+b)^{1/6}]/4$
3. (A) 1.297 (B) 0.034 28 (C) 1.587×10^{18}

Exercise A-3

All letters represent positive real numbers unless otherwise stated.

A *In Problems 1–12 find rational representations for each, if they exist.*

1. $16^{1/2}$ 2. $64^{1/3}$ 3. $16^{3/2}$

Appendix A-3 Rational Exponents

4. $16^{3/4}$
5. $-36^{1/2}$
6. $32^{3/5}$
7. $(-36)^{1/2}$
8. $(-32)^{3/5}$
9. $(\frac{4}{25})^{3/2}$
10. $(\frac{8}{27})^{2/3}$
11. $9^{-3/2}$
12. $8^{-2/3}$

Simplify Problems 13–36 and express answers using positive exponents only.

13. $y^{1/5}y^{2/5}$
14. $x^{1/4}x^{3/4}$
15. $d^{2/3}d^{-1/3}$
16. $x^{1/4}x^{-3/4}$
17. $(y^{-8})^{1/16}$
18. $(x^{-2/3})^{-6}$
19. $(8x^3y^{-6})^{1/3}$
20. $(4u^{-2}v^4)^{1/2}$

B 21. $\left(\dfrac{a^{-3}}{b^4}\right)^{1/12}$
22. $\left(\dfrac{m^{-2/3}}{n^{-1/2}}\right)^{-6}$

23. $\left(\dfrac{4x^{-2}}{y^4}\right)^{-1/2}$
24. $\left(\dfrac{w^4}{9x^{-2}}\right)^{-1/2}$

25. $\left(\dfrac{8a^{-4}b^3}{27a^2b^{-3}}\right)^{1/3}$
26. $\left(\dfrac{25x^5y^{-1}}{16x^{-3}y^{-5}}\right)^{1/2}$

27. $\left(\dfrac{9}{25}x^2y^{-1/3}\right)^{-3/2}$
28. $\left(\dfrac{27}{8}x^{-3}y^{1/2}\right)^{-4/3}$

29. $\dfrac{8x^{-1/3}}{12x^{1/4}}$
30. $\dfrac{6a^{3/4}}{15a^{-1/3}}$

31. $\left(\dfrac{a^{2/3}b^{-1/2}}{a^{1/2}b^{1/2}}\right)^2$
32. $\left(\dfrac{x^{-1/3}y^{1/2}}{x^{-1/4}y^{1/3}}\right)^6$

33. $\left(\dfrac{9x^{1/3}x^{1/2}}{x^{-1/6}}\right)^{1/2}$
34. $\left(\dfrac{8y^{1/3}y^{-1/4}}{y^{-1/12}}\right)^2$

35. $(125x^{1/2}y^{-1/3})^{-2/3}(x^{1/3}y^{-2/3})$
36. $(9x^{1/3}y^{-1/2})^{3/2}(x^{-1/3}y^{1/4})$

C In Problems 37–40, m and n represent positive integers. Simplify and express answers using positive exponents.

37. $(a^{3/n}b^{3/m})^{1/3}$
38. $(a^{n/2}b^{n/3})^{1/n}$
39. $(x^{m/4}y^{n/3})^{-12}$
40. $(a^{m/3}b^{n/2})^{-6}$

41. Find a real value of x such that
    (A) $(x^2)^{1/2} \neq x$ (B) $(x^2)^{1/2} = x$

42. Find a real value of x such that
    (A) $(x^2)^{1/2} \neq -x$ (B) $(x^2)^{1/2} = -x$

CALCULATOR PROBLEMS In Problems 43–50 evaluate to four significant digits using a hand calculator. (Refer to the instruction book for your calculator to see how exponential forms are evaluated.)

43. $15^{5/4}$
44. $22^{3/2}$

45. $103^{-3/4}$
46. $827^{-3/8}$
47. $2.876^{8/5}$
48. $37.09^{7/3}$
49. $(0.000\ 000\ 077\ 35)^{-2/7}$
50. $(491{,}300{,}000{,}000)^{7/4}$

Appendix A-4 Radicals

- From Rational Exponents to Radicals
- Properties of Radicals
- Changing Radical Forms
- Simplifying $\sqrt[n]{x^n}$ for All Real x

From Rational Exponents to Radicals

For n a natural number greater than 1 and b any real number except b negative when n is even, we define

$$\text{nTH ROOT OF } b: \quad \sqrt[n]{b} = b^{1/n}$$

The symbol $\sqrt{}$ is called a **radical**, n is called the **index**, and b is called the **radicand**. (Note that, if n = 2, we write $\sqrt{b}$ in place of $\sqrt[2]{b}$.) There are occasions when it is more convenient to work with radicals than with rational exponents, or vice versa. It is often an advantage to be able to shift back and forth between the two forms. The following relationships are useful in this regard:

$$b^{m/n} = (b^m)^{1/n} = \sqrt[n]{b^m}$$
$$b^{m/n} = (b^{1/n})^m = (\sqrt[n]{b})^m$$

where b is not negative for n even.

[*Note:* Unless stated to the contrary, all variables in the rest of the discussion represent positive real numbers.]

Appendix A-4 Radicals

EXAMPLE 1 From rational exponent form to radical form.

(A) $x^{1/7} = \sqrt[7]{x}$

(B) $(3u^2v^3)^{3/5} = \sqrt[5]{(3u^2v^3)^3}$ or $(\sqrt[5]{3u^2v^3})^3$ The first is usually preferred.

(C) $y^{-2/3} = \dfrac{1}{y^{2/3}} = \dfrac{1}{\sqrt[3]{y^2}}$ or $\sqrt[3]{y^{-2}}$ or $\sqrt[3]{\dfrac{1}{y^2}}$

From radical form to rational exponent form.

(D) $\sqrt[5]{6} = 6^{1/5}$ (E) $-\sqrt[3]{x^2} = -x^{2/3}$ (F) $\sqrt{x^2+y^2} = (x^2+y^2)^{1/2}$

[Note: $(x^2+y^2)^{1/2} \neq x+y$ (Why?)]

PROBLEM 1 Convert to radical form.

(A) $u^{1/5}$ (B) $(6x^2y^5)^{2/9}$ (C) $(3xy)^{-3/5}$

Convert to rational exponent form.

(D) $\sqrt[4]{9u}$ (E) $-\sqrt[7]{(2x)^4}$ (F) $\sqrt[3]{x^3+y^3}$

■ **Properties of Radicals**

Changing and simplifying radical expressions are aided by the introduction of several properties of radicals that follow directly from exponent properties considered earlier.

Properties of Radicals

For k, n, and m natural numbers 2 or larger, and x and y positive real numbers:

1. $\sqrt[n]{x^n} = x$ $\sqrt[3]{x^3} = x$
2. $\sqrt[n]{xy} = \sqrt[n]{x}\sqrt[n]{y}$ $\sqrt[5]{xy} = \sqrt[5]{x}\sqrt[5]{y}$
3. $\sqrt[n]{\dfrac{x}{y}} = \dfrac{\sqrt[n]{x}}{\sqrt[n]{y}}$ $\sqrt[4]{\dfrac{x}{y}} = \dfrac{\sqrt[4]{x}}{\sqrt[4]{y}}$
4. $\sqrt[kn]{x^{km}} = \sqrt[n]{x^m}$ $\sqrt[12]{x^8} = \sqrt[4\cdot3]{x^{4\cdot2}} = \sqrt[3]{x^2}$
5. $\sqrt[m]{\sqrt[n]{x}} = \sqrt[mn]{x}$ $\sqrt[4]{\sqrt[3]{x}} = \sqrt[12]{x}$

Can you supply the reasons for the proofs of properties 1, 2, and 4? (Proofs of properties 3 and 5 are asked for in Exercise A-4, Problems 71 and 72.)

Proofs of 1, 2, and 4

1. $\sqrt[n]{x^n} = (x^n)^{1/n} = x^{n/n} = x$

2. $\sqrt[n]{xy} = (xy)^{1/n} = x^{1/n}y^{1/n} = \sqrt[n]{x}\sqrt[n]{y}$

4. $\sqrt[kn]{x^{km}} = (x^{km})^{1/kn} = x^{km/kn} = x^{m/n} = (x^m)^{1/n} = \sqrt[n]{x^m}$

EXAMPLE 2

(A) $\sqrt[5]{(3x^2y)^5} = 3x^2y$

(B) $\sqrt{10}\sqrt{5} = \sqrt{50} = \sqrt{25 \cdot 2} = \sqrt{25}\sqrt{2} = 5\sqrt{2}$

(C) $\sqrt[3]{\dfrac{x}{27}} = \dfrac{\sqrt[3]{x}}{\sqrt[3]{27}} = \dfrac{\sqrt[3]{x}}{3}$ or $\dfrac{1}{3}\sqrt[3]{x}$

(D) $\sqrt[6]{x^4} \boxed{= \sqrt[2\cdot3]{x^{2\cdot2}}} = \sqrt[3]{x^2}$

(E) $\sqrt[3]{\sqrt[4]{x}} \boxed{= \sqrt[3\cdot4]{x}} = \sqrt[12]{x}$

PROBLEM 2 Simplify as in Example 2.

(A) $\sqrt[7]{(u^2+v^2)^7}$ (B) $\sqrt{6}\sqrt{2}$ (C) $\sqrt[3]{\dfrac{x^2}{8}}$

(D) $\sqrt[8]{y^6}$ (E) $\sqrt[5]{\sqrt{x}}$

■ **Changing Radical Forms**

The properties of radicals provide us with the means of changing algebraic expressions containing radicals to a variety of equivalent forms. One form that is often useful is the simplest radical form. An algebraic expression that contains radicals is said to be in the **simplest radical form** if all four of the conditions listed in the box are satisfied.

Simplest Radical Form

1. A radicand (the expression within the radical sign) contains no factor to a power greater than or equal to the index of the radical. ($\sqrt{x^5}$ violates this condition.)
2. The power of the radicand and the index of the radical have no common factor other than 1. ($\sqrt[6]{x^4}$ violates this condition.)
3. No radical appears in a denominator. ($y/\sqrt{x}$ violates this condition.)
4. No fraction appears within a radical. ($\sqrt{\tfrac{3}{5}}$ violates this condition.)

Appendix A-4 Radicals

EXAMPLE 3 Express radicals in simplest radical form.

(A) $\sqrt[3]{54} = \sqrt[3]{3^3 \cdot 2}$ $\qquad$ $54 = 27 \cdot 2 = 3^3 \cdot 2$
  $= \sqrt[3]{3^3}\sqrt[3]{2}$ $\qquad$ Condition 1 is not met.
  $= 3\sqrt[3]{2}$

(B) $\sqrt{12x^3y^5z^2} = \sqrt{(4x^2y^4z^2)(3xy)}$ $\qquad$ Condition 1 is not met.
  $\boxed{= \sqrt{(2xy^2z)^2(3xy)}}$
  $\boxed{= \sqrt{(2xy^2z)^2}\sqrt{3xy}}$
  $= 2xy^2z\sqrt{3xy}$

(C) $\sqrt[6]{16x^4y^2} = \sqrt[6]{(4x^2y)^2}$ $\qquad$ Condition 2 is not met.
  $= \sqrt[2\cdot3]{(4x^2y)^{2\cdot 1}}$
  $= \sqrt[3]{4x^2y}$

(D) $\sqrt[3]{\sqrt{27}} = \sqrt[6]{27}$ $\qquad$ Condition 2 is not met.
  $= \sqrt[3\cdot 2]{3^{3\cdot 1}} = \sqrt{3}$

PROBLEM 3 Express in simplest radical form.

(A) $\sqrt[3]{16}$ $\qquad$ (B) $\sqrt{18x^5y^2z^3}$ $\qquad$ (C) $\sqrt[9]{8x^6y^3}$ $\qquad$ (D) $\sqrt{\sqrt[3]{4}}$

Eliminating a radical from a denominator or a fraction from within a radical is referred to as **rationalizing denominators**.

EXAMPLE 4 Rationalize denominators.

(A) $\dfrac{3}{\sqrt{5}} = \dfrac{3}{\sqrt{5}} \cdot \dfrac{\sqrt{5}}{\sqrt{5}}$ $\qquad$ Condition 3 is not met.

  $\boxed{= \dfrac{3\sqrt{5}}{\sqrt{5^2}}}$

  $= \dfrac{3\sqrt{5}}{5}$ or $\tfrac{3}{5}\sqrt{5}$

(B) $\dfrac{6x^2}{\sqrt[3]{9x}} = \dfrac{6x^2}{\sqrt[3]{9x}} \cdot \dfrac{\sqrt[3]{3x^2}}{\sqrt[3]{3x^2}}$ $\qquad$ Condition 3 is not met.

  $\boxed{= \dfrac{6x^2\sqrt[3]{3x^2}}{\sqrt[3]{3^3x^3}}}$

  $= \dfrac{6x^2\sqrt[3]{3x^2}}{3x} = 2x\sqrt[3]{3x^2}$

(C) $\sqrt[3]{\dfrac{2a^2}{3b^2}} = \sqrt[3]{\dfrac{2a^2}{3b^2} \cdot \dfrac{3^2 b}{3^2 b}}$ Condition 4 is not met.

$= \sqrt[3]{\dfrac{18a^2 b}{3^3 b^3}}$

$= \dfrac{\sqrt[3]{18a^2 b}}{\sqrt[3]{3^3 b^3}} = \dfrac{\sqrt[3]{18a^2 b}}{3b}$

PROBLEM 4 Write in simplest radical form by rationalizing denominators.

(A) $\dfrac{6}{\sqrt{2x}}$ (B) $\dfrac{10x^3}{\sqrt[3]{2x^2}}$ (C) $\sqrt[3]{\dfrac{3y^2}{2x^4}}$

Simplest radical forms are often useful in simplifying more involved algebraic expressions that contain radicals. However, there are many situations in which the simplest radical form may not be the most useful form. For example, in finding a decimal approximation for $\sqrt{\tfrac{2}{5}}$ using a hand calculator, it is probably easier to evaluate $\sqrt{\tfrac{2}{5}}$ directly rather than expressing it in the simplest radical form $\sqrt{10}/5$ first.

EXAMPLE 5 (A) $\sqrt[3]{6x^2 y} \sqrt[3]{4x^5 y^2} = \sqrt[3]{(6x^2 y)(4x^5 y^2)}$

$= \sqrt[3]{24 x^7 y^3}$
$= \sqrt[3]{(8x^6 y^3)(3x)}$
$= \sqrt[3]{(2x^2 y)^3 (3x)}$
$= \sqrt[3]{(2x^2 y)^3} \sqrt[3]{3x}$
$= 2x^2 y \sqrt[3]{3x}$

(B) $\sqrt{2u^5 v} \sqrt[3]{4u^4 v^2} = \sqrt[3 \cdot 2]{(2u^5 v)^3} \sqrt[2 \cdot 3]{(4u^4 v^2)^2}$
$= \sqrt[6]{2^3 u^{15} v^3} \sqrt[6]{2^4 u^8 v^4}$
$= \sqrt[6]{2^7 u^{23} v^7}$
$= \sqrt[6]{(2^6 u^{18} v^6)(2u^5 v)}$
$= \sqrt[6]{(2u^3 v)^6 (2u^5 v)}$
$= 2u^3 v \sqrt[6]{2u^5 v}$

[Note: Index numbers must be the same for both radicals before they can be multiplied (see property 2 on page A17).]

Appendix A-4 Radicals

PROBLEM 5 Multiply (as in Example 5) and express the answer in simplest radical form.

(A) $\sqrt[4]{27a^3b^3}\sqrt[4]{3a^5b^3}$ (B) $\sqrt[4]{8x^7y^3}\sqrt{2x^3y}$

■ **Simplifying $\sqrt[n]{x^n}$ for All Real x**

We digress for a moment to recall that the absolute value of a real number is defined as follows:

Absolute Value

For all real numbers a, the **absolute value** of a, denoted by $|a|$, is given by

$$|a| = \begin{cases} a & \text{if } a \text{ is positive} \\ 0 & \\ -a & \text{if } a \text{ is negative} \end{cases}$$

$|3| = 3$
$|0| = 0$
$|-3| = -(-3) = 3$

Now let us return to the first property of radicals:

$\sqrt[n]{x^n} = x$ for x a positive number

What happens if we do not restrict x to positive numbers? For example, does

$$\sqrt{x^2} = x$$

for *all* real numbers x? The answer is no. Testing the equation for $x = 2$ and $x = -2$, we see that

$$\begin{array}{cc} x = 2 & x = -2 \\ \sqrt{2^2} \stackrel{?}{=} 2 & \sqrt{(-2)^2} \stackrel{?}{=} -2 \\ \sqrt{4} \stackrel{?}{=} 2 & \sqrt{4} \stackrel{?}{=} -2 \\ 2 \stackrel{\checkmark}{=} 2 & 2 \neq -2 \end{array}$$

Thus, if x is negative, we must write

$$\sqrt{x^2} = -x$$

then both sides will represent the same positive number.

In summary, for x any real number,

$$\sqrt{x^2} = \begin{cases} x & \text{if } x \text{ is positive} \\ 0 & \text{if } x \text{ is } 0 \\ -x & \text{if } x \text{ is negative} \end{cases}$$

And we see that $\sqrt{x^2}$ and $|x|$ are the same; thus,

For any real number x,

$$\sqrt{x^2} = |x|^*$$

Now let us turn to $\sqrt[3]{x^3}$. Here we do not have the same problem as earlier.

For any real number x,

$$\sqrt[3]{x^3} = x$$

Evaluate both sides for $x = 2$ and for $x = -2$.

If asked to simplify $\sqrt[3]{x^3} + \sqrt{x^2}$, many students would write

$$\sqrt[3]{x^3} + \sqrt{x^2} = x + x \;\boxed{= (1+1)x} \;= 2x$$

and not think any more about it. But if we evaluate both sides for $x = -2$, we find that

$$\sqrt[3]{(-2)^3} + \sqrt{(-2)^2} = \sqrt[3]{-8} + \sqrt{4} = -2 + 2 = 0 \quad \text{Left side}$$

and

$$2(-2) = -4 \quad \text{Right side}$$

Both sides are not equal! What is wrong? When x is not restricted to positive real values—that is, when x is allowed to take on any real value—then we should write

$$\sqrt[3]{x^3} + \sqrt{x^2} = x + |x|$$

Then the left side will equal the right side for all real numbers.

EXAMPLE 6 For x a positive number,

$$\sqrt[3]{x^3} + \sqrt{x^2} = x + |x| = x + x \;\boxed{= (1+1)x} \;= 2x \quad |x| = x \text{ if } x \text{ is positive.}$$

---

*This relationship is also of use when a programmable calculator without an $|x|$ function key is used.

Appendix A-4 Radicals

For x a negative number,

$$\sqrt[3]{x^3} + \sqrt{x^2} = x + |x| = x + (-x) = 0 \quad |x| = -x \text{ if } x \text{ is negative.}$$

PROBLEM 6 Given $4\sqrt[3]{x^3} + \sqrt{x^2}$, simplify.

(A) For x a positive number   (B) For x a negative number

Following the same kind of reasoning, we can obtain the more general result:

For x any real number and n a positive integer 2 or larger,

$$\sqrt[n]{x^n} = \begin{cases} |x| & \text{if } n \text{ is even} \\ x & \text{if } n \text{ is odd} \end{cases}$$

Answers to Matched Problems

1. (A) $\sqrt[5]{u}$ (B) $\sqrt[9]{(6x^2y^5)^2}$ or $(\sqrt[9]{6x^2y^5})^2$ (C) $1/\sqrt[5]{(3xy)^3}$
   (D) $(9u)^{1/4}$ (E) $-(2x)^{4/7}$ (F) $(x^3 + y^3)^{1/3}$ (not $x + y$)
2. (A) $u^2 + v^2$ (B) $2\sqrt{3}$ (C) $(\sqrt[3]{x^2})/2$ or $\tfrac{1}{2}\sqrt[3]{x^2}$
   (D) $\sqrt[4]{y^3}$ (E) $\sqrt[10]{x}$
3. (A) $2\sqrt[3]{2}$ (B) $3x^2yz\sqrt{2xz}$ (C) $\sqrt[3]{2x^2y}$ (D) $\sqrt[3]{2}$
4. (A) $(3\sqrt{2x})/x$ (B) $5x^2\sqrt[3]{4x}$ (C) $(\sqrt[3]{12x^2y^2})/2x^2$
5. (A) $3a^2b\sqrt[4]{b^2} = 3a^2b\sqrt{b}$ (B) $2x^3y\sqrt[4]{2xy}$
6. (A) $5x$ (B) $3x$

Exercise A-4 ■ Unless stated to the contrary, all letters and radicands represent positive real numbers.

A Change to radical form. Do not simplify.

1. $m^{2/3}$
2. $n^{4/5}$
3. $6x^{3/5}$
4. $7y^{2/5}$
5. $(4xy^3)^{2/5}$
6. $(7x^2y)^{5/7}$
7. $(x + y)^{1/2}$
8. $x^{1/2} + y^{1/2}$

Change to rational exponent form. Do not simplify.

9. $\sqrt[5]{b}$
10. $\sqrt{c}$
11. $5\sqrt[4]{x^3}$
12. $7m\sqrt[5]{n^2}$
13. $\sqrt[5]{(2x^2y)^3}$
14. $\sqrt[9]{(3m^4n)^2}$
15. $\sqrt[3]{x} + \sqrt[3]{y}$
16. $\sqrt[3]{x + y}$

Simplify and write in simplest radical form.

17. $\sqrt[3]{-8}$
18. $\sqrt[3]{-27}$
19. $\sqrt{9x^8y^4}$

20. $\sqrt{16m^4y^8}$
21. $\sqrt[4]{16m^4n^8}$
22. $\sqrt[5]{32a^{15}b^{10}}$
23. $\sqrt{8a^3b^5}$
24. $\sqrt{27m^2n^7}$
25. $\sqrt[3]{2^4x^4y^7}$
26. $\sqrt[4]{2^4x^5y^8}$
27. $\sqrt[4]{m^2}$
28. $\sqrt[10]{n^6}$
29. $\sqrt[5]{\sqrt[3]{xy}}$
30. $\sqrt{\sqrt[4]{5x}}$
31. $\sqrt[3]{9x^2}\sqrt[3]{9x}$
32. $\sqrt{2x}\sqrt{8xy}$
33. $1/\sqrt{5}$
34. $1/\sqrt{7}$
35. $6/\sqrt[3]{3}$
36. $2/\sqrt[3]{2}$
37. $\sqrt{6x/7y}$
38. $\sqrt{3a/2b}$

B Simplify and write in simplest radical form.

39. $x\sqrt[5]{3^6x^7y^{11}}$
40. $2a\sqrt[3]{8a^8b^{13}}$
41. $\dfrac{\sqrt[4]{32m^7n^9}}{2mn}$
42. $\dfrac{\sqrt[5]{32u^{12}v^8}}{uv}$
43. $\sqrt[6]{a^4(b-a)^2}$
44. $\sqrt[8]{3^6(u+v)^6}$
45. $\sqrt[3]{\sqrt[4]{a^9b^3}}$
46. $\sqrt{\sqrt[6]{x^8y^6}}$
47. $\sqrt[3]{2x^2y^4}\sqrt[3]{3x^5y}$
48. $\sqrt[4]{4m^5n}\sqrt[4]{6m^3n^4}$
49. $\dfrac{\sqrt{2m}\sqrt{5}}{\sqrt{20m}}$
50. $\dfrac{\sqrt{6}\sqrt{8c}}{\sqrt{18c}}$
51. $\dfrac{4a^3b^2}{\sqrt[3]{2ab^2}}$
52. $\dfrac{8x^3y^5}{\sqrt[3]{4x^2y}}$
53. $\sqrt[4]{\dfrac{3y^3}{4x^3}}$
54. $\sqrt[5]{\dfrac{4x^2}{16y^3}}$
55. $\sqrt[3]{\dfrac{2a^4}{9b^5}}$
56. $\sqrt[3]{\dfrac{3y^5}{4x^4}}$

In Problems 57–62 rationalize the numerators; that is, perform operations on the algebraic expression that eliminate radicals from the numerator. (This is a particularly useful operation in some problems in calculus.)

57. $\dfrac{\sqrt{2x}}{4}$
58. $\dfrac{\sqrt{3y}}{\sqrt{2x}}$
59. $\dfrac{\sqrt[3]{4m}}{2m^2}$
60. $\dfrac{\sqrt[3]{3x^2}}{6xy}$
61. $\sqrt[4]{8x^3}$
62. $\sqrt[5]{8y^3}$

C Simplify Problems 63–70 and express answers in simplest radical form.

63. $\sqrt[3]{\sqrt[5]{\sqrt{32a^{15}b^{10}}}}$
64. $\sqrt{\sqrt[4]{\sqrt[3]{8x^6y^9}}}$
65. $\sqrt[3]{8\sqrt{16x^6y^4}}$
66. $\sqrt[4]{16x^4\sqrt[3]{16x^{24}y^4}}$
67. $\sqrt{2a^5b^3}\sqrt[3]{16a^7b^7}$
68. $\sqrt[3]{3x^2y^2}\sqrt[4]{3x^3y^2}$
69. $\dfrac{\sqrt[3]{4a^2b^2}}{\sqrt{2ab}}$
70. $\dfrac{\sqrt{2x}}{\sqrt[3]{x}}$

71. Show that: $\sqrt[n]{\dfrac{x}{y}} = \dfrac{\sqrt[n]{x}}{\sqrt[n]{y}}$ **72.** Show that: $\sqrt[m]{\sqrt[n]{x}} = \sqrt[mn]{x}$

Simplify Problems 73–78 for (A) x a positive number and (B) x a negative number.

73. $4\sqrt[3]{x^3} + 2\sqrt{x^2}$ **74.** $7\sqrt[3]{x^3} + 2\sqrt{x^2}$ **75.** $\sqrt[5]{x^5} + \sqrt[4]{x^4}$

76. $\sqrt[7]{x^7} + \sqrt[8]{x^8}$ **77.** $2\sqrt[4]{x^4} + 3\sqrt[5]{x^5}$ **78.** $3\sqrt[6]{x^6} + 4\sqrt[9]{x^9}$

CALCULATOR PROBLEMS

Evaluate to four significant digits using a hand calculator. (Read the instruction booklet accompanying your calculator for the process required to evaluate $\sqrt[n]{x}$.)

79. $\sqrt{0.049\ 375}$ **80.** $\sqrt{306.721}$

81. $\sqrt[5]{27.0635}$ **82.** $\sqrt[8]{0.070\ 144}$

83. $\sqrt[7]{0.000\ 000\ 008\ 066}$ **84.** $\sqrt[12]{6{,}423{,}000{,}000{,}000}$

85. $\sqrt[3]{7} + \sqrt[3]{7}$ **86.** $\sqrt[5]{4} + \sqrt[5]{4}$

87. $\sqrt[3]{\sqrt[4]{2}}$ and $\sqrt[12]{2}$ **88.** $\sqrt[3]{\sqrt{5}}$ and $\sqrt[6]{5}$

89. $\dfrac{1}{\sqrt[3]{4}}$ and $\dfrac{\sqrt[3]{2}}{2}$ **90.** $\dfrac{1}{\sqrt[3]{5}}$ and $\dfrac{\sqrt[3]{25}}{5}$

Appendix A-5 Algebraic Expressions, Basic Operations

- Algebraic Expressions and Polynomials
- Addition and Subtraction
- Multiplication
- Combined Operations

In this appendix we will review the basic operations of addition, subtraction, and multiplication on algebraic expressions. For ease of reference, we start by repeating the special subsection on polynomials from Section 1-5.

- **Algebraic Expressions and Polynomials**

Algebraic expressions are formed by using constants and variables and the algebraic operations of addition, subtraction, multiplication, division, and the taking of roots. For example,

$$\sqrt[3]{x^3 - 2x + 1} \qquad \dfrac{x-5}{x^2 + 2x - 5} \qquad (3x^{-5} - 2x^{-3})^{2/3}$$

are all algebraic expressions. An algebraic expression involving only the operations of addition, subtraction, and multiplication on variables and constants, such as $x^3 - 2x^2 + 5x - 1$, is called a **polynomial**.

Polynomial in x

A **polynomial in x** is an algebraic expression of the form

$$a_n x^n + a_{n-1} x^{n-1} + \cdots + a_1 x + a_0$$

where the coefficients $a_0, a_1, \ldots, a_n$ are real numbers and n is a nonnegative integer.

Of course, we may consider polynomials in more than one variable. A polynomial in the two variables x and y is an algebraic expression formed by adding terms of the form $ax^m y^n$, where a is a real number and m and n are nonnegative integers. For example,

$$3x^3 - \sqrt{2} x^2 y + xy - \tfrac{1}{2} xy^2 + y^3 + 2x - 3$$

is a polynomial in two variables. Polynomials in three and more variables are defined in a similar way.

Polynomial forms are encountered frequently in mathematics, and for their more efficient study it is useful to classify them according to their degree. If a term in a polynomial has only one variable as a factor, then the **degree of that term** is the power of the variable. If two or more variables are present in a term as factors, then the **degree of the term** is the sum of the powers of the variables. The **degree of a polynomial** is the degree of the nonzero term with the highest degree in the polynomial. Any nonzero constant is defined to be a **polynomial of degree 0**. The number 0 is also a polynomial but is not assigned a degree.

EXAMPLE 1 (A) Polynomials in one variable:

$$x^2 - 3x + 2 \qquad 6x^3 - \sqrt{2} x - \tfrac{1}{3}$$

(B) Polynomials in several variables:

$$3x^2 - 2xy + y^2 \qquad 4x^3 y^2 - \sqrt{3} xy^2 z^5$$

(C) Nonpolynomials:

$$\sqrt{2x} - \frac{3}{x} + 5 \qquad \frac{x^2 - 3x + 2}{x - 3} \qquad \sqrt{x^2 - 3x + 1}$$

(D) The degree of the first term in $6x^3 - \sqrt{2} x - \tfrac{1}{3}$ is 3, the second term 1, the third term 0, and the whole polynomial 3.

(E) The degree of the first term in $4x^3y^2 - \sqrt{3}xy^2$ is 5, the second 3, and the whole polynomial 5.

PROBLEM 1 (A) Which of the following are polynomials?

$$3x^2 - 2x + 1 \qquad \sqrt{x-3} \qquad x^2 - 2xy + y^2 \qquad \frac{x-1}{x^2+2}$$

(B) Given the polynomial $3x^5 - 6x^3 + 5$, what is the degree of the first term? The second term? The whole polynomial?

(C) Given the polynomial $6x^4y^2 - 3xy^3$, what is the degree of the first term? The second term? The whole polynomial?

In addition to classifying polynomials by degree, we also call a single-term polynomial a **monomial**, a two-term polynomial a **binomial**, and a three-term polynomial a **trinomial**.

■ Addition and Subtraction

We now turn to the addition and subtraction of polynomials and other algebraic expressions. All letters in the following discussion and examples represent real numbers; hence, all the properties of real numbers we have discussed apply. We now add one more property that we will find useful—namely, that multiplication distributes over subtraction:

$a(b - c) = (b - c)a = ab - ac$

This is easy to see as follows:

$$\begin{aligned} a(b-c) &= a[b + (-c)] \\ &= ab + a(-c) \\ &= ab + [-(ac)] \\ &= ab - ac \end{aligned}$$

Can you supply the reasons for each of these steps?

EXAMPLE 2 (A) Add $5x^3 - 2x^2 + x - 3$ and $7x^3 + 5x^2 + 9$.

Solution

$(5x^3 - 2x^2 + x - 3) + (7x^3 + 5x^2 + 9)$

$= 1(5x^3 - 2x^2 + x - 3) + 1(7x^3 + 5x^2 + 9)$

$= 5x^3 - 2x^2 + x - 3 + 7x^3 + 5x^2 + 9$

$= 5x^3 + 7x^3 + 5x^2 - 2x^2 + x + 6$

$= (5 + 7)x^3 + (5 - 2)x^2 + x + 6$

$= 12x^3 + 3x^2 + x + 6$

By placing 1's in front of the parentheses, we can use the distributive property to "clear" the parentheses. Notice how the commutative and associative properties are also used in this example.

(B) Subtract $5x^3 - 2x^2 + x - 3$ from $7x^3 + 5x^2 + 9$.

Solution $(7x^3 + 5x^2 + 9) - (5x^3 - 2x^2 + x - 3)$ Subtracting a number is the same as adding its negative.

$= 1(7x^3 + 5x^2 + 9) + (-1)(5x^3 - 2x^2 + x - 3)$

$= 7x^3 + 5x^2 + 9 - 5x^3 + 2x^2 - x + 3$ The negative of a number can be obtained by multiplying it by -1.

$= 7x^3 - 5x^3 + 5x^2 + 2x^2 - x + 12$

$= (7 - 5)x^3 + (5 + 2)x^2 - x + 12$

$= 2x^3 + 7x^2 - x + 12$

PROBLEM 2 Given the polynomials $x^3 - 7x + 2$ and $4x^3 - x^2 + x - 1$.

(A) Add them. (B) Subtract the first from the second.

EXAMPLE 3 (A) Add $3\sqrt{x} + 5\sqrt{y} + 2$ and $\sqrt{x} + \sqrt[3]{y} - 4$.

Solution $(3\sqrt{x} + 5\sqrt{y} + 2) + (\sqrt{x} + \sqrt[3]{y} - 4)$ These are not polynomials.

$= 3\sqrt{x} + 5\sqrt{y} + 2 + \sqrt{x} + \sqrt[3]{y} - 4$

$= 3\sqrt{x} + \sqrt{x} + 5\sqrt{y} + \sqrt[3]{y} - 2$

$= (3 + 1)\sqrt{x} + 5\sqrt{y} + \sqrt[3]{y} - 2$

$= 4\sqrt{x} + 5\sqrt{y} + \sqrt[3]{y} - 2$

(B) Subtract $4x^{2/3} - x^{1/3} + 2$ from $3x^{2/3} + 2x^{1/3} - 8$.

Solution $(3x^{2/3} + 2x^{1/3} - 8) - (4x^{2/3} - x^{1/3} + 2)$ These are not polynomials.

$= 3x^{2/3} + 2x^{1/3} - 8 - 4x^{2/3} + x^{1/3} - 2$

$= 3x^{2/3} - 4x^{2/3} + 2x^{1/3} + x^{1/3} - 10$

$= (3 - 4)x^{2/3} + (2 + 1)x^{1/3} - 10$

$= -x^{2/3} + 3x^{1/3} - 10$

PROBLEM 3 (A) Add $\left(2\sqrt[4]{m} + 3\sqrt{pq} - 3\right)$ and $\left(\sqrt[3]{m} - \sqrt{pq} + 5\right)$.

(B) Subtract the first algebraic expression from the second in part (A).

■ **Multiplication**

Multiplication of algebraic expressions involves the extensive use of distributive properties for real numbers, as well as other real number properties.

EXAMPLE 4 Multiply $(2x - 3)(3x^2 - 2x + 3)$.

Appendix A-5 Algebraic Expressions, Basic Operations

Solution $(2x - 3)(3x^2 - 2x + 3)$

$= 2x(3x^2 - 2x + 3) - 3(3x^2 - 2x + 3)$

$= 6x^3 - 4x^2 + 6x - 9x^2 + 6x - 9$

$= 6x^3 - 13x^2 + 12x - 9$

or

$3x^2 - 2x + 3$
$2x - 3$
$\overline{6x^3 - 4x^2 + 6x \phantom{{}-9}}$
$ - 9x^2 + 6x - 9$
$\overline{6x^3 - 13x^2 + 12x - 9}$

PROBLEM 4 Multiply $(2x - 3)(2x^2 + 3x - 2)$.

Certain types of binomial products occur so frequently that it is useful to note formulas for them.

Special Products

$(ax + b)(cx + d) = acx^2 + (ad + bc)x + bd$
$(A - B)(A + B) = A^2 - B^2$
$(A + B)^2 = A^2 + 2AB + B^2$
$(A - B)^2 = A^2 - 2AB + B^2$

EXAMPLE 5 (A) $(2x - 3y)(5x + 2y) = 10x^2 - 11xy - 6y^2$

(B) $(\sqrt{2} - \sqrt{3})(\sqrt{2} + \sqrt{3}) = (\sqrt{2})^2 - (\sqrt{3})^2$

$\phantom{(B) (\sqrt{2} - \sqrt{3})(\sqrt{2} + \sqrt{3})} = 2 - 3 = -1$

(C) $(3x^{1/2} - 2y^{1/2})^2 = 9x - 12x^{1/2}y^{1/2} + 4y$

(D) $(3\sqrt{x} + 2\sqrt{y})(2\sqrt{x} - \sqrt{y}) = 6x + \sqrt{xy} - 2y$

PROBLEM 5 Multiply and simplify.

(A) $(5u + 3v)(4u - v)$ (B) $(\sqrt{x} - \sqrt{y})(\sqrt{x} + \sqrt{y})$
(C) $(2^{1/2} - 3^{1/2})^2$ (D) $(4\sqrt{a} - \sqrt{b})(3\sqrt{a} + 2\sqrt{b})$

Combined Operations

We complete this appendix by considering several examples that combine operations discussed in this and preceding appendixes.

EXAMPLE 6 (A) $\sqrt{2}(\sqrt{10} - 3) - 4(\sqrt{45} + \sqrt{2})$ Multiply, transform terms to simplest radical form, and combine terms where possible.
$= \sqrt{20} - 3\sqrt{2} - 4\sqrt{45} - 4\sqrt{2}$
$= \sqrt{4 \cdot 5} - 3\sqrt{2} - 4\sqrt{9 \cdot 5} - 4\sqrt{2}$
$= 2\sqrt{5} - 3\sqrt{2} - 12\sqrt{5} - 4\sqrt{2}$
$= -10\sqrt{5} - 7\sqrt{2}$

(B) $3x - \{5 - 3[x - x(3 - x)]\} = 3x - \{5 - 3[x - 3x + x^2]\}$
$= 3x - \{5 - 3x + 9x - 3x^2\}$
$= 3x - 5 + 3x - 9x + 3x^2$
$= 3x^2 - 3x - 5$

(C) $(\sqrt[3]{m} + \sqrt[3]{n^2})(\sqrt[3]{m^2} - \sqrt[3]{n}) = \sqrt[3]{m^3} - \sqrt[3]{mn} + \sqrt[3]{m^2n^2} - \sqrt[3]{n^3}$
$= m - \sqrt[3]{mn} + \sqrt[3]{m^2n^2} - n$

PROBLEM 6 Multiply and simplify.

(A) $\sqrt{10}(\sqrt{5} + \sqrt{2}) + \sqrt{6}(\sqrt{3} - \sqrt{2})$ (B) $2t - \{7 - 2[t - t(4 + t)]\}$
(C) $(\sqrt[3]{x^2} - \sqrt[3]{y^2})(\sqrt[3]{x} + \sqrt[3]{y})$

Answers to Matched Problems

1. (A) $3x^2 - 2x + 1$, $x^2 - 2xy + y^2$ (B) 5, 3, 5
   (C) 6, 4, 6
2. (A) $5x^3 - x^2 - 6x + 1$ (B) $3x^3 - x^2 + 8x - 3$
3. (A) $2\sqrt[4]{m} + \sqrt[3]{m} + 2\sqrt{pq} + 2$ (B) $-2\sqrt[4]{m} + \sqrt[3]{m} - 4\sqrt{pq} + 8$
4. $4x^3 - 13x + 6$
5. (A) $20u^2 + 7uv - 3v^2$ (B) $x - y$
   (C) $5 - 2 \cdot 2^{1/2} \cdot 3^{1/2}$ or $5 - 2(6)^{1/2}$ (D) $12a + 5\sqrt{ab} - 2b$
6. (A) $8\sqrt{2} + 2\sqrt{5} - 2\sqrt{3}$ (B) $-2t^2 - 4t - 7$
   (C) $x + \sqrt[3]{x^2y} - \sqrt[3]{xy^2} - y$

Exercise A-5 ■ Unless stated to the contrary, express all answers in simplest radical form with positive exponents. All variables are positive real numbers.

A Given the polynomials: $2x^3 - 3x^2 + x + 5$, $2x^2 + x - 1$, and $3x - 2$.

1. What is the degree of the first?
2. What is the degree of the second?

3. Add the first and second.
4. Add the second and third.
5. Subtract the second from the first.
6. Subtract the third from the second.
7. Multiply the first and third.
8. Multiply the second and third.

In Problems 9–28 perform the indicated operations and simplify.

9. $2(x-1) + 3(2x-3) - (4x-5)$
10. $2(u-1) - (3u+2) - 2(2u-3)$
11. $2y - 3y[4 - 2(y-1)]$
12. $4a - 2a[5 - 3(a+2)]$
13. $\sqrt{5} - 2\sqrt{3} + 3\sqrt{5}$
14. $3\sqrt{2} - 2\sqrt{3} - \sqrt{2}$
15. $2\sqrt[3]{a} + 3\sqrt[3]{a} - \sqrt[4]{a}$
16. $4\sqrt[3]{y} - \sqrt[3]{y} + 2\sqrt{y}$
17. $(3x - 2y)(3x + 2y)$
18. $(4m + 3n)(4m - 3n)$
19. $(\sqrt{c} - \sqrt{d})(\sqrt{c} + \sqrt{d})$
20. $(x^{1/2} + y^{1/2})(x^{1/2} - y^{1/2})$
21. $(4x - y)^2$
22. $(3u + 4v)^2$
23. $(x^{1/2} + y^{1/2})^2$
24. $(\sqrt{y} + \sqrt{z})^2$
25. $(\sqrt{c} - \sqrt{d})^2$
26. $(x^{1/2} - y^{1/2})^2$
27. $(a + b)(a^2 - ab + b^2)$
28. $(a - b)(a^2 + ab + b^2)$

B In Problems 29–54 perform the indicated operations and simplify (or vice versa).

29. $2x - 3\{x + 2[x - (x+5)] + 1\}$
30. $m - \{m - [m - (m-1)]\}$
31. $(2x^2 + x - 2)(x^2 - 3x + 5)$
32. $(x^2 - 2xy + y^2)(x^2 + 2xy + y^2)$
33. $(2x - 1)^2 - (3x + 2)(3x - 2)$
34. $(3a - b)(3a + b) - (2a - 3b)^2$
35. $(2m - n)^3$
36. $(x - 2y)^3$
37. $\sqrt{8} - \sqrt{20} + 4\sqrt{2}$
38. $3\sqrt{3} - \sqrt{12} + \sqrt{24}$
39. $2\sqrt{12x} - 3\sqrt{27x} + \sqrt{3x}$
40. $\sqrt[3]{8a} - 2\sqrt[3]{27a} + \sqrt{4a}$
41. $(a\sqrt{2} - b\sqrt{5})(a\sqrt{2} + b\sqrt{5})$
42. $(x + y\sqrt{3})(x - y\sqrt{3})$
43. $(3\sqrt{x} - \sqrt{y})^2$
44. $(5\sqrt{x} + 2)(2\sqrt{x} - 3)$
45. $(2x^{1/2} + y^{1/2})(x^{1/2} + y^{1/2})$
46. $(3u^{1/2} - 2)(2u^{1/2} + 4)$
47. $2x^{1/3}(3x^{2/3} - x^6)$
48. $3m^{3/4}(4m^{1/4} - 2m^8)$
49. $\sqrt{3 - \sqrt{8}}\sqrt{3 + \sqrt{8}}$
50. $\sqrt{4 - \sqrt{10}}\sqrt{4 + \sqrt{10}}$
51. $\sqrt{ab}\left(\sqrt{\dfrac{a}{b}} + \sqrt{\dfrac{b}{a}}\right)$
52. $(xy)^{1/3}\left[\left(\dfrac{x^2}{y}\right)^{1/3} - \left(\dfrac{y^2}{x}\right)^{1/3}\right]$
53. $(\sqrt{x+h} - \sqrt{x})(\sqrt{x+h} + \sqrt{x})$
54. $[(u+k)^{1/2} - u^{1/2}][(u+k)^{1/2} + u^{1/2}]$

In Problems 55–58 evaluate each polynomial for the indicated value.

55. $x^2 + 2x - 5$ for $x = -1 + \sqrt{6}$
56. $m^2 - 4m + 1$ for $m = 2 + \sqrt{3}$
57. $u^2 - 6u + 3$ for $u = 3 - \sqrt{2}$
58. $y^2 - 3y - 2$ for $y = 1 - \sqrt{3}$

C Simplify.

59. $(x^{-1/2} - y^{-1/2})^2$
60. $(a^{-1/2} + 3b^{-1/2})(2a^{-1/2} - b^{-1/2})$
61. $(\sqrt[3]{t} - \sqrt[3]{x})(\sqrt[3]{t^2} + \sqrt[3]{tx} + \sqrt[3]{x^2})$
62. $(\sqrt[3]{t} + \sqrt[3]{x})(\sqrt[3]{t^2} - \sqrt[3]{tx} + \sqrt[3]{x^2})$
63. $[3(x + 3)^{1/2} + 2][2(x + 3)^{1/2} - 3]$
64. $[2(x - 2)^{1/2} - 3][(x - 2)^{1/2} + 3]$
65. $(\sqrt[3]{x} - \sqrt[3]{y^2})(\sqrt[3]{x^2} + 2\sqrt[3]{y})$
66. $(\sqrt[5]{y^2} - \sqrt[5]{z^3})(\sqrt[5]{y^3} + \sqrt[5]{z^2})$

APPLICATIONS

67. *Geometric.* The width of a rectangle is 5 centimeters less than its length. If x represents the length, write an algebraic expression in terms of x that represents the perimeter of the rectangle. Simplify the expression.

68. *Geometric.* The length of a rectangle is 8 meters more than its width. If x represents the width of the rectangle, write an algebraic expression in terms of x that represents its area. Change the expression to a form without parentheses.

69. *Coin problem.* A parking meter contains nickels, dimes, and quarters. There are five fewer dimes than nickels, and two more quarters than dimes. If x represents the number of nickels, write an algebraic expression in terms of x that represents the value of all the coins in the meter in cents. Simplify the expression.

70. *Coin problem.* A vending machine contains dimes and quarters only. There are four more dimes than quarters. If x represents the number of quarters, write an algebraic expression in terms of x that represents the value of all the coins in the vending machine. Simplify the expression.

CALCULATOR PROBLEMS

Evaluate each to four significant digits using a hand calculator.

71. $(2\sqrt{5} - \sqrt{2})^2 - (24 - 4\sqrt{10})$
72. $(\sqrt{3} - \sqrt{5})^2 - (8 - 2\sqrt{15})$
73. $\sqrt{28 - 10\sqrt{3}} - (5 - \sqrt{3})$
74. $(\sqrt{2} + \sqrt{3}) - \sqrt{5 + 2\sqrt{6}}$

Appendix A-6 Factoring

- Factoring—What Does It Mean?
- Common Factors

Appendix A-6 Factoring

- Factoring by Grouping
- Special Factoring Formulas

- Factoring—What Does It Mean?

If a number is written as the product of other numbers, then each number in the product is called a **factor** of the original number. Similarly, if an algebraic expression is written as the product of other algebraic expressions, then each algebraic expression in the product is called a **factor** of the original algebraic expression.

$$30 = 2 \cdot 3 \cdot 5 \qquad \text{2, 3, and 5 are factors of 30.}$$
$$x^2 - 4 = (x - 2)(x + 2) \qquad (x-2) \text{ and } (x+2) \text{ are factors of } x^2 - 4.$$

The process of writing a number or algebraic expression as the product of other numbers or algebraic expressions is called **factoring**. We start our discussion of factoring with the positive integers.

An integer such as 30 can be represented in a factored form in many ways. The products

$$6 \cdot 5 \qquad (\tfrac{1}{2})(10)(6) \qquad 15 \cdot 2 \qquad \sqrt{15} \cdot \sqrt{60}$$

all yield 30. A particularly useful way of factoring positive integers greater than 1 is in terms of prime numbers.

Prime and Composite Numbers

A positive integer greater than 1 is **prime** if its only positive integer factors are itself and 1. A positive integer greater than 1 that is not prime is called a **composite number**. The integer 1 is neither prime nor composite.

PRIME NUMBERS: 2, 3, 5, 7, 11, 13, . . .

COMPOSITE NUMBERS: 4, 6, 8, 9, 10, 12, . . .

A composite integer greater than 1 is said to be **factored completely** if it is represented as a product of prime factors. The only factoring of 30 given that meets this condition is $30 = 2 \cdot 3 \cdot 5$.

EXAMPLE 1 Write 60 in a completely factored form.

Solution $60 = 6 \cdot 10 = 2 \cdot 3 \cdot 2 \cdot 5 = 2^2 \cdot 3 \cdot 5$

or

$60 = 5 \cdot 12 = 5 \cdot 4 \cdot 3 = 2^2 \cdot 3 \cdot 5$

or

$$60 = 2 \cdot 30 = 2 \cdot 2 \cdot 15 = 2^2 \cdot 3 \cdot 5$$

Notice in Example 1 that we end up with the same prime factors for 60 irrespective of how we progress through the factoring process. This illustrates a basic property of integers:

Fundamental Theorem of Arithmetic

Each positive integer greater than 1 is either prime or has, except for the order of factors, a unique set of prime factors.

PROBLEM 1 Write 180 in a completely factored form.

We can also talk about writing polynomials in a completely factored form. The following polynomials are written in a factored form:

$$x^2 - 9 = (x - 3)(x + 3)$$
$$2x^3 - 4x = 2x(x - \sqrt{2})(x + \sqrt{2})$$
$$2x^4 - 15x^2 - 27 = (x^2 - 9)(2x^2 + 3)$$
$$x^2 + 3x + \tfrac{9}{4} = (x + \tfrac{3}{2})^2$$

But which are in a completely factored form? Paralleling our discussion with prime numbers, we define a **prime polynomial** as follows:

Prime Polynomials

A polynomial is said to be **prime** relative to a given set of numbers if: (1) it has coefficients from that set, and (2) it cannot be written as a product of two polynomials of positive degree having coefficients from that set.

For example, $x^2 - 2$ is prime relative to the integers but is not prime relative to the real numbers [since $x^2 - 2 = (x - \sqrt{2})(x + \sqrt{2})$]. A nonprime polynomial is said to be **factored completely relative to a given set of numbers** if it is represented as a product of prime polynomials relative to that set of numbers.

Appendix A-6 Factoring

Writing polynomials in a completely factored form is often a difficult task. But accomplishing it can lead to the simplification of certain algebraic expressions and to the solutions of certain types of equations.

■ **Common Factors**

The distributive property for real numbers is behind the process of factoring out common factors.

EXAMPLE 2 Take out all factors common to all terms.

(A) $2x^3y - 8x^2y^2 - 6xy^3$ (B) $2x(3x - 2) - 7(3x - 2)$

Solutions (A) $2x^3y - 8x^2y^2 - 6xy^3 \;\vert\; = \mathbf{(2xy)}x^2 - \mathbf{(2xy)}4xy - \mathbf{(2xy)}3y^2$

$= \mathbf{2xy}(x^2 - 4xy - 3y^2)$

(B) $2x(3x - 2) - 7\,3x - 2) \;\vert\; = 2x\mathbf{(3x - 2)} - 7\mathbf{(3x - 2)}$

$= (2x - 7)\mathbf{(3x - 2)}$

PROBLEM 2 Take out all factors common to all terms.

(A) $3x^3y - 6x^2y^2 - 3xy^3$ (B) $3y(2y + 5) + 2(2y + 5)$

■ **Factoring by Grouping**

Some polynomials can be factored by first grouping terms. We can then complete the factoring by removing common factors.

EXAMPLE 3 Factor completely, relative to the integers, by grouping.

(A) $3x^2 - 6x + 4x - 8$ (B) $2y^2 - 2y - y + 1$
(C) $4x^2 + 8xy - xy - 2y^2$ (D) $2ac - 2ad - bc + bd$

Solutions (A) $3x^2 - 6x + 4x - 8$ Group the first two and last two terms.

$= (3x^2 - 6x) + (4x - 8)$ Remove common factors from each group.

$= 3x\,\mathbf{(x - 2)} + 4\,\mathbf{(x - 2)}$ The common factor $(x - 2)$ can be taken out.

$= (3x + 4)\,\mathbf{(x - 2)}$ The factoring is complete.

(B) $2y^2 - 2y - y + 1$

$= (2y^2 - 2y) - (y - 1)$ Be careful of sign errors here.

$= 2y(y - 1) - 1(y - 1)$

$= (2y - 1)(y - 1)$

(C) $4x^2 + 8xy - xy - 2y^2 = (4x^2 + 8xy) - (xy + 2y^2)$
$= 4x(x + 2y) - y(x + 2y)$
$= (4x - y)(x + 2y)$

(D) $2ac - 2ad - bc + bd = (2ac - 2ad) - (bc - bd)$
$= 2a(c - d) - b(c - d)$
$= (2a - b)(c - d)$

PROBLEM 3 Factor completely, relative to the integers, by grouping.

(A) $2x^2 + 6x + 5x + 15$ (B) $3m^2 + 3m - m - 1$
(C) $6u^2 - 3uv - 2uv + v^2$ (D) $6wy + 3wz - 2xy - xz$

■ **Special Factoring Formulas**

Several factoring formulas are worthwhile observing, since they show us how to factor certain polynomial forms that occur often.

Special Factoring Formulas

1. $u^2 + (a + b)u + ab = (u + a)(u + b)$
2. $acu^2 + (ad + bc)u + bd = (au + b)(cu + d)$
3. $a^2u^2 + 2abuv + b^2v^2 = (au + bv)^2$ PERFECT SQUARE
4. $u^2 - v^2 = (u - v)(u + v)$ DIFFERENCE OF TWO SQUARES
5. $u^3 - v^3 = (u - v)(u^2 + uv + v^2)$ DIFFERENCE OF TWO CUBES
6. $u^3 + v^3 = (u + v)(u^2 - uv + v^2)$ SUM OF TWO CUBES

The formulas in the box can be established by multiplying the factors on the right.

EXAMPLE 4 Factor completely relative to the integers.

(A) $x^2 - 5x - 6$ (B) $6x^2 - 5x - 4$ (C) $x^2 + 6xy + 9y^2$
(D) $9x^2 - 4y^2$ (E) $8m^3 - 1$ (F) $x^3 + y^3z^3$

Solutions
(A) $x^2 - 5x - 6 = (x - 6)(x + 1)$
(B) $6x^2 - 5x - 4 = (3x - 4)(2x + 1)$
(C) $x^2 + 6xy + 9y^2 = (x + 3y)^2$
(D) $9x^2 - 4y^2 = (3x - 2y)(3x + 2y)$

Appendix A-6 Factoring

(E) $8m^3 - 1$ $\boxed{\begin{aligned}&= (2m)^3 - 1^3\\&= (2m-1)[(2m)^2 + (2m)(1) + 1^2]\end{aligned}}$

$= (2m-1)(4m^2 + 2m + 1)$

(F) $x^3 + y^3z^3$ $\boxed{= x^3 + (yz)^3}$

$= (x + yz)(x^2 - xyz + y^2z^2)$

PROBLEM 4   Factor completely relative to the integers.

(A) $x^2 + 7x - 8$ (B) $4m^2 - 4mn - 3n^2$
(C) $4m^2 - 12mn + 9n^2$ (D) $x^2 - 16y^2$
(E) $z^3 - 1$ (F) $m^3 + n^3$

We complete this appendix by considering factoring that involves combinations of the preceding techniques as well as a few additional ones. Generally speaking, **when asked to factor a polynomial, we first take out all factors common to all terms, if they are present, and then proceed as above until all factors are prime.**

EXAMPLE 5   Factor completely relative to the integers.

(A) $18x^3 - 8x$ (B) $x^2 - 6x + 9 - y^2$
(C) $4m^3n - 2m^2n^2 + 2mn^3$ (D) $2t^4 - 16t$
(E) $2y^4 - 5y^2 - 12$ (F) $x^4 + x^2 + 1$

Solutions

(A) $18x^3 - 8x = 2x(9x^2 - 4)$
$= 2x(3x - 2)(3x + 2)$

(B) $x^2 - 6x + 9 - y^2$ Group the first three terms.
$= (x^2 - 6x + 9) - y^2$ Factor $x^2 - 6x + 9$.
$= (x - 3)^2 - y^2$ Difference of two squares
$= [(x - 3) - y][(x - 3) + y]$
$= (x - 3 - y)(x - 3 + y)$

(C) $4m^3n - 2m^2n^2 + 2mn^3 = 2mn(2m^2 - mn + n^2)$

(D) $2t^4 - 16t = 2t(t^3 - 8)$
$= 2t(t - 2)(t^2 + 2t + 4)$

(E) $2y^4 - 5y^2 - 12 = (2y^2 + 3)(y^2 - 4)$
$= (2y^2 + 3)(y - 2)(y + 2)$

(F) $x^4 + x^2 + 1$
$= (x^4 + 2x^2 + 1) - x^2$
$= (x^2 + 1)^2 - x^2$
$= [(x^2 + 1) - x][(x^2 + 1) + x]$
$= (x^2 - x + 1)(x^2 + x + 1)$

Proceeding as in (E) does not work; this involves a "trick." Adding and subtracting x^2 lead to a difference of two squares form.

PROBLEM 5 Factor completely relative to the integers.

(A) $3x^3 - 48x$ (B) $x^2 - y^2 - 4y - 4$
(C) $3u^4 - 3u^3v - 9u^2v^2$ (D) $3m^4 - 24mn^3$
(E) $3x^4 - 5x^2 + 2$ (F) $x^4 + x^2 + 25$

Remark: It should be noted that if one writes a polynomial with integer coefficients at random, then the resulting polynomial is more likely to be prime than not prime; that is, it most likely will not have polynomial factors of positive degree relative to the integers. If it does, however, the results may be very useful, as was pointed out earlier.

Answers to Matched Problems

1. $2^2 \cdot 3^2 \cdot 5$
2. (A) $3xy(x^2 - 2xy - y^2)$ (B) $(3y + 2)(2y + 5)$
3. (A) $(2x + 5)(x + 3)$ (B) $(3m - 1)(m + 1)$
    (C) $(3u - v)(2u - v)$ (D) $(3w - x)(2y + z)$
4. (A) $(x + 8)(x - 1)$ (B) $(2m - 3n)(2m + n)$
    (C) $(2m - 3n)^2$ (D) $(x - 4y)(x + 4y)$
    (E) $(z - 1)(z^2 + z + 1)$ (F) $(m + n)(m^2 - mn + n^2)$
5. (A) $3x(x - 4)(x + 4)$ (B) $(x - y - 2)(x + y + 2)$
    (C) $3u^2(u^2 - uv - 3v^2)$ (D) $3m(m - 2n)(m^2 + 2mn + 4n^2)$
    (E) $(3x^2 - 2)(x - 1)(x + 1)$ (F) $(x^2 - 3x + 5)(x^2 + 3x + 5)$

Exercise A-6

A   *Factor out all factors common to all terms.*

1. $6x^4 - 8x^3 - 2x^2$ 2. $6m^4 - 9m^3 - 3m^2$
3. $10x^3y + 20x^2y^2 - 15xy^3$ 4. $8u^3v - 6u^2v^2 + 4uv^3$
5. $5x(x + 1) - 3(x + 1)$ 6. $7m(2m - 3) + 5(2m - 3)$
7. $2w(y - 2z) - x(y - 2z)$ 8. $a(3c + d) - 4b(3c + d)$

Factor completely relative to the integers. Start by grouping the first two and last two terms.

9. $x^2 - 2x + 3x - 6$ 10. $2y^2 - 6y + 5y - 15$
11. $6m^2 + 10m - 3m - 5$ 12. $5x^2 - 40x - x + 8$

13. $2x^2 - 4xy - 3xy + 6y^2$
14. $3a^2 - 12ab - 2ab + 8b^2$
15. $8ac - 4ad - 6bc + 3bd$
16. $3pr + 6ps - qr - 2qs$

Factor completely relative to the integers. If a polynomial is prime relative to the integers, say so.

17. $2x^2 + 5x - 3$
18. $3y^2 - y - 2$
19. $x^2 - 4xy - 12y^2$
20. $u^2 - 2uv - 15v^2$
21. $x^2 + x - 4$
22. $m^2 - 6m - 3$
23. $25m^2 - 16n^2$
24. $w^2x^2 - y^2$
25. $x^2 + 10xy + 25y^2$
26. $9m^2 - 6mn + n^2$
27. $u^2 + 81$
28. $y^2 + 16$
29. $6x^2 + 48x + 72$
30. $4z^2 - 28z + 48$
31. $2y^3 - 22y^2 + 48y$
32. $2x^4 - 24x^3 + 40x^2$
33. $16x^2y - 8xy + y$
34. $4xy^2 - 12xy + 9x$
35. $6s^2 + 7st - 3t^2$
36. $6m^2 - mn - 12n^2$
37. $x^3y - 9xy^3$
38. $4u^3v - uv^3$
39. $3m^3 - 6m^2 + 15m$
40. $2x^3 - 2x^2 + 8x$
41. $m^3 + n^3$
42. $r^3 - t^3$
43. $c^3 - 1$
44. $a^3 + 1$

B Factor completely relative to the integers. In polynomials involving more than three terms, try grouping the terms in various combinations as a first step. If a polynomial is prime relative to the integers, say so.

45. $(a - b)^2 - 4(c - d)^2$
46. $(x + 2)^2 - 9y^2$
47. $2am - 3an + 2bm - 3bn$
48. $15ac - 20ad + 3bc - 4bd$
49. $3x^2 - 2xy - 4y^2$
50. $5u^2 + 4uv - 2v^2$
51. $x^3 - 3x^2 - 9x + 27$
52. $x^3 - x^2 - x + 1$
53. $a^3 - 2a^2 - a + 2$
54. $t^3 - 2t^2 + t - 2$
55. $4(A + B)^2 - 5(A + B) - 6$
56. $6(x - y)^2 + 23(x - y) - 4$
57. $m^4 - n^4$
58. $y^4 - 3y^2 - 4$
59. $s^4t^4 - 8st$
60. $27a^2 + a^5b^3$
61. $m^2 + 2mn + n^2 - m - n$
62. $y^2 - 2xy + x^2 - y + x$

Factor the following algebraic expressions by using the factoring formulas discussed in this appendix.

63. $\dfrac{4}{x^2} - \dfrac{9}{y^2}$
64. $u^2 - \dfrac{4}{v^2}$
65. $x^2 + 3x + \dfrac{9}{4}$
66. $x^2 - 5x + \dfrac{25}{4}$

67. $\dfrac{1}{x^3} - 8$ 68. $27 + \dfrac{8}{t^3}$

69. $x^{2/3} - 3x^{1/3} + 2$ 70. $9x^{2/3} - 4y^{2/5}$

71. $4x^{-2} - y^{-4}$ 72. $2y^{-2} - y^{-1} - 3$

C   Factor completely to the integers. In polynomials that involve more than three terms, try grouping the terms in various combinations as a first step. Also try the "trick" illustrated in Example 5F where appropriate.

73. $18a^3 - 8a(x^2 + 8x + 16)$

74. $25(4x^2 - 12xy + 9y^2) - 9a^2b^2$

75. $x^4 + 2x^2 + 1 - x^2$ 76. $a^4 + 2a^2b^2 + b^4 - a^2b^2$

77. $16x^4 + 4x^2 + 1$ 78. $a^4 + a^2b^2 + b^4$

79. $x^6 - 1$ 80. $a^6 - b^6$

81. $x^4 + 6x^2y^2 + 25y^4$ 82. $4a^4 + 8a^2b^2 + 9b^4$

Appendix A-7   Fractions

- Multiplication and Division
- Addition and Subtraction
- Complex Fractions

Algebraic fractions represent quotients, and for those replacements of the variables by real numbers that result in the quotient of real numbers, division by 0 excluded, the properties of real fractions discussed in Section 1-2 apply. In particular, we will very frequently be using the **fundamental principle of fractions**:

$$\frac{ak}{bk} = \frac{a}{b} \qquad b, k \neq 0$$

Using this principle from left to right to eliminate all common factors from a numerator and a denominator of a given fraction is referred to as **reducing a fraction to lowest terms**. We are actually dividing the numerator and denominator by the same nonzero common factor.

Using the principle from right to left—that is, multiplying a numerator and a denominator by the same nonzero factor—is referred to as **raising a fraction to higher terms**. We will use the principle in both directions in the material that follows.

A particular type of algebraic fraction, the quotient of two polynomials, is called a **rational expression**. We say that a rational expression is reduced to lowest terms if the numerator and denominator do not have

Appendix A-7 Fractions

any prime factors in common. (Unless stated to the contrary, *prime* will mean relative to the integers.)

EXAMPLE 1 Reduce each rational expression to lowest terms.

(A) $\dfrac{x^2 - 6x + 9}{x^2 - 9} = \dfrac{(x - 3)^2}{(x - 3)(x + 3)}$ Factor numerator and denominator completely. Divide numerator and denominator by $(x - 3)$—a valid operation as long as $x \neq 3$ and $x \neq -3$ as well.

$= \dfrac{x - 3}{x + 3}$

(B) $\dfrac{x^3 - 1}{x^2 - 1} = \dfrac{(x - 1)(x^2 + x + 1)}{(x - 1)(x + 1)}$

$= \dfrac{x^2 + x + 1}{x + 1}$

[*Note:* Throughout our work on fractions, we will always assume without specific statement that variables are restricted to avoid division by 0.]

PROBLEM 1 Reduce to lowest terms.

(A) $\dfrac{6x^2 + x - 2}{2x^2 + x - 1}$ (B) $\dfrac{x^4 - 8x}{3x^3 - 2x^2 - 8x}$

■ **Multiplication and Division**

Since in each algebraic fraction we restrict variable replacements to real numbers that produce real fractions, multiplication and division of algebraic fractions follow the rules for multiplying and dividing fractions in real numbers; that is (excluding division by 0),

$$\dfrac{a}{b} \cdot \dfrac{c}{d} = \dfrac{ac}{bd} \qquad \dfrac{a}{b} \div \dfrac{c}{d} = \dfrac{a}{b} \cdot \dfrac{d}{c}$$

EXAMPLE 2 (A) $\dfrac{10x^3y}{3xy + 9y} \cdot \dfrac{x^2 - 9}{4x^2 - 12x}$ Factor numerators and denominators; then divide any numerator and any denominator with a like common factor (this can be done either before or after multiplying numerators or denominators).

$= \dfrac{\overset{5x^2}{\cancel{10x^3y}}}{\underset{3 \cdot 1}{\cancel{3y}(x + 3)}} \cdot \dfrac{\overset{1 \cdot 1}{\cancel{(x - 3)}\cancel{(x + 3)}}}{\underset{2 \cdot 1}{\cancel{4x}\cancel{(x - 3)}}}$

$= \dfrac{5x^2}{6}$

(B) $\dfrac{4-2x}{4} \div (x-2) = \dfrac{\overset{1}{\cancel{2}}(2-x)}{\underset{2}{\cancel{4}}} \cdot \dfrac{1}{x-2}$ $x-2$ is the same as $\dfrac{x-2}{1}$.

$= \dfrac{2-x}{2(x-2)} = \dfrac{\overset{-1}{\cancel{-(x-2)}}}{\underset{1}{\cancel{2(x-2)}}}$ $b - a = -(a - b)$, a useful change in some problems.

$= -\dfrac{1}{2}$

(C) $\dfrac{2x^3 - 2x^2y + 2xy^2}{x^3y - xy^3} \div \dfrac{x^3 + y^3}{x^2 + 2xy + y^2}$

$= \dfrac{\overset{2}{\cancel{2x}}(\overset{1}{\cancel{x^2 - xy + y^2}})}{\underset{y}{\cancel{xy}}(x+y)(x-y)} \cdot \dfrac{\overset{1}{\cancel{(x+y)^2}}}{\underset{1}{\cancel{(x+y)}}(\underset{1}{\cancel{x^2 - xy + y^2}})}$

$= \dfrac{2}{y(x-y)}$

PROBLEM 2   Perform the indicated operations and reduce to lowest terms.

(A) $\dfrac{12x^2y^3}{2xy^2 + 6xy} \cdot \dfrac{y^2 + 6y + 9}{3y^3 + 9y^2}$ (B) $(4-x) \div \dfrac{x^2 - 16}{5}$

(C) $\dfrac{m^3 + n^3}{2m^2 + mn - n^2} \div \dfrac{m^3n - m^2n^2 + mn^3}{2m^3n^2 - m^2n^3}$

We will now use the fundamental principle of fractions to rationalize denominators and numerators in fractional expressions that involve radicals.

EXAMPLE 3   (A) Rationalize the denominator: $\dfrac{\sqrt{x} - \sqrt{y}}{\sqrt{x} + \sqrt{y}}$

Solution $\dfrac{\sqrt{x} - \sqrt{y}}{\sqrt{x} + \sqrt{y}} = \dfrac{\sqrt{x} - \sqrt{y}}{\sqrt{x} + \sqrt{y}} \cdot \dfrac{\sqrt{x} - \sqrt{y}}{\sqrt{x} - \sqrt{y}}$ Multiplying numerator and denominator by $(\sqrt{x} - \sqrt{y})$ eliminates radicals from the denominator, since $(a+b)(a-b) = a^2 - b^2$.

$= \dfrac{x - 2\sqrt{xy} + y}{x - y}$

(B) Rationalize the numerator: $\dfrac{\sqrt{x+h} - \sqrt{x}}{h}$

Solution

$$\frac{\sqrt{x+h}-\sqrt{x}}{h} = \frac{\sqrt{x+h}-\sqrt{x}}{h} \cdot \frac{\sqrt{x+h}+\sqrt{x}}{\sqrt{x+h}+\sqrt{x}}$$

$$= \frac{x \times h - x}{h(\sqrt{x+h}+\sqrt{x})}$$

$$= \frac{h}{h(\sqrt{x+h}+\sqrt{x})} = \frac{1}{\sqrt{x+h}+\sqrt{x}}$$

PROBLEM 3 (A) Rationalize the denominator: $\dfrac{\sqrt{m}+\sqrt{n}}{\sqrt{m}-\sqrt{n}}$

(B) Rationalize the numerator: $\dfrac{\sqrt{3+h}-\sqrt{3}}{h}$

■ Addition and Subtraction

Because in each algebraic fraction we restrict variable replacements to real numbers that produce real fractions, addition and subtraction of algebraic fractions follow the rules for adding and subtracting fractions in real numbers; that is (excluding division by 0),

$$\frac{a}{b}+\frac{c}{b}=\frac{a+c}{b} \qquad \frac{a}{b}-\frac{c}{b}=\frac{a-c}{b}$$

Thus, we add algebraic fractions, if their denominators are the same, by adding or subtracting their numerators and placing the result over the common denominator. If the denominators are not the same, we raise the fractions to higher terms, using the fundamental principle of fractions to obtain common denominators, and then proceed as described.

Even though any common denominator will do, the problem will generally become less involved if the least common denominator (LCD)

The Least Common Denominator

The LCD of two or more rational expressions is found as follows:

1. Factor each denominator completely.
2. Form a product that contains each different factor from all denominators to the highest power it occurs in any one denominator. This product is the LCD.

is used. Often, the LCD is obvious, but if it is not, proceed as described in the box to find it.

EXAMPLE 4 Combine into a single fraction and reduce to lowest terms.

(A) $\dfrac{3}{10} + \dfrac{5}{6} - \dfrac{11}{45}$ (B) $\dfrac{4}{9x} - \dfrac{5x}{6y^2} + 1$

(C) $\dfrac{x+3}{x^2 - 6x + 9} - \dfrac{x+2}{x^2 - 9} - \dfrac{5}{3-x}$

Solution (A) To find the LCD, factor each denominator completely.

$$\left.\begin{array}{l} 10 = 2 \cdot 5 \\ 6 = 2 \cdot 3 \\ 45 = 3^2 \cdot 5 \end{array}\right\} \text{LCD} = 2 \cdot 3^2 \cdot 5 = 90$$

Now use the fundamental principle of fractions to make each denominator 90.

$$\dfrac{3}{10} + \dfrac{5}{6} - \dfrac{11}{45} = \dfrac{9 \cdot 3}{9 \cdot 10} + \dfrac{15 \cdot 5}{15 \cdot 6} - \dfrac{2 \cdot 11}{2 \cdot 45}$$

$$= \dfrac{27}{90} + \dfrac{75}{90} - \dfrac{22}{90}$$

$$= \dfrac{27 + 75 - 22}{90} = \dfrac{80}{90} = \dfrac{8}{9}$$

(B) $\left.\begin{array}{l} 9x = 3^2 x \\ 6y^2 = 2 \cdot 3y^2 \end{array}\right\} \text{LCD} = 2 \cdot 3^2 xy^2 = 18xy^2$

$$\dfrac{4}{9x} - \dfrac{5x}{6y^2} + 1 = \dfrac{2y^2 \cdot 4}{2y^2 \cdot 9x} - \dfrac{3x \cdot 5x}{3x \cdot 6y^2} + \dfrac{18xy^2}{18xy^2}$$

$$= \dfrac{8y^2 - 15x^2 + 18xy^2}{18xy^2}$$

(C) $\dfrac{x+3}{x^2 - 6x + 9} - \dfrac{x+2}{x^2 - 9} - \dfrac{5}{3-x} = \dfrac{x+3}{(x-3)^2} - \dfrac{x+2}{(x-3)(x+3)} + \dfrac{5}{x-3}$

Note: $-\dfrac{5}{3-x} = -\dfrac{5}{-(x-3)} = \dfrac{5}{x-3}$ We have again used the fact that $(a-b) = -(b-a)$.

Appendix A-7 Fractions

The LCD $= (x-3)^2(x+3)$. Thus,

$$\frac{(x+3)^2}{(x-3)^2(x+3)} - \frac{(x-3)(x+2)}{(x-3)^2(x+3)} + \frac{5(x-3)(x+3)}{(x-3)^2(x+3)}$$

$$= \frac{(x^2+6x+9)-(x^2-x-6)+5(x^2-9)}{(x-3)^2(x+3)}$$

$$= \frac{x^2+6x+9-x^2+x+6+5x^2-45}{(x-3)^2(x+3)}$$

$$= \frac{5x^2+7x-30}{(x-3)^2(x+3)}$$

PROBLEM 4 Combine into a single fraction and reduce to lowest terms.

(A) $\dfrac{5}{28} - \dfrac{1}{10} + \dfrac{6}{35}$

(B) $\dfrac{1}{4x^2} - \dfrac{2x+1}{3x^3} + \dfrac{3}{12x}$

(C) $\dfrac{y-3}{y^2-4} - \dfrac{y+2}{y^2-4y+4} - \dfrac{2}{2-y}$

■ **Complex Fractions**

A fractional form with fractions in its numerator, denominator, or both is called a **complex fraction**. It is often necessary to represent a complex fraction as a **simple fraction**—that is (in all cases we will consider), as the quotient of two polynomials. The process does not involve any new concepts. It is a matter of applying old concepts and processes in the right sequence. We will illustrate two approaches to the problem, each with its own merits, depending on the particular problem under consideration. One of the methods makes very effective use of the fundamental principle of fractions:

$$\frac{a}{b} = \frac{ka}{kb} \qquad b, k \neq 0$$

EXAMPLE 5 Express as a simple fraction: $\dfrac{\dfrac{y}{x^2} - \dfrac{x}{y^2}}{\dfrac{y}{x} - \dfrac{x}{y}}$

Solution *Method 1:* Multiply the numerator and denominator by the LCD of all fractions in the numerator and denominator—in this case, x^2y^2.

Appendixes

$$\frac{x^2y^2\left(\dfrac{y}{x^2} - \dfrac{x}{y^2}\right)}{x^2y^2\left(\dfrac{y}{x} - \dfrac{x}{y}\right)} = \frac{y^3 - x^3}{xy^3 - x^3y} = \frac{\cancel{(y-x)}(y^2 + xy + x^2)}{xy\cancel{(y-x)}(y+x)}$$

$$= \frac{y^2 + xy + x^2}{xy(y+x)} \quad \text{or} \quad \frac{x^2 + xy + y^2}{xy(x+y)}$$

Method 2: Write the numerator and denominator as single fractions. Then treat as a quotient.

$$\frac{\dfrac{y}{x^2} - \dfrac{x}{y^2}}{\dfrac{y}{x} - \dfrac{x}{y}} = \frac{\dfrac{y^3 - x^3}{x^2y^2}}{\dfrac{y^2 - x^2}{xy}} = \frac{y^3 - x^3}{x^2y^2} \div \frac{y^2 - x^2}{xy}$$

$$= \frac{\cancel{(y-x)}(y^2 + xy + x^2)}{\underset{xy}{x^2y^2}} \cdot \frac{\cancel{xy}}{\cancel{(y-x)}(y+x)}$$

$$= \frac{x^2 + xy + y^2}{xy(x+y)}$$

PROBLEM 5 Express as a simple fraction reduced to lowest terms. Use the two methods described in Example 5.

$$\frac{\dfrac{a}{b} - \dfrac{b}{a}}{\dfrac{a}{b} + 2 + \dfrac{b}{a}}$$

EXAMPLE 6 Express as simple fractions reduced to lowest terms.

(A) $\displaystyle \frac{x^{-2} - y^{-2}}{x^{-1} + y^{-1}} = \frac{\dfrac{1}{x^2} - \dfrac{1}{y^2}}{\dfrac{1}{x} + \dfrac{1}{y}} = \frac{x^2y^2\left(\dfrac{1}{x^2} - \dfrac{1}{y^2}\right)}{x^2y^2\left(\dfrac{1}{x} + \dfrac{1}{y}\right)}$

$$= \frac{y^2 - x^2}{xy^2 + x^2y} = \frac{(y-x)\cancel{(y+x)}}{xy\cancel{(y+x)}}$$

$$= \frac{y - x}{xy}$$

Appendix A-7 Fractions

(B) $2 - \dfrac{1}{2 - \dfrac{2}{2 + \dfrac{1}{x}}} = 2 - \dfrac{1}{2 - \dfrac{x \cdot 2}{x\left(2 + \dfrac{1}{x}\right)}}$ Write $\dfrac{2}{2 + \dfrac{1}{x}}$ as a simple fraction first.

$= 2 - \dfrac{1}{2 - \dfrac{2x}{2x + 1}}$

$= 2 - \dfrac{(2x + 1) \cdot 1}{(2x + 1)\left(2 - \dfrac{2x}{2x + 1}\right)}$

$= 2 - \dfrac{2x + 1}{4x + 2 - 2x}$

$= 2 - \dfrac{2x + 1}{2x + 2} = \dfrac{4x + 4 - 2x - 1}{2x + 2}$

$= \dfrac{2x + 3}{2x + 2}$

PROBLEM 6 Express as simple fractions reduced to lowest terms.

(A) $\dfrac{x - x^{-1}}{1 - x^{-2}}$ (B) $2 - \dfrac{1}{2 - \dfrac{1}{1 - \dfrac{1}{x}}}$

Answers to Matched Problems

1. (A) $(3x + 2)/(x + 1)$ (B) $(x^2 + 2x + 4)/(3x + 4)$
2. (A) $2x$ (B) $-5/(x + 4)$ (C) mn
3. (A) $(m + 2\sqrt{mn} + n)/(m - n)$ (B) $1/(\sqrt{3 + h} + \sqrt{3})$
4. (A) $\tfrac{1}{4}$ (B) $(3x^2 - 5x - 4)/12x^3$
   (C) $(2y^2 - 9y - 6)/[(y - 2)^2(y + 2)]$
5. $(a - b)/(a + b)$
6. (A) x (B) $(x - 3)/(x - 2)$

Exercise A-7 ■ **A** *Perform the indicated operations and reduce to lowest terms. Represent all complex fractions as simple fractions reduced to lowest terms.*

1. $\left(\dfrac{d^5}{3a} \div \dfrac{d^2}{6a^2}\right) \cdot \dfrac{a}{4d^3}$ 2. $\dfrac{d^5}{3a} \div \left(\dfrac{d^2}{6a^2} \cdot \dfrac{a}{4d^3}\right)$

3. $\dfrac{2y}{18} - \dfrac{-1}{28} - \dfrac{y}{42}$
4. $\dfrac{x^2}{12} + \dfrac{x}{18} - \dfrac{1}{30}$

5. $\dfrac{3x+8}{4x^2} - \dfrac{2x-1}{x^3} - \dfrac{5}{8x}$
6. $\dfrac{4m-3}{18m^3} + \dfrac{3}{4m} - \dfrac{2m-1}{6m^2}$

7. $\dfrac{2x^2+7x+3}{4x^2-1} \div (x+3)$
8. $\dfrac{x^2-9}{x^2-3x} \div (x^2 - x - 12)$

9. $\dfrac{m+n}{m^2-n^2} \div \dfrac{m^2-mn}{m^2-2mn+n^2}$
10. $\dfrac{x^2-6x+9}{x^2-x-6} \div \dfrac{x^2+2x-15}{x^2+2x}$

11. $\dfrac{1}{a^2-b^2} + \dfrac{1}{a^2+2ab+b^2}$
12. $\dfrac{3}{x^2-1} - \dfrac{2}{x^2-2x+1}$

13. $m - 3 - \dfrac{m-1}{m-2}$
14. $\dfrac{x+1}{x-1} - 1$

15. $\dfrac{5}{x-3} - \dfrac{2}{3-x}$
16. $\dfrac{3}{a-1} - \dfrac{2}{1-a}$

17. $\dfrac{2}{y+3} - \dfrac{1}{y-3} + \dfrac{2y}{y^2-9}$
18. $\dfrac{2x}{x^2-y^2} + \dfrac{1}{x+y} - \dfrac{1}{x-y}$

19. $\dfrac{1 - \dfrac{y^2}{x^2}}{1 - \dfrac{y}{x}}$
20. $\dfrac{1 + \dfrac{3}{x}}{x - \dfrac{9}{x}}$

21. $\dfrac{m^{-1}+1}{m+1}$
22. $\dfrac{m^{-2}-1}{m^{-1}+1}$

B
23. $\dfrac{y}{y^2-y-2} - \dfrac{1}{y^2+5y-14} - \dfrac{2}{y^2+8y+7}$

24. $\dfrac{x^2}{x^2+2x+1} + \dfrac{x-1}{3x+3} - \dfrac{1}{6}$

25. $\dfrac{9-m^2}{m^2+5m+6} \cdot \dfrac{m+2}{m-3}$
26. $\dfrac{2-x}{2x+x^2} \cdot \dfrac{x^2+4x+4}{x^2-4}$

27. $\dfrac{x+7}{ax-bx} + \dfrac{y+9}{by-ay}$
28. $\dfrac{c+2}{5c-5} - \dfrac{c-2}{3c-3} + \dfrac{c}{1-c}$

29. $\dfrac{x^2-16}{2x^2+10x+8} \div \dfrac{x^2-13x+36}{x^3+1}$

30. $\left(\dfrac{x^3-y^3}{y^3} \cdot \dfrac{y}{x-y}\right) \div \dfrac{x^2+xy+y^2}{y^2}$

31. $\dfrac{x^2-xy}{xy+y^2} \div \left(\dfrac{x^2-y^2}{x^2+2xy+y^2} \div \dfrac{x^2-2xy+y^2}{x^2y+xy^2}\right)$

32. $\left(\dfrac{x^2 - xy}{xy + y^2} \div \dfrac{x^2 - y^2}{x^2 + 2xy + y^2}\right) \div \dfrac{x^2 - 2xy + y^2}{x^2y + xy^2}$

33. $\left(\dfrac{x}{x^2 - 16} - \dfrac{1}{x + 4}\right) \div \dfrac{4}{x + 4}$

34. $\left(\dfrac{3}{x - 2} - \dfrac{1}{x + 1}\right) \div \dfrac{x + 4}{x - 2}$

35. $\dfrac{x^{-1} + y^{-1}}{x + y}$

36. $\dfrac{c - d}{c^{-1} - d^{-1}}$

37. $\dfrac{1 + \dfrac{2}{x} - \dfrac{15}{x^2}}{1 + \dfrac{4}{x} - \dfrac{5}{x^2}}$

38. $\dfrac{\dfrac{x}{y} - 2 + \dfrac{y}{x}}{\dfrac{x}{y} - \dfrac{y}{x}}$

39. $\dfrac{xy^{-2} - yx^{-2}}{y^{-1} - x^{-1}}$

40. $\dfrac{b^{-2} - c^{-2}}{b^{-3} - c^{-3}}$

41. $\dfrac{y - \dfrac{y^2}{y - x}}{1 + \dfrac{x^2}{y^2 - x^2}}$

42. $\dfrac{\dfrac{s^2}{s - t} - s}{\dfrac{t^2}{s - t} + t}$

43. $2 - \dfrac{1}{1 - \dfrac{2}{a + 2}}$

44. $1 - \dfrac{1}{1 - \dfrac{1}{1 - \dfrac{1}{x}}}$

45. $\left(\dfrac{-b + \sqrt{b^2 - 4ac}}{2a}\right)\left(\dfrac{-b - \sqrt{b^2 - 4ac}}{2a}\right)$

46. $\dfrac{-b + \sqrt{b^2 - 4ac}}{2a} + \dfrac{-b - \sqrt{b^2 - 4ac}}{2a}$

Rationalize denominators. [Hint: $(a - b)(a + b) = a^2 - b^2$.]

47. $\dfrac{3 - \sqrt{a}}{\sqrt{a} - 2}$

48. $\dfrac{2 + \sqrt{x}}{\sqrt{x} - 3}$

49. $\dfrac{2\sqrt{5} - 3\sqrt{2}}{5\sqrt{5} + 2\sqrt{2}}$

50. $\dfrac{3\sqrt{2} + 2\sqrt{3}}{2\sqrt{2} - 3\sqrt{3}}$

51. $\dfrac{x^2}{\sqrt{x^2 + 9} - 3}$

52. $\dfrac{-y^2}{2 - \sqrt{y^2 + 4}}$

Rationalize numerators.

53. $\dfrac{\sqrt{t} - \sqrt{x}}{t - x}$

54. $\dfrac{\sqrt{x} - \sqrt{y}}{\sqrt{x} + \sqrt{y}}$

55. $\dfrac{\sqrt{x + h} - \sqrt{x}}{h}$

56. $\dfrac{\sqrt{2 + h} + \sqrt{2}}{h}$

C Combine into single terms. First write each radical in simplest radical form.

57. $\sqrt{\dfrac{3xy}{2}} + \sqrt{\dfrac{2xy}{3}}$

58. $\sqrt{\dfrac{5a}{8}} - \sqrt{\dfrac{2a}{5}}$

59. $\sqrt{50} - \dfrac{2}{\sqrt{2}} + \sqrt{\dfrac{9}{2}}$

60. $\sqrt{\dfrac{1}{3}} - \dfrac{2}{\sqrt{3}} + \sqrt{12}$

Write as simple fractions.

61. $\left(\dfrac{x^{-1}}{x^{-1} - y^{-1}}\right)^{-1}$

62. $\left[\dfrac{u^{-2} - v^{-2}}{(u^{-1} - v^{-1})^2}\right]^{-1}$

63. $1 - \dfrac{1}{1 - \dfrac{1}{1 - \dfrac{1}{1 - \dfrac{1}{x}}}}$

64. $1 + \dfrac{1}{1 + \dfrac{1}{1 + \dfrac{1}{1 + x}}}$

Rationalize the numerators. [Hint: $(a - b)(a^2 + ab + b^2) = a^3 - b^3$.]

65. $\dfrac{\sqrt[3]{t} - \sqrt[3]{x}}{t - x}$

66. $\dfrac{\sqrt[3]{x + h} - \sqrt[3]{x}}{h}$

Appendix A-8 Table Evaluation of Trigonometric Functions

- Evaluation for $0 < x < \pi/2$ and $0° < \theta < 90°$
- Evaluation for All x and All θ

- **Evaluation for $0 < x < \pi/2$ and $0° < \theta < 90°$**

Tables V and VI (following this appendix) give values of the trigonometric functions for angles in radian measure (or for real numbers) and for angles in degree measure. Table V includes values from 0 to 1.60 (from 0 to a little more than $\pi/2$) in increments of 0.01. Table VI includes values from 0° to 90° in increments of 10′.

EXAMPLE 1 Use Table V or VI to find the approximate values of:

(A) sin 37°20′   (B) cos 63°42′   (C) tan 1.35   (D) csc 0.3453

Solution (A) Read the value directly from Table VI:

$\sin 37°20′ = 0.6065$

(B) Round 63°42′ to the nearest Table VI value, 63°40′. The latter is found in the right-hand column and cosine is read up from the bottom.

Thus,

$$\cos 63°42' \approx \cos 63°40' = 0.4436$$

(C) Read the value directly from Table V:

$$\tan 1.35 = 4.455$$

(D) Round 0.3453 to the nearest Table V value, 0.35, and read from the table:

$$\csc 0.3453 \approx \csc 0.35 = 2.916$$

PROBLEM 1 Use Table V or VI to find the approximate values for:

(A) $\cot 8°50'$ (B) $\tan 72°27'$ (C) $\cos 0.84$ (D) $\sec 1.406\ 82$

■ **Evaluation for All *x* and All θ**

How can tables restricted to first-quadrant values be used to approximate values of the trigonometric functions for any angle or real number? Reference angles and triangles and the basic definitions of the functions are at the heart of the solution to this problem. (In the discussion that follows, we assume that no angle or real number is associated with an angle that is coterminal with a coordinate axis. For such angles, all trigonometric functions, if they are defined, are easily evaluated without a table or a calculator.)

For a given angle θ, in degree or radian measure, we find a reference triangle and reference angle as follows:

Reference Triangle and Angle

1. To form a **reference triangle** for θ, drop a perpendicular from a point $P(a, b)$ on the terminal side of θ to the horizontal axis.
2. The **reference angle** α is the acute angle (always taken positive) between the terminal side of θ and the horizontal axis.

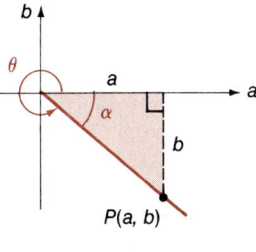

$(a, b) \neq (0, 0)$

EXAMPLE 2 Find a reference triangle and a reference angle for:

(A) 212°23′   (B) 2.03   (C) −6.83

Solution   Draw angles in standard position and form reference triangles; then compute reference angles.

(A) α = 212°23′ − 180°
    = 32°23′

(B) α = 3.14 − 2.03
    = 1.11 radians

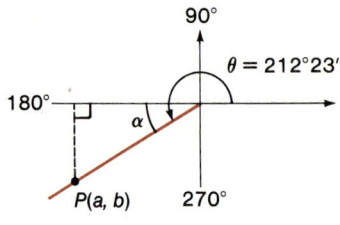
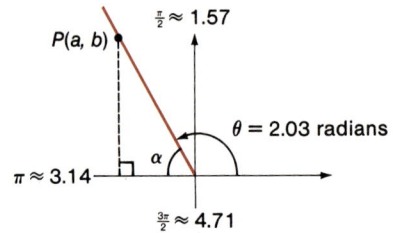

(C) α = 6.83 − 6.28
    = 0.55 radian

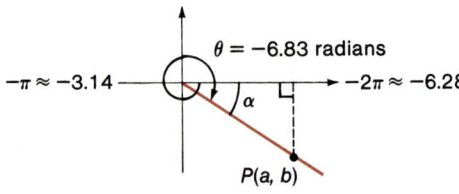

PROBLEM 2 Find reference angles for:

(A) 306°25′   (B) −2.34   (C) 6.56

Theorem 1 shows how to use reference angles in evaluating trigonometric functions.

THEOREM 1   Let α be the reference angle corresponding to a given angle θ. Then:

$$\sin \theta = (\pm) \sin \alpha \qquad \csc \theta = (\pm) \csc \alpha$$
$$\cos \theta = (\pm) \cos \alpha \qquad \sec \theta = (\pm) \sec \alpha$$
$$\tan \theta = (\pm) \tan \alpha \qquad \cot \theta = (\pm) \cot \alpha$$

where the sign on the right of each equation is chosen to be the same as the sign of the particular function in the quadrant that contains the terminal side of θ.

We will prove a small part of Theorem 1 to illustrate how it can be proved in general. Suppose the terminal side of θ is in quadrant II. We locate the reference triangle and the reference angle, and also place the reference triangle in the first quadrant with α in the standard position (Fig. 1). Let (a, b) be a point on the terminal side of α in the first quadrant, excluding the origin. Then (−a, b) is on the terminal side of θ. Thus,

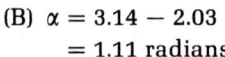

FIGURE 1

using the definitions of the trigonometric functions, we have

$$\sin\theta = \frac{b}{R} = \sin\alpha$$

$$\cos\theta = \frac{-a}{R} = -\frac{a}{R} = -\cos\alpha$$

$$\tan\theta = \frac{b}{-a} = -\frac{b}{a} = -\tan\alpha$$

and so on.

EXAMPLE 3 Use Theorem 1 and Table V or VI to find:

(A) $\cos 222°20'$ (B) $\tan(-4.32)$

Solution (A) *Step 1.* Find reference angle α.

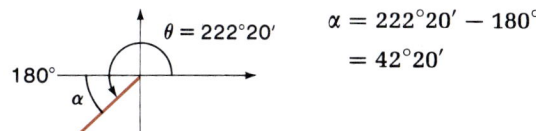

$$\alpha = 222°20' - 180°$$
$$= 42°20'$$

Step 2. Determine the sign of the cosine in the third quadrant. It is negative, since a in $P(a, b)$ is negative in the third quadrant.

Step 3. Use Theorem 1 and Table VI.

$$\cos 222°20' = -\cos 42°20'$$
$$= -0.7392$$

(B) Find reference angle α.

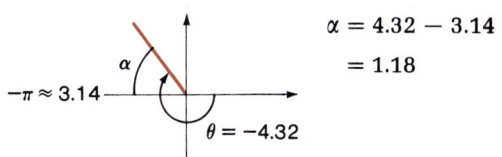

$$\alpha = 4.32 - 3.14$$
$$= 1.18$$

The terminal side of θ is in the second quadrant and the tangent function is negative in the second quadrant. Hence, using Theorem 1 and Table V, we obtain

$$\tan(-4.32) = -\tan 1.18$$
$$= -2.427$$

PROBLEM 3 Use Theorem 1 and Table V or VI to find:

(A) $\sin(-3.64)$ (B) $\cos(-53°40')$

Answers to Matched Problems
1. (A) 6.435 (B) 3.172 (C) 0.6675 (D) 6.246
2. (A) 53°35' (B) 0.80 radian (C) 0.28 radian

3. (A) sin(−3.64) = sin 0.50 = 0.4794
   (B) cos(−53°40′) = cos 53°40′ = 0.5925

Exercise A-8

A *Find the following using the nearest table values in Table V or VI.*

1. sin 12°10′
2. cos 33°30′
3. tan 78°27′
4. cot 53°52′
5. sec 80°17′
6. csc 69°47′
7. cos 0.35
8. sin 0.08
9. cot 1.0356
10. tan 0.3043
11. csc 1.038
12. sec 0.0832

Find the reference angle α for each. Use π = 3.14 where necessary.

13. 195°
14. 230°
15. 340°
16. 280°
17. −55°
18. −78°
19. −340°
20. −260°
21. 165°40′
22. 195°10′
23. 3
24. 2
25. 7
26. 5
27. −4.53
28. −1.55
29. −1.72
30. −5.37

B *Select the appropriate sign on the right side of each equation to make the equation correct according to Theorem 1.*

31. cos 170° = (±) cos 10°
32. sin 195° = (±) sin 15°
33. cot 230° = (±) cot 50°
34. tan 150° = (±) tan 30°
35. csc 342°20′ = (±) csc 17°40′
36. sec 108°50′ = (±) sec 71°10′
37. tan 4.08 = (±) tan 0.94
38. sin 2.93 = (±) sin 0.21
39. cos(−1.37) = (±) cos 1.37
40. sin(−2.89) = (±) sin 0.25

Use Theorem 1 and Table V or VI to evaluate each of the following to three significant digits. Use π = 3.14 where necessary.

41. cos 153°
42. sin 192°
43. cot 340°
44. tan 281°
45. sec(−35°)
46. csc(−35°)
47. sin(−172°30′)
48. tan(−210°40′)
49. cos 253°10′
50. csc 322°40′
51. cos 3
52. sin 5
53. sec 4
54. tan 6
55. cot(−2)
56. sin(−2.34)
57. cos 2.32
58. tan(−2.13)
59. cot 4.15
60. sec(−7.23)

C *Use Theorem 1 and Table V or VI to evaluate to four significant digits. Use π = 3.14 where necessary.*

61. sin 503°20′
62. cos(−423°10′)
63. tan 8.25
64. cot 10.56
65. sec(−9.03)
66. csc(−13.28)

Tables

I   Values of e^x and e^{-x} (0.00 to 3.00)
II   Common Logarithms
III   Natural Logarithms ($\ln x = \log_e x$)
IV   Logarithms of Factorial n
V   Trigonometric Functions—Radians or Real Numbers
VI   Trigonometric Functions—Degrees and Minutes

TABLE I Values of e^x and e^{-x} (0.00 to 3.00)

x	e^x	e^{-x}	x	e^x	e^{-x}	x	e^x	e^{-x}
0.00	1.0000	1.000 00	0.50	1.6487	0.606 53	1.00	2.7183	0.367 88
0.01	1.0101	0.990 05	0.51	1.6653	0.600 50	1.01	2.7456	0.364 22
0.02	1.0202	0.980 20	0.52	1.6820	0.594 52	1.02	2.7732	0.360 59
0.03	1.0305	0.970 45	0.53	1.6989	0.588 60	1.03	2.8011	0.357 01
0.04	1.0408	0.960 79	0.54	1.7160	0.582 75	1.04	2.8292	0.353 45
0.05	1.0513	0.951 23	0.55	1.7333	0.576 95	1.05	2.8577	0.349 94
0.06	1.0618	0.941 76	0.56	1.7507	0.571 21	1.06	2.8864	0.346 46
0.07	1.0725	0.932 39	0.57	1.7683	0.565 53	1.07	2.9154	0.343 01
0.08	1.0833	0.923 12	0.58	1.7860	0.559 90	1.08	2.9447	0.339 60
0.09	1.0942	0.913 93	0.59	1.8040	0.554 33	1.09	2.9743	0.336 22
0.10	1.1052	0.904 84	0.60	1.8221	0.548 81	1.10	3.0042	0.332 87
0.11	1.1163	0.895 83	0.61	1.8404	0.543 35	1.11	3.0344	0.329 56
0.12	1.1275	0.886 92	0.62	1.8589	0.537 94	1.12	3.0649	0.326 28
0.13	1.1388	0.878 10	0.63	1.8776	0.532 59	1.13	3.0957	0.323 03
0.14	1.1503	0.869 36	0.64	1.8965	0.527 29	1.14	3.1268	0.319 82
0.15	1.1618	0.860 71	0.65	1.9155	0.522 05	1.15	3.1582	0.316 64
0.16	1.1735	0.852 14	0.66	1.9348	0.516 85	1.16	3.1899	0.313 49
0.17	1.1853	0.843 66	0.67	1.9542	0.511 71	1.17	3.2220	0.310 37
0.18	1.1972	0.835 27	0.68	1.9739	0.506 62	1.18	3.2544	0.307 28
0.19	1.2092	0.826 96	0.69	1.9937	0.501 58	1.19	3.2871	0.304 22
0.20	1.2214	0.818 73	0.70	2.0138	0.496 59	1.20	3.3201	0.301 19
0.21	1.2337	0.810 58	0.71	2.0340	0.491 64	1.21	3.3535	0.298 20
0.22	1.2461	0.802 52	0.72	2.0544	0.486 75	1.22	3.3872	0.295 23
0.23	1.2586	0.794 53	0.73	2.0751	0.481 91	1.23	3.4212	0.292 29
0.24	1.2712	0.786 63	0.74	2.0959	0.477 11	1.24	3.4556	0.289 38
0.25	1.2840	0.778 80	0.75	2.1170	0.472 37	1.25	3.4903	0.286 50
0.26	1.2969	0.771 05	0.76	2.1383	0.467 67	1.26	3.5254	0.283 65
0.27	1.3100	0.763 38	0.77	2.1598	0.463 01	1.27	3.5609	0.280 83
0.28	1.3231	0.755 78	0.78	2.1815	0.458 41	1.28	3.5966	0.278 04
0.29	1.3364	0.748 26	0.79	2.2034	0.453 84	1.29	3.6328	0.275 27
0.30	1.3499	0.740 82	0.80	2.2255	0.449 33	1.30	3.6693	0.272 53
0.31	1.3634	0.733 45	0.81	2.2479	0.444 86	1.31	3.7062	0.269 82
0.32	1.3771	0.726 15	0.82	2.2705	0.440 43	1.32	3.7434	0.267 14
0.33	1.3910	0.718 92	0.83	2.2933	0.436 05	1.33	3.7810	0.264 48
0.34	1.4049	0.711 77	0.84	2.3164	0.431 71	1.34	3.8190	0.261 85
0.35	1.4191	0.704 69	0.85	2.3396	0.427 41	1.35	3.8574	0.259 24
0.36	1.4333	0.697 68	0.86	2.3632	0.423 16	1.36	3.8962	0.256 66
0.37	1.4477	0.690 73	0.87	2.3869	0.418 95	1.37	3.9354	0.254 11
0.38	1.4623	0.683 86	0.88	2.4109	0.414 78	1.38	3.9749	0.251 58
0.39	1.4770	0.677 06	0.89	2.4351	0.410 66	1.39	4.0149	0.249 08
0.40	1.4918	0.670 32	0.90	2.4596	0.406 57	1.40	4.0552	0.246 60
0.41	1.5068	0.663 65	0.91	2.4843	0.402 52	1.41	4.0960	0.244 14
0.42	1.5220	0.657 05	0.92	2.5093	0.398 52	1.42	4.1371	0.241 71
0.43	1.5373	0.650 51	0.93	2.5345	0.394 55	1.43	4.1787	0.239 31
0.44	1.5527	0.644 04	0.94	2.5600	0.390 63	1.44	4.2207	0.236 93
0.45	1.5683	0.637 63	0.95	2.5857	0.386 74	1.45	4.2631	0.234 57
0.46	1.5841	0.631 28	0.96	2.6117	0.382 89	1.46	4.3060	0.232 24
0.47	1.6000	0.625 00	0.97	2.6379	0.379 08	1.47	4.3492	0.229 93
0.48	1.6161	0.618 78	0.98	2.6645	0.375 31	1.48	4.3939	0.227 64
0.49	1.6323	0.612 63	0.99	2.6912	0.371 58	1.49	4.4371	0.225 37
0.50	1.6487	0.606 53	1.00	2.7183	0.367 88	1.50	4.4817	0.223 13

x	e^x	e^{-x}	x	e^x	e^{-x}	x	e^x	e^{-x}
1.50	4.4817	0.223 13	2.00	7.3891	0.135 34	2.50	12.182	0.082 085
1.51	4.5267	0.220 91	2.01	7.4633	0.133 99	2.51	12.305	0.081 268
1.52	4.5722	0.218 71	2.02	7.5383	0.132 66	2.52	12.429	0.080 460
1.53	4.6182	0.216 54	2.03	7.6141	0.131 34	2.53	12.554	0.079 659
1.54	4.6646	0.214 38	2.04	7.6906	0.130 03	2.54	12.680	0.078 866
1.55	4.7115	0.212 25	2.05	7.7679	0.128 73	2.55	12.807	0.078 082
1.56	4.7588	0.210 14	2.06	7.8460	0.127 45	2.56	12.936	0.077 305
1.57	4.8066	0.208 05	2.07	7.9248	0.126 19	2.57	13.066	0.076 536
1.58	4.8550	0.205 98	2.08	8.0045	0.124 93	2.58	13.197	0.075 774
1.59	4.9037	0.203 93	2.09	8.0849	0.123 69	2.59	13.330	0.075 020
1.60	4.9530	0.201 90	2.10	8.1662	0.122 46	2.60	13.464	0.074 274
1.61	5.0028	0.199 89	2.11	8.2482	0.121 24	2.61	13.599	0.073 535
1.62	5.0531	0.197 90	2.12	8.3311	0.120 03	2.62	13.736	0.072 803
1.63	5.1039	0.195 93	2.13	8.4149	0.118 84	2.63	13.874	0.072 078
1.64	5.1552	0.193 98	2.14	8.4994	0.117 65	2.64	14.013	0.071 361
1.65	5.2070	0.192 05	2.15	8.5849	0.116 48	2.65	14.154	0.070 651
1.66	5.2593	0.190 14	2.16	8.6711	0.115 33	2.66	14.296	0.069 948
1.67	5.3122	0.188 25	2.17	8.7583	0.114 18	2.67	14.440	0.069 252
1.68	5.3656	0.186 37	2.18	8.8463	0.113 04	2.68	14.585	0.068 563
1.69	5.4195	0.184 52	2.19	8.9352	0.111 92	2.69	14.732	0.067 881
1.70	5.4739	0.182 68	2.20	9.0250	0.110 80	2.70	14.880	0.067 206
1.71	5.5290	0.180 87	2.21	9.1157	0.109 70	2.71	15.029	0.066 537
1.72	5.5845	0.179 07	2.22	9.2073	0.108 61	2.72	15.180	0.065 875
1.73	5.6407	0.177 28	2.23	9.2999	0.107 53	2.73	15.333	0.065 219
1.74	5.6973	0.175 52	2.24	9.3933	0.106 46	2.74	15.487	0.064 570
1.75	5.7546	0.173 77	2.25	9.4877	0.105 40	2.75	15.643	0.063 928
1.76	5.8124	0.172 04	2.26	9.5831	0.104 35	2.76	15.800	0.063 292
1.77	5.8709	0.170 33	2.27	9.6794	0.103 31	2.77	15.959	0.062 662
1.78	5.9299	0.168 64	2.28	9.7767	0.102 28	2.78	16.119	0.062 039
1.79	5.9895	0.166 96	2.29	9.8749	0.101 27	2.79	16.281	0.061 421
1.80	6.0496	0.165 30	2.30	9.9742	0.100 26	2.80	16.445	0.060 810
1.81	6.1104	0.163 65	2.31	10.074	0.099 261	2.81	16.610	0.060 205
1.82	6.1719	0.162 03	2.32	10.176	0.098 274	2.82	16.777	0.059 606
1.83	6.2339	0.160 41	2.33	10.278	0.097 296	2.83	16.945	0.059 013
1.84	6.2965	0.158 82	2.34	10.381	0.096 328	2.84	17.116	0.058 426
1.85	6.3598	0.157 24	2.35	10.486	0.095 369	2.85	17.288	0.057 844
1.86	6.4237	0.155 67	2.36	10.591	0.094 420	2.86	17.462	0.057 269
1.87	6.4883	0.154 12	2.37	10.697	0.093 481	2.87	17.637	0.056 699
1.88	6.5535	0.152 59	2.38	10.805	0.092 551	2.88	17.814	0.056 135
1.89	6.6194	0.151 07	2.39	10.913	0.091 630	2.89	17.993	0.055 576
1.90	6.6859	0.149 57	2.40	11.023	0.090 718	2.90	18.174	0.055 023
1.91	6.7531	0.148 08	2.41	11.134	0.089 815	2.91	18.357	0.054 476
1.92	6.8210	0.146 61	2.42	11.246	0.088 922	2.92	18.541	0.053 934
1.93	6.8895	0.145 15	2.43	11.359	0.088 037	2.93	18.728	0.053 397
1.94	6.9588	0.143 70	2.44	11.473	0.087 161	2.94	18.916	0.052 866
1.95	7.0287	0.142 27	2.45	11.588	0.086 294	2.95	19.106	0.052 340
1.96	7.0993	0.140 86	2.46	11.705	0.085 435	2.96	19.298	0.051 819
1.97	7.1707	0.139 46	2.47	11.822	0.084 585	2.97	19.492	0.051 303
1.98	7.2427	0.138 07	2.48	11.941	0.083 743	2.98	19.688	0.050 793
1.99	7.3155	0.136 70	2.49	12.061	0.082 910	2.99	19.886	0.050 287
2.00	7.3891	0.135 34	2.50	12.182	0.082 085	3.00	20.086	0.049 787

TABLE II Common Logarithms

x	0	1	2	3	4	5	6	7	8	9
1.0	0.0000	0.004321	0.008600	0.01284	0.01703	0.02119	0.02531	0.02938	0.03342	0.03743
1.1	0.04139	0.04532	0.04922	0.05308	0.05690	0.06070	0.06446	0.06819	0.07188	0.07555
1.2	0.07918	0.08279	0.08636	0.08991	0.09342	0.09691	0.1004	0.1038	0.1072	0.1106
1.3	0.1139	0.1173	0.1206	0.1239	0.1271	0.1303	0.1335	0.1367	0.1399	0.1430
1.4	0.1461	0.1492	0.1523	0.1553	0.1584	0.1614	0.1644	0.1673	0.1703	0.1732
1.5	0.1761	0.1790	0.1818	0.1847	0.1875	0.1903	0.1931	0.1959	0.1987	0.2014
1.6	0.2041	0.2068	0.2095	0.2122	0.2148	0.2175	0.2201	0.2227	0.2253	0.2279
1.7	0.2304	0.2330	0.2355	0.2380	0.2405	0.2430	0.2455	0.2480	0.2504	0.2529
1.8	0.2553	0.2577	0.2601	0 2625	0.2648	0.2673	0.2695	0.2718	0.2742	0.2765
1.9	0.2788	0.2810	0.2833	0.2856	0.2878	0.2900	0.2923	0.2945	0.2967	0.2989
2.0	0.3010	0.3032	0.3054	0.3075	0.3096	0.3118	0.3139	0.3160	0.3181	0.3201
2.1	0.3222	0.3243	0.3263	0.3284	0.3304	0.3324	0.3345	0.3365	0.3385	0.3404
2.2	0.3424	0.3444	0.3464	0.3483	0.3502	0.3522	0.3541	0.3560	0.3579	0.3598
2.3	0.3617	0.3636	0.3655	0.3674	0.3692	0.3711	0.3729	0.3747	0.3766	0.3784
2.4	0.3802	0.3820	0.3838	0.3856	0.3874	0.3892	0.3909	0.3927	0.3945	0.3962
2.5	0.3979	0.3997	0.4014	0.4031	0.4048	0.4065	0.4082	0.4099	0.4116	0.4133
2.6	0.4150	0.4166	0.4183	0.4200	0.4216	0.4232	0.4249	0.4265	0.4281	0.4298
2.7	0.4314	0.4330	0.4346	0.4362	0.4378	0.4393	0.4409	0.4425	0.4440	0.4456
2.8	0.4472	0.4487	0.4502	0.4518	0.4533	0.4548	0.4564	0.4579	0.4594	0.4609
2.9	0.4624	0.4639	0.4654	0.4669	0.4683	0.4698	0.4713	0.4728	0.4742	0.4757
3.0	0.4771	0.4786	0.4800	0.4814	0.4829	0.4843	0.4857	0.4871	0.4886	0.4900
3.1	0.4914	0.4928	0.4942	0.4955	0.4969	0.4983	0.4997	0.5011	0.5024	0.5038
3.2	0.5051	0.5065	0.5079	0.5092	0.5105	0.5119	0.5132	0.5145	0.5159	0.5172
3.3	0.5185	0.5198	0.5211	0.5224	0.5237	0.5250	0.5263	0.5276	0.5289	0.5302
3.4	0.5315	0.5328	0.5340	0.5353	0.5366	0.5378	0.5391	0.5403	0.5416	0.5428
3.5	0.5441	0.5453	0.5465	0.5478	0.5490	0.5502	0.5514	0.5527	0.5539	0.5551
3.6	0.5563	0.5575	0.5587	0.5599	0.5611	0.5623	0.5635	0.5647	0.5658	0.5670
3.7	0.5682	0.5694	0.5705	0.5717	0.5729	0.5740	0.5752	0.5763	0.5775	0.5786
3.8	0.5798	0.5809	0.5821	0.5832	0.5843	0.5855	0.5866	0.5877	0.5888	0.5899
3.9	0.5911	0.5922	0.5933	0.5944	0.5955	0.5966	0.5977	0.5988	0.5999	0.6010
4.0	0.6021	0.6031	0.6042	0.6053	0.6064	0.6075	0.6085	0.6096	0.6107	0.6117
4.1	0.6128	0.6138	0.6149	0.6160	0.6170	0.6180	0.6191	0.6201	0.6212	0.6222
4.2	0.6232	0.6243	0.6253	0.6263	0.6274	0.6284	0.6294	0.6304	0.6314	0.6325
4.3	0.6335	0.6345	0.6355	0.6365	0.6375	0.6385	0.6395	0.6405	0.6415	0.6425
4.4	0.6435	0.6444	0.6454	0.6464	0.6474	0.6484	0.6493	0.6503	0.6513	0.6522
4.5	0.6532	0.6542	0.6551	0.6561	0.6571	0.6580	0.6590	0.6599	0.6609	0.6618
4.6	0.6628	0.6637	0.6646	0.6656	0.6665	0.6675	0.6684	0.6693	0.6702	0.6712
4.7	0.6721	0.6730	0.6739	0.6749	0.6758	0.6767	0.6776	0.6785	0.6794	0.6803
4.8	0.6812	0.6821	0.6830	0.6839	0.6848	0.6857	0.6866	0.6875	0.6884	0.6893
4.9	0.6902	0.6911	0.6920	0.6928	0.6937	0.6946	0.6955	0.6964	0.6972	0.6981
5.0	0.6990	0.6998	0.7007	0.7016	0.7024	0.7033	0.7042	0.7050	0.7059	0.7067
5.1	0.7076	0.7084	0.7093	0.7101	0.7110	0.7118	0.7126	0.7135	0.7143	0.7152
5.2	0.7160	0.7168	0.7177	0.7185	0.7193	0.7202	0.7210	0.7218	0.7226	0.7235
5.3	0.7243	0.7251	0.7259	0.7267	0.7275	0.7284	0.7292	0.7300	0.7308	0.7316
5.4	0.7324	0.7332	0.7340	0.7348	0.7356	0.7364	0.7372	0.7380	0.7388	0.7396

x	0	1	2	3	4	5	6	7	8	9
5.5	0.7404	0.7412	0.7419	0.7427	0.7435	0.7443	0.7451	0.7459	0.7466	0.7474
5.6	0.7482	0.7490	0.7497	0.7505	0.7513	0.7520	0.7528	0.7536	0.7543	0.7551
5.7	0.7559	0.7566	0.7574	0.7582	0.7589	0.7597	0.7604	0.7612	0.7619	0.7627
5.8	0.7634	0.7642	0.7649	0.7657	0.7664	0.7672	0.7679	0.7686	0.7694	0.7701
5.9	0.7709	0.7716	0.7723	0.7731	0.7738	0.7745	0.7752	0.7760	0.7767	0.7774
6.0	0.7782	0.7789	0.7796	0.7803	0.7810	0.7818	0.7825	0.7832	0.7839	0.7846
6.1	0.7853	0.7860	0.7868	0.7875	0.7882	0.7889	0.7896	0.7903	0.7910	0.7917
6.2	0.7924	0.7931	0.7938	0.7945	0.7952	0.7959	0.7966	0.7973	0.7980	0.7987
6.3	0.7993	0.8000	0.8007	0.8014	0.8021	0.8028	0.8035	0.8041	0.8048	0.8055
6.4	0.8062	0.8069	0.8075	0.8082	0.8089	0.8096	0.8102	0.8109	0.8116	0.8122
6.5	0.8129	0.8136	0.8142	0.8149	0.8156	0.8162	0.8169	0.8176	0.8182	0.8189
6.6	0.8195	0.8202	0.8209	0.8215	0.8222	0.8228	0.8235	0.8241	0.8248	0.8254
6.7	0.8261	0.8267	0.8274	0.8280	0.8287	0.8293	0.8299	0.8306	0.8312	0.8319
6.8	0.8325	0.8331	0.8338	0.8344	0.8351	0.8357	0.8363	0.8370	0.8376	0.8382
6.9	0.8388	0.8395	0.8401	0.8407	0.8414	0.8420	0.8426	0.8432	0.8439	0.8445
7.0	0.8451	0.8457	0.8463	0.8470	0.8476	0.8482	0.8488	0.8494	0.8500	0.8506
7.1	0.8513	0.8519	0.8525	0.8531	0.8537	0.8543	0.8549	0.8555	0.8561	0.8567
7.2	0.8573	0.8579	0.8585	0.8591	0.8597	0.8603	0.8609	0.8615	0.8621	0.8627
7.3	0.8633	0.8639	0.8645	0.8651	0.8657	0.8663	0.8669	0.8675	0.8681	0.8686
7.4	0.8692	0.8698	0.8704	0.8710	0.8716	0.8722	0.8727	0.8733	0.8739	0.8745
7.5	0.8751	0.8756	0.8762	0.8768	0.8774	0.8779	0.8785	0.8791	0.8797	0.8802
7.6	0.8808	0.8814	0.8820	0.8825	0.8831	0.8837	0.8842	0.8848	0.8854	0.8859
7.7	0.8865	0.8871	0.8876	0.8882	0.8887	0.8893	0.8899	0.8904	0.8910	0.8915
7.8	0.8921	0.8927	0.8932	0.8938	0.8943	0.8949	0.8954	0.8960	0.8965	0.8971
7.9	0.8976	0.8982	0.8987	0.8993	0.8998	0.9004	0.9009	0.9015	0.9020	0.9025
8.0	0.9031	0.9036	0.9042	0.9047	0.9053	0.9058	0.9063	0.9069	0.9074	0.9079
8.1	0.9085	0.9090	0.9096	0.9101	0.9106	0.9112	0.9117	0.9122	0.9128	0.9133
8.2	0.9138	0.9143	0.9149	0.9154	0.9159	0.9165	0.9170	0.9175	0.9180	0.9186
8.3	0.9191	0.9196	0.9201	0.9206	0.9212	0.9217	0.9222	0.9227	0.9232	0.9238
8.4	0.9243	0.9248	0.9253	0.9258	0.9263	0.9269	0.9274	0.9279	0.9284	0.9289
8.5	0.9294	0.9299	0.9304	0.9309	0.9315	0.9320	0.9325	0.9330	0.9335	0.9340
8.6	0.9345	0.9350	0.9355	0.9360	0.9365	0.9370	0.9375	0.9380	0.9385	0.9390
8.7	0.9395	0.9400	0.9405	0.9410	0.9415	0.9420	0.9425	0.9430	0.9435	0.9440
8.8	0.9445	0.9450	0.9455	0.9460	0.9465	0.9469	0.9474	0.9479	0.9484	0.9489
8.9	0.9494	0.9499	0.9504	0.9509	0.9513	0.9518	0.9523	0.9528	0.9533	0.9538
9.0	0.9542	0.9547	0.9552	0.9557	0.9562	0.9566	0.9571	0.9576	0.9581	0.9586
9.1	0.9590	0.9595	0.9600	0.9605	0.9609	0.9614	0.9619	0.9624	0.9628	0.9633
9.2	0.9638	0.9643	0.9647	0.9652	0.9657	0.9661	0.9666	0.9671	0.9675	0.9680
9.3	0.9685	0.9689	0.9694	0.9699	0.9703	0.9708	0.9713	0.9717	0.9722	0.9727
9.4	0.9731	0.9736	0.9741	0.9745	0.9750	0.9754	0.9759	0.9763	0.9768	0.9773
9.5	0.9777	0.9782	0.9786	0.9791	0.9795	0.9800	0.9805	0.9809	0.9814	0.9818
9.6	0.9823	0.9827	0.9832	0.9836	0.9841	0.9845	0.9850	0.9854	0.9859	0.9863
9.7	0.9868	0.9872	0.9877	0.9881	0.9886	0.9890	0.9894	0.9899	0.9903	0.9908
9.8	0.9912	0.9917	0.9921	0.9926	0.9930	0.9934	0.9939	0.9943	0.9948	0.9952
9.9	0.9956	0.9961	0.9965	0.9969	0.9974	0.9978	0.9983	0.9987	0.9991	0.9996

TABLE III Natural Logarithms (ln x = $\log_e x$)

$\ln 10 = 2.3026$ $\quad$ $6 \ln 10 = 13.8155$
$2 \ln 10 = 4.6052$ $\quad$ $7 \ln 10 = 16.1181$
$3 \ln 10 = 6.9078$ $\quad$ $8 \ln 10 = 18.4207$
$4 \ln 10 = 9.2103$ $\quad$ $9 \ln 10 = 20.7233$
$5 \ln 10 = 11.5130$ $\quad$ $10 \ln 10 = 23.0259$

Note: $\ln 35{,}200 = \ln (3.52 \times 10^4) = \ln 3.52 + 4 \ln 10$
$\ln 0.008\ 64 = \ln (8.64 \times 10^{-3}) = \ln 8.64 - 3 \ln 10$

x	.00	.01	.02	.03	.04	.05	.06	.07	.08	.09
1.0	0.0000	0.0100	0.0198	0.0296	0.0392	0.0488	0.0583	0.0677	0.0770	0.0862
1.1	0.0953	0.1044	0.1133	0.1222	0.1310	0.1398	0.1484	0.1570	0.1655	0.1740
1.2	0.1823	0.1906	0.1989	0.2070	0.2151	0.2231	0.2311	0.2390	0.2469	0.2546
1.3	0.2624	0.2700	0.2776	0.2852	0.2927	0.3001	0.3075	0.3148	0.3221	0.3293
1.4	0.3365	0.3436	0.3507	0.3577	0.3646	0.3716	0.3784	0.3853	0.3920	0.3988
1.5	0.4055	0.4121	0.4187	0.4253	0.4318	0.4383	0.4447	0.4511	0.4574	0.4637
1.6	0.4700	0.4762	0.4824	0.4886	0.4947	0.5008	0.5068	0.5128	0.5188	0.5247
1.7	0.5306	0.5365	0.5423	0.5481	0.5539	0.5596	0.5653	0.5710	0.5766	0.5822
1.8	0.5878	0.5933	0.5988	0.6043	0.6098	0.6152	0.6206	0.6259	0.6313	0.6366
1.9	0.6419	0.6471	0.6523	0.6575	0.6627	0.6678	0.6729	0.6780	0.6831	0.6881
2.0	0.6931	0.6981	0.7031	0.7080	0.7129	0.7178	0.7227	0.7275	0.7324	0.7372
2.1	0.7419	0.7467	0.7514	0.7561	0.7608	0.7655	0.7701	0.7747	0.7793	0.7839
2.2	0.7885	0.7930	0.7975	0.8020	0.8065	0.8109	0.8154	0.8198	0.8242	0.8286
2.3	0.8329	0.8372	0.8416	0.8459	0.8502	0.8544	0.8587	0.8629	0.8671	0.8713
2.4	0.8755	0.8796	0.8838	0.8879	0.8920	0.8961	0.9002	0.9042	0.9083	0.9123
2.5	0.9163	0.9203	0.9243	0.9282	0.9322	0.9361	0.9400	0.9439	0.9478	0.9517
2.6	0.9555	0.9594	0.9632	0.9670	0.9708	0.9746	0.9783	0.9821	0.9858	0.9895
2.7	0.9933	0.9969	1.0006	1.0043	1.0080	1.0116	1.0152	1.0188	1.0225	1.0260
2.8	1.0296	1.0332	1.0367	1.0403	1.0438	1.0473	1.0508	1.0543	1.0578	1.0613
2.9	1.0647	1.0682	1.0716	1.0750	1.0784	1.0818	1.0852	1.0886	1.0919	1.0953
3.0	1.0986	1.1019	1.1053	1.1086	1.1119	1.1151	1.1184	1.1217	1.1249	1.1282
3.1	1.1314	1.1346	1.1378	1.1410	1.1442	1.1474	1.1506	1.1537	1.1569	1.1600
3.2	1.1632	1.1663	1.1694	1.1725	1.1756	1.1787	1.1817	1.1848	1.1878	1.1909
3.3	1.1939	1.1969	1.2000	1.2030	1.2060	1.2090	1.2119	1.2149	1.2179	1.2208
3.4	1.2238	1.2267	1.2296	1.2326	1.2355	1.2384	1.2413	1.2442	1.2470	1.2499
3.5	1.2528	1.2556	1.2585	1.2613	1.2641	1.2669	1.2698	1.2726	1.2754	1.2782
3.6	1.2809	1.2837	1.2865	1.2892	1.2920	1.2947	1.2975	1.3002	1.3029	1.3056
3.7	1.3083	1.3110	1.3137	1.3164	1.3191	1.3218	1.3244	1.3271	1.3297	1.3324
3.8	1.3350	1.3376	1.3403	1.3429	1.3455	1.3481	1.3507	1.3533	1.3558	1.3584
3.9	1.3610	1.3635	1.3661	1.3686	1.3712	1.3737	1.3762	1.3788	1.3813	1.3838
4.0	1.3863	1.3888	1.3913	1.3938	1.3962	1.3987	1.4012	1.4036	1.4061	1.4085
4.1	1.4110	1.4134	1.4159	1.4183	1.4207	1.4231	1.4255	1.4279	1.4303	1.4327
4.2	1.4351	1.4375	1.4398	1.4422	1.4446	1.4469	1.4493	1.4516	1.4540	1.4563
4.3	1.4586	1.4609	1.4633	1.4656	1.4679	1.4702	1.4725	1.4748	1.4770	1.4793
4.4	1.4816	1.4839	1.4861	1.4884	1.4907	1.4929	1.4951	1.4974	1.4996	1.5019
4.5	1.5041	1.5063	1.5085	1.5107	1.5129	1.5151	1.5173	1.5195	1.5217	1.5239
4.6	1.5261	1.5282	1.5304	1.5326	1.5347	1.5369	1.5390	1.5412	1.5433	1.5454
4.7	1.5476	1.5497	1.5518	1.5539	1.5560	1.5581	1.5602	1.5623	1.5644	1.5665
4.8	1.5686	1.5707	1.5728	1.5748	1.5769	1.5790	1.5810	1.5831	1.5851	1.5872
4.9	1.5892	1.5913	1.5933	1.5953	1.5974	1.5994	1.6014	1.6034	1.6054	1.6074
5.0	1.6094	1.6114	1.6134	1.6154	1.6174	1.6194	1.6214	1.6233	1.6253	1.6273
5.1	1.6292	1.6312	1.6332	1.6351	1.6371	1.6390	1.6409	1.6429	1.6448	1.6467
5.2	1.6487	1.6506	1.6525	1.6544	1.6563	1.6582	1.6601	1.6620	1.6639	1.6658
5.3	1.6677	1.6696	1.6715	1.6734	1.6752	1.6771	1.6790	1.6808	1.6827	1.6845
5.4	1.6864	1.6882	1.6901	1.6919	1.6938	1.6956	1.6974	1.6993	1.7011	1.7029

x	.00	.01	.02	.03	.04	.05	.06	.07	.08	.09
5.5	1.7047	1.7066	1.7084	1.7102	1.7120	1.7138	1.7156	1.7174	1.7192	1.7210
5.6	1.7228	1.7246	1.7263	1.7281	1.7299	1.7317	1.7334	1.7352	1.7370	1.7387
5.7	1.7405	1.7422	1.7440	1.7457	1.7475	1.7492	1.7509	1.7527	1.7544	1.7561
5.8	1.7579	1.7596	1.7613	1.7630	1.7647	1.7664	1.7681	1.7699	1.7716	1.7733
5.9	1.7750	1.7766	1.7783	1.7800	1.7817	1.7834	1.7851	1.7867	1.7884	1.7901
6.0	1.7918	1.7934	1.7951	1.7967	1.7984	1.8001	1.8017	1.8034	1.8050	1.8066
6.1	1.8083	1.8099	1.8116	1.8132	1.8148	1.8165	1.8181	1.8197	1.8213	1.8229
6.2	1.8245	1.8262	1.8278	1.8294	1.8310	1.8326	1.8342	1.8358	1.8374	1.8390
6.3	1.8405	1.8421	1.8437	1.8453	1.8469	1.8485	1.8500	1.8516	1.8532	1.8547
6.4	1.8563	1.8579	1.8594	1.8610	1.8625	1.8641	1.8656	1.8672	1.8687	1.8703
6.5	1.8718	1.8733	1.8749	1.8764	1.8779	1.8795	1.8810	1.8825	1.8840	1.8856
6.6	1.8871	1.8886	1.8901	1.8916	1.8931	1.8946	1.8961	1.8976	1.8991	1.9006
6.7	1.9021	1.9036	1.9051	1.9066	1.9081	1.9095	1.9110	1.9125	1.9140	1.9155
6.8	1.9169	1.9184	1.9199	1.9213	1.9228	1.9242	1.9257	1.9272	1.9286	1.9301
6.9	1.9315	1.9330	1.9344	1.9359	1.9373	1.9387	1.9402	1.9416	1.9430	1.9445
7.0	1.9459	1.9473	1.9488	1.9502	1.9516	1.9530	1.9544	1.9559	1.9573	1.9587
7.1	1.9601	1.9615	1.9629	1.9643	1.9657	1.9671	1.9685	1.9699	1.9713	1.9727
7.2	1.9741	1.9755	1.9769	1.9782	1.9796	1.9810	1.9824	1.9838	1.9851	1.9865
7.3	1.9879	1.9892	1.9906	1.9920	1.9933	1.9947	1.9961	1.9974	1.9988	2.0001
7.4	2.0015	2.0028	2.0042	2.0055	2.0069	2.0082	2.0096	2.0109	2.0122	2.0136
7.5	2.0149	2.0162	2.0176	2.0189	2.0202	2.0215	2.0229	2.0242	2.0255	2.0268
7.6	2.0281	2.0295	2.0308	2.0321	2.0334	2.0347	2.0360	2.0373	2.0386	2.0399
7.7	2.0412	2.0425	2.0438	2.0451	2.0464	2.0477	2.0490	2.0503	2.0516	2.0528
7.8	2.0541	2.0554	2.0567	2.0580	2.0592	2.0605	2.0618	2.0631	2.0643	2.0656
7.9	2.0669	2.0681	2.0694	2.0707	2.0719	2.0732	2.0744	2.0757	2.0769	2.0782
8.0	2.0794	2.0807	2.0819	2.0832	2.0844	2.0857	2.0869	2.0882	2.0894	2.0906
8.1	2.0919	2.0931	2.0943	2.0956	2.0968	2.0980	2.0992	2.1005	2.1017	2.1029
8.2	2.1041	2.1054	2.1066	2.1078	2.1090	2.1102	2.1114	2.1126	2.1138	2.1150
8.3	2.1163	2.1175	2.1187	2.1199	2.1211	2.1223	2.1235	2.1247	2.1258	2.1270
8.4	2.1282	2.1294	2.1306	2.1318	2.1330	2.1342	2.1353	2.1365	2.1377	2.1389
8.5	2.1401	2.1412	2.1424	2.1436	2.1448	2.1459	2.1471	2.1483	2.1494	2.1506
8.6	2.1518	2.1529	2.1541	2.1552	2.1564	2.1576	2.1587	2.1599	2.1610	2.1622
8.7	2.1633	2.1645	2.1656	2.1668	2.1679	2.1691	2.1702	2.1713	2.1725	2.1736
8.8	2.1748	2.1759	2.1770	2.1782	2.1793	2.1804	2.1815	2.1827	2.1838	2.1849
8.9	2.1861	2.1872	2.1883	2.1894	2.1905	2.1917	2.1928	2.1939	2.1950	2.1961
9.0	2.1972	2.1983	2.1994	2.2006	2.2017	2.2028	2.2039	2.2050	2.2061	2.2072
9.1	2.2083	2.2094	2.2105	2.2116	2.2127	2.2138	2.2148	2.2159	2.2170	2.2181
9.2	2.2192	2.2203	2.2214	2.2225	2.2235	2.2246	2.2257	2.2268	2.2279	2.2289
9.3	2.2300	2.2311	2.2322	2.2332	2.2343	2.2354	2.2364	2.2375	2.2386	2.2396
9.4	2.2407	2.2418	2.2428	2.2439	2.2450	2.2460	2.2471	2.2481	2.2492	2.2502
9.5	2.2513	2.2523	2.2534	2.2544	2.2555	2.2565	2.2576	2.2586	2.2597	2.2607
9.6	2.2618	2.2628	2.2638	2.2649	2.2659	2.2670	2.2680	2.2690	2.2701	2.2711
9.7	2.2721	2.2732	2.2742	2.2752	2.2762	2.2773	2.2783	2.2793	2.2803	2.2814
9.8	2.2824	2.2834	2.2844	2.2854	2.2865	2.2875	2.2885	2.2895	2.2905	2.2915
9.9	2.2925	2.2935	2.2946	2.2956	2.2966	2.2976	2.2986	2.2996	2.3006	2.3016

TABLE IV Logarithms of Factorial n

n	log $n!$	n	log $n!$	n	log $n!$	n	log $n!$
		50	64.483 07	100	157.970 00	150	262.756 89
1	0.000 00	51	66.190 64	101	159.974 32	151	264.935 87
2	0.301 03	52	67.906 65	102	161.982 93	152	267.117 71
3	0.778 15	53	69.630 92	103	163.995 76	153	269.302 41
4	1.380 21	54	71.363 32	104	166.012 80	154	271.489 93
5	2.079 18	55	73.103 68	105	168.033 99	155	273.680 26
6	2.857 33	56	74.851 87	106	170.059 29	156	275.873 38
7	3.702 43	57	76.607 74	107	172.086 67	157	278.069 28
8	4.605 52	58	78.371 17	108	174.122 10	158	280.267 94
9	5.559 76	59	80.142 02	109	176.159 52	159	282.469 34
10	6.559 76	60	81.920 17	110	178.200 92	160	284.673 46
11	7.601 16	61	83.705 50	111	180.246 24	161	286.880 28
12	8.680 34	62	85.497 90	112	182.295 46	162	289.089 80
13	9.794 28	63	87.297 24	113	184.348 54	163	291.301 98
14	10.940 41	64	89.103 42	114	186.405 44	164	293.516 83
15	12.116 50	65	90.916 33	115	188.466 14	165	295.734 31
16	13.320 62	66	92.735 87	116	190.530 60	166	297.954 42
17	14.551 07	67	94.561 95	117	192.598 78	167	300.177 14
18	15.806 34	68	96.394 46	118	194.670 67	168	302.402 45
19	17.085 09	69	98.233 31	119	196.746 21	169	304.630 33
20	18.386 12	70	100.078 40	120	198.825 39	170	306.860 78
21	19.708 34	71	101.929 66	121	200.908 18	171	309.093 78
22	21.050 77	72	103.787 00	122	202.994 54	172	311.329 31
23	22.412 49	73	105.650 32	123	205.084 44	173	313.567 35
24	23.792 71	74	107.519 55	124	207.177 87	174	315.807 90
25	25.190 65	75	109.394 61	125	209.274 78	175	318.050 94
26	26.605 62	76	111.275 43	126	211.375 15	176	320.296 45
27	28.036 98	77	113.161 92	127	213.478 95	177	322.544 43
28	29.484 14	78	115.054 01	128	215.586 16	178	324.794 85
29	30.946 54	79	116.951 64	129	217.696 75	179	327.047 70
30	32.423 66	80	118.854 73	130	219.810 69	180	329.302 97
31	33.915 02	81	120.763 21	131	221.927 96	181	331.560 65
32	35.420 17	82	122.677 03	132	224.048 54	182	333.820 72
33	36.938 69	83	124.596 10	133	226.172 39	183	336.083 17
34	38.470 16	84	126.520 38	134	228.299 49	184	338.347 99
35	40.014 23	85	128.449 80	135	230.429 83	185	340.615 16
36	41.570 54	86	130.384 30	136	232.563 37	186	342.884 68
37	43.138 74	87	132.323 82	137	234.700 09	187	345.156 52
38	44.718 52	88	134.268 30	138	236.839 97	188	347.430 67
39	46.309 59	89	136.217 69	139	238.982 98	189	349.707 14
40	47.911 65	90	138.171 94	140	241.129 11	190	351.985 89
41	49.524 43	91	140.130 98	141	242.278 33	191	354.266 92
42	51.147 68	92	142.094 76	142	245.430 62	192	356.550 22
43	52.781 15	93	144.063 25	143	247.585 95	193	358.835 78
44	54.424 60	94	146.036 38	144	249.744 32	194	361.123 58
45	56.077 81	95	148.014 10	145	251.905 68	195	363.413 62
46	57.740 57	96	149.996 37	146	254.070 04	196	365.705 87
47	59.412 67	97	151.983 14	147	256.237 35	197	368.000 34
48	61.093 91	98	153.974 37	148	258.407 62	198	370.297 01
49	62.784 10	99	155.970 00	149	260.580 80	199	372.595 86

TABLE V Trigonometric Functions—Radians or Real Numbers

	sin	cos	tan	cot	sec	csc
.00	.0000	1.0000	.0000	—	1.000	—
.01	.0100	1.0000	.0100	99.997	1.000	100.00
.02	.0200	.9998	.0200	49.993	1.000	50.00
.03	.0300	.9996	.0300	33.323	1.000	33.34
.04	.0400	.9992	.0400	24.987	1.001	25.01
.05	.0500	.9988	.0500	19.983	1.001	20.01
.06	.0600	.9982	.0601	16.647	1.002	16.68
.07	.0699	.9976	.0701	14.262	1.002	14.30
.08	.0799	.9968	.0802	12.473	1.003	12.51
.09	.0899	.9960	.0902	11.081	1.004	11.13
.10	.0998	.9950	.1003	9.967	1.005	10.02
.11	.1098	.9940	.1104	9.054	1.006	9.109
.12	.1197	.9928	.1206	8.293	1.007	8.353
.13	.1296	.9916	.1307	7.649	1.009	7.714
.14	.1395	.9902	.1409	7.096	1.010	7.166
.15	.1494	.9888	.1511	6.617	1.011	6.692
.16	.1593	.9872	.1614	6.197	1.013	6.277
.17	.1692	.9856	.1717	5.826	1.015	5.911
.18	.1790	.9838	.1820	5.495	1.016	5.586
.19	.1889	.9820	.1923	5.200	1.018	5.295
.20	.1987	.9801	.2027	4.933	1.020	5.033
.21	.2085	.9780	.2131	4.692	1.022	4.797
.22	.2182	.9759	.2236	4.472	1.025	4.582
.23	.2280	.9737	.2341	4.271	1.027	4.386
.24	.2377	.9713	.2447	4.086	1.030	4.207
.25	.2474	.9689	.2553	3.916	1.032	4.042
.26	.2571	.9664	.2660	3.759	1.035	3.890
.27	.2667	.9638	.2768	3.613	1.038	3.749
.28	.2764	.9611	.2876	3.478	1.041	3.619
.29	.2860	.9582	.2984	3.351	1.044	3.497
.30	.2955	.9553	.3093	3.233	1.047	3.384
.31	.3051	.9523	.3203	3.122	1.050	3.278
.32	.3146	.9492	.3314	3.018	1.053	3.179
.33	.3240	.9460	.3425	2.920	1.057	3.086
.34	.3335	.9428	.3537	2.827	1.061	2.999
.35	.3429	.9394	.3650	2.740	1.065	2.916
.36	.3523	.9359	.3764	2.657	1.068	2.839
.37	.3616	.9323	.3879	2.578	1.073	2.765
.38	.3709	.9287	.3994	2.504	1.077	2.696
.39	.3802	.9249	.4111	2.433	1.081	2.630
	sin	cos	tan	cot	sec	csc

TABLE V (*continued*)

	sin	cos	tan	cot	sec	csc
.40	.3894	.9211	.4228	2.365	1.086	2.568
.41	.3986	.9171	.4346	2.301	1.090	2.509
.42	.4078	.9131	.4466	2.239	1.095	2.452
.43	.4169	.9090	.4586	2.180	1.100	2.399
.44	.4259	.9048	.4708	2.124	1.105	2.348
.45	.4350	.9004	.4831	2.070	1.111	2.299
.46	.4439	.8961	.4954	2.018	1.116	2.253
.47	.4529	.8916	.5080	1.969	1.122	2.208
.48	.4618	.8870	.5206	1.921	1.127	2.166
.49	.4706	.8823	.5334	1.875	1.133	2.125
.50	.4794	.8776	.5463	1.830	1.139	2.086
.51	.4882	.8727	.5594	1.788	1.146	2.048
.52	.4969	.8678	.5726	1.747	1.152	2.013
.53	.5055	.8628	.5859	1.707	1.159	1.978
.54	.5141	.8577	.5994	1.668	1.166	1.945
.55	.5227	.8525	.6131	1.631	1.173	1.913
.56	.5312	.8473	.6269	1.595	1.180	1.883
.57	.5396	.8419	.6410	1.560	1.188	1.853
.58	.5480	.8365	.6552	1.526	1.196	1.825
.59	.5564	.8309	.6696	1.494	1.203	1.797
.60	.5646	.8253	.6841	1.462	1.212	1.771
.61	.5729	.8196	.6989	1.431	1.220	1.746
.62	.5810	.8139	.7139	1.401	1.229	1.721
.63	.5891	.8080	.7291	1.372	1.238	1.697
.64	.5972	.8021	.7445	1.343	1.247	1.674
.65	.6052	.7961	.7602	1.315	1.256	1.652
.66	.6131	.7900	.7761	1.288	1.266	1.631
.67	.6210	.7838	.7923	1.262	1.276	1.610
.68	.6288	.7776	.8087	1.237	1.286	1.590
.69	.6365	.7712	.8253	1.212	1.297	1.571
.70	.6442	.7648	.8423	1.187	1.307	1.552
.71	.6518	.7584	.8595	1.163	1.319	1.534
.72	.6594	.7518	.8771	1.140	1.330	1.517
.73	.6669	.7452	.8949	1.117	1.342	1.500
.74	.6743	.7385	.9131	1.095	1.354	1.483
.75	.6816	.7317	.9316	1.073	1.367	1.467
.76	.6889	.7248	.9505	1.052	1.380	1.452
.77	.6961	.7179	.9697	1.031	1.393	1.437
.78	.7033	.7109	.9893	1.011	1.407	1.422
.79	.7104	.7038	1.009	.9908	1.421	1.408
	sin	cos	tan	cot	sec	csc

	sin	cos	tan	cot	sec	csc
.80	.7174	.6967	1.030	.9712	1.435	1.394
.81	.7243	.6895	1.050	.9520	1.450	1.381
.82	.7311	.6822	1.072	.9331	1.466	1.368
.83	.7379	.6749	1.093	.9146	1.482	1.355
.84	.7446	.6675	1.116	.8964	1.498	1.343
.85	.7513	.6600	1.138	.8785	1.515	1.331
.86	.7578	.6524	1.162	.8609	1.533	1.320
.87	.7643	.6448	1.185	.8437	1.551	1.308
.88	.7707	.6372	1.210	.8267	1.569	1.297
.89	.7771	.6294	1.235	.8100	1.589	1.287
.90	.7833	.6216	1.260	.7936	1.609	1.277
.91	.7895	.6137	1.286	.7774	1.629	1.267
.92	.7956	.6058	1.313	.7615	1.651	1.257
.93	.8016	.5978	1.341	.7458	1.673	1.247
.94	.8076	.5898	1.369	.7303	1.696	1.238
.95	.8134	.5817	1.398	.7151	1.719	1.229
.96	.8192	.5735	1.428	.7001	1.744	1.221
.97	.8249	.5653	1.459	.6853	1.769	1.212
.98	.8305	.5570	1.491	.6707	1.795	1.204
.99	.8360	.5487	1.524	.6563	1.823	1.196
1.00	.8415	.5403	1.557	.6421	1.851	1.188
1.01	.8468	.5319	1.592	.6281	1.880	1.181
1.02	.8521	.5234	1.628	.6142	1.911	1.174
1.03	.8573	.5148	1.665	.6005	1.942	1.166
1.04	.8624	.5062	1.704	.5870	1.975	1.160
1.05	.8674	.4976	1.743	.5736	2.010	1.153
1.06	.8724	.4889	1.784	.5604	2.046	1.146
1.07	.8772	.4801	1.827	.5473	2.083	1.140
1.08	.8820	.4713	1.871	.5344	2.122	1.134
1.09	.8866	.4625	1.917	.5216	2.162	1.128
1.10	.8912	.4536	1.965	.5090	2.205	1.122
1.11	.8957	.4447	2.014	.4964	2.249	1.116
1.12	.9001	.4357	2.066	.4840	2.295	1.111
1.13	.9044	.4267	2.120	.4718	2.344	1.106
1.14	.9086	.4176	2.176	.4596	2.395	1.101
1.15	.9128	.4085	2.234	.4475	2.448	1.096
1.16	.9168	.3993	2.296	.4356	2.504	1.091
1.17	.9208	.3902	2.360	.4237	2.563	1.086
1.18	.9246	.3809	2.427	.4120	2.625	1.082
1.19	.9284	.3717	2.498	.4003	2.691	1.077
	sin	cos	tan	cot	sec	csc

TABLE V (*continued*)

	sin	cos	tan	cot	sec	csc
1.20	.9320	.3624	2.572	.3888	2.760	1.073
1.21	.9356	.3530	2.650	.3773	2.833	1.069
1.22	.9391	.3436	2.733	.3659	2.910	1.065
1.23	.9425	.3342	2.820	.3546	2.992	1.061
1.24	.9458	.3248	2.912	.3434	3.079	1.057
1.25	.9490	.3153	3.010	.3323	3.171	1.054
1.26	.9521	.3058	3.113	.3212	3.270	1.050
1.27	.9551	.2963	3.224	.3102	3.375	1.047
1.28	.9580	.2867	3.341	.2993	3.488	1.044
1.29	.9608	.2771	3.467	.2884	3.609	1.041
1.30	.9636	.2675	3.602	.2776	3.738	1.038
1.31	.9662	.2579	3.747	.2669	3.878	1.035
1.32	.9687	.2482	3.903	.2562	4.029	1.032
1.33	.9711	.2385	4.072	.2456	4.193	1.030
1.34	.9735	.2288	4.256	.2350	4.372	1.027
1.35	.9757	.2190	4.455	.2245	4.566	1.025
1.36	.9779	.2092	4.673	.2140	4.779	1.023
1.37	.9799	.1994	4.913	.2035	5.014	1.021
1.38	.9819	.1896	5.177	.1931	5.273	1.018
1.39	.9837	.1798	5.471	.1828	5.561	1.017
1.40	.9854	.1700	5.798	.1725	5.883	1.015
1.41	.9871	.1601	6.165	.1622	6.246	1.013
1.42	.9887	.1502	6.581	.1519	6.657	1.011
1.43	.9901	.1403	7.055	.1417	7.126	1.010
1.44	.9915	.1304	7.602	.1315	7.667	1.009
1.45	.9927	.1205	8.238	.1214	8.299	1.007
1.46	.9939	.1106	8.989	.1113	9.044	1.006
1.47	.9949	.1006	9.887	.1011	9.938	1.005
1.48	.9959	.0907	10.983	.0910	11.029	1.004
1.49	.9967	.0807	12.350	.0810	12.390	1.003
1.50	.9975	.0707	14.101	.0709	14.137	1.003
1.51	.9982	.0608	16.428	.0609	16.458	1.002
1.52	.9987	.0508	19.670	.0508	19.695	1.001
1.53	.9992	.0408	24.498	.0408	24.519	1.001
1.54	.9995	.0308	32.461	.0308	32.476	1.000
1.55	.9998	.0208	48.078	.0208	48.089	1.000
1.56	.9999	.0108	92.620	.0108	92.626	1.000
1.57	1.0000	.0008	1,255.8	.0008	1,255.8	1.000
1.58	1.0000	−.0092	−108.65	−.0092	−108.65	1.000
1.59	.9998	−.0192	−52.067	−.0192	−52.08	1.000
1.60	.9996	−.0292	−34.233	−.0292	−34.25	1.000
	sin	cos	tan	cot	sec	csc

TABLE VI Trigonometric Functions—Degrees and Minutes

↱	sin	cos	tan	cot	sec	csc	
0°00′	0.0000	1.000	0.0000	—	1.000	—	90°00′
10′	0.0029	1.000	0.0029	343.8	1.000	343.8	89°50′
20′	0.0058	1.000	0.0058	171.9	1.000	171.9	40′
30′	0.0087	1.000	0.0087	114.6	1.000	114.6	30′
40′	0.0116	0.9999	0.0116	85.94	1.000	85.95	20′
0°50′	0.0145	0.9999	0.0145	68.75	1.000	68.76	10′
1°00′	0.0175	0.9998	0.0175	57.29	1.000	57.30	89°00′
10′	0.0204	0.9998	0.0204	49.10	1.000	49.11	88°50′
20′	0.0233	0.9997	0.0233	42.96	1.000	42.98	40′
30′	0.0262	0.9997	0.0262	38.19	1.000	38.20	30′
40′	0.0291	0.9996	0.0291	34.37	1.000	34.38	20′
1°50′	0.0320	0.9995	0.0320	31.24	1.001	31.26	10′
2°00′	0.0349	0.9994	0.0349	28.64	1.001	28.65	88°00′
10′	0.0378	0.9993	0.0378	26.43	1.001	26.45	87°50′
20′	0.0407	0.9992	0.0407	24.54	1.001	24.56	40′
30′	0.0436	0.9990	0.0437	22.90	1.001	22.93	30′
40′	0.0465	0.9989	0.0466	21.47	1.001	21.49	20′
2°50′	0.0494	0.9988	0.0495	20.21	1.001	20.23	10′
3°00′	0.0523	0.9986	0.0524	19.08	1.001	19.11	87°00′
10′	0.0552	0.9985	0.0553	18.07	1.002	18.10	86°50′
20′	0.0581	0.9983	0.0582	17.17	1.002	17.20	40′
30′	0.0610	0.9981	0.0612	16.35	1.002	16.38	30′
40′	0.0640	0.9980	0.0641	15.60	1.002	15.64	20′
3°50′	0.0669	0.9978	0.0670	14.92	1.002	14.96	10′
4°00′	0.0698	0.9976	0.0699	14.30	1.002	14.34	86°00′
10′	0.0727	0.9974	0.0729	13.73	1.003	13.76	85°50′
20′	0.0756	0.9971	0.0758	13.20	1.003	13.23	40′
30′	0.0785	0.9969	0.0787	12.71	1.003	12.75	30′
40′	0.0814	0.9967	0.0816	12.25	1.003	12.29	20′
4°50′	0.0843	0.9964	0.0846	11.83	1.004	11.87	10′
5°00′	0.0872	0.9962	0.0875	11.43	1.004	11.47	85°00′
10′	0.0901	0.9959	0.0904	11.06	1.004	11.10	84°50′
20′	0.0929	0.9957	0.0934	10.71	1.004	10.76	40′
30′	0.0958	0.9954	0.0963	10.39	1.005	10.43	30′
40′	0.0987	0.9951	0.0992	10.08	1.005	10.13	20′
5°50′	0.1016	0.9948	0.1022	9.788	1.005	9.839	10′
6°00′	0.1045	0.9945	0.1051	9.514	1.006	9.567	84°00′
10′	0.1074	0.9942	0.1080	9.255	1.006	9.309	83°50′
20′	0.1103	0.9939	0.1110	9.010	1.006	9.065	40′
30′	0.1132	0.9936	0.1139	8.777	1.006	8.834	30′
40′	0.1161	0.9932	0.1169	8.556	1.007	8.614	20′
6°50′	0.1190	0.9929	0.1198	8.345	1.007	8.405	10′
7°00′	0.1219	0.9925	0.1228	8.144	1.008	8.206	83°00′
	cos	sin	cot	tan	csc	sec	↵

TABLE VI (*continued*)

	sin	cos	tan	cot	sec	csc	
7°00′	0.1219	0.9925	0.1228	8.144	1.008	8.206	83°00′
10′	0.1248	0.9922	0.1257	7.953	1.008	8.016	82°50′
20′	0.1276	0.9918	0.1287	7.770	1.008	7.834	40′
30′	0.1305	0.9914	0.1317	7.596	1.009	7.661	30′
40′	0.1334	0.9911	0.1346	7.429	1.009	7.496	20′
7°50′	0.1363	0.9907	0.1376	7.269	1.009	7.337	10′
8°00′	0.1392	0.9903	0.1405	7.115	1.010	7.185	82°00′
10′	0.1421	0.9899	0.1435	6.968	1.010	7.040	81°50′
20′	0.1449	0.9894	0.1465	6.827	1.011	6.900	40′
30′	0.1478	0.9890	0.1495	6.691	1.011	6.765	30′
40′	0.1507	0.9886	0.1524	6.561	1.012	6.636	20′
8°50′	0.1536	0.9881	0.1554	6.435	1.012	6.512	10′
9°00′	0.1564	0.9877	0.1584	6.314	1.012	6.392	81°00′
10′	0.1593	0.9872	0.1614	6.197	1.013	6.277	80°50′
20′	0.1622	0.9868	0.1644	6.084	1.013	6.166	40′
30′	0.1650	0.9863	0.1673	5.976	1.014	6.059	30′
40′	0.1679	0.9858	0.1703	5.871	1.014	5.955	20′
9°50′	0.1708	0.9853	0.1733	5.769	1.015	5.855	10′
10°00′	0.1736	0.9848	0.1763	5.671	1.015	5.759	80°00′
10′	0.1765	0.9843	0.1793	5.576	1.016	5.665	79°50′
20′	0.1794	0.9838	0.1823	5.485	1.016	5.575	40′
30′	0.1822	0.9833	0.1853	5.396	1.017	5.487	30′
40′	0.1851	0.9827	0.1883	5.309	1.018	5.403	20′
10°50′	0.1880	0.9822	0.1914	5.226	1.018	5.320	10′
11°00′	0.1908	0.9816	0.1944	5.145	1.019	5.241	79°00′
10′	0.1937	0.9811	0.1974	5.066	1.019	5.164	78°50′
20′	0.1965	0.9805	0.2004	4.989	1.020	5.089	40′
30′	0.1994	0.9799	0.2035	4.915	1.020	5.016	30′
40′	0.2022	0.9793	0.2065	4.843	1.021	4.945	20′
11°50′	0.2051	0.9787	0.2095	4.773	1.022	4.876	10′
12°00′	0.2079	0.9781	0.2126	4.705	1.022	4.810	78°00′
10′	0.2108	0.9775	0.2156	4.638	1.023	4.745	77°50′
20′	0.2136	0.9769	0.2186	4.574	1.024	4.682	40′
30′	0.2164	0.9763	0.2217	4.511	1.024	4.620	30′
40′	0.2193	0.9757	0.2247	4.449	1.025	4.560	20′
12°50′	0.2221	0.9750	0.2278	4.390	1.026	4.502	10′
13°00′	0.2250	0.9744	0.2309	4.331	1.026	4.445	77°00′
10′	0.2278	0.9737	0.2339	4.275	1.027	4.390	76°50′
20′	0.2306	0.9730	0.2370	4.219	1.028	4.336	40′
30′	0.2334	0.9724	0.2401	4.165	1.028	4.284	30′
40′	0.2363	0.9717	0.2432	4.113	1.029	4.232	20′
13°50′	0.2391	0.9710	0.2462	4.061	1.030	4.182	10′
14°00′	0.2419	0.9703	0.2493	4.011	1.031	4.134	76°00′
	cos	sin	cot	tan	csc	sec	

→	sin	cos	tan	cot	sec	csc	
14°00′	0.2419	0.9703	0.2493	4.011	1.031	4.134	76°00′
10′	0.2447	0.9696	0.2524	3.962	1.031	4.086	75°50′
20′	0.2476	0.9689	0.2555	3.914	1.032	4.039	40′
30′	0.2504	0.9681	0.2586	3.867	1.033	3.994	30′
40′	0.2532	0.9674	0.2617	3.821	1.034	3.950	20′
14°50′	0.2560	0.9667	0.2648	3.776	1.034	3.906	10′
15°00′	0.2588	0.9659	0.2679	3.732	1.035	3.864	75°00′
10′	0.2616	0.9652	0.2711	3.689	1.036	3.822	74°50′
20′	0.2644	0.9644	0.2742	3.647	1.037	3.782	40′
30′	0.2672	0.9636	0.2773	3.606	1.038	3.742	30′
40′	0.2700	0.9628	0.2805	3.566	1.039	3.703	20′
15°50′	0.2728	0.9621	0.2836	3.526	1.039	3.665	10′
16°00′	0.2756	0.9613	0.2867	3.487	1.040	3.628	74°00′
10′	0.2784	0.9605	0.2899	3.450	1.041	3.592	73°50′
20′	0.2812	0.9596	0.2931	3.412	1.042	3.556	40′
30′	0.2840	0.9588	0.2962	3.376	1.043	3.521	30′
40′	0.2868	0.9580	0.2994	3.340	1.044	3.487	20′
16°50′	0.2896	0.9572	0.3026	3.305	1.045	3.453	10′
17°00′	0.2924	0.9563	0.3057	3.271	1.046	3.420	73°00′
10′	0.2952	0.9555	0.3089	3.237	1.047	3.388	72°50′
20′	0.2979	0.9546	0.3121	3.204	1.048	3.356	40′
30′	0.3007	0.9537	0.3153	3.172	1.049	3.326	30′
40′	0.3035	0.9528	0.3185	3.140	1.049	3.295	20′
17°50′	0.3062	0.9520	0.3217	3.108	1.050	3.265	10′
18°00′	0.3090	0.9511	0.3249	3.078	1.051	3.236	72°00′
10′	0.3118	0.9502	0.3281	3.047	1.052	3.207	71°50′
20′	0.3145	0.9492	0.3314	3.018	1.053	3.179	40′
30′	0.3173	0.9483	0.3346	2.989	1.054	3.152	30′
40′	0.3201	0.9474	0.3378	2.960	1.056	3.124	20′
18°50′	0.3228	0.9465	0.3411	2.932	1.057	3.098	10′
19°00′	0.3256	0.9455	0.3443	2.904	1.058	3.072	71°00′
10′	0.3283	0.9446	0.3476	2.877	1.059	3.046	70°50′
20′	0.3311	0.9436	0.3508	2.850	1.060	3.021	40′
30′	0.3338	0.9426	0.3541	2.824	1.061	2.996	30′
40′	0.3365	0.9417	0.3574	2.798	1.062	2.971	20′
19°50′	0.3393	0.9407	0.3607	2.773	1.063	2.947	10′
20°00′	0.3420	0.9397	0.3640	2.747	1.064	2.924	70°00′
10′	0.3448	0.9387	0.3673	2.723	1.065	2.901	69°50′
20′	0.3475	0.9377	0.3706	2.699	1.066	2.878	40′
30′	0.3502	0.9367	0.3739	2.675	1.068	2.855	30′
40′	0.3529	0.9356	0.3772	2.651	1.069	2.833	20′
20°50′	0.3557	0.9346	0.3805	2.628	1.070	2.812	10′
21°00′	0.3584	0.9336	0.3839	2.605	1.071	2.790	69°00′
	cos	sin	cot	tan	csc	sec	←

TABLE VI (*continued*)

	sin	cos	tan	cot	sec	csc	
21°00′	0.3584	0.9336	0.3839	2.605	1.071	2.790	69°00′
10′	0.3611	0.9325	0.3872	2.583	1.072	2.769	68°50′
20′	0.3638	0.9315	0.3906	2.560	1.074	2.749	40′
30′	0.3665	0.9304	0.3939	2.539	1.075	2.729	30′
40′	0.3692	0.9293	0.3973	2.517	1.076	2.709	20′
21°50′	0.3719	0.9283	0.4006	2.496	1.077	2.689	10′
22°00′	0.3746	0.9272	0.4040	2.475	1.079	2.669	68°00′
10′	0.3773	0.9261	0.4074	2.455	1.080	2.650	67°50′
20′	0.3800	0.9250	0.4108	2.434	1.081	2.632	40′
30′	0.3827	0.9239	0.4142	2.414	1.082	2.613	30′
40′	0.3854	0.9228	0.4176	2.394	1.084	2.595	20′
22°50′	0.3881	0.9216	0.4210	2.375	1.085	2.577	10′
23°00′	0.3907	0.9205	0.4245	2.356	1.086	2.559	67°00′
10′	0.3934	0.9194	0.4279	2.337	1.088	2.542	66°50′
20′	0.3961	0.9182	0.4314	2.318	1.089	2.525	40′
30′	0.3987	0.9171	0.4348	2.300	1.090	2.508	30′
40′	0.4014	0.9159	0.4383	2.282	1.092	2.491	20′
23°50′	0.4041	0.9147	0.4417	2.264	1.093	2.475	10′
24°00′	0.4067	0.9135	0.4452	2.246	1.095	2.459	66°00′
10′	0.4094	0.9124	0.4487	2.229	1.096	2.443	65°50′
20′	0.4120	0.9112	0.4522	2.211	1.097	2.427	40′
30′	0.4147	0.9100	0.4557	2.194	1.099	2.411	30′
40′	0.4173	0.9088	0.4592	2.177	1.100	2.396	20′
24°50′	0.4200	0.9075	0.4628	2.161	1.102	2.381	10′
25°00′	0.4226	0.9063	0.4663	2.145	1.103	2.366	65°00′
10′	0.4253	0.9051	0.4699	2.128	1.105	2.352	64°50′
20′	0.4279	0.9038	0.4734	2.112	1.106	2.337	40′
30′	0.4305	0.9026	0.4770	2.097	1.108	2.323	30′
40′	0.4331	0.9013	0.4806	2.081	1.109	2.309	20′
25°50′	0.4358	0.9001	0.4841	2.066	1.111	2.295	10′
26°00′	0.4384	0.8988	0.4877	2.050	1.113	2.281	64°00′
10′	0.4410	0.8975	0.4913	2.035	1.114	2.268	63°50′
20′	0.4436	0.8962	0.4950	2.020	1.116	2.254	40′
30′	0.4462	0.8949	0.4986	2.006	1.117	2.241	30′
40′	0.4488	0.8936	0.5022	1.991	1.119	2.228	20′
26°50′	0.4514	0.8923	0.5059	1.977	1.121	2.215	10′
27°00′	0.4540	0.8910	0.5095	1.963	1.122	2.203	63°00′
10′	0.4566	0.8897	0.5132	1.949	1.124	2.190	62°50′
20′	0.4592	0.8884	0.5169	1.935	1.126	2.178	40′
30′	0.4617	0.8870	0.5206	1.921	1.127	2.166	30′
40′	0.4643	0.8857	0.5243	1.907	1.129	2.154	20′
27°50′	0.4669	0.8843	0.5280	1.894	1.131	2.142	10′
28°00′	0.4695	0.8829	0.5317	1.881	1.133	2.130	62°00′
	cos	sin	cot	tan	csc	sec	

A70

↓	sin	cos	tan	cot	sec	csc	
28°00′	0.4695	0.8829	0.5317	1.881	1.133	2.130	62°00′
10′	0.4720	0.8816	0.5354	1.868	1.134	2.118	61°50′
20′	0.4746	0.8802	0.5392	1.855	1.136	2.107	40′
30′	0.4772	0.8788	0.5430	1.842	1.138	2.096	30′
40′	0.4797	0.8774	0.5467	1.829	1.140	2.085	20′
28°50′	0.4823	0.8760	0.5505	1.816	1.142	2.074	10′
29°00′	0.4848	0.8746	0.5543	1.804	1.143	2.063	61°00′
10′	0.4874	0.8732	0.5581	1.792	1.145	2.052	60°50′
20′	0.4899	0.8718	0.5619	1.780	1.147	2.041	40′
30′	0.4924	0.8704	0.5658	1.767	1.149	2.031	30′
40′	0.4950	0.8689	0.5696	1.756	1.151	2.020	20′
29°50′	0.4975	0.8675	0.5735	1.744	1.153	2.010	10′
30°00′	0.5000	0.8660	0.5774	1.732	1.155	2.000	60°00′
10′	0.5025	0.8646	0.5812	1.720	1.157	1.990	59°50′
20′	0.5050	0.8631	0.5851	1.709	1.159	1.980	40′
30′	0.5075	0.8616	0.5890	1.698	1.161	1.970	30′
40′	0.5100	0.8601	0.5930	1.686	1.163	1.961	20′
30°50′	0.5125	0.8587	0.5969	1.675	1.165	1.951	10′
31°00′	0.5150	0.8572	0.6009	1.664	1.167	1.942	59°00′
10′	0.5175	0.8557	0.6048	1.653	1.169	1.932	58°50′
20′	0.5200	0.8542	0.6088	1.643	1.171	1.923	40′
30′	0.5225	0.8526	0.6128	1.632	1.173	1.914	30′
40′	0.5250	0.8511	0.6168	1.621	1.175	1.905	20′
31°50′	0.5275	0.8496	0.6208	1.611	1.177	1.896	10′
32°00′	0.5299	0.8480	0.6249	1.600	1.179	1.887	58°00′
10′	0.5324	0.8465	0.6289	1.590	1.181	1.878	57°50′
20′	0.5348	0.8450	0.6330	1.580	1.184	1.870	40′
30′	0.5373	0.8434	0.6371	1.570	1.186	1.861	30′
40′	0.5398	0.8418	0.6412	1.560	1.188	1.853	20′
32°50′	0.5422	0.8403	0.6453	1.550	1.190	1.844	10′
33°00′	0.5446	0.8387	0.6494	1.540	1.192	1.836	57°00′
10′	0.5471	0.8371	0.6536	1.530	1.195	1.828	56°50′
20′	0.5495	0.8355	0.6577	1.520	1.197	1.820	40′
30′	0.5519	0.8339	0.6619	1.511	1.199	1.812	30′
40′	0.5544	0.8323	0.6661	1.501	1.202	1.804	20′
33°50′	0.5568	0.8307	0.6703	1.492	1.204	1.796	10′
34°00′	0.5592	0.8290	0.6745	1.483	1.206	1.788	56°00′
10′	0.5616	0.8274	0.6787	1.473	1.209	1.781	55°50′
20′	0.5640	0.8258	0.6830	1.464	1.211	1.773	40′
30′	0.5664	0.8241	0.6873	1.455	1.213	1.766	30′
40′	0.5688	0.8225	0.6916	1.446	1.216	1.758	20′
34°50′	0.5712	0.8208	0.6959	1.437	1.218	1.751	10′
35°00′	0.5736	0.8192	0.7002	1.428	1.221	1.743	55°00′
	cos	sin	cot	tan	csc	sec	↵

TABLE VI (*continued*)

↓→	sin	cos	tan	cot	sec	csc	
35°00′	0.5736	0.8192	0.7002	1.428	1.221	1.743	55°00′
10′	0.5760	0.8175	0.7046	1.419	1.223	1.736	54°50′
20′	0.5783	0.8158	0.7089	1.411	1.226	1.729	40′
30′	0.5807	0.8141	0.7133	1.402	1.228	1.722	30′
40′	0.5831	0.8124	0.7177	1.393	1.231	1.715	20′
35°50′	0.5854	0.8107	0.7221	1.385	1.233	1.708	10′
36°00′	0.5878	0.8090	0.7265	1.376	1.236	1.701	54°00′
10′	0.5901	0.8073	0.7310	1.368	1.239	1.695	53°50′
20′	0.5925	0.8056	0.7355	1.360	1.241	1.688	40′
30′	0.5948	0.8039	0.7400	1.351	1.244	1.681	30′
40′	0.5972	0.8021	0.7445	1.343	1.247	1.675	20′
36°50′	0.5995	0.8004	0.7490	1.335	1.249	1.668	10′
37°00′	0.6018	0.7986	0.7536	1.327	1.252	1.662	53°00′
10′	0.6041	0.7969	0.7581	1.319	1.255	1.655	52°50′
20′	0.6065	0.7951	0.7627	1.311	1.258	1.649	40′
30′	0.6088	0.7934	0.7673	1.303	1.260	1.643	30′
40′	0.6111	0.7916	0.7720	1.295	1.263	1.636	20′
37°50′	0.6134	0.7898	0.7766	1.288	1.266	1.630	10′
38°00′	0.6157	0.7880	0.7813	1.280	1.269	1.624	52°00′
10′	0.6180	0.7862	0.7860	1.272	1.272	1.618	51°50′
20′	0.6202	0.7844	0.7907	1.265	1.275	1.612	40′
30′	0.6225	0.7826	0.7954	1.257	1.278	1.606	30′
40′	0.6248	0.7808	0.8002	1.250	1.281	1.601	20′
38°50′	0.6271	0.7790	0.8050	1.242	1.284	1.595	10′
39°00′	0.6293	0.7771	0.8098	1.235	1.287	1.589	51°00′
10′	0.6316	0.7753	0.8146	1.228	1.290	1.583	50°50′
20′	0.6338	0.7735	0.8195	1.220	1.293	1.578	40′
30′	0.6361	0.7716	0.8243	1.213	1.296	1.572	30′
40′	0.6383	0.7698	0.8292	1.206	1.299	1.567	20′
39°50′	0.6406	0.7679	0.8342	1.199	1.302	1.561	10′
40°00′	0.6428	0.7660	0.8391	1.192	1.305	1.556	50°00′
10′	0.6450	0.7642	0.8441	1.185	1.309	1.550	49°50′
20′	0.6472	0.7623	0.8491	1.178	1.312	1.545	40′
30′	0.6494	0.7604	0.8541	1.171	1.315	1.540	30′
40′	0.6517	0.7585	0.8591	1.164	1.318	1.535	20′
40°50′	0.6539	0.7566	0.8642	1.157	1.322	1.529	10′
41°00′	0.6561	0.7547	0.8693	1.150	1.325	1.524	49°00′
10′	0.6583	0.7528	0.8744	1.144	1.328	1.519	48°50′
20′	0.6604	0.7509	0.8796	1.137	1.332	1.514	40′
30′	0.6626	0.7490	0.8847	1.130	1.335	1.509	30′
40′	0.6648	0.7470	0.8899	1.124	1.339	1.504	20′
41°50′	0.6670	0.7451	0.8952	1.117	1.342	1.499	10′
42°00′	0.6691	0.7431	0.9004	1.111	1.346	1.494	48°00′
	cos	sin	cot	tan	csc	sec	↵

↱	sin	cos	tan	cot	sec	csc	
42°00′	0.6691	0.7431	0.9004	1.111	1.346	1.494	48°00′
10′	0.6713	0.7412	0.9057	1.104	1.349	1.490	47°50′
20′	0.6734	0.7392	0.9110	1.098	1.353	1.485	40′
30′	0.6756	0.7373	0.9163	1.091	1.356	1.480	30′
40′	0.6777	0.7353	0.9217	1.085	1.360	1.476	20′
42°50′	0.6799	0.7333	0.9271	1.079	1.364	1.471	10′
43°00′	0.6820	0.7314	0.9325	1.072	1.367	1.466	47°00′
10′	0.6841	0.7294	0.9380	1.066	1.371	1.462	46°50′
20′	0.6862	0.7274	0.9435	1.060	1.375	1.457	40′
30′	0.6884	0.7254	0.9490	1.054	1.379	1.453	30′
40′	0.6905	0.7234	0.9545	1.048	1.382	1.448	20′
43°50′	0.6926	0.7214	0.9601	1.042	1.386	1.444	10′
44°00′	0.6947	0.7193	0.9657	1.036	1.390	1.440	46°00′
10′	0.6967	0.7173	0.9713	1.030	1.394	1.435	45°50′
20′	0.6988	0.7153	0.9770	1.024	1.398	1.431	40′
30′	0.7009	0.7133	0.9827	1.018	1.402	1.427	30′
40′	0.7030	0.7112	0.9884	1.012	1.406	1.423	20′
44°50′	0.7050	0.7092	0.9942	1.006	1.410	1.418	10′
45°00′	0.7071	0.7071	1.000	1.000	1.414	1.414	45°00′
	cos	sin	cot	tan	csc	sec	↵

Answers

Chapter 1   Exercise 1-1

1. T   **3.** T   **5.** T   **7.** T   **9.** $\{3, 4, 5, 6, 7\}$ **11.** $\{5\}$ **13.** $\varnothing$ **15.** $\{8\}$ **17.** $\{-7, 7\}$
19. $\{2, 3, 5, 7\}$ **21.** $\{-2, 1, 2\}$ **23.** 100   **25.** 61   **27.** 80   **29.** 41   **31.** 89   **33.** 20
35. (A) $\{2, 3, 4, 5, 6, 7\}$; (B) $\{2, 3, 4, 5, 6, 7\}$ **37.** $\{1, 2, 3, 4, 6\}$ **39.** Yes   **41.** No   **43.** No
45. They are all different: $\varnothing$ is the empty set, $\{0\}$ has one element 0, and $\{\varnothing\}$ has one element $\varnothing$.   **47.** 850   **49.** 150
51. 130   **53.** 850   **55.** 150   **57.** 130   **59.** $AB-, AB+$ **61.** $A-, AB-, B-, A+, AB+, B+$ **63.** $O-$
65. $A-, AB-$

Exercise 1-2

1. Commutative   **3.** Associative   **5.** Distributive   **7.** Negatives   **9.** Def. subtraction   **11.** Negatives
13. Identity   **15.** Def. division   **17.** Inverse   **19.** Distributive   **21.** Distributive   **23.** Zero
25. Commutative   **27.** Associative   **29.** Distributive   **31.** Distributive   **33.** Negatives   **35.** Yes
37. (A) T; (B) F; (C) T   **39.** $\frac{3}{5}$ and -1.43 are two examples of infinitely many.
41. (A) Z, Q, R; (B) Q, R; (C) R; (D) Q, R
43. (B) is false, since, for example, $5 - 3 \neq 3 - 5$; (D) is false, since, for example, $9 \div 3 \neq 3 \div 9$.   **45.** $\frac{1}{11}$
47.
$$\begin{array}{r} 23 \\ \underline{12} \\ 46 \\ \underline{230} \\ 276 \end{array} \quad \begin{aligned} 23 \cdot 12 &= 23(2 + 10) \\ &= 23 \cdot 2 + 23 \cdot 10 \\ &= 46 + 230 \\ &= 276 \end{aligned}$$
49. (A) $0.888\ 888\ 88\ldots$; (B) $0.272\ 727\ 27\ldots$; (C) $2.236\ 067\ 97\ldots$; (D) $1.375\ 000\ 00\ldots$

Exercise 1-3

1. $-8 \leq x \leq 7$ **3.** $-6 \leq x < 6$
5. $x \geq -6$ **7.** $(-2, 6]$
9. $(-7, 8)$ **11.** $(-\infty, -2]$
13. 9   **15.** $m > 3$ or $(3, \infty)$ **17.** $\frac{11}{2} = 5.5$
19. $B \geq -4$ or $[-4, \infty)$ **21.** 8   **23.** No solution   **25.** $-2 < t \leq 3$ or $(-2, 3]$
27. $\frac{8}{5}$ **29.** $q < -14$ or $(-\infty, -14)$ **31.** 2   **33.** -4
35. $x \geq 4.5$ or $[4.5, \infty)$ **37.** $-30 \leq x < 18$ or $[-30, 18)$

Answers

39. $-8 \leq x < -3$ or $[-8, -3)$ **41.** $d = (a_n - a_1)/(n - 1)$ **43.** $f = d_1 d_2/(d_1 + d_2)$
45. $x = (5y + 3)/(2 - 3y)$ **47.** Positive
51. $m - n - p = 0$ and division by 0 is not defined **53.** $8,000 \leq h \leq 20,000$ or $[8,000, 20,000]$ **55.** $x > 600$
57. $90 **59.** 30 liters of 20% solution and 70 liters of 80% solution
61. 10 AM; 24 miles **63.** 90 miles **65.** $5\frac{5}{11}$ minutes after 1 PM

Exercise 1-4

1. $\sqrt{5}$ **3.** 4 **5.** $5 - \sqrt{5}$ **7.** $5 - \sqrt{5}$ **9.** 12 **11.** 12 **13.** 9 **15.** 4 **17.** 4 **19.** 9
21. $x = \pm 7$ **23.** $-7 \leq x \leq 7$
25. $x \leq -7$ or $x \geq 7$ **27.** $y = 2$ or 8
29. $2 < y < 8$ **31.** $y < 2$ or $y > 8$
33. $u = -11$ or -5 **35.** $-11 \leq u \leq -5$
37. $u \leq -11$ or $u \geq -5$ **39.** $x = -4, \frac{4}{3}$ **41.** $-\frac{9}{5} \leq x \leq 3$
43. $y < 3$ or $y > 5$ **45.** $t = -\frac{4}{5}, \frac{18}{5}$ **47.** $-\frac{5}{7} < u < \frac{23}{7}$ **49.** $x \leq -6$ or $x \geq 9$ **51.** $-35 < C < -\frac{5}{9}$
53. $x \geq 5$ **55.** $x \leq -8$ **57.** $x \geq -\frac{3}{4}$ **59.** $x \leq \frac{2}{5}$
61. Case 1: $a = b$; $|b - a| = |0| = 0$; $|a - b| = |0| = 0$
   Case 2: $a > b$; $|b - a| = -(b - a) = a - b$
   $|a - b| = a - b$
   Case 3: $b > a$; $|b - a| = b - a$
   $|a - b| = -(a - b) = b - a$

Exercise 1-5

1. $-3 < x < 4$
$(-3, 4)$

3. $x \leq -3$ or $x \geq 4$
$(-\infty, -3] \cup [4, \infty)$

5. $-5 < x < 2$
$(-5, 2)$

7. $x < 3$ or $x > 7$
$(-\infty, 3) \cup (7, \infty)$

9. $0 \leq x \leq 8$
$[0, 8]$

11. $-5 \leq x \leq 0$
$[-5, 0]$

13. $x < -2$ or $x > 2$
$(-\infty, -2) \cup (2, \infty)$

15. True for all real numbers.
Graph: Whole real number line

17. $-1 \leq x \leq 1$ or $x \geq 5$
$[-1, 1] \cup [5, \infty]$

19. $x < -5$ or $3 < x < 5$
$(-\infty, -5) \cup (3, 5)$

21. $-4 < x \leq 2$
$(-4, 2]$

23. $-5 \leq x \leq 0$ or $x > 3$
$[-5, 0] \cup (3, \infty)$

25. $x \leq -4$ or $x > 1$
$(-\infty, -4] \cup (1, \infty)$

27. $x < 0$ or $x > \frac{1}{4}$
$(-\infty, 0) \cup (\frac{1}{4}, \infty)$

29. $x < -3$ or $x \geq 3$
$(-\infty, -3) \cup [3, \infty)$

31. $-4 < x \leq \frac{3}{2}$
$(-4, \frac{3}{2}]$

33. $-1 < x < 2$ or $x \geq 5$
$(-1, 2) \cup [5, \infty)$

35. No solutions

37. $x > 4$
$(4, \infty)$

39. $-2 \leq x \leq -\frac{1}{2}$ or $\frac{1}{2} \leq x \leq 2$
$[-2, -\frac{1}{2}] \cup [\frac{1}{2}, 2]$

41. $-2 \leq x \leq 2$
$[-2, 2]$

Exercise 1-6

1. $7 + 5i$ **3.** $5 + 3i$ **5.** $2 + 4i$ **7.** $5 + 9i$ **9.** $4 - 3i$ **11.** -24 or $-24 + 0i$ **13.** $-12 - 6i$
15. $15 - 3i$ **17.** $-4 - 33i$ **19.** 65 or $65 + 0i$ **21.** $\frac{2}{5} - \frac{1}{5}i$ **23.** $\frac{3}{13} + \frac{11}{13}i$ **25.** $5 + 3i$ **27.** $7 - 5i$
29. $-3 + 2i$ **31.** $8 + 25i$ **33.** $\frac{5}{7} - \frac{2}{7}i$ **35.** $\frac{2}{13} + \frac{3}{13}i$ **37.** $-\frac{2}{5}i$ or $0 - \frac{2}{5}i$ **39.** $\frac{3}{2} - \frac{1}{2}i$ **41.** $-6i$ or $0 - 6i$
43. 0 or $0 + 0i$ **45.** $-1, -i, 1, i, -1, -i, 1$ **47.** $x = 3, y = -2$ **49.** $(a + c) + (b + d)i$
51. $a^2 + b^2$ or $(a^2 + b^2) + 0i$ **53.** $(ac - bd) + (ad + bc)i$ **55.** $i^{4k} = (i^4)^k = (i^2 \cdot i^2)^k = [(-1)(-1)]^k = 1^k = 1$
57. (1) Def. addition, (2) property of real numbers, (3) def. addition

Exercise 1-7

1. $0, 2$ **3.** $-8, \frac{1}{2}$ **5.** $\frac{3}{2}, 4$ **7.** $-\frac{4}{3}, \frac{1}{2}$ **9.** ± 5 **11.** $\pm 5i$ **13.** $\pm 2\sqrt{3}$ **15.** $\pm \frac{4}{3}$ **17.** $\pm \frac{5}{2}i$
19. $-2, -8$ **21.** $3 \pm 2i$ **23.** $5 \pm 2\sqrt{7}$ **25.** $(-1 \pm \sqrt{5})/2$ **27.** $2 \pm 2i$ **29.** $(2 \pm \sqrt{2})/2$
31. $(-3 \pm \sqrt{17})/4$ **33.** $\frac{1}{5} \pm \frac{3}{5}i$ **35.** $3 \pm 2\sqrt{3}$ **37.** $(3 \pm \sqrt{3})/2$ **39.** $(1 \pm \sqrt{7})/3$ **41.** $(-m \pm \sqrt{m^2 - 4n})/2$
43. $-\frac{5}{4}, \frac{2}{3}$ **45.** $(3 \pm \sqrt{5})/2$ **47.** $(3 \pm \sqrt{13})/2$ **49.** $-\frac{4}{7}, 0$ **51.** $-\frac{1}{2}, 2$ **53.** $2 \pm 2i$ **55.** $-50, 2$
57. $(-5 \pm \sqrt{57})/2$ **59.** $(-3 \pm \sqrt{57})/4$ **61.** $t = \sqrt{2s/g}$ **63.** $I = (E + \sqrt{E^2 - 4RP})/2R$
65. $[(-b + \sqrt{b^2 - 4ac})/2a] \times [(-b - \sqrt{b^2 - 4ac})/2a] = [b^2 - (b^2 - 4ac)]/4a^2 = c/a$
67. Substitute $r_1 = c/r_2 a$ into $r_1 + r_2 = -b/a$ to obtain $c/r_2 a + r_2 = -b/a$ or $ar_2^2 + br_2 + c = 0$. Similarly, substitute $r_2 = c/r_1 a$ into $r_1 + r_2 = -b/a$ to obtain after simplification $ar_1^2 + br_1 + c = 0$.
69. The $\pm$ in front still yields the same two numbers even if a is negative. **71.** $1.35, 0.48$ **73.** $-1.05, 0.63$
75. Has real solutions, since discriminant is positive **77.** Has no real solutions, since discriminant is negative
79. $8, 13$ **81.** $12, 14$ **83.** 5.12 by 3.12 inches **85.** 179 miles **87.** 20% **89.** 5 and 12 miles/hour
91. 13.09 and 8.09 hours **93.** 70 miles/hour **95.** 50 miles/hour

Exercise 1-8

1. 22 **3.** 8 **5.** No solution **7.** $0, 4$ **9.** $\pm 2, \pm \sqrt{2}i$ **11.** $-\sqrt[3]{5}, \sqrt[3]{2}$ **13.** $\frac{1}{8}, -8$
15. $-2, 3, \frac{1}{2} \pm \sqrt{7}/2i$ **17.** No solution **19.** 1 **21.** 2 **23.** $-\frac{3}{4}, \frac{1}{5}$ **25.** $\pm 1, \pm 3$ **27.** $1, 16$
29. $2, 3, 7, 8$ **31.** -2 **33.** $9, 16$ **35.** $4, 81$

Exercise 1-9 Chapter Review

1. (A) $\{1, 3, 4, 5\}$; (B) $\{3, 5\}$; (C) $\emptyset$; (D) $\{3, 4, 5\}$; (E) $\{2, 4, 6\}$ (1-1)
2. (A) F; (B) T; (C) T; (D) T; (E) T; (F) T (1-1)   3. $\{-\frac{2}{3}, \frac{5}{2}\}$ (1-1, 1-7)
4. Substitution property (1-3)   5. 21 (1-3)   6. $\frac{30}{11}$ (1-3)   7. $x \geq 1$ (1-3)
   $[1, \infty)$
   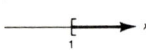

8. $-14 < y < -4$ (1-4)
   $(-14, -4)$
9. $-1 \leq x \leq 4$ (1-4)
   $[-1, 4]$
10. $-5 < x < 4$ (1-5)
    $(-5, 4)$

11. $x \leq -3$ or $x \geq 7$ (1-5)
    $(-\infty, -3] \cup [7, \infty)$
12. (A) $3 - 6i$; (B) $15 + 3i$; (C) $2 + i$ (1-6)   13. $\pm\sqrt{\frac{7}{2}}$ or $\pm\sqrt{\frac{14}{2}}$ (1-7)

14. $0, 2$ (1-7)   15. $\frac{1}{2}, 3$ (1-7)   16. $-\frac{1}{2} \pm (\sqrt{3}/2)i$ (1-7)   17. $(3 \pm \sqrt{33})/4$ (1-7)   18. $2, 3$ (1-8)
19. (A) $H = \{x | 10x + 11 = 6/x\}$; (B) $H = \{-\frac{3}{2}, \frac{2}{5}\}$ (1-1, 1-7)
20. (A) $\{-1, 0, 1, 2, 3\}$; (B) $\{1\}$; (C) No; (D) Yes; (E) No; (F) No (1-1, 1-2)
21. $\emptyset$ (each of the others is a set containing one element) (1-1)   22. Def. subtraction (1-2)   23. Commutative (1-2)
24. Distributive (1-2)   25. Associative (1-2)   26. Negatives (1-2)   27. Identity (1-2)
28. (A) T; (B) F (1-2)   29. 0 and -3 are two examples of infinitely many. (1-2)   30. -15 (1-3)
31. No solution (1-3)   32. $x \geq -19$ (1-3)   33. $x < 2$ or $x > \frac{10}{3}$ (1-4)   34. $x < 0$ or $x > \frac{1}{2}$ (1-5)
    $[-19, \infty)$ $(-\infty, 2) \cup (\frac{10}{3}, \infty)$ $(-\infty, 0) \cup (\frac{1}{2}, \infty)$

35. $x \leq 1$ or $3 < x < 4$ (1-5)   36. (A) 6; (B) 6 (1-4)   37. (A) $5 + 4i$; (B) $-i$ (1-6)
    $(-\infty, 1] \cup (3, 4)$

38. (A) $-1 + i$; (B) $\frac{4}{13} - \frac{7}{13}i$; (C) $\frac{5}{2} - 2i$ (1-6)   39. $(-5 \pm \sqrt{5})/2$ (1-7)   40. $1 \pm i\sqrt{2}$ (1-7)
41. $(1 \pm \sqrt{43})/3$ (1-7)   42. $-\frac{27}{8}, 64$ (1-7)   43. $\pm 2, \pm 3i$ (1-8)   44. $\frac{9}{4}, 3$ (1-8)   45. $M = P/(1 - dt)$ (1-8)
46. $I = (E \pm \sqrt{E^2 - 4PR})/2R$ (1-8)   47. $A \cap B$ (1-1)   48. (A) T; (B) F; (C) T (1-1)
49. $\frac{6}{11}$ (1-2)   50. 1 (1-6)   51. No solution (1-5)   52. Set of all real numbers (1-7)
53. $x \leq -4$ or $-2 \leq x < 0$ or $0 < x \leq 2$ or $x \geq 4$; $(-\infty, -4] \cup [-2, 0) \cup (0, 2] \cup [4, \infty)$ (1-4)   54. $9, 25$ (1-8)
55. 450 milliliters (1-3)   56. (A) 925; (B) 75 (1-1)   57. $\frac{5}{3}$ or $-\frac{3}{5}$ (1-7)
58. (A) 2,000 and 8,000; (B) 5,000 (1-7)
59. $x = (13 \pm \sqrt{45})/2$ thousand, or approximately 3,146 and 9,854 (1-7)

Practice Test: Chapter 1

1. B (1-1)   2. C (1-1)
3. (A) Commutative; (B) Distributive; (C) Def. subtraction; (D) Identity; (E) Associative (1-2)
4. (A) T; (B) F (1-2)   5. $\frac{8}{5} - \frac{4}{5}i$ (1-6)   6. $x \leq 18$ (1-3)
   $(-\infty, 18]$

7. $-1 < x < 4$ (1-4)
$(-1, 4)$

8. $x \leq -1$ or $x \geq 7$ (1-4)
$(-\infty, -1] \cup [7, \infty)$

9. $0 \leq x \leq 3$ or $x > 4$ (1-5)
$[0, 3] \cup (4, \infty)$

10. No solution (1-3)   **11.** $2 \pm i\sqrt{3}$ (1-7)   **12.** $-3 \pm \sqrt{2}$ (1-7)   **13.** $-1, 32$ (1-8)
14. $\pm 2i, \pm\sqrt{1/2}$ or $\pm\sqrt{2}/2$ (1-8)   **15.** 5 (0 is extraneous) (1-8)   **16.** 2 miles/hour (1-7)
17. 20 milliliters of 30% solution, 30 milliliters of 80% solution (1-3)

Chapter 2   Exercise 2-1

1.   **3.** Symmetric with respect to the y axis

5. Symmetric with respect to the x axis   **7.** Symmetric with respect to the x axis, y axis, and origin   **9.** $\sqrt{145}$

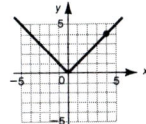

11. $\sqrt{68}$ **13.** $x^2 + y^2 = 49$ **15.** $(x - 2)^2 + (y - 3)^2 = 36$ **17.** $(x + 4)^2 + (y - 1)^2 = 7$
19. $(x + 3)^2 + (y + 4)^2 = 2$ **21.** Symmetric with respect to the x axis   **23.** Symmetric with respect to the y axis

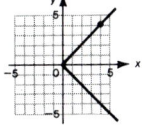
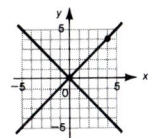

25. Symmetric with respect to the x axis, y axis, and origin   **27.** Symmetric with respect to the x axis, y axis, and origin

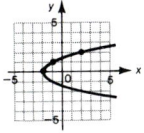
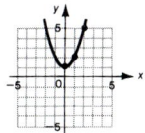

29. Symmetric with respect to the origin   **31.** A right triangle   **33.** $x = -3, 7$ **35.** Center: (3, 5); radius = 7

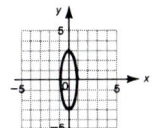
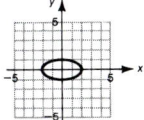

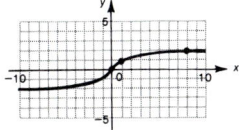

37. Center: $(-4, 2)$; radius $= \sqrt{7}$ **39.** $(x-3)^2 + (y-2)^2 = 49$, $C(3, 2)$, $r = 7$
41. $(x+4)^2 + (y-3)^2 = 17$, $C(-4, 3)$, $r = \sqrt{17}$ **43.** Symmetric with respect to the y axis

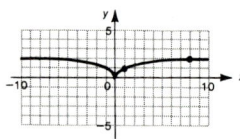

45. Symmetric with respect to the origin   **47.**   **49.** $5x + 3y = -2$

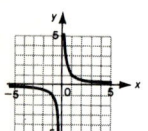   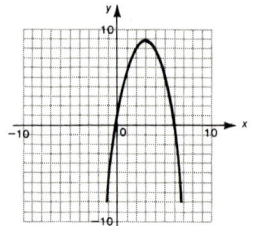

51. $(x-4)^2 + (y-2)^2 = 34$ **53.** 18.11   **55.** Symmetric with respect to the y axis

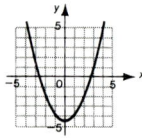

57. Symmetric with respect to the y axis   **59.** Symmetric with respect to the y axis

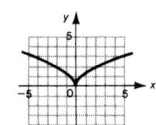

Exercise 2-2

1. A function   **3.** Not a function   **5.** A function   **7.** A function   **9.** Not a function   **11.** Not a function
13. -8 **15.** -6 **17.** 7   **19.** 26   **21.** $-\frac{10}{9}$ **23.** 3   **25.** A function   **27.** A function
29. Not a function   **31.** A function   **33.** Not a function   **35.** A function   **37.** A function   **39.** A function
41. A function   **43.** Domain $= \{2, 4\}$; range $= \{-2, 0, 2, 4\}$; not a function
45. Domain $= \{-4, 0, 4\}$; range $= \{1\}$; a function   **47.** Domain $= \{0, 1, 4\}$; range $= \{-2, -1, 0, 1, 2\}$; not a function
49. Domain $= \{-2, 2\}$; range $= \{2, 6\}$; a function   **51.** Domain $= R$ **53.** Domain $= R$ **55.** Domain: $x \geq -2$
57. Domain: $-9 \leq x \leq 9$ **59.** Domain: All R except 2   **61.** Domain: $x \leq -3$ or $x > 2$ **63.** $-6 - h$ **65.** h
67. $2a + h$ **69.** $2x + h$ **71.** $3x^2 + 3xh + h^2$ **73.** $C(x) = 10 + 0.12x$ **75.** $A(r) = \pi r^2$
77. $V(x) = x(8-2x)(12-2x)$; domain: $0 < x < 4$
79. (A) 0, 16, 64, 144; (B) $64 + 16h$; (C) 64; this number appears to be the speed of the object at the end of 2 seconds.

Exercise 2-3

1. f **3.** f **5.** (A) None; (B) $[0, \infty)$; (C) $(-\infty, 0]$
7. (A) $(-\infty, -2)$; (B) $[-2, -1]$, $[1, \infty)$; (C) $[-1, 1)$ **9.** f, g, and p
11. q; discontinuous at $x = -2$ and $x = 1$ **13.** Odd   **15.** Even   **17.** Neither   **19.** Even   **21.** Neither

23.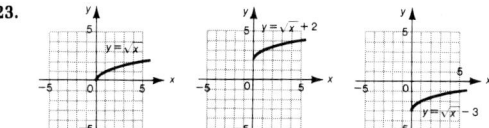

25.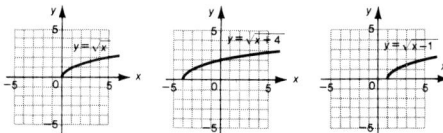

27.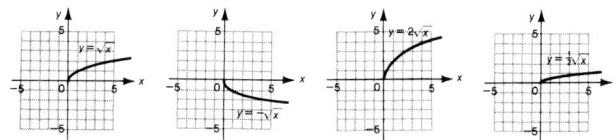

29. It is the same as the graph of $y = x^2$ reflected with respect to the x axis and shifted to the left two units.

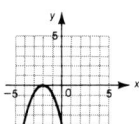

31. It is the same as the graph of $y = |x|$ reflected with respect to the x axis and shifted to the left two units.

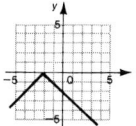

33. It is the same as the graph of $y = -\sqrt{x}$ reflected with respect to the x axis and shifted to the right one unit.

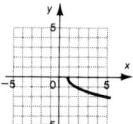

35. Even

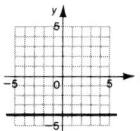

37.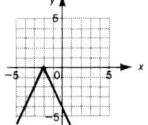

39.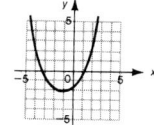

41. Even; discontinuous at $x = 0$

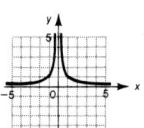

43. Odd; discontinuous at $x = 0$

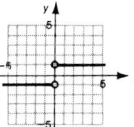

45. Discontinuous at $x = 2$

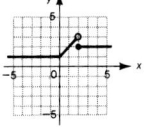

47. Discontinuous at $x = 0$

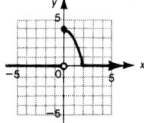

49. Discontinuous at the integers

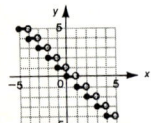

51. Discontinuous at the integers

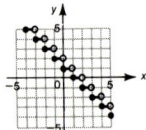

53. Even

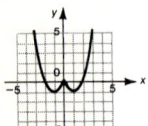

55. Odd

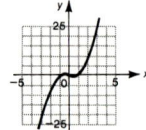

57. Discontinuous at the integers

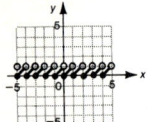

Exercise 2-4

1. Slope = 2

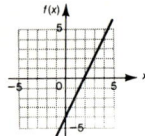

3. Slope = $-\frac{3}{5}$

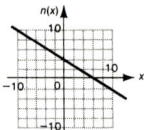

5. Slope = $-\frac{3}{4}$

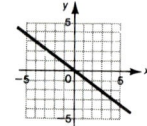

7. Slope = $\frac{2}{3}$

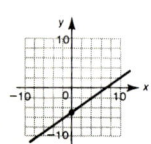

9. Slope = $\frac{4}{5}$

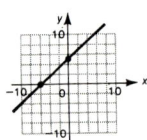

11. Slope = $\frac{3}{2}$

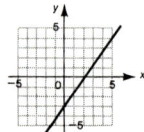

13. Slope = 2

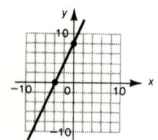

15. Slope not defined

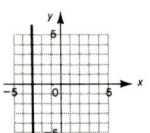

17. Slope = 0

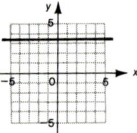

19. $y = -3x + 4$ **21.** $y = -\frac{2}{5}x + 2$ **23.** $y = 5$ **25.** $y = -2x + 8$ **27.** $y = -\frac{4}{3}x + \frac{8}{3}$ **29.** $y = 4$
31. $x = 4$ **33.** $y = -\frac{1}{3}x + 2$ **35.** $y = \frac{3}{4}x + 3$ **37.** $3x - y = -13$ **39.** $3x - y = 9$ **41.** $x = 2$

43. $x = 3$ **45.** $3x - 2y = 15$ **47.** $3x - y = 4$ **49.**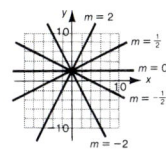

51. Slope $AB = -\frac{3}{4}$ = Slope DC **53.** (Slope AB)(Slope BC) = $(-\frac{3}{4})(\frac{4}{3}) = -1$ **55.** $6x + 8y = -9$ **57.** $g(x) = -\frac{2}{3}x$
59. $f(x) = \frac{1}{2}x + \frac{7}{2}$
61. (A) $s = f(w) = w/10$ **63.** $0.2x + 0.1y = 20$
(B) $f(15) = 1.5$ inches, $f(30) = 3$ inches
(C) Slope $= \frac{1}{10}$
(D)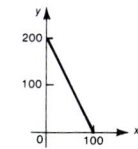

Exercise 2-5

1. 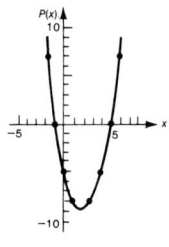   **3.** 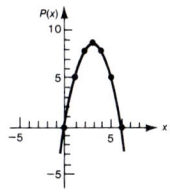   **5.** Min $f(x) = f(3) = 2$
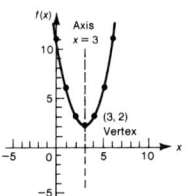

7. Max $f(x) = f(-3) = -2$ **9.** Min $f(x) = f(-3) = 2$ **11.** Max $f(x) = f(3) = 3$
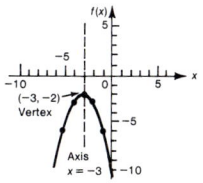   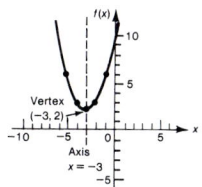   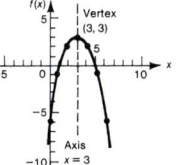

13. 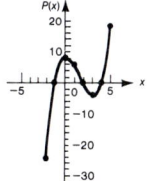   **15.** 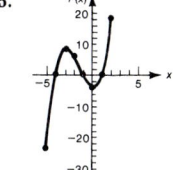   **17.** Min $f(x) = f(-2) = -2$
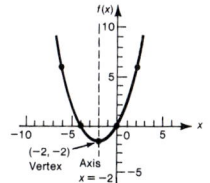

19. Max $f(x) = f(-2) = 6$

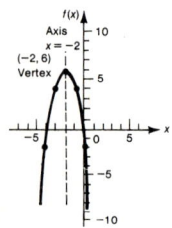

21.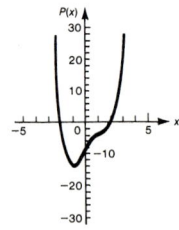

23.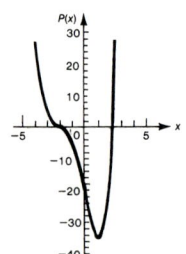

25. (A) $A(x) = 50x - x^2$
    (B) Domain: $0 < x < 50$
    (C)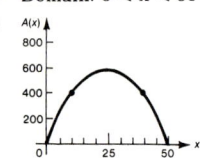
    (D) 25 by 25 feet

27. (A) $V(x) = (12 - 2x)(8 - 2x)x$
        $= 4x^3 - 40x^2 + 96x$
    (B) Domain: $0 < x < 4$. [Note: At $x = 0$ and $x = 4$, we have zero volume.]
    (C)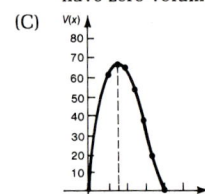
    (D) Max $V(x) \approx V(1.5) \approx 67.5$ cubic inches
        A 1.5-inch square should be cut from each corner.

Exercise 2-6

1.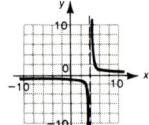

3.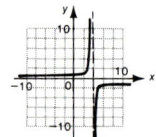

5.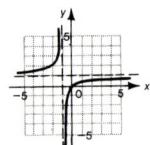

7.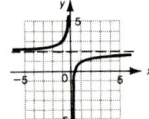

9.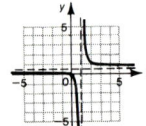

11.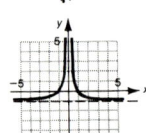

13.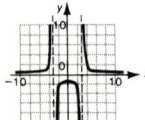

15.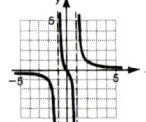

17.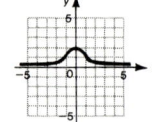

19.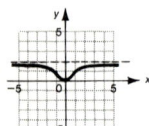

21.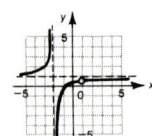

Exercise 2-7

1. $(f \circ g)(x) = (x^2 - x + 1)^7$, $(g \circ f)(x) = x^{14} - x^7 + 1$ 3. $(f \circ g)(x) = \sqrt{2x + 5}$, $(g \circ f)(x) = 2\sqrt{x} + 5$
5. $(f \circ g)(x) = (1 - x^2)^{1/2}$, $(g \circ f)(x) = 1 - x$ 7. $(f \circ g)(x) = |3x - 2|$, $(g \circ f)(x) = 3|x| - 2$
9. $(f \circ g)(x) = (x^3 - 4)^{2/3}$, $(g \circ f)(x) = x^2 - 4$ 11. $(f \circ g)(x) = 7$, $(g \circ f)(x) = -7$
13. Both H and H^{-1} are functions.   15. Both g and g^{-1} are functions.   17. p is a function

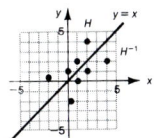   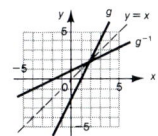   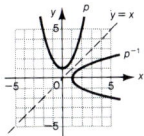

19. (A) Domain of $f = R = $ Range of f^{-1}, Range of $f = R = $ Domain of f^{-1}; (B) $(x + 2)/2$; (C) 3; (D) 2; (E) x
21. (A) Domain of $f = R = $ Range of f^{-1}, Range of $f = R = $ Domain of f^{-1}; (B) $(x + 5)/3$; (C) 3; (D) 2; (E) x
23. Not one to one
25. (A) Domain of $f = [0, \infty) = $ Range of f^{-1}, Range of $f = [1, \infty) = $ Domain of f^{-1}; (B) $\sqrt{x - 1}$; (C) $\sqrt{3}$; (D) 2;
    (E) x 27. Not one to one
29. (A) Domain of $f = [1, \infty) = $ Range of f^{-1}, Range of $f = [0, \infty) = $ Domain of f^{-1}; (B) $\sqrt{x} + 1$; (C) 3; (D) 2; (E) x
31. Domain of $f \circ g = R$, Domain of $g \circ f = R$ 33. Domain of $f \circ g = [-\frac{5}{2}, \infty)$, Domain of $g \circ f = [0, \infty)$
35. Domain of $f \circ g = [-1, 1]$, Domain of $g \circ f = [0, \infty)$ 37. Domain of $f \circ g = R$, Domain of $g \circ f = R$
39. Domain of $f \circ g = R$, Domain of $g \circ f = R$ 41. Domain of $f \circ g = R$, Domain of $g \circ f = R$
43. (A) x; (B) $4 - x$; (C) x; (D) $f^{-1}(x) = f(x)$ 45. (A) x;   (B) $1/x$; (C) x; (D) $f^{-1}(x) = f(x)$
47. $g(x) = (2x + 1)/(x - 1)$ [Note: $g = f^{-1}$.]   49. $(f^{-1} \circ f)(x) = x = I(x)$ for all x in X
51. $(f \circ g)(R) = 2\pi R$, a formula for the circumference of a circle in terms of R

Exercise 2-8

1. $F = kv^2$ 3. $f = k\sqrt{T}$ 5. $y = k/\sqrt{x}$ 7. $t = k/T$ 9. $R = kSTV$ 11. $V = khr^2$ 13. 4   15. $9\sqrt{3}$
17. $U = k(ab/c^3)$ 19. $L = k(wh^2/l)$ 21. -12 23. 83 pounds   25. 20 amperes
27. The new horsepower must be eight times the old.   29. No effect   31. $t^2 = kd^3$ 33. 1.47 hours (approx.)
35. 20 days   37. Quadrupled   39. 540 pounds   41. (A) $\Delta S = kS$; (B) 10 ounces; (C) 8 candlepower
43. 32 times per second   45. $N = k(F/d)$ 47. 1.2 miles/second   49. 20 days
51. The volume is increased by a factor of 8.

Exercise 2-9   Chapter Review

1. (A) Reflected across x axis;   (B) Shifted down three units;   (C) Shifted left three units   (2-3)
2. (A) $\sqrt{45}$;   (B) $-\frac{1}{2}$;   (C) 2   (2-1, 2-4)   3. Vertical: $x = -3$, slope not defined; horizontal: $y = 4$, slope $= 0$ (2-4)
4. 16   (2-2)   5. 1   (2-2)   6. 3   (2-2)   7. $-2a - h$ (2-2)
8. $(f \circ g)(x) = 17 - 3x^2$; $(g \circ f)(x) = 4 - (3x + 5)^2 = -21 - 30x - 9x^2$ (2-7)   9. (A) $(x + 1)/4$;   (B) 2;   (C) x (2-7)
10. Slope $= -\frac{3}{2}$ (2-4)   11. $2x + 3y = 12$ (2-4)   12. (2-6)

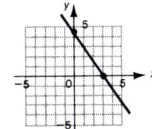   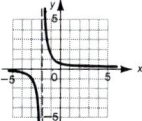

13. (A) $y = k(x/z)$;   (B) $\frac{4}{3}$ (2-8)   14. $x - 2y = 2$ (2-4)   15. $(x - 3)^2 + (y + 2)^2 = 4$ (2-1)
16. Decreasing   (2-3, 2-4)   17. It is symmetric with respect to all three.   (2-1)   18. (A) f;   (B) g;   (C) h (2-3)

AA12   Answers

19. (2-3) 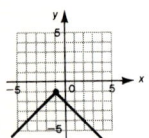   **20.** (A) $y = -2x - 3$; (B) $y = \frac{1}{2}x + 2$ (2-4)   **21.** Min $f(x) = f(3) = -4$ (2-5)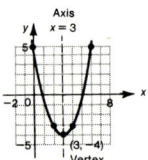

22. (A) No; (B) x intercept = 1, y intercept = $-\frac{1}{2}$; (C) $x = -2$; (D) Horizontal asymptote: $y = 1$, vertical asymptote: $x = -2$; (E) Below: $-2 < x < 1$, above: $x < -2$ and for $x > 1$ (F) (2-6)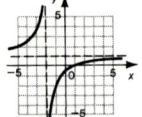

23. Center: $(3, -4)$, radius = 5 (2-1)   **24.** $P(x) = [(x - 2)x - 5]x + 6$ (2-5)   **25.** All but B (2-7)

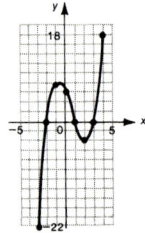

26. (A) Domain of $f = \{-2, -1, 0, 1, 2\} = $ Range of f^{-1}, Range of $f = \{\frac{1}{4}, \frac{1}{2}, 1, 2, 4\} = $ Domain of f^{-1}
(B)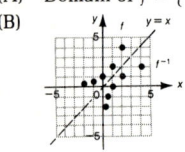

(C) Both f and f^{-1} are functions. (2-2, 2-3, 2-7)
27. (A) Domain of $f = [0, \infty) = $ Range of f^{-1}; Range of $f = [-1, \infty) = $ Domain of f^{-1}; (B) $\sqrt{x + 1}$; (C) 2; (D) 4; (E) x (2-7)   **28.** $t = kwd/p$ (2-8)
29. (A) Horizontal: $y = \frac{3}{2}$, vertical: $x = -\frac{3}{2}$; (B) horizontal: $y = 0$, vertical: $x = -2, x = 3$ (2-6)
30. f is discontinuous at $x = -\frac{3}{2}$ and g is discontinuous at $x = -2$ and $x = 3$. (2-6)
31. (A) $(3x + 2)/(x - 1)$; (B) $\frac{11}{2}$; (C) x (2-7)   **32.** $x - y = 3$; a line (2-1, 2-4)
33. Above for $(-\infty, -2)$, below for $(-2, \infty)$ (2-6)   **34.** The force is doubled. (2-8)

Practice Test: Chapter 2

1. (A) $3x + 2y = -6$; (B) $\sqrt{52}$ (2-1, 2-4)   **2.** (A) $y = -\frac{3}{4}x + 1$; (B) $y = \frac{4}{3}x - \frac{22}{3}$ (2-4)
3. Center: $(-2, 3)$, radius = 4 (2-1)   **4.** (A) -1; (B) $-4a - 2h$ (2-2)
5. (A) g and h; (B) f; (C) f (2-3, 2-7)   **6.** Min $f(x) = f(3) = 2$, vertex: $(3, 2)$ (2-5)
7. (A) No; (B) At $x = -1$; (C) x intercept = 1, y intercept = $-\frac{1}{2}$;
(D) Horizontal asymptote: $y = \frac{1}{2}$, vertical asymptote: $x = -1$ (2-6)
8. Above for $x < -1$ and for $x > 1$, below for $-1 < x < 1$ (2-6)   **9.** (2-6)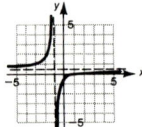

Answers   AA13

10. (A) $(f \circ g)(x) = \sqrt{|x|} - 8$, $(g \circ f)(x) = |\sqrt{x} - 8|$; (B) Domain of $f \circ g = R$, Domain of $g \circ f = [0, \infty)$ (2-7)
11. (A) $(x + 7)/3$; (B) 4; (C) x; (D) Increasing (2-3, 2-7)
12. (A) Domain of $f = [1, \infty) = $ Range of f^{-1}, Range of $f = [0, \infty) = $ Domain of f^{-1}
    (B) Both f and f^{-1} are functions.

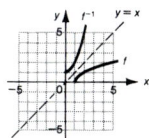

(C) $f^{-1}(x) = x^2 + 1$, $x \geq 0$ (2-7)
13. (A) $H = K(n/m^2)$; (B) 18; (C) H is cut in half. (2-8)
14. It is the same as the graph of g shifted to the right two units and down one unit; then turned upside down. (2-3)
15. (A) $f(x) = [(x - 3)x - 1]x + 3$
    (B) (2-5)

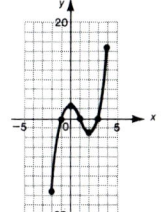

Chapter 3   Exercise 3-2

1. $2m + 1$ 3. $4x - 5$, $R = 11$ 5. $x^2 + x + 1$ 7. $2y^2 - 5y + 13$, $R = -27$
9. $(x^2 + 3x - 7)/(x - 2) = x + 5 + [3/(x - 2)]$ 11. $(4x^2 + 10x - 9)/(x + 3) = 4x - 2 - [3/(x + 3)]$
13. $(2x^3 - 3x + 1)/(x - 2) = 2x^2 + 4x + 5 + [11/(x - 2)]$ 15. $3x^3 - 3x^2 + 3x - 4$ 17. $x^4 - x^3 + x^2 - x + 1$
19. $2x^2 - 2x - 3$, $R = -2$ 21. $2x^3 - 3x^2 - x - 5$, $R = -10$ 23. $4x^3 - 6x - 2$, $R = 2$ 25. $4x^2 - 2x - 4$
27. $x^2 - 1.7x + 2.49$, $R = -0.253$ 29. $3x^2 + 0.6x + 2.12$, $R = -3.576$ 31. $2x^2 - 3x + 2$, $R = 0$
33. $2x^2 + 1.6x - 2.72$, $R = -2.976$ 35. $2x^2 - 7.8x + 2.92$, $R = 1.912$ 37. $x^2 + (-3 + i)x - 3i$
39. (A) In both cases, the coefficient of x is a_2, the constant term is $a_2 r + a_1$, and the remainder is $(a_2 r + a_1)r + a_0$;
    (B) The remainder expanded is $a_2 r^2 + a_1 r + a_0 = P(r)$.

Exercise 3-3

1. 4   3. 3   5. -6 7. $3, -5$ 9. $-\frac{1}{2}, 8, -2$ 11. Yes   13. Yes   15. -3 17. 0.427
19.   21.   23. $-4, -8, 1$ 25. $\frac{1}{8}, -\frac{3}{5}, -4$

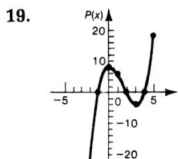

27. $P(x) = \left[x - \left(\frac{3 + \sqrt{5}}{2}\right)\right]\left[x - \left(\frac{3 - \sqrt{5}}{2}\right)\right]$ 29. $P(x) = [x - (3 + i)][x - (3 - i)]$ 31. Yes   33. No

35. 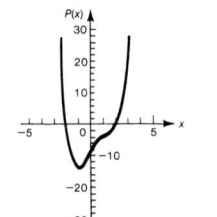 **37.**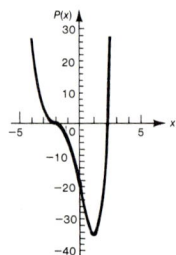

39. $P(2) = 10$. Both methods involve exactly the same operations on the same numbers.
41. (A) $P(r) = (a_2 r + a_1)r + a_0$; (B) $P(r) = R = (a_2 r + a_1)r + a_0$. Both are the same.

Exercise 3-4

1. -8 (multiplicity 3), 6 (multiplicity 2); degree of $P(x)$ is 5   **3.** -4 (multiplicity 3), 3 (multiplicity 2), -1; degree of $P(x)$ is 6
5. $P(x) = (x - 3)^2(x + 4)$, degree 3   **7.** $P(x) = (x + 7)^3(x - \frac{2}{3})(x + 5)$, degree 5
9. $P(x) = [x - (2 - 3i)][x - (2 + 3i)](x + 4)^2$, degree 4   **11.** One real and two complex, or three real
13. Six complex, two real and four complex, four real and two complex, or six real   **15.** $P(x) = (x + 4)^2(x + 1)$
17. $P(x) = (x - 1)(x + 1)(x - i)(x + i)$ **19.** $P(x) = (2x - 1)[x - (4 + 5i)][x - (4 - 5i)]$
21. Four complex, two complex and two real, or four real
23. Six complex, two real and four complex, four real and two complex, or six real   **25.** $x^2 - 8x + 41$
27. $x^2 - 2ax + (a^2 + b^2)$ **29.** $3 + i, -1$ **31.** $5i, 3$ **33.** (A) 3; (B) $-\frac{1}{2} - (\sqrt{3}/2)i, -\frac{1}{2} + (\sqrt{3}/2)i$ **35.** $n, 1$
37. No, since $P(x)$ is not a polynomial with real coefficients (the coefficient of x is the complex number $2i$).

Exercise 3-5

1. One positive, one negative   **3.** No positive, two or no negative   **5.** One or three positive, no negative
7. UB = 2, LB = -1 **9.** UB = 2, LB = -3 **11.** UB = 1, LB = -2 **13.** $P(3) = -2, P(4) = 2$
15. $P(-3) = -13, P(-2) = 3$ **17.** $Q(1) = 4, Q(2) = -1$
19. (A) Zero or two positive, one negative; (B) UB = 3, LB = -3;
  (C) 1 is a zero and one zero in each interval $(-3, -2)$ and $(2, 3)$
21. (A) One positive, zero or two negative; (B) UB = 3, LB = -2; (C) One real zero in interval (2, 3)
23. (A) One positive, one or three negative; (B) UB = 2, LB = -5;
  (C) Each interval contains exactly one real zero: $(-4, -3), (-2, -1) (-1, 0)$, and $(1, 2)$.
25. (A) One or three positive, zero or two negative; (B) UB = 2, LB = -2; (C) At least one real zero in interval (1, 2)
27. By Descartes' rule of signs $P(x)$ has one positive and one negative real zero. By the fundamental theorem of algebra, $P(x)$ has four zeros, and 0 is not a zero. Therefore, the other two zeros must be complex.
29. By Descartes' rule of signs, $P(x)$ has no positive or negative zeros. Since $P(0) = 0, P(-1) < 0$, and $P(1) > 0$, the graph crosses the x axis only at the origin.

Exercise 3-6

1. (A) $\pm 1, \pm 2, \pm 3, \pm 6$; (B) $-2, 1, 3$ **3.** (A) $\pm 1, \pm 2, \pm 4, \pm\frac{1}{3}, \pm\frac{2}{3}, \pm\frac{4}{3}$; (B) 2 (double zero), $-\frac{1}{3}$
5. (A) $\pm 1, \pm 3, \pm\frac{1}{2}, \pm\frac{3}{2}, \pm\frac{1}{3}, \pm\frac{1}{4}, \pm\frac{3}{4}, \pm\frac{1}{6}, \pm\frac{1}{12}$; (B) $-\frac{1}{2}, \frac{1}{3}, \frac{3}{2}$ **7.** (A) $\pm 1, \pm 2, \pm 4, \pm\frac{1}{3}, \pm\frac{2}{3}, \pm\frac{4}{3}$; (B) $-\frac{1}{3}$
9. (A) $\pm 1, \pm 2, \pm 3, \pm 6$; (B) No rational zeros   **11.** (A) $\pm 1, \pm 2, \pm 4, \pm 8$; (B) ± 2
13. (A) $\pm 1, \pm 2, \pm 3, \pm 6, \pm\frac{1}{3}, \pm\frac{2}{3}$; (B) $-\frac{1}{3}, 2$ **15.** $\frac{1}{2}, 1 \pm \sqrt{2}$ **17.** -2 (double root), $\pm\sqrt{5}$ **19.** $\pm 1, \frac{3}{2}, \pm i$
21. $P(x) = (x + 2)(2x - 1)(3x + 2)$ **23.** $P(x) = (x + 4)\left[x - \left(1 + \sqrt{2}\right)\right]\left[x - \left(1 - \sqrt{2}\right)\right]$
25. $\sqrt{6}$ is a zero of $P(x) = x^2 - 6$, but $P(x)$ has no rational zeros.
27. $\sqrt[3]{5}$ is a zero of $P(x) = x^3 - 5$, but $P(x)$ has no rational zeros.

Answers

29. Inequality notation: $2 - \sqrt{3} \le x \le 2 + \sqrt{3}$, interval notation: $[2 - \sqrt{3},\ 2 + \sqrt{3}]$
31. Inequality notation: $-3 \le x \le \frac{1}{2}$ or $x \ge 2$, interval notation: $[-3, \frac{1}{2}] \cup [2, \infty)$

Exercise 3-7

1. 4.9   **3.** 0.7   **5.** $-1, 2.1$ **7.** $-\frac{1}{2}, 3, 1.9$
9. One irrational root is in each of the following intervals: $[-1, 0], [0, 1], [4, 5]$; largest ≈ 4.87
11. $P(2) = 2, P(2.5) = -0.1875, P(3) = 2$ **13.** 1.8 inches   **15.** 0.4 foot

Exercise 3-8

1. $A = 2, B = 5$ **3.** $A = 7, B = -2$ **5.** $A = 1, B = 2, C = 3$ **7.** $A = 2, B = 1, C = 3$
9. $A = 0, B = 2, C = 2, D = -3$ **11.** $\dfrac{3}{x-4} - \dfrac{4}{x+2}$ **13.** $\dfrac{3}{3x+4} - \dfrac{1}{2x-3}$ **15.** $\dfrac{2}{x} - \dfrac{1}{x-3} - \dfrac{3}{(x-3)^2}$
17. $\dfrac{2}{x} + \dfrac{3x-1}{x^2+2x+3}$ **19.** $\dfrac{2x}{x^2+2} + \dfrac{3x+5}{(x^2+2)^2}$ **21.** $\dfrac{x-2+3}{x-2} - \dfrac{2}{x-3}$ **23.** $\dfrac{2}{x-3} + \dfrac{2x+5}{x^2+3x+3}$
25. $\dfrac{2}{x-4} - \dfrac{1}{x+3} + \dfrac{3}{(x+3)^2}$ **27.** $\dfrac{2}{x-2} - \dfrac{3}{(x-2)^2} - \dfrac{2x}{x^2-x+1}$ **29.** $\dfrac{x+2+1}{2x-1} - \dfrac{2}{x+2} + \dfrac{x-1}{2x^2-x+1}$

Exercise 3-9 Chapter Review

1. $2x^3 + 3x^2 - 1 = (x+2)(2x^2 - x + 2) - 5$ (3-2)   **2.** $P(3) = -8$ (3-3)   **3.** $2, -4, -1$ (3-3)   **4.** $1 - i$ (3-4)
5. (A) Zero or two positive, one negative; (B) Zero positive, one negative (3-5)   **6.** LB: -2 and -1, UB: 4 (3-5)
7. $P(1) = -5$ and $P(2) = 1$ are of opposite sign. (3-5)   **8.** $\pm 1, \pm 2, \pm 3, \pm 6$ (3-6)   **9.** $-1, 2, 3$ (3-6)
10. $\dfrac{2}{x-3} + \dfrac{5}{x+2}$ (3-8)   **11.** $P(x) = (x - \frac{2}{3})(3x^2 - 6x - 3) - 5$ (3-2)   **12.** -4 (3-3)
13. $P(x) = [x - (1 + \sqrt{2})][x - (1 - \sqrt{2})]$ (3-3)   **14.** Yes, since $P(-1) = (-1)^{25} + 1 = 0$ (3-3)
15. (A) Zero or two positive, zero or two negative; (B) LB: -3, UB: 4; (C) -2 is a zero; $(-1, 0), (0, 1)$, and $(3, 4)$ each contains a zero. (3-3)   **16.** $-2, -\frac{1}{2}, 4$ (3-6)   **17.** $P(x) = 2(x+2)(x+\frac{1}{2})(x-4)$ (3-3)   **18.** No rational zeros (3-6)
19. $\frac{1}{2}, \dfrac{1 + \sqrt{3}i}{2}, \dfrac{1 - \sqrt{3}i}{2}$ (3-6, 3-7)   **20.** $P(x) = 2(x - \frac{1}{2})\left(x - \dfrac{1 + \sqrt{3}i}{2}\right)\left(x - \dfrac{1 - \sqrt{3}i}{2}\right)$ (3-3)   **21.** 1.4 (3-7)
22. $\dfrac{1}{x} - \dfrac{2}{x-2} + \dfrac{3}{(x-2)^2}$ (3-8)   **23.** $\dfrac{3}{x} + \dfrac{2x-1}{2x^2-3x+3}$ (3-8)
24. $P(x) = [x - (1 + i)][x^2 + (1 + i)x + (3 + 2i)] + (3 + 5i)$ (3-2)   **25.** $P(x) = (x + \frac{1}{2})^2(x+3)(x-1)^3$, degree 6 (3-3)
26. $P(x) = (x+5)[x - (2 - 3i)][x - (2 + 3i)]$, degree 3 (3-3)   **27.** $\frac{1}{2}, 3.0$ (3-6, 3-7)   **28.** $\dfrac{2}{x-3} - \dfrac{3}{x} + \dfrac{x-1}{x^2+1}$ (3-8)

Practice Test: Chapter 3

1. $Q(x) = 8x^3 - 12x^2 - 16x - 8, R = 5, P(\frac{1}{4}) = R = 5$ (3-3)   **2.** Since $P(-1) = 0, x - (-1) = x + 1$ is a factor of $P(x)$. (3-3)
3. (A) $\pm 1, \pm 2, \pm 3, \pm 6, \pm \frac{1}{2}, \pm \frac{3}{2}$; (B) Two or no positive real zeros and one negative real zero (3-5, 3-6)
4. (A) $[-1, 0], [1, 2], [2, 3]$; (B) LB: -1, UB: 4 (3-5)   **5.** $\frac{3}{2}, 1 \pm \sqrt{3}$ (3-6)
6. $P(x) = 2(x - \frac{3}{2})[x - (1 + \sqrt{3})][x - (1 - \sqrt{3})]$ (3-3, 3-6)

Answers

7. Since $P(x) = x^3 + 3x - 5$ has one variation in sign and $P(-x)$ has no variations in sign, by Descartes' rule of sign $P(x)$ has no negative real zeros and exactly one positive real zero. (3-5)

8. 1.2 (3-7)   9. $\dfrac{2}{x+2} - \dfrac{1}{x-1} + \dfrac{3}{(x-1)^2}$ (3-8)   10. $\dfrac{1}{x+1} + \dfrac{2x}{x^2+4}$ (3-8)

Chapter 4 Exercise 4-1

1.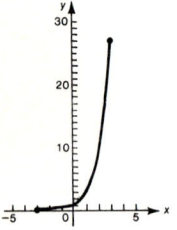

3.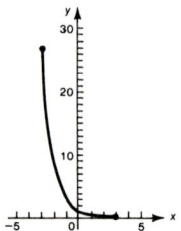

5.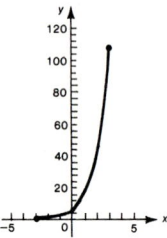

7.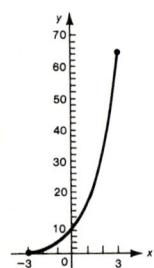

9.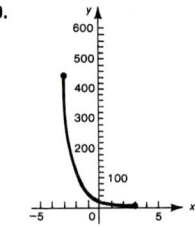

11.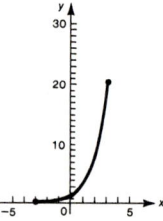

13.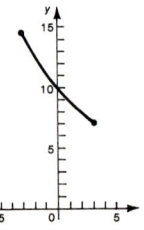

15.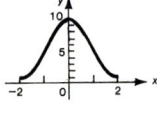

17.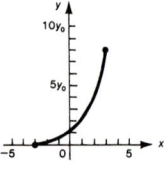

19.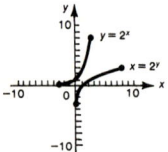

21.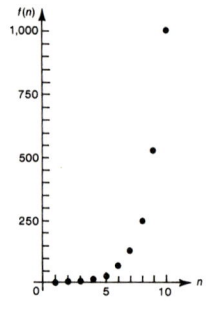

23.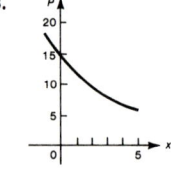

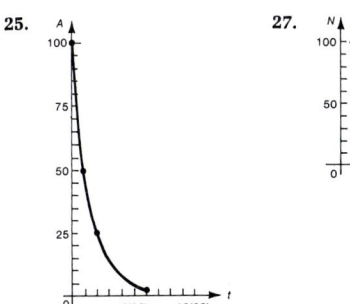

25.

27.

Exercise 4-2

1. $9 = 3^2$ 3. $81 = 3^4$ 5. $1{,}000 = 10^3$ 7. $1 = e^0$ 9. $\log_8 64 = 2$ 11. $\log_{10} 10{,}000 = 4$ 13. $\log_v u = x$
15. $\log_{27} 9 = \tfrac{2}{3}$ 17. 5   19. -4 21. 2   23. 3   25. $x = 4$ 27. $y = 2$ 29. $b = 4$
31. $0.001 = 10^{-3}$ 33. $3 = 81^{1/4}$ 35. $16 = (\tfrac{1}{2})^{-4}$ 37. $N = a^e$ 39. $\log_{10} 0.01 = -2$ 41. $\log_e 1 = 0$
43. $\log_2 (\tfrac{1}{8}) = -3$ 45. $\log_{81} (\tfrac{1}{3}) = -\tfrac{1}{4}$ 47. $\log_{49} 7 = \tfrac{1}{2}$ 49. u 51. $\tfrac{1}{2}$ 53. $\tfrac{3}{2}$ 55. 0   57. $\tfrac{3}{2}$
59. $x = 2$ 61. $y = -2$ 63. $b = 100$ 65. $b =$ Any positive real number except 1
67. Domain of f is the set of all real numbers; the range of f is $\{1\}$. The domain of f^{-1} is $\{1\}$; the range of f^{-1} is the set of all real numbers. No, f^{-1} is not.
69. (A)   (B) Domain of f is the set of real numbers; range of f is the set of positive real numbers. The domain of f is the range of f^{-1} and the range of f is the domain of f^{-1}.
   (C) f^{-1} is called the logarithmic function with base 10.

71. $f^{-1}(x) = [1 + \log_5 (x - 4)]/3$ 73. $g^{-1}(x) = (b^{x/3} + 2)/5$

Exercise 4-3

1. $\log_b u + \log_b v$ 3. $\log_b A - \log_b B$ 5. $5 \log_b u$ 7. $\tfrac{3}{5} \log_b N$ 9. $\tfrac{1}{2} \log_b Q$ 11. $\log_b u + \log_b v + \log_b w$
13. $\log_b AB$ 15. $\log_b X/Y$ 17. $\log_b wx/y$ 19. 3.40   21. -0.92 23. 3.30   25. $2 \log_b u + 7 \log_b v$
27. $-\log_b a$ 29. $\tfrac{1}{3} \log_b N - 2 \log_b p - 3 \log_b q$ 31. $\tfrac{1}{4}(2 \log_b x + 3 \log_b y - \tfrac{1}{2} \log_b z)$ 33. $\log_b x^2/y$
35. $\log_b x^3 y^2/z^4$ 37. $\log_b \sqrt[3]{x^2 y^3}$ 39. 2.02   41. 0.23   43. -0.05 45. 8   47. $y = cb^{-kt}$
49. Let $u = \log_b M$ and $v = \log_b N$; then $M = b^u$ and $N = b^v$. Thus, $\log_b M/N = \log_b b^u/b^v = \log_b b^{u-v} = u - v = \log_b M - \log_b N$.
51. $MN = b^{\log_b M} b^{\log_b N} = b^{\log_b M + \log_b N}$; hence, by definition of logarithm $\log_b MN = \log_b M + \log_b N$.

Exercise 4-4

1. 4.9177   3. -2.8419 5. 3.7623   7. -2.5128 9. 0.8627   11. 3.3096   13. -1.3840 15. 200,800
17. 0.000 6648   19. 47.73   21. 0.6760   23. 5.4843   25. -2.3215 27. 2.74×10^7 29. 1.58×10^{-5}
31. 4.959   33. 7.861   35. 3.301   37. 3.6776   39. -1.6094 41. -1.7372 43. 0.8544   45. 13.3114
47. -5.7541 49. 12.725   51. -25.715 53. 1.1709×10^{32} 55. 4.2672×10^{-7}

Exercise 4-5

1. 1.46   3. 0.321   5. 1.29   7. 3.50   9. 1.80   11. 2.07   13. 20   15. 5   17. 14.2   19. -1.83
21. 11.7   23. 5   25. $1, e^2, e^{-2}$ 27. $x = e^e$ 29. 100, 0.1   31. $x = -(1/k) \ln (I/I_0)$ 33. $I = I_0 10^{N/10}$
35. $t = (-L/R) \ln [1 - (RI/E)]$
37. Inequality sign should have been reversed when both sides were multiplied by $\log \frac{1}{2}$, a negative quantity.
39. 5 years to the nearest year   41. Approx. 3.8 hours   43. Approx. 35 years   45. Approx. 28 years
47. Divide both sides by I_0, take logs of both sides, and then multiply both sides by 10.   49. 95 feet, 489 feet

Exercise 4-6   Chapter Review

1. $n = \log_{10} m$ (4-2)   2. $x = 10^y$ (4-2)   3. 8 (4-2)   4. 5 (4-2)   5. 3 (4-2)   6. 1.24 (4-5)
7. 11.9 (4-5)   8. 900 (4-3, 4-5)   9. 5 (4-3, 4-5)   10. $y = e^x$ (4-2)   11. $y = \ln x$ (4-2)   12. -2 (4-2)
13. $\frac{1}{3}$ (4-2)   14. 64 (4-2)   15. e (4-2)   16. 33 (4-2)   17. 1 (4-2)   18. 2.32 (4-5)   19. 3.92 (4-5)
20. 92.1 (4-5)   21. 300 (4-3, 4-5)   22. 2 (4-3, 4-5)   23. $1, 10^3, 10^{-3}$ (4-3, 4-5)   24. 10^e (4-3, 4-5)
25. 1.95 (4-4)   26. $y = ce^{-5t}$ (4-3, 4-5)   27. Domain $f = (0, \infty) = $ Range f^{-1}, Range $f = R = $ Domain f^{-1} (4-2)

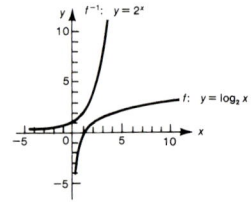

28. If $\log_1 x = y$, then we would have to have $1^y = x$; that is, $1 = x$ for arbitrary positive x, which is impossible. (4-2)
29. Let $u = \log_b M$ and $v = \log_b N$; then $M = b^u$ and $N = b^v$. Thus, $\log_b (M/N) = \log_b (b^u/b^v) = \log_b b^{u-v} = u - v = \log_b M - \log_b N$. (4-3)
30. 23.4 years (4-5)   31. 23.1 years (4-5)   32. 37,100 years (4-5)   33. $I = I_0 e^{-kx}$ (4-3, 4-5)
34. $n = -\log [1 - (Pi/R)]/\log (1 + i)$ (4-3, 4-5)

Practice Test: Chapter 4

1. $y = e^{-x^2}$ (4-2)   2. $\log_b [(x^{1/2} y^3)/z^2]$ (4-3)   3. 1.63 (4-5)   4. 2.11 (4-5)   5. 11,200 (4-5)
6. -3 (4-2)   7. 8 (4-2)   8. $\frac{1}{10}$ (4-2)   9. 6 (4-3, 4-5)   10. e^{10} (4-3, 4-5)
11. $f^{-1}(x) = \ln x$, Domain of $f = R = $ Range of f^{-1}, Range of $f = (0, \infty) = $ Domain of f^{-1} (4-2)   12. 0.837 (4-4)
13. 9.24 years (4-5)   14. $r = (1/t) \ln (A/P)$ (4-3, 4-5)   15. $n = -\log [1 - (Pi/R)]/\log (1 + i)$ (4-3, 4-5)

Chapter 5   Exercise 5-2

1. $(-1, 0)$ 3. $(1, 0)$ 5. $(-1, 0)$ 7. $(0, -1)$ 9. $(0, -1)$ 11. $(0, -1)$ 13. $(1/\sqrt{2}, 1/\sqrt{2})$
15. $(\sqrt{3}/2, 1/2)$ 17. $(1/2, -\sqrt{3}/2)$ 19. $(-1/2, \sqrt{3}/2)$ 21. $(-1/\sqrt{2}, -1/\sqrt{2})$ 23. $(1/2, -\sqrt{3}/2)$
25. $(-1/\sqrt{2}, -1/\sqrt{2})$ 27. $(\sqrt{3}/2, -1/2)$ 29. Negative   31. Negative   33. Negative   35. Positive
37. Positive   39. Negative   41. T   43. F   45. T   47. $x = -7\pi/4, \pi/4$ 49. $x = -4\pi/3, 2\pi/3$
51. $x = -5\pi/6, 7\pi/6$ 53. $x = \pi/4 + 2k\pi$, k any integer

Exercise 5-3

1. (A) a; (B) $1/b$; (C) a/b; (D) $1/a$; (E) b/a; (F) b 3. 1 5. $\frac{1}{2}$ 7. 1 9. $\sqrt{3}$
11. Not defined 13. 1 15. $\sqrt{2}$ 17. 1 19. Not defined 21. Quadrant II or III 23. Quadrant I or II
25. Quadrant II or IV 27. -0.6573 29. -14.60 31. 1.000 33. -1 35. 0 37. $1/\sqrt{2}$
39. Not defined 41. $-1/\sqrt{3}$ 43. $\sqrt{3}/2$ 45. $-1/\sqrt{2}$ 47. $-\frac{1}{2}$ 49. 2 51. 1 53. $\sqrt{2}$
55. $-\sqrt{3}/2$ 57. None 59. $\pi/2, 3\pi/2$ 61. $\pi/2, 3\pi/2$
63. (A) 0 to 1; (B) 1 to 0; (C) 0 to -1; (D) -1 to 0 65. 0.8138 67. 0.5290 69. $\frac{1}{2}$ 71. $\sqrt{3}$
73. -5 75. $\sin x = -\sqrt{3}/2, \tan x = -\sqrt{3}, \cot x = -1/\sqrt{3}, \csc x = -2/\sqrt{3}, \sec x = 2$
77. $\cos x = -1/\sqrt{2}, \tan x = 1, \cot x = 1, \csc x = -\sqrt{2}, \sec x = -\sqrt{2}$
79. $\cot x = 1/\sqrt{3}, \sin x = -\sqrt{3}/2, \cos x = -\frac{1}{2}, \csc x = -2/\sqrt{3}, \sec x = -2$ 81. π 83. $5\pi/6$ 85. $5\pi/6$
87. (A) Identity (5); (B) Identity (9); (C) Identity (1) 89. 75 square meters 91. $12\sqrt{3} \approx 20.78$ square inches

Exercise 5-4

1. $40°$ 3. $270°$ 5. $405°$ 7. 6 9. 2.5 11. $\pi/4$ 13. $3\pi/2$ 15. $13\pi/6$
17. $\pi/6, \pi/3, \pi/2, 2\pi/3, 5\pi/6, \pi$ 19. $-\pi/4, -\pi/2, -3\pi/4, -\pi$ 21. $60°, 120°, 180°, 240°, 300°, 360°$
23. $-90°, -180°, -270°, -360°$ 25. 5.859° 27. 354.141° 29. 1.117 31. 1.892 33. 0.234
35. 53.29° 37. 64.74° 39. -134.65 41. Quadrant III 43. Quadrant II 45. Quadrant III
47. Quadrantal angle 49. Quadrant IV 51. Quadrant IV 53. Quadrant II 55. Quadrantal angle
57. Quadrant II 59. Quadrant III 61. Coterminal 63. Coterminal 65. Coterminal 67. Not coterminal
69. Coterminal 71. Coterminal 73. 12 75. 200

Exercise 5-5

1. $\sin \theta = \frac{4}{5}, \cos \theta = \frac{3}{5}, \tan \theta = \frac{4}{3}, \csc \theta = \frac{5}{4}, \sec \theta = \frac{5}{3}, \cot \theta = \frac{3}{4}$
3. $\sin \theta = -1/\sqrt{5}, \cos \theta = -2/\sqrt{5}, \tan \theta = \frac{1}{2}, \csc \theta = -\sqrt{5}, \sec \theta = -\sqrt{5}/2, \cot \theta = 2$
5. $\sin \theta = \sqrt{3}/2, \cos \theta = -\frac{1}{2}, \tan \theta = -\sqrt{3}, \csc \theta = 2/\sqrt{3}, \sec \theta = -2, \cot \theta = -1/\sqrt{3}$ 7. 0.4226 9. -1.573
11. 0.8439 13. -0.3363 15. 0.9174 17. 1.009 19. 0 21. $\sqrt{3}$ 23. $1/\sqrt{2}$ 25. $\sqrt{2}$
27. Not defined 29. Not defined 31. $-\frac{1}{2}$ 33. 0 35. $-1/\sqrt{3}$ 37. $\sqrt{3}/2$ 39. $1/\sqrt{2}$ 41. 2
43. 2 45. 0 47. $-\sqrt{3}$ 49. $-\sqrt{3}/2$ 51. Not defined 53. $-\frac{1}{2}$ 55. $120°$ or $2\pi/3$ radians
57. $210°$ or $7\pi/6$ radians 59. $240°$ or $4\pi/3$ radians 61. $\cos \theta = -\frac{4}{5}, \tan \theta = -\frac{3}{4}, \csc \theta = \frac{5}{3}, \sec \theta = -\frac{5}{4}, \cot \theta = -\frac{4}{3}$
63. $\sin \theta = -\frac{2}{3}, \tan \theta = 2/\sqrt{5}, \csc \theta = -\frac{3}{2}, \sec \theta = -3/\sqrt{5}, \cot \theta = \sqrt{5}/2$
65. $\sin \theta = -\sqrt{2}/\sqrt{3}, \cos \theta = 1/\sqrt{3}, \csc \theta = -\sqrt{3}/\sqrt{2}, \sec \theta = \sqrt{3}, \cot \theta = -1/\sqrt{2}$ 67. Secant and tangent

Exercise 5-6

1.

FUNCTION	VALUE AT x				
	0	$\pi/2$	π	$3\pi/2$	2π
$\sin x$	0	1	0	-1	0
$\cos x$	1	0	-1	0	1
$\tan x$	0	Not defined	0	Not defined	0
$\cot x$	Not defined	0	Not defined	0	Not defined
$\sec x$	1	Not defined	-1	Not defined	1
$\csc x$	Not defined	1	Not defined	-1	Not defined

3.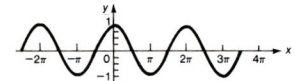

5.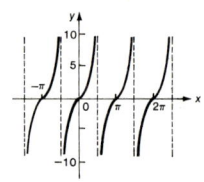

7.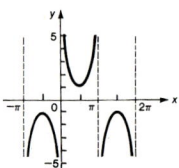

Exercise 5-7

1. $A = 3$, $P = 2\pi$

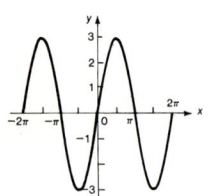

3. $A = \frac{1}{2}$, $P = 2\pi$

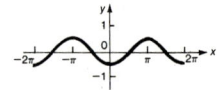

5. $A = 1$, $P = 2\pi/3$

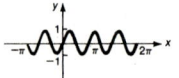

7. $A = 1$, $P = 4\pi$

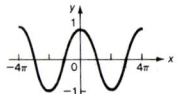

9. $A = 1$, $P = 2$

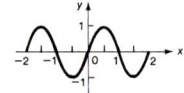

11. $A = 3$, $P = \pi$

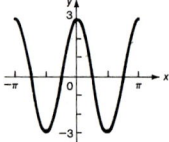

13. $A = \frac{1}{2}$, $P = 1$

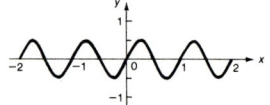

15. $A = 3$, $P = 4\pi$

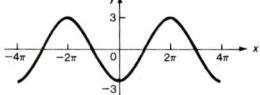

17. $A = 1$, $P = 2\pi$, phase shift $= \pi/2$ left

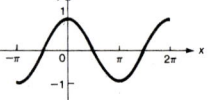

19. $A = \frac{1}{2}$, $P = 2\pi$, phase shift $= \pi/4$ right

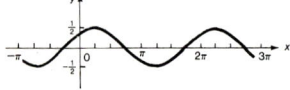

21. $A = 1$, $P = 2$, phase shift $= 1$ right

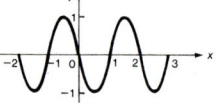

23. $A = 3$, $P = 2$, phase shift $= \frac{1}{2}$ left

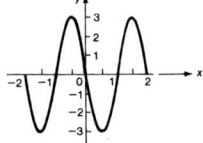

25. $A = 4$, $P = \pi$, phase shift $= \pi/2$ right

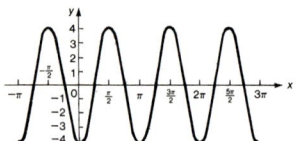

27. $A = 3.5$, $P = 4$, phase shift $= 0.5$ left

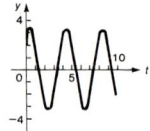

29. $A = 50$, $P = 1$, phase shift $= 0.25$ right

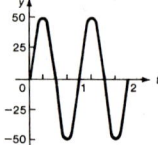

31. $P = \pi/2$, phase shift: none

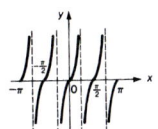

33. $P = \pi/2$, phase shift $= \pi/2$ left

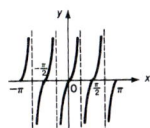

35. $A = \frac{1}{3}$, $P = \pi/4$

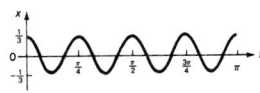

37. $A = 3$, $P = \frac{1}{3}$

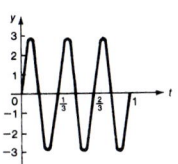

Exercise 5-8

1. 0 **3.** $\pi/6$ **5.** $\pi/4$ **7.** $\pi/3$ **9.** $\pi/6$ **11.** 0 **13.** 1.155 **15.** 1.548 **17.** Not defined
19. $2\pi/3$ **21.** $-\pi/4$ **23.** $-\pi/3$ **25.** $5\pi/6$ **27.** -0.6 **29.** -1.5 **31.** $\sqrt{3}$ **33.** $\sqrt{2}/2$ **35.** $-1/\sqrt{3}$
37. -1.328 **39.** 1.001 **41.** -1.029 **43.** 2.456 **45.** $\sqrt{1-x^2}$ **47.** $x/\sqrt{1-x^2}$
49. $f^{-1}(x) = 1 + \sin^{-1}[(x-3)/5]$, $-2 \leq x \leq 8$

Exercise 5-9 Chapter Review

1. (A) $(\sqrt{3}/2, 1/2)$; (B) $(1/\sqrt{2}, 1/\sqrt{2})$ (5-2) **2.** 7.5 centimeters (5-4)
3. (A) III, IV; (B) II, III; (C) II, IV (5-3) **4.** (A) $-\frac{3}{5}$; (B) $\frac{5}{4}$; (C) $-\frac{4}{3}$ (5-5)
5. (5-5)

θ degrees	θ radians	$\sin\theta$	$\cos\theta$	$\tan\theta$	$\csc\theta$	$\sec\theta$	$\cot\theta$
0°	0	0	1	0	ND*	1	ND
30°	$\pi/6$	1/2	$\sqrt{3}/2$	$1/\sqrt{3}$	2	$2/\sqrt{3}$	$\sqrt{3}$
45°	$\pi/4$	$1/\sqrt{2}$	$1/\sqrt{2}$	1	$\sqrt{2}$	$\sqrt{2}$	1
60°	$\pi/3$	$\sqrt{3}/2$	1/2	$\sqrt{3}$	$2/\sqrt{3}$	2	$1/\sqrt{3}$
90°	$\pi/2$	1	0	ND	1	ND	0
180°	π	0	-1	0	ND	-1	ND
270°	$3\pi/2$	-1	0	ND	-1	ND	0
360°	2π	0	1	0	ND	1	ND

* ND = Not defined

6. (A) 2π; (B) 2π; (C) π (5-6)
7. (A) Domain $= R$, range $= [-1, 1]$; (B) Domain is the set of all real numbers except $x = \dfrac{2k+1}{2}\pi$, k an integer, range $= R$ (5-6)
8. (5-6) 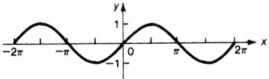 **9.** (5-6)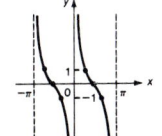

Answers

10. (A) $(-1/2, -\sqrt{3}/2)$; (B) $(0, -1)$ (5-2)   **11.** (A) II; (B) Quadrantal; (C) III (5-4)
12. (A) and (C) (5-4)   **13.** (B) and (C) (5-5)   **14.** (A) $\pi/2, 3\pi/2$; (B) $0, \pi$; (C) $0, \pi$ (5-3, 5-5)
15. 0 (5-3, 5-5)   **16.** Not defined (5-5)   **17.** 0 (5-8)   **18.** $-1/\sqrt{2}$ or $-\sqrt{2}/2$ (5-3, 5-5)   **19.** $\pi/4$ (5-8)
20. $-2/\sqrt{3}$ or $-2\sqrt{3}/3$ (5-5)   **21.** $\pi/3$ (5-8)   **22.** $-\frac{1}{2}$ (5-5)   **23.** $-\pi/4$ (5-8)
24. $-1/\sqrt{3}$ or $-\sqrt{3}/3$ (5-3, 5-5)   **25.** $-\pi/6$ (5-8)   **26.** $5\pi/6$ (5-8)   **27.** 0.33 (5-8)   **28.** $-\sqrt{2}$ (5-8)
29. $\sqrt{3}/2$ (5-8)   **30.** 0.4431 (5-5)   **31.** -15.17 (5-5)   **32.** -2.077 (5-3, 5-5)   **33.** -0.9750 (5-8)
34. Not defined (5-8)   **35.** 1.557 (5-8)   **36.** 1.095 (5-8)   **37.** Not defined (5-8)
38. (A) $-2\pi, -\pi, 0, \pi, 2\pi$; (B) $-3\pi/2, -\pi/2, \pi/2, 3\pi/2$ (5-3, 5-5)   **39.** (A) All integers; (B) All integers (5-6)
40. $A = 2, P = 2$ (5-7)   **41.** (5-7)

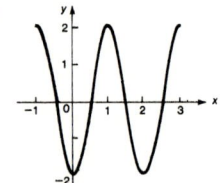
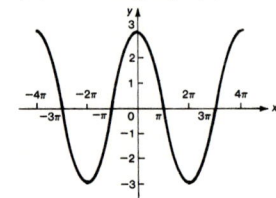

42. $A = 2, P = 4$, phase shift $= \frac{1}{2}$ right (5-7)
43. $y = \cos^{-1} x = \arccos x$, domain $= [-1, 1]$, range $= [0, \pi]$ (5-8)

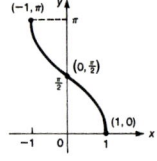

44. (A) $2\pi/3$; (B) $5\pi/4$ (5-3, 5-5)
45. $y = \sec x$ (5-6)   **46.** $y = \tan^{-1} x = \arctan x$, domain $= R$, range $= (-\pi/2, \pi/2)$ (5-8)

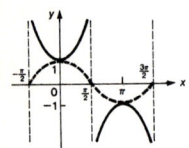
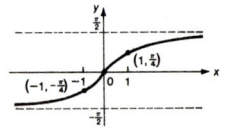

47. $P = 1$; phase shift $= \frac{1}{2}$ left (5-7)   **48.** (A) Odd; (B) Even; (C) Odd (5-3)   **49.** 2.490 (5-8)
50. $1/\sqrt{1-x^2}$ (5-8)   **51.** (A) $\frac{4}{5}$; (B) $\frac{3}{4}$; (C) $-\frac{4}{5}$ (5-3)

Practice Test: Chapter 5

1. (B) (5-4)   **2.** (C) (5-3, 5-6)   **3.** (A) $\frac{3}{5}$; (B) $-\frac{4}{3}$; (C) $\frac{5}{4}$ (5-3)   **4.** $-\sqrt{3}/2$ (5-3, 5-5)
5. $-\sqrt{3}/2$ (5-3, 5-5)   **6.** π (5-8)   **7.** $-1/\sqrt{3}$ or $-\sqrt{3}/3$ (5-8)   **8.** Not defined (5-3, 5-5)
9. $1/\sqrt{3}$ or $\sqrt{3}/3$ (5-3, 5-5)   **10.** 25 (5-8)   **11.** $5\pi/6$ (5-8)   **12.** -75.18 (5-3, 5-5)   **13.** 13.80 (5-8)
14. -0.2314 (5-5)   **15.** -2.695 (5-5)
16. $A = 2, P = 2$ (5-7)   **17.** $A = 1, P = 4\pi$, phase shift $= \pi$ left (5-7)

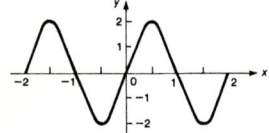
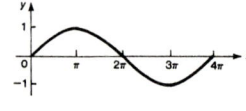

18. $y = \sin^{-1} x = \arcsin x$, domain $= [-1, 1]$, range $= [-\pi/2, \pi/2]$ (5-8)

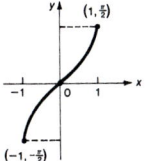

19. $-30°, -150°$ (5-5)   **20.** $1/\sqrt{1-x^2}$ (5-8)

Chapter 6

Exercise 6-2

73. III, IV   **75.** I, II   **77.** All quadrants   **79.** I, IV

Exercise 6-3

13. $(\sqrt{2}/2)(\sin x - \cos x)$ **15.** $-\cos x$ **17.** $(1 - \tan x)/(1 + \tan x)$ **19.** $(\sqrt{3}+1)/(2\sqrt{2})$ **21.** $(\sqrt{3}+1)/(2\sqrt{2})$
23. $\sqrt{3}/2$ **25.** $\sqrt{3}$ **27.** $\sin(x-y) = (-2 - 5\sqrt{3})/12$, $\tan(x+y) = (-2 + 5\sqrt{3})/(\sqrt{5}+2\sqrt{15})$
29. $\sin(x-y) = (-4\sqrt{2}-1)/(3\sqrt{5})$, $\tan(x+y) = (1-4\sqrt{2})/(2+2\sqrt{2})$
43. $\frac{24}{25}$ **45.** $-\frac{1}{2}$ **47.** $xy + (\sqrt{1-x^2})(\sqrt{1-y^2})$ **53.** $0.6115, 0.6115; -1.155, -1.155$
55. $0.9756, 0.9756; -0.4895, -0.4895$

Exercise 6-4

1. $1 = 1$ **3.** $\sqrt{3} = \sqrt{3}$ **5.** $1 = 1$ **7.** $\sqrt{2-\sqrt{2}}/2$ **9.** $\sqrt{2-\sqrt{2}}/2$
29. $\sin 2x = \frac{24}{25}, \cos 2x = \frac{7}{25}, \tan 2x = \frac{24}{7}$ **31.** $\sin 2x = -\frac{24}{25}, \cos 2x = \frac{7}{25}, \tan 2x = -\frac{24}{7}$
33. $\sin 2x = -\frac{120}{169}, \cos 2x = \frac{119}{169}, \tan 2x = -\frac{120}{119}$ **35.** $\sin(x/2) = \sqrt{3}/3, \cos(x/2) = \sqrt{6}/3, \tan(x/2) = \sqrt{2}/2$
37. $\sin(x/2) = \sqrt{\frac{3+2\sqrt{2}}{6}}, \cos(x/2) = -\sqrt{\frac{3-2\sqrt{3}}{6}}, \tan(x/2) = -3 - 2\sqrt{2}$
39. $\sin(x/2) = -2\sqrt{5}/5, \cos(x/2) = \sqrt{5}/5, \tan(x/2) = -2$ **41.** $\sin\theta = 2\sqrt{5}/5, \cos\theta = \sqrt{5}/5$ **47.** $-7/25$
49. $-24/7$ **51.** $\sqrt{5}/5$ **53.** (A) $-0.723\ 35 = -0.723\ 35$; (B) $-0.588\ 21 = -0.588\ 21$
55. (A) $-3.2518 = -3.2518$; (B) $0.892\ 79 = 0.892\ 79$

Exercise 6-5

1. $\frac{1}{2}\cos 12A + \frac{1}{2}\cos 2A$ **3.** $\frac{1}{2}\sin 5\theta + \frac{1}{2}\sin\theta$ **5.** $2\cos 6\theta \cos\theta$ **7.** $-2\cos 3u \sin 2u$
9. $(2-\sqrt{3})/4$ **11.** $\frac{1}{4}$ **13.** $\sqrt{2}/2$ **15.** $\sqrt{2}/2$ **31.** (A) $0.198\ 53 = 0.198\ 53$; (B) $1.5918 = 1.5918$
33. (A) $0.572\ 85 = 0.572\ 85$; (B) $1.8186 = 1.8186$

Exercise 6-6

1. π **3.** $225°, 315°$ **5.** $\pi/6, 5\pi/6, 7\pi/6, 11\pi/6$ **7.** $0°, 90°, 180°$ **9.** $\pi/6, \pi/2, 5\pi/6, 3\pi/2$
11. $2k\pi + 0.3023, (2k+1)\pi - 0.3023$, k any integer   **13.** $2k\pi \pm 2.660$, k any integer   **15.** $k\pi + 1.494$, k any integer
17. $2k\pi \pm 1.687$, k any integer   **19.** $k\pi - 0.2800$, k any integer   **21.** $\pi/6, \pi/2, 5\pi/6$ **23.** $\pi/6, 5\pi/6, 7\pi/6, 11\pi/6$
25. $k\pi + \pi/4$, k any integer   **27.** $\pi/6, 5\pi/6$ **29.** $k\pi, 2k\pi \pm \pi/3$, k any integer   **31.** $180°$ **33.** $104.5°$
35. $0.9987, 5.284$ **37.** $2k\pi \pm 0.9987$, k any integer   **39.** $2k\pi \pm \pi/3, 2k\pi$, k any integer   **41.** 0 **43.** 0
45. $0, \pi/2, \pi$ **47.** $0, \pi/12, 5\pi/12$ **49.** $0.002\ 613$ second   **51.** $64.1°$ **53.** $(r, \theta) = (1, 30°), (1, 150°)$

Exercise 6-7   Chapter Review

5. $135°, 225°$ (6-6)   **6.** $0, \pi, \pi/4, 5\pi/4$ (6-6)   **7.** $2k\pi + 0.7878, (2k+1)\pi - 0.7878$, k any integer (6-6)
8. $k\pi - 1.526$, k any integer (6-6)   **15.** (A) $3\sqrt{10}/10$; (B) $-\frac{24}{25}$; (C) $\frac{24}{7}$ (6-4)   **16.** $2\cos 2x \sin x$ (6-5)
17. $\cos 7x + \cos 3x$ (6-5)   **18.** $\pi/3, 2\pi/3, 4\pi/3, 5\pi/3$ (6-6)   **19.** $0°, 120°$ (6-6)
20. $k\pi, 2k\pi \pm \pi/6$, k any integer (6-6)   **21.** $1.178, 2.749$ (6-6)   **22.** $-\frac{24}{25}$ (6-4)   **23.** $\frac{24}{25}$ (6-3)
24. $0, \pi/3, 2\pi/3$ (6-6)   **25.** $\pi/12, 5\pi/12, 0, \pi/3$ (6-6)   **26.** $0, \pi/2$ (6-6)

Practice Test: Chapter 6

4. $\cos x$ (6-3)   **5.** (A) $3/\sqrt{10}$ or $3\sqrt{10}/10$; (B) $\frac{7}{25}$ (6-4)   **6.** $\frac{1}{2}\sin 10x + \frac{1}{2}\sin 4x$ (6-5)
7. $k\pi, 2k\pi + \pi/6, (2k+1)\pi - \pi/6$, k any integer (6-6)   **8.** $120°, 240°$ (6-6)   **9.** $0, 2\pi/3$ (6-6)
10. $0.6259, 2.516$ (6-6)

Chapter 7   Exercise 7-1

1. $\alpha = 72°10', a = 3.28, b = 1.06$ **3.** $\alpha = 46°40', b = 116, c = 169$ **5.** $\beta = 67°0', b = 127, c = 138$
7. $\beta = 36°48', a = 31.84, c = 39.76$ **9.** $\alpha = 35°20', \beta = 54°40', c = 10.4$ **11.** $\alpha = 37°30', \beta = 52°30', a = 7.67$
13. $\alpha = 48°40', \beta = 41°20', b = 2.12$ **15.** 218 feet   **17.** 127.5 feet   **19.** 2,292 feet   **21.** 2,225 miles
23. $44°$ **25.** 0.5 mile   **27.** 86.2 square centimeters   **29.** 0.77 meter

Exercise 7-2

1. $\alpha = 67°20', a = 55.1, c = 58.8$ **3.** $\alpha = 98°, b = 4.32, c = 7.62$ **5.** $\beta = 135°15', \gamma = 18°55', c = 48.36$
7. $\alpha = 52°, \gamma = 96°20', c = 15.1$ **9.** $\alpha = 141°, \alpha' = 39°, \beta = 9°, \beta' = 111°, b = 13, b' = 75$ **11.** No triangle
13. $26.997 \approx 26.999$ **17.** 4.06 miles, 2.47 miles   **19.** 353 feet   **21.** 4.42×10^7 kilometers, 2.36×10^8 kilometers

Exercise 7-3

1. $a = 6.00, \beta = 65°0', \gamma = 64°20'$ **3.** $c = 14.0, \alpha = 20°40', \beta = 39°0'$ **5.** $\alpha = 23°30', \beta = 92°30', \gamma = 64°0'$
7. $\alpha = 22°20', \beta = 131°30', \gamma = 26°10'$ **9.** $c^2 = a^2 + b^2 - 2ab \cos 90° = c^2 = a^2 + b^2$, since $\cos 90° = 0$
11. $-0.87 = -0.87$ **13.** 5.81 feet   **15.** 121 miles   **17.** 74.1 meters

Exercise 7-4

1. $H = 28$ pounds, $V = 10$ pounds   **3.** $H = 6.8$ knots, $V = 19$ knots   **5.** 1,700 miles north; 1,000 miles east
7. $M_3 = 69$ pounds, $\beta = 16°$ **9.** $M_3 = 22$ miles/hour, $\beta = 6°$ **11.** Magnitude = 13 miles/hour, direction = $252°$
13. 349 pounds   **15.** Left cable: 518 pounds; right cable: 390 pounds
17. For AB, a compression of 9,050 pounds; for BC a tension of 7,540 pounds   **19.** $350.8°$, 247 miles/hour
21. $51.0°$; 3.83 knots

Exercise 7-5

7. 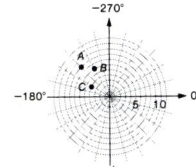 **9.** 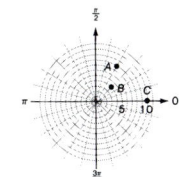 **11.**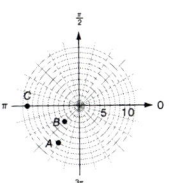

13. 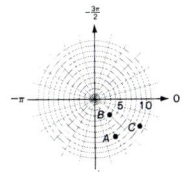 **15.**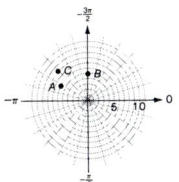

17. $(7\sqrt{2}/2, 7\sqrt{2}/2)$ **19.** $(8, 0)$ **21.** $(-3\sqrt{3}, -3)$ **23.** $(-4\sqrt{3}, -4)$ **25.** $(2\sqrt{2}, 2\sqrt{2})$ **27.** $(-5/2, 5\sqrt{3}/2)$
29. $(6, \pi/3)$ **31.** $(12, 5\pi/6)$ **33.** $(10, 7\pi/4)$ **35.** $(10, \pi)$
37. 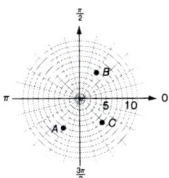 **39.**

(37 and 39 are separate polar plots shown side by side)

41. $r = 5 \sin \theta$ **43.** $r(3 \cos \theta - 5 \sin \theta) = -2$ **45.** $\tan \theta = 1$ or $\theta = \pi/4$ **47.** $3x - 4y = -1$
49. $x^2 + y^2 = -2y$ **51.** $y = x$ **53.** $(r \sin \theta - 3)^2 = 4r^2$

Exercise 7-6

1. 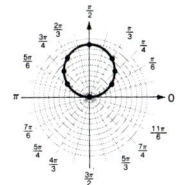 **3.** 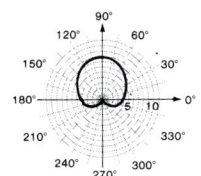 **5.**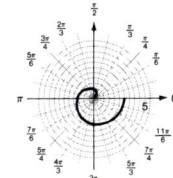

7. 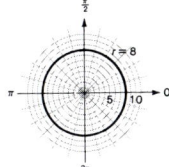 **9.** 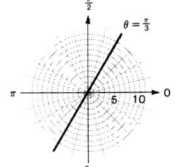 **11.**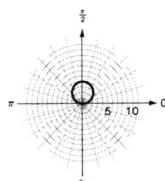

13. 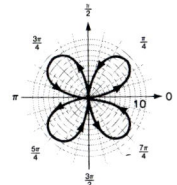 **15.** 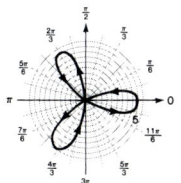 **17.**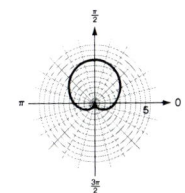

19.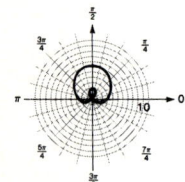
21.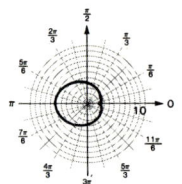
23.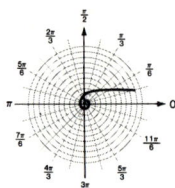

25.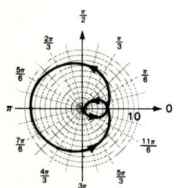
27.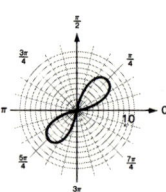
29. $(\sqrt{2}, \pi/4)$
Note: The pole has no ordered pairs that simultaneously satisfy both equations.
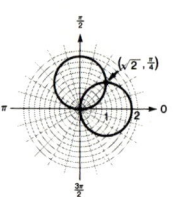

Exercise 7-7

1. 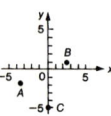 **3.** (graph) **5.** (graph)

7. (graph) **9.** $z_1 = 2 \text{ cis}(\pi/6)$, $z_2 = \sqrt{2} \text{ cis}(3\pi/4)$, $z_3 = 2 \text{ cis}(\pi/2)$ 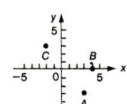 **11.** $z_1 = 2 \text{ cis } 300°$, $z_2 = 2 \text{ cis } 30°$, $z_3 = 4 \text{ cis } 0°$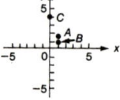

13. $z_1 = 1 + i\sqrt{3}$, $z_2 = 1 + i$, $z_3 = 3i$

15. $z_1 = \dfrac{5}{2} - \dfrac{5\sqrt{3}}{2}i$, $z_2 = i$, $z_3 = -\dfrac{3\sqrt{3}}{2} - \dfrac{3}{2}i$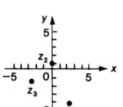

17. $z_1 = 1.83 + i1.39$, $z_2 = 1.56 + i3.02$

19. $z_1 z_2 = 14 \text{ cis } 113°$, $z_1/z_2 = 3.5 \text{ cis } 51°$ **21.** $z_1 z_2 = 20 \text{ cis } 135°$, $z_1/z_2 = 1.25 \text{ cis}(-31°)$
23. $(1 + i)^2 = (1 + i)(1 + i) = 2i$, $(\sqrt{2} \text{ cis } 45°)^2 = (\sqrt{2})^2 \text{ cis }(45°) = 2 \text{ cis } 90°$
25. $[1 \text{ cis}(\pi/3)]^2 = 1^2 \text{ cis}(2\pi/3) = \cos(2\pi/3) + i \sin(\pi/3) = -\tfrac{1}{2} + i\sqrt{3}/2$

27. $(r_1 \text{ cis } \theta_1)/(r_2 \text{ cis } \theta_2) = R \text{ cis } \theta \Leftrightarrow r_1 \text{ cis } \theta_1 = (r_2 \text{ cis } \theta_2)(R \text{ cis } \theta) = r_2 R \text{ cis}(\theta_2 + \theta) \Leftrightarrow r_1 = r_2 R \text{ or } R = r_1/r_2 \text{ and } \theta_2 + \theta = \theta_1 + 2n\pi \text{ or } \theta = \theta_1 - \theta_2 + 2n\pi, n \in Z$

Exercise 7-8

1. $8 \text{ cis } 90° = 8i$ **3.** $8 \text{ cis } 60° = 4 + i4\sqrt{3}$ **5.** $-128 - i128\sqrt{3}$ **7.** 1 **9.** $16\sqrt{3} + 16i$ **11.** $-\frac{1}{8} + \frac{1}{8}i$
13. $64 + i64\sqrt{3}$ **15.** -8 **17.** $w_1 = 2 \text{ cis } 10°, w_2 = 2 \text{ cis } 130°, w_3 = 2 \text{ cis } 250°$
19. $w_1 = \sqrt{2} \text{ cis } 10°, w_2 = \sqrt{2} \text{ cis } 70°, w_3 = \sqrt{2} \text{ cis } 130°, w_4 = \sqrt{2} \text{ cis } 190°, w_5 = \sqrt{2} \text{ cis } 250°, w_6 = \sqrt{2} \text{ cis } 310°$
21. $w_1 = 2^{1/10} \text{ cis } 27°, w_2 = 2^{1/10} \text{ cis } 99°, w_3 = 2^{1/10} \text{ cis } 171°, w_4 = 2^{1/10} \text{ cis } 243°, w_5 = 2^{1/10} \text{ cis } 315°$
23. $w_1 = 2 \text{ cis } 0°, w_2 = 2 \text{ cis } 120°, w_3 = 2 \text{ cis } 240°$
25. $w_1 = \text{cis } 15°, w_2 = \text{cis } 75°, w_3 = \text{cis } 135°, w_4 = \text{cis } 195°, w_5 = \text{cis } 255°, w_6 = \text{cis } 315°$
27. $w_1 = 2^{1/6} \text{ cis } 35°, w_2 = 2^{1/6} \text{ cis } 95°, w_3 = 2^{1/6} \text{ cis } 155°, w_4 = 2^{1/6} \text{ cis } 215°, w_5 = 2^{1/6} \text{ cis } 275°, w_6 = 2^{1/6} \text{ cis } 335°$
29. $[r^{1/n} \text{cis}(\theta/n)]^n = r^{n/n} \text{ cis } n(\theta/n) = r \text{ cis } \theta = z$
31. $x_1 = 2 \text{ cis } 0°, x_2 = 2 \text{ cis } 72°, x_3 = 2 \text{ cis } 144°, x_4 = 2 \text{ cis } 216°, x_5 = 2 \text{ cis } 288°$
33. $x_1 = \text{cis } 36°, x_2 = \text{cis } 108°, x_3 = \text{cis } 180°, x_4 = \text{cis } 252°, x_5 = \text{cis } 324°$
35. $P(x) = (x - 2i)(x + 2i)[x - (-\sqrt{3} + i)][x - (-\sqrt{3} - i)][x - (\sqrt{3} + i)][x - (\sqrt{3} - i)]$

Exercise 7-9   Chapter Review

1. $\beta = 35°40', a = 1.62, b = 1.17$ (7-1)   **2.** $\alpha = 39°40', \beta = 20°10', a = 7.38$ (7-2)   **3.** $a = 2.67$ (7-3)
4. (7-5)   **5.** (7-5)   **6.** (7-5)

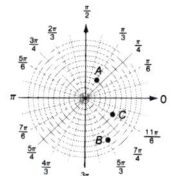
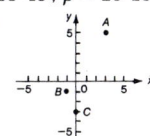
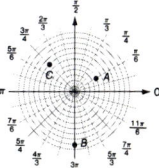

7. 1 (7-8)   **8.** $8 + i8\sqrt{3}$ (7-8)   **9.** (7-6)

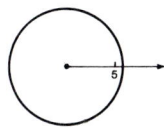

10. $22°$, 11 pounds (7-4)   **11.** 28.3 (7-3)
12. $\beta = 83°40', \gamma = 58°40', b = 73.5$ or $\gamma' = 121°20', \beta' = 21°0', b' = 26.5$ (7-2)
13. $a = 10.6, \beta = 19°0', \gamma = 40°30'$ (7-2, 7-3)   **14.** $15°$, 177 miles/hour (7-2, 7-3)
15. (7-5)   **16.** (7-6)

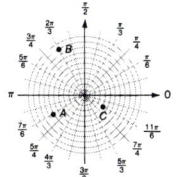
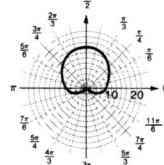

17. $z_1 = \sqrt{2}\operatorname{cis}(3\pi/4)$, $z_2 = 2\operatorname{cis}(2\pi/3)$, $z_3 = 5\operatorname{cis}0$ (7-7) **18.** $z_1 = 1+i$, $z_2 = -3\sqrt{3}/2 - \frac{3}{2}i$, $z_3 = 1 - i\sqrt{3}$ (7-7)
19. $z_1 z_2 = 12\operatorname{cis}124°$, $z_1/z_2 = 3\operatorname{cis}60°$ (7-7) **20.** $-1/32 + (\sqrt{3}/32)i$ (7-8)
21. $h = 0.819$ (7-1, 7-2) **22.** $86°$, 464 miles/hour (7-4)
23. (7-6) **24.** $r = 9\sin\theta$ (7-5)
25. $x^2 + y^2 - 5y = 0$ (7-5)

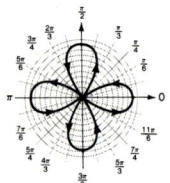

26. $-1, 1, -i, i, \sqrt{2}/2 + i(\sqrt{2}/2), \sqrt{2}/2 - i(\sqrt{2}/2), -\sqrt{2}/2 + (\sqrt{2}/2)i - \sqrt{2}/2 - (\sqrt{2}/2)i$ (7-8)

Practice Test: Chapter 7

1. $a = 5.80$, $c = 9.09$, $\beta = 50°20'$ (7-1) **2.** $b = 5.82$, $\alpha = 44°40'$, $\gamma = 80°20'$ (7-2, 7-3) **3.** (7-5)

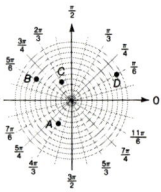

4. (A) $2\operatorname{cis}300°$; (B) $2\sqrt{3} - 2i$ (7-7) **5.** (A) $32\operatorname{cis}44°$; (B) $2\operatorname{cis}6°$ (7-7)
6. $\sqrt{3}/2 + \frac{1}{2}i, -\sqrt{3}/2 + \frac{1}{2}i, -i$ (7-8) **7.** $29°$, 7.3 pounds (7-2, 7-3, 7-4)
8. (7-6) **9.** (7-6) **10.** $x^2 + y^2 - 8x = 0$ (7-5)

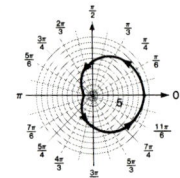

Chapter 8

Exercise 8-2

1. 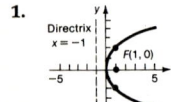 **3.** 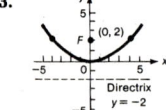 **5.** 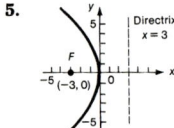 **7.** 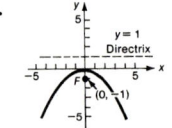 **9.**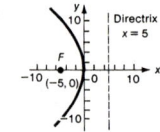

11. 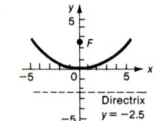 **13.** $(9.75, 0)$ **15.** $(0, -26.25)$ **17.** $(-19.25, 0)$ **19.** $x^2 = 8y$ **21.** $y^2 = -12x$

23. $x^2 = -4y$ **25.** $x^2 - 4x + 4y - 8 = 0$ **27.** $x^2 = -200y$

Exercise 8-3

1. Foci: $F'(-\sqrt{21}, 0)$, $F(\sqrt{21}, 0)$
Major axis length = 10
Minor axis length = 4

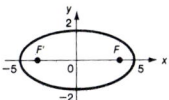

3. Foci: $F'(0, -\sqrt{21})$, $F(0, \sqrt{21})$
Major axis length = 10
Minor axis length = 4

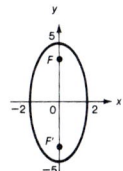

5. Foci: $F'(-\sqrt{8}, 0)$, $F(\sqrt{8}, 0)$
Major axis length = 6
Minor axis length = 2

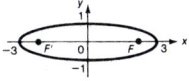

7. Foci: $F'(0, -4)$, $F(0, 4)$
Major axis length = 10
Minor axis length = 6

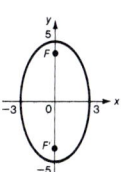

9. Foci: $F'(0, -\sqrt{6})$, $F(0, \sqrt{6})$
Major axis length = $2\sqrt{12} \approx 6.93$
Minor axis length = $2\sqrt{6} \approx 4.90$

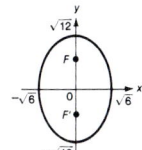

11. Foci: $F'(-\sqrt{3}, 0)$, $F(\sqrt{3}, 0)$
Major axis length = $2\sqrt{7} \approx 5.29$
Minor axis length = 4

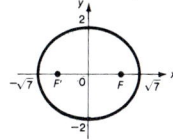

13. $\dfrac{x^2}{16} + \dfrac{y^2}{9} = 1$
15. $\dfrac{x^2}{64} + \dfrac{y^2}{121} = 1$
17. $\dfrac{x^2}{64} + \dfrac{y^2}{28} = 1$
19. $\dfrac{x^2}{100} + \dfrac{y^2}{170} = 1$
21. $\dfrac{x^2}{16} + \dfrac{y^2}{12} = 1$; ellipse

23. $\dfrac{x^2}{400} + \dfrac{y^2}{144} = 1$; 7.94 feet

Exercise 8-4

1. Foci: $F'(-\sqrt{13}, 0)$, $F(\sqrt{13}, 0)$
Transverse axis length = 6
Conjugate axis length = 4

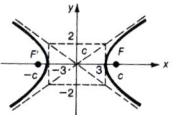

3. Foci: $F'(0, -\sqrt{13})$, $F(0, \sqrt{13})$
Transverse axis length = 4
Conjugate axis length = 6

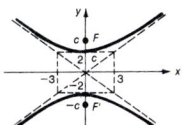

5. Foci: $F'(-\sqrt{20}, 0)$, $F(\sqrt{20}, 0)$
Transverse axis length = 4
Conjugate axis length = 8

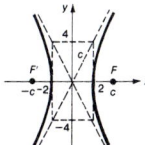

7. Foci: $F'(0, -5)$, $F(0, 5)$
Transverse axis length = 8
Conjugate axis length = 6

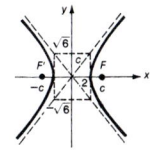

9. Foci: $F'(-\sqrt{10}, 0)$, $F(\sqrt{10}, 0)$
Transverse axis length = 4
Conjugate axis length = $2\sqrt{6} \approx 4.90$

11. Foci: $F'(0, -\sqrt{11})$, $F(0, \sqrt{11})$
Transverse axis length = 4
Conjugate axis length = $2\sqrt{7} \approx 5.29$

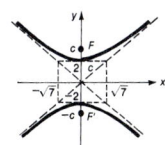

AA30   Answers

13. $\dfrac{x^2}{49} - \dfrac{y^2}{25} = 1$ **15.** $\dfrac{y^2}{144} - \dfrac{x^2}{81} = 1$ **17.** $\dfrac{x^2}{81} - \dfrac{y^2}{40} = 1$ **19.** $\dfrac{y^2}{151} - \dfrac{x^2}{49} = 1$ **21.** $\dfrac{x^2}{4} - \dfrac{y^2}{5} = 1$; hyperbola

23. $\dfrac{y^2}{16} - \dfrac{x^2}{8} = 1$; 5.38 feet above vertex

Exercise 8-5

1. (A) $x' = x - 3$, $y' = y - 5$, $(h, k) = (3, 5)$; (B) $x'^2 + y'^2 = 81$; (C) Circle

3. (A) $x' = x + 7$, $y' = y - 4$, $(h, k) = (-7, 4)$; (B) $\dfrac{x'^2}{9} + \dfrac{y'^2}{16} = 1$; (C) Ellipse

5. (A) $x' = x - 4$, $y' = y + 9$, $(h, k) = (4, -9)$; (B) $y'^2 = 16x'$; (C) Parabola

7. (A) $x' = x + 8$, $y' = y + 3$, $(h, k) = (-8, -3)$; (B) $\dfrac{x'^2}{12} + \dfrac{y'^2}{8} = 1$; (C) Ellipse

9. (A) $\dfrac{x'^2}{9} - \dfrac{y'^2}{16} = 1$; hyperbola; (B) $x' = x - 3$, $y' = y + 2$, $(h, k) = (3, -2)$

11. (A) $\dfrac{x'^2}{5} + \dfrac{y'^2}{6} = 1$; ellipse; (B) $x' = x + 5$, $y' = y + 7$, $(h, k) = (-5, -7)$

13. (A) $x'^2 = -24y'$; parabola; (B) $x' = x + 6$, $y' = y - 4$, $(h, k) = (-6, 4)$

15. $\dfrac{x'^2}{9} + \dfrac{y'^2}{4} = 1$; ellipse   **17.** $x'^2 = -8y'$; parabola   **19.** $x'^2 + y'^2 = 16$; circle   **21.** $\dfrac{y'^2}{9} - \dfrac{x'^2}{16} = 1$; hyperbola

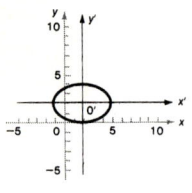 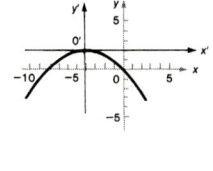 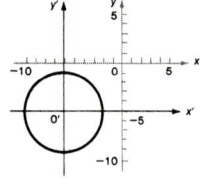 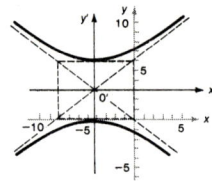

23. $F'(-\sqrt{5} + 2, 2)$, $F(\sqrt{5} + 2, 2)$ **25.** $F(-4, 0)$ **27.** $F'(-4, -2)$, $F(-4, 8)$

Exercise 8-6

1. $x'^2 + y'^2 = 49$; circle   **3.** $\dfrac{x'^2}{4} + \dfrac{y'^2}{20} = 1$; ellipse   **5.** $\dfrac{y'^2}{4} - \dfrac{x'^2}{12} = 1$; hyperbola   **7.** $\dfrac{x'^2}{9} + \dfrac{y'^2}{4} = 1$; ellipse

$\theta = 45°$ $\theta \approx 63.43°$

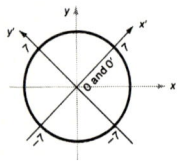 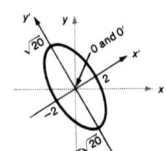 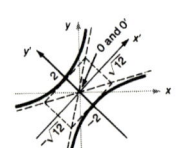 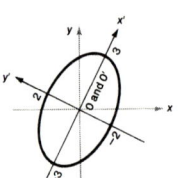

9. $y'^2 = 8x'$; parabola
$\theta = 30°$

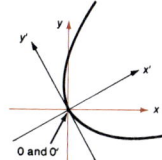

11. $\theta = 60°$
$x'^2 - 2\sqrt{3}x' + 2y' - 1 = 0$
Translate $0'$ to $A'(\sqrt{3}, 2)$
$x''^2 = -2y''$; parabola

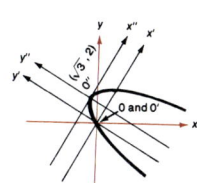

Exercise 8-7

1. $y = -2x - 2$; straight line   3. $y = -2x - 2$, $x \leq 0$; a ray (part of a straight line)   5. $y = -\frac{2}{3}x$; straight line

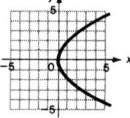 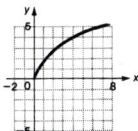 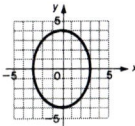

7. $y^2 = 4x$; parabola   9. $y^2 = 4x$, $y \geq 0$; parabola (upper half)   11. $\dfrac{x^2}{9} + \dfrac{y^2}{16} = 1$; ellipse

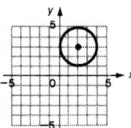 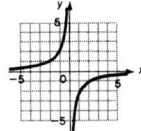 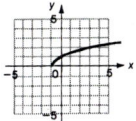

13. $(x - 2)^2 + (y - 3)^2 = 4$; circle   15. $y = -\dfrac{2}{x}$; hyperbola   17. $y^2 = (x + 1)$, $y \geq 0$, $x \geq -1$; parabola (upper half)

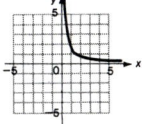 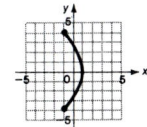 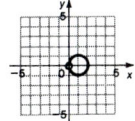

19. $y = 1/x$, $x > 0$; hyperbola (one branch)   21. $y^2 = -8(x - 1)$, $-1 \leq x \leq 1$; part of a parabola   23. $x^2 + y^2 = 2x$, $x \neq 0$ or $(x - 1)^2 + y^2 = 1$, $x \neq 0$; circle (note hole at origin)

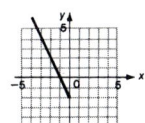 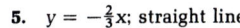  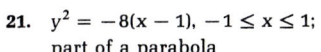 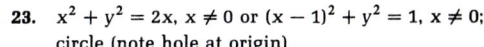

AA32   Answers

25.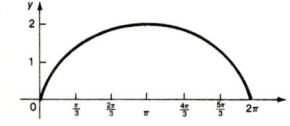

27. (A) (4.8, −1.5);
(B) $x^2 + y^2 = 25$; circle

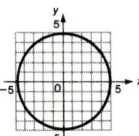

29. (A) 44.194 seconds; (B) 31,249.877 feet, 5.919 miles; (C) 7,812.500 feet

Exercise 8-8 Chapter Review

1. Foci: $F'(-4, 0)$, $F(4, 0)$
Major axis length = 10
Minor axis length = 6   (8-3)

2. (8-2)

3. Foci: $F'(0, -\sqrt{34})$, $F(0, \sqrt{34})$
Transverse axis length = 6
Conjugate axis length = 10   (8-4)

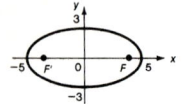   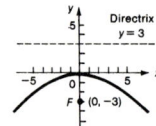   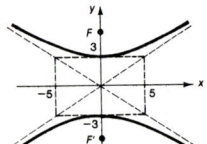

4. (A) $\dfrac{y'^2}{25} - \dfrac{x'^2}{4} = 1$; hyperbola; (B) $x' = x - 4$, $y' = y + 2$, $(h, k) = (4, -2)$ (8-5)

5. (A) $x'^2 = -12y'$; parabola; (B) $x' = x + 5$, $y' = y + 4$, $(h, k) = (-5, -4)$ (8-5)

6. (A) $\dfrac{x'^2}{9} + \dfrac{y'^2}{16} = 1$; ellipse; (B) $x' = x - 6$, $y' = y - 4$, $(h, k) = (6, 4)$ (8-5)

7. $\dfrac{x'^2}{20} + \dfrac{y'^2}{4} = 1$; ellipse   (8-6)

8. $y = \tfrac{1}{2}x + 1$, $x \leq 0$;
a ray (part of a straight line)   (8-7)

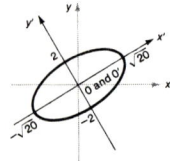

9. $y^2 = -x$ (8-2)   **10.** $\dfrac{x^2}{9} + \dfrac{y^2}{25} = 1$ (8-3)   **11.** $\dfrac{y^2}{9} - \dfrac{x^2}{16} = 1$ (8-4)

12. $\dfrac{x'^2}{4} + \dfrac{y'^2}{16} = 1$; ellipse   (8-5)   **13.** $x'^2 = 8y'$; parabola   (8-5)   **14.** $\dfrac{x'^2}{9} - \dfrac{y'^2}{4} = 1$; hyperbola   (8-5)

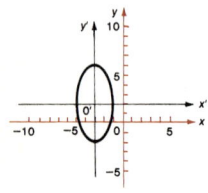   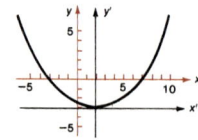   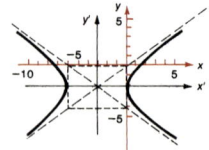

15. $\dfrac{y'^2}{9} - \dfrac{x'^2}{4} = 1$; hyperbola; $\theta = 45°$ (8-6)   **16.** $\dfrac{(x+2)^2}{4} + \dfrac{(y-3)^2}{16} = 1$; ellipse (8-3)

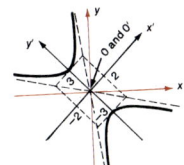

17. $(y-4)^2 = -8(x-4)$ or $y^2 - 8y + 8x - 16 = 0$ (8-2)   **18.** $\dfrac{x^2}{4} - \dfrac{y^2}{12} = 1$; hyperbola (8-4)

19. $\dfrac{x^2}{36} + \dfrac{y^2}{20} = 1$; ellipse (8-3)   **20.** $F'(-3, -\sqrt{12} + 2)$, $F(-3, \sqrt{12} + 2)$ (8-3, 8-5)

21. $F(2, -1)$ (8-2, 8-3)   **22.** $F'(-\sqrt{13} - 3, -2)$, $F(\sqrt{13} - 3, -2)$ (8-4, 8-5)

23. $y = \dfrac{1}{x}$, $x > 0$; hyperbola (one branch) (8-7)

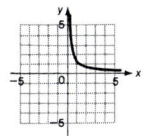

Practice Test: Chapter 8

1. Parabola (8-2)   **2.** Focus: $(-1, 0)$; directrix: $x = 1$ (8-2)   **3.** Hyperbola (8-4)

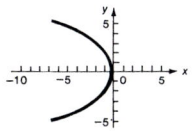
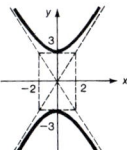

4. $F'(0, -\sqrt{13})$, $F(0, \sqrt{13})$; conjugate axis length is 4 (8-4)   **5.** $\dfrac{x'^2}{16} + \dfrac{y'^2}{4} = 1$; ellipse (8-5)

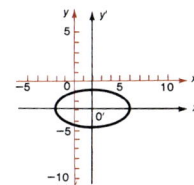

AA34 Answers

6. $F'(-\sqrt{12}+2, -3)$, $F(\sqrt{12}+2, -3)$; major axis length is 8   **7.** $y = -2x + 2$; straight line (8-7)
(8-3, 8-5)

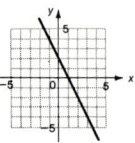

8. $\dfrac{x^2}{16} + \dfrac{(y-3)^2}{9} = 1$; ellipse   **9.** $\dfrac{x^2}{16} - \dfrac{y^2}{4} = 1$ **10.** $x = \dfrac{\sqrt{3}}{2}x' - \dfrac{1}{2}y'$, $y = \dfrac{1}{2}x' + \dfrac{\sqrt{3}}{2}y'$

(8-3, 8-7)   (8-4)   (8-6)

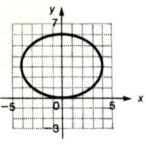

Chapter 9   Exercise 9-1

1. $(2, -3)$ **3.** No solution (parallel lines)   **5.** $(6, 2)$ **7.** $(2, -1)$ **9.** $(2, -1)$ **11.** $(5, 2)$
13. Infinitely many solutions (dependent)   **15.** $(1, -\tfrac{2}{3})$ **17.** No solution (inconsistent)   **19.** $(2{,}500, 200)$
21. $(1, 0.2)$ **23.** $(3, -1, -2)$ **25.** $(1, -1, 2)$ **27.** $(-2, 1, 0)$ **29.** Infinitely many solutions (dependent)
31. 35 20¢ stamps, 12 15¢ stamps   **33.** 40 milliliters of 50% solution and 60 milliliters of 80% solution
35. $3\tfrac{1}{3}$ grams of 18-carat and $6\tfrac{2}{3}$ grams of 12-carat   **37.** 80 grams of mix A, 60 grams of mix B
39. 40 seconds; 24 seconds; 120 miles   **41.** 1,200 style A; 800 style B; 2,000 style C
43. 60 grams of mix A, 50 grams of mix B, 40 grams of mix C

Exercise 9-2

1. $\begin{bmatrix} 4 & -6 & | & -8 \\ 1 & -3 & | & 2 \end{bmatrix}$ **3.** $\begin{bmatrix} -4 & 12 & | & -8 \\ 4 & -6 & | & -8 \end{bmatrix}$ **5.** $\begin{bmatrix} 1 & -3 & | & 2 \\ 8 & -12 & | & -16 \end{bmatrix}$ **7.** $\begin{bmatrix} 1 & -3 & | & 2 \\ 0 & 6 & | & -16 \end{bmatrix}$
9. $\begin{bmatrix} 1 & -3 & | & 2 \\ 2 & 0 & | & -12 \end{bmatrix}$ **11.** $\begin{bmatrix} 1 & -3 & | & 2 \\ 3 & -3 & | & -10 \end{bmatrix}$ **13.** $x_1 = 3, x_2 = 2$ **15.** $x_1 = 3, x_2 = 1$ **17.** $x_1 = 2, x_2 = 1$
19. $x_1 = 2, x_2 = 4$ **21.** No solution   **23.** $x_1 = 1, x_2 = 4$
25. Infinitely many solutions: $x_2 = s$, $x_1 = 2s - 3$ for any real number s
27. Infinitely many solutions: $x_2 = s$, $x_1 = \tfrac{1}{2}s + \tfrac{1}{2}$ for any real number s **29.** $x_1 = 2, x_2 = -1$ **31.** $x_1 = 2, x_2 = -1$
33. $x_1 = 1.1, x_2 = 0.3$

Exercise 9-3

1. Yes 3. No 5. No 7. Yes 9. $x_1 = -2, x_2 = 3, x_3 = 0$ 11. $x_1 = 2t + 3$
$x_2 = -t - 5$
$x_3 = t$
t any real number 13. No solution

15. $x_1 = 2s + 3t - 5$
$x_2 = s$
$x_3 = -3t + 2$
$x_4 = t$
s and t any real numbers

17. $\begin{bmatrix} 1 & 0 & | & -7 \\ 0 & 1 & | & 3 \end{bmatrix}$

19. $\begin{bmatrix} 1 & 0 & 0 & | & -5 \\ 0 & 1 & 0 & | & 4 \\ 0 & 0 & 1 & | & -2 \end{bmatrix}$

21. $\begin{bmatrix} 1 & 0 & 2 & | & -\frac{5}{3} \\ 0 & 1 & -2 & | & \frac{1}{3} \\ 0 & 0 & 0 & | & 0 \end{bmatrix}$

23. $x_1 = -2, x_2 = 3, x_3 = 1$ 25. $x_1 = 0, x_2 = -2, x_3 = 2$ 27. $x_1 = 2t + 3$
$x_2 = t - 2$
$x_3 = t$
t any real number 29. $x_1 = (-4t - 4)/7$
$x_2 = (5t + 5)/7$
$x_3 = t$
t any real number

31. $x_1 = -1, x_2 = 2$ 33. No solution 35. No solution 37. $x_1 = 0, x_2 = 2, x_3 = -3$
39. $x_1 = 1, x_2 = -2, x_3 = 1$ 41. $x_1 = 2s - 3t + 3$
$x_2 = s + 2t + 2$
$x_3 = s$
$x_4 = t$
s and t any real numbers

43. 20 one-person boats, 220 two-person boats, 100 four-person boats
45. $(t - 80)$ one-person boats, $(-2t + 420)$ two-person boats, t four-person boats, $80 \le t \le 210$, t an integer
47. No solution; no production schedule will use all the work-hours in all departments.
49. 8 ounces food A, 2 ounces food B, 4 ounces food C 51. No solution
53. 8 ounces food A, $(-2t + 10)$ ounces food B, t ounces food C, $0 \le t \le 5$
55. Company A: 10 hours; company B: 15 hours

Exercise 9-4

1. $(-12, 5), (-12, -5)$ 3. $(2, 4), (-2, -4)$ 5. $(5i, -5i), (-5i, 5i)$ 7. $(3 + 4i, -1 + 2i), (3 - 4i, -1 - 2i)$
9. $(2, 4), (2, -4), (-2, 4), (-2, -4)$ 11. $(1, 3), (1, -3), (-1, 3), (-1, -3)$
13. $\left(-1 + \sqrt{3}i, 1 + \sqrt{3}i\right), \left(-1 - \sqrt{3}i, 1 - \sqrt{3}i\right)$ 15. $(0, -1), (-4, -3)$ 17. $(2, 2i), (2, -2i), (-2, 2i), (-2, -2i)$
19. $\left(2, \sqrt{2}\right), \left(2, -\sqrt{2}\right), (-1, i), (-1, -i)$ 21. $(2, 1), (-2, 1), (2i, 3), (-2i, 3)$ 23. $(2, 1), (-2, -1), (i, -2i), (-i, 2i)$
25. $(2, 2), (-2, -2), \left(\sqrt{2}, -\sqrt{2}\right), \left(-\sqrt{2}, \sqrt{2}\right)$ 27. $(-3, 1), (3, -1), (-i, i), (i, -i)$
29. $\left(\tfrac{1}{2} + \left(\sqrt{3}/2\right)i, \tfrac{1}{2} - \left(\sqrt{3}/2\right)i\right), \left(\tfrac{1}{2} - \left(\sqrt{3}/2\right)i, \tfrac{1}{2} + \left(\sqrt{3}/2\right)i\right)$ 31. 12 by 5 inches 33. 20¢

Exercise 9-5

1.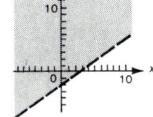
3.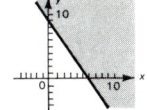
5.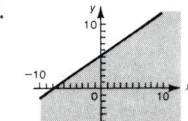

Answers

7. 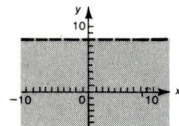   **9.** 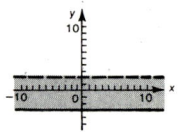   **11.**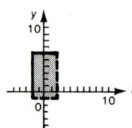

13. 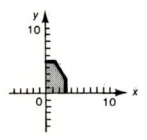   **15.** 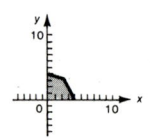   **17.**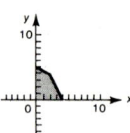

19. 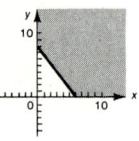   **21.** 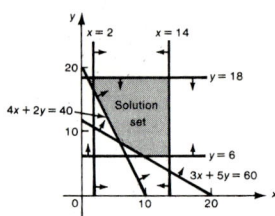   **23.** $6x + 4y \leq 108$
$x + y \leq 24$
$x \geq 0$
$y \geq 0$

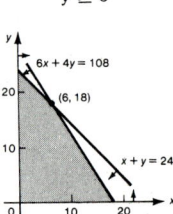

Exercise 9-6   Chapter Review

1. (2, 3) (9-1)   **2.** No solution (inconsistent) (9-1)
3. Infinitely many solutions: $(x, (4x + 8)/3)$ for any real number x (9-1)   **4.** (2, 1, −1) (9-1)
5. $(1, -1), (\frac{7}{5}, -\frac{1}{5})$ (9-4)   **6.** (1, 3), (1, −3), (−1, 3), (−1, −3) (9-4)
7. (9-1)   **8.** (9-5)

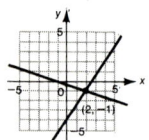

9. $x_1 = -1, x_2 = 3$ (9-2, 9-3)   **10.** $x_1 = -1, x_2 = 2, x_3 = 1$ (9-3)   **11.** $x_1 = 2, x_2 = 1, x_3 = -1$ (9-3)
12. Infinitely many solutions: $x_1 = -5t - 12, x_2 = 3t + 7, x_3 = t$, t any real number (9-3)   **13.** No solution (9-3)
14. Infinitely many solutions: $x_1 = -\frac{3}{7}t - \frac{4}{7}, x_2 = \frac{5}{7}t + \frac{9}{7}, x_3 = t$, t any real number (9-3)
15. $(2, \sqrt{2}), (2, -\sqrt{2}), (-1, i), (-1, -i)$ (9-4)   **16.** $(2, 2), (-2, -2), (\sqrt{2}, -\sqrt{2}), (-\sqrt{2}, \sqrt{2})$ (9-4)
17. (9-5)   **18.** $x_1 = 1,000, x_2 = 4,000, x_3 = 2,000$ (9-3)

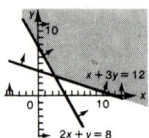

19. $(2, 2), (-2, -2), (\frac{4}{7}\sqrt{7}, -\frac{2}{7}\sqrt{7}), (-\frac{4}{7}\sqrt{7}, \frac{2}{7}\sqrt{7})$ (9-4)   **20.** $48\frac{1}{2}$-pound packages, $72\frac{1}{3}$-pound packages (9-1, 9-3)
21. 6 by 8 meters (9-1, 9-3)   **22.** 40 grams mix A, 60 grams mix B, 30 grams mix C (9-1, 9-3)

Answers

Practice Test: Chapter 9

1. (9-1)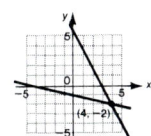

2. (9-5)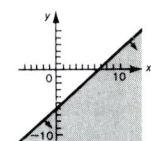

3. (9-5)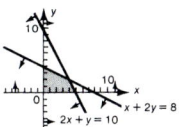

4. $(1 + i, 2i)$ $(1 - i, -2i)$ (9-4) **5.** $(1, 3), (1, -3), (-1, 3), (-1, -3)$ (9-4)
6. $(0, 2\sqrt{2}), (0, -2\sqrt{2}), (\sqrt{2}, \sqrt{2}), (-\sqrt{2}, -\sqrt{2})$ (9-4) **7.** $x_1 = 3, x_2 = -1$ (9-2, 9-3)
8. $x_1 = 2, x_2 = -2, x_3 = 1$ (9-3) **9.** No solution (9-3)
10. $x_1 = 2t - 3, x_2 = -5t + 1, x_3 = t$, t any real number (9-3)

Chapter 10   Exercise 10-1

1. $2 \times 2, 1 \times 4$ **3.** 2 **5.** $\begin{bmatrix} 0 & 0 \\ 0 & 0 \end{bmatrix}$ **7.** C, D **9.** A, B **11.** $\begin{bmatrix} -1 & 0 \\ 5 & -3 \end{bmatrix}$ **13.** $\begin{bmatrix} -2 \\ 3 \\ 0 \end{bmatrix}$ **15.** $\begin{bmatrix} -1 \\ 6 \\ 5 \end{bmatrix}$

17. $\begin{bmatrix} -15 & 5 \\ 10 & -15 \end{bmatrix}$ **19.** $[0\ 1\ 4]$ **21.** $\begin{bmatrix} 250 & 360 \\ 40 & 350 \end{bmatrix}$ **23.** $\begin{bmatrix} -1 & 0 \\ 0 & 6 \\ -1 & -1 \end{bmatrix}$ **25.** $a = -1, b = 1, c = 3, d = -5$

27. Guitar Banjo
$\begin{bmatrix} \$33 & \$26 \\ \$57 & \$77 \end{bmatrix}$ Materials
Labor

29. $\begin{bmatrix} 135 & 282 & 50 \\ 55 & 258 & 155 \end{bmatrix} \cdot \begin{bmatrix} 0.14 & 0.30 & 0.05 \\ 0.06 & 0.28 & 0.17 \end{bmatrix}$

Exercise 10-2

1. 10 **3.** -1 **5.** $[12\ 13]$ **7.** $\begin{bmatrix} 5 \\ -3 \end{bmatrix}$ **9.** $\begin{bmatrix} 2 & 4 \\ 1 & -5 \end{bmatrix}$ **11.** $\begin{bmatrix} 10 & 1 \\ -2 & 10 \end{bmatrix}$ **13.** 6 **15.** 15

17. $\begin{bmatrix} 0 & 9 \\ 5 & -4 \end{bmatrix}$ **19.** $\begin{bmatrix} 5 & 8 & -5 \\ -1 & -3 & 2 \\ -2 & 8 & -6 \end{bmatrix}$ **21.** $[11]$ **23.** $\begin{bmatrix} 3 & -2 & -4 \\ 6 & -4 & -8 \\ -9 & 6 & 12 \end{bmatrix}$ **25.** $AB = \begin{bmatrix} 5 & 7 \\ 2 & 3 \end{bmatrix}, BA = \begin{bmatrix} 1 & 3 \\ 2 & 7 \end{bmatrix}$

27. Both sides equal $\begin{bmatrix} 0 & 12 \\ 1 & 5 \end{bmatrix}$ **29.** (A) $9 per boat (B) $[1.5\ 1.2\ 0.4] \cdot \begin{bmatrix} 7 \\ 10 \\ 4 \end{bmatrix} = \$24.10$ (C) 3×2

(D) $\begin{array}{c} \\ \\ \\ \end{array}\begin{matrix} \text{I} & \text{II} \\ \end{matrix}$
$\begin{bmatrix} \$9.00 & \$11.00 \\ \$14.10 & \$17.20 \\ \$19.80 & \$24.10 \end{bmatrix}$ One-person
Two-person
Four-person
Labor costs per boat at each plant

31. (A) $2,025 (B) $[2{,}000\ 800\ 8{,}000] \cdot \begin{bmatrix} \$0.40 \\ \$0.75 \\ \$0.25 \end{bmatrix} = \$3{,}400$ (C) $\begin{bmatrix} \$2{,}025 \\ \$3{,}400 \end{bmatrix}$ Berkeley
Oakland
Cost per town

(D) Telephone House Letter
$[3{,}000\quad 1{,}300\quad 13{,}000]$
Number of each type of contact made

Exercise 10-3

1. $\begin{bmatrix} 2 & -3 \\ 4 & 5 \end{bmatrix}$ 3. $\begin{bmatrix} -2 & 1 & 3 \\ 2 & 4 & -2 \\ 5 & 1 & 0 \end{bmatrix}$ 9. $x_1 = -8, x_2 = 2$ 11. $x_1 = 0, x_2 = 4$ 13. $\begin{bmatrix} 3 & -2 \\ -1 & 1 \end{bmatrix}$

15. $\begin{bmatrix} 7 & -3 \\ -2 & 1 \end{bmatrix}$ 17. $\begin{bmatrix} 7 & 6 & -3 \\ 2 & 2 & -1 \\ -6 & -5 & 3 \end{bmatrix}$ 19. $\frac{1}{2}\begin{bmatrix} 3 & -1 & -1 \\ -1 & 1 & 1 \\ -3 & 1 & 3 \end{bmatrix}$ 21. (A) $x_1 = -3, x_2 = 2$
(B) $x_1 = -1, x_2 = 2$
(C) $x_1 = -8, x_2 = 3$

23. (A) $x_1 = 17, x_2 = -5$ 25. (A) $x_1 = 1, x_2 = 0, x_3 = 0$ 27. (A) $x_1 = 1, x_2 = 1, x_3 = 3$
(B) $x_1 = 7, x_2 = -2$ (B) $x_1 = -1, x_2 = 0, x_3 = 1$ (B) $x_1 = -1, x_2 = 1, x_3 = -1$
(C) $x_1 = 24, x_2 = -7$ (C) $x_1 = -1, x_2 = -1, x_3 = 1$ (C) $x_1 = 5, x_2 = -1, x_3 = -5$

35. Concert 1: 6,000 $4 tickets and 4,000 $8 tickets; concert 2: 5,000 $4 tickets and 5,000 $8 tickets; concert 3: 3,000 $4 tickets and 7,000 $8 tickets
37. Diet 1: 60 ounces mix A and 80 ounces mix B; diet 2: 20 ounces mix A and 60 ounces mix B; diet 3: 0 ounces mix A and 100 ounces mix B

Exercise 10-4

1. 8   3. -20 5. -0.88 7. $\begin{vmatrix} a_{22} & a_{23} \\ a_{32} & a_{33} \end{vmatrix}$ 9. $\begin{vmatrix} a_{11} & a_{12} \\ a_{31} & a_{32} \end{vmatrix}$ 11. $(-1)^{1+1}\begin{vmatrix} a_{22} & a_{23} \\ a_{32} & a_{33} \end{vmatrix}$ 13. $(-1)^{2+3}\begin{vmatrix} a_{11} & a_{12} \\ a_{31} & a_{32} \end{vmatrix}$

15. $\begin{vmatrix} 1 & -2 \\ -4 & 8 \end{vmatrix}$ 17. $\begin{vmatrix} -2 & 0 \\ 5 & -2 \end{vmatrix}$ 19. $(-1)^{1+1}\begin{vmatrix} 1 & -2 \\ -4 & 8 \end{vmatrix} = 0$ 21. $(-1)^{3+2}\begin{vmatrix} -2 & 0 \\ 5 & -2 \end{vmatrix} = -4$ 23. 10

25. -21 27. -40 29. $(-1)^{1+1}\begin{vmatrix} a_{22} & a_{23} & a_{24} \\ a_{32} & a_{33} & a_{34} \\ a_{42} & a_{43} & a_{44} \end{vmatrix}$ 31. $(-1)^{4+3}\begin{vmatrix} a_{11} & a_{12} & a_{14} \\ a_{21} & a_{22} & a_{24} \\ a_{31} & a_{32} & a_{34} \end{vmatrix}$ 33. 22   35. -12

37. 0   39. 6   41. 60   43. $\begin{vmatrix} a & b \\ ka & kb \end{vmatrix} = akb - kab = 0$ 45. $\begin{vmatrix} a & b \\ c & d \end{vmatrix} = ad - cb = ad - bc = \begin{vmatrix} a & c \\ b & d \end{vmatrix}$

49. $49 = (-7)(-7)$

Exercise 10-5

1. Theorem 2   3. Theorem 2   5. Theorem 3   7. Theorem 4   9. Theorem 6   11. $x = 0$ 13. $x = 5$
15. 25   17. -12 19. Theorem 2   21. Theorem 3   23. Theorem 6   25. $x = 5, y = 0$
27. $x = -3, y = 10$ 29. -28 31. 106   33. 0   35. 6   37. 14
39. Expand the left member of the equation using minors.   41. Expand both members of the equation and compare.
43. This follows from Theorem 5.
45. Expand the determinant about the first row to obtain $(y_1 - y_2)x - (x_1 - x_2)y + (x_1y_2 - x_2y_1) = 0$; then show that the two points satisfy this linear equation.
47. If the determinant is 0, then the area of the triangle formed by the three points is 0. The only way that this can happen is if the three points are on the same line; that is, the points are collinear.

Exercise 10-6

1. $x = 5, y = -2$ 3. $x = 1, y = -1$ 5. $x = -1, y = 1$ 7. $x = 2, y = -2, z = -1$ 9. $x = 2, y = -1, z = 2$
11. $x = 2, y = -3, z = -1$ 13. $x = 1, y = -1, z = 2$
15. Since $D = 0$, the system has either no solution or infinitely many. Since $x = 0, y = 0, z = 0$ is a solution, the second case must hold.

Exercise 10-7 Chapter Review

1. $\begin{bmatrix} 3 & 3 \\ 4 & 2 \end{bmatrix}$ (10-1) 2. Not defined (10-1) 3. $\begin{bmatrix} -3 & 0 \\ 1 & -1 \end{bmatrix}$ (10-1) 4. $\begin{bmatrix} 4 & 3 \\ 7 & 4 \end{bmatrix}$ (10-2) 5. Not defined (10-2)
6. $\begin{bmatrix} 5 \\ 5 \end{bmatrix}$ (10-2) 7. $\begin{bmatrix} 2 & 3 \\ 4 & 6 \end{bmatrix}$ (10-2) 8. 8 (a real number) (10-2) 9. Not defined (10-2) 10. $\begin{bmatrix} 3 & -2 \\ -4 & 3 \end{bmatrix}$ (10-3)
11. (A) $x_1 = -1, x_2 = 3$; (B) $x_1 = 1, x_2 = 2$; (C) $x_1 = 8, x_2 = -10$ (10-3) 12. -17 (10-4) 13. 0 (10-4)
14. $x = 2, y = -1$ (10-6) 15. Not defined (10-1) 16. $\begin{bmatrix} 10 & -8 \\ 4 & 6 \end{bmatrix}$ (10-1, 10-2) 17. $\begin{bmatrix} -2 & 8 \\ 8 & 6 \end{bmatrix}$ (10-1, 10-2)
18. 9 (a real number) (10-2) 19. [9] (a matrix) (10-2) 20. $\begin{bmatrix} 10 & -5 & 1 \\ -1 & -4 & -5 \\ 1 & -7 & -2 \end{bmatrix}$ (10-1, 10-2)
21. $\begin{bmatrix} -\frac{5}{2} & 2 & -\frac{1}{2} \\ 1 & -1 & 1 \\ \frac{1}{2} & 0 & -\frac{1}{2} \end{bmatrix}$ or $\frac{1}{2}\begin{bmatrix} -5 & 4 & -1 \\ 2 & -2 & 2 \\ 1 & 0 & -1 \end{bmatrix}$ (10-3)
22. (A) $x_1 = 2, x_2 = 1, x_3 = -1$; (B) $x_1 = 1, x_2 = -2, x_3 = 1$; (C) $x_1 = -1, x_2 = 2, x_3 = -2$ (10-3)
23. $-\frac{11}{12}$ (10-4) 24. 35 (10-4, 10-5) 25. $y = \frac{10}{5} = 2$ (10-6) 26. $\begin{bmatrix} -\frac{11}{12} & -\frac{1}{12} & 5 \\ \frac{10}{12} & \frac{2}{12} & -4 \\ \frac{1}{12} & -\frac{1}{12} & 0 \end{bmatrix}$ or $\frac{1}{12}\begin{bmatrix} -11 & -1 & 60 \\ 10 & 2 & -48 \\ 1 & -1 & 0 \end{bmatrix}$ (10-3)
27. $x_1 = 1{,}000, x_2 = 4{,}000, x_3 = 2{,}000$ (10-3) 28. 42 (10-4, 10-5)
29. $\begin{vmatrix} u + kv & v \\ w + kx & x \end{vmatrix} = (u + kv)x - (w + kx)v = ux + kvx - wv - kvx = ux - wv = \begin{vmatrix} u & v \\ w & x \end{vmatrix}$ (10-4, 10-5)

Practice Test: Chapter 10

1. 0 (a real number) (10-2) 2. $\begin{bmatrix} 7 & -8 & 6 \\ -4 & 5 & -3 \end{bmatrix}$ (10-2) 3. Not defined (10-2)
4. $\begin{bmatrix} 6 & -12 & -12 \\ 5 & 4 & 7 \\ 0 & -4 & -4 \end{bmatrix}$ (10-1, 10-2) 5. Not defined (10-1) 6. $A^{-1} = \begin{bmatrix} -4 & 3 \\ -3 & 2 \end{bmatrix}$ (10-3)
7. (A) $x_1 = -17, x_2 = -12$; (B) $x_1 = 11, x_2 = 8$ (10-3) 8. 20 (10-4, 10-5) 9. $z = \frac{70}{14} = 5$ (10-6)
10. Since $D = 0$, we conclude (from Cramer's rule) that the system has either no solution or infinitely many. Since (0, 0, 0) is a solution, the first case does not hold; therefore, there must be infinitely many solutions. (10-6)

Chapter 11   Exercise 11-1

1. $-1, 0, 1, 2$ 3. $0, \frac{1}{3}, \frac{1}{2}, \frac{3}{5}$ 5. $4, -8, 16, -32$ 7. 6 9. $\frac{99}{101}$ 11. $S_5 = 1 + 2 + 3 + 4 + 5$
13. $S_3 = \frac{1}{10} + \frac{1}{100} + \frac{1}{1{,}000}$ 15. $S_4 = -1 + 1 - 1 + 1$ 17. $1, -4, 9, -16, 25$ 19. $0.3, 0.33, 0.333, 0.3333, 0.333\,33$
21. $1, -\frac{1}{2}, \frac{1}{4}, -\frac{1}{8}, \frac{1}{16}$ 23. $7, 3, -1, -5, -9$ 25. $4, 1, \frac{1}{4}, \frac{1}{16}, \frac{1}{64}$ 27. $a_n = n + 3$ 29. $a_n = 3n$
31. $a_n = n/(n + 1)$ 33. $a_n = (-1)^{n+1}$ 35. $a_n = (-2)^n$ 37. $a_n = x^n/n$ 39. $S_4 = \frac{4}{1} - \frac{8}{2} + \frac{16}{3} - \frac{32}{4}$
41. $S_3 = x^2 + (x^3/2) + (x^4/3)$ 43. $S_5 = x - (x^2/2) + (x^3/3) - (x^4/4) + (x^5/5)$ 45. $S_4 = \sum_{k=1}^{4} k^2$ 47. $S_5 = \sum_{k=1}^{5} \frac{1}{2^k}$
49. $S_n = \sum_{k=1}^{n} \frac{1}{k^2}$ 51. $S_n = \sum_{k=1}^{n} (-1)^{k+1} k^2$ 55. (A) 3, 1.83, 1.46, 1.415; (B) $\sqrt{2} = 1.414$; (C) 1, 1.5, 1.417, 1.414
57. Series approx. of $e^{0.2} = 1.221\,400\,0$, calculator value of $e^{0.2} = 1.221\,402\,8$

Exercise 11-2

1. Fails at $n = 2$ 3. Fails at $n = 3$ 5. $P_1: 2 = 2 \cdot 1^2$; $P_2: 2 + 6 = 2 \cdot 2^2$; $P_3: 2 + 6 + 10 = 2 \cdot 3^2$
7. $P_1: a^5 a = a^{5+1}$; $P_2: a^5 a^2 = a^5(a^1 a) = (a^5 a)a = a^6 a = a^7 = a^{5+2}$; $P_3: a^5 a^3 = a^5(a^2 a) = a^5(a^1 a)a = [(a^5 a)a]a = a^8 = a^{5+3}$
9. $P_1: 9^1 - 1 = 8$ is divisible by 4; $P_2: 9^2 - 1 = 80$ is divisible by 4; $P_3: 9^3 - 1 = 728$ is divisible by 4
11. $P_k: 2 + 6 + 10 + \cdots + (4k - 2) = 2k^2$; $P_{k+1}: 2 + 6 + 10 + \cdots + (4k - 2) + (4k + 2) = 2(k + 1)^2$
13. $P_k: a^5 a^k = a^{5+k}$; $P_{k+1}: a^5 a^{k+1} = a^{5+k+1}$ 15. $P_k: 9^k - 1 = 4r$; $P_{k+1}: 9^{k+1} - 1 = 4s$; $r, s \in N$
39. Formula: $2 + 4 + 6 + \cdots + 2n = n(n + 1)$ 41. $1 + 2 + 3 + \cdots + (n - 1) = n(n - 1)/2$, $n \geq 2$
47. $3^4 + 4^4 + 5^4 + 6^4 \neq 7^4$

Exercise 11-3

1. (B) $d = -0.5$; 5.5, 5;   (C) $d = -5$; $-26, -31$ 3. $a_2 = -1, a_3 = 3, a_4 = 7$ 5. $a_{15} = 67, S_{11} = 242$
7. $S_{21} = 861$ 9. $a_{15} = -21$ 11. $d = 6, a_{101} = 603$ 13. $S_{40} = 200$ 15. $a_{11} = 2, S_{11} = \frac{77}{6}$ 17. $a_1 = 1$
19. $S_{51} = 4{,}131$ 21. $-1{,}071$ 23. $4{,}446$ 27. $a_n = -3 + (n - 1)3$
29. (A) 336 feet;   (B) 1,936 feet;   (C) $16t^2$ 33. Hint: $y = x + d$, $z = x + 2d$ 35. $x = -1, y = 2$

Exercise 11-4

1. (A) $r = -2$; $-16, 32$;   (D) $r = \frac{1}{3}; \frac{1}{54}, \frac{1}{162}$ 3. $a_2 = 3, a_3 = -\frac{3}{2}, a_4 = \frac{3}{4}$ 5. $a_{10} = \frac{1}{243}$ 7. $S_7 = 3{,}279$
9. $r = 0.398$ 11. $S_{10} = -1{,}705$ 13. $a_2 = 6, a_3 = 4$ 15. $S_7 = 547$ 17. $\frac{1{,}023}{1{,}024}$ 19. $x = 2\sqrt{3}$
21. $S_\infty = \frac{9}{2}$ 23. No sum   25. $S_\infty = \frac{8}{5}$ 27. $\frac{7}{9}$ 29. $\frac{6}{11}$ 31. $3\frac{8}{37}$ or $\frac{119}{37}$
33. $A = P(1 + r)^n$; approx. 12 years   35. 900   37. $a_n = (-2)(-3)^{n-1}$ 41. \$4,000,000

Exercise 11-5

1. \$700 per year; \$115,500   3. 0.0015 pound/square inch   5. 1,250,000   7. $A = A_0 2^{2t}$ 9. $r = 10^{-0.4} = 0.398$
11. 2   13. $\$9.223 \times 10^{16}$; $\$1.845 \times 10^{17}$

Exercise 11-6

1. 720   3. 20   5. 720   7. 15   9. 1   11. 28   13. $9!/8!$ 15. $8!/5!$ 17. 126   19. 6   21. 1
23. 2,380   25. $m^3 + 3m^2 n + 3mn^2 + n^3$ 27. $8x^3 - 36x^2 y + 54xy^2 - 27y^3$ 29. $x^4 - 8x^3 + 24x^2 - 32x + 16$
31. $m^4 + 12m^3 n + 54m^2 n^2 + 108mn^3 + 81n^4$ 33. $32x^5 - 80x^4 y + 80x^3 y^2 - 40x^2 y^3 + 10xy^4 - y^5$
35. $m^6 + 12m^5 n + 60m^4 n^2 + 160m^3 n^3 + 240m^2 n^4 + 192mn^5 + 64n^6$ 37. $5{,}005 u^9 v^6$ 39. $264 m^2 n^{10}$ 41. $924 w^6$
43. $-48{,}384 x^3 y^5$ 45. 1.1046   47. $\binom{n}{r} = \frac{n!}{r!(n-r)!} = \frac{n!}{(n-r)![n-(n-r)]!} = \binom{n}{n-r}$
49. $\binom{k}{r-1} + \binom{k}{r} = \frac{k!}{(r-1)!(k-r+1)!} + \frac{k!}{r!(k-r)!} = \frac{rk! + (k+1-r)k!}{r!(k+1-r)!} = \frac{(k+1)!}{r!(k+1-r)!} = \binom{k+1}{r}$
51. $\binom{k}{k} = \frac{k!}{k!(k-k)!} = 1 = \frac{(k+1)!}{(k+1)![(k+1)-(k+1)]!} = \binom{k+1}{k+1}$

Exercise 11-7 Chapter Review

1. (A) Geometric;   (B) Arithmetic;   (C) Arithmetic;   (D) Neither;   (E) Geometric   (11-3, 11-4)
2. (A) 5, 7, 9, 11;   (B) $a_{10} = 23$;   (C) $S_{10} = 140$ (11-1, 11-3, 11-4)
3. (A) 16, 8, 4, 2;   (B) $a_{10} = \frac{1}{32}$;   (C) $S_{10} = 31\frac{31}{32}$ (11-1, 11-3, 11-4)
4. (A) $-8, -5, -2, 1$;   (B) $a_{10} = 19$;   (C) $S_{10} = 55$ (11-1, 11-3, 11-4)
5. (A) $-1, 2, -4, 8$;   (B) $a_{10} = 512$;   (C) $S_{10} = 341$ (11-1, 11-3, 11-4)   6. $S_\infty = 32$ (11-4)   7. 720   (11-6)

Answers

8. $20 \cdot 21 \cdot 22 = 9{,}240$ *(11-6)* 9. 21 *(11-6)*
10. $P_1: 5 = 1^2 + 4 \cdot 1; P_2: 5 + 7 = 2^2 + 4 \cdot 2; P_3: 5 + 7 + 9 = 3^2 + 4 \cdot 3$ *(11-2)*
11. $P_1: 2 = 2^{1+1} - 2; P_2: 2 + 4 = 2^{2+1} - 2; P_3: 2 + 4 + 8 = 2^{3+1} - 2$ *(11-2)*
12. $P_1: 49^1 - 1 = 48$ is divisible by 6; $P_2: 49^2 - 1 = 2{,}400$ is divisible by 6; $P_3: 49^3 - 1 = 117{,}648$ is divisible by 6 *(11-2)*
13. $P_k: 5 + 7 + 9 + \cdots + (2k+3) = k^2 + 4k; P_{k+1}: 5 + 7 + 9 + \cdots + (2k+3) + (2k+5) = (k+1)^2 + 4(k+1)$ *(11-2)*
14. $P_k: 2 + 4 + 8 + \cdots + 2^k = 2^{k+1} - 2; P_{k+1}: 2 + 4 + 8 + \cdots + 2^k + 2^{k+1} = 2^{k+2} - 2$ *(11-2)*
15. $P_k: 49^k - 1 = 6r$ for some integer r; $P_{k+1}: 49^{k+1} - 1 = 6s$ for some integer s *(11-2)*
16. $S_{10} = -6 - 4 - 2 + 0 + 2 + 4 + 6 + 8 + 10 + 12 = 30$ *(11-1, 11-3, 11-4)*
17. $S_7 = 8 + 4 + 2 + 1 + \frac{1}{2} + \frac{1}{4} + \frac{1}{8} = 15\frac{7}{8}$ *(11-1, 11-3, 11-4)* 18. $S = \frac{81}{5}$ *(11-4)*
19. $S_n = \sum_{k=1}^{n} \frac{(-1)^{k+1}}{3^k}, S_\infty = \frac{1}{4}$ *(11-4)* 20. $d = 3, a_5 = 25$ *(11-3)* 21. $\frac{8}{11}$ *(11-4)* 22. 190 *(11-6)*
23. 1,820 *(11-6)* 24. 1 *(11-6)* 25. $x^5 - 5x^4y + 10x^3y^2 - 10x^2y^3 + 5xy^4 - y^5$ *(11-6)* 26. $-1{,}760x^3y^9$ *(11-6)*
30. $49g/2$ feet; $625g/2$ feet *(11-3)* 31. $x^6 + 6ix^5 - 15x^4 - 20ix^3 + 15x^2 + 6ix - 1$ *(11-6)*

Practice Test: Chapter 11

1. (A) Neither; (B) Geometric, $r = -\frac{1}{4}$; (C) Arithmetic, $d = -4$ *(11-3, 11-4)*
2. 64, 60, 56, 52; $a_{51} = -136$; $S_{31} = 124$ *(11-3)* 3. 32, 16, 8, 4; $a_{10} = \frac{1}{16}$; $S_{10} = \frac{1{,}023}{16}$ *(11-4)*
4. $x^2 + 4x^5 + 9x^8 + 16x^{11}$ *(11-1)* 5. $\sum_{k=1}^{4} (-1)^{k+1}(\frac{2}{3})^k$ *(11-1)* 6. $\frac{2}{111}$ *(11-4)*
7. Write $P_n: 3 + 3^2 + 3^3 + \cdots + 3^n = (3^{n+1} - 3)/2$
    $S = \{n \in N \mid P_n \text{ is true}\}$
    Part 1. Show $1 \in S$. $3 = (3^2 - 3)/2$ $\therefore 1 \in S$
    Part 2. Show that S is inductive.
    $$3 + 3^2 + 3^3 + \cdots + 3^k = (3^{k+1} - 3)/2$$
    $$3 + 3^2 + 3^3 + \cdots + 3^k + 3^{k+1} = (3^{k+1} - 3)/2 + 3^{k+1}$$
    $$= (3 \cdot 3^{k+1} - 3)/2 = (3^{k+2} - 3)/2$$
    Thus, $k \in S \Rightarrow k + 1 \in S$, and S is inductive. $\therefore S = N$ *(11-2)*
8. (A) 455; (B) 253 *(11-6)* 9. $81x^4 - 108x^3y + 54x^2y^2 - 12xy^3 + y^4$ *(11-6)* 10. $-4{,}032x^4$ *(11-6)*

Appendixes

Exercise A-2

1. 1 3. $6x^9$ 5. $9x^6/y^4$ 7. $(a^4b^{12})/(c^8d^4)$ 9. 10^{17} 11. $(2x)/y^2$ 13. n^2 15. 4×10^8
17. 3.225×10^7 19. 8.5×10^{-2} 21. 7.29×10^{-8} 23. 0.005 25. 26,900,000 27. 0.000 000 000 59
29. $3y^4/2$ 31. x^{12}/y^8 33. $4x^8/y^6$ 35. $w^{12}/(u^{20}v^4)$ 37. $(27y^3)/(2x^3)$ 39. $1/(x+y)^2$ 41. $2(a+2b)^5$
43. $5/(u - v + w)^3$ 45. 1.3×10^{25} pounds
47. 10^8 or 100 million, 10^{10} or 10 billion; 6×10^9 or 6 billion, 6×10^{11} or 600 billion 49. 6.65×10^{-17}
51. 1.54×10^{12} 53. 1.0295×10^{11} 55. -4.3647×10^{-18} 57. 9.4697×10^{29}

Exercise A-3

1. 4 3. 64 5. -6 7. Not a real number 9. $\frac{8}{125}$ 11. $\frac{1}{27}$ 13. $y^{3/5}$ 15. $d^{1/3}$ 17. $1/y^{1/2}$
19. $2x/y^2$ 21. $1/(a^{1/4}b^{1/3})$ 23. $xy^2/2$ 25. $2b^2/3a^2$ 27. $125y^{1/2}/27x^3$ 29. $2/3x^{7/12}$ 31. $a^{1/3}/b^2$
33. $3x^{1/2}$ 35. $1/25y^{2/9}$ 37. $a^{1/n}b^{1/m}$ 39. $1/(x^{3m}y^{4n})$ 41. (A) $x = -2$, for example; (B) $x = 2$, for example
43. 29.52 45. 0.03093 47. 5.421 49. 107.6

Exercise A-4

1. $\sqrt[3]{m^2}$ or $(\sqrt[3]{m})^2$ (first preferred) 3. $6\sqrt[5]{x^3}$ (not $\sqrt[5]{6x^3}$) 5. $\sqrt[5]{(4xy^3)^2}$ 7. $\sqrt{x+y}$ 9. $b^{1/5}$ 11. $5x^{3/4}$
13. $(2x^2y)^{3/5}$ 15. $x^{1/3} + y^{1/3}$ 17. -2 19. $3x^4y^2$ 21. $2mn^2$ 23. $2ab^2\sqrt{2ab}$ 25. $2xy^2\sqrt[3]{2xy}$

27. $\sqrt{m}$ 29. $\sqrt[12]{xy}$ 31. $3x\sqrt[3]{3}$ 33. $\sqrt{5}/5$ 35. $2\sqrt[3]{9}$ 37. $\sqrt{42xy}/7y$ or $(1/7y)\sqrt{42xy}$ 39. $3x^2y^2\sqrt[5]{3x^2y}$
41. $n\sqrt[5]{2m^3n}$ 43. $\sqrt[3]{a^2(b-a)}$ 45. $\sqrt[4]{a^3b}$ 47. $x^2y\sqrt[3]{6xy^2}$ 49. $\sqrt{2}/2$ or $\frac{1}{2}\sqrt{2}$ 51. $2a^2b\sqrt[3]{4a^2b}$
53. $\sqrt[3]{12xy^3}/2x$ or $(1/2x)\sqrt[3]{12xy^3}$ 55. $(a\sqrt[3]{6ab})/3b^2$ or $(a/3b^2)\sqrt[3]{6ab}$ 57. $x/2\sqrt{2x}$ 59. $1/m\sqrt[3]{2m^2}$ 61. $2x/\sqrt[4]{2x}$
63. $\sqrt[5]{2a^3b^2}$ 65. $2x\sqrt[3]{4y^2}$ 67. $2a^4b^3\sqrt[6]{2^5a^5b^5}$ 69. $\sqrt[6]{2ab}$ 71. $\sqrt[n]{x/y} = (x/y)^{1/n} = (x^{1/n})/(y^{1/n}) = \sqrt[n]{x}/\sqrt[n]{y}$
73. (A) 6x; (B) 2x 75. (A) 2x; (B) 0 77. (A) 5x; (B) x 79. 0.2222 81. 1.934 83. 0.069 79
85. 2.073 87. Both are 1.059. 89. Both are 0.6300.

Exercise A-5

1. 3 3. $2x^3 - x^2 + 2x + 4$ 5. $2x^3 - 5x^2 + 6$ 7. $6x^4 - 13x^3 + 9x^2 + 13x - 10$ 9. $4x - 6$
11. $6y^2 - 16y$ 13. $4\sqrt{5} - 2\sqrt{3}$ 15. $5\sqrt[3]{a} - \sqrt[3]{a}$ 17. $9x^2 - 4y^2$ 19. $c - d$ 21. $16x^2 - 8xy + y^2$
23. $x + 2x^{1/2}y^{1/2} + y$ 25. $c - 2\sqrt{cd} + d$ 27. $a^3 + b^3$ 29. $-x + 27$ 31. $2x^4 - 5x^3 + 5x^2 + 11x - 10$
33. $-5x^2 - 4x + 5$ 35. $8m^3 - 12m^2n + 6mn^2 - n^3$ 37. $6\sqrt{2} - 2\sqrt{5}$ 39. $-4\sqrt{3x}$ 41. $2a^2 - 5b^2$
43. $9x - 6\sqrt{xy} + y$ 45. $2x + 3x^{1/2}y^{1/2} + y$ 47. $6x - 2x^{19/3}$ 49. 1 51. $a + b$ 53. h 55. 0
57. -4 59. $1/x - 2/(x^{1/2}y^{1/2}) + 1/y$ 61. $t - x$ 63. $6x + 12 - 5(x + 3)^{1/2}$ 65. $x + 2\sqrt[3]{xy} - \sqrt[3]{x^2y^2} - 2y$
67. Perimeter $= 2x + 2(x - 5) = 4x - 10$ 69. Value $= 5x + 10(x - 5) + 25(x - 3) = 40x - 125$ 71. -2.000 73. 0

Exercise A-6

1. $2x^2(3x^2 - 4x - 1)$ 3. $5xy(2x^2 + 4xy - 3y^2)$ 5. $(5x - 3)(x + 1)$ 7. $(2w - x)(y - 2z)$ 9. $(x + 3)(x - 2)$
11. $(2m - 1)(3m + 5)$ 13. $(2x - 3y)(x - 2y)$ 15. $(4a - 3b)(2c - d)$ 17. $(2x - 1)(x + 3)$ 19. $(x - 6y)(x + 2y)$
21. Prime 23. $(5m - 4n)(5m + 4n)$ 25. $(x + 5y)^2$ 27. Prime 29. $6(x + 2)(x + 6)$ 31. $2y(y - 3)(y - 8)$
33. $y(4x - 1)^2$ 35. $(3s - t)(2s + 3t)$ 37. $xy(x - 3y)(x + 3y)$ 39. $3m(m^2 - 2m + 5)$
41. $(m + n)(m^2 - mn + n^2)$ 43. $(c - 1)(c^2 + c + 1)$ 45. $[(a - b) - 2(c - d)][(a - b) + 2(c - d)]$
47. $(2m - 3n)(a + b)$ 49. Prime 51. $(x + 3)(x - 3)^2$ 53. $(a - 2)(a + 1)(a - 1)$
55. $[4(A + B) + 3][(A + B) - 2]$ 57. $(m - n)(m + n)(m^2 + n^2)$ 59. $st(st - 2)(s^2t^2 + 2st + 4)$
61. $(m + n)(m + n - 1)$ 63. $(2/x - 3/y)(2/x + 3/y)$ 65. $(x + \frac{3}{2})^2$ 67. $(1/x - 2)(1/x^2 + 2/x + 4)$
69. $(x^{1/3} - 2)(x^{1/3} - 1)$ 71. $(2x^{-1} - y^{-2})(2x^{-1} + y^{-2})$ 73. $2a[3a - 2(x + 4)][3a + 2(x + 4)]$
75. $(x^2 - x + 1)(x^2 + x + 1)$ 77. $(4x^2 + 2x + 1)(4x^2 - 2x + 1)$
79. $(x^3 - 1)(x^3 + 1) = (x - 1)(x^2 + x + 1)(x + 1)(x^2 - x + 1)$ 81. $(x^2 - 2xy + 5y^2)(x^2 + 2xy + 5y^2)$

Exercise A-7

1. $a^2/2$ 3. $(22y + 9)/252$ 5. $(x^2 + 8)/8x^3$ 7. $1/(2x - 1)$ 9. $1/m$ 11. $2a/[(a + b)^2(a - b)]$
13. $(m^2 - 6m + 7)/(m - 2)$ 15. $7/(x - 3)$ 17. $3/(y + 3)$ 19. $(x + y)/x$ 21. $1/m$
23. $(y + 3)/[(y - 2)(y + 7)]$ 25. -1 27. $(7y - 9x)/[xy(a - b)]$ 29. $(x^2 - x + 1)/[2(x - 9)]$
31. $[(x - y)^2]/[y^2(x + y)]$ 33. $1/(x - 4)$ 35. $1/xy$ 37. $(x - 3)/(x - 1)$ 39. $(x^2 + xy + y^2)/xy$
41. $[-x(x + y)]/y$ 43. $(a - 2)/a$ 45. c/a 47. $(6 + \sqrt{a} - a)/(a - 4)$ 49. $(62 - 19\sqrt{10})/117$
51. $\sqrt{x^2 + 9} + 3$ 53. $1/(\sqrt{t} + \sqrt{x})$ 55. $1/(\sqrt{x + h} + \sqrt{x})$ 57. $\frac{5}{6}\sqrt{6xy}$ 59. $\frac{11}{2}\sqrt{2}$ 61. $(y - x)/y$
63. $(x - 1)/x$ 65. $1/(\sqrt[3]{t^2} + \sqrt[3]{tx} + \sqrt[3]{x^2})$

Exercise A-8

1. 0.2108 3. 4.915 5. 5.955 7. 0.9394 9. 0.5870 11. 1.160 13. 15° 15. 20° 17. 55°
19. 20° 21. 14°20' 23. 0.14 radian 25. 0.72 radian 27. 1.39 radians 29. 1.42 radians 31. —
33. + 35. — 37. + 39. + 41. -0.891 43. -2.75 45. 1.22 47. -0.131 49. -0.290
51. -0.990 53. -1.53 55. 0.460 57. -0.682 59. 0.628 61. 0.5972 63. -2.360 65. -1.081

Index

Abscissa, 83
Absolute value
　definition of, 35
　distance, 36
　radical form, A21
Addition
　fractions, 16, A43
　polynomials, A27–A28
　vectors, 412
Addition and subtraction identities, 358–362
Additive identity, 13
Additive inverse, 13
Algebraic expression, 42, A25
Amplitude, 326, 327, 330
Angles
　complementary, 292
　coterminal, 291
　definition of, 290–291
　　initial side, 291
　　terminal side, 291
　　vertex, 290
　degree measure, 292
　　minutes and seconds, 292
　grad unit, 301
　initial side, 291
　negative, 291
　obtuse, 292
　positive, 291
　quadrantal, 291
　radian measure, 293
　right, 292
　sides of, 290
　special angles (30°, 60°, 45°), 309
　standard position, 291
　straight, 292
　supplementary, 292
　terminal side, 291
　vertex, 290
Arithmetic progression (see Arithmetic sequence)
Arithmetic sequence
　common difference, 620

Arithmetic sequence (continued)
　definition of, 620
　nth term formula, 620–621
　sum formulas, 622
Associative property, 12
Asymptotes, 144–147, 469–471
Augmented matrix, 518–519

Basic identities, 282–285, 351
Binomial, 43, A27
Binomial formula, 635–638
Blood types, 9

Cartesian coordinate system
　abscissa, 83
　coordinate axes, 82–83
　coordinates, 83
　distance formula, 90
　fundamental theorem of analytic geometry, 84
　midpoint formula, 95
　ordered pair, 83
　ordinate, 83
　origin, 83
　polar–rectangular relationships, 422–424
　quadrants, 83
Catenary, 454
Change-of-base formula (logarithms), 255
Circle
　center, 90
　definition of, 90
　equations of, 91
　radius, 90
Circular functions (see also Trigonometric functions)
　basic identities, 282–285
　calculator evaluation, 285–286
　definition of, 279
　exact values, 280–281
　sign properties, 282
Circular point, 271

Cofactor (determinant), 580
Cofunction, 359–360, 362
Collinear, 123
Combinations, 635
Common denominator, A43
　least common denominator, A43
Common logarithms, 249–254
Common multiple, 22
Commutative property, 11
Complement of a set, 5
Complementary angles, 292
Completing the square, 60
Complex fraction, A45–A47
Complex numbers, 50–55, 432–443
　absolute value, 434
　addition, 52
　complex plane, 433
　　imaginary axis, 433
　　real axis, 433
　conjugate of, 51
　definition of, 51
　equality, 52
　history of, 50
　imaginary unit, 51
　multiplication, 52
　polar (trigonometric) form, 433–434
　　argument, 434
　　DeMoivre's theorem, 439, 441
　　division, 436–437
　　modulus (absolute value), 434
　　multiplication, 436–437
　　nth root theorem, 440
　properties, 52
　and radicals, 54–55
　rectangular form, 433
　zero, 51
Component (vectors), 412
Composite functions, 155–156
Composite number, A33
Conditional equation, 350
Cone
　axis, 448
　definition of, 448

I-1

Cone (*continued*)
  nappes, 448
  vertex, 448
Conic sections, 448–490
  eccentricity, 476
Conjugate (complex number), 51
Conjugate axis (hyperbola), 470–471
Constant, 3
Constant function, 113
Continuity
  interval, 115
  point, 114
Coordinates
  Cartesian coordinate system, 82–83
  real line, 11
Coterminal angles, 291
Counterexample, 611
Cramer's rule, 592–596
Cube root, A10
Cycloid, 494–496

Decreasing function, 113
Degree measure of angles, 292
Degree of a polynomial, 42
DeMoivre's theorem, 439, 441
Denominator, 16
Dependent variable, 99
Descartes' rule of signs, 200–201
Determinant
  cofactor, 580
  Cramer's rule, 592–596
  function, 578
  minor, 580
  order n, 578
  properties, 585–588
  second-order, 578–579
  third-order, 579–582
Difference of two cubes, A36
Difference of two squares, A36
Directrix, 449
Discontinuous, 115
Discriminant (quadratic formula), 63
Disjoint sets, 5
Distance
  Cartesian coordinate system, 89–90
  real line, 36
Distance formula, 90
Distributive property, 14
Division
  algebraic long, 184
  algorithm, 188
  complex numbers, 436–437
  fractions, 16

Division (*continued*)
  radicals, A19, A42
  real numbers, 15
  synthetic, 184–186
Domain, 97
Dot product (matrices), 558
Double-angle identities, 367–369

e, 233
Eccentricity, 476
Element of a set, 2
Ellipse
  applications, 463
  center, 456, 460
  definition of, 456
  drawing, 456
  eccentricity, 476
  equations of, 457–460
  foci, 456, 460
  graphing, 460–461
  major axis, 456, 458, 460
  minor axis, 456, 458, 460
  vertex, 456
Empty set, 3
Equality
  definition of, 19
  not equal to, 20
  properties of, 20, 21
Equations
  and absolute value, 37–40
  conditional, 350
  equivalent, 21
  exponential, 257–260
  linear, 20–24
  logarithmic, 260–261
  matrix, 570–572
  polynomial, 181–218
  quadratic, 57–71
  reducible to quadratic, 71–74
  replacement set, 21
  root, 21
  solution, 21
  solution set, 21
  trigonometric, 379–385

Equivalent equations, 21
Equivalent inequalities, 26
Even function, 111–112
Exponent
  base, A3
  integer, A3–A6
  negative integer, A3

Exponent (*continued*)
  positive integer, A3
  properties of, A3, A4
  rational, A10–A14
  scientific notation, A6–A8
  zero, A3
Exponential functions
  definition of, 230
  domain, 230
  e, 233
  graphs, 231–233
  properties, 235
  range, 231
Extraneous solutions, 72

Factor, A33
Factor identities, 376–378
Factor theorem, 191
Factorial, 634–635
Factoring
  common factors, A35
  difference of two cubes, A36
  difference of two squares, A36
  grouping, A35–A36
  nested, 136–137
  special formulas, A36
  square, A36
  sum of two cubes, A36
Fermat's last theorem, 616
First-degree equation (*see* Linear equations)
Focus (parabola), 449, 452
Force, 414
Fractions
  addition, 16, A43
  complex, A45–A47
  denominator, 16
  division, 16, A41
  equal, 16
  fundamental principle of, 16, A40
  least common denominator (LCD), A43
  multiplication, 16, A41
  numerator, 16
  raising to higher terms, A40
  rational expression, A40
  real number properties, 16
  reducing to lowest terms, A40
  simple, A45
  subtraction, 16, A43
Functions
  circular, 279–286
  composite, 155–156
  constant, 113

Functions (continued)
  continuous, 114–115, 136
  decreasing, 113
  dependent variable, 99
  determinant, 578
  domain, 97
  even, 111–112
  exponential, 230–236
  factorial, 634–635
  graph of, 110–111
  greatest integer, 114
  history of, 105–106
  increasing, 113
  independent variable, 99
  input, 99
  inverse relations and functions, 156–162
  inverse trigonometric functions, 334–341
  linear (see Linear relations and functions)
  logarithmic, 238–255
  map, 101
  notation, 101–105
  odd, 111–112
  output, 99
  periodic, 316
  polynomial, 135–137 (see also Polynomial functions)
  quadratic, 138–141
  range, 97
  rational, 143–153 (see also Rational functions)
  rule definition, 97
  sequence, 604
  set definition, 97
  trigonometric, 299–301
  vertical line test, 99
  wrapping, 270–277
Fundamental theorem of algebra, 194
Fundamental theorem of analytic geometry, 84

Gauss–Jordan elimination, 524–531
Geometric progression (see Geometric sequence)
Geometric sequence
  common ratio, 625
  definition of, 625
  infinite sum formula, 628
  nth term formula, 628
  sum formulas, 627
Goldbach's problem, 616

Grad unit (angle measure), 301
Graphing
  asymptotes, 144–147, 469–471
  Cartesian coordinate system, 82
  circles, 90–92
  continuity, 114–115
  contraction, 119
  expansion, 119
  exponential functions, 231–233
  horizontal shift, 118
  parabola, 449–453
  point-by-point, 84–85
  polar graphs, 426–431
  polynomial functions 135–141, 190–191
  rational functions, 143–153
  rectangular coordinate system, 82
  reflection, 119
  symmetry
    origin, 86
    tests for, 87
    x axis, 86
    y axis, 86
  translation, 116–118
  trigonometric functions, 315–331
  vertical shift, 117
  x intercept, 130
  y intercept, 130
Greater than, 24
Greatest integer function, 114

Half-angle identities, 369–373
Half-planes, 540
Harmonic analysis, 325
Hyperbola
  applications, 474–475
  asymptotes, 469–471
  center, 466, 471
  conjugate axis, 470, 471
  definition of, 466
  drawing, 466–467
  eccentricity, 476
  equations of, 467–471
  foci, 466, 471
  graphing, 471–473
  transverse axis, 466, 471
  vertices, 466, 471
Hypotenuse, 394

Identities
  addition and subtraction, 358–362
  basic, 282–285, 351

Identities (continued)
  cofunction, 359–360, 362
  definition of, 350
  double-angle, 367–369
  factor, 376–378
  half-angle, 369–373
  for negatives, 283, 351
  product, 375–376
  Pythagorean, 283, 351
  quotient, 283, 351
  reciprocal, 283, 351
  steps in proving, 353
Imaginary axis, 443
Imaginary numbers (see Complex numbers)
Increasing function, 113
Independent variable, 99
Inequalities (see also Inequality relation)
  and absolute value, 37–40
  equivalent, 26
  first-degree, one variable, 26–31
  first-degree, two variables, 540–543
  interval notation, 25
  polynomial, 43–47
  rational, 47–49
  solution, 26
  solution set, 26
Inequality relation
  greater than, 24
  less than, 24
  properties, 26
  trichotomy property, 24
Input, 99
Integer exponent, A3–A6
Integers, 10
Intersection of sets, 5
Interval notation, 25
Inverse relations and functions, 156–162
  domain, 157
  one-to-one correspondence, 158–162
  range, 157
Inverse trigonometric functions, 334–341
Irrational numbers, 10

Law of cosines, 406–410
Law of sines, 399–404
Least common denominator (LCD), A43
Least common multiple (LCM), 22
Less than, 24
Linear equations
  applications, 28–30
  solving, 20–23

Linear inequalities
  applications, 28–30
  solving, 26–28
Linear relations and functions
  function, 122
  graphs, 123
  slope, 124–125
Lines
  equations of
    intercept form, 130
    point–slope form, 127
    slope–intercept form, 129
  horizontal, 124, 126
  parallel, 131
  perpendicular, 131
  slope, 124–125
  vertical, 124, 126
Logarithmic functions
  change-of-base formula, 255
  common: calculator evaluation, 249–251
  common: table evaluation, 251–254
  definition of, 239
  history of, 248–249
  logarithmic–exponential identities, 242
  natural: calculator evaluation, 249–251
  natural: table evaluation, 254
  properties of, 244, 246

Map, 101
Mathematical induction, 610–615
Matrix
  addition, 553
  augmented, 518–519
    row-equivalent, 519
  column, 518, 552
  definition of, 518, 552
  dimension, 552
  element, 518
  equal, 552
  equations, 570–572
  Gauss–Jordan elimination, 524, 525–531
  identity, 565
  inverse, 565–570
  main diagonal, 565
  multiplication
    by a number, 554
    dot product, 558
    matrix, 559–561
  negative of, 553

Matrix (continued)
  properties of, 553, 561
  reduced form, 525
  row, 518, 552
  row equivalent, 519
  square, 552, 565
  subscript notation, 518
  subtraction, 554
  zero, 553
Member of a set, 2
Midpoint formula, 95
Minor (determinant) 580
Mollweide's equation, 405
Monomial, 43, A27
Multiplication
  complex numbers, 436–437
  fractions, 16
  polynomials, 195
  radicals,
Multiplicative identity, 13
Multiplicative inverse, 13

n factorial, 634–635
nth root theorem, 440
Nappes of a cone, 448
Natural logarithm, 249–254
Natural numbers, 10
  composite, A33
  prime, A33
Negative of a number, 15
Null set, 3
Numbers
  complex, 50–55, 432–443
  integers, 10
  irrational, 10
  natural, 10
  rational, 10
  real, 10
Numerator, 16

Oblique triangle, 399
Obtuse angle, 292
Odd function, 111–112
One-to-one correspondence, 158–159
Ordered pair, 83
Ordinate, 83
Origin, 11, 83, 422
Output, 99

Parabola
  applications, 453
  axis, 138, 449

Parabola (continued)
  definition of, 449
  directrix, 449, 452
  drawing, 450
  eccentricity, 476
  equations, of 138, 452
  focus, 449, 452
  graph, 138, 448, 449–453
  vertex, 138, 140, 449, 452
Parameter, 491, 509
Parametric equations, 491–496
  cycloid, 494–496
  plane curve, 492
  projectile motion, 493–494
Partial fraction decomposition, 218–224
Pascal's triangle, 640
Periodic functions, 316
Phase shift, 329–330
Plane curve, 492
Polar coordinate system, 420–432
  graphing, 426–431
  origin, 420
  pole, 420
  polar axis, 420
  polar–rectangular relationships, 422
Polynomial
  addition, A27–A28
  binomial, 43, A27
  completely factored form, A34
  definition of, 42, A26
  degree of, 42, A26
  factoring, A32–A38
  monomial, 43, A27
  multiplication, A28–A29
  prime, A34
  subtraction, A27–A28
  trinomial, 43, A27
Polynomial functions
  bounding real zeros, 201–203
  complex zeros, 196–198
  definition of, 183
  Descartes' rule of signs, 200–201
  factor theorem, 191
  first-degree, 122
  fundamental theorem of algebra, 194
  graphing, 135–141, 190–191
  location theorem, 204
  multiplicity (of zeros), 195
  n zero theorem, 195
  quadratic, 138–141
  rational zeros
    strategy for finding, 208
    theorem, 206

Index

Polynomial functions (*continued*)
  real zeros
    strategy for finding, 214
    successive approximation, 215
  remainder theorem, 189
  zero of, 183
Prime number, A33
Principal nth root, A12
Production identities, 375–376
Projectile motion, 493–494
Pythagorean identities, 283, 351
Pythagorean theorem, 65

Quadrantal angle, 291
Quadratic equations
  applications 63–66
  solution by completing the square, 59–61
  solution by factoring, 57–58
  solution by quadratic formula, 61–63
  solution by square root, 58–59
  standard form, 57
Quadratic formula, 62
  discriminant, 63
Quadratic function
  definition of, 138
  graph, 138–141
    axis, 138, 140
    vertex, 138, 140
Quotient identities, 283, 251

Radian measure of angles, 293
Radicals
  absolute value, A21
  index, A16
  nth root, A16
  properties of, A17
  radical sign, A16
  radicand, A16
  rationalizing denominators, A19, A42
  simplest radical form, A18
Range, 97
Rational exponent, A10–A14
Rational expression, A40
Rational functions
  asymptotes, 144–147
  definition of, 143
  graphing, 143–153
  partial fraction decomposition, 284–290
Rational numbers, 10
Rational zeros theorem, 206
Rationalizing denominators, A19, A42

Real number line, 11
  coordinate, 11
  origin, 11
Real numbers, 10–17
  additive identity, 13
  additive inverse, 13
  associative property, 12
  basic properties, 11–14
  commutative property, 11
  distributive property, 14
  division, 15
  fraction properties, 16
  identities, 13
  integers, 10
  inverses, 13
  irrational numbers, 10
  multiplicative identity, 13
  multiplicative inverse, 13
  natural numbers, 10
  negative of, 15
  rational numbers, 10
  real line, 11
  reciprocal, 15
  subtraction, 15
  zero properties, 15
Real plane, 82
Reciprocal, 15
Reciprocal identities, 283, 351
Rectangular coordinate system (*see* Cartesian coordinate system)
Recursion formula, 605
Reference angle, 308
Reference triangle, 304, 308
Reflection, 119
Relation
  domain, 97
  range, 97
  rule definition, 97
  set definition, 97
Remainder theorem, 189
Replacement set, 21
Resultant (vectors), 412
Right angle, 292
Right triangle, 394–397
  Pythagorean theorem, 65
  ratios, 394
Root of an equation, 21
Roots
  cube root, A10
  nth root, 496, A11
  principal nth root, A12
  square root, A10
Rotation of axes, 484–490

Scalar, 412
Scientific notation, A6–A8
Sequence
  arithmetic, 619–623
  definition of, 604
  finite, 605
  geometric, 625–629
  infinite, 605
  nth term formula, 620
  recursion formula, 605
Series
  arithmetic, 621–623
  finite, 606
  geometric, 625–629
  infinite, 606
  nth term formula, 626
  summation notation, 606
Sets, 2–6
  complement, 5
  definition, 2
  disjoint, 5
  element of, 2
  empty, 3
  equal, 3
  inductive, 612
  intersection, 5
  member of, 2
  notation, 2–3
    listing method, 3
    rule method, 3
  null, 3
  replacement set, 21
  subset, 3–4
  union, 4–5
  universal, 5
  well defined, 2
Significant digits, 395, A1–A2
Simple fraction, A45
Simple harmonic motion, 325
Simplest radical form, A18
Slope, 124–125
Solution, 21, 24
Solution set, 21, 24
Square root, A10
Static equilibrium, 415
Step function, 114
Subset, 3–4
Subtraction
  fractions, 16, A43
  polynomials, A27–A28
  real numbers, 15
Sum of two cubes, A36
Summation notation, 606

Supplementary angles, 292
Symmetry, 86–87
Synthetic division, 184–186
Systems involving nonlinear equations, 536–538
Systems of linear equations
  consistent, 509
  dependent, 509
  equivalent systems, 508
  Gauss–Jordan elimination, 524, 525–531
  inconsistent, 509
  three variables
    solution by elimination, 509–512
  two variables
    solution by elimination using addition, 507–509
    solution by elimination using substitution, 507
    solution by graphing, 505–506
  solution set, 504
Systems of linear inequalities, 540–545

Translation of axes, 477–483
Transverse axis (hyperbola), 466, 471
Triangles
  law of cosines, 406–410
  law of sines, 399–404
  oblique triangles, 399
  right triangles, 394–397
Trichotomy property, 24
Trigonometric functions
  amplitude, 326, 327, 330
  calculator evaluation, 301

Trigonometric functions (*continued*)
  circular functions (*see* Circular functions)
  definition of, 299–301, 303–307
  exact values, 307–313
  graphs of, 315–331
  identities (*see* Identities)
  inverse, 334–341
  law of cosines, 406–410
  law of sines, 399–404
  period, 316, 318, 319, 322, 323, 327, 330
  phase shift, 329–331
  reference angle, 308
  reference triangle, 304, 308
  right triangle solutions, 394–397
  table evaluation, A50–A53
Trinomial, 43, A27

Union of sets, 5
Universal set, 5

Variable, 3
Variation
  combined, 168–170
  direct, 165–166
  inverse, 166–167
  joint, 167
Vectors
  components, 412, 415
  direction, 412
  equal, 412
  equivalent, 412
  force, 414

Vectors (*continued*)
  geometric, 412
  magnitude, 412
  resultant, 412
  scalar quantities, 412
  static equilibrium, 415
  sum, 412
  vector quantities, 412
  velocity, 413
Velocity, 413
Vertex
  angle, 290
  cone, 448
  ellipse, 456
  hyperbola, 466, 471
  parabola, 138, 140

Well-ordering principle, 612
Wrapping function, 270–277
  circular point, 271
  property, 277

x intercept, 130

y intercept, 130

zero
  complex number, 51
  matrix, 553
  real number properties, 15

Chapter-Opening Photo Sources

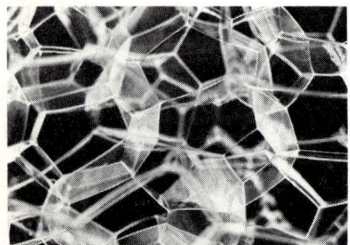

Chapter 1 Soap bubbles
Photo © by Peter Pearce.

Chapter 2 Feather
Photo © by Anne Monk.

Chapter 3 Fractured rock
Photo © by William C. Ferguson.

Chapter 4 Christmas tree worm
Photo © by S. K. Webster, Monterey Bay Aquarium/BPS.

Chapter 5 Old log
Photo © by Glenn R. Steiner.

Chapter 6 Plant
Photo © by Anne Monk.

Chapter 7 Crystal
Photo © by Anne Monk.

Chapter 8 Madrone bark.
Photo © by R. Humbert, Stanford University/BPS.

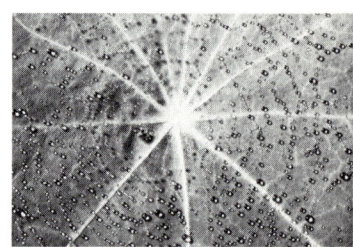

Chapter 9 Leaf with rain drops
Photo © by Anne Monk.

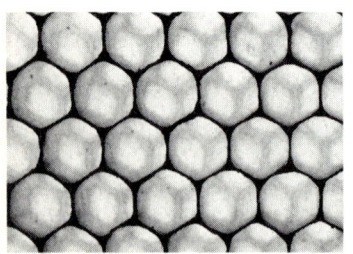

Chapter 10 Honeycomb
Photo © by Peter Pearce.

Chapter 11 Shell
Photo © by Anne Monk.

TRIGONOMETRIC IDENTITIES (6-2 – 6-5)

RECIPROCAL IDENTITIES

$$\csc x = \frac{1}{\sin x} \qquad \sec x = \frac{1}{\cos x} \qquad \cot x = \frac{1}{\tan x}$$

QUOTIENT IDENTITIES

$$\tan x = \frac{\sin x}{\cos x} \qquad \cot x = \frac{\cos x}{\sin x}$$

IDENTITIES FOR NEGATIVES

$$\sin(-x) = -\sin x \qquad \cos(-x) = \cos x \qquad \tan(-x) = -\tan x$$

PYTHAGOREAN IDENTITIES

$$\sin^2 x + \cos^2 x = 1 \qquad \tan^2 x + 1 = \sec^2 x$$
$$1 + \cot^2 x = \csc^2 x$$

SUM IDENTITIES

$$\sin(x + y) = \sin x \cos y + \cos x \sin y$$
$$\cos(x + y) = \cos x \cos y - \sin x \sin y$$
$$\tan(x + y) = \frac{\tan x + \tan y}{1 - \tan x \tan y}$$

DIFFERENCE IDENTITIES

$$\sin(x - y) = \sin x \cos y - \cos x \sin y$$
$$\cos(x - y) = \cos x \cos y + \sin x \sin y$$
$$\tan(x - y) = \frac{\tan x - \tan y}{1 + \tan x \tan y}$$

COFUNCTION IDENTITIES

$$\sin\left(\frac{\pi}{2} - y\right) = \cos y \qquad \tan\left(\frac{\pi}{2} - y\right) = \cot y$$
$$\sin(90° - \theta) = \cos\theta \qquad \tan(90° - \theta) = \cot\theta$$
$$\sec\left(\frac{\pi}{2} - y\right) = \csc y$$
$$\sec(90° - \theta) = \csc\theta$$

PRODUCT IDENTITIES

$$\sin x \cos y = \frac{1}{2}[\sin(x + y) + \sin(x - y)]$$
$$\cos x \sin y = \frac{1}{2}[\sin(x + y) - \sin(x - y)]$$
$$\sin x \sin y = \frac{1}{2}[\cos(x - y) - \cos(x + y)]$$
$$\cos x \cos y = \frac{1}{2}[\cos(x + y) + \cos(x - y)]$$

FACTOR IDENTITIES

$$\sin x + \sin y = 2 \sin\frac{x + y}{2} \cos\frac{x - y}{2}$$
$$\sin x - \sin y = 2 \cos\frac{x + y}{2} \sin\frac{x - y}{2}$$
$$\cos x + \cos y = 2 \cos\frac{x + y}{2} \cos\frac{x - y}{2}$$
$$\cos x - \cos y = -2 \sin\frac{x + y}{2} \sin\frac{x - y}{2}$$

DOUBLE-ANGLE IDENTITIES

$$\sin 2x = 2 \sin x \cos x$$
$$\cos 2x = \begin{cases} \cos^2 x - \sin^2 x \\ 1 - 2\sin^2 x \\ 2\cos^2 x - 1 \end{cases}$$
$$\tan 2x = \frac{2 \tan x}{1 - \tan^2 x} = \frac{2 \cot x}{\cot^2 x - 1} = \frac{2}{\cot x - \tan x}$$

HALF-ANGLE IDENTITIES

$$\sin\frac{x}{2} = \pm\sqrt{\frac{1 - \cos x}{2}}$$
$$\cos\frac{x}{2} = \pm\sqrt{\frac{1 + \cos x}{2}}$$

Signs are determined by quadrant in which $x/2$ lies

$$\tan\frac{x}{2} = \frac{1 - \cos x}{\sin x} = \frac{\sin x}{1 + \cos x} = \pm\sqrt{\frac{1 - \cos x}{1 + \cos x}}$$

DEGREES AND RADIANS (5-4)

$$\frac{\theta°}{180°} = \frac{\theta}{\pi \text{ rad}} \qquad \begin{array}{l} \theta° \text{ in degrees} \\ \theta \text{ in radians} \end{array}$$

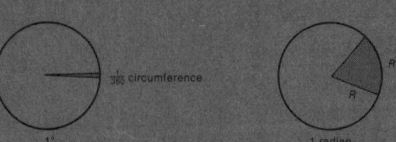

SPECIAL TRIANGLES (5-5)

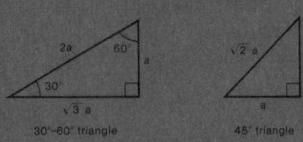